Automatic Transmissions

Other Texts by the Same Authors:

Automotive Principles and Service
Study Guide and Workbook for Automotive Principles and Service
Instructor's Guide for Automotive Principles and Service
Diesel Fundamentals
Instructor's Guide for Diesel Fundamentals
Automotive Engines
Automotive Steering, Suspension, and Braking Systems
Automotive Electronics and Engine Performance
Automotive Drive Trains

Automatic Transmissions

Frank J. Thiessen
Davis N. Dales

Reston Publishing Company, Inc.
A Prentice-Hall Company
Reston, Virginia

Library of Congress Cataloging in Publication Data

Thiessen, Frank.
Automatic transmissions.

1. Automobiles—Transmission devices, Automatic.
2. Automobiles—Transmission devices, Automatic—Maintenance and repair. I. Dales, Davis. II. Title.
TL263.T44 1983 629.2′446 82–25032
ISBN 0–8359–0328–1
ISBN 0–8359–0327–3 (pbk.)

10 9 8 7 6 5 4 3 2 1

Printed in the United States of America

Contents

Preface

This text is designed to help the student in the study of the purpose, construction, operation, and servicing of automatic transmissions.

Chapter 1 deals briefly with the development of automatic transmissions. The progress that has been made in improving the design and operation of automatic transmissions in a relatively short period of time is one example of the rapid changes taking place in automotive technology. Chapter 1 also acquaints the student with the major components and systems required in a modern automatic transmission. A general concept of an automatic transmission is required for a better understanding of each component.

Chapter 2 discusses the purpose, construction, and operation of the various types of torque converters used in present-day automatic transmissions.

Chapter 3 deals with the mechanical gear train of the automatic transmission. The discussion progresses from basic gear principles to simple planetary gear operation and then to the Simpson and Ravigneaux compound planetary gears. Purpose, construction, and operation of each of these gear systems is covered. Power flow in all gear positions including overdrive is covered.

In order to operate properly in an automatic transmission, the planetary gears must be connected to and disconnected from the torque converter while different members of the gear train are being held during the various operating modes. Chapter 4 deals with the various devices used for driving and holding planetary gear components.

Chapter 5 introduces the basic principles of hydraulic system operation and applies these principles to the components of the automatic transmission hydraulic control system.

Chapter 6 covers in more detail the purpose, construction, and operation of the many different components that make up the hydraulic control system. Included are the hydraulic schematics of many of these components that explain how hydraulic pressure is started, stopped, and regulated in order to operate the driving and holding devices covered in Chapter 4.

Chapter 7 is a collection of the hydraulic schematics of various makes and models of transmissions which can be used to follow the regulation and flow of hydraulic pressures during various operating modes. These schematics must be studied circuit by circuit and system by system to be useful in consolidating an understanding of hydraulic control system operation and problem diagnosis.

Automatic transmission diagnosis and service and transmission overhaul are covered in Chapters 8 and 9. Diagnosing, testing, servicing, and overhaul methods are generally similar regardless of the make or model of the transmission. Differences in step-by-step procedures always exist and are best dealt with by consulting manufacturer's service manuals and bulletins, which should always be reviewed for possible changes to earlier information on parts, procedures, and specifications.

This text is designed to deal with the principles of

construction operation and service rather than with the minor design differences that are constantly changing. The technician who understands the principles of the various planetary gear systems, the hydraulic control system components, the torque converter, and other transmission components will be able to apply that knowledge to any automatic transmission since these principles are used in all automatic transmissions regardless of make or model.

Appropriate training models and shop activities must also be available to the student to provide reinforcement to the knowledge gained. A set of Performance Objectives is provided at the beginning of each Chapter and a Performance Evaluation section is included at the end of each chapter. Evaluation of performance is simply a method of determining whether stated objectives have been achieved. Worksheets and Self-Checks are included in each chapter to assist in achieving objectives.

Many opportunities for employment and business opportunities exist in the automatic transmission service industry. The knowledgeable and ambitious student can achieve satisfaction and ample financial returns in this field.

CAUTION: Proper service and repair is important to the reliable operation of all motor vehicles. The service procedures recommended and described in this text are effective methods of performing service operations. Some of these service operations require the use of tools specially designed for the purpose. The special tools should be used as recommended in the manufacturers' service manuals.

It is important to note that this text contains various general precautions, which should be read carefully in order to minimize the risk of personal injury or damage to the vehicle resulting from improper service methods. It also is important to understand that these general precautions are not exhaustive. The authors could not possibly know, evaluate, and advise the service trade of all conceivable ways in which service might be carried out, or of the possible hazardous consequences of each method. Accordingly, anyone who uses any given service procedure or tool must first satisfy himself thoroughly that neither his safety nor the safety of the vehicle will be jeopardized by the service method he selects.

Automobiles contain many parts dimensioned in the metric system as well as in the customary system. Many fasteners are metric and are very close in dimension to familiar customary fasteners in the inch system. It is important to note that, during any maintenance procedures, replacement fasteners must have the same measurements and strength as those removed, whether metric or customary. (Numbers on the heads of metric bolts and on surfaces of metric nuts indicate their strength. Customary bolts use radial lines for this purpose, while most customary nuts do not have strength markings.) Mismatched or incorrect fasteners can result in vehicle damage or malfunction, and possibly personal injury. Therefore, fasteners removed from the car should be saved for re-use in the same locations whenever possible. Where the fasteners are not satisfactory for re-use, care should be taken to select a replacement that matches the original. For information on any specific make or model, refer to the appropriate service manual.

Acknowledgments

The authors have made every effort to give proper credit to the sources of illustrations and other materials, and would be grateful for information from readers on errors or omissions in this regard. The authors are grateful to the various vehicle manufacturers and tool and equipment manufacturers for their kind assistance in producing this text.

The authors are especially grateful to Linda Zuk for the fine work done on the design and layout of this text.

A very special thank you to Ford Motor Co., General Motors Corporation, Chrysler Corporation, and American Motors Corporation for their kind cooperation in the development of this text.

Automatic Transmissions

Chapter 1

Performance Objectives

After thorough study of this chapter, you should be able to perform these tasks:

1. State the need for the automatic transmission.
2. Describe early automatic transmissions briefly.
3. Name the most common early automatic transmissions.
4. Describe briefly the basic function of the five major components of the automatic transmission.
5. Complete the worksheet for this chapter.
6. Complete the self-check at the end of this chapter with at least 80 percent accuracy.
7. Evaluate your performance based on the performance evaluation at the end of this chapter.

Transmission Development and Basic Function

PART 1-1 INTRODUCTION

This chapter deals briefly with the history and development of the automatic transmission. Early automatic transmissions and recent developments are dealt with to provide some background for the rest of the text.

Automatic transmissions are used in passenger cars, light and heavy trucks, buses, and off-road equipment. They are used in rear-wheel-drive, front-wheel-drive, and four-wheel-drive vehicles.

A combined automatic transmission, final drive, and differential is used on front-wheel-drive vehicles and on some rear-wheel-drive vehicles. This type of unit is called a transaxle.

The basic functions of the major components common to automatic transmissions are also covered in this chapter.

PART 1-2 THE NEED FOR AN AUTOMATIC TRANSMISSION

Purpose of a Transmission

A transmission is required to provide a means of selecting a forward or reverse direction of vehicle movement. A number of speed and torque ratio combinations are also required to start the vehicle into motion and to keep it moving with the best combina-

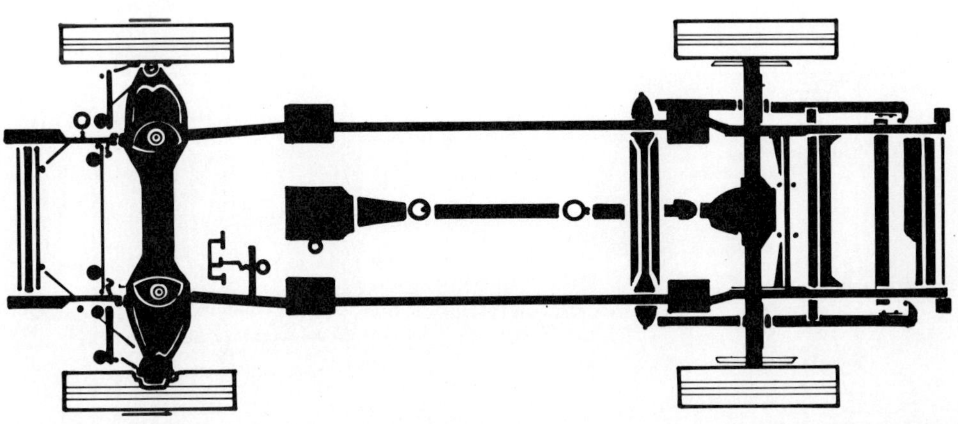

Figure 1-1. Drive train layout for rear-wheel-drive vehicle with engine in front.

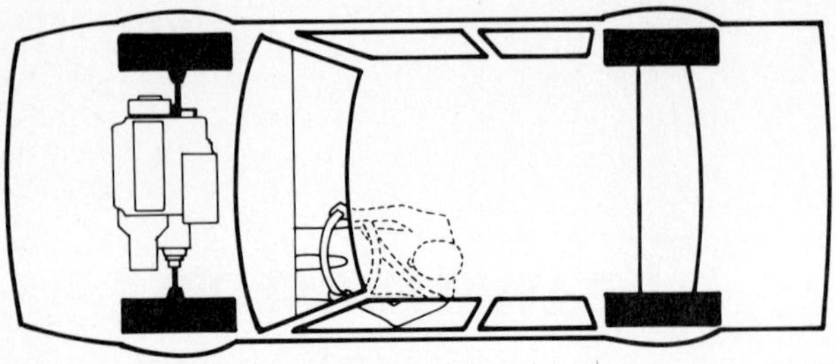

Figure 1-2. Drive train layout for front-wheel-drive vehicle with transverse engine and transaxle.

tion of power, speed, and economy desired by the operator.

A clutch is needed to allow the driver to select different gears while the engine is running. The clutch connects or disconnects the engine to or from the transmission. With the transmission in gear, the clutch allows gradual application of the vehicle load to the engine without stalling the engine.

The clutch allows the transmission to be shifted into reverse or into forward low gear while the engine is running and the vehicle is stationary. Once the vehicle is moving, the clutch is depressed and the next "higher" gear is selected and so on until the transmission is in high or in overdrive.

This combination of a clutch and manual transmission allows the driver to shift "up" or "down" as vehicle load and driver preference may require. This method of shifting gears requires a combination of skill and manipulation to prevent damage to the vehicle.

The automatic transmission performs most of the tasks of the clutch and manual transmission completely automatically. The need for an automatic transmission arose from the desire to eliminate (as much as possible) the need for the driver to upshift and downshift the transmission. The automatic transmission reduced the skills required to operate the vehicle and allowed the driver to concentrate more on the other aspects of driving, making the vehicle more convenient to operate and perhaps a bit safer.

PART 1-3 EARLY AUTOMATIC TRANSMISSIONS

Before the truly automatic transmission, several developments provided the first automatic shift systems. One of these was the introduction of the planetary gear automatic overdrive by Chrysler in 1934. This unit was attached to the rear of the manual transmission and provided automatic upshifts and downshifts from third-speed direct drive to overdrive by means of a governor-controlled electric-solenoid shift mechanism. A dash-mounted control allowed the driver to determine whether the unit would be effective or locked out.

In 1938 Chrysler introduced its Fluid Drive unit. A fluid coupling was placed between the engine crankshaft and the conventional dry-disc friction clutch and standard transmission. This unit allowed the engine to idle with the transmission in gear and the clutch applied. The fluid coupling provided a smooth power application that could not be achieved with the mechanical drive.

The transmission used with this arrangement provided two speed ranges selected by the shift lever on the steering column—low and high—with two speeds in each range. Low and high speed in each range were shifted automatically but controlled by the driver's use of the accelerator pedal. A vacuum motor was used in early models to do the shifting while later designs used a hydraulic pump and hydraulic servo for this purpose.

Although planetary gear transmissions were used

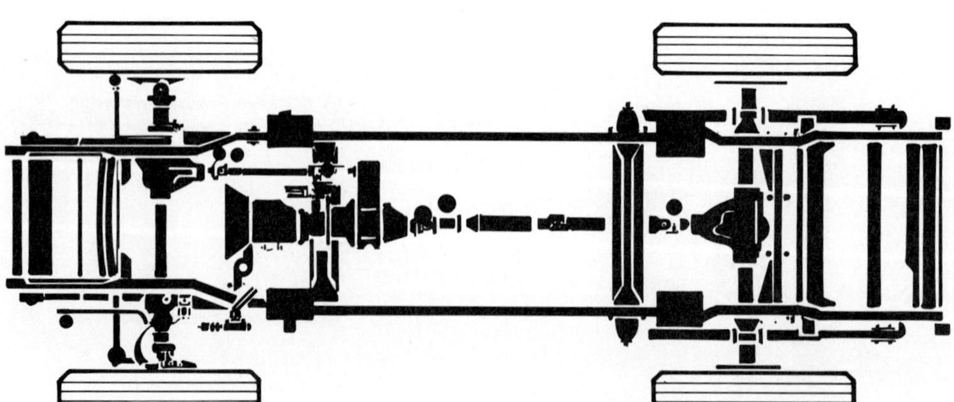

Figure 1-3. Drive train layout for four-wheel-drive vehicle with engine in front.

in early Ford and Cadillac automobiles, they were manually controlled. A semi-automatic transmission using planetary gears was used by Oldsmobile in 1938 models. However, this unit still used some conventional gears as well. A hydraulic servo operated band and multiple-disc clutch controlled planetary gear operation.

Early General Motors Transmissions

General Motors (GM) automatic transmission development took several different tracks following different vehicle product lines, although transmission availability later crossed product lines. Three general areas of transmission development took place. These are the Hydra-Matic series of transmissions used in Oldsmobile, Cadillac, and Pontiac products; the Dynaflow series of transmissions used by Buick; and the Powerglide series used in Chevrolets and the smaller Pontiacs. The Hydra-Matic transmission was also used in other makes such as early Lincoln, Nash, Kaiser, and Fraser automobiles.

General Motors Hydra-Matic Transmissions

In 1940 Oldsmobile made available the first fully automatic four-speed planetary gear transmission. This unit used a fluid coupling, three sets of planetary gears, two multiple-disc clutches, two bands, and a hydraulic control system to operate these devices. A mechanically operated anchor controlled the reverse planetary gearset to provide reverse. This transmission was named the Hydra-Matic transmission.

Later versions of the Hydra-Matic transmission saw the reverse anchor replaced by a cone-type clutch. The front clutch was replaced by a fluid coupling, which could be filled or emptied as required; the two bands were replaced by sprag-type one-way clutches; and a low band was used. A separate park gear was also added. This transmission was used on some General Motors products through the 1964 model year.

A three-speed automatic transmission known as the Roto Hydra-Matic appeared in 1961. This unit used two planetary gearsets and a fluid coupling with a torque multiplier. Two multiple-disc clutches, a cone clutch, a sprag-type one-way clutch, and a band controlled the planetary gears to provide three forward speeds, neutral, and reverse. A smaller version of this unit used on smaller cars used a band instead of the sprag clutch with the rear planetary unit. The Roto Hydra-Matic transmission was last used on some General Motors products in 1964.

Redesigned versions of the Hydra-Matic series of General Motors transmissions adopted the use of a Simpson planetary gearset and a torque converter.

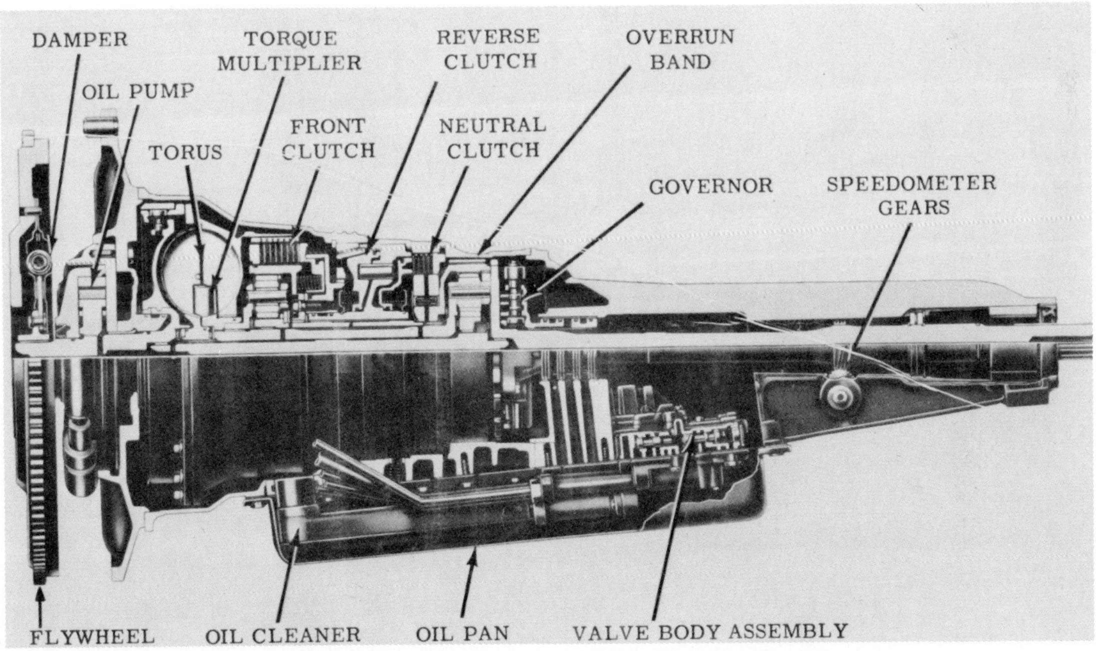

Figure 1-4. Early 1960s Oldsmobile Hydra-Matic transmission. (*Courtesy of General Motors Corporation*)

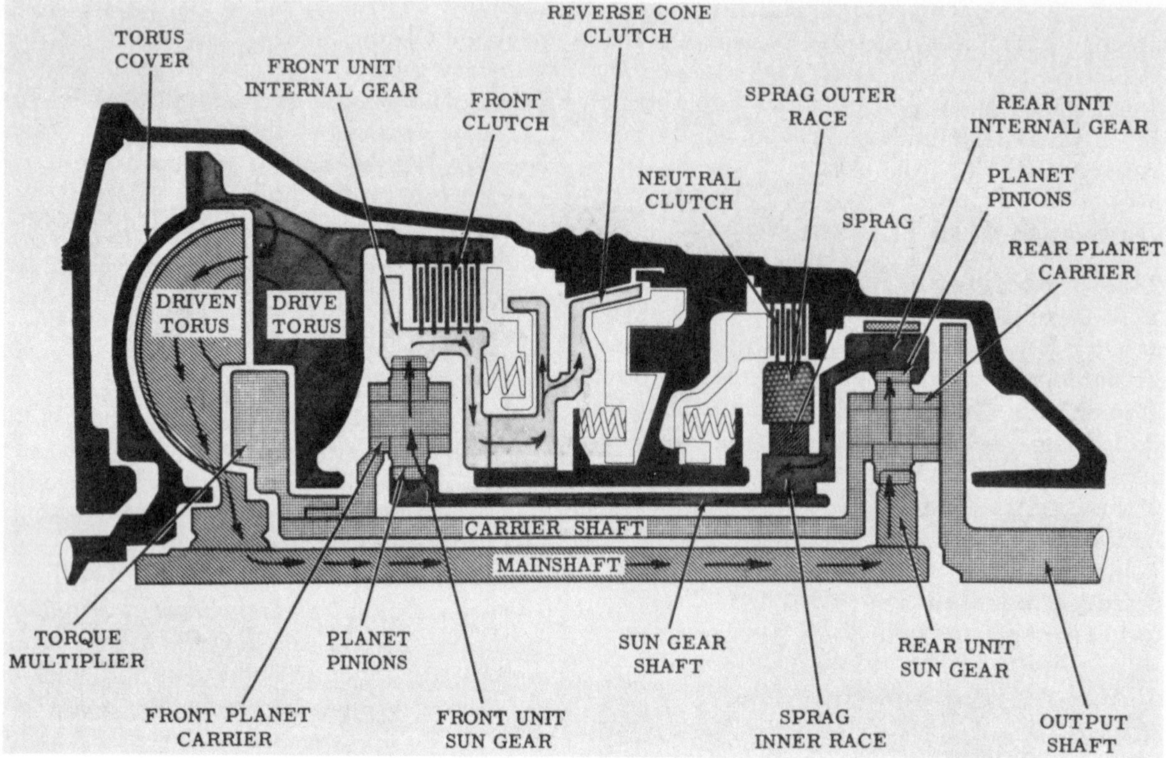

Figure 1-5. Hydra-Matic in neutral. (*Courtesy of General Motors Corporation*)

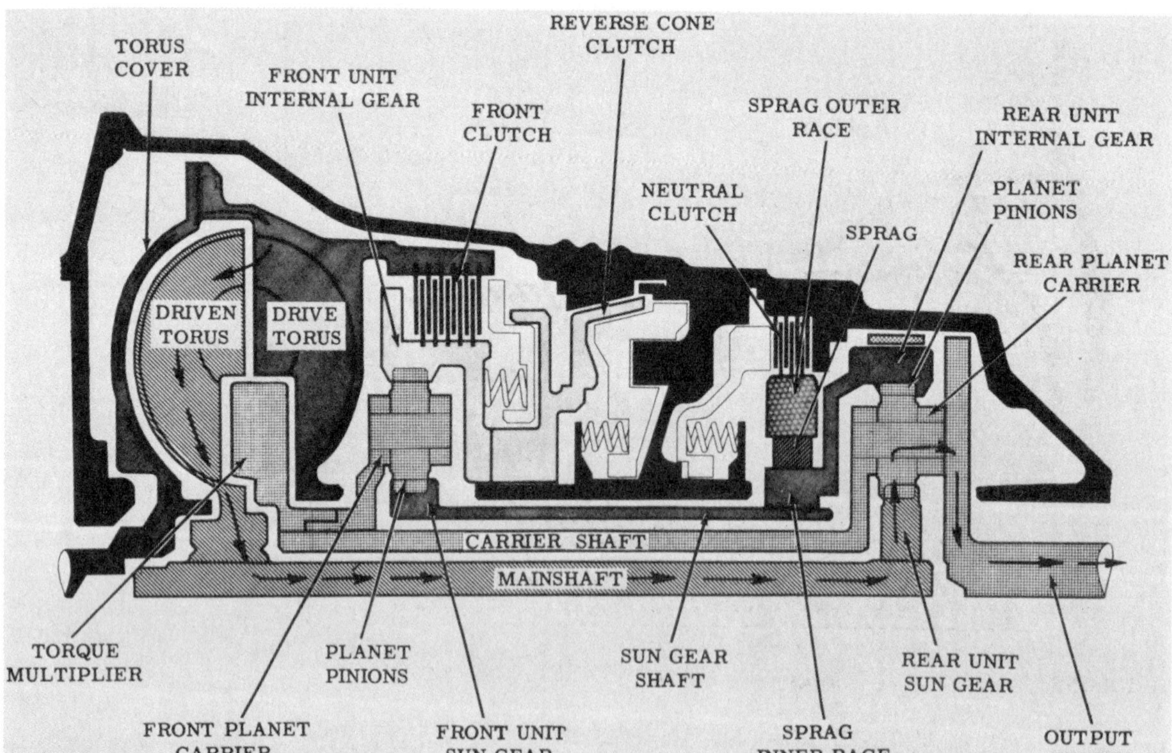

Figure 1-6. First gear. (*Courtesy of General Motors Corporation*)

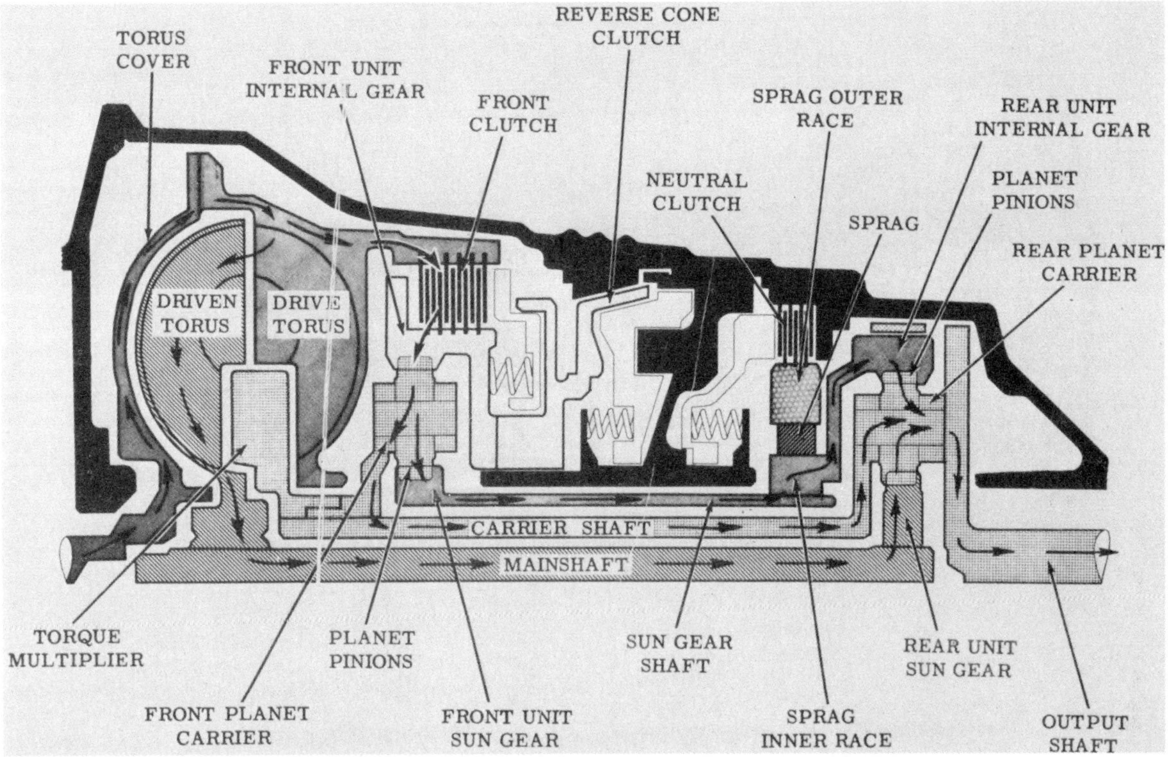

Figure 1-7. Second gear. (*Courtesy of General Motors Corporation*)

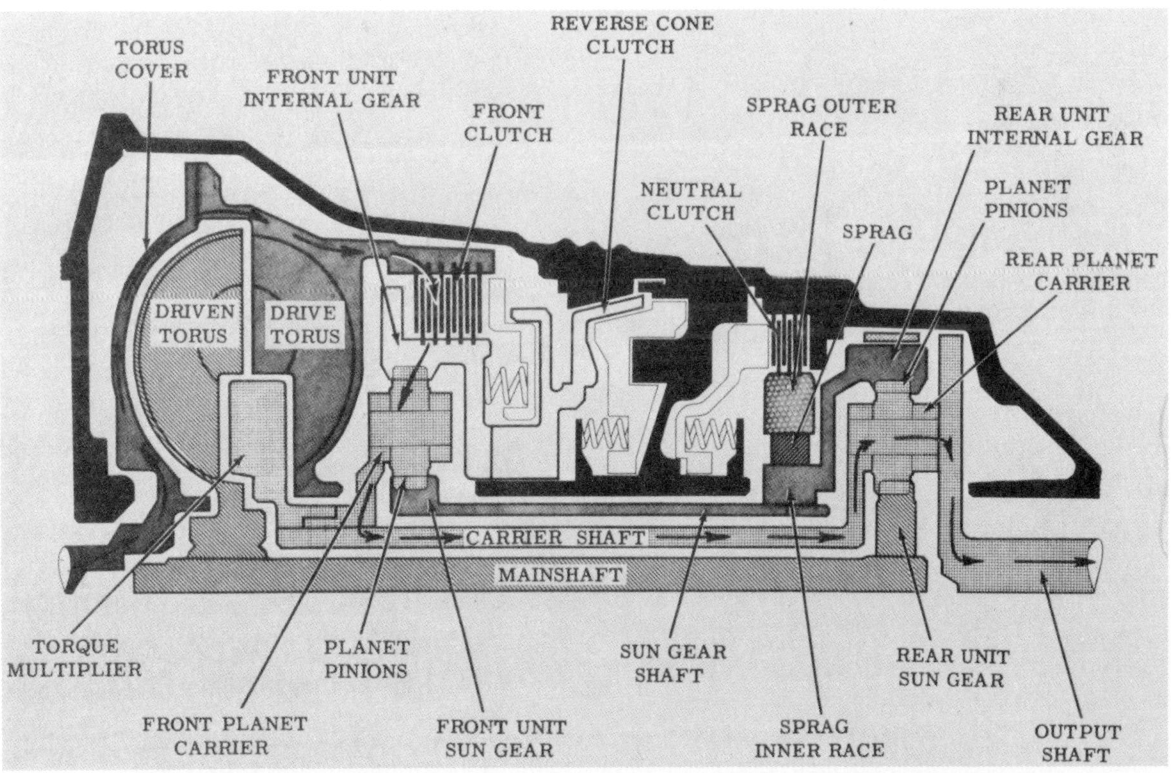

Figure 1-8. Third gear. (*Courtesy of General Motors Corporation*)

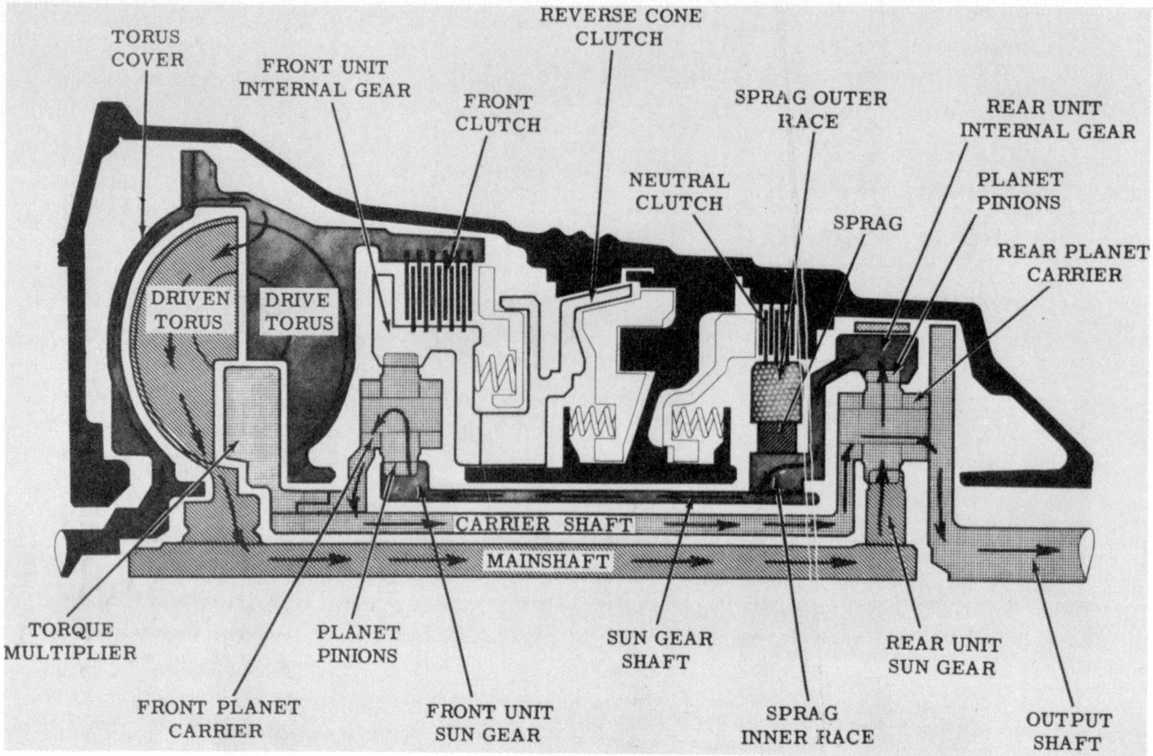

Figure 1-9. Reverse. (*Courtesy of General Motors Corporation*)

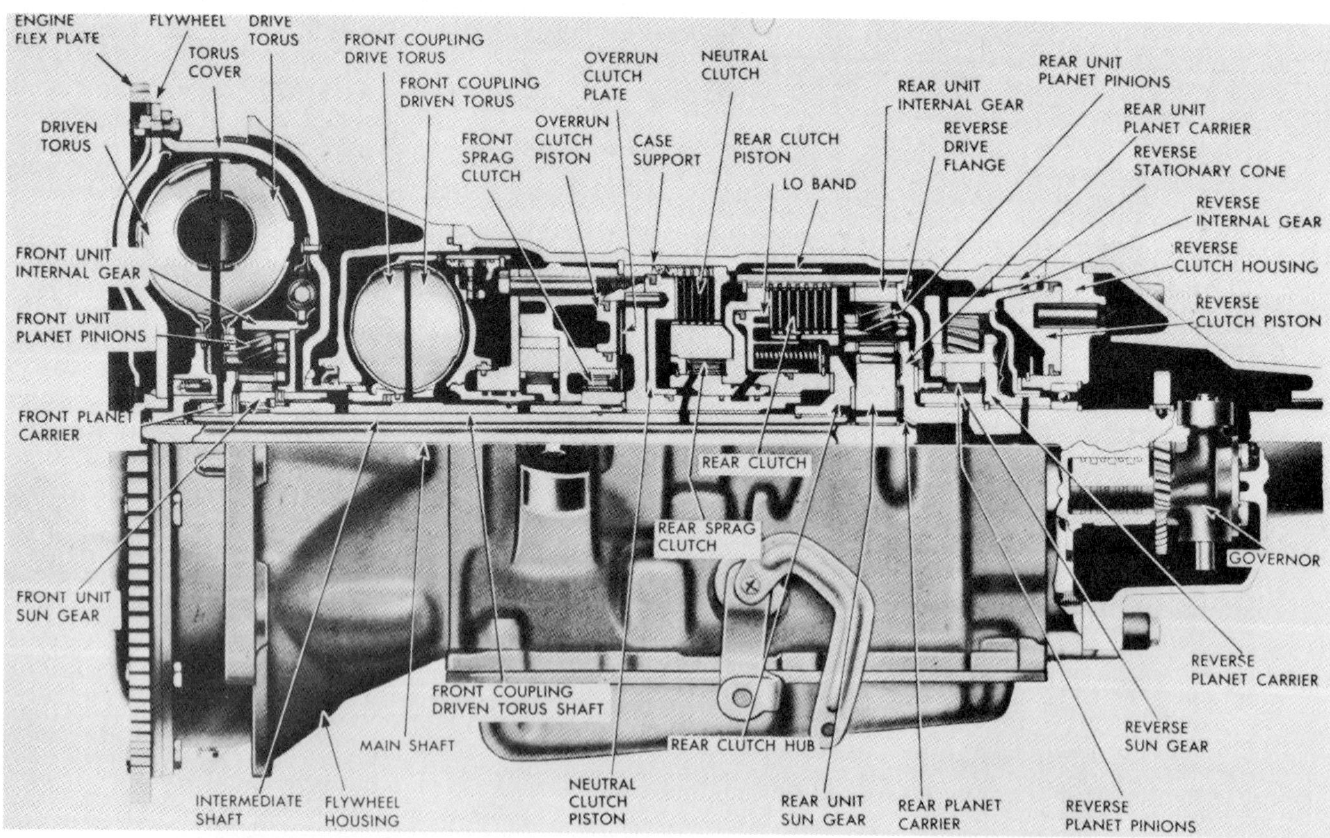

Figure 1-10. Cross-sectional view of dual-coupling four-speed Hydra-Matic transmission used in Cadillacs, Oldsmobiles, and Pontiacs in the early 1960s. (*Courtesy of General Motors Corporation*)

The Super Turbine 400 transmission featured a variable-pitch stator in its torque converter. This provided a wider range of torque converter ratios than the fixed-vane stator type of torque converter. This transmission became available on Buicks in 1964.

The Turbo Hydra-Matic 300 series of transmissions with a three-member fixed-vane stator torque converter and Simpson planetary gear system become available in 1964 and was used on many larger-engine General Motors automobiles. Both the Super Turbine 400 and the Turbo Hydra-Matic 400 transmissions were fully automatic three-speed transmissions and enjoyed great popularity for many years.

General Motors Dynaflow Transmissions

The Buick Division of General Motors made the first Dynaflow transmission available in 1948. This transmission used a five-element torque converter and a Ravigneaux planetary gearset to provide two forward speeds and reverse. Two bands and one multiple-disc clutch controlled the planetary gears. No automatic shifting was provided. The driver could select neutral, drive, low, or reverse.

The torque converter consisted of a primary and a secondary pump, a primary and a secondary reactor, and a turbine. The Ravigneaux gearset consisted of two sun gears, one carrier with two sets of pinions, and one ring gear.

Over the years the Dynaflow torque converter underwent many design variations, including a four-element converter with a planetary splitter gear and a five-element converter with a variable-pitch stator.

The two-speed Ravigneaux gearset was used throughout the series of Dynaflow transmissions, which ended with the 1963 version.

General Motors Powerglide Transmissions

The Chevrolet Powerglide appeared in 1950 as a smaller version of the Buick Dynaflow transmission. The same five-element torque converter and Ravigneaux gearset were used. The cast-iron case was replaced with the aluminum-case Powerglide in 1962. The three-element torque converter replaced the five-element converter in 1953.

In 1957 the Turboglide version of the Powerglide design became available. A five-element variable-

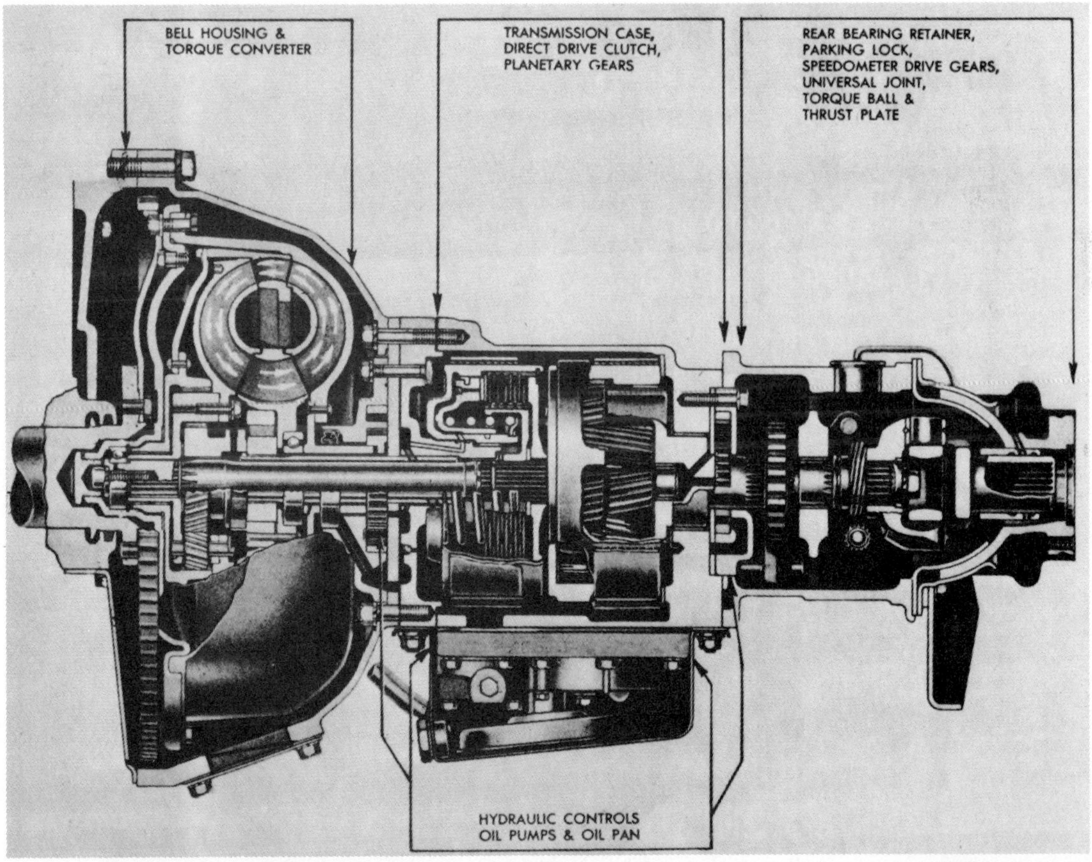

Figure 1-11. Buick Dynaflow two-speed transmission used in early 1960s. *(Courtesy of General Motors Corporation)*

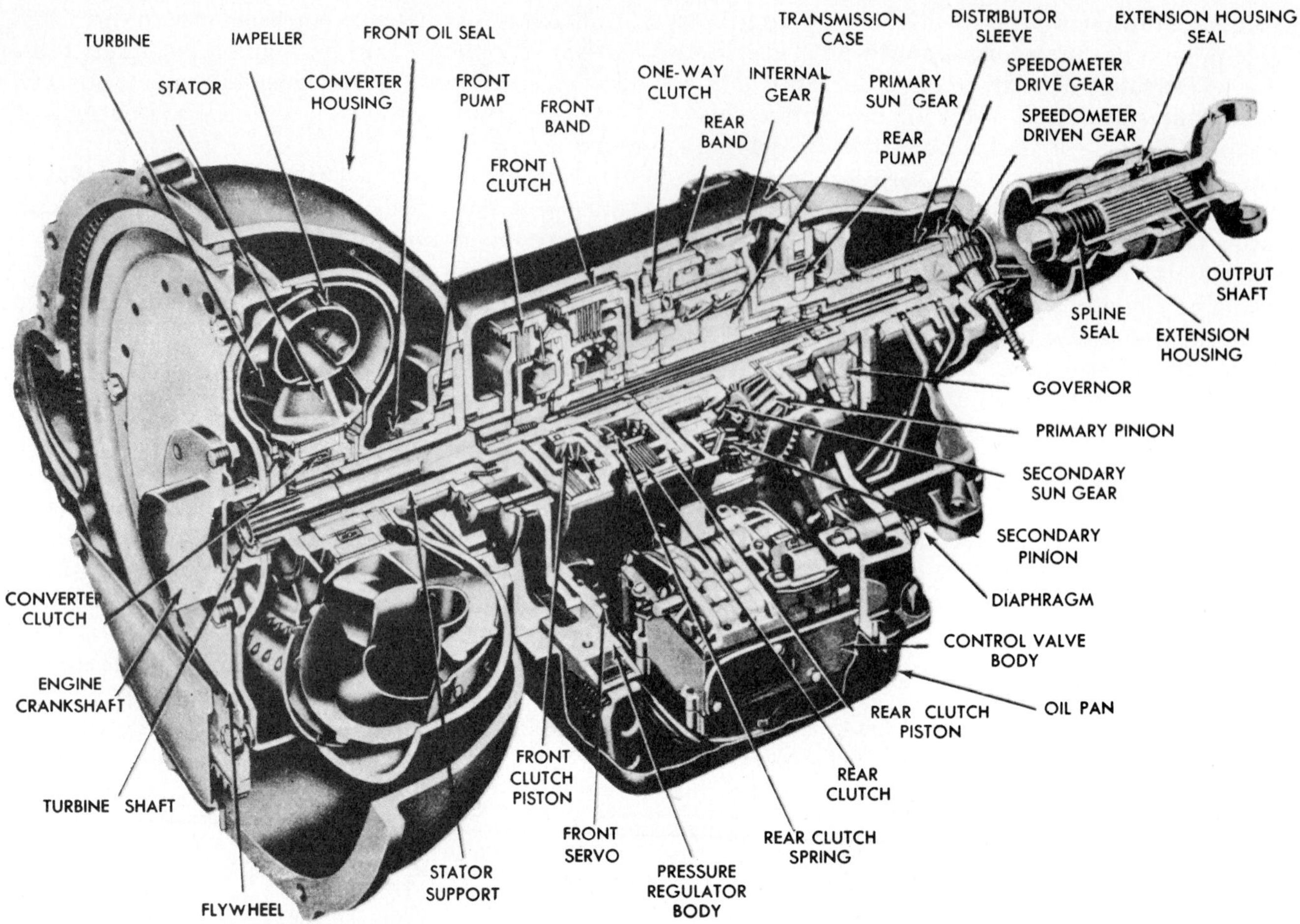

Figure 1-12. Typical Cruise-O-Matic three-speed transmission used in early 1960s by the Ford Motor Company. (*Courtesy of Ford Motor Company*)

pitch stator type of torque converter was used with this transmission. Turboglide transmission production ended in 1961.

Other versions of this two-speed transmission design included the Super Turbine 300, available from 1964 to 1969; the Jetaway, from 1964 to 1969; and the various Pontiac two-speed transmissions.

The last year for Powerglide transmission production was 1973, after which General Motors three-speed automatic transmissions took over.

Ford Motor Company Transmissions

The first automatic transmission used by Ford appeared in 1951. This was a three-speed transmission with a Ravigneaux planetary gearset and a three-element torque converter. The transmission was produced by Borg-Warner. This gearset is still used in many Ford transmissions.

The gearset was controlled by two bands and two multiple-disc clutches. This transmission was known as the Fordomatic or Mercomatic transmission, depending on the application. (Ford Motor Company's early Lincolns used the General Motors Hydra-Matic transmission.)

The Simpson planetary gearset was used in Ford's C4 and C6 automatic transmissions, which came on the scene in 1964 and 1966, respectively. Both these transmissions had three-element torque converters and an aluminum case and provided three forward speeds.

A two-speed version of the Fordomatic-Mercomatic transmission was used briefly on some smaller Ford products in 1963 and 1964.

Later versions of Ford's automatic transmissions include models using the Simpson gearset as well as the Ravigneaux gearset.

Chrysler Corporation Transmissions

Aside from Chrysler's early fluid coupling or torque converter and standard transmission combinations,

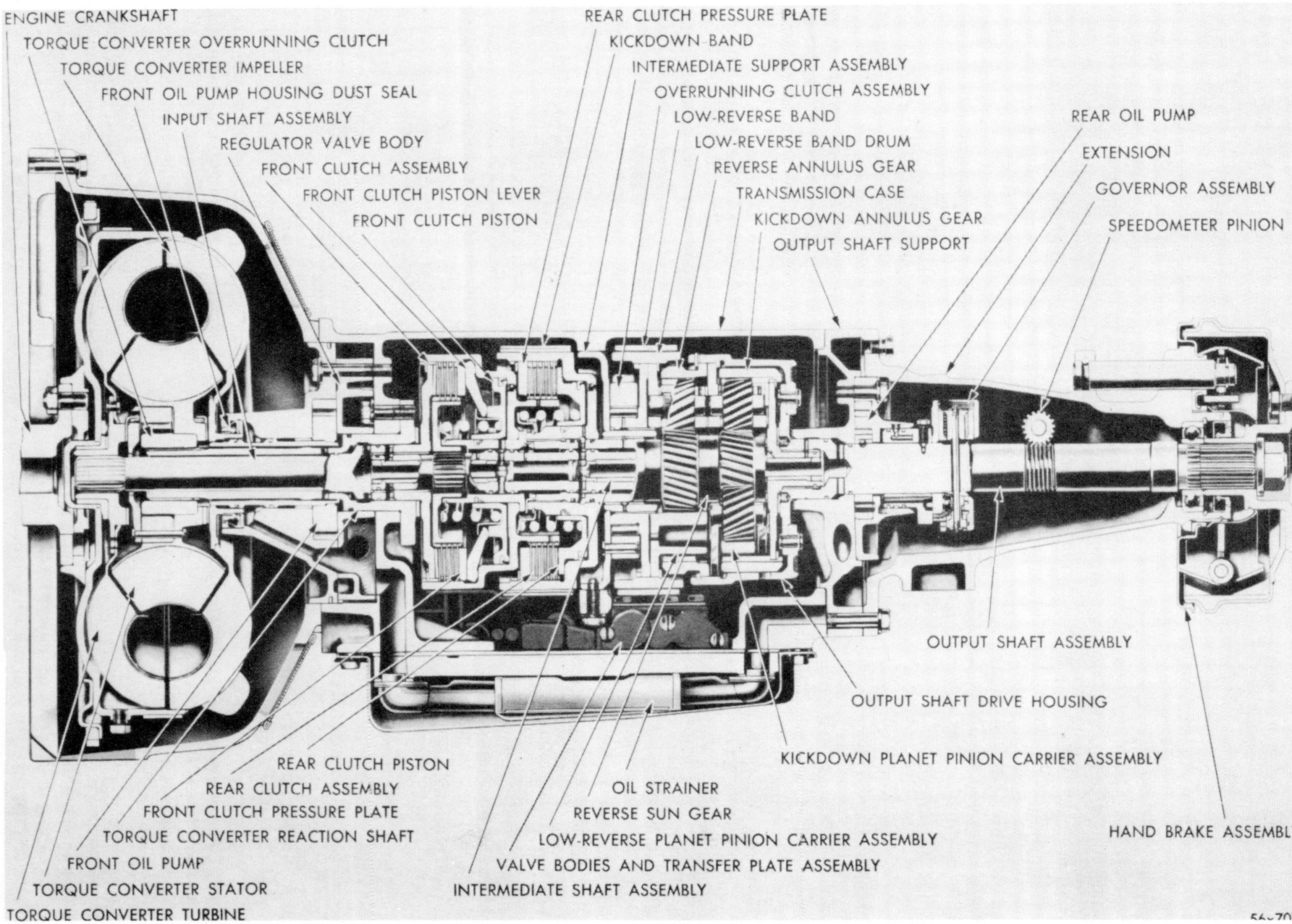

Figure 1-13. Early 1960s TorqueFlite three-speed automatic transmission by Chrysler. (*Courtesy of Chrysler Corporation*)

the first truly automatic Chrysler transmission came on the scene in 1953. This was the Powerflite transmission.

The Powerflite transmission provided two forward speeds and reverse. This was accomplished with a four-element torque converter and two planetary gearsets. Two bands and one multiple-disc clutch controlled planetary gear operation.

In 1956 the three-element torque converter was adopted and the three-speed TorqueFlite automatic transmission became available. Two bands, two multiple-disc clutches, and a one-way clutch controlled the Simpson planetary gearset. This arrangement is still the basis for present-day TorqueFlite automatic transmissions.

American Motors Transmissions

American Motors has relied heavily on other manufacturers to supply it with automatic transmissions for its vehicles. Rambler originally (1953) used the General Motors Hydra-Matic transmission.

Later the Flash-O-Matic and Shift Command versions of the Borg-Warner Mercomatic and Fordomatic transmissions were used. Since the early 1970s the Torque-Command version of Chrysler's TorqueFlite transmission has been used by American Motors vehicles.

Other Automatic Transmissions

The Ultramatic transmission used on the 1949 Packard automobile was equipped with a four-element torque converter. A unique feature of this torque converter was its direct-drive lockup clutch, which was operated by a hydraulic piston. A Ravigneaux two-speed gearset provided low and high ranges. Two bands and one multiple-disc clutch were used to control gearset operation.

The Studebaker automatic transmission of 1950 was a Borg-Warner transmission. A three-element torque converter and a three-speed Ravigneaux planetary gearset provided the necessary drive ranges. The torque converter was equipped with a direct-drive lockup clutch.

A unique feature of this transmission was an anti-rollback device, which prevented the vehicle from rolling back when stopped at a stop sign on an upgrade as long as the transmission was in the D position with the engine running.

Packard and Studebaker, however, did not survive much past the middle of the 20th century.

PART 1-4 RECENT DEVELOPMENTS

Torque Converters

For many years, car manufacturers have used the three-member torque converter as standard for the automatic transmissions in their passenger cars. Some are equipped with a lockup clutch to provide direct drive through the converter, thereby improving fuel economy. Without the lockup clutch–direct-drive feature, the torque converter is only approximately 90 percent efficient in the coupling stage.

Another method of improving converter efficiency is to use a splitter gear built into the converter. This arrangement provides for some mechanical drive to take place through the converter, thereby improving efficiency to approximately 99 percent.

Still another method of improving efficiency through the converter is to provide two power outputs from the converter—one from the turbine through the turbine shaft and the other from the converter housing (internally) through a spring-dampened plate to a direct-drive shaft. This arrangement provides several combinations of mechanical and hydraulic drive to the transmission gear train, depending on selector lever position and on the gear (reverse, first, second, third, or fourth) of the transmission. In reverse, for example, only hydraulic drive is provided by the converter to the gear train. In fourth-gear overdrive, only mechanical drive is provided, and in third there is a 40 percent–60 percent split between hydraulic and mechanical drive through the converter.

For more information on the construction and operation of the different types of torque converters, see Chapter 2.

Planetary Gears

Passenger-car automatic transmissions have adopted the use of two types of planetary gear systems for the actual gearbox. These two compound planetary gearsets are known as the Simpson gearset and the Ravigneaux gearset. Some automatic transmissions may be equipped with additional planetary gears in the converter to improve converter efficiency or to function as an overdrive unit.

The Simpson planetary gearset consists of two simple planetary gears operating separately or together, depending on the transmission gear selected. The Ravigneaux planetary gearset consists of two separate sun gears, one carrier with two sets of pinions, and one ring or internal gear.

For more information on the construction and

operation of planetary gear systems, see Chapters 2 and 4.

Transmission Design

Automobile design is a major factor in transmission design. The front engine–rear wheel drive automobile has a transmission and torque converter attached to the rear of the engine. A drive shaft transmits engine torque and motion to the rear axle, which houses the differential. The differential splits drive to the two drive wheels through the differential and drive axles, which may be enclosed or open.

The front engine–front wheel drive automobile has a transmission-torque converter assembly that includes the final drive and differential in one assembly. Open drive axles or halfshafts transmit drive from the differential to the front drive wheels. This type of unit is commonly referred to as a transaxle, which is a contraction of the two words *transmission* and *axle*. Another version of the transaxle similar in basic design is the one used in rear engine–rear wheel drive automobiles.

Similarities

Regardless of the type of automatic transmission, the basic function and operation of each of their components is similar in all of them. Torque converter principles and operation are always the same whether in a transaxle, front-wheel-drive, or a rear-wheel drive automobile.

Planetary gears operate on the same principles regardless of transmission design or vehicle make. Hydraulic systems and valving use the same principles to achieve the desired results in every make and model of automatic transmission.

Once these principles have been mastered, they are easily applied to any make or model of transmission. Understanding changes in design then becomes relatively simple. These same principles apply equally well to domestic automobiles and to foreign or import vehicles.

PART 1-5 MAJOR COMPONENTS AND THEIR BASIC FUNCTIONS

The study of the automatic transmission is simplified considerably when the major components and systems are first considered separately. When this is done, the purpose and function of each of the major components are more easily understood. This also facilitates comprehending how the automatic transmission functions as a whole. Let's consider the basic function of each of these major components and systems one by one.

The Torque Converter

The torque converter is a hydrodynamic unit situated between the engine and the transmission gear train. The term *hydrodynamic* basically means that the torque converter is a unit of fluid (hydro) in motion (dynamic). In other words, motion is transmitted from the engine to the transmission by means of a fluid rather than by a mechanical connection. It can also be described as a fluid coupling since engine crankshaft rotating force is coupled to the transmission by means of hydraulic fluid. But it is more than just a fluid coupling, hence comes the name *torque converter*—a unit that converts or changes engine torque output.

The purpose of the torque converter is to:

1. Transmit engine torque from the engine to the transmission;
2. Multiply engine torque (convert engine torque to an increased value) and reduce speed as required under certain conditions;
3. Absorb drive train and engine torsional vibrations;
4. Act as an automatic clutch;
5. Drive the transmission's hydraulic pump;
6. Smooth engine operation by acting as the engine flywheel; and
7. Provide the means for mounting the starter ring gear in some cases; in others the ring gear is mounted on the converter drive plate.

The torque converter obviously has a big job to do. Chapter 2 explains how the torque converter does all these jobs and what can happen to it so that it sometimes cannot do everything that it was designed to do.

The Gear Train

The gear train consists primarily of the planetary gears and shafts, which connect the torque converter to the drive shaft or final drive. The gear train in most automatic transmissions is connected to or disconnected from the torque converter by means of hydraulically operated clutches. Other transmissions may have a direct mechanical connection between the torque converter and the gear train (more about this in Chapter 3).

The purpose of the gear train in the automatic transmission is to:

1. Provide a number of forward speeds to match driving conditions and the driver's desire (lower speed gears provide decreased speed and increased torque for heavier loads; higher speed gears provide less torque and increased speed for fuel economy);

2. Provide a reverse gear with speed reduction to move the vehicle in reverse;

3. Provide a neutral gear position to allow the engine to run when the vehicle is stationary (without causing unnecessary torque converter action);

4. Provide a park position, which locks the gear train output shaft and drive wheels into a stationary position to prevent the vehicle from moving when parked.

All of this is accomplished by means of the hydraulic control system and the manual control system. Chapter 3 explains the various designs of planetary gear trains used in automatic transmissions and how each of these gear trains works.

The Hydraulic Control System

A hydraulic system is a method of transmitting force and motion using a fluid as the medium. A basic hydraulic system requires a pump, a valve, a servo piston, and some kind of fluid. The pump puts the fluid in motion and transfers pump piston force to the fluid. The fluid force acts on the servo piston by applying a force or pressure against it. The servo piston will move when sufficient force (pressure) is applied to it. The valve prevents fluid from returning to the pump on the pump intake stroke. This is a simple check valve. A second valve may be used to release or apply pressure when desired. A simple hydraulic system such as this is used in the common hydraulic jack.

The hydraulic control system in the automatic transmission uses the same hydraulic principles to control the planetary gear system. The basic system includes the fluid, an engine-driven pump, a number of valves, and the servo pistons that operate the clutches and bands. Some of the valves are controlled by the driver of the vehicle while others operate automatically.

The hydraulic control system in the automatic transmission has several tasks to perform. These are:

1. To provide hydraulic fluid for torque converter operation;

2. To provide hydraulic pressure to operate the clutches and bands that control planetary gear operation;

3. To provide lubrication for all torque converter and transmission parts; and

4. To provide cooling to the torque converter and transmission.

How the hydraulic system does each of these jobs is explained in Chapters 5, 6, and 7. Proper operation of the transmission's hydraulic control system is the key to troublefree automatic transmission operation.

Manual Controls

How the automatic transmission functions depends to a large extent on the driver of the vehicle or automobile. Two controls that the driver may use will determine whether and when the transmission will upshift or downshift and whether the vehicle will move forward or in reverse. These two controls are (1) the selector lever—either floor-mounted (console shift) or steering-column–mounted—and (2) the accelerator pedal.

By moving the selector lever, the driver can place the transmission in the park position, reverse position, neutral position, or one of two or three forward operating ranges, depending on transmission design. Selecting the manual low position will allow the vehicle to operate in low gear only; no upshifts will take place. On most vehicles this is the lever position farthest to the rear on console shift systems and the lowest level position on column shift systems.

Selecting the next position forward (on the console shift) or upward (on the column shift) on most three-speed automatic transmissions will allow the transmission to be operated only in first and second gears on most designs or only in second gear on other designs. Selecting the next position (to the right of or behind the N or neutral position) will allow the transmission to operate in all three forward speeds on many designs with automatic 1–2 and 2–3 upshifts and 3–2, 2–1, and 3–1 downshifts. Other designs may operate in three or four forward gears with a 2–3 and 3–4 automatic upshifts or 1–2, 2–3, and 3–4 automatic upshifts and downshifts from 4–3, 3–2, 2–1, or similar combinations.

The other control that the driver may use is the accelerator pedal. Although the accelerator pedal controls the speed of the engine, it also has a direct effect on automatic transmission operation. *Light throttle* and *heavy throttle* or *light acceleration* and *heavy acceleration* are terms used to describe how the driver moves the accelerator pedal or gas pedal, as it is sometimes called in spite of diesel engines being used in some vehicles.

With the transmission lever in the D or drive position and with light acceleration, the transmission will upshift automatically at relatively low road speeds. Under heavier or more rapid acceleration, the transmission will upshift later or at higher road speeds.

Once the transmission is in high gear and cruising along at normal road speeds, the transmission may be downshifted by suddenly depressing the accelerator pedal to one of two positions—to detent or through detent. The transmission will downshift from 3 to 2 or from 3 to 1, depending on accelerator operation, vehicle road speed, and transmission design.

For more detail on these manual controls, which are in fact part of the hydraulic control system, see Chapters 6 and 7.

The Transmission Case

The transmission case contains all of the internal parts of the automatic transmission. This includes the torque converter, planetary gear system, and most of the hydraulic control system.

The transmission case usually consists of (1) the bell housing, which contains the torque converter; (2) the case, which contains the planetary gear system and most of the hydraulic control system; (3) the oil pan, which contains a reservoir of fluid; and (4) the extension housing, which contains the output shaft and related parts. In the case of front-wheel-drive vehicles usually no extension housing is needed but the case may also contain the final drive and differential assembly.

Since all the control linkages and the input and output shafts must enter the case at some point, a number of oil seals are used at these points to contain the hydraulic fluid and prevent the entry of dirt or other foreign material. The case is also vented to the atmosphere to allow air to enter and exit as required.

Older transmission cases were made of cast iron and were quite heavy. Modern transmission cases are made of lighter aluminum alloys.

Part 1–6 Worksheet

1. In your own words write why the automatic transmission was developed.
2. Name three early General Motors automatic transmission types, one early Ford automatic transmission, one early Chrysler automatic transmission, one early American Motors automatic transmission, and two other early automatic transmissions.
3. Go to your local public library or your school library to find some information on early automatic transmission development. List the title of each book, the author, the publisher, and the pages on this subject. Note anything unique and interesting.
4. Describe the basic function of each of the major components of the automatic transmission as outlined in this chapter.
5. Complete the self-check at the end of this chapter with at least 80 percent accuracy.
6. Read the performance evaluation section and decide whether you have met the performance objectives given at the beginning of this chapter.

PART 1-7 SELF-CHECK

1. A combined transmission, final drive, and differential unit is called a ____________.
2. Why is a clutch needed in an automobile?
3. The earliest planetary gear transmissions were manually controlled. True or false?
4. Early General Motors automatic transmissions were known as ____________ transmissions.
5. The Super Turbine 400 and Super Turbine 300 GM transmissions featured a ____________ ____________ stator.
6. The Dynaflow transmission was used on Cadillac automobiles. True or false?
7. How many forward speeds did the General Motors Powerglide transmission feature?
8. Ford's automatic transmissions use two different types of planetary gearsets. What are they called?
9. Ford's early Lincolns used a General Motors automatic transmission. True or false?
10. Chrysler's Powerflite and Powerglide transmissions came on the scene in 1953. True or false?
11. How many forward speeds does the TorqueFlite transmission have?
12. Name two automobiles no longer being produced that also featured automatic transmissions.
13. Variations of the three-member torque converter design include those with a lockup ____________ ____________ and those with a ____________ ____________ gear.
14. List six functions of a torque converter.
15. What is the purpose of the planetary gear train?
16. List four functions of the hydraulic control system.
17. What two manual controls over the automatic transmission can the driver use?

PART 1-8 PERFORMANCE EVALUATION

The degree to which the student is asked to fulfill the requirements of stated performance objectives is usually determined by course requirements and the class instructor.

The student, however, should answer the following questions based on performance required.

1. Did I state the need for the automatic transmission?
2. Did I describe early automatic transmissions briefly?
3. Did I name the most common early automatic transmissions?
4. Did I briefly describe the basic functions of the five major components of the automatic transmission?
5. Did I complete the worksheet for this chapter?
6. Did I complete the self-check for this chapter with at least 80 percent accuracy?

Chapter 2

Torque Converters

Performance Objectives

After a thorough study of this chapter and the appropriate training models, you should be able to perform these tasks:

1. State the six basic functions of the torque converter.
2. Describe basic three-member torque converter construction.
3. Describe the operation of the three-member torque converter, including torque multiplication, coupling stage, acceleration, and deceleration.
4. Describe hydraulic clutch torque converter lockup operation.
5. Describe centrifugal clutch torque converter lockup operation.
6. Describe splitter gear torque converter operation.
7. Describe torque converter wear and damage, including causes and effects.
8. Complete the worksheet for this chapter.
9. Complete the self-check for this chapter with at least 80 percent accuracy.
10. Evaluate your performance by using the performance evaluation section of this chapter.

PART 2-1 INTRODUCTION

This chapter deals with the purpose, construction, and operation of the three-member torque converter as used in automotive-type automatic transmissions. Included is a discussion of such features as the hydraulic lockup clutch, the centrifugal lockup clutch, and the planetary splitter gear used in some torque converters. The causes and effects of torque converter wear and damage are also covered.

A good understanding of the purpose, construction, and operation of the torque converter is necessary for the modern transmission technician in order to diagnose torque converter problems intelligently as well as to service a defective torque converter properly. This requires taking advantage of every learning opportunity available in the text and in class.

PART 2-2 PURPOSE AND FUNCTION

The torque converter is a type of fluid coupling located between the engine and the transmission. Sometimes referred to as a hydrodynamic unit, (meaning fluid in motion), the torque converter transmits engine crankshaft torque and motion (rotation) to the input shaft of the transmission through the medium of a fluid. Let's examine what the torque converter must do to provide proper vehicle operation.

The purpose of the torque converter is to:

1. Transfer engine torque,
2. Multiply engine torque,
3. Dampen torsional vibrations,
4. Act as an automatic clutch,
5. Drive the transmission hydraulic pump,
6. Act as the engine flywheel, and
7. Provide the means for mounting the starter gear.

Transferring Engine Torque

Torque is a twisting or turning force. Torque imparted by the engine's pistons and connecting rods causes the engine crankshaft to rotate. The torque of the rotating crankshaft is transmitted or transferred to the input shaft of the transmission by the torque converter. If sufficient engine torque is produced and transmitted to overcome the load of the stationary vehicle, the vehicle will start to move; if in motion, it will continue moving.

At first glance it would seem that all the torque produced by the engine is being transmitted to the transmission. However, this is not the case. Some loss of torque occurs in the process. This is discussed in Part 2–3.

Multiplying Engine Torque

Torque is multiplied or increased by the torque converter under certain operating conditions, which are discussed in Part 2–3. Torque input to the converter can be multiplied in the converter to provide an increased torque output. There is, however, a corresponding decrease in rotating speed of the torque converter output as compared to input speed.

The torque converter is basically an engine-driven pump throwing fluid against the vanes of the torque converter turbine. The engine-driven pump part of the torque converter is known by several other names such as drive member, drive torus, and impeller. The turbine is also known as the driven member or driven torus. However, to make this unit a torque converter rather than just a fluid coupling, a third member is required: the stator.

The stator is required to maintain control of fluid flow returning from the turbine to the impeller. Without it fluid flow from the turbine would act against impeller rotation, which would reduce the efficiency of the unit during certain operating conditions.

Dampening Torsional Vibrations

Every time any of the engine's cylinders fires, the engine crankshaft attempts to speed up the rotating speed of the torque converter. This results in torsional pulsations being transmitted to the torque converter. If undampened, these torsional pulsations are transmitted through the drive train, transmission, and differential. The shock of continuously repeated pulsations is hard on gears and other drive train parts.

Since the standard three-member torque converter (without direct-drive capability) relies entirely on fluid for torque transfer, torsional vibrations are dampened and absorbed by the fluid in the torque converter. Torque converters with direct-drive capabilities are equipped with torsional springs, which dampen torsional vibrations during direct-drive operation.

Acting as an Automatic Clutch

The torque converter allows repeated stop-and-go vehicle operation without moving the gearshift lever from neutral to drive and back again every time the vehicle comes to a stop.

This automatic clutch-operating feature of the torque converter is possible since the impeller (drive member) and turbine (driven member) can turn independently. There is no mechanical drive connection between the two. With the transmission selector lever in the drive position, the driver can apply the brakes to stop the vehicle with the engine still running. This means that the engine crankshaft and the impeller are still turning while the turbine and transmission input shaft have stopped.

Although the impeller is still turning and transmitting engine torque through the fluid to the turbine, sufficient torque is not being produced to force the turbine and transmission input shaft to turn. As the driver releases the brakes and speeds up the engine, increasing fluid velocity increases torque transfer and causes the turbine to turn and the vehicle to move once again.

Driving the Hydraulic Pump

Since hydraulic pressure and force are used to operate the clutches and bands in the automatic transmission, a hydraulic pump is required to provide the fluid pressure. This hydraulic pressure is required whether the vehicle is stationary or in motion. This means that the hydraulic pump must provide hydraulic pressure whenever the engine is running or whether the vehicle is moving or stopped. This is achieved by having the torque converter drive the hydraulic pump by means of a direct mechanical connection.

The torque converter is attached to the engine crankshaft by means of a flex plate or drive plate. The drive plate is bolted to a flange on the crankshaft. The torque converter is bolted to the flex plate. The rear hub of the torque converter is designed to engage the hydraulic pump by means of lugs and notches or by flat sections on the hub and pump rotor.

Consequently whenever the engine is running, the hydraulic pump provides hydraulic pressure to the transmission and torque converter.

Acting as the Engine's Flywheel

A flywheel is a heavy metal disc attached to the engine's crankshaft. The heavy mass of metal is required to help smooth engine operation. It also provides the means for mounting the starter ring gear and provides one of the driving friction surfaces for the clutch on manual-transmission-equipped vehicles.

On automatic-transmission-equipped vehicles the torque converter serves this purpose. The torque converter filled with fluid and the flex plate or drive plate provide the mass needed to smooth the engine's power impulses. The starter ring gear may be welded directly to the torque converter, or it may be welded to the flex plate.

PART 2-3 THREE-MEMBER TORQUE CONVERTER

Since the two-member fluid coupling and the four- and five-member torque converters have not been in use in passenger cars and light trucks for many years, this discussion will be confined to the common three-member torque converter and the various direct-drive features of torque converters.

Major Components

The three-member torque converter consists of (1) the impeller, (2) the turbine, and (3) the stator assembly. These three members are enclosed in a heavy, steel, fluid-filled case or cover. This assembly is bolted to a flex plate (converter drive plate), which is bolted to the engine crankshaft. The torque converter is supported and centered at the front by a hub, which fits into a close-fitting recess in the engine crankshaft. The rear hub of the converter is supported by a bushing mounted in the hydraulic pump housing.

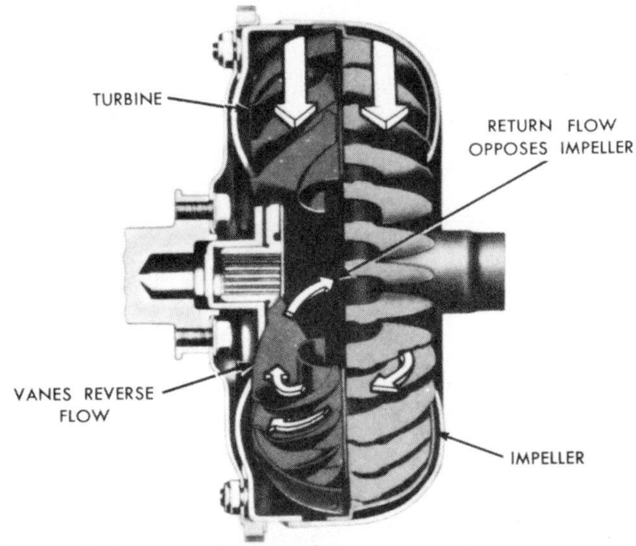

Figure 2-1. Fluid flow from impeller to turbine (turbine is splined to transmission input shaft). (*Courtesy of Ford Motor Co. of Canada Ltd.*)

The Impeller

The impeller is the drive member of the torque converter. Its job is to put the fluid in motion and to direct that fluid against the vanes of the turbine. It is in fact a centrifugal type of fluid pump.

The impeller consists of a series of vanes arranged radially and attached to the inside of the rear half of the converter cover. The cover and impeller rotate with the engine crankshaft.

With the engine running, the rotating impeller

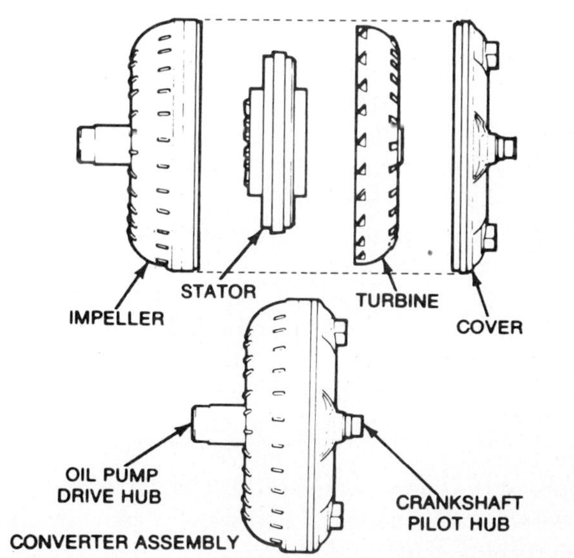

Figure 2-2. Torque converter components (*Courtesy of Ford Motor Co. of Canada Ltd.*)

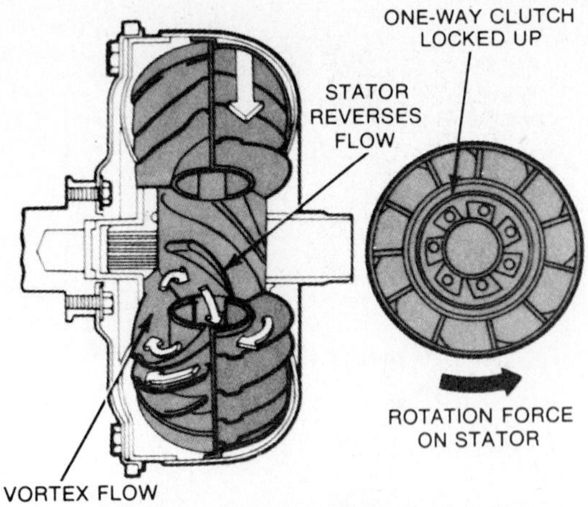

Figure 2-3. Fluid flow from turbine is redirected by stator vanes (during torque multiplication) to same direction as impeller. This aids impeller rotation. This flow is called vortex flow. (*Courtesy of Ford Motor Co. of Canada Ltd.*)

causes the fluid to rotate with the impeller vanes. At the same time due to centrifugal force acting on the fluid, the fluid flows away from the axis or center of rotation. Because of the curved design of the converter cover and the guide ring, the fluid is directed forward to the turbine vanes at the same time as it is being forced to rotate in a clockwise direction.

The Turbine

The turbine is the driven member of the torque converter. The turbine is situated inside the torque converter cover but is free to turn independently of the impeller and cover.

The turbine is designed to absorb the force of the moving fluid and convert it to rotary motion. The turbine is splined to the transmission input shaft, which extends from the transmission into the torque converter.

Therefore when sufficient fluid force (caused by the impeller) is applied to the turbine, the turbine and transmission input shaft (turbine shaft) will start to turn. If the transmission is in a forward or reverse gear position, the vehicle will start to move. If the transmission is in park or neutral, only the parts mechanically connected to the turbine shaft will turn with the turbine and the vehicle will not move.

The turbine vanes are arranged radially around the axis. The vanes are curved in order to absorb more of the energy of the fluid. The vanes can be considered as extensions of a series of levers attached to the turbine shaft.

The curved shape of the turbine shell (like half of a doughnut ring) and the smaller split guide ring confine the fast-moving fluid to passages between the vanes, the split guide ring, and the turbine outer shell. This is similar to fluid flow in the impeller. The difference is that fluid flow in the impeller moves away from the center of rotation whereas fluid flow between the turbine vanes moves toward the center of rotation.

The Stator Assembly (Reactor)

The stator and one-way clutch assembly are positioned between the impeller and the turbine. The stator consists of a series of curved vanes arranged radially around the one-way clutch assembly. The one-way clutch allows the stator to rotate in the same direction (clockwise) as the impeller and turbine but prevents rotation in the opposite direction.

The stator is smaller in diameter than the turbine and impeller. The fluid passages between the stator vanes are positioned to accept fluid flow from the inner openings of the turbine and direct that fluid back to the impeller in the same direction as impeller rotation. Without the stator, fluid leaving the turbine would oppose impeller rotation.

The stator and one-way clutch assembly are mounted on a tubular extension of the transmission pump. This extension is part of the pump housing and is bolted to the transmission case. The tubular extension known as the reaction shaft or stator shaft extends into the torque converter. Its external splines fit into the internally splined inner race of the one-way clutch.

When fluid flow strikes the stator vanes in a direction that tries to turn the stator backward or counterclockwise, the one-way clutch locks up and prevents the stator from turning. When fluid flow strikes the stator vanes in a direction that tries to turn the stator clockwise, the one-way clutch unlocks and overruns or freewheels, allowing the stator to turn in a forward or clockwise direction.

Torque Multiplication

As stated, as soon as the engine starts, the impeller and torque converter case turn at crankshaft speed. The driver moves the selector lever into position to put the transmission in gear.

The rotating impeller causes the fluid in the converter to flow in a rotary direction. This is called rotary flow. Centrifugal force also causes the fluid to flow away from the center of the rotating impeller and exit from the impeller at the openings around its outer edges.

The fluid exiting from the impeller enters the turbine at the openings around its outer edges and is

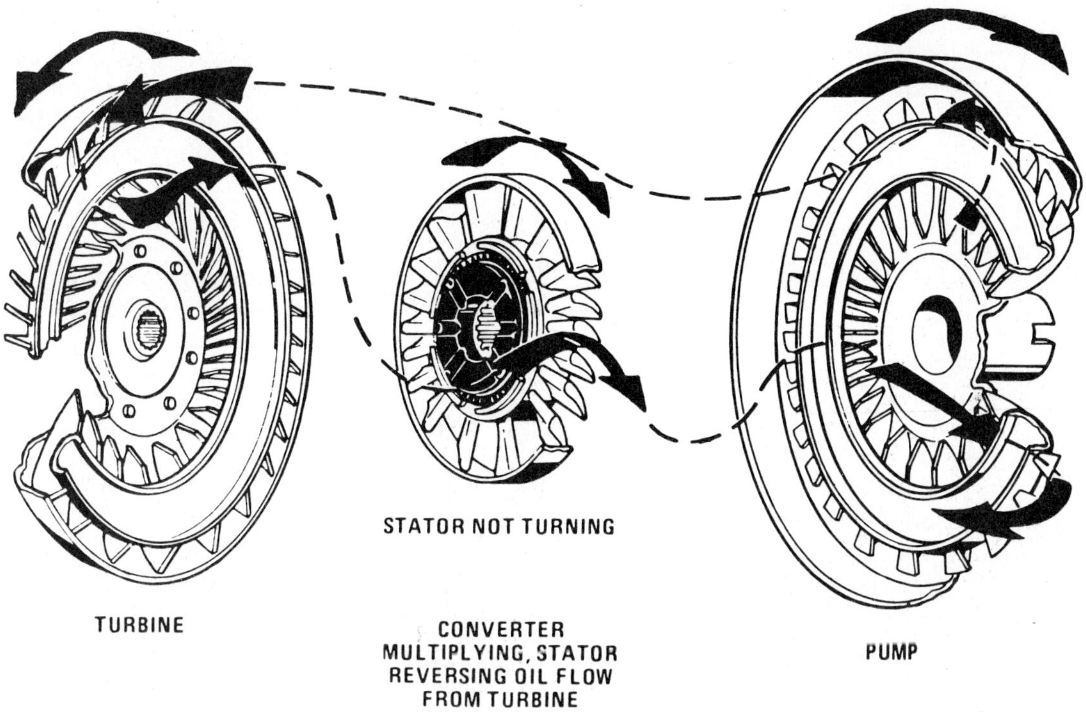

Figure 2-4. (*Courtesy of General Motors Corporation*)

forced to travel in a curved direction (due to curved turbine vanes) and exit at the inner openings of the turbine. Fluid leaving the impeller in a clockwise direction strikes the turbine vanes and attempts to turn the turbine in a clockwise direction.

Fluid leaving the turbine exits in a counterclockwise direction due to the curved turbine vanes. This counterclockwise-moving fluid strikes the stator vanes attempting to turn the stator backward (counterclockwise). This causes the one-way clutch to

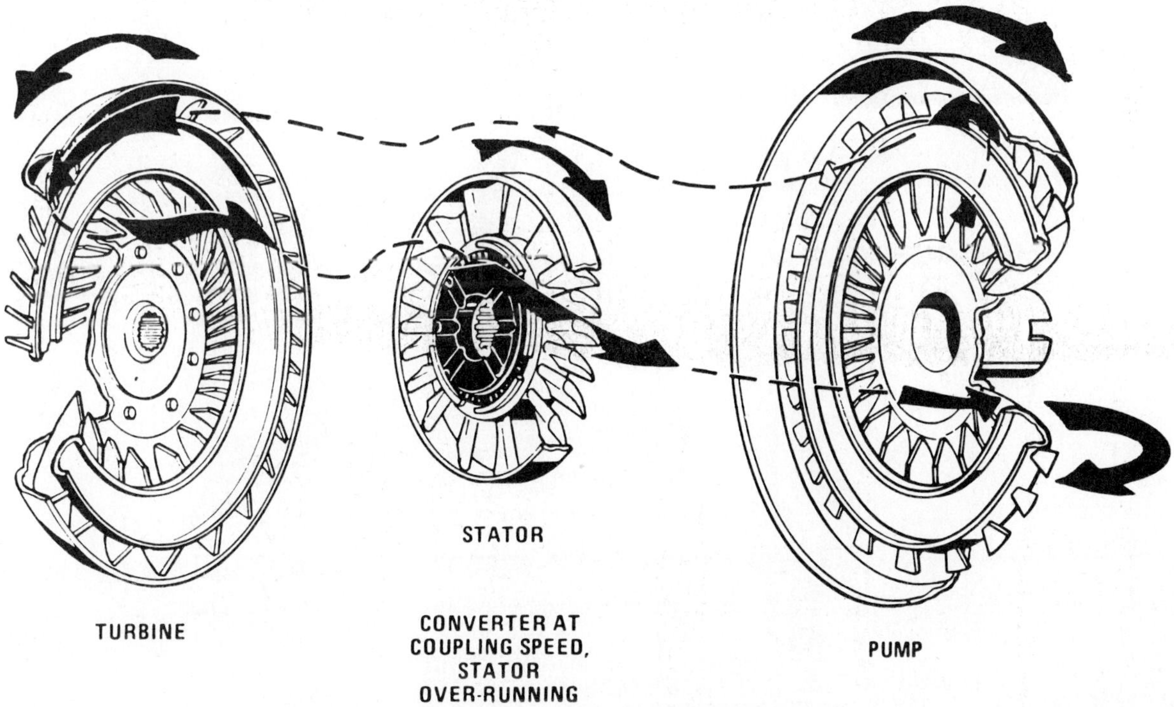

Figure 2-5. (*Courtesy of General Motors Corporation*)

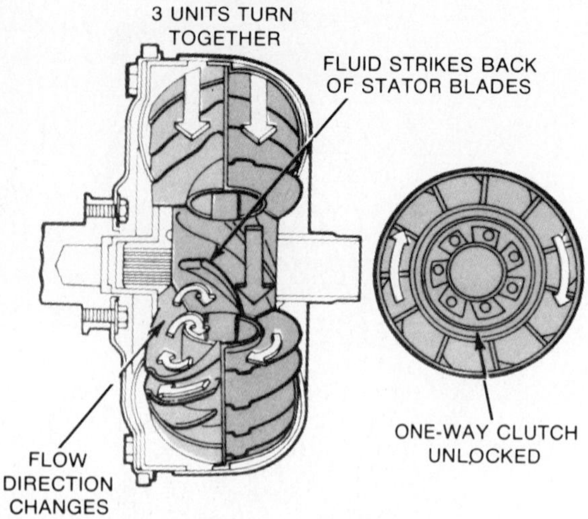

Figure 2-6. As the turbine approaches impeller speed, fluid from the turbine strikes the back side of the stator vanes, causing the entire assembly to rotate as a unit. (*Courtesy of Ford Motor Co. of Canada Ltd.*)

lock up and prevent stator rotation. The stator vanes being curved redirect the fluid into a clockwise direction before it enters the inner openings of the impeller to repeat the circulation from the impeller to the turbine, then to the stator and back to the impeller on a continuous basis. This is called vortex flow.

Since considerable energy is left in the fluid exiting from the stator and since it is moving in a clockwise direction as it enters the impeller, it in fact aids impeller rotation. This is torque multiplication.

As the driver speeds up the engine, impeller rotation and fluid flow are speeded up. When the fluid leaving the impeller strikes the vanes of the turbine with sufficient force, the turbine and turbine shaft start to turn, causing the drive wheels to turn and move the vehicle.

Under light acceleration relatively low torque is delivered to the drive wheels. Under heavy acceleration high torque is delivered to the drive wheels to accelerate the vehicle rapidly or to move heavy loads such as trailers.

Fluid flow caused by the combination of rotary and vortex flow results in a rotating corkscrew pattern in the torque converter.

Torque Converter Stall

With the transmission in gear, the engine running and the vehicle stationary, the condition in the torque converter is known as stall. Stall can also be described as the condition in the torque converter when the impeller is turning and the turbine is stationary or stalled. Stall can occur at any throttle position.

Stall speed refers to the speed of the engine when the transmission is in gear, the vehicle is held stationary by a very firm application of the brakes, and the engine is operated at wide open throttle.

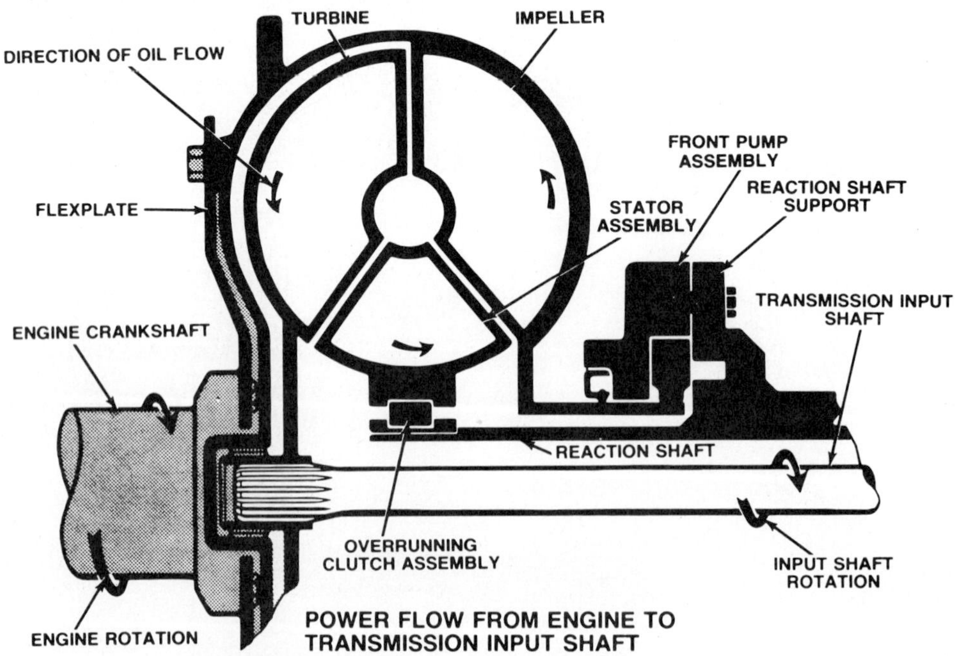

Figure 2-7. (*Courtesy of Chrysler Corporation*)

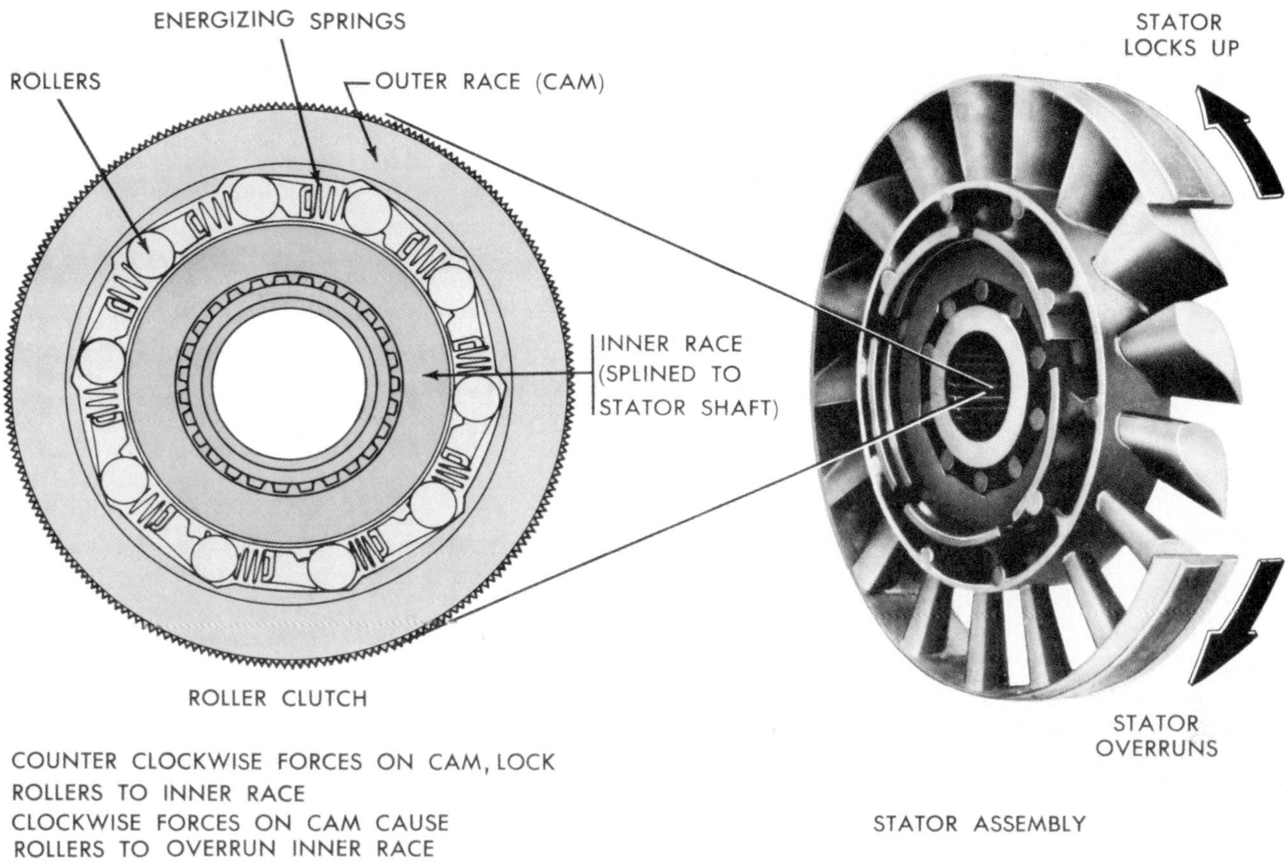

Figure 2-8. (*Courtesy of General Motors Corporation*)

Caution: Never attempt this unless you are fully experienced and aware of all the consequences of this procedure. Never attempt this on any vehicle unless the vehicle manufacturer recommends this procedure. Never perform this procedure under any conditions other than those recommended by the vehicle manufacturer.

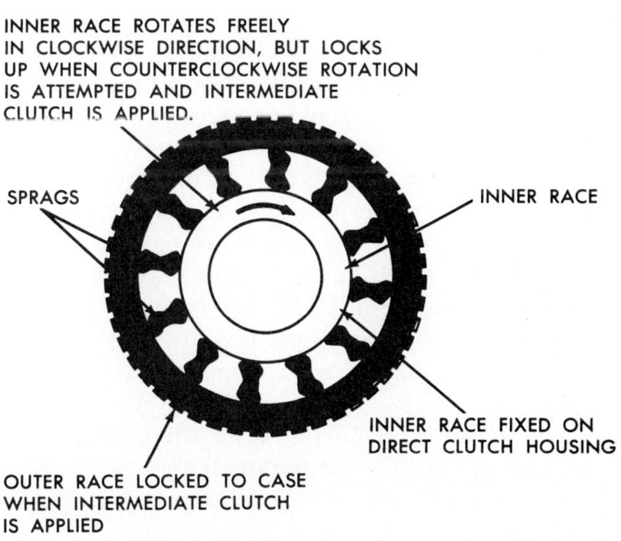

Figure 2-9. Sprag clutch operation. (*Courtesy of General Motors Corporation*)

Serious damage can result if all precautions and conditions are not followed. Stall testing is covered in Chapter 9.

At stall, vortex flow is at its maximum while rotary flow is minimal. As turbine speed increases in relation to impeller speed, vortex flow decreases and rotary flow increases.

Torque Converter Ratios

Two ratios should be considered in any torque converter ratio or gear ratio discussion. One is the ratio of torque increase and the other is the ratio of speed decrease.

At or near stall, the torque increase through the converter is at its maximum. At the same time, speed reduction through the converter will also be at its maximum.

During maximum torque multiplication, turbine speed is approximately 40 percent of impeller speed or a 2.5:1 speed reduction. This means that the impeller makes 2.5 revolutions to every single revolution of the turbine. As turbine speed increases, this ratio gradually changes until the turbine reaches approximately 90 percent of impeller speed. At this point, torque multiplication has ceased and the converter is in its coupling stage.

When this stage is reached, the stator turns with the impeller and turbine at approximately the same speed or at a 0.9:1 turbine-to-impeller ratio.

Coupling Stage

As the speed of the turbine approaches that of the impeller, vortex flow has decreased considerably. However, rotary flow has increased proportionately.

The combined effect of decreased vortex flow and increased rotary flow results in a change in the direction or angle of the fluid exiting from the turbine. The angle of this fluid flow has in fact changed enough to cause it to strike the opposite (rear) side of the stator vanes to that during torque multiplication. This results in the stator turning in a forward or clockwise direction with the impeller and turbine, which means that torque multiplication has ceased, the converter has reached the coupling stage, and it now acts as a simple fluid coupling.

Converter Efficiency

The three-member torque converter without any lockup device or mechanical drive feature is only about 90 percent efficient when in the coupling stage. Some loss of efficiency results from slippage between the impeller and the turbine since the medium used to transfer torque is a fluid. Another factor resulting in only a minor loss at the coupling stage is friction.

Converter efficiency is reduced as the distance between the impeller and the turbine increases. Fluid forces on the turbine tend to try to push it away from the impeller. This eventually results in some wear on the turbine thrust surfaces and an increased clearance between the turbine and impeller. This wear is measurable and is known as turbine end play. See Chapter 9 on how to measure turbine end play.

Converter Cooling

The torque converter generates a lot of heat when it is multiplying torque. The tremendous velocity of the fluid and the shearing action that takes place between the impeller and turbine cause considerable friction of the fluid, which produces heat.

Fluid is constantly circulating from the transmission to the torque converter and back again so that the heated fluid carries the heat to the transmission internal parts. If some of this heat is not removed, the fluid overheats and consequently the seals and clutch discs in the transmission do also. This results in early failure of the transmission.

Some method of cooling is therefore needed. Two methods are used: the air cooling system and the liquid cooling system.

Some air-cooled systems use externally finned converters and vented converter housings. The fins are attached in a manner that causes air to circulate through the housing from the inlet vent opening to the outlet vent opening.

Another method also uses a vented converter housing and a shroud attached to the rear of the converter. The shroud is mounted with a space between the converter and shroud for air circulation. The rotating converter and shroud act as a centrifugal air pump, taking in cooler air and expelling the heated air.

The liquid-cooled system circulates fluid through a heat exchanger located in the coolest of the two radiator tanks. In the vertical flow radiator it is in the lower tank, and in the cross flow radiator it could be in either of the side tanks, depending on coolant flow direction in the radiator. Heat from the fluid is transferred to the liquid coolant in the radiator, where it is cooled by the engine's cooling system and returned to the transmission hydraulic system from the heat exchanger in the radiator.

Converter Capacity

The torque capacity of the torque converter must match the engine's torque capability. If the torque capacity of the converter is too low, slippage, overheating, and too high an engine operating speed will result. Using a torque converter too high in torque capacity would result in too low an engine operating speed for efficiency and performance.

Factors that affect the performance and torque capacity of the torque converter are converter diameter and stator vane angle. Larger diameter torque converters have greater torque capacity; smaller diameter converters have lower torque capacity.

Stator vane angles that are more nearly parallel to the direction of fluid flow exiting from the turbine result in a higher torque capacity. Stator vanes that cause a considerable change in fluid flow direction and therefore velocity result in a lower torque capacity converter.

A low torque capacity converter provides better vehicle acceleration over a wider range of vehicle speed but results in higher fuel consumption since the coupling stage is not reached until a higher vehicle speed is achieved.

A high torque capacity converter results in less rapid acceleration and an earlier coupling point of the converter. Performance suffers but fuel economy is improved.

A low torque capacity converter is sometimes referred to as a high stall converter. This means that the stall speed of the engine during a stall test is fairly high. A high torque capacity converter on the other hand may be called a low stall converter.

From this discussion it can be seen that vehicle manufacturers design their torque converters to match engine torque capacity and vehicle performance and fuel economy objectives. Torque converter design is a careful compromise between acceptable performance and fuel economy.

Variable-Pitch Converter

As stated, one of the design factors affecting converter capacity and performance is stator vane angle. The fixed stator vane design is obviously a compromise between performance and economy. It is not possible to have the best performance and the best economy from fixed vane stator design.

To overcome this situation, some torque converters were designed with a variable-pitch (angle) stator. The stator vanes were mounted to allow them to pivot. A hydraulic piston connected to the vanes by small connecting rods caused the vanes to open or close, depending on operating conditions.

Moving the piston to close the stator vanes resulted in changing the exit angle of the fluid to aid impeller rotation better. This position also reduced the exit openings of the stator. Since the impeller was still trying to move the same volume of fluid through the stator, the velocity of the fluid was increased, resulting in greater torque multiplication, higher engine speed, and more rapid acceleration. As the stator angle changed to the more open position, torque multiplication and acceleration were reduced.

This design was used with General Motors two-speed transmissions until 1967. The increased number of gear ratios available in the three-speed transmissions eliminated the need for the variable-pitch stator design.

PART 2-4 LOCKUP TORQUE CONVERTERS

The lockup design feature used in many torque converters provides a direct mechanical drive connection between the engine and the transmission. This feature provides better torque converter efficiency during certain operating conditions, which in turn results in better fuel economy.

Vehicle manufacturers use several different methods to provide some means of mechanical drive through the torque converter. A centrifugal clutch and damper assembly is used by one manufacturer to lock the turbine to the converter cover, thereby providing a mechanical drive connection.

A hydraulically operated disc-type clutch and damper is another method used by several manufacturers to connect the turbine to the torque converter cover. A third method is to incorporate a simple planetary gearset into the converter to provide varying degrees of mechanical drive under various operating conditions. A fourth method is to use a direct drive shaft and damper assembly splined to the converter cover in addition to the usual turbine shaft.

It should be remembered that the torque converter is the unit that transmits engine torque from the engine crankshaft to the transmission input shaft or shafts. With this in mind it is easier to understand the various designs of torque converters and the different methods used to provide some means of mechanical power transmission through the converter in addition to the normal hydrodynamic method.

Centrifugal Clutch Lockup Converter

This unit is similar to the conventional three-member torque converter. It has the impeller, turbine, and stator assembly, which operate in the usual manner described earlier. The added features are the centrifugal clutch and damper assembly, which includes a coasting one-way clutch.

As seen from the illustrations the converter clutch consists of a series of sliding friction shoes arranged around the circumference of the damper assembly. As the speed of the turbine increases, the friction shoes slide outward due to increased centrifugal force. When the friction shoes are fully applied against the converter cover, power flow is from the converter cover to the damper clutch, through the torsional springs to the turbine hub and turbine shaft, thereby providing mechanical drive through the converter. Torsional shocks are dampened by the torsional springs.

The centrifugal clutch is designed to provide some slippage as torque demand from the engine increases. Consequently when the vehicle is under load, there may be a split between mechanical drive and hydraulic drive through the converter.

The factors determining whether the centrifugal

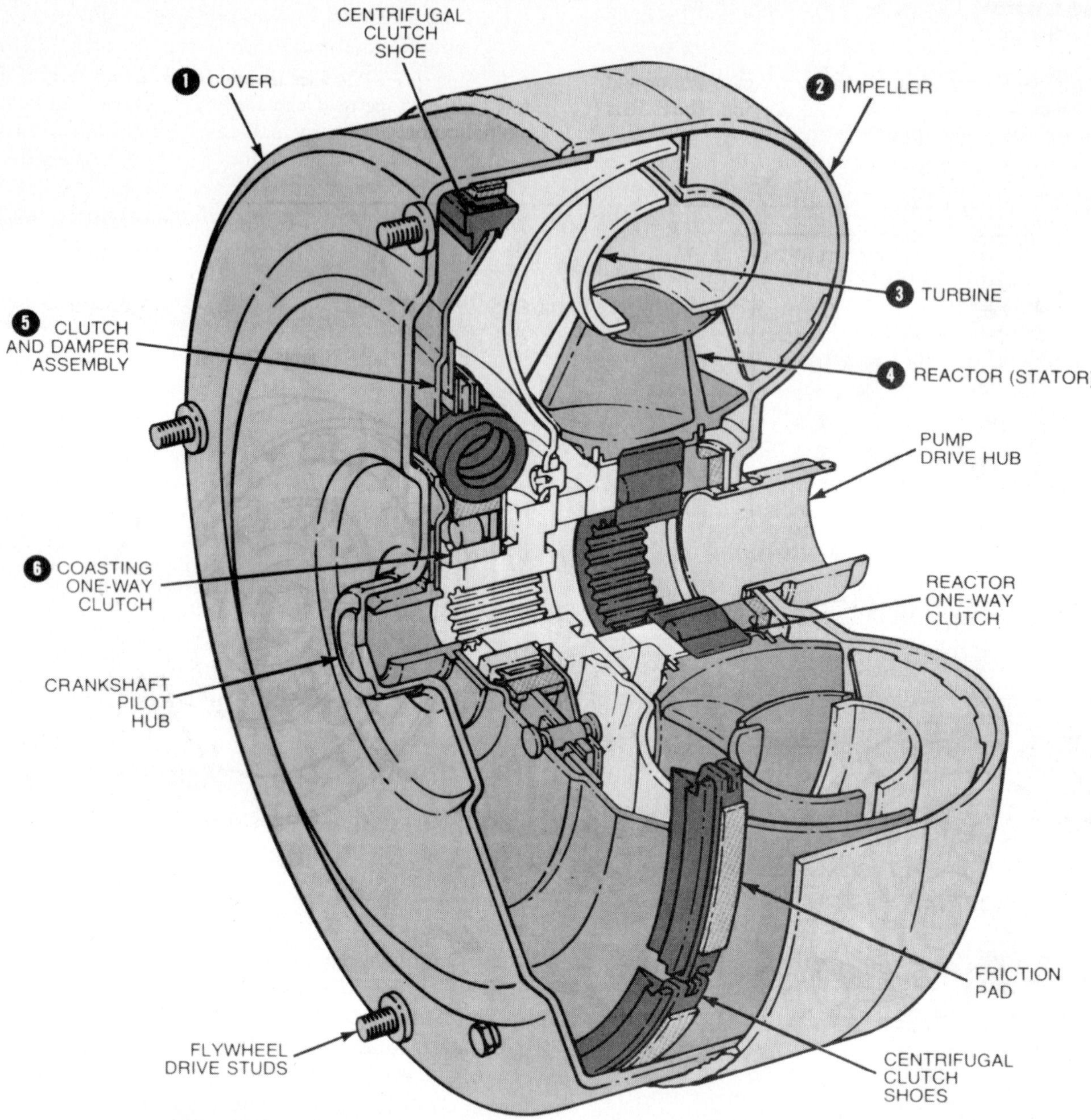

Figure 2-10. Cutaway view of centrifugal clutch lockup converter. (*Courtesy of Ford Motor Company*)

clutch will be applied and the degree of application are rotating speed and vehicle load. During mechanical drive the clutch and damper assembly drive the turbine and turbine shaft through the one-way clutch assembly. During coasting conditions the one-way clutch freewheels or overruns.

This unit is used with Ford's C5 automatic transmissions.

Hydraulically Controlled Lockup Clutch Converters

This type of torque converter is also similar in design and operation to the conventional three-member torque converter. The difference is in the addition of a hydraulically operated single-plate direct-drive clutch.

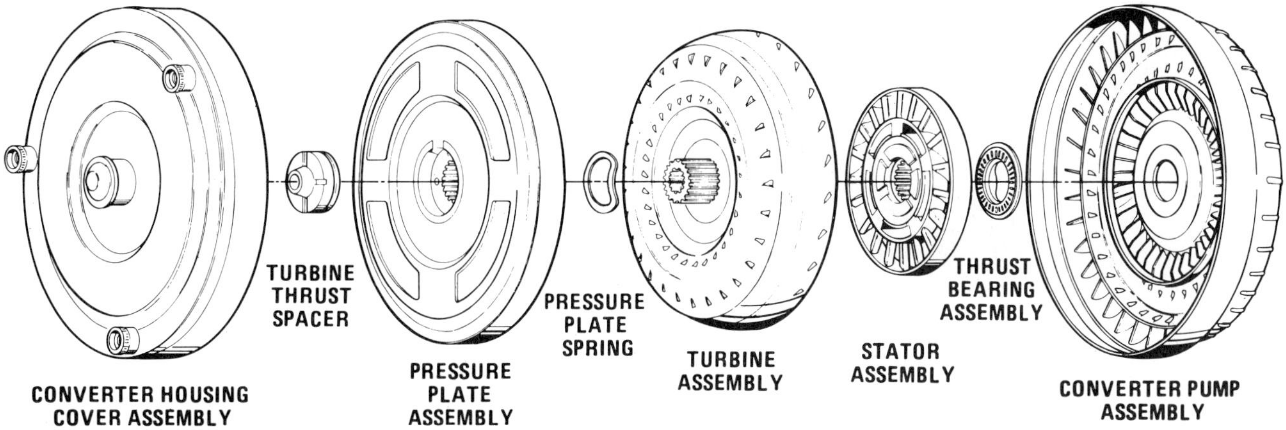

Figure 2-11. Exploded view of friction disc lockup clutch torque converter. (*Courtesy of General Motors Corporation*)

The basic function of the General Motors lockup converter and the Chrysler lockup converter (also used by American Motors) is similar. Both designs use a clutch apply piston or pressure plate assembly, which is splined to the turbine. This unit is equipped with a series of torsional springs through which torque is transmitted from the clutch to the turbine shaft. The purpose of the torsional springs is to dampen the effects of engine power impulses and the shock of clutch lockup.

The clutch piston or pressure plate is applied and released by hydraulic pressure from the transmission hydraulic system. Clutch friction material is attached to the inside of the converter cover. When applied, mechanical power flow is from the converter cover through the clutch to the torsional springs, the pressure plate hub, and the turbine shaft.

The Chrysler lockup converter operates as any three-element converter in all gears except third or direct. In third gear the clutch is applied at approximately 25 to 40 mph (40 to 64 km/h), depending on vehicle engine size and drive axle ratio. On a forced downshift (kickdown), whether part throttle or full throttle, the torque converter clutch releases; normal hydrodynamic torque converter operation transmits torque to the transmission planetary gears.

The General Motors torque converter clutch may apply in second, third, and fourth gears or in third or fourth on four-speed transmissions (depending on vehicle model) and in third gear only on three-speed

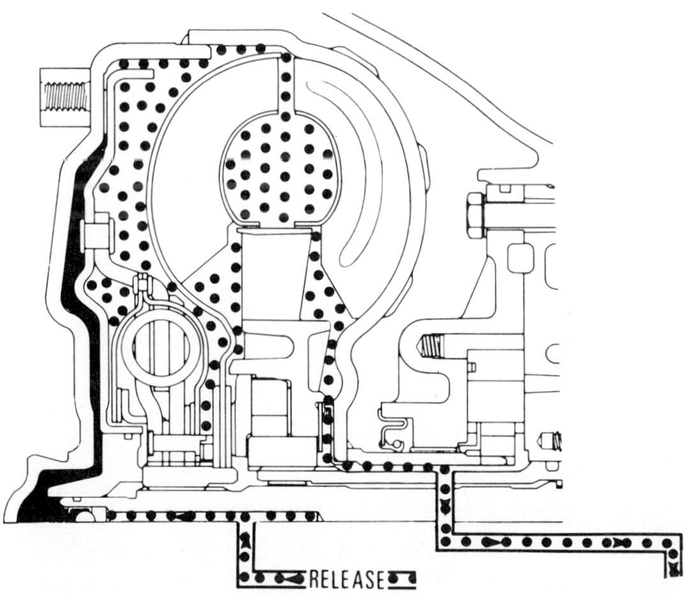

Figure 2-12. Fluid flow in converter clutch release position. (*Courtesy of General Motors Corporation*)

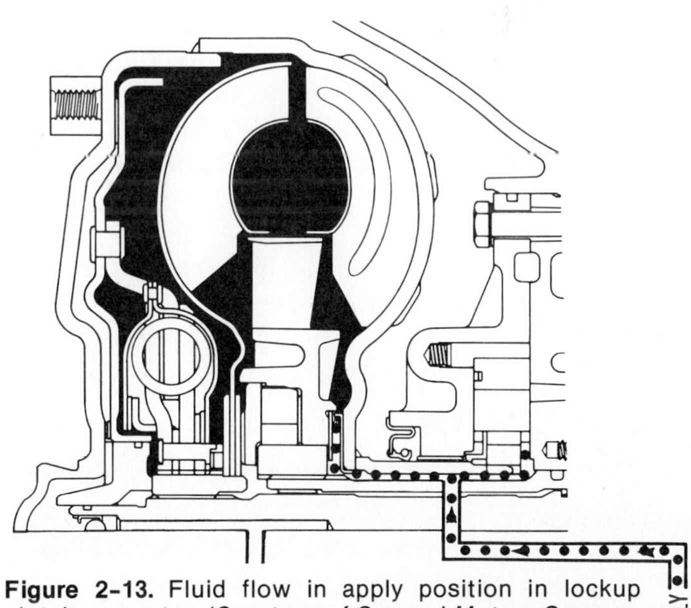

Figure 2-13. Fluid flow in apply position in lockup clutch converter. (*Courtesy of General Motors Corporation*)

transmissions. A number of factors other than gear position determine whether the converter clutch will be applied or released.

On vehicles not equipped with Computer Command Control the following items control the converter clutch:

1. Brake release switch. To avoid stalling the engine when braking, any time the brakes are applied the converter clutch is released.
2. Thermal vacuum valve. It prevents the converter clutch from applying until the engine coolant temperature has reached 130°F.
3. Engine vacuum switch. It releases the converter clutch when engine vacuum drops to approximately 1.5 to 3 inches of mercury (5 to 10 kPa) during moderate acceleration, prior to a part-throttle to detent downshift.
4. Vacuum delay valve. This slows the vacuum switch response to vacuum changes.
5. Ported vacuum. The source of vacuum to vacuum switch, it opens the vacuum switch to release the clutch during a closed throttle coastdown.

Vehicles equipped with Computer Command Control have the following items controlling the converter clutch:

1. Brake release switch. To avoid stalling the engine when braking, any time the brakes are applied the converter clutch is released.
2. Electronic control module. This energizes and grounds transmission electrical system.
3. Vehicle speed sensor. This sends vehicle speed information to the electronic control module.
4. Throttle position sensor. It sends throttle position information to the electronic control module.
5. Vacuum sensor. This sends engine vacuum (load) information to the electronic control module.

More detailed information on how the hydraulic control system affects converter clutch operation is provided in Chapters 7 and 8.

Planetary Splitter Gear Torque Converter

The Ford ATX automatic transaxle uses a torque converter equipped with a simple planetary gearset to provide mechanical drive through the converter. This unit is sometimes referred to as a splitter gear since it splits or divides engine torque transmission between mechanical and hydraulic means in second and third gears.

In second gear the turbine supplies 62 percent of the torque hydraulically; 38 percent is transmitted mechanically. In third gear the planetary gear mechanical torque output is increased to 93 percent with only 7 percent torque transmitted hydraulically.

Operation of this unit is dependent on the interaction between the torque converter and the planetary gear train in the transmission. Understanding the operation of this unit may require studying planetary gear operation and power flow, covered in Chapters 3 and 4.

Construction of this converter includes a torsional spring damper assembly riveted to the inside of the converter cover. The planetary ring gear (internal gear) is splined into the damper assembly so that these parts always turn at engine speed. The planet pinion carrier is splined to the transmission intermediate shaft and the pinions are, of course, in mesh with the ring gear. The sun gear, which is in mesh with the planet pinions, is splined to the turbine hub and to the turbine shaft.

Power flow and operation are as follows.

First Gear

1. The engine drives the impeller clockwise.
2. The impeller drives the turbine clockwise hydraulically.
3. The turbine drives the turbine shaft clockwise.
4. The turbine shaft (transmission input shaft) drives the planetary gears in the transmission.

This is normal three-member hydrodynamic torque converter operation and, of course, includes the stator, as described earlier in converter operation.

Second Gear

1. The converter cover mechanically drives the impeller and the converter ring gear clockwise.
2. The impeller hydraulically drives the turbine clockwise.
3. Hydraulic input torque is transmitted through the converter planetary unit to the intermediate shaft.
4. Mechanical input torque is transmitted from the damper and ring gear through the planetary unit, to the intermediate shaft.
5. The intermediate shaft drives the planetary gears in the transmission.

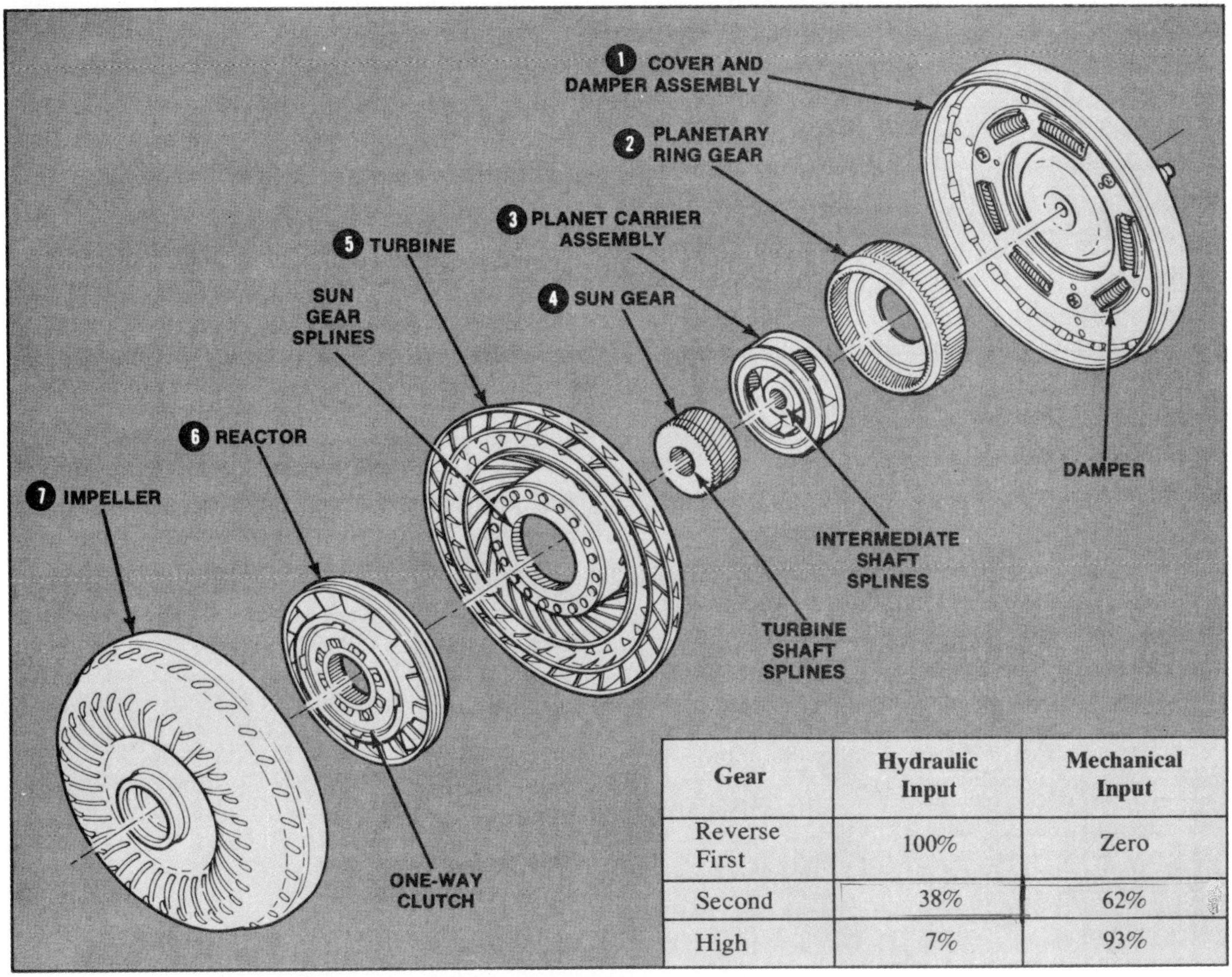

Gear	Hydraulic Input	Mechanical Input
Reverse First	100%	Zero
Second	38%	62%
High	7%	93%

Figure 2-14A. Exploded view of splitter gear type of torque converter used in Ford transaxle. (*Courtesy of Ford Motor Company*)

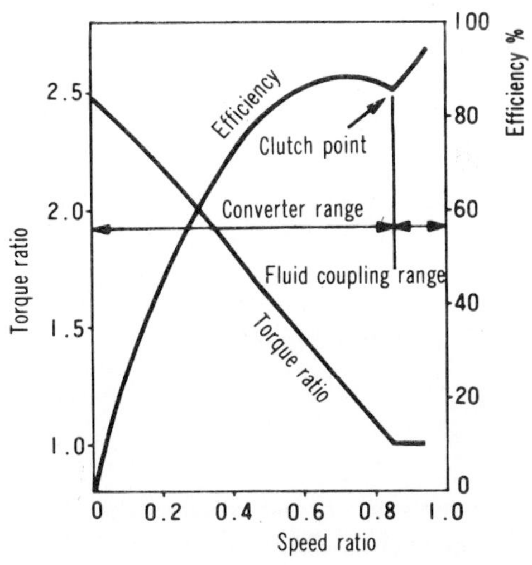

Figure 2-14B. Typical torque converter performance chart. (*Courtesy of Chrysler Corporation*)

6. The turbine input shaft also turns with the turbine but does not transmit any power. The one-way clutch overruns.
7. This provides for 62 percent of torque transmitted hydraulically and 38 percent transmitted mechanically.

Third Gear

1. The converter cover mechanically drives the impeller and the converter ring gear clockwise.
2. The impeller hydraulically drives the turbine clockwise.
3. Hydraulic input torque is transmitted through the converter planetary unit to the intermediate shaft.
4. Mechanical input torque is transmitted from the damper and ring gear through the planetary unit to the intermediate shaft.
5. The intermediate shaft drives the planetary gears in the transmission.
6. Presuming the converter to be in the coupling phase, the turbine shaft will be turning at essentially the same speed.
7. The low-reverse sun gear thus is driven along with the ring gear of the compound planetary unit in the transmission.
8. This provides for 7 percent of hydraulic transmission of torque and 93 percent of torque transmitted mechanically.

Mechanical transmission of torque through the torque converter is dampened by the torsional springs located between the converter cover and planetary ring gear. The degree of mechanical transmission of torque is dependent on gear position and the interaction between the transmission planetary gear train and the torque converter.

Ford Automatic Overdrive Torque Converter

The torque converter used with the Ford automatic overdrive transmission is a conventional three-member torque converter with a damper assembly attached to the inside of the converter cover. It has the

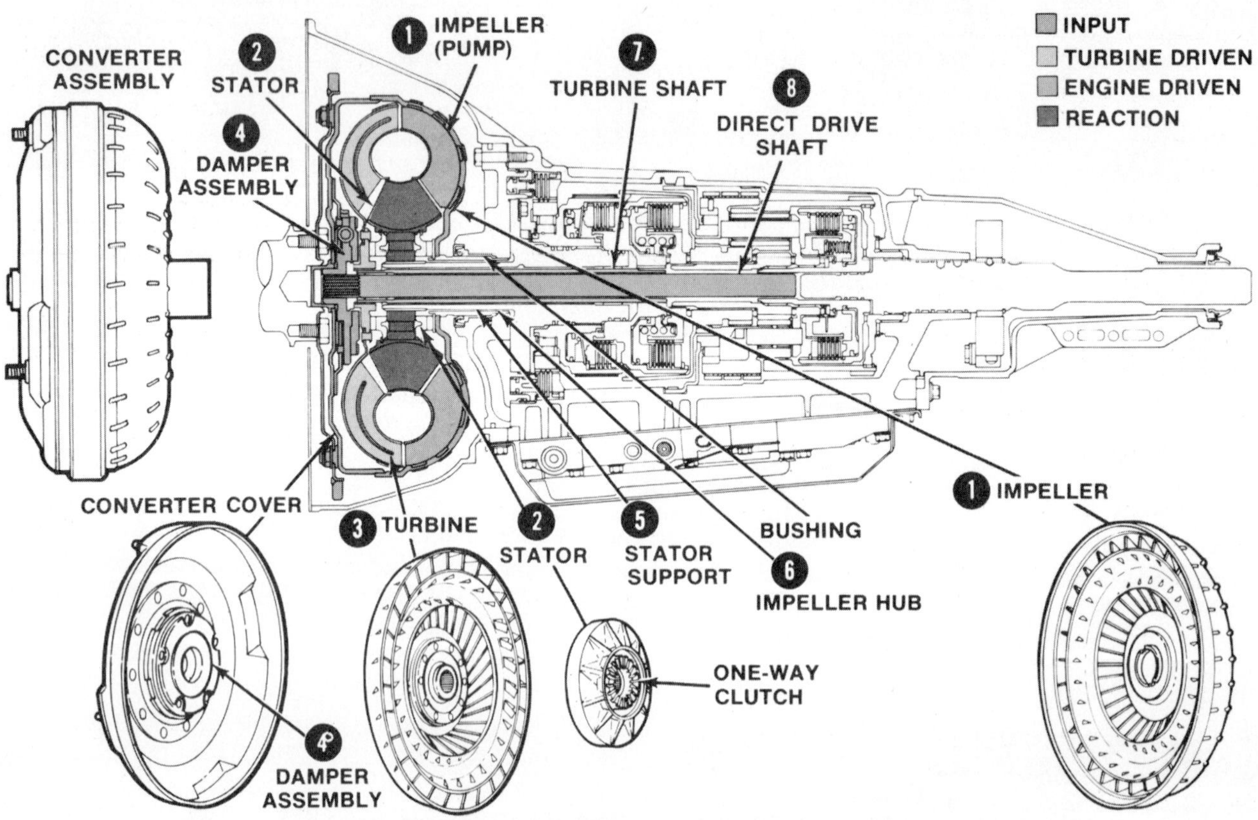

Figure 2-15. Ford automatic overdrive transmission torque converter power flow. Note direct drive shaft. (*Courtesy of Ford Motor Company*)

usual turbine shaft power output and an additional direct drive converter output shaft.

In operation this unit has the usual turbine-driven turbine shaft (transmission input shaft) for hydraulic transmission of torque. In reverse, first, and second gears, hydraulic drive only through the turbine and turbine shaft transmits engine torque to the transmission.

In third gear, torque is split between the turbine shaft (40 percent) and a direct-drive shaft splined to the hub of the torsional damper in the converter cover (60 percent).

In fourth gear (overdrive), the direct-drive shaft only is transmitting engine torque mechanically. Torsional pulsations are dampened by the damper assembly.

Action by the various clutches and band in the transmission (determined by gear position) cause the converter to deliver torque as described. Chapter 3 details the planetary gear operation in the transmission.

PART 2-5 TORQUE CONVERTER WEAR AND DAMAGE

Under ideal operating conditions the torque converter and automatic transmission have a service life of approximately 100,000 miles (160,000 km) or even more. Ideal conditions, however, do not always exist. Consequently many automatic transmissions require servicing long before such mileage is reached.

Vehicle drivers do not always operate the transmission in an ideal manner. Sometimes the transmission is put into drive or reverse when the engine is running too fast. At times when the vehicle is stuck or on a slippery surface, repeated shifts from reverse to drive and drive to reverse are made too suddenly and at too high an engine speed.

Engines do not always remain properly tuned especially when regular tuneups and required routine maintenance are neglected. Faulty engine performance, too high an engine idle speed, or too fast idle speed has a detrimental effect on transmission service life. A faulty engine cooling system, damaged or restricted transmission oil cooler lines, or overloaded engine and transmission (incorrect trailer towing for example) can seriously affect the life expectancy of the automatic transmission and the torque converter. Listed here are some of the possible torque converter conditions resulting from both normal and abnormal operating conditions and their probable causes.

This information is of a general nature and is given simply to show what can happen to a torque converter and how the driver perceives the problem and the probable cause; it should not be used as a diagnostic guide. Diagnosis, testing, inspection, measurement, and correction of torque converter problems are covered in Chapters 8 and 9.

Causes given here assume no manufacturing or assembly defects and do not list other conditions that may result in similar operating problems.

Problem	Cause
1. Noise, rattle, or vibration	A. Attaching bolts loose or worn.
	B. Worn front hub.
	C. Internal wear and interference between stator and turbine, turbine and cover.
	D. Loose turbine or impeller vanes.
	E. Flex plate bent.
	F. Damaged inspection cover causing interference with converter.
	G. Loose cooling fins.
2. Poor acceleration	A. Low fluid level in transmission.
	B. Stator freewheels in both directions.

Part 2–6 Worksheet

1. Examine a cutaway model (if available) of a torque converter; name all its parts.
2. At the request of your instructor, describe the operation of a three-member torque converter either orally or in writing.
3. List the four methods of mechanical torque transmission through the torque converter used by different manufacturers.
4. If you are licensed to drive a car and can have the use of a car, operate the car away from traffic and identify torque multiplication operation. Also identify torque converter lockup if you have access to a vehicle equipped with a lockup converter. In other words, get the feel of torque converter operation while you are driving the vehicle. Report your findings to your instructor.
5. If cutaway models of the different converter designs are available, examine them and note the differences in construction.

PART 2-7 SELF-CHECK

1. What is the purpose of a torque converter? (List all its functions.)
2. The impeller drives the ____________ and the turbine drives the ____________.
3. Define the terms *rotary flow* and *vortex flow.*
4. The torque converter is usually bolted to the ____________.
5. The stator assembly is splined to the ____________.
6. During torque multiplication the stator one-way clutch allows the stator to freewheel. True or false?
7. During the coupling stage the stator turns in a ____________ direction.
8. Define the term *torque converter stall.*
9. Explain the difference between a *low stall* and a *high stall* converter.
10. What factors determine torque converter torque capacity?
11. List four methods of obtaining mechanical drive through the torque converter.
12. Explain the relationship between speed and torque as it applies to torque converters.
13. What is the purpose of the torsional springs in a lockup converter?
14. One type of torque converter uses planetary gears in its design. What is the purpose of these planetary gears?
15. Define the term *transaxle.* List the causes and effects of converter wear.

PART 2-8 PERFORMANCE EVALUATION

The following statements can be used to determine the ability of the student to perform the tasks required in this chapter. Generally evaluating one's performance means determining whether stated objectives have been achieved and to what degree they have been achieved. The level of performance required from the student will be determined by the school and the instructor.

After thorough study of this chapter and the appropriate training models, the student should be able to:

1. State the six basic functions of the torque converter.
2. Describe basic three-member torque converter construction.
3. Describe the operation of the three-member torque converter, including torque multiplication, coupling stage, acceleration, and deceleration.
4. Describe hydraulic clutch torque converter lockup operation (in converter only).
5. Describe centrifugal clutch torque converter lockup operation.
6. Describe splitter gear torque converter operation.
7. Describe torque converter wear and damage, including causes and effects.
8. Complete the worksheet for this chapter.
9. Complete the self-check for this chapter with at least 80 percent accuracy.
10. Evaluate performance based on this performance evaluation section.

Chapter 3

The Gear Train

Performance Objectives

After thorough study of this chapter and the appropriate training models, you should be able to perform these tasks:

1. State the purpose of gears in an automatic transmission.
2. Describe the principles of basic gear operation, including gear terminology, the lever principle, direction of rotation, torque-speed relationship, and basic gear construction.
3. Name the components of a simple planetary gearset; describe its basic construction features, operating principles, and ratios.
4. Name the components of the Simpson planetary gearset; describe its basic construction and operation in both three-speed and four-speed arrangements.
5. Name the components of the Ravigneaux planetary gearset; describe its basic construction and operation in both three-speed and four-speed arrangements.
6. Describe normal and abnormal planetary gear wear and damage.
7. State the causes and effects of planetary gear wear and damage.
8. Identify by name the most common domestic and import automatic transmissions and state which type of planetary gear system is used by each.
9. Complete the worksheet as required.
10. Complete the self-check with at least 80 percent accuracy.
11. Evaluate your performance by using the performance evaluation section at the end of this chapter.

PART 3-1 INTRODUCTION

This chapter deals with the purpose of gears in an automatic transmission, basic gear principles, terminology and operation, and the torque-speed relationship of gear combinations. Also discussed in this chapter are the design, construction, operation, and wear of simple planetary gears and Simpson and Ravigneaux compound planetary gears. The type of planetary gear system used in the various domestic and import automatic transmissions is also covered.

The laws of planetary gear operation apply to all automatic transmissions regardless of make or model. A clear understanding of these operating principles is essential to understanding the mechanical and hydraulic devices used to control the operation of planetary gears. This kind of understanding is also required for any intelligent and accurate diagnosis of automatic transmission problems.

PART 3-2 PRINCIPLES OF GEAR OPERATION

Gear Terminology

What is a gear? Simply stated, a gear is a series of levers.

Gears are normally mounted on a shaft. The shaft may drive the gear, the gear may drive the shaft, or the gear may be free to turn on the shaft and simply act as an idler gear, depending on application.

Gears can be used to (1) transfer torque un-

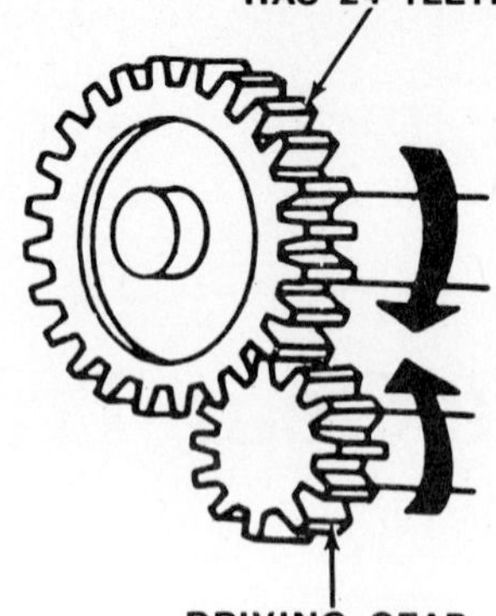

Figure 3-1. Small gear driving large gear.

changed, (2) multiply torque and decrease speed, and (3) increase speed and decrease torque. In order to see how gears are used to do these things, it is necessary to define some common terms used to describe gear construction and operation.

Drive Gear (Input)

A drive gear is one that drives another gear or causes another gear to turn.

Driven Gear (Output)

A driven gear is one that is driven or forced to turn by a drive gear, a shaft, or other rotating device.

Rotation and Direction of Rotation

Rotation is simply a term used to describe the fact that a gear, shaft, or other device is turning. The direction of rotation is described by comparing it to the rotation of the hands of a clock. All clocks that have hands rotate them in the same direction.

In order to tell what time it is, one must face the front of the clock. As time progresses, the hands of the clock move in a clockwise (CW) direction. This can also be described as turning in a right-hand direction (as with right-hand threads on a bolt or nut). Rotation in the opposite direction is described as counterclockwise (CCW) rotation or turning to the left (as with left-hand threads on a bolt or nut).

Figure 3-2. Two-to-one gear ratio.

When considering the power flow through the planetary gear system of an automatic transmission, it is important to have everyone use the same definition of terminology. Otherwise we are really not speaking the same technical language, and no one will be properly understood.

As with a clock, to determine the direction of rotation of gears in a transmission, one must face the front of the transmission. The front of the transmission is the end attached to the engine, whether mounted in a transverse (crosswise) or longitudinal manner. The direction of rotation of all transmission gears and shafts is therefore determined by viewing the end of the transmission that is attached to the engine.

Another term for clockwise rotation is *forward rotation.* Other terms used for counterclockwise rotation are *backward rotation, turning backward,* or *turning in a reverse direction.*

Speed of Rotation

The term *speed* is used to describe the rotating frequency of gears or shafts. When used to describe engine speed, it refers to the number of turns or revolutions that the engine crankshaft makes in one minute stated in revolutions per minute (rpm).

When used to describe the rotation of transmission parts, the word *speed* is used in a more general way to compare the rotating speed of one part to another. Terms used to describe speed comparisons include *faster, slower, increased speed, decreased speed, direct drive* (same speed) or *one to one,* and *overdrive* (output faster than input).

Ratios

For an accurate description of speed comparison, the term *gear ratio* is used. In automatic applications the accepted practice is to compare the speed of the input—drive gear or shaft—to the speed of the output—driven gear or shaft. All comparisons (ratios) are stated as the number of turns (revolutions) of the input or drive member in order to cause the output or driven member to make one turn. For example, a small gear driving a gear twice as large would result in a ratio of 2:1. The figure for the input, or drive member, is stated first and compared to one turn of the output, or driven member.

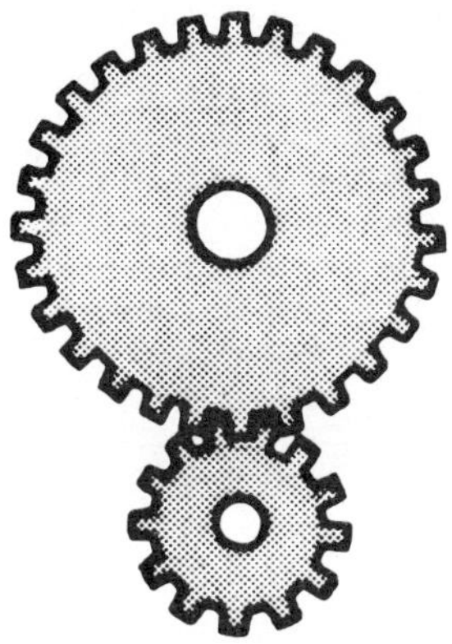

LARGE GEAR
DRIVING SMALL
GEAR =
MORE SPEED
BUT LESS TORQUE

Figure 3-3.

To calculate the ratio of a two-gear drive combination, simply count the number of teeth on the driven gear and divide by the number of teeth on the drive gear and compare to 1 $\frac{\text{(driven:1)}}{\text{(drive)}}$. For example, if a 20-tooth gear drives a 40-tooth driven gear, the ratio can be calculated as follows:

$$\frac{40:1}{20}$$

or 2/1:1

or 2.0:1

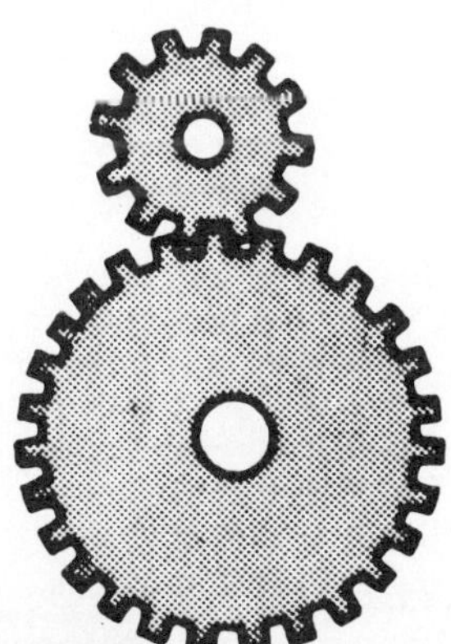

SMALL GEAR
DRIVING LARGER
GEAR =
LESS SPEED
BUT MORE TORQUE

Figure 3-4.

which is the proper way to state the ratio. This causes a speed reduction and a torque increase.

On the other hand, if a 40-tooth drive gear drives a 20-tooth driven gear, the ratio is calculated as follows:

$$\frac{20:1}{40}$$

or 1/2:1

or 0.5:1

which is a torque reduction and a speed increase (overdrive).

Upon closer examination it can be seen that there is a definite relationship between speed and torque ratios. Using gears to double the torque output, for example, reduces the speed to half the input speed. Stated in ratios it would be done this way: speed ratio of 2:1; torque ratio of 0.5 to 1. It can also be seen that using gears to increase speed reduces torque output. A 0.5:1 speed ratio doubles the output speed but results in a 50 percent torque decrease or torque ratio of 2:1. Ignoring the relatively minor friction losses of gears and shafts, the relationship of speed and torque output is inversely proportional.

Axis (of Rotation)

The axis is the center line about which a gear or part rotates.

External Gears

An external gear is one that has the gear teeth arranged around the outer circumference.

Internal Gear

An internal gear is a ring-type gear with the teeth arranged around the inner circumference of the ring.

Straight Spur Gears

A straight spur gear has teeth that are parallel to the axis of rotation of the gear.

Helical Gears

A helical geal has gear teeth at an angle to the axis of rotation of the gear. Advantages of helical gears include the fact that more than one tooth is doing the driving at all times, which is not the case with straight spur gears. Helical gears also run more quietly since they create a wiping action as they engage and disengage the teeth on another gear. A disadvantage is that helical teeth on a gear cause the gear to move fore or aft on a shaft, depending on the direction of the angle of the gear teeth. This axial thrust must be absorbed

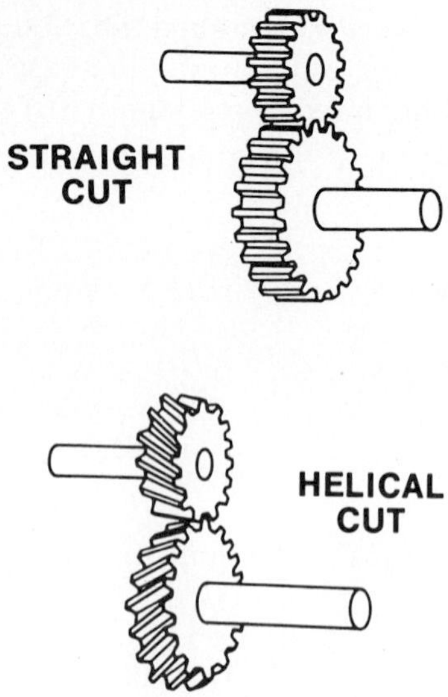

Figure 3-5.

by thrust washers and other transmission gears, shafts, or the transmission case.

Drive and Coast

The drive side of gear teeth is the side in contact with teeth of another gear while torque is being applied. This is the side of gear teeth that is subject to the most wear. The coast side of gear teeth is the opposite side from the drive side.

Constant Mesh

The term *constant mesh* refers to gears that remain in mesh with each other and are not engaged or disengaged from each other by transmission or driver action when shifting gears.

Backlash

Backlash is the term used to describe the amount that one gear is able to move when the gear with which it is in mesh is held stationary. All gears have some backlash to allow for expansion of metal due to heat and for proper lubrication. Excessive backlash is an indication of gear tooth wear.

Bottoming

The condition known as bottoming involves the teeth of one gear touching the lowest point between teeth of

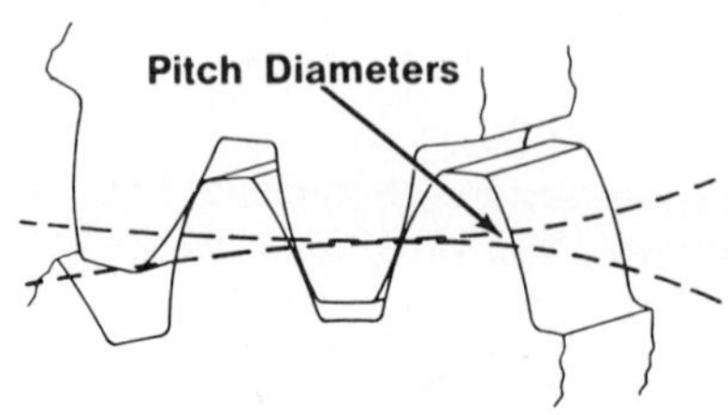

Figure 3-6.

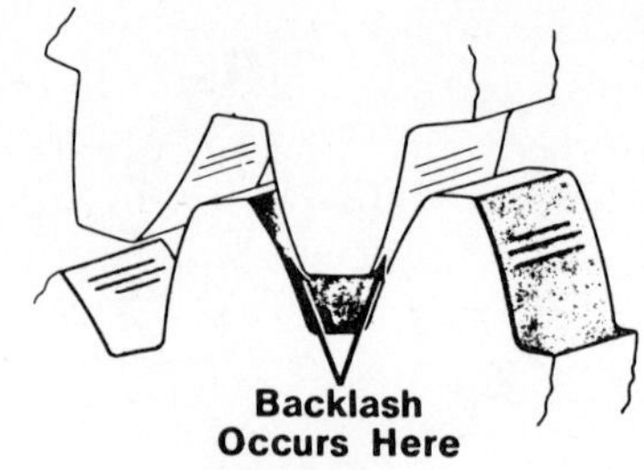

Figure 3-7.

a mating gear. Bottoming does not occur in a two-gear drive combination but can occur in multiple-gear drive combinations. A simple two-gear drive combination always tends to force the two gears apart; therefore bottoming cannot occur in this arrangement.

Climbing

Climbing occurs when there is so much wear in gears, bearings, and shafts that the gears move apart sufficiently to cause the apex (or point) of the teeth on one gear to climb over the apex of the teeth on another gear with which it is meshed. This results in a loss of drive until other teeth are engaged; this causes rapid destruction of the gears.

PART 3-3 SIMPLE PLANETARY GEARS

Planetary gears are compact, are in constant mesh, and rotate around a common center. This makes them ideal for use in an automatic transmission. The

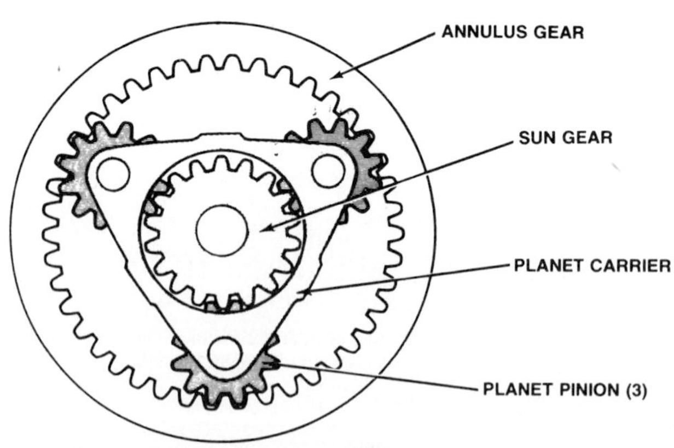

SIMPLE PLANETARY GEARSET

Figure 3-8. Planetary gear components. Annulus gear is also called an *internal* gear or a *ring* gear. Planet carrier is also called a *cage.* Planetary gears transmit torque when one of the three members is the drive member, one is the held member, and the other is the driven member. (*Courtesy of Chrysler Corporation*)

laws of planetary gear operation allow a smooth transition from one gear ratio to another without shifting gears in and out of mesh with each other.

The word *planetary* in the term *planetary gears* relates to the fact that their operation can be compared to the function of the solar system. The earth, a planet, rotates on its own axis, which causes the day-night cycle. The earth also orbits the sun, which causes the seasonal and yearly cycles. Planetary gears operate in a similar manner.

Basic Components and Construction

A simple planetary gearset consists of a sun gear, a set of planet pinions mounted in a pinion carrier or cage, and a ring gear (also known as an internal gear or annulus gear).

The sun gear has external teeth, which are in constant mesh with the external teeth of the planet pinions. The planet pinion teeth are also in constant mesh with the internal teeth of the ring gear.

The simple planetary gearset can be viewed as consisting of three major components: the sun gear, the planet pinion carrier, and the ring gear. The action of any one of the three members has a direct effect on the action of the other two.

Simple Planetary Gear Operation

The simple planetary gearset is arranged so that the sun gear is able to turn on its axis, the ring gear is able to turn on its axis, and the carrier is able to turn on its axis. All three units can rotate around a common axis or center. The planet pinions can turn on their axes, and when the carrier is turning, the pinions orbit the sun gear.

This planetary gearset can be utilized in a number of ways, depending on the particular arrangement used. In every application of planetary gears there must be at least one drive member and one held (reaction or stationary) member or two drive members for the other (third) member to become the driven (output) member. Obviously several different arrangements of this principle are possible, even though some of them are not practical for use in an automotive transmission.

The manufacturer has the option to design a transmission that uses any one or any two of the three planetary gear members as drive members and use the third member as the output member. Study Figures 3–9 to 3–14, using a model planetary gearset from an automatic transmission to demonstrate the results of each possible drive combination.

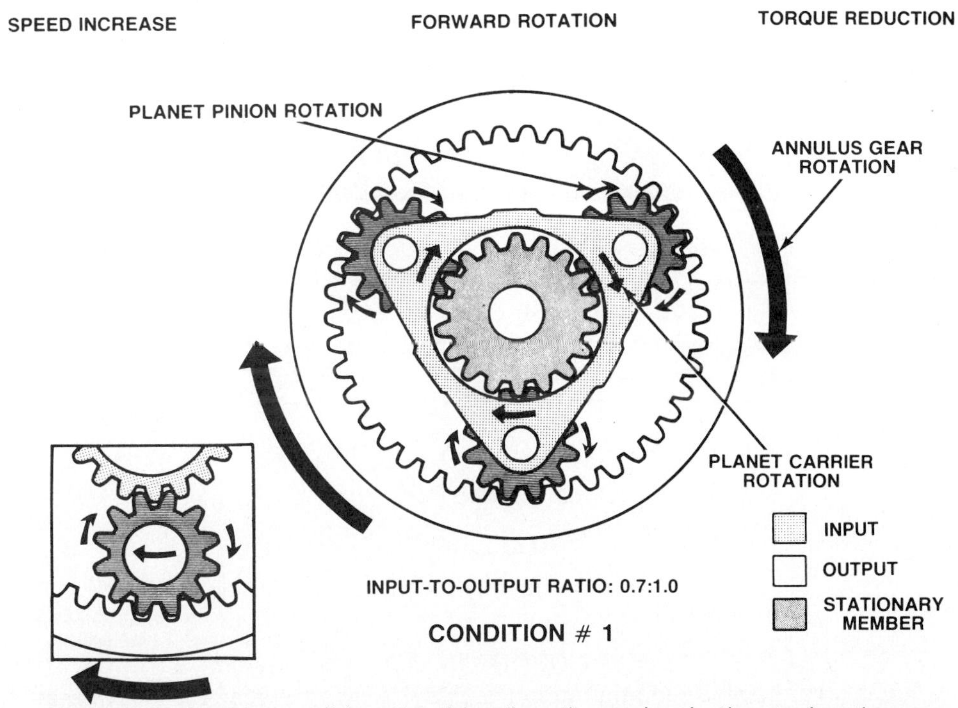

Figure 3-9. In this condition the drive (input) member is the carrier, the sun gear is being held (stationary), and the ring (annulus) gear is the driven member providing an overdrive. In this condition the pinions "walk" around the stationary sun gear, forcing the carrier to rotate. This arrangement is used to achieve overdrive. (*Courtesy of Chrysler Corporation*)

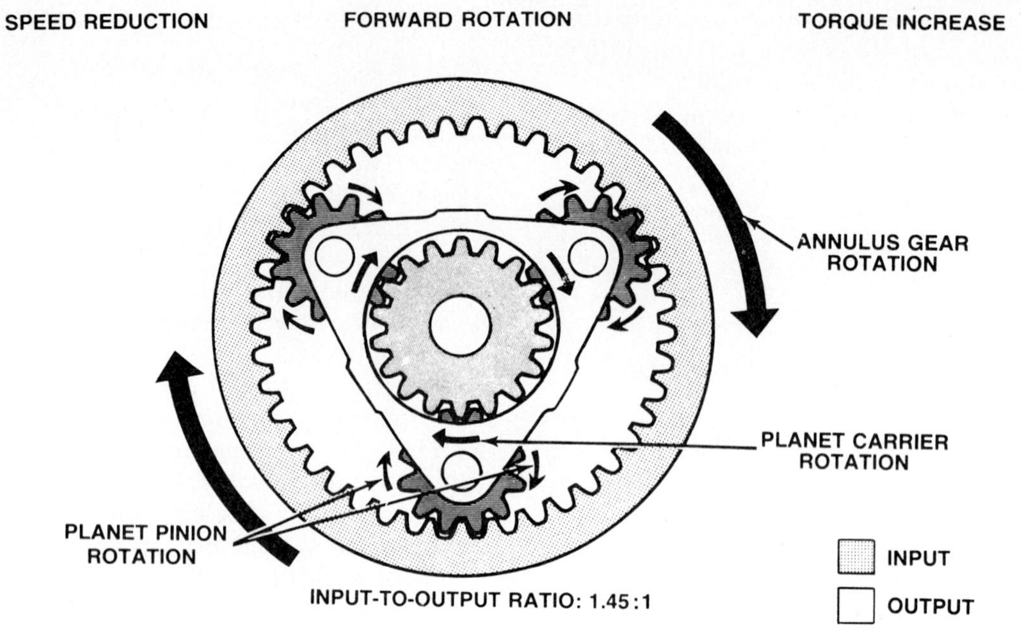

CONDITION # 2

Figure 3-10. In this condition the drive member is the ring gear, the sun gear is held, and the carrier is the driven member providing a gear reduction. In this condition the pinions "walk" around the stationary sun gear, forcing the carrier to rotate. This arrangement is used in automatic transmissions to achieve second gear. (*Courtesy of Chrysler Corporation*)

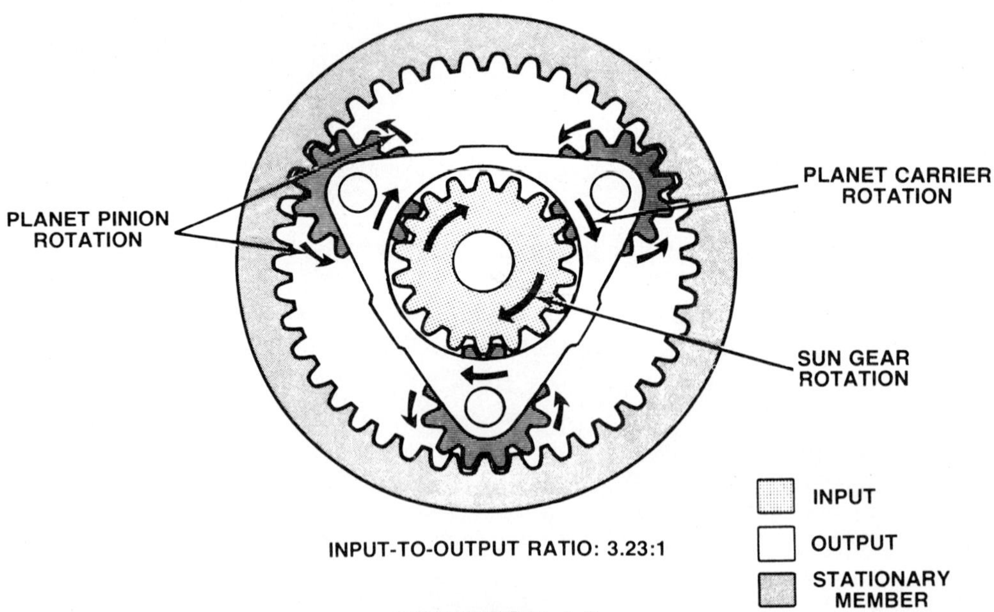

CONDITION # 3

Figure 3-11. In this condition the sun gear is the drive member, the ring gear is held, and the carrier is the driven member. In this condition the pinions "walk" around the stationary ring gear, forcing the carrier to rotate. This arrangement is not used in automatic transmissions. (*Courtesy of Chrysler Corporation*)

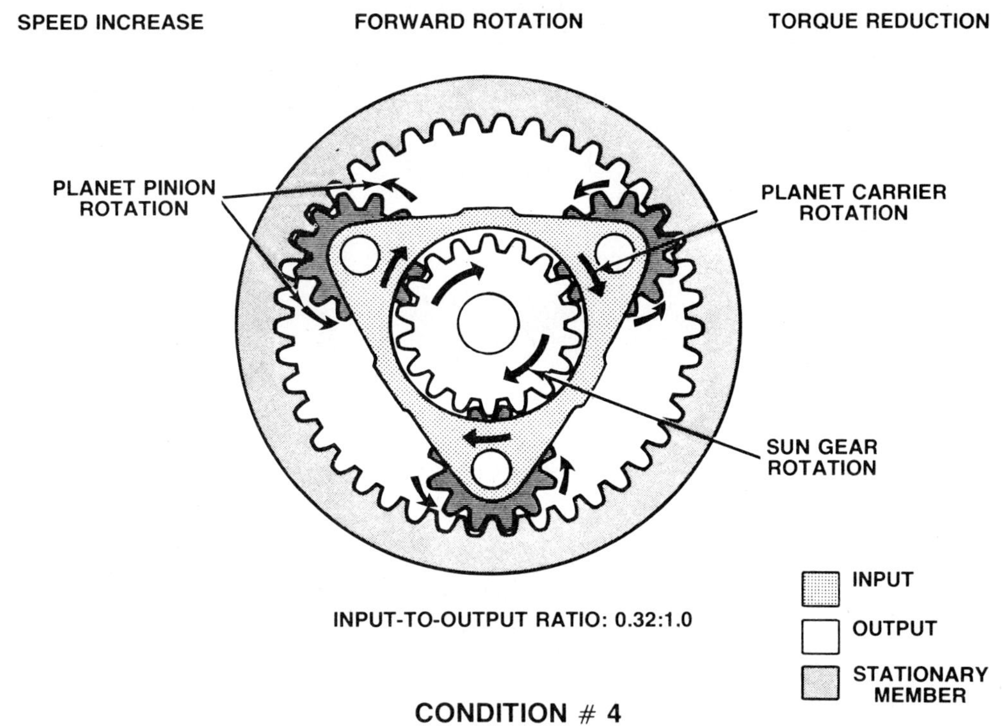

Figure 3-12. The carrier is the input, the ring gear is held, and the sun gear is the output member. In this condition the pinions "walk" around the stationary ring gear, forcing the carrier to rotate. This arrangement is not used in automatic transmissions. (*Courtesy of Chrysler Corporation*)

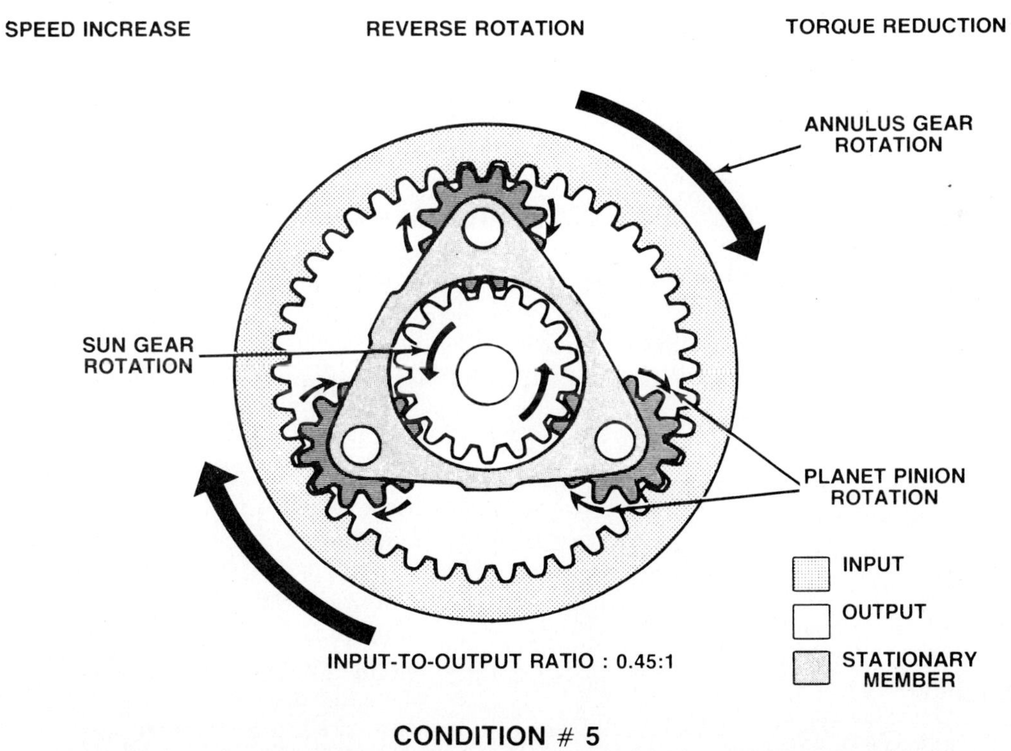

Figure 3-13. Here the ring gear is the drive gear, the carrier is held, and the sun gear is the driven gear. In this case the pinion gears simply act as idlers, reversing the direction of rotation. Since this is an overdrive, it is not practical to use this arrangement for reverse. (*Courtesy of Chrysler Corporation*)

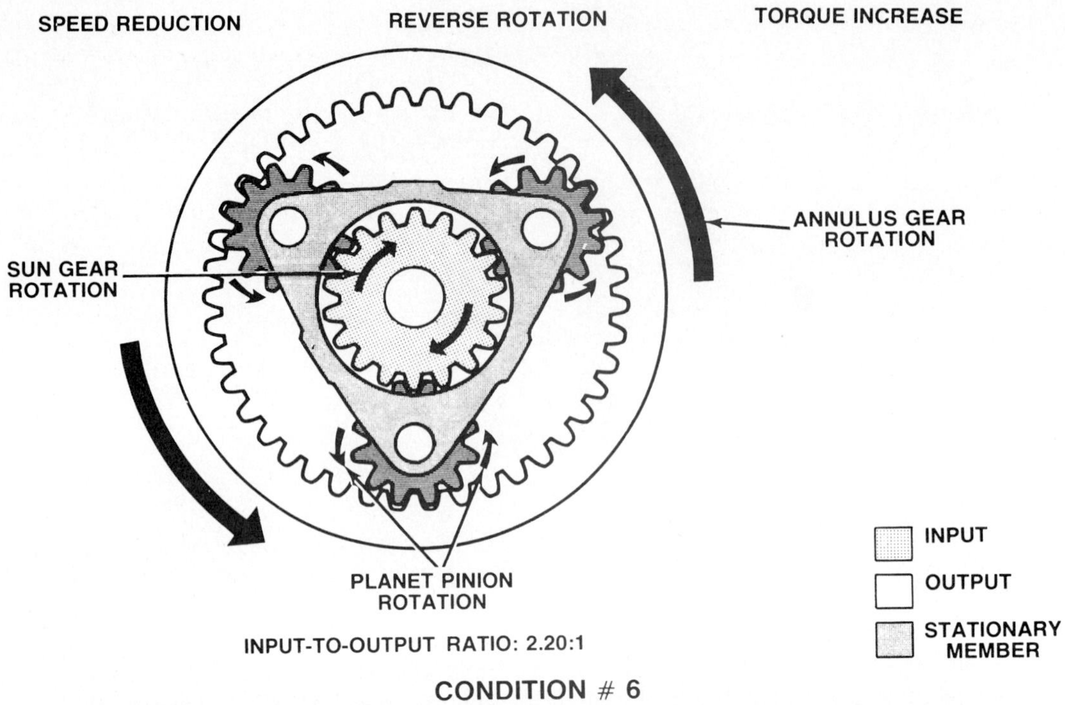

Figure 3-14. Here the sun gear is the drive gear, the carrier is held, and the ring gear is the driven gear. In this condition the pinion gears simply act as idlers, reversing the direction of rotation. This arrangement is commonly used to achieve reverse in automatic transmissions. (*Courtesy of Chrysler Corporation*)

Here are the laws that apply to *simple* planetary gear operation.

1. When the carrier is the input or drive member, the result is forward overdrive.
2. When the carrier is the output or driven member, the result is forward gear reduction.
3. When the carrier is the held member, the result is always reverse.
4. When the sun gear is the input member, the result is gear reduction.
5. When the sun gear is the output member, the result is overdrive.
6. When there is only one drive member and no held member, the result is neutral.
7. When any two members are locked together as drive members, the result is direct drive.

When simple planetary gear operation is fully understood, compound planetary gear operation is much easier to follow.

Simple Planetary Gear Ratios

A gear ratio is the number of turns that a drive gear makes while the driven gear makes one turn. In planetary gear terms, it is the number of revolutions made by the drive member while the driven member makes one revolution.

As mentioned, when a small gear drives a larger gear, the result is a gear reduction. On the other hand when a large gear drives a smaller gear, the result is overdrive. From this it may appear that relative physical size is the determining factor in gear ratios. While this is the case with standard externally toothed gear transmissions, this is not the case with planetary gears.

Operating a model simple planetary gearset by hand through all the possible drive variations will prove that even though the ring gear appears to be the largest of the three planetary gear members, this is not the case with ratios. Operating the gearset with careful observation will soon prove that the sun gear is the smallest, the ring gear is larger, and the carrier is the largest of the three planetary gear members.

The ratio of a simple planetary gearset can be calculated by counting the number of teeth on the driven member and dividing by the number of teeth on the drive member. The number of teeth on the pinions must always be ignored since they serve only as idlers. However, since there are no teeth on the carrier itself, an effective number must be given to the carrier. This is calculated by adding the number of teeth on the sun gear to the number of teeth on the ring gear:

$$S + R = C \text{ (effective number)}$$

Assuming that we have a planetary gearset with 30 teeth on the sun gear and 70 teeth on the ring gear, the effective number for the carrier is 70 + 30 = 100. With these numbers it is easy to calculate the various ratios possible through this gearset as follows: the number of teeth on the driven gear divided by the number of teeth on the drive gear. For example, with the sun gear as the driving gear and the ring gear as the output or driven gear, the ratio is R/S, 70/30:1, or 2.33:1. With the ring gear driving and the carrier being driven the ratio would be C/R, 100/70:1, or 1.43:1 when rounded off. As noted, the figure 100 for the carrier was obtained by adding the number of teeth on the ring gear to the number of teeth on the sun gear, 70 + 30 = 100.

A thorough understanding of the various ratios possible in a simple planetary gearset will help in understanding the operation of the compound planetary gears used in automatic transmissions and how speed and torque changes are achieved without going into complex compound gear ratio formulas.

PART 3-4 SIMPSON PLANETARY GEARS

The Simpson compound planetary gear system is commonly used in many automatic transmissions, domestic and import. This gearset operates similarly in most transmission applications.

Operation of this gearset is discussed on the basis that it is possible to connect or disconnect one or more planetary gear members to or from the transmission input shaft and that it is possible to hold (prevent from turning) one or more planetary gear members without describing how the driving and holding devices operate. Operation of driving and holding devices is described in Chapter 4. A planetary gear train can be used to increase or decrease torque, increase or decrease speed, reverse the direction of rotation, or act as a coupling to provide direct drive.

The Simpson three-speed planetary gear system is used in Chrysler's TorqueFlite series of transmissions and in the American Motors Torque Command version of the TorqueFlite. It is also used in Ford's C series of transmissions C-3, C-4, C-5, and C-6 and Ford's Jatco. General Motors' series 125, 200, 250, 325, 350, and 400 also use the Simpson gearset. A number of European and Japanese automobiles also use the Simpson type of planetary gear system in their transmissions.

The operation of the Simpson planetary gear train is described here as it is used in such transmissions as Ford's C-6, C-4, and C-3, Chrysler's TorqueFlite, and General Motors' Turbo Hydra-Matics.

The addition of a simple planetary gearset between the torque converter and the Simpson planetary gear train as used in the General Motors 325-4L to obtain overdrive is explained as well as the Turbo Hydra-Matic THM 700-R4 overdrive transmission gear train.

Components and Contruction (Simpson Gear Train)

The Simpson compound planetary gearset consists of two simple planetary gearsets in tandem, one behind the other. The gears in the two sets are the identical size in some transmissions and different sizes in others, depending on the gear ratios desired by the manufacturer.

Gears are tough, hard, long-lasting, highly finished steel as are the pinion shafts. Carriers are usually die-cast aluminum or stamped steel.

Multiple-disc clutches are used to connect or disconnect planetary gear drive members with the transmission input shaft. Multiple-disc clutches, brake bands, or one-way roller clutches are used to prevent certain planetary gear members from rotating under certain conditions. Operation of these devices is described in Chapter 4.

Common Simpson Gearset Operation

Neutral and Park Position

When the gearshift selector lever is in the neutral position, no clutches or bands are applied, and therefore no power is transmitted from the torque converter to the planetary gears. This is also the condition when the selector lever is in the park position. In the park position, however, a toothed lever engages the park gear, which is splined to the transmission output shaft. This effectively locks the output shaft to the transmission case, which also prevents the drive wheels of the car from turning.

Reverse Position

With the selector lever in the reverse position, hydraulic pressure applies the reverse and high clutch, which connects the sun gear to the turbine shaft. At the same time hydraulic pressure also applies the reverse band (or reverse holding clutch,

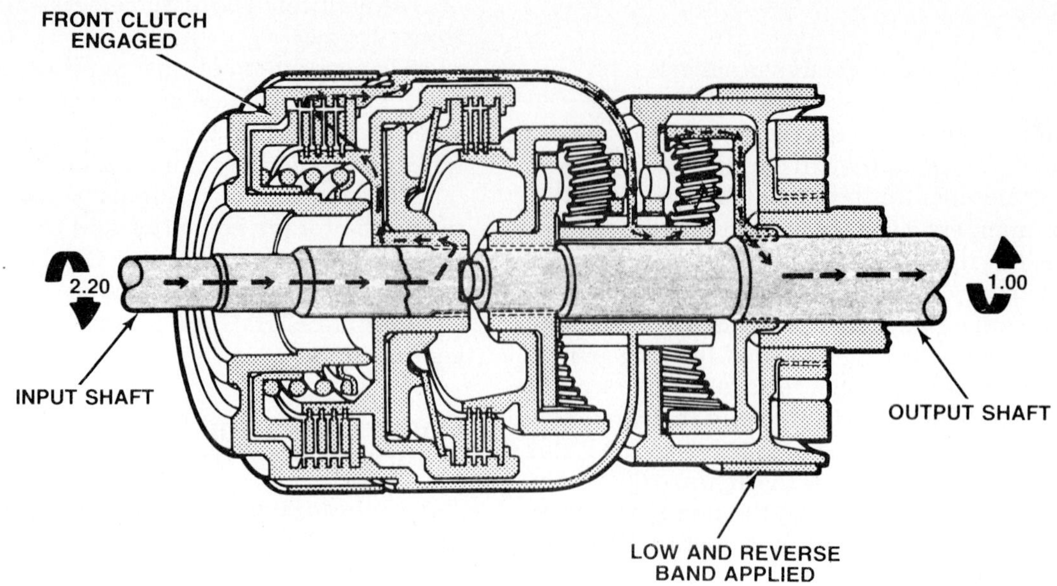

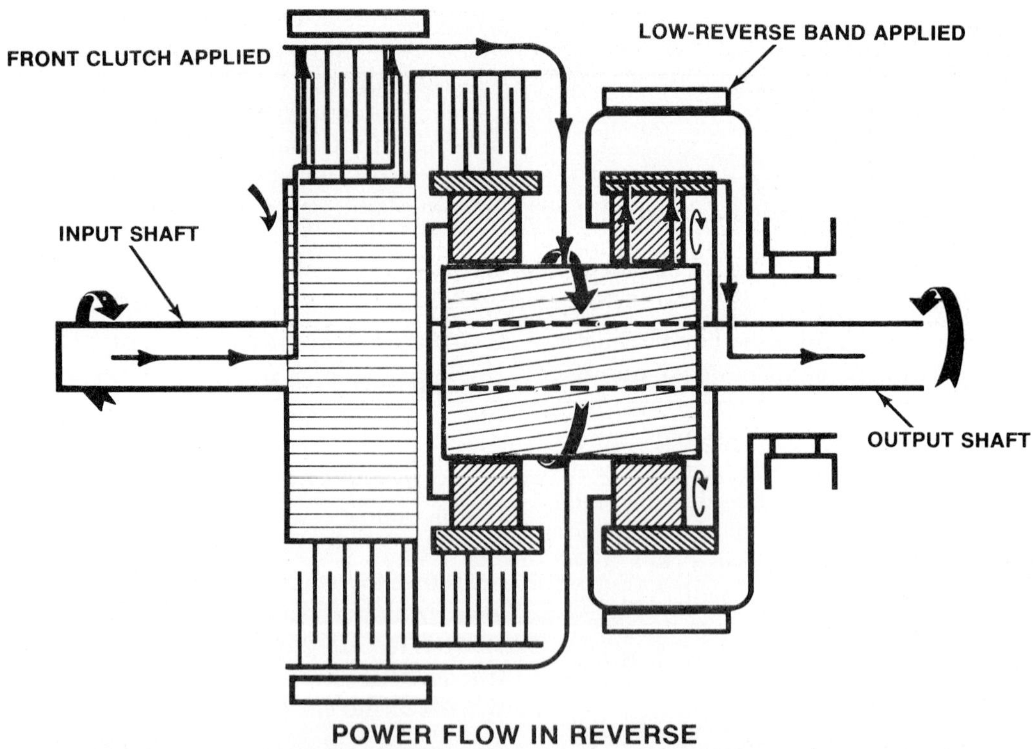

POWER FLOW IN REVERSE

Figure 3-15. In reverse only the rear half of the gearset is in use. The front clutch is applied driving the sun gear. The low and reverse band is applied holding the rear carrier. (*Courtesy of Chrysler Corporation*)

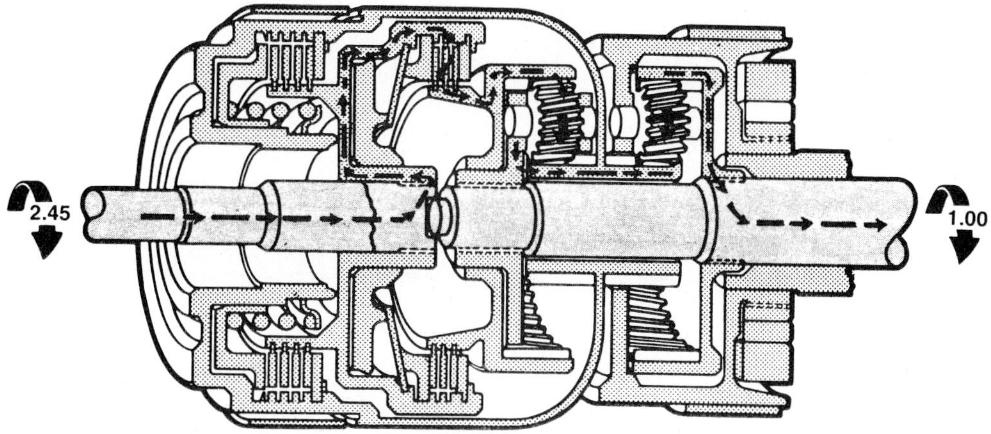

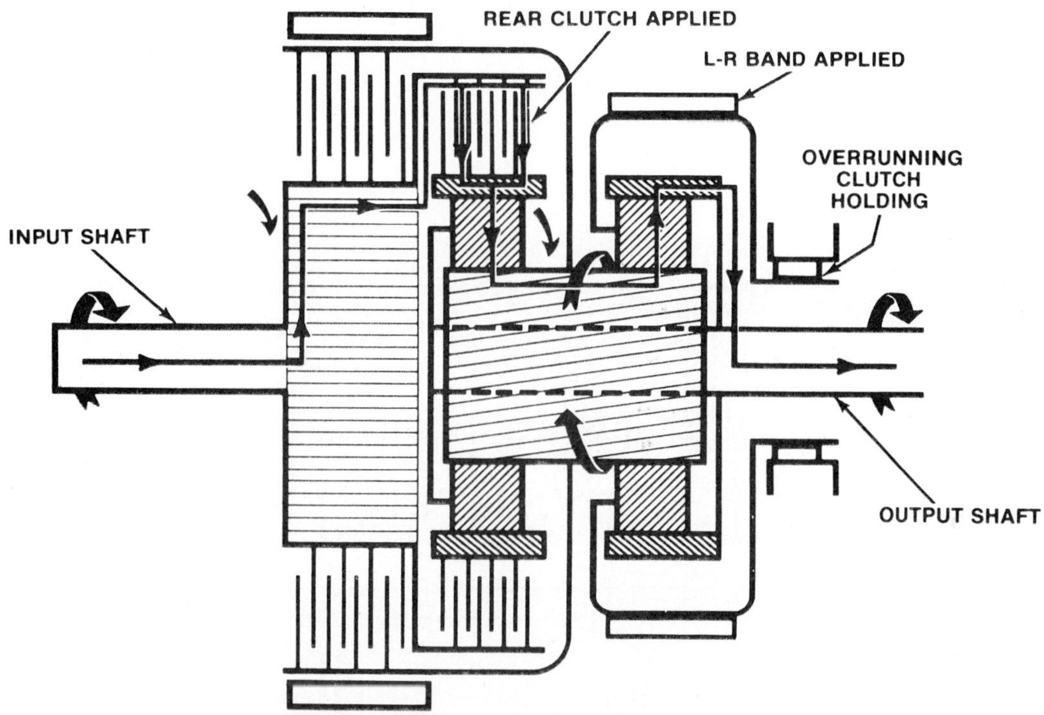

POWER FLOW IN DRIVE BREAKAWAY OR MANUAL LOW

Figure 3–16. In drive low, the low and reverse band is not applied. in manually selected low, the l and r band is applied for extra holding power and to provide engine braking during deceleration. (Some transmissions use a low and reverse holding clutch instead of a low and reverse band.) (*Courtesy of Chrysler Corporation*)

depending on transmission design) to lock the reverse carrier effectively to the transmission case.

Engine power through the torque converter drives the transmission input (turbine) shaft in a clockwise direction. Since the reverse and high clutch is applied, this drives the sun gear in a clockwise direction. The sun gear drives the planet pinions in a counterclockwise direction. Since the carrier is prevented from turning by the reverse band or clutch, the pinions drive the output ring gear in a counterclockwise direction. The output ring gear is splined to the transmission output shaft, which is forced to turn in a

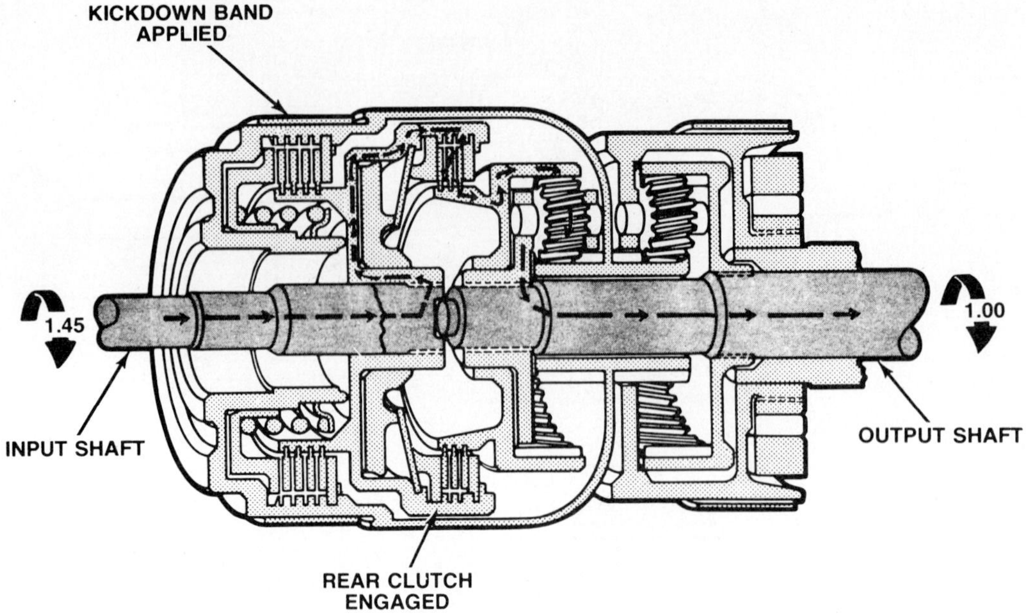

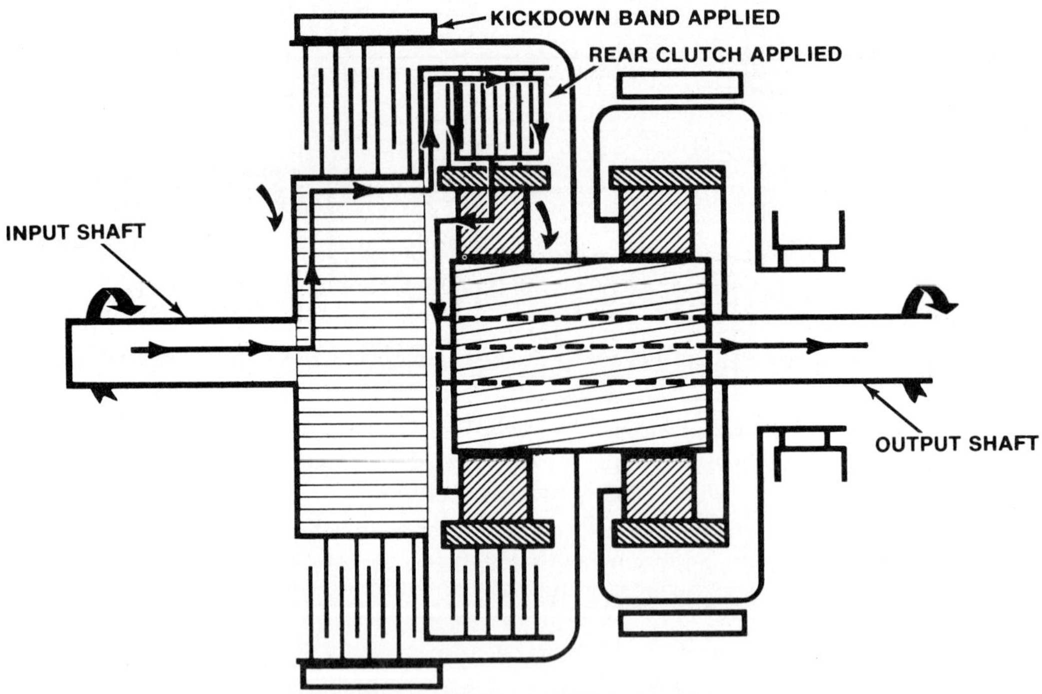

POWER FLOW IN DRIVE SECOND

Figure 3-17. When the transmission has shifted from low to second (with lever in D position), the kickdown band (intermediate band) is applied and the rear clutch is still applied as it was in low. (*Courtesy of Chrysler Corporation*)

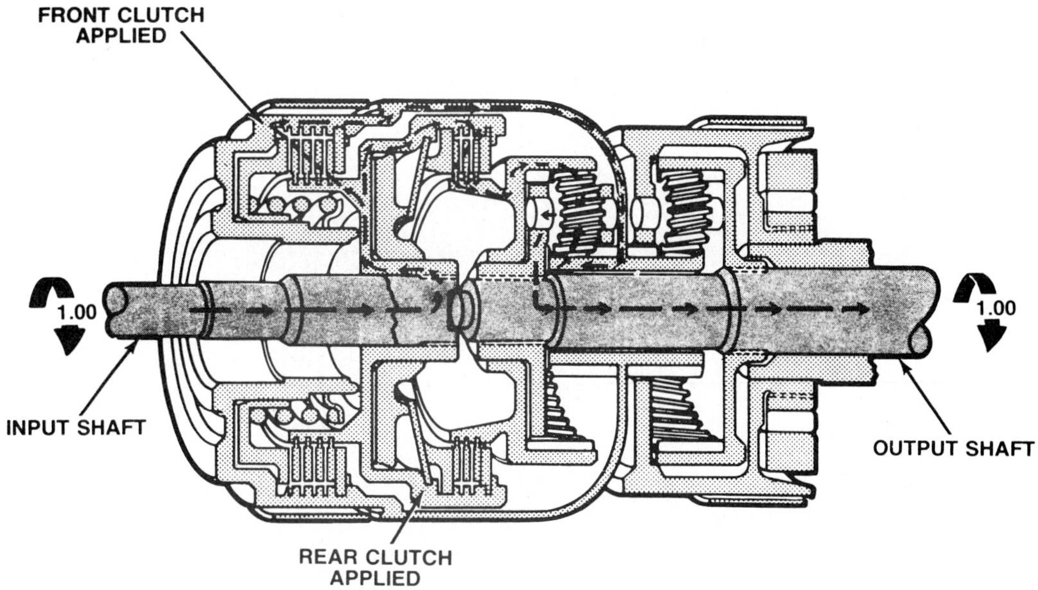

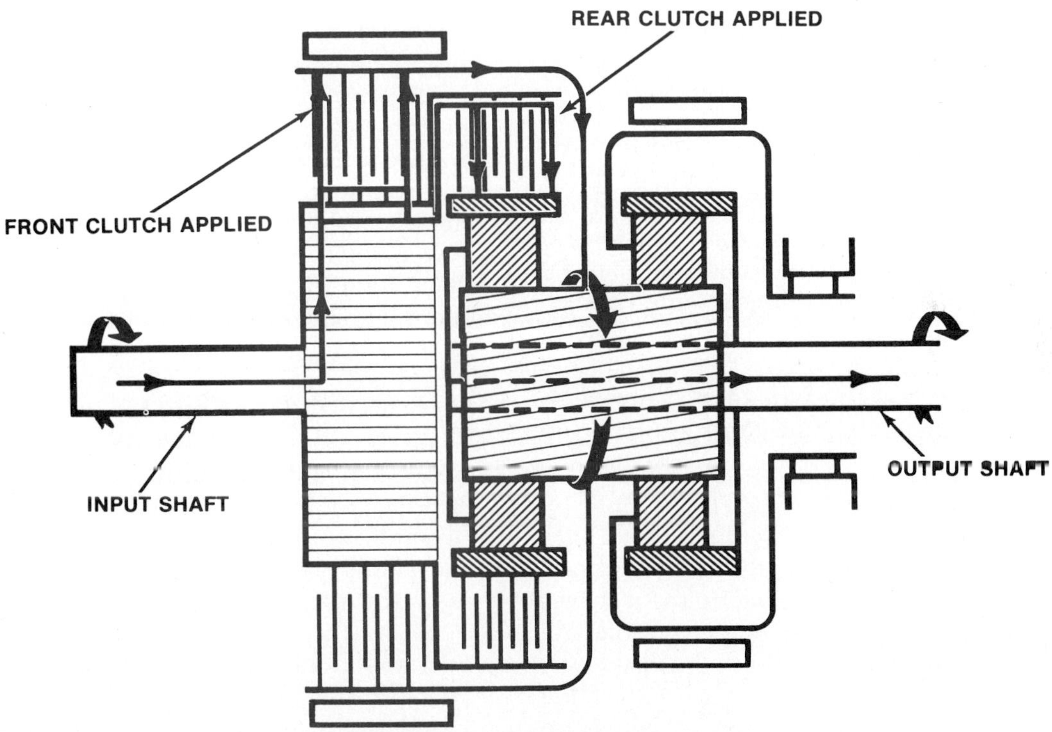

POWER FLOW IN DIRECT DRIVE

Figure 3-18. When the transmission shifts automatically from second to third, the intermediate band releases and the front clutch is applied. The rear clutch is also still applied providing direct drive. (The entire gear train is locked together and rotates as a unit.) (*Courtesy of Chrysler Corporation*)

LOW (D) (Breakaway) 2.45:1	LOW (1) (Manual) 2.45:1	SECOND 1.45:1	DIRECT 1.0:1	REVERSE 2.20:1
REAR CLUTCH DRIVES FRONT ANNULUS GEAR	REAR CLUTCH DRIVES FRONT ANNULUS GEAR	REAR CLUTCH DRIVES FRONT ANNULUS GEAR	REAR CLUTCH DRIVES FRONT ANNULUS GEAR	FRONT CLUTCH DRIVES SUN GEAR
OVERRUNNING CLUTCH HOLDS REAR PLANET CARRIER	LOW AND REVERSE BAND HOLDS REAR PLANET CARRIER	KICKDOWN BAND HOLDS SUN GEAR	FRONT CLUTCH DRIVES SUN GEAR	LOW AND REVERSE BAND HOLDS REAR PLANET CARRIER

Figure 3-19. Summary of clutch and band applications in all gear positions. *(Courtesy of Chrysler Corporation)*

reverse direction, turning the drive wheels in reverse. Reverse gear ratio is generally approximately 2.20:1 to 2.45:1, depending on transmission make and model as well as engine application.

In reverse only one half of the Simpson gearset is in actual use; the other half of the gearset is also turning, but is not transmitting power.

Drive Position

When the selector lever is placed in the drive position in the three-speed automatic transmission, the transmission starts off in first (low) gear. When the vehicle reaches the required road speed, the transmission automatically shifts to second (intermediate) gear. As road speed increases further to the required speed, the transmission shifts automatically from second to third (direct).

The power flow in each of the three gears—first (low), second (intermediate), and third (direct)—follows.

First Gear (Low)

In first gear the forward clutch is applied. This drives the ring gear of the forward half of the planetary gearset clockwise. The clockwise-turning ring gear drives the pinions clockwise, which in turn drive the sun gear in a counterclockwise direction. Since the forward sun gear extends into the reverse half of the planetary gearset, the counterclockwise-turning sun gear drives the pinions in the reverse half of the planetary gears in a clockwise direction. The pinions drive the reverse ring gear and output shaft in a clockwise direction, causing the drive wheels to turn in a forward direction.

All of this could not happen, however, unless the planet pinion carrier in the reverse half of the gearset is prevented from turning in a reverse (counterclockwise) direction. It tries to turn backward since the vehicle load places a resistance to rotation on the output shaft and reverse ring gear. Backward turning of the reverse carrier is prevented by a one-way roller clutch, which allows clockwise rotation of the carrier but prevents counterclockwise rotation.

Although the low and reverse band (or clutch) can also prevent rear carrier rotation, it is not applied in drive low. The one-way roller clutch provides instant release (overrun) when the transmission upshifts automatically to second and provides instant holding power when the transmission automatically downshifts from third or second to first. A hydraulically operated band or clutch could not do this as quickly or as smoothly.

Second Gear (Intermediate)

When the vehicle reaches a high enough road speed (about 7 mph [11 km/h] to 12 mph [19km/h] at light throttle), the hydraulic system in the transmission causes the transmission to shift to second gear. In second gear the forward clutch remains on, just as in first gear, still driving the forward ring gear clockwise. However, the intermediate band (kickdown band or intermediate clutch in some transmissions) is applied. This prevents the sun gear from turning.

Since the forward ring gear is still turning clockwise, it drives the pinions in a clockwise direction. However, since the sun gear cannot turn, the clockwise turning pinions are forced to "walk" around the stationary sun gear, taking the pinion shafts and carrier with them in a clockwise direction. Since the forward carrier is splined to the front end of the transmission output shaft, the shaft is forced to turn clockwise, driving the drive wheels in a forward direction in second gear.

In second gear only the forward half of the planetary gearset is in actual use to transmit power. The reverse half of the gearset is also turning but is not involved in transmitting any power. The second gear ratio is approximately 1.45:1. In second gear the reverse pinion carrier is turning in a forward direction, which causes the one-way clutch to overrun.

Third Gear (Direct)

As the vehicle's road speed continues to increase to about 12 mph (19 km/h) to 20 mph (32 km/h) at light throttle, the hydraulic system causes the transmission to shift automatically from second to third.

In third or direct, the forward clutch remains on, still driving the forward ring gear clockwise just as in first and second gears. In third, however, the intermediate band (or clutch) must be released and the direct clutch (reverse and high clutch) applied. The direct clutch connects the sun gear to the transmission input shaft. Since the forward ring gear is also connected to the input shaft through the forward clutch, there are now two planetary gear members connected to the input (turbine) shaft.

Connecting two planetary gear members to the input shaft causes the entire gear train to turn as a single locked-up unit. There is no individual rotation of gears, and the transmission input shaft and output shaft turn at the same speed, or 1:1 direct drive.

Manually Selected Low (1 or L) Position

When the selector lever is placed in the manual low (1 or L, depending on the make and model of the transmission) position, the gear train operates exactly as in the drive low range described earlier but with two exceptions.

First, the low and reverse band (or clutch) is applied to hold the reverse carrier. This increases the holding power on the reverse carrier and also provides engine braking during deceleration. Increased holding power is desirable for pulling heavy loads in low gear. Engine braking is also desirable.

Second, the automatic hydraulic shift circuit is locked out in manual low so that no automatic upshift takes place. The transmission starts in low gear and stays in low unless the selector lever is shifted to the 2 or D position.

Manually Selected Second (2 or S) Position

With the selector lever in the 2 position on the quadrant (S for super on some transmissions) the transmission starts in low gear just as it does in the D position and shifts automatically to second gear when the vehicle reaches the required road speed for a 1–2 upshift. The transmission then stays in second gear unless the driver downshifts by depressing the accelerator, when vehicle speed drops to the automatic 2–1 downshift speed, or when the selector lever is moved to the D or L position. (Some transmissions use the letter *I* for intermediate or second.)

First- and second-gear power flow is exactly as described for the D selector position. However, there is no upshift from second to third because the selector lever position has locked out the 2–3 automatic upshift hydraulic circuit in the transmission.

Some three-speed automatic transmissions start in second gear when the 2 position is selected. In other words, when the second or 2 position is selected with the shift lever, the transmission operates in second gear only with no upshift or downshift. The position of the shift lever locks out the first- and third-gear hydraulic circuits in the transmission.

THM 325-4L Four-Speed Overdrive

The THM 325-4L Turbo Hydra-Matic transmission is a four-speed overdrive transmission with a three-element lockup torque converter for use in front-wheel-drive vehicles. It consists of the common Simpson compound planetary gear drive train chain driven from the overdrive planetary gearset and torque converter.

The advantages of overdrive include reduced engine speed (in relation to road speed) and therefore less fuel consumption and less wear on engine parts and accessories.

The lockup torque converter and Simpson gear train operate as described earlier. The addition of a simple planetary gearset between the torque converter and the Simpson gear train allows the transmission input shaft to be driven at either turbine shaft speed or at overdrive speed, which is 0.67:1 (turbine shaft to transmission input shaft ratio). The overdrive planetary gear unit is controlled by two multiple-disc clutches and a one-way roller clutch.

Overdrive is achieved when the carrier is driving, the sun gear is held, and the ring gear is the driven member. The ring gear has external teeth to engage the drive link or chain.

The overdrive carrier is splined to the turbine shaft of the converter. The carrier is attached to the outer race of the one-way roller clutch. The sun gear is attached to the inner race of the roller clutch.

Because of the resistance to rotation of the ring gear (because of vehicle load), the carrier and pinions try to turn the sun gear faster than the carrier. This causes the one-way clutch to lock up, turning the car-

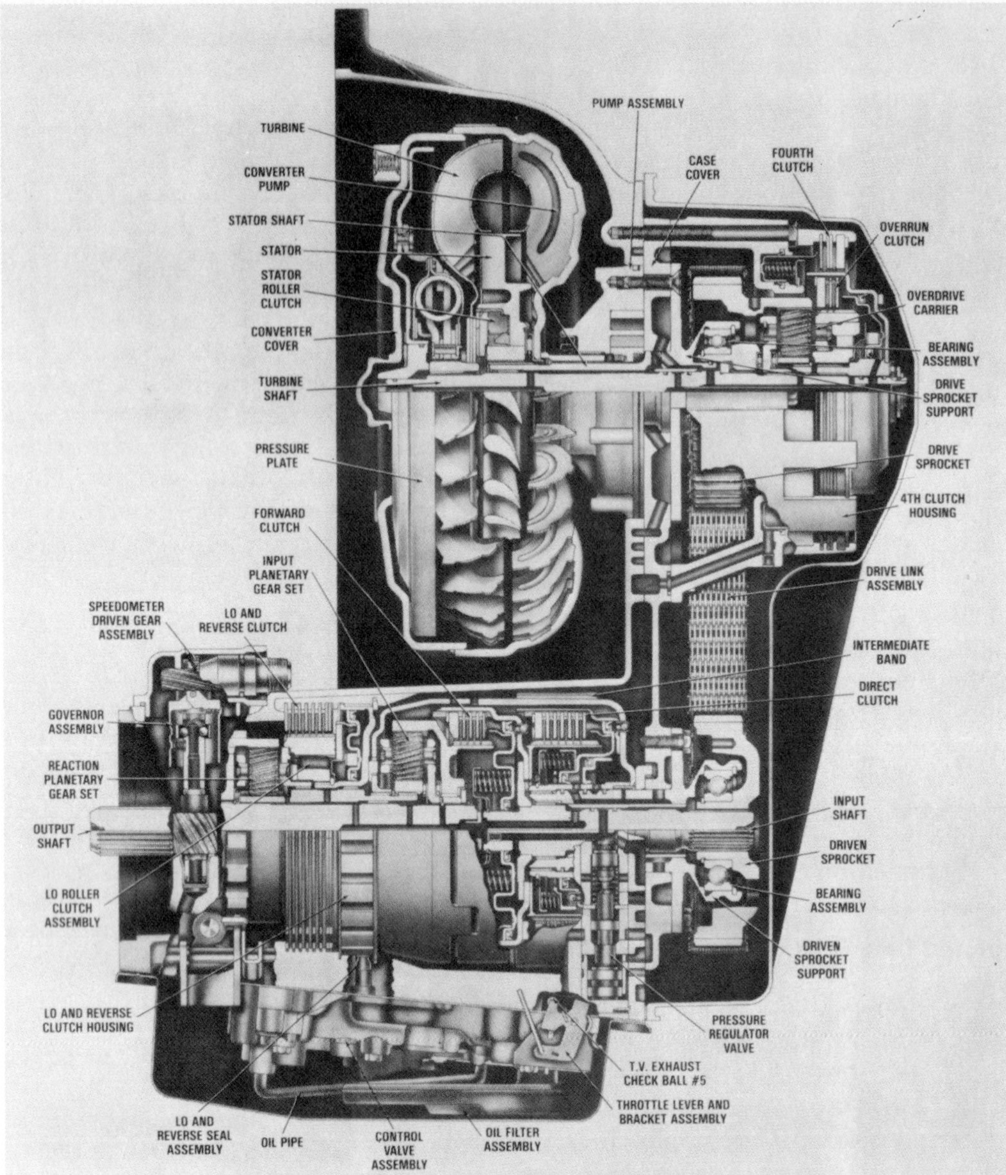

Figure 3-20. General Motors 325-4L four-speed overdrive transmission for front-wheel drive. (*Courtesy of General Motors Corporation*)

rier, sun gear, and ring gear all at the same speed. This is the condition any time engine torque is being applied. When the driver releases the accelerator pedal, the vehicle drive wheels try to drive the engine through the transmission and overdrive unit.

However, under these conditions the one-way clutch releases and overruns, allowing the sun gear to turn faster than the carrier. This results in no engine braking when the selector lever is in the drive position. In manually selected first, second, or third, the overrun clutch applies, locking the sun gear to the carrier, which prevents the one-way clutch from overrunning. This allows engine compression to be used for engine braking when going downhill.

Overdrive occurs with the selector lever in the drive (D) position when the transmission has shifted into fourth gear (overdrive). In fourth-gear overdrive the fourth clutch applies to hold the overdrive sun gear. The torque converter drives the carrier, which in turn drives the overdrive ring gear at an increased speed.

The clockwise-turning carrier forces the pinions to walk around the stationary sun gear, which causes the ring gear to turn faster than the carrier.

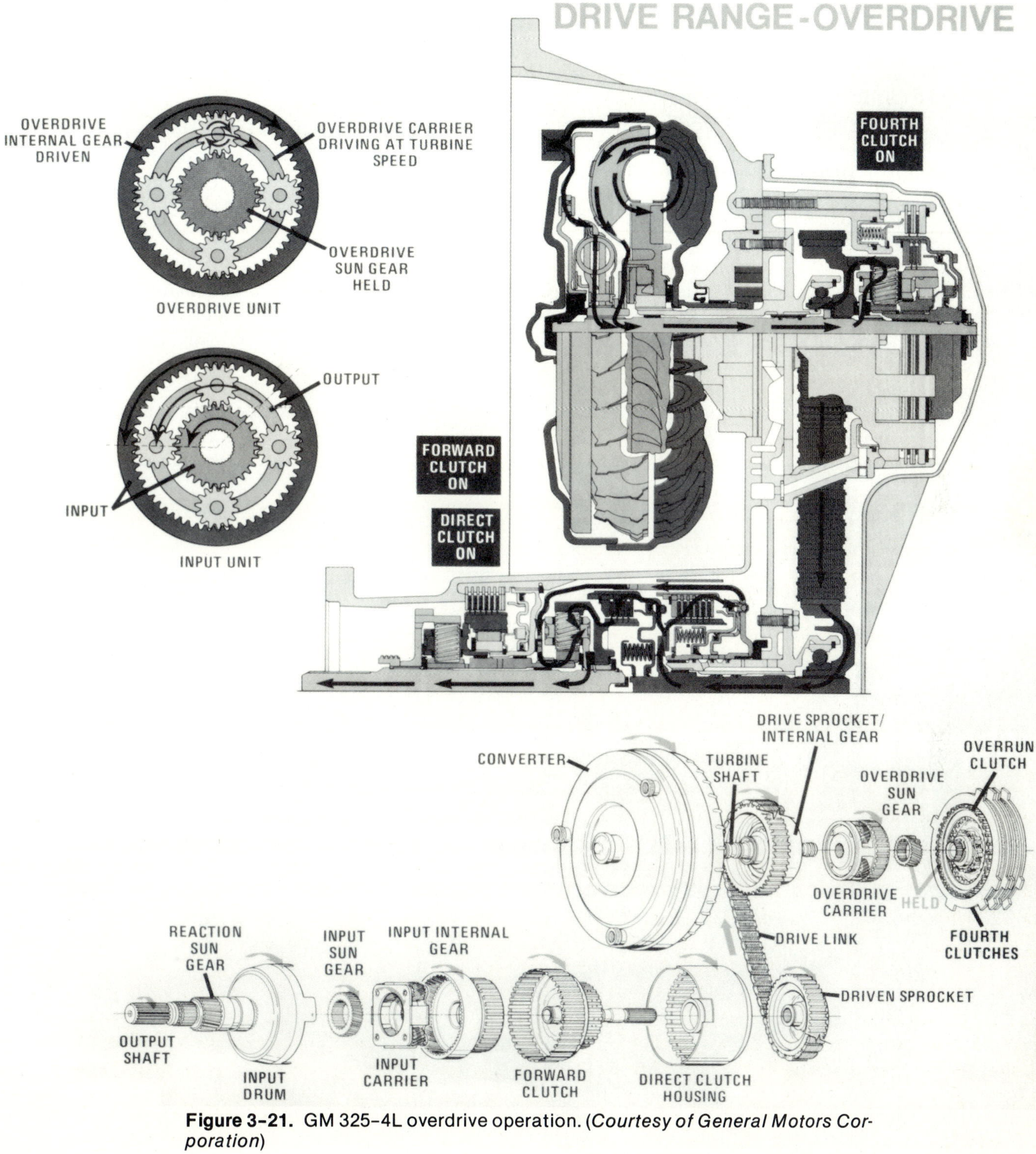

Figure 3-21. GM 325-4L overdrive operation. (*Courtesy of General Motors Corporation*)

Torque converter lockup occurs in drive-range second, third, and fourth only.

THM 700-R4 Four-Speed Overdrive

The THM 700-R4 Turbo Hydra-Matic transmission is a four-speed overdrive unit with a lockup torque converter. The planetary gear train resembles the Simpson gearset but operates differently. The torque converter locks up in drive-range second, third and fourth only.

The planetary gearset consists of two simple planetary gearsets similar to the Simpson component gear train. However, in the 700-R4 transmission the two sun gears are able to operate independently as necessary, which is not the case with the common Simpson gearset. In this transmission five multiple-disc clutches, two roller clutches, and one band control operation of the gearset.

Overdrive is achieved when the selector lever is in the drive range and the transmission has upshifted into fourth-gear overdrive. The rear or reaction planetary gearset produces overdrive when the 2–4 band prevents the reaction sun gear from turning.

The 3–4 clutch is applied to drive the front ring gear. The front ring gear is splined to the rear carrier, which is now driven at turbine speed.

With the 2–4 band on, the rear sun gear cannot turn. This forces the pinions in the rear carrier to walk around the stationary sun gear and causes the rear ring gear to turn faster than the carrier. The rear ring gear is splined to the transmission output shaft. This produces an overdrive ratio between turbine shaft and transmission output shaft of approximately 0.70:1.

First Gear

In first gear the forward clutch and input roller clutch are applied to drive the input or front sun gear clockwise. The input or front ring gear is prevented from turning in a counterclockwise direction by a one-way low roller clutch.

Driving the sun gear clockwise causes the pinions in the front carrier to walk around the stationary front ring gear. This forces the carrier and output

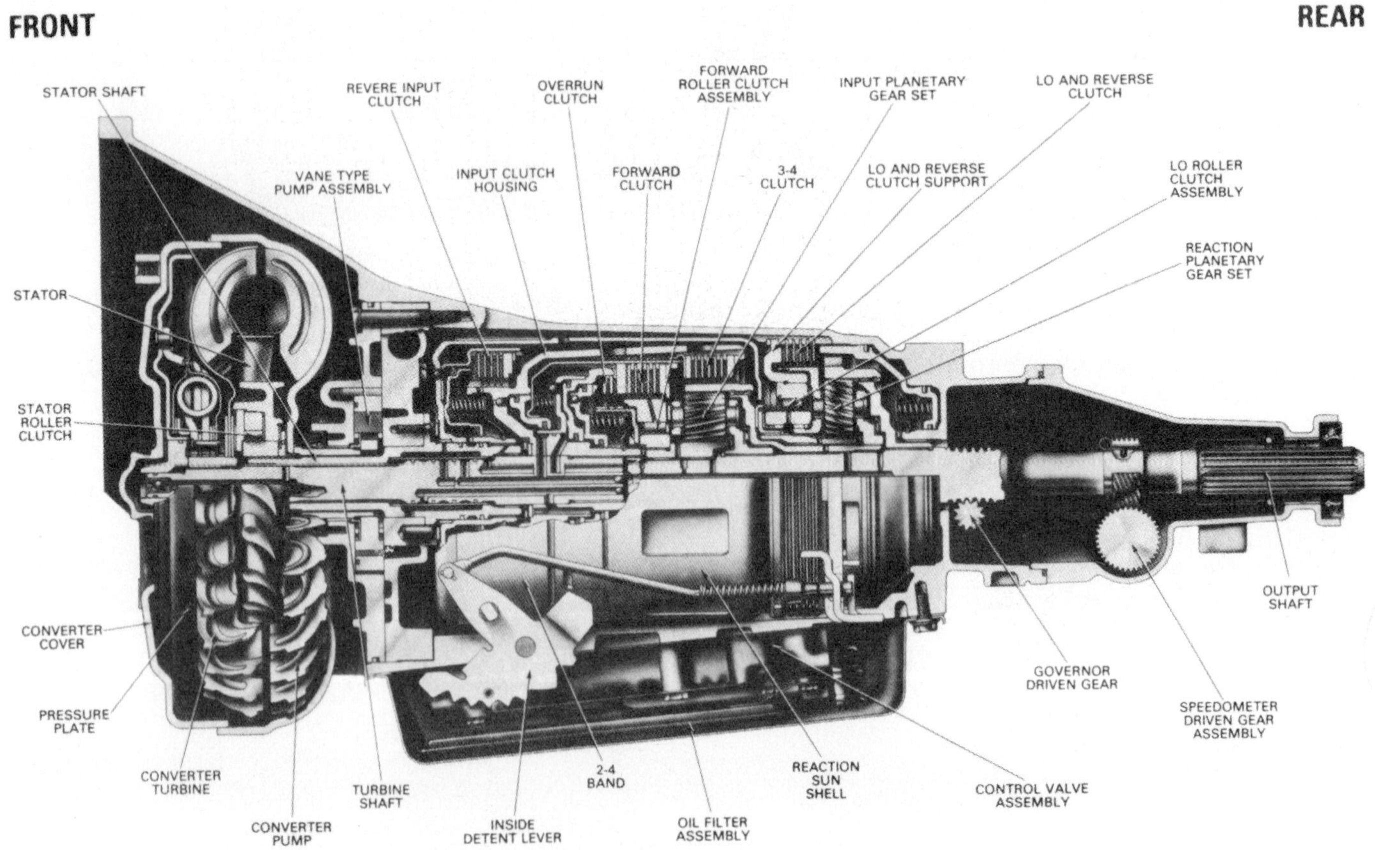

Figure 3-22. GM 700–R4 four-speed overdrive transmission. (*Courtesy of General Motors Corporation*)

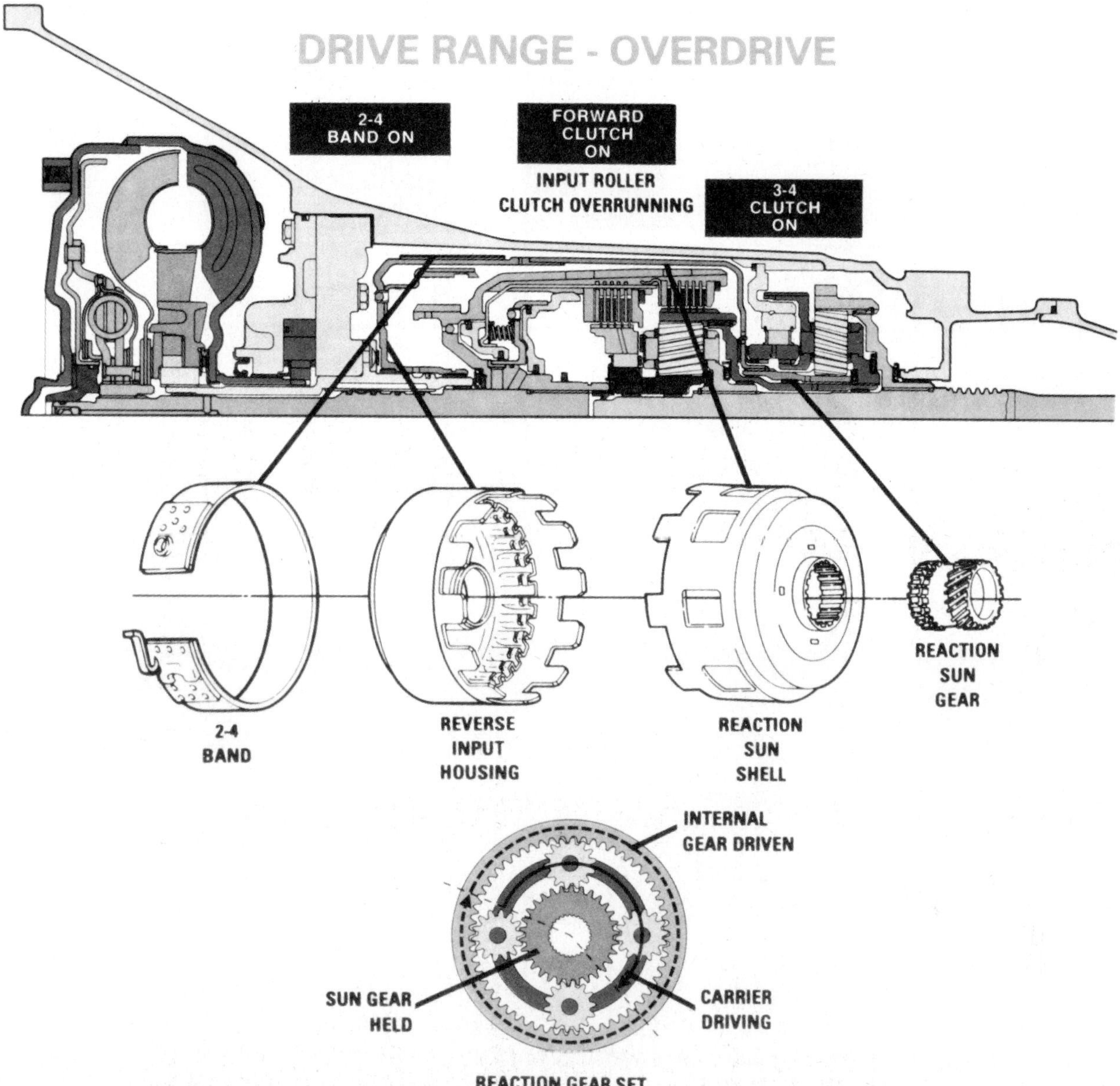

Figure 3-23. GM 700-R4 overdrive operation. (*Courtesy of General Motors Corporation*)

shaft (to which it is splined) to turn at reduced speed for low gear. This is the condition when the transmission is in low gear and the selector lever is in the drive, 3, or 2 position. This condition does not provide any engine braking since the low roller clutch will overrun on deceleration.

In the manually selected low or 1 position the only difference is that the low and reverse multiple-disc clutch is applied to prevent the front ring gear from turning and the low roller clutch from overrunning. This position provides engine braking. No automatic upshifts occur since all the hydraulic upshift circuits are locked out in this selector lever position.

Second Gear

To shift from first to second gear, the 2–4 band is applied to hold the reaction or rear sun gear. The forward clutch is still on, which still drives the front or input sun gear through the input roller clutch in a clockwise direction.

In low gear the front or input ring gear was sta-

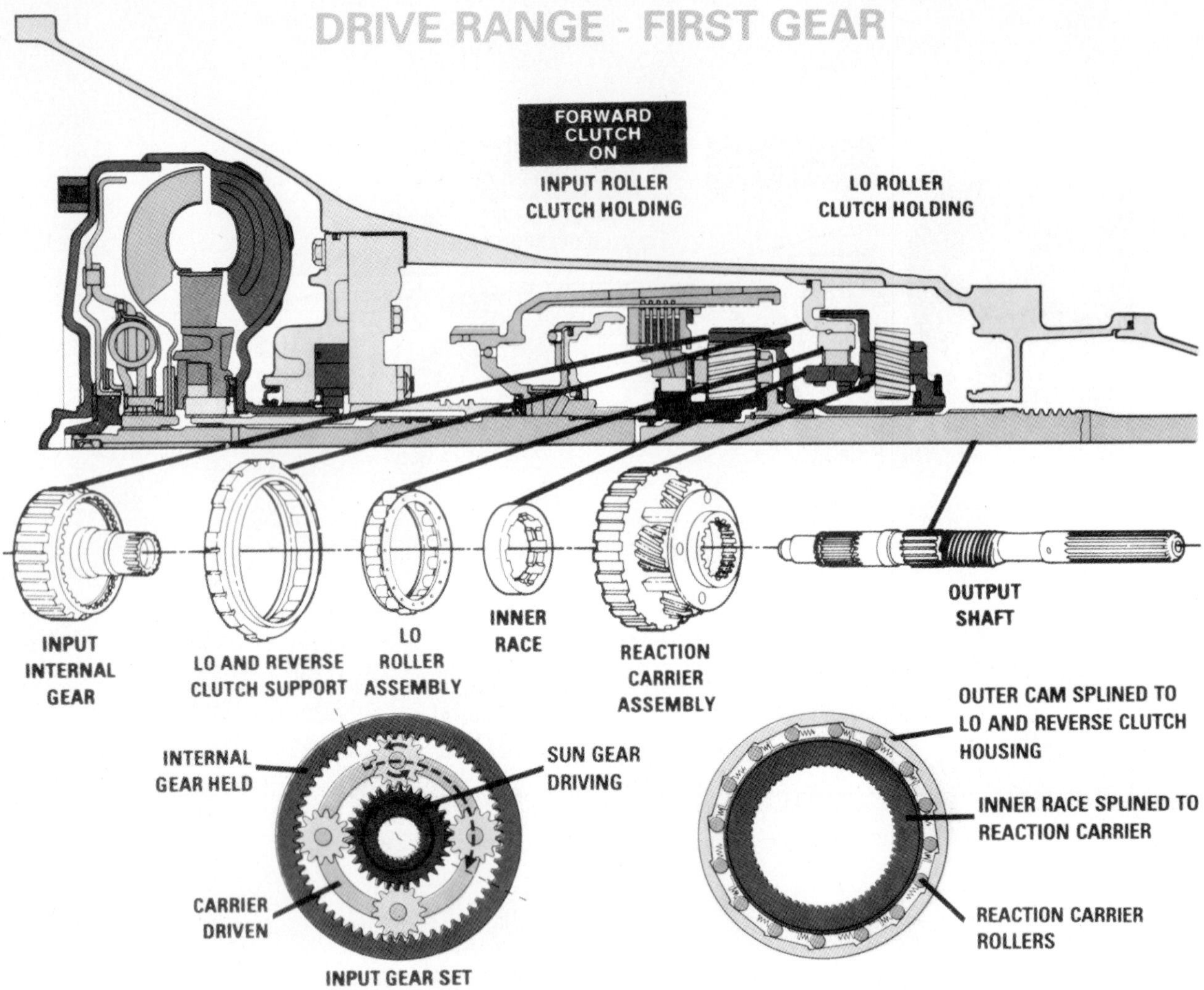

Figure 3-24. 700-R4. (*Courtesy of General Motors Corporation*)

tionary, providing a 3.06:1 sun-gear-to-carrier reduction.

In second gear the front or input ring gear is forced to turn in a forward or clockwise direction since the rear sun gear is being held by the 2–4 band. Holding the rear sun gear, which was turning counterclockwise in low gear, causes the front or input ring gear to turn in a forward to clockwise direction. This increases the speed of the input carrier and output shaft to a second gear speed of approximately 1.63:1. This arrangement does not provide any engine braking since the input roller clutch overruns when decelerating.

In manually selected second the transmission starts in low and automatically upshifts to second. Further upshifts are prevented.

However, in manual second the overrun clutch is applied, locking the input sun gear to the turbine shaft and effectively eliminating any overrun of the one-way clutch on deceleration. This provides engine braking in manually selected second gear.

Third Gear

To shift into third gear the 2–4 band is released and the 3–4 clutch is applied.

Applying the 3–4 clutch connects the front or input ring gear to the turbine shaft. Since the forward clutch and roller clutch are still on, the input sun gear is also being driven by the turbine shaft. Driving these two members of the planetary gearset causes the front carrier and output shaft to be driven at turbine shaft speed. This is direct drive, or 1:1. Engine braking is not possible since the input roller clutch will overrun when decelerating.

In the manually selected third position the trans-

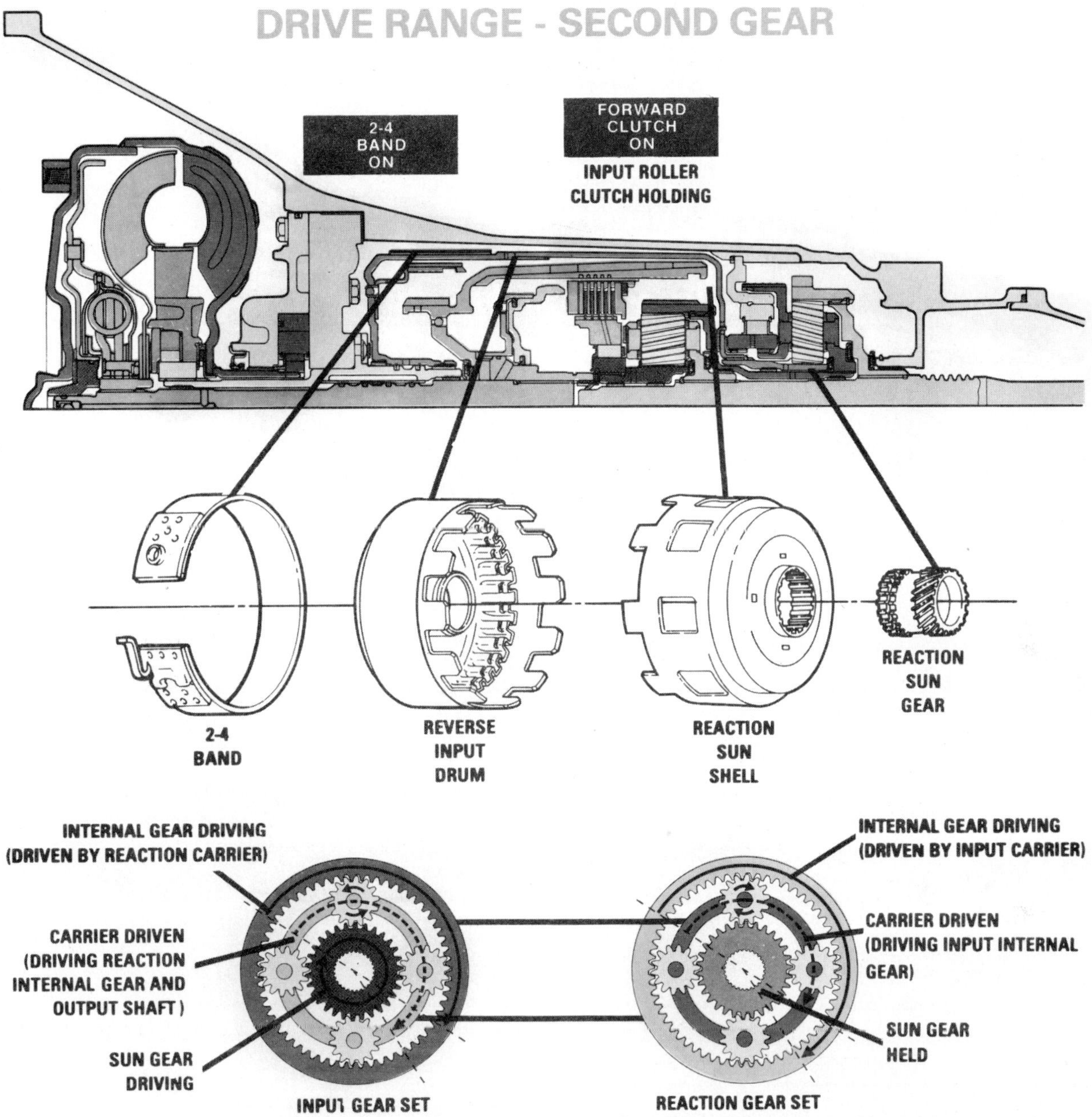

Figure 3-25. GM 700-R4. (*Courtesy of General Motors Corporation*)

mission shifts from first to second and from second to third as it does in the drive range. However, no 3–4 upshift takes place. In addition, the overrun clutch is applied just as in manual second. This prevents the input roller clutch from overrunning when decelerating, thereby providing engine braking.

Reverse

Reverse is provided in the 700-R4 transmission by applying the reverse input clutch and the low and reverse holding clutch. Applying the reverse input clutch causes the turbine shaft to drive the rear or reaction sun gear in a clockwise direction. Applying the low and reverse holding clutch prevents the reaction carrier from turning.

The clockwise-turning sun gear drives the pinions counterclockwise. The pinions drive the ring gear and output shaft (to which it is splined) in a counterclockwise or reverse direction at a reduction of about 2.30:1.

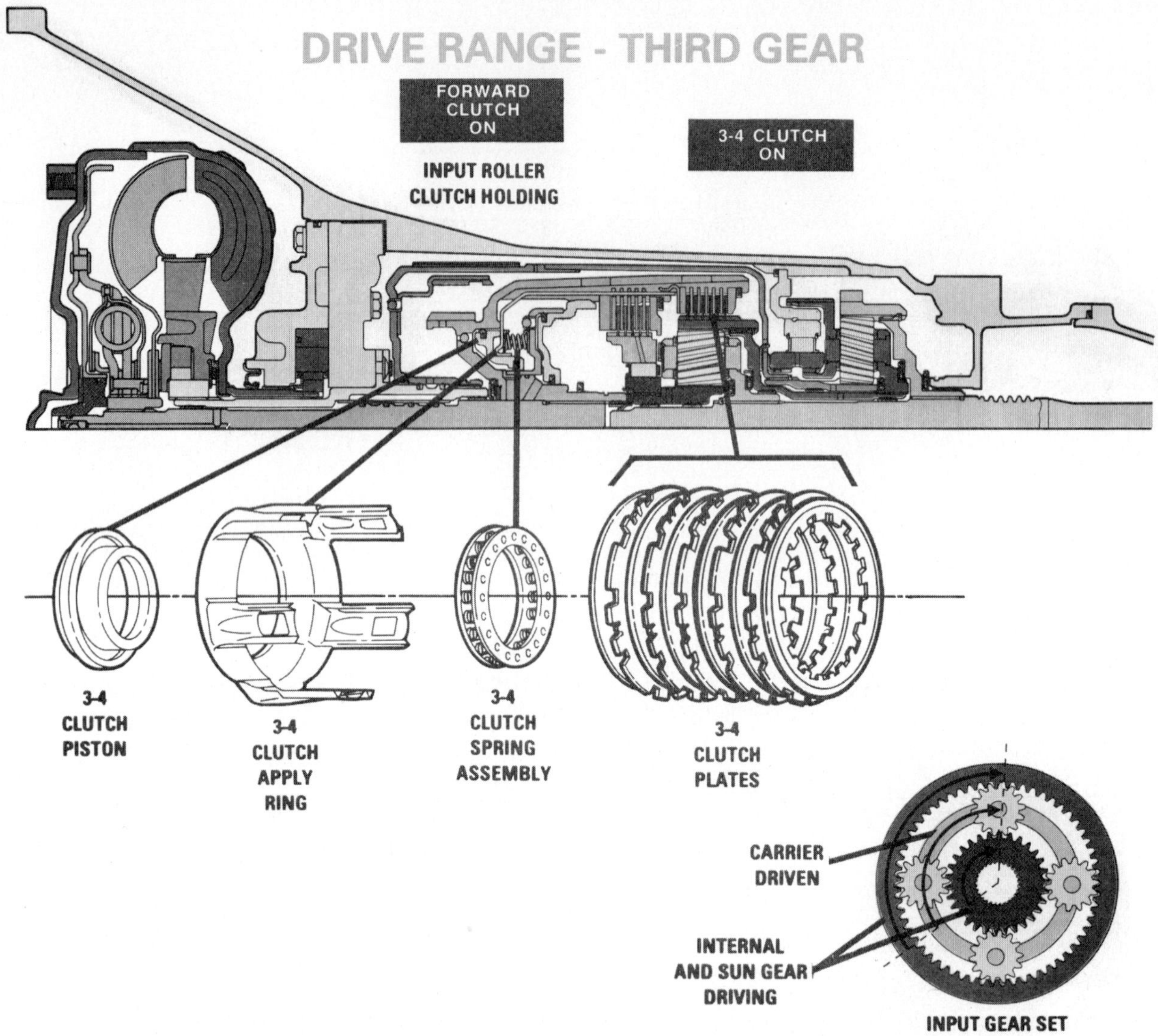

Figure 3-26. GM 700-R4. (*Courtesy of General Motors Corporation*)

PART 3-5 RAVIGNEAUX PLANETARY GEARS

The Ravigneaux compound planetary gear train consists of two sun gears, one planet pinion carrier containing two sets of pinions, and one ring gear. The discussion of this gearset is confined to the three-speed and four-speed overdrive arrangements and does not include the older two-speed version.

The Ravigneaux planetary gear system is used in several domestic and import automatic transmissions. Among these are Ford's FMX, automatic overdrive, and ATX transaxle; General Motors' (MD3) 180; BMW's 3HP22; Mitsubishi's Borg-Warner; Volvo's Borg-Warner; and others.

This discussion will use the Ford FMX three-speed, Ford automatic overdrive four-speed, and ATX transaxle arrangements to explain the operation of this gearset.

Components and Construction (FMX)

In the three-speed arrangement of this planetary gearset there are two sun gears, one behind the other and each free to turn separately. The front sun gear is slightly larger than the rear sun gear.

The rear sun gear is in mesh with one set of short pinions. These short pinions are in mesh with a set of long pinions mounted in a single carrier. The long pinions are in mesh with the front or larger sun gear and also with the ring gear or internal gear. This completes the physical arrangement of these gears.

The ring gear also has external teeth on it, which

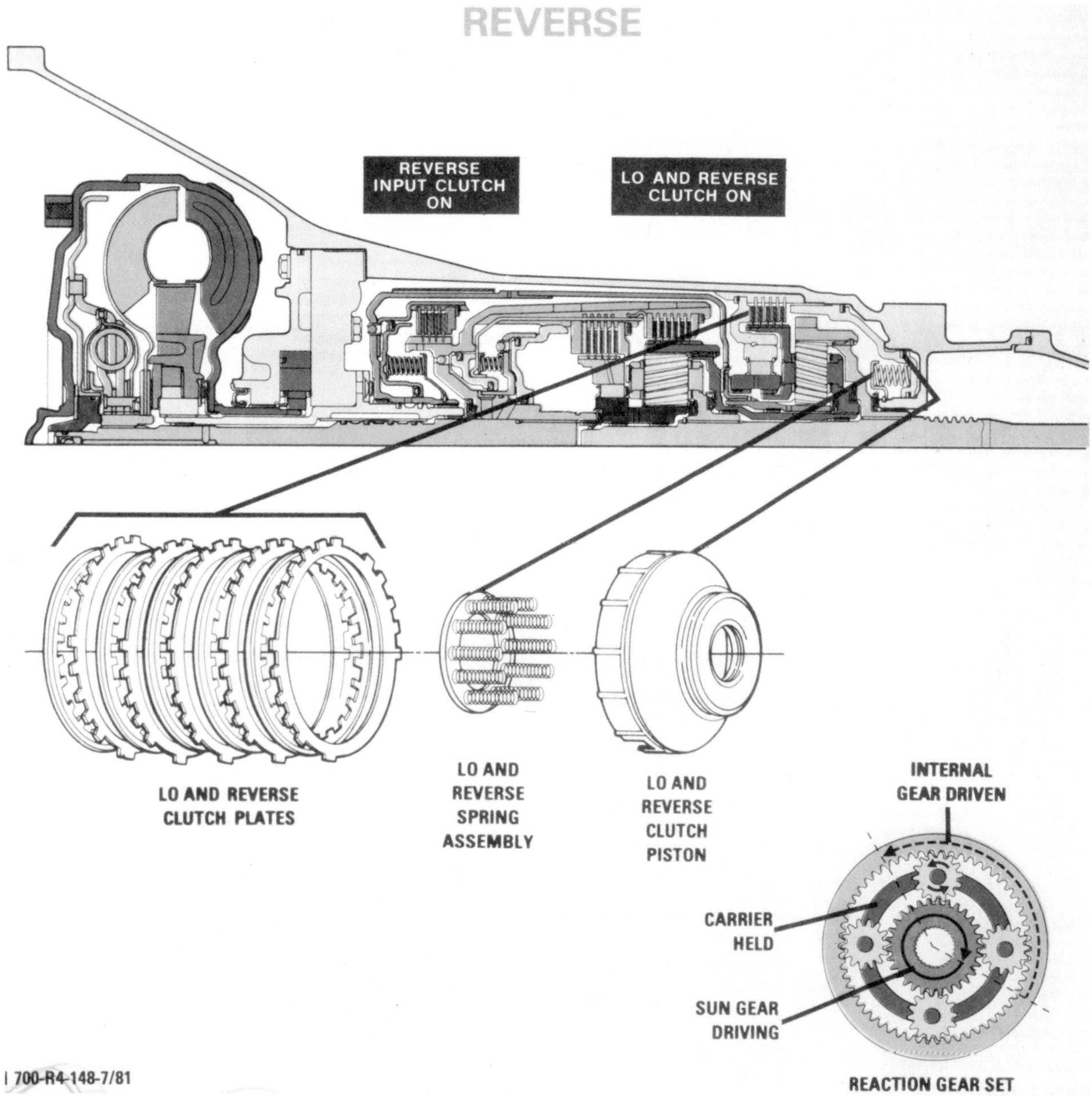

Figure 3-27. GM 700-R4. (*Courtesy of General Motors Corporation*)

are used to provide the park position for this gearset. All gears have helical teeth for quiet operation and are highly finished high-grade steel for toughness and durability.

The gearset is controlled by two multiple-disc clutches, two bands, and a one-way clutch.

Operation

The selector lever has six positions: park, reverse, neutral, drive, 2, and 1.

In park, the ring gear and output shaft are held stationary by a park pawl, which engages the external teeth on the ring gear. The park pawl is anchored to the transmission case.

In reverse, the rear (reverse and high) clutch is applied to drive the front or secondary sun gear in a clockwise direction. The secondary sun gear drives the secondary (long) pinions counterclockwise. The counterclockwise-turning secondary pinions drive the ring gear and output shaft in a counterclockwise direction for reverse. The low and reverse holding clutch is applied to prevent the carrier from turning in a forward direction. It tries to do this since there is a resistance to motion of the ring gear and output shaft due to vehicle load.

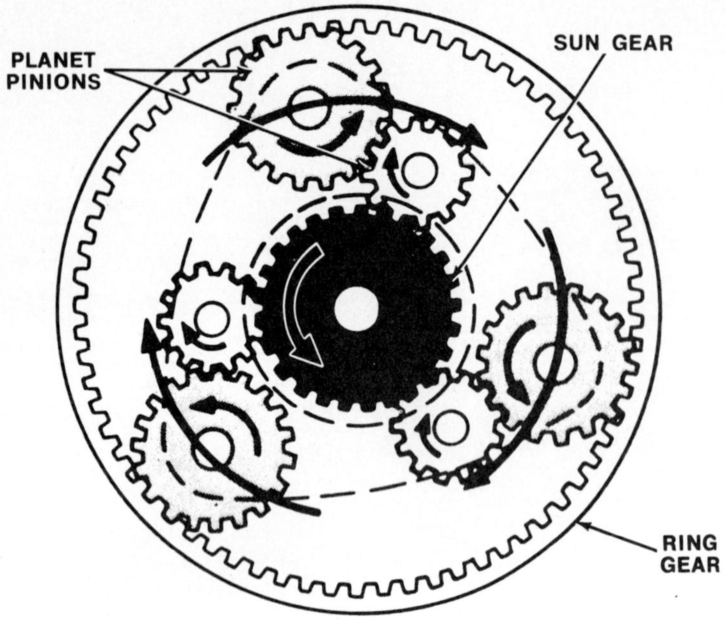

Figure 3-28. Ravigneaux single-carrier, dual-pinion, dual-sun-gear, single-ring-gear, compound-planetary-gearset used in Ford FMX and automatic overdrive transmissions.

In neutral no clutches or bands are applied.

In the drive position, the transmission starts in low and shifts automatically to second and to third.

Low Gear

In low gear the front or forward clutch is applied to drive the primary or smaller sun gear clockwise. The primary sun gear drives the primary (short) pinions in a counterclockwise direction. The primary pinions drive the secondary pinions in a clockwise direction, causing the ring gear and output shaft to turn forward (clockwise).

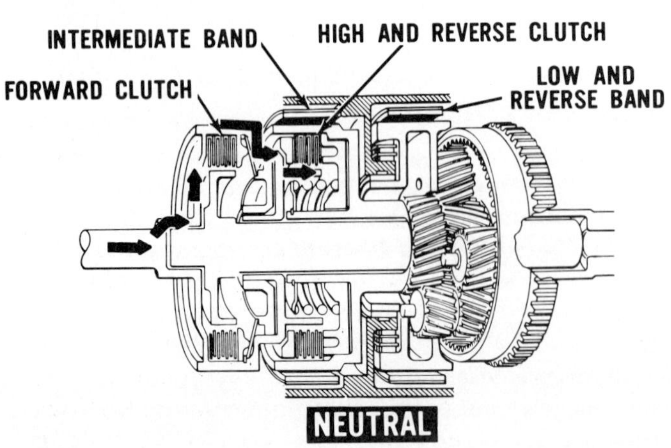

Figure 3-29. In neutral, no clutches or bands are applied; therefore, no power is transmitted to the output shaft in this Ravigneaux three-speed gearset. (*Courtesy of Ford Motor Company*)

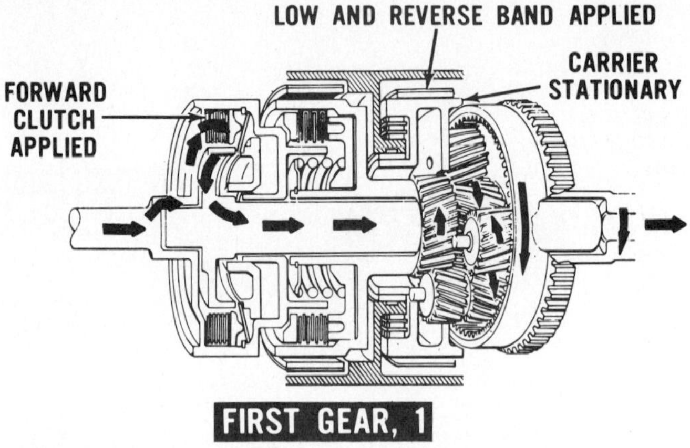

Figure 3-30. In first gear at selector level position 1, the pinion carrier is held against rotation by the low and reverse band. (*Courtesy of Ford Motor Company*)

Vehicle load impacts a resistance to output shaft and ring gear rotation. This results in the carrier trying to turn backward (counterclockwise). This is prevented by the one-way clutch, which is mounted between the carrier and a center support attached to the transmission case.

During the application of engine torque, the one-way clutch locks up, preventing carrier rotation. During deceleration the one-way clutch freewheels or overruns, resulting in no engine braking.

When manual low is selected, the low and reverse band is applied to lock the carrier effectively to the transmission case. This provides for vehicle retard by engine braking. No automatic upshifts occur in the manually selected low gear position.

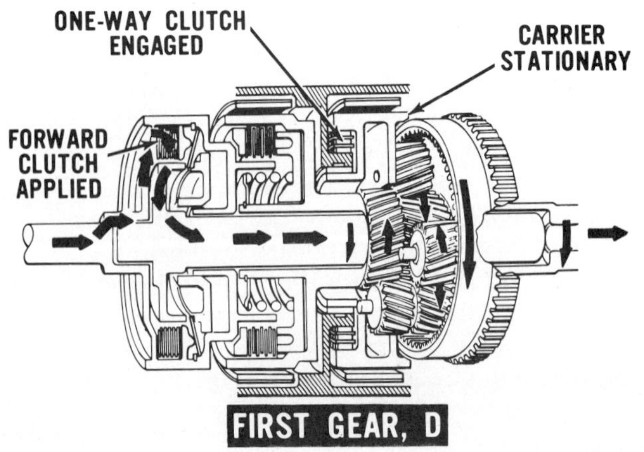

Figure 3-31. In first gear at selector lever position D, the pinion carrier is held against rotation by the one-way clutch instead of the low and reverse band. First gear in D is the only gear that uses the one-way clutch. (*Courtesy of Ford Motor Company*)

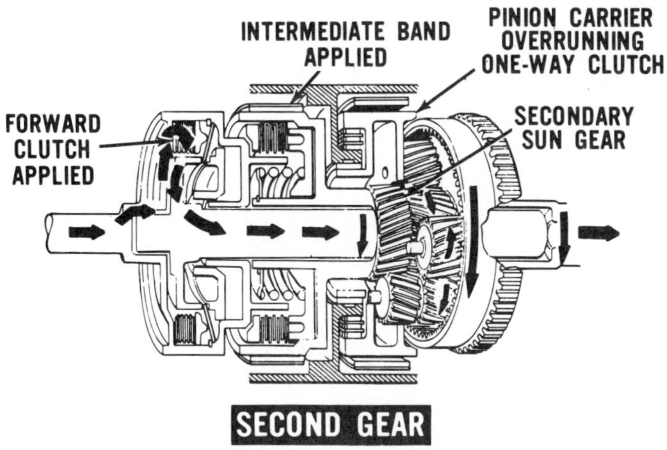

Figure 3-32. Second-gear ratio is obtained by driving the primary sun gear and holding the secondary sun gear. The primary pinions drive the secondary pinions, causing them to "walk" around the secondary sun gear and to carry the internal gear and output shaft around with them. (*Courtesy of Ford Motor Company*)

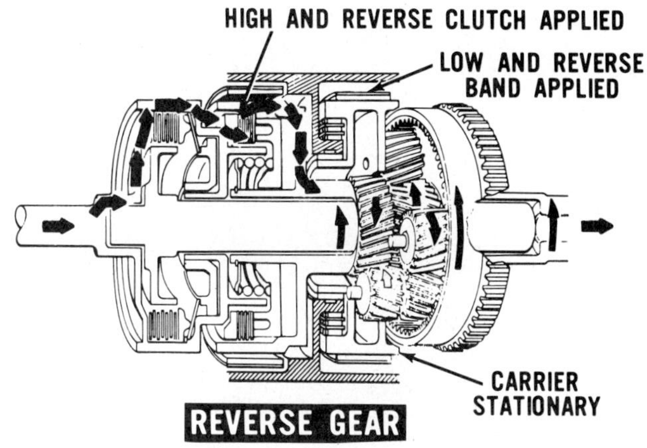

Figure 3-34. Reverse gear is obtained by driving the secondary sun gear and holding the pinion carrier. The secondary pinions drive the internal gear in the reverse direction. The primary sun gear and the primary pinions rotate freely and have no effect on the gear train. (*Courtesy of Ford Motor Company*)

Second Gear

When the transmission shifts to second gear, the intermediate (front) band is applied. Applying the intermediate band holds the secondary (front) sun gear stationary. In first gear the secondary sun gear was allowed to turn backward (counterclockwise). In second gear the intermediate band prevents this. The result is that the secondary pinions are forced to walk around the stationary secondary sun gear. This causes the carrier to be turned clockwise.

In first gear the carrier was stationary and the ring gear and output shaft were driven by secondary pinion rotation only. In second gear additional ring gear speed results from the forward turning carrier while the pinions are still turning on their axes just as they were in first gear.

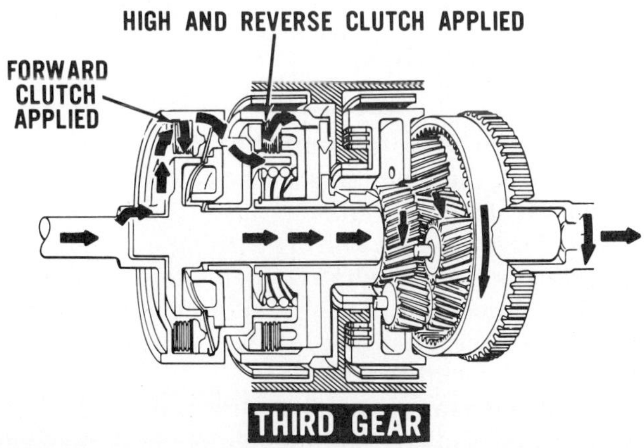

Figure 3-33. In third gear, the primary and secondary sun gears are locked together and driven as a unit. Therefore, the pinions cannot rotate and the entire planetary train revolves as a unit, which causes the output shaft to rotate at the same speed as the turbine shaft. (*Courtesy of Ford Motor Company*)

As long as engine torque is applied, the reaction on the secondary sun gear attempts to turn it backward. The intermediate band prevents this. On deceleration the reaction on the secondary sun gear is opposite; however, the intermediate band still holding the secondary sun gear provides for vehicle retard by engine braking.

In manually selected second the transmission starts in low and shifts to second automatically. No upshift to third occurs. The transmission operates in second as just described.

Ford Automatic Overdrive (Four-Speed)

The Ford automatic overdrive four-speed transmission uses four multiple-disc clutches, two one-way clutches, two bands, and two input shafts to control a Ravigneaux-type planetary gear system. A tubular turbine shaft is used, allowing a direct-drive shaft to be located inside the turbine shaft. The direct-drive shaft is spring-dampened in the torque converter as described in Chapter 2.

Selector lever positions are park, reverse, neutral, drive, 3, and 1. The drive position is the overdrive position, and the 3 position locks out the overdrive.

In the park position the park pawl locks the ring gear and output shaft to the transmission case. The

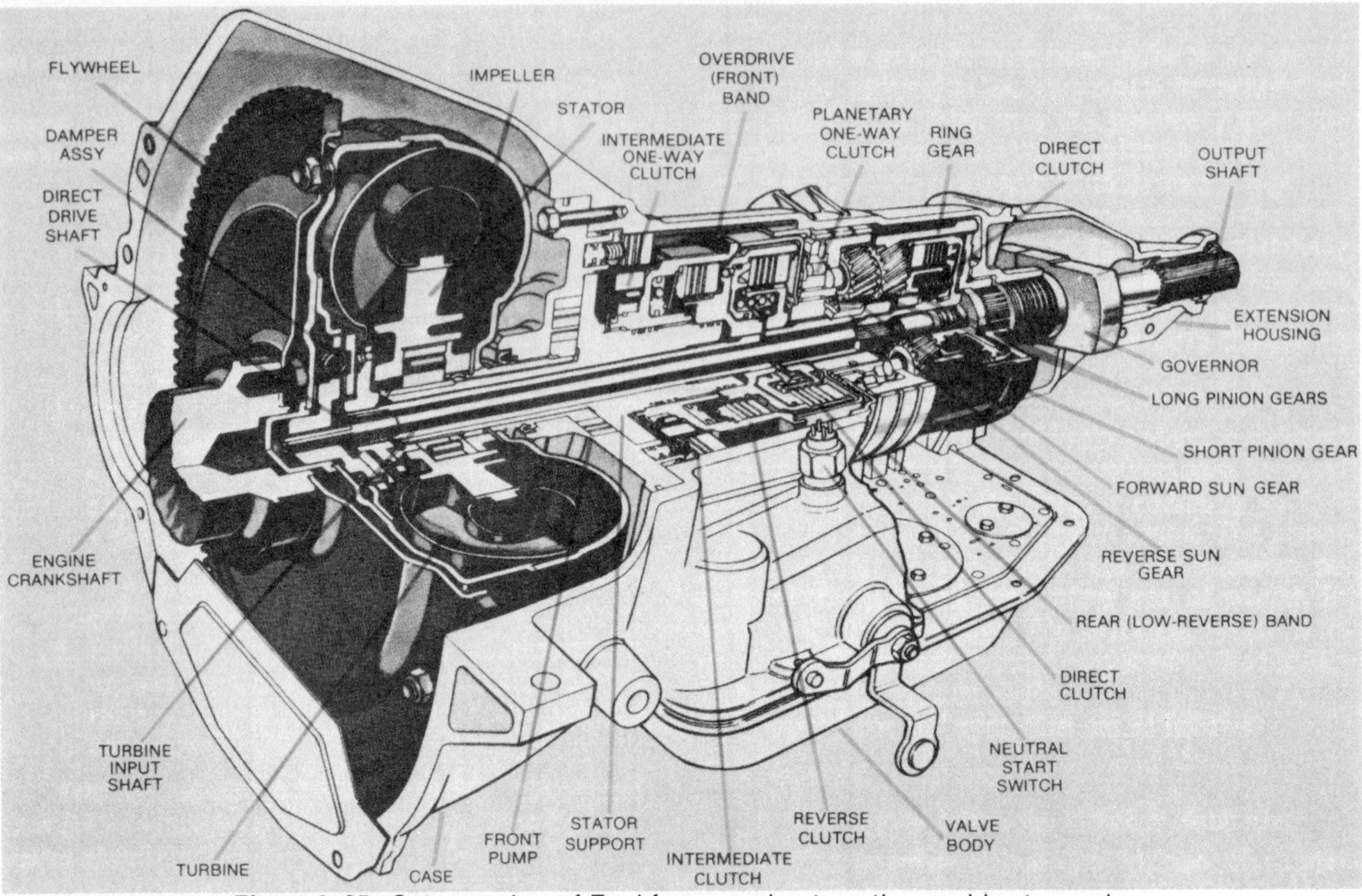

Figure 3-35. Cutaway view of Ford four-speed automatic overdrive transmission with Ravigneaux planetary gearset. (*Courtesy of Ford Motor Company*)

low and reverse band is applied in park but no drive condition is possible since no clutches are applied.

In reverse the reverse clutch is applied to drive the reverse (front, larger) sun gear clockwise. The low and reverse band holds the carrier to prevent it from turning forward. This allows the clockwise-turning reverse sun gear to drive the long pinions counterclockwise. The long pinions drive the ring gear and output shaft in reverse (counterclockwise).

In neutral no clutches or bands are applied.

Low Gear

In low gear the forward clutch is applied to cause the turbine shaft to drive the forward (rear or smaller) sun gear in a clockwise direction. This drives the short pinions in a counterclockwise direction. The short pinions drive the long pinions in a clockwise direction, which drive the ring gear and output shaft forward or clockwise for low gear.

Resistance to motion of the ring gear creates a reaction on the carrier, which attempts to turn backward. This is prevented by a one-way clutch just as in the FMX three-speed transmission.

On deceleration the one-way clutch overruns, providing no engine braking.

In manual low the low and reverse band is applied to prevent carrier rotation and provide engine braking on deceleration.

Second Gear

In second gear the intermediate holding clutch is applied to lock the outer race of the intermediate one-way clutch to the transmission case. This prevents the reverse (larger) sun gear from turning counterclockwise. This is the case during engine torque application. During deceleration the intermediate one-way clutch overruns, resulting in a freewheeling–no engine braking condition.

In second gear, then, the reverse sun gear is held by the intermediate holding clutch and one-way clutch. Since the forward clutch is still applied, drive is from the turbine shaft to the forward clutch to the forward (small) sun gear driving it clockwise. The forward sun gear drives the short pinions counterclockwise. The short pinions drive the long pinions clockwise. The long pinions are forced to walk

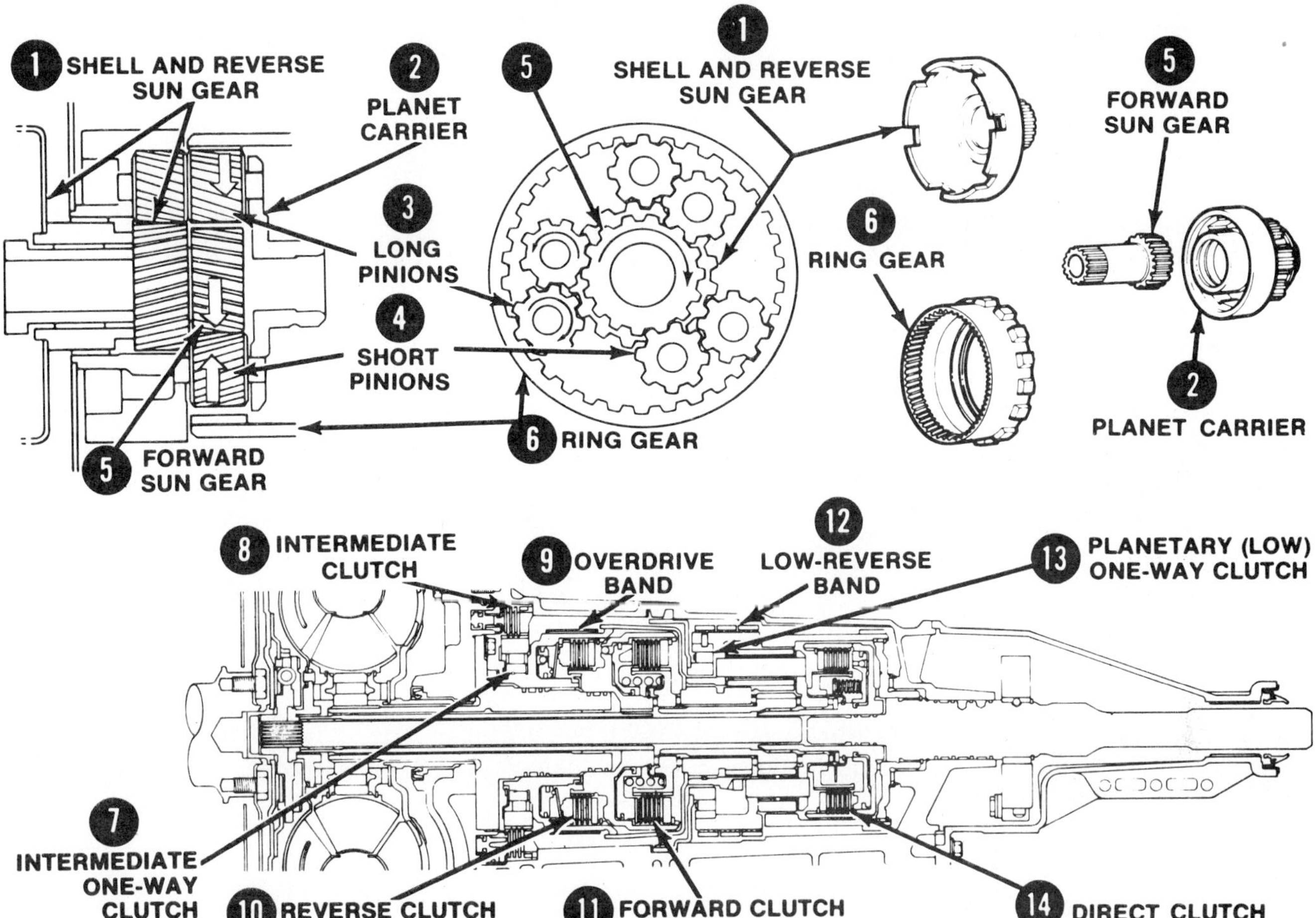

Figure 3-36. Ford automatic overdrive gear train. (*Courtesy of Ford Motor Company*)

around the stationary reverse sun gear, causing the carrier to rotate and driving the ring gear and output shaft at an increased speed compared to low gear.

The intermediate holding clutch remains applied in third and fourth gear but has no effect since the entire gear train is turning in a clockwise direction in third and fourth. The fact that the intermediate clutch remains applied in third and fourth results in smoother 3–2 downshifts.

There is no manually selected second gear in this transmission. However, if the selector lever is moved to the 1 or manual low position when car speed is above 25 mph (40 km/h), the transmission will operate in second gear until vehicle speed drops below that speed.

Third Gear (Direct)

To upshift from second gear to third gear, the direct clutch must be applied. The direct clutch is driven by the direct-drive shaft (not the turbine shaft) and when applied drives the carrier at engine speed.

Since the forward clutch is still applied, driving the forward sun gear, and the carrier is driven through the direct clutch, the entire gear train is forced to turn at engine speed, or 1:1. The direct clutch transmits about 60 percent of the drive while the forward clutch transmits the other 40 percent.

However, although the gear train is locked up and rotates as a unit for all practical purposes, there is some converter slip in direct drive, causing the planet pinions to turn on their shafts to a very small degree. As mentioned, the intermediate holding clutch remains applied but has no effect on third-gear operation since the entire gear train is turning clockwise and the intermediate one-way clutch therefore overruns.

Fourth Gear (Overdrive)

As the transmission automatically shifts from third gear direct to fourth gear overdrive, the overdrive band is applied to hold the reverse sun gear stationary. At the same time the forward clutch is released while the direct-drive clutch is still left applied. The transmission is now driven through the direct-drive

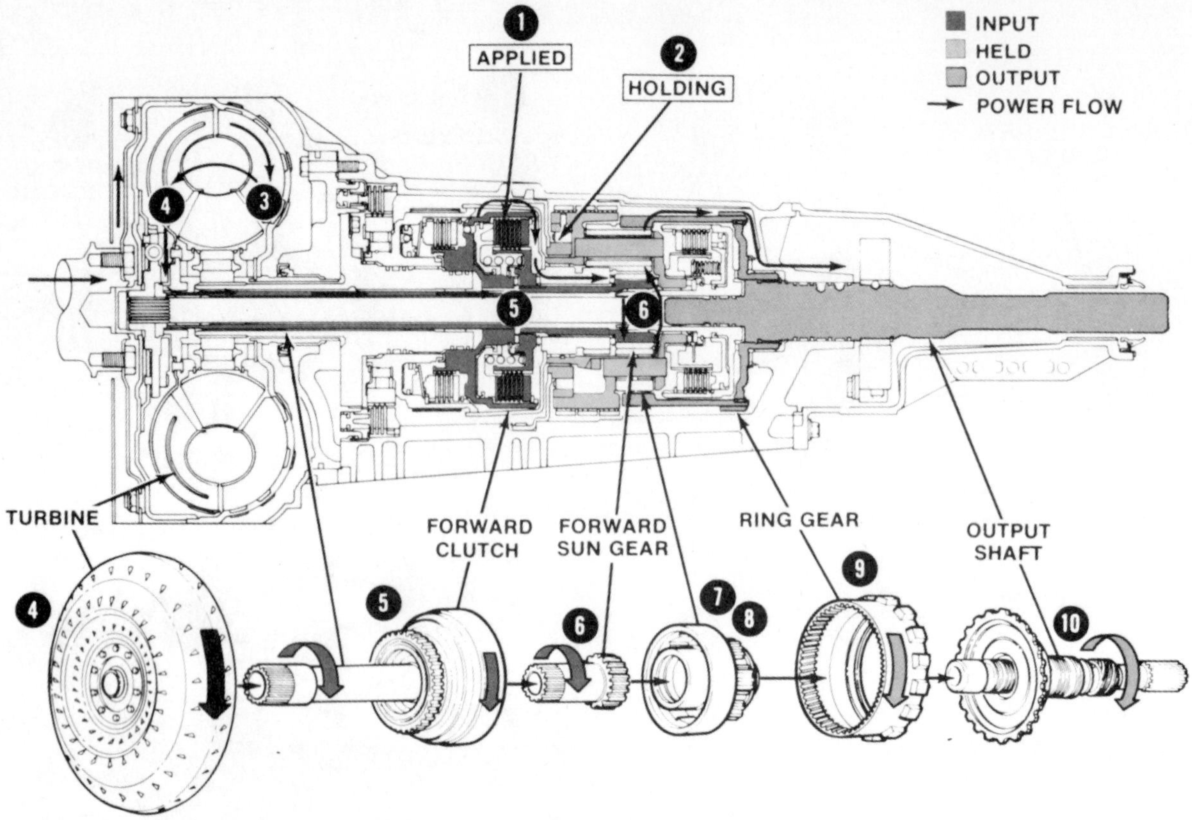

Figure 3-37. Ford AOD first-gear power flow. (*Courtesy of Ford Motor Company*)

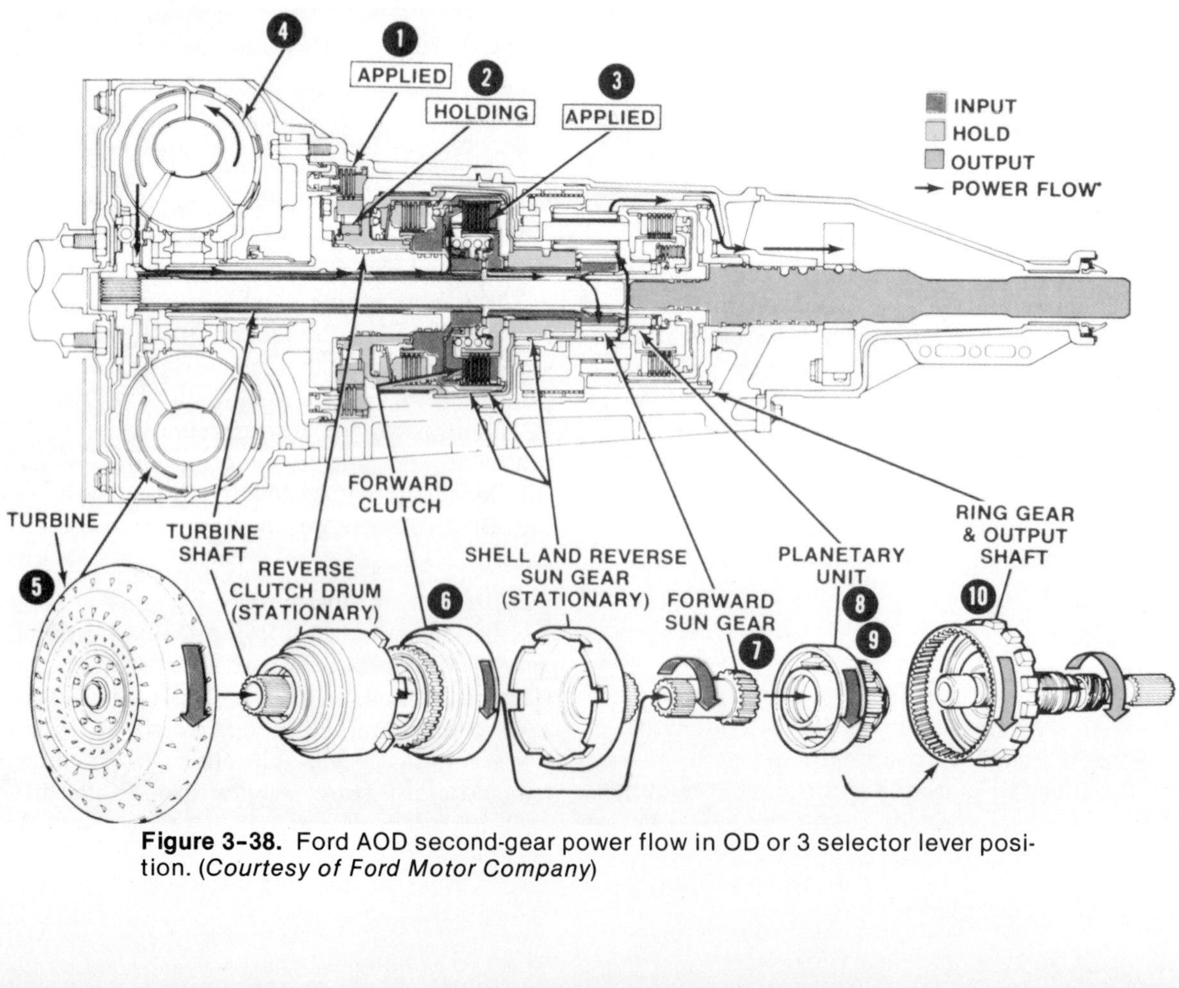

Figure 3-38. Ford AOD second-gear power flow in OD or 3 selector lever position. (*Courtesy of Ford Motor Company*)

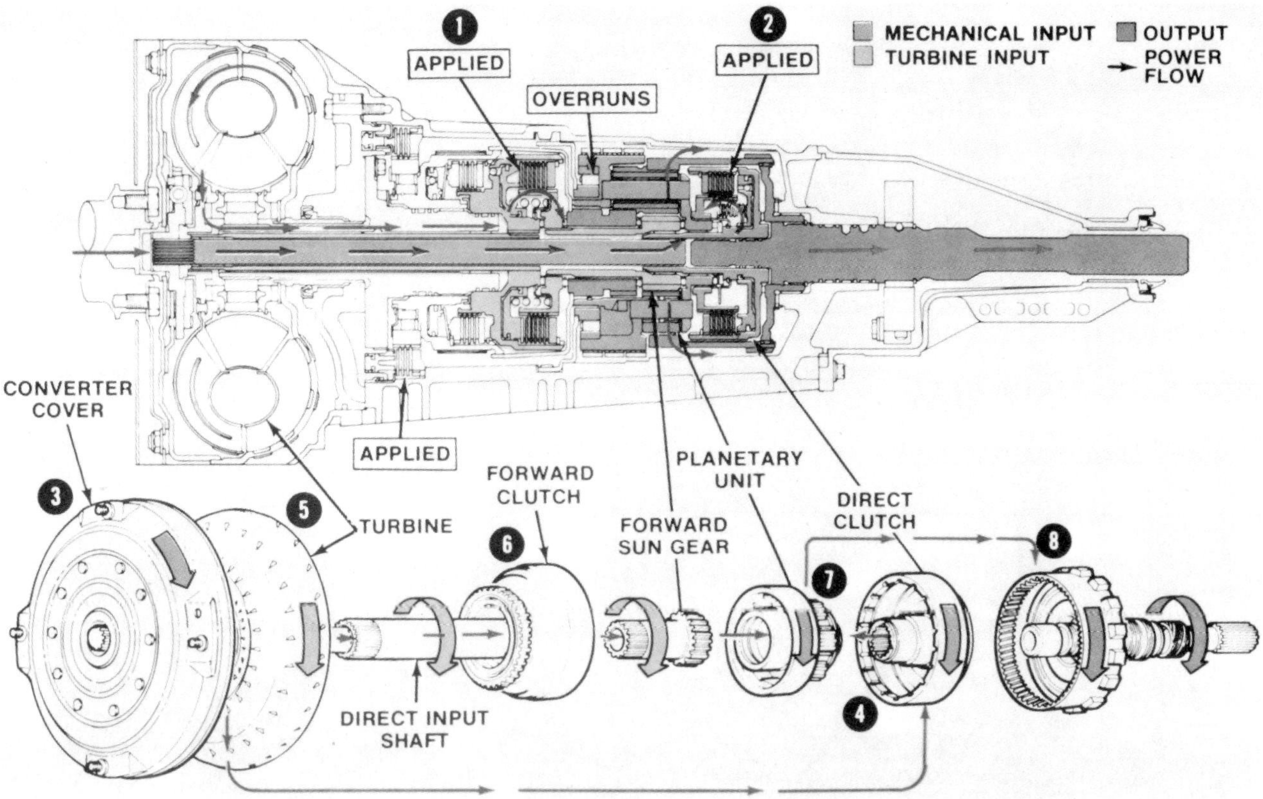

Figure 3-39. Ford AOD third-gear (direct-drive) power flow. (*Courtesy of Ford Motor Company*)

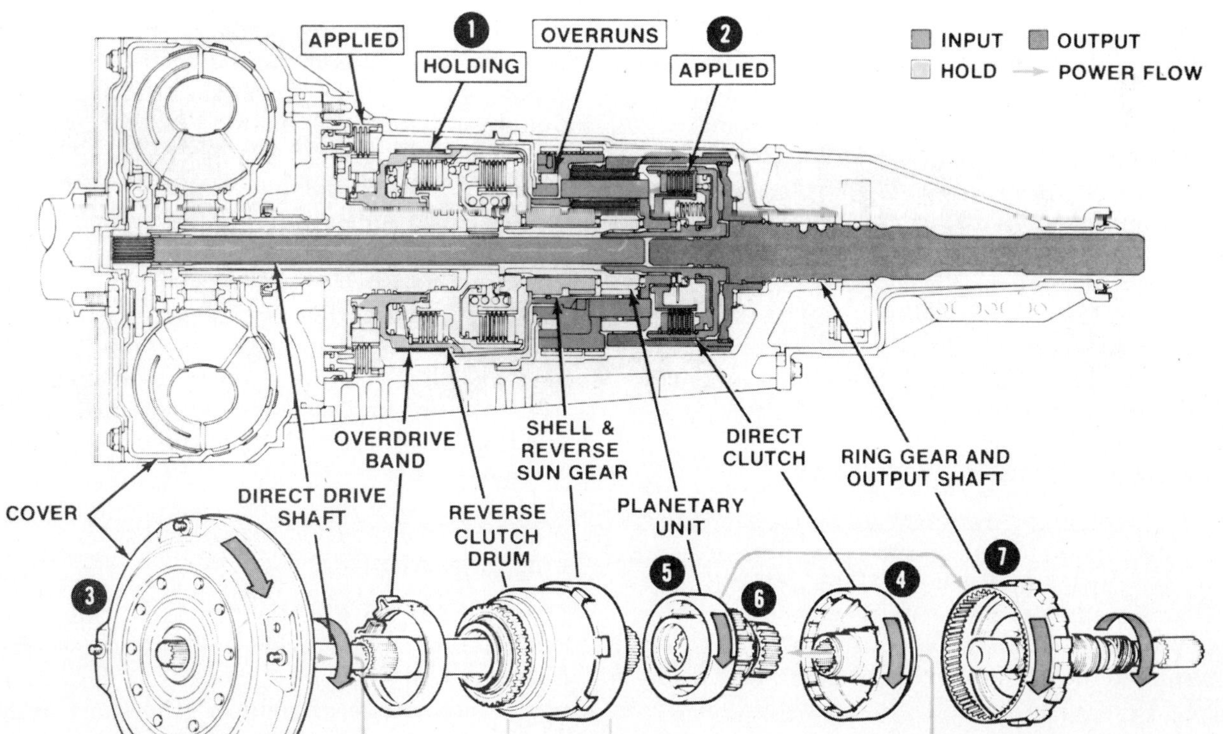

Figure 3-40. Fourth-gear (overdrive) power flow in Ford AOD transmission. (*Courtesy of Ford Motor Company*)

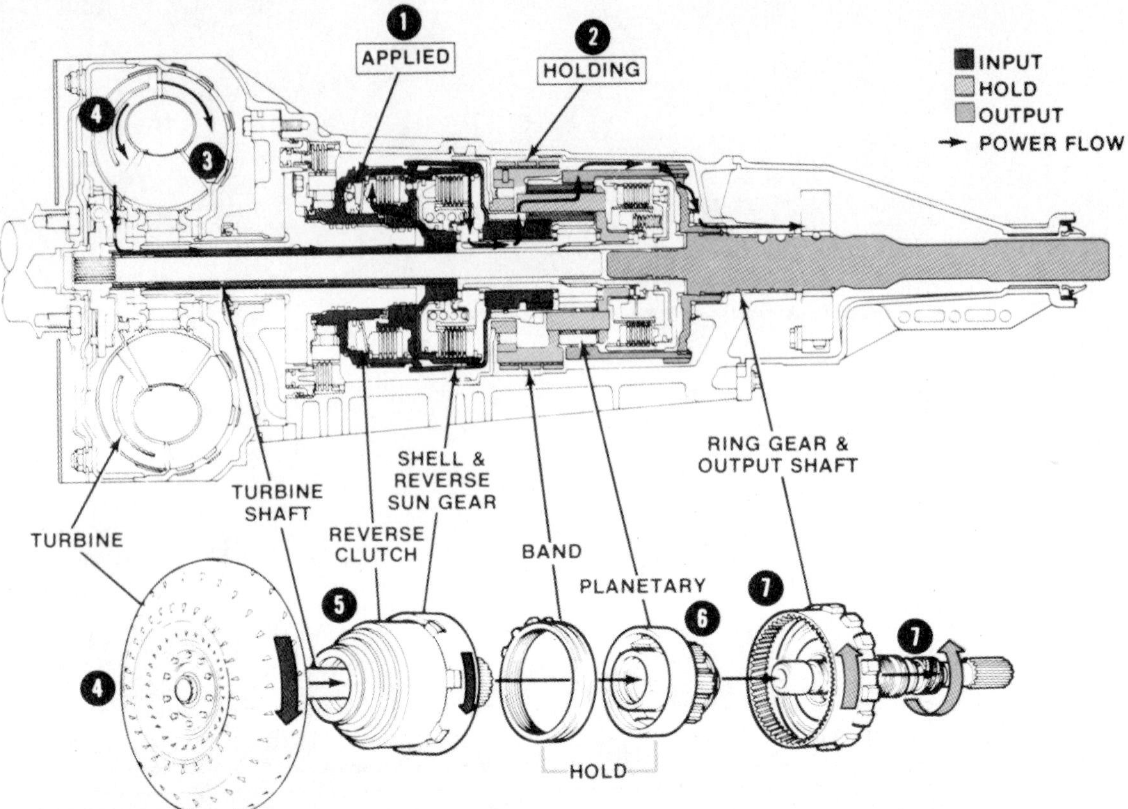

Figure 3-41. Ford AOD reverse-gear power flow. (*Courtesy of Ford Motor Company*)

shaft only, bypassing the turbine and turbine shaft and thereby eliminating converter slip.

Power flow is from the direct-drive shaft, through the direct clutch, driving the carrier clockwise. Since the overdrive band is holding the reverse sun gear stationary, carrier rotation causes the long pinions to walk around the reverse sun gear. This causes the pinions to turn on their shafts, forcing the ring gear and output shaft to turn faster than the carrier in overdrive. The result is an overdrive ratio of 0.667:1. The intermediate clutch though still applied has no effect on overdrive operation since all gear components are turning clockwise and the intermediate one-way clutch therefore overruns.

Ford ATX Automatic Transaxle

Ford's ATX automatic transaxle is a three-speed automatic transmission and final drive assembly used on front-wheel-drive Ford Escort and Mercury Lynx cars.

A unique three-element torque converter with a simple planetary splitter gear is used to transmit engine torque to the Ravigneaux gear train in the transmission. As described in Chapter 2 this torque converter has two output shafts: the turbine shaft, which transmits torque converter hydraulic torque output, and the intermediate shaft, which transmits engine torque hydraulically and mechanically through a torsional damper assembly. Both shafts are tubular, with the intermediate shaft inside the turbine shaft. A third shaft inside the intermediate shaft is used to drive the transmission hydraulic pump, which is located at the opposite end of the transmission from the torque converter. The pump drive shaft is splined to the torque converter cover. The turbine shaft is splined to the sun gear in the converter. The sun gear is splined to the turbine. The intermediate shaft is splined to the splitter gear carrier in the converter. The ring gear is splined to the converter damper and cover assembly.

ATX Gear Train Components and Operation

The ATX Ravigneaux planetary gear train consists of two sun gears, one carrier, and one ring gear: the reverse sun gear and a larger forward sun gear; the carrier; and two sets of pinions—the short pinions in

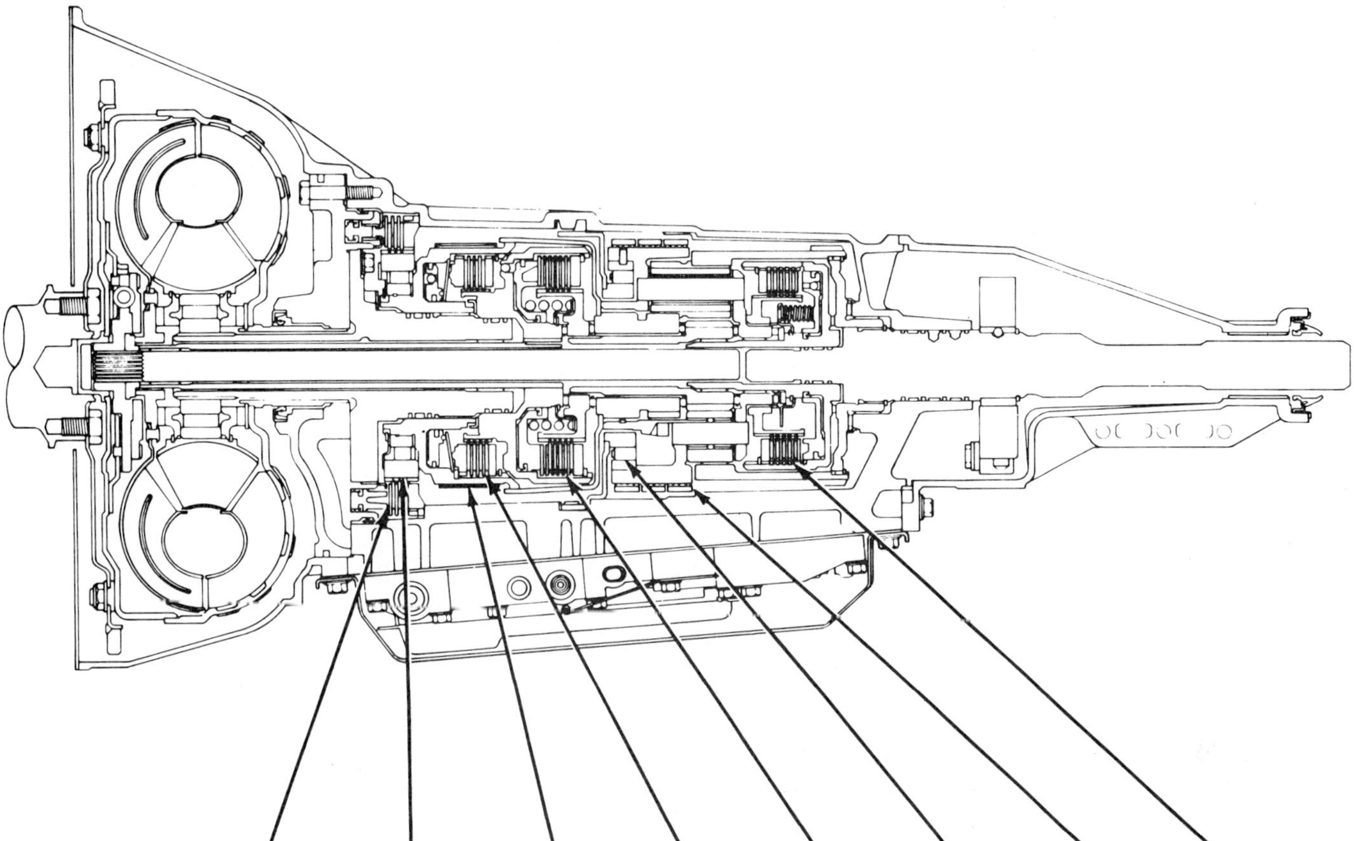

Gear	**Intermediate Friction Clutch**	**Intermediate Roller Clutch**	**Overdrive Band**	**Reverse Clutch**	**Forward Clutch**	**Planetary (Low) One-Way Clutch**	**Low and Reverse Band**	**Direct Clutch**	*** Gear Ratio**
1 Range Low					**APPLIED**	HOLDING	**APPLIED**		2.4-to-1
O/D and 3 Range Low					**APPLIED**	**HOLDING**			2.4-to-1
Second (Intermediate)	**APPLIED**	**HOLDING**			**APPLIED**	OVERRUNS			1.467-to-1
Third (Direct)	APPLIED	OVERRUNS			**APPLIED**	OVERRUNS		**APPLIED**	1-to-1
Fourth (Overdrive)	APPLIED		**APPLIED**			OVERRUNS		**APPLIED**	0.667-to-1
Reverse				**APPLIED**			**APPLIED**		2-to-1

*Not including torque converter reduction in 1st, Second and Reverse.

Figure 3-42. Summary of clutch and band operation in Ford AOD automatic transmission. (*Courtesy of Ford Motor Company*)

mesh with the low and reverse sun gear and the long pinions; the long pinions in mesh with the short pinions, the forward sun gear, and the ring gear.

This gearset is controlled by a one-way clutch, three multiple-disc clutches, and a band. The input members of this gear train are the low and reverse sun gear and the ring gear. The gear train members that can be held are the ring gear and the forward sun gear. The carrier serves as the output member and is splined to the final drive gear.

The clutches and band act as follows.

- One-way clutch allows the turbine shaft to drive the reverse sun gear.
- Direct clutch allows the turbine shaft to drive the reverse sun gear without any freewheeling.
- Intermediate clutch allows the intermediate shaft to drive the ring gear.
- Reverse clutch allows the ring gear to be held stationary.

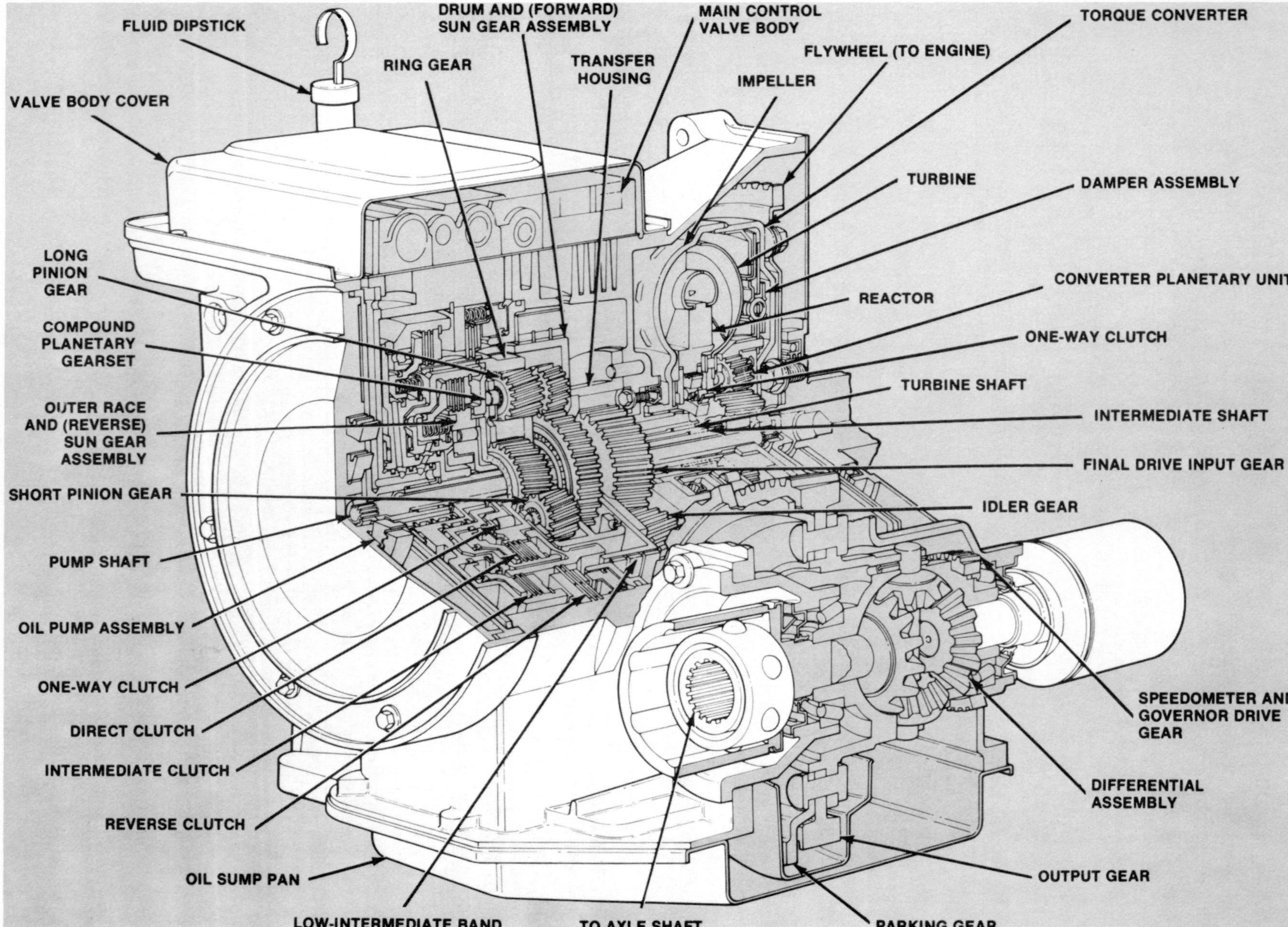

Figure 3-43. Cutaway view of Ford ATX transaxle including final drive and differential gears. (*Courtesy of Ford Motor Company*)

- Band allows the forward sun gear to be held stationary.

Selector lever positions for this transmission are:

P—park,

R—reverse,

N—neutral,

D—drive,

2—intermediate,

1—manual low.

- Park. All transmission components are in neutral with clutches and band released. Although the turbine is driving the low-reverse sun gear, no torque is being transmitted through the transmission since there is no reaction or held member. The differential case is locked to the transmission case by a park pawl, preventing drive wheel movement.
- Reverse. The ring gear is held stationary by applying the reverse clutch. Drive is from the turbine shaft, the direct clutch that is applied, and the low and reverse sun gear in a clockwise direction. The low and reverse sun gear drives the short pinions counterclockwise, which in turn drive the long pinions clockwise. The clockwise-turning long pinions are forced to walk around the stationary ring gear, forcing the carrier and output to turn counterclockwise in reverse reduction of 1.97:1. Reverse gear drive through the torque converter is 100 percent hydraulic from the turbine and turbine shaft only.
- Neutral. Neutral is the same as park as far as action in the transmission gear train is concerned; however, in neutral the park pawl is not engaged and the vehicle is able to roll freely.
- Drive. In the drive range the transmission starts in low and shifts automatically to second and then to third or direct.
- In drive low the band is applied to hold the forward sun gear stationary. Drive is from the turbine shaft through the one-way clutch to the low and reverse sun gear in a clockwise direction. The low and reverse sun gear drives the short pinions counterclockwise, which drive the long pinions clockwise. The clockwise-turning long pinions are forced to walk around the stationary forward sun gear in a clockwise direction taking the carrier and output forward in a first-gear reduction of 2.79:1 ratio. All torque in first gear is transmitted hydraulically to the turbine shaft and low-reverse sun gear. On deceleration, while the transmission is in drive low, the one-way clutch will overrun, which results in no engine braking.
- Manual Low. In manually selected low gear the only difference from drive low is that the direct clutch is applied to connect the low-reverse sun gear solidly to the turbine shaft. Power flow is exactly the same as in drive low but freewheeling does not occur and engine braking on deceleration takes place.
- Intermediate (drive range). In intermediate the band remains applied just as in drive low to hold the forward sun gear stationary. The intermediate clutch applies to drive the planetary ring gear.

Power flow is now split in the converter. The turbine provides the hydraulic drive torque to the intermediate shaft since the turbine is splined to the converter sun gear. The converter cover provides the mechanical drive to the intermediate shaft since the ring gear is splined to the damper assembly in the converter cover. This means that two members of the splitter gear are being driven—one hydraulically and the other mechanically—driving the third member, the carrier, at relatively the same speed. There is, of course, some hydraulic slip in the converter. Torque input to the intermediate shaft in second gear is 62 percent hydraulic and 38 percent mechanical.

To continue the power flow in the transmission gear train, the intermediate shaft drives the intermediate clutch and planetary ring gear clockwise. This drives the long pinions clockwise and forces them to walk around the stationary forward sun gear. This causes the output carrier to turn forward in a reduction ratio of 1.61:1.

- Manual second. In manually selected second gear, power flow is exactly as in first and second in the drive range. However, no upshift to third is possible since the 2–3 upshift circuit is locked out.
- Third (Direct). In third or direct drive, the band is released, the intermediate clutch remains applied, and the direct clutch is applied. Applying the direct clutch allows the turbine shaft to drive the low and reverse sun gear. The intermediate shaft is still driving the ring gear through the intermediate clutch. Driving these two planetary members forces the carrier to

turn at the same speed. For all practical purposes the entire planetary unit is locked together and turns as a unit in direct drive in a virtual 1:1 ratio.

There is, of course, still some hydraulic slip in the converter, allowing some minor rotation of pinions on their axes. This is insignificant, however, since input torque in direct is 93 percent mechanical and only 7 percent hydraulic, resulting in virtually 100 percent mechanical drive if we consider a 10 percent slip in the 7 percent hydraulic torque input. Since the direct clutch is applied, the one-way clutch cannot overrun, allowing engine braking in third gear.

PART 3-6 PLANETARY GEAR WEAR AND DAMAGE

Planetary gears are very reliable and have a long service life. Nevertheless extended mileage will eventually result in some wear.

Angled teeth of helical gears cause axial force to wear thrust surfaces. Gear teeth themselves will eventually wear, especially on the drive sides of the teeth. Pinion bearings or bushings and shafts are subject to wear as are sun gear bushings.

A combination of gear tooth wear on sun gears, pinions, and internal gears as well as pinion bearing and pinion shaft wear, if severe, can cause gear teeth to attempt to climb over each other. This results in tremendous forces being exerted on the entire gearset and may cause cracked pinion carriers, ring gears, or sun gears. Transmission failure is the obvious result, with the only solution being complete overhaul or replacement.

Causes of premature wear and damage include low fluid levels and overheating of clutches and bands with consequent overheating and damage to planetary gear components. Overheating can be caused by a fluid level that is too low or too high. It can also be caused by driver abuse of the transmission and by overloading.

Part 3–7 Worksheet

1. Using the appropriate training model or other available gears, demonstrate your ability to calculate gear ratios of various gear combinations.
2. Using a simple planetary gearset (a training model or half of a Simpson planetary gearset from an automatic transmission disassembled earlier), demonstrate all the possible drive conditions (six drive conditions plus neutral).
3. Using a Simpson planetary gearset, demonstrate your ability to follow and explain all the power flow combinations in a three-speed automatic transmission.
4. Using a Ravigneaux planetary gearset, demonstrate your ability to describe and explain all the power flow combinations as used in a three-speed automatic transmission.
5. Using a simple planetary gearset, demonstrate how overdrive is achieved in a THM 325-4L automatic transmission.
6. Using a Ravigneaux gearset, demonstrate how overdrive is achieved in the Ford automatic overdrive transmission.
7. Using a simple planetary gearset, demonstrate how overdrive is achieved in a THM 700-R4 automatic transmission.

PART 3-8 SELF-CHECK

1. Define the following terms as they apply to gears:
   - *Drive gear*
   - *Driven gear*
   - *Clockwise*
   - *Counterclockwise*
   - *Rpm*
   - *Ratio*
   - *Axis*
   - *External gear*
   - *Internal gear*
   - *Straight spur gear*
   - *Helical spur gear*
   - *Drive side*
   - *Coast side*
   - *Constant mesh*
   - *Backlash*
   - *Bottoming*
   - *Climbing*
2. How is reverse obtained in a simple planetary gearset?
3. State seven laws which can be used to describe all possible uses of a simple planetary gearset.
4. What is the effective number for the carrier in a planetary gearset with a 30-tooth sun gear and a 70-tooth ring gear?
5. What is the gear ratio when the sun gear is the input member and the carrier is the output member in a gearset as described in question 4?
6. How is reverse obtained in a Simpson compound planetary gear train?
7. Trace the power flow in second gear in the Simpson gear train by naming all major parts from the turbine shaft to the output shaft.
8. Name the major components of the Ravigneaux planetary gear train.
9. In the three-speed Ravigneaux gear train, direct drive is obtained by driving the ____________ and the ____________.
10. Why is it possible to leave the intermediate holding clutch applied in third and fourth gears in Ford's automatic overdrive transmission?
11. A fluid level that is too high or too low in a transmission may cause overheating. True or false?
12. What is the purpose of the automatic transmission gear train?

PART 3-9 PERFORMANCE EVALUATION

The following statements can be used to evaluate a student's performance of the tasks required to master the content of this chapter. This can be done by the student or by the instructor.

The minimum level of performance required will be determined by the school and the instructor.

After sufficient study of this chapter and the appropriate training models, you should be able to perform these tasks.

1. State the purpose of gears in an automatic transmission.
2. Describe the principles of basic gear operation, including gear terminology, the lever principle, direction of rotation, torque-speed relationship, and basic gear construction.
3. Name the components of a simple planetary gearset; describe its basic construction features, operating principles, and ratios.
4. Name the components of the Simpson planetary gearset; describe its basic construction and operation in both three-speed and four-speed arrangements.
5. Name the components of the Ravigneaux planetary gearset; describe its basic construction and operation in both three-speed and four-speed arrangements.
6. Describe normal and abnormal planetary gear wear and damage.
7. State the causes and effects of planetary gear wear and damage.
8. Identify by name the most common domestic and import automatic transmissions and state which type of planetary gear system is used by each.
9. Complete the worksheet as required.
10. Complete the self-check with at least 80 percent accuracy.
11. Evaluate performance based on this performance evaluation section.

Chapter 4

Driving and Holding Devices

Performance Objectives

After sufficient study of this chapter and the approriate training models, the student should be able to perform the following tasks.

1. State the purpose of a driving device in an automatic transmission.
2. Describe the basic construction and operation of multiple-disc driving clutches, including apply and release.
3. Describe the basic construction and operation of multiple-disc holding clutches, including apply and release.
4. State the purpose of a holding device in an automatic transmission.
5. Describe the basic construction and operation of roller-type and sprag-type one-way holding clutches.
6. Describe the basic construction and operation of the band and servo in an automatic transmission, including apply and release.
7. List the different types of multiple-disc clutch plates and band types used in automatic transmission and describe their operating characteristics.
8. Describe the effects of insufficient hydraulic apply pressure, wear, and excessive or insufficient clearance in multiple-disc driving and holding clutches.
9. Describe the effects of inoperative one-way driving or holding clutches.
10. Describe the effects of insufficient hydraulic apply pressure and excessive or insufficient clearances in band and servo operation.
11. Describe the factors that determine multiple-disc clutch torque capacity.

PART 4-1 INTRODUCTION

This chapter deals with the purpose, construction, and operation (both normal and abnormal) of multiple-disc driving and holding clutches, bands and servos, and one-way driving and holding clutches. The different types of construction used in these devices are also discussed.

PART 4-2 MULTIPLE-DISC CLUTCHES

Automatic transmissions use multiple-disc–type clutches to connect a planetary gear member to the transmission input shaft. Many automatic transmissions also employ one or more multiple-disc clutches as holding devices to prevent the rotation of one or more planetary gear members.

The simplest Simpson compound three-speed planetary gear arrangement uses two multiple-disc clutches to connect the input shaft to the planetary gear system. One of these clutches is called the forward clutch since it is applied in all three forward speeds; the other is known as the reverse and high clutch since it is applied in high gear and reverse only. Some shop manuals refer to these clutches as the rear clutch and the front clutch because of physical position in the transmission.

The forward clutch in this arrangement connects the ring gear of the forward half of the planetary gearset to the transmission input shaft when it is applied.

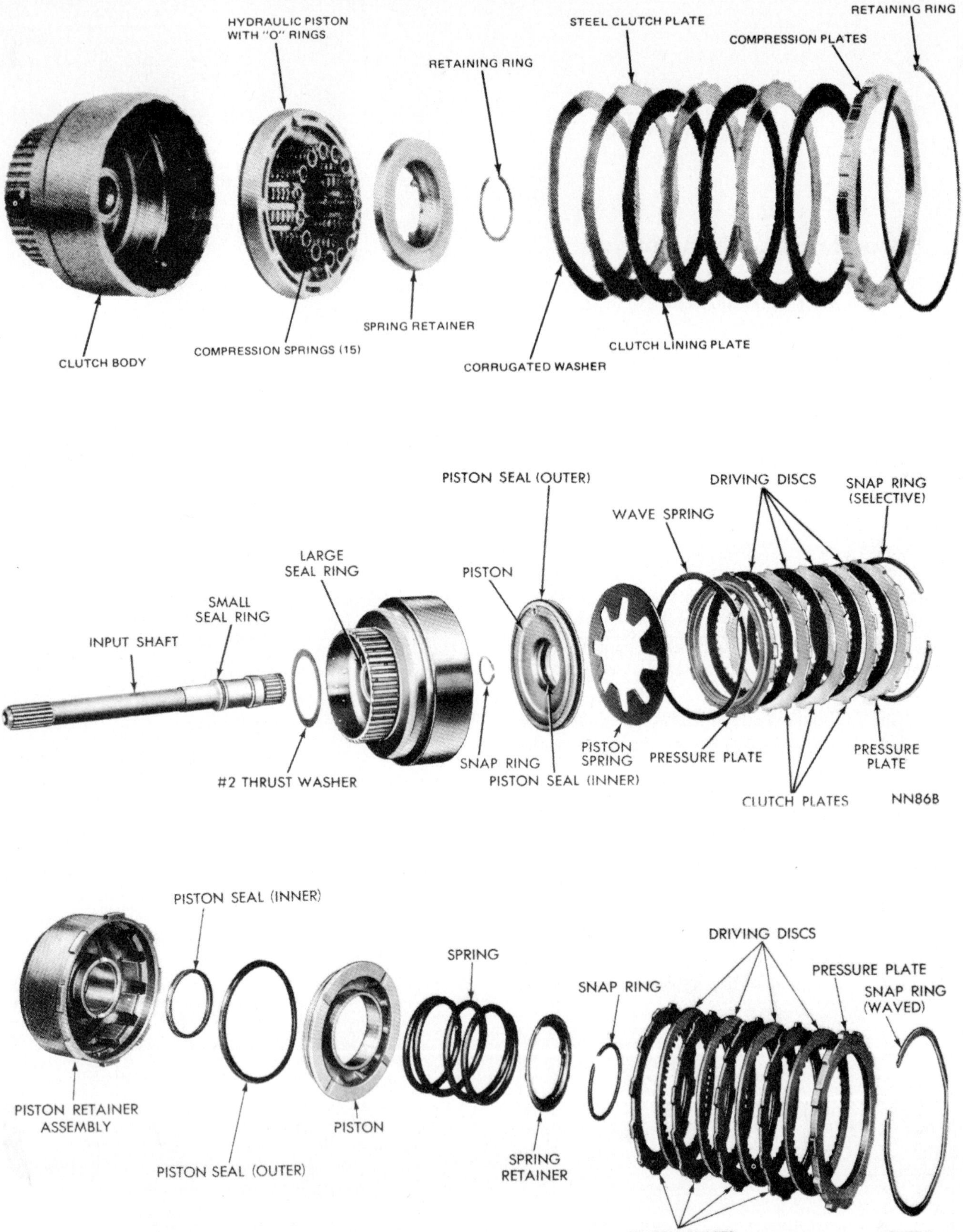

Figure 4–1. Typical multiple-disc drive clutches. Note difference in release springs. (*Courtesy of Ford Motor Co. of Canada Ltd. and Chrysler Corporation*)

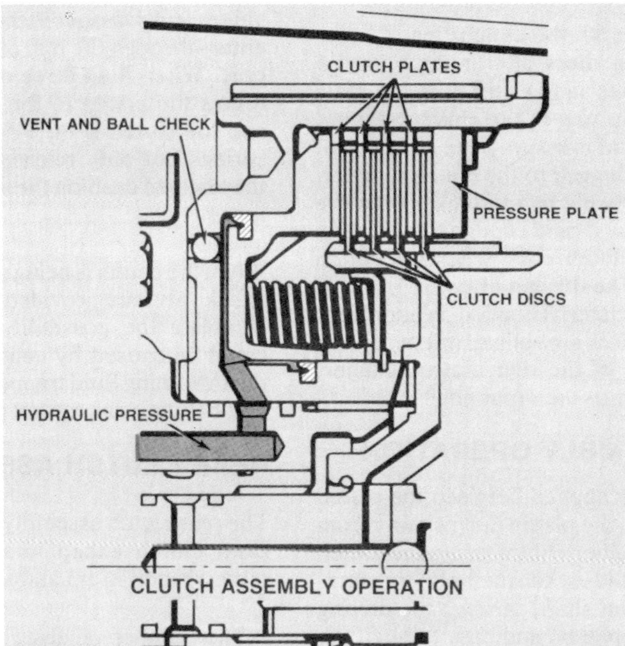

Figure 4-2. Hydraulic pressure acting on clutch piston squeezes clutch plates and discs together against pressure plate to apply clutch. (*Courtesy of Chrysler Corporation*)

The reverse and high clutch connects the common sun gear to the input shaft when the clutch is applied.

The transmission input shaft (turbine shaft), the reverse and high (front) clutch hub, and the forward (rear) clutch cylinder (drum) rotate as an assembly. The clutch pack consists of an alternately stacked arrangement of externally toothed or tanged steel plates and internally toothed friction surface plates. The external tangs of the steel plates engage grooves in the clutch drum and therefore rotate with the drum. The internal teeth of the friction plates engage splines or grooves on the clutch hub and therefore rotate with the hub.

When a clutch is applied, the hub, steel plates, friction plates, and drum rotate as a unit. When the clutch is released, one set of plates may be stationary while the other set is turning or the steel plates and friction plates in the same clutch may be turning in opposite directions, depending on planetary gear operation and the drive range in which the transmission is operating. Precise clutch pack clearance is critical to proper clutch apply and release.

A hydraulic piston in the clutch cylinder applies the clutch while a pressure plate and retaining ring on the other side of the clutch pack provide the reaction. Synthetic rubber seals between the piston and cylinder prevent fluid leakage and loss of apply pressure.

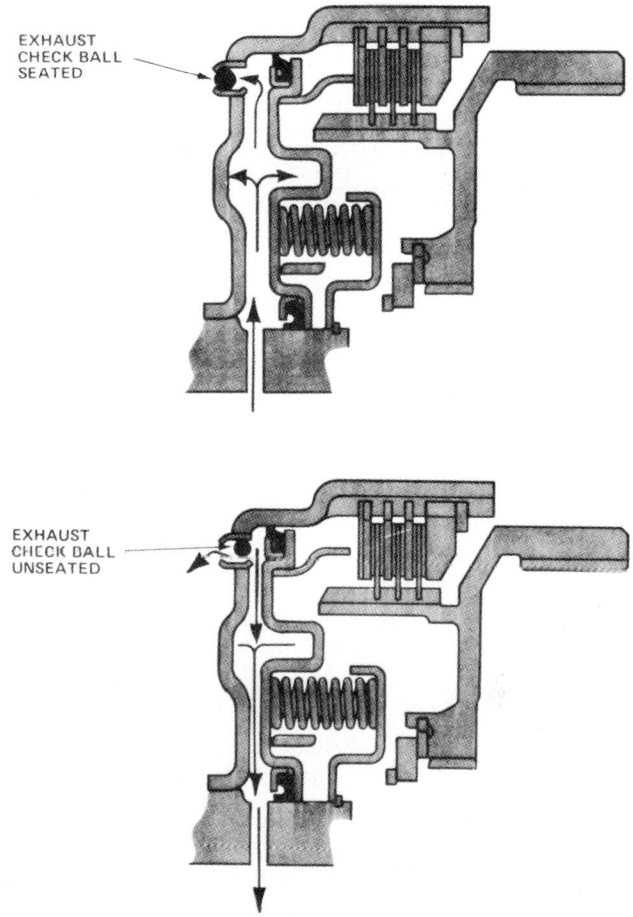

Figure 4-3. Clutch check ball seats when hydraulic pressure applies clutch, preventing escape of fluid. When hydraulic pressure is released, centrifugal force unseats check ball and causes fluid to be exhausted through check ball port. This prevents partial clutch application during clutch release. (*Courtesy of General Motors Corporation*)

A dished-plate–type Belleville spring, a heavy coil spring, or a series of small coil springs is used to return the piston to the release position when hydraulic apply force is removed. A Belleville spring is used only in the forward clutch since fast release is required only in the reverse and high clutch during a part throttle or full throttle downshift. In many cases the clutch pack also includes a snap ring type of wave spring or in some cases one or more dished steel plates to cushion clutch application for smoother shifts.

A ball type of check valve is located in the clutch piston. The check ball controls a fluid exhaust passage. Fast clutch release is required to provide smooth rapid shifts and prevent unwanted friction in the clutch. The rapidly spinning clutch assembly deposits residual fluid between the piston and cylinder base circumference, which attempts to continue clutch apply. Centrifugal force unseats the fluid exhaust check ball and allows fluid to exhaust, preventing this condition. When the clutch is once again applied, hydraulic pressure seats the check ball and seals off the exhaust orifice.

Clutch Plates

Steel clutch discs are stamped from steel stock and are finished with a dull surface to facilitate proper break-in with new friction discs. After proper break-in, steel discs become polished.

Friction discs are steel plates with friction material bonded to both sides. Friction material may be organic or metallic, depending on friction characteristics required.

Metallic material is usually a sintered or powdered copper and graphite mix with bonding agents. Semimetallic materials include asbestos powders. Nonmetallic organic friction materials are normally a paper type sometimes combined with asbestos and the proper bonding agents.

The friction characteristics of these materials are modified by the type of fluid used and by grooving of the friction discs. The slip-lockup characteristics of multiple-disc clutches are critical to shift quality. Due to the nature of the frictional characteristics of the clutch plates, new and old plates should never be mixed, since glazing could occur.

Clutch Torque Capacity

The maximum torque capacity of the multiple-disc clutch is determined by the following factors:

1. Clutch diameter
2. Frictional surface area (number of plates)
3. Hydraulic pressure
4. Surface area acted on by hydraulic pressure
5. In the case of a Belleville-spring–type clutch, the mechanical advantage gained by the lever action of the spring

The diameter of the clutch is limited by the room available in the transmission case. The frictional surface area can be increased or decreased by the number of plates used in the clutch pack. Increasing hydraulic apply pressure will, of course, increase the torque capacity of a clutch and also will result in harsher shifts. The surface area against which the hydraulic pressure acts (piston size) is also limited by the available room in the clutch cylinder.

Engine size and vehicle design are factors that influence the design and balance of torque capacity factors utilized by the vehicle manufacturer.

Multiple-Disc Holding Clutch

Many transmission designs utilize multiple-disc holding clutches in place of bands and servos. The clutch pack and piston are located in the same bore whereas a band requires a separate servo to operate the band. The multiple-disc holding clutch uses the transmission case as the clutch cylinder.

A hydraulically applied and spring-released piston is used. Steel and friction plates stacked alternately comprise the clutch pack. The externally tanged steel plates engage corresponding grooves in the transmission case. The internally toothed friction plates engage grooves on a drum attached to the planetary gear member to be held. A heavy pressure plate and snap ring absorb clutch apply reaction pressure. A wave spring may be used to help cushion clutch application.

Typical examples of transmissions using multiple-disc clutches as holding devices are the General Motors Turbo Hydra-Matics and the Ford C6 transmissions.

PART 4-3 BANDS AND SERVOS

Many automatic transmission designs use the hydraulic-servo–operated wraparound band as a holding device to control planetary gear operation. The band does not increase the diameter of the gear train as much as a multiple-disc clutch does but requires room for a servo to operate the band.

Hydraulic Servo

A hydraulic servo converts hydraulic force to mechanical force. The servo consists of a piston operating in a bore in the transmission case or in a separate servo cylinder bolted to the transmission case. A piston rod is used to act directly on the band or through a lever and link assembly. The lever usually is used to increase apply force. Reaction force is applied to an anchor at the other end of the band. The anchor is attached to or part of the transmission case.

The piston in a low and reverse band servo is usually single acting. Hydraulic apply force is available on one side of the piston with a spring on the other side of the piston to return the piston to the release position when apply pressure is removed. A piston plug in the servo piston may be used to cushion servo apply. The two piston return springs are located between the piston and a land on the piston plug and

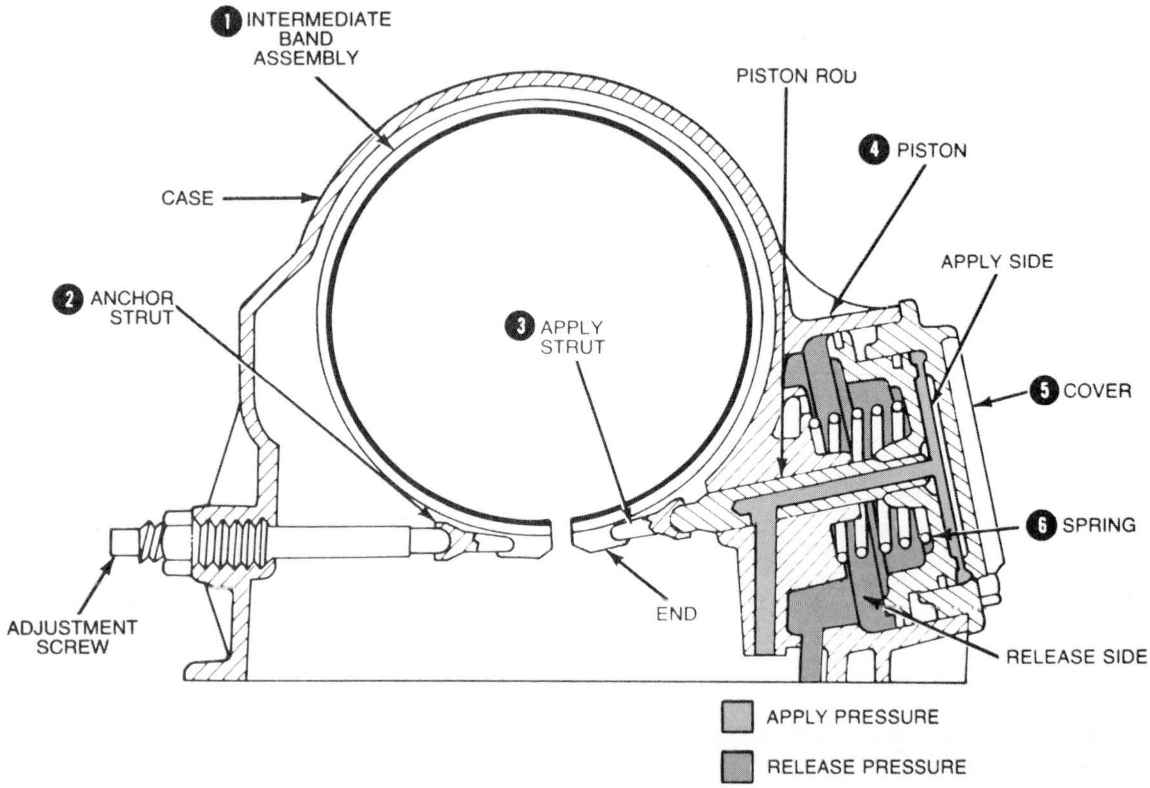

Figure 4–4. Arrangement of band, servo, and anchor in transmission case (typical). Note apply and release hydraulic pressures. (*Courtesy of Ford Motor Company*)

between the piston and spring retainer. Band adjustment may be provided at the lever or the anchor to maintain proper operating clearance between the band and the drum.

The intermediate servo usually has a two-land piston and is double acting. The servo bore and piston are stepped—the small end of the piston operating in the smaller diameter bore and the larger diameter of the piston in the larger bore. A piston rod and return spring are located inside the piston. A piston return spring is used between the piston and piston rod guide, which is held in the bore by means of a snap ring.

Hydraulic pressure applied between the piston

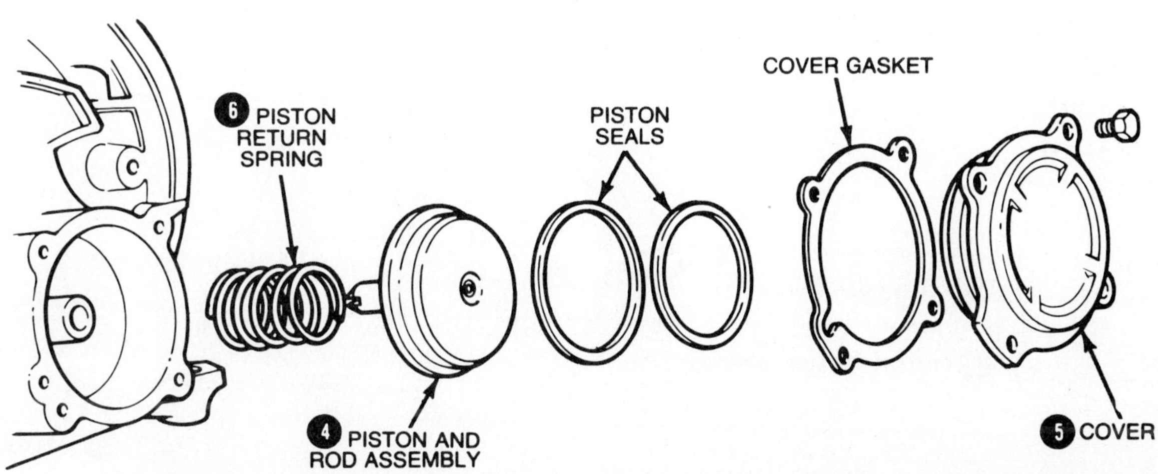

Figure 4–5. Exploded view of intermediate band servo components (typical). (*Courtesy of Ford Motor Company*)

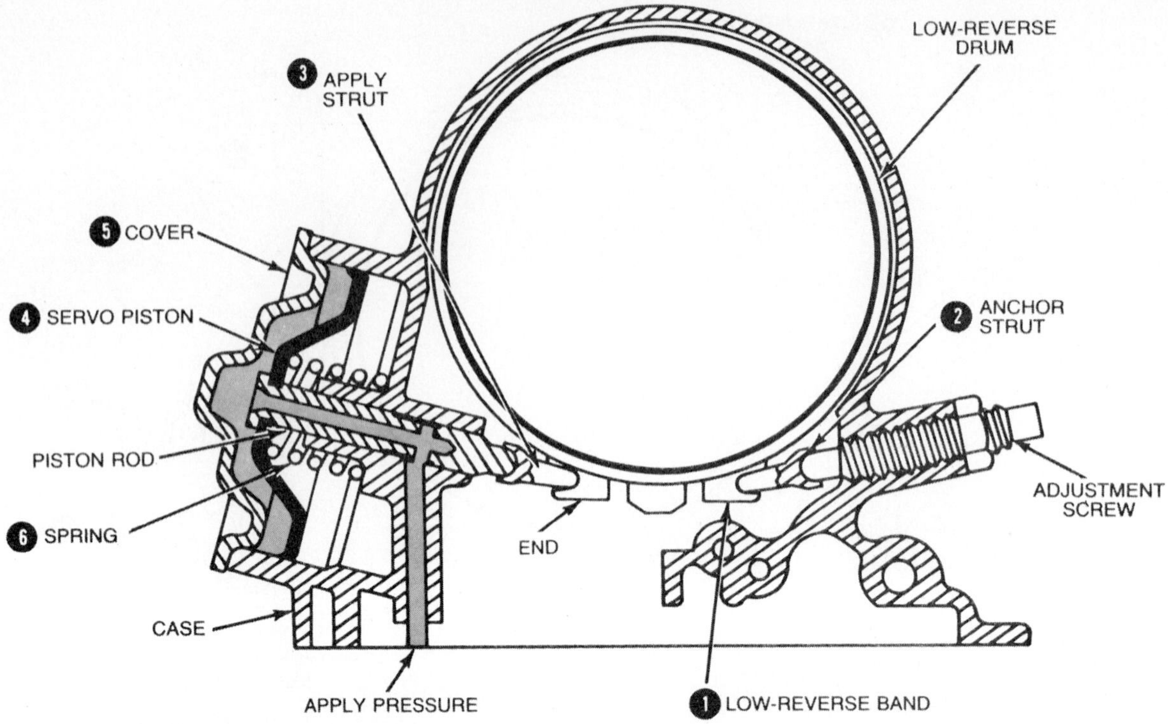

Figure 4-6. Low and reverse band servo operation. (*Courtesy of Ford Motor Company*)

and piston bore applies the intermediate band on a 1–2 upshift or a 3–2 downshift. Spring pressure returns the piston to the release position on a 2–1 downshift. On a 2–3 upshift, hydraulic pressure introduced to the release side of the piston plus spring pressure forces the piston back against apply pressure to release the band for third gear. During a 3–2 downshift, intermediate servo release pressure is dumped. Since apply pressure is still present, the intermediate band applies for second-gear operation.

During a part-throttle or full-throttle 3–2 downshift, it is essential that the intermediate band be applied immediately as the direct clutch is released in order to prevent excessive engine speed increase. The intermediate servo piston rod may be direct acting on the band or it may act on the band through a lever and link or strut. Reaction force is absorbed by the other end of the band through an anchor in the transmission case.

Proper operating clearance may be adjusted on

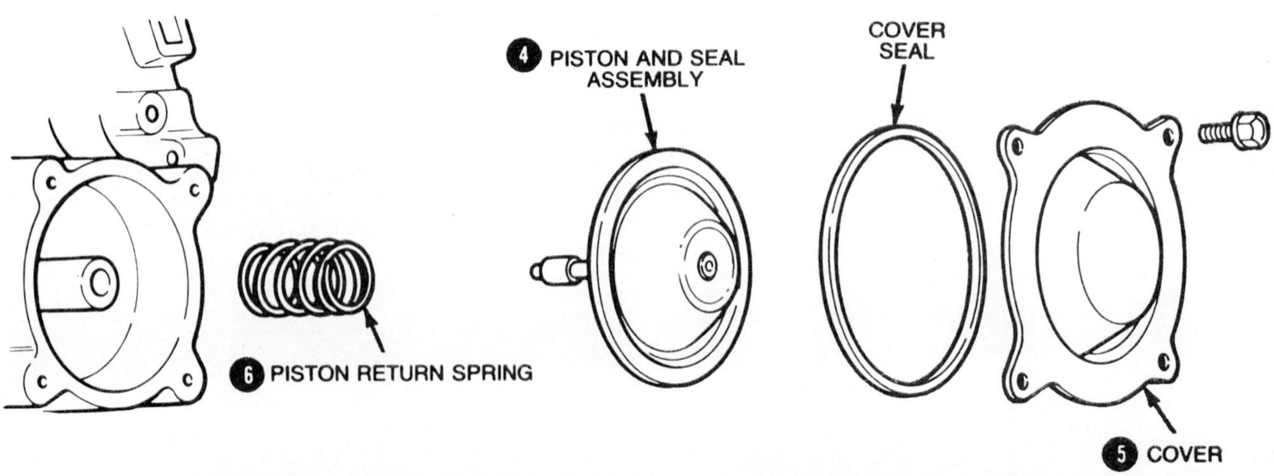

Figure 4-7. Low and reverse servo components (typical). (*Courtesy of Ford Motor Company*)

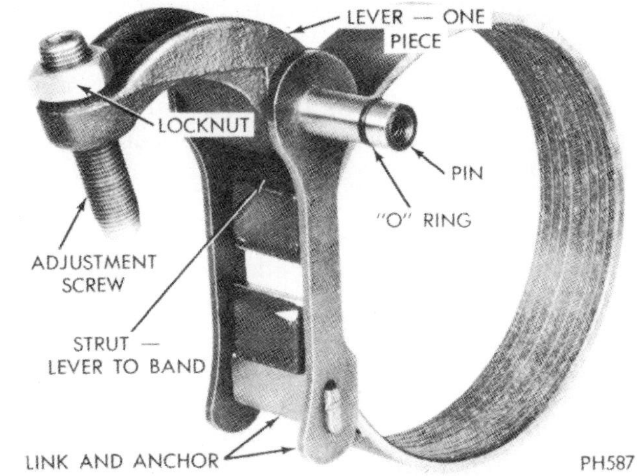

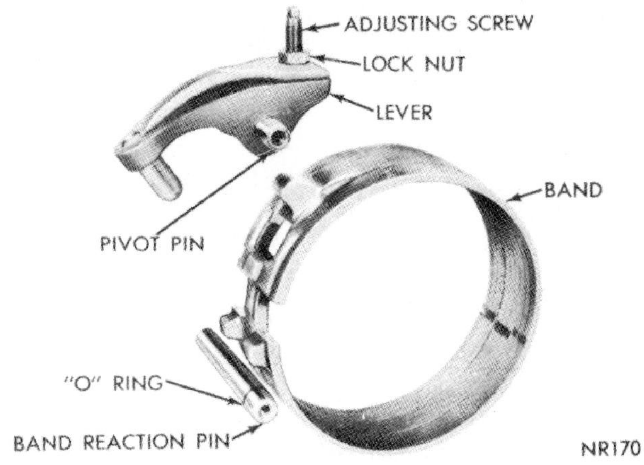

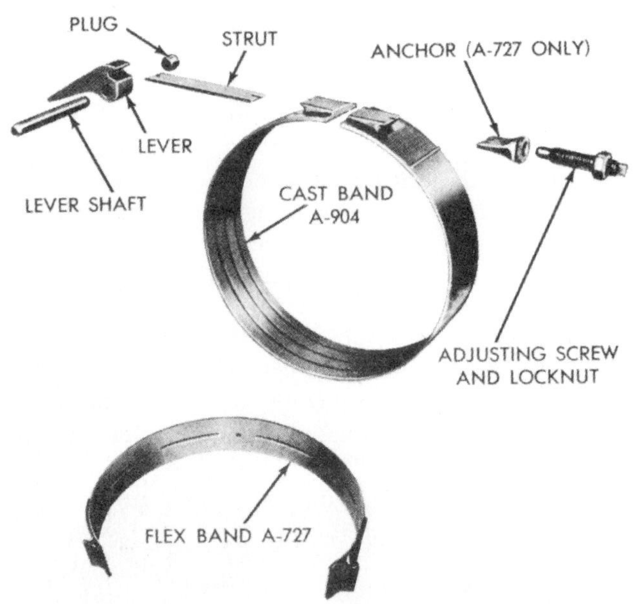

Figure 4-8. Several types of bands used in automatic transmissions. Bands are lined with friction material ranging from fiber to sintered metal. (*Courtesy of Chrysler Corporation*)

some transmissions by means of an anchor screw adjustment. Other transmissions may use selective length band actuating rods or may have no adjustment. Proper band-to-drum clearance is required to prevent drag on the drum when the band is in the released position and to assure full band application before the servo piston bottoms in its bore to prevent slippage.

Bands

Transmission bands may be of the rigid or flexible type. The rigid type is also flexible to the extent that it will contract during band application and expand on band release. The flexible type when removed from the transmission does not retain its circular shape to fit the drum. Bands are either single wrap or double wrap. The double-wrap design increases the effects of self-energization when the band is applied. Self-energizing is the tendency of a band to tighten around a drum during drum rotation.

The double-wrap band design is used for low and reverse where greater holding force is required due to the high torque of low and reverse gear operation. Bands have either a relatively hard metallic friction lining or a softer organic friction material.

The metallic materials are generally more abrasive than the organic materials. For this reason they are usually used only on the low and reverse band since this band is usually applied to a static or slow-turning drum.

The softer organic materials, usually a paper-based compound, are used on bands required to operate on fast-turning drums. Metallic lined bands used on these drums would result in much more rapid drum wear.

Drum material is also a factor in the slip-lockup characteristics of band operation. The proper combination of drum material, drum surface finish, and band frictional material is designed to provide the required slip-lockup characteristics for smooth positive shifts without causing premature band and drum wear or glazing.

Band frictional linings are usually grooved to control the escape of transmission fluid. This helps in providing the desired band application characteristics.

To ensure proper operation of the bands, the fluid specified by the transmission or vehicle manufacturer must be used. Using incorrect fluid results in a change of the coefficient of friction of all friction devices (clutches and bands) and consequent changes in shift characteristics. Harsher, more severe shifts or "mushy" delayed shifts can be the result of using improper fluid in the transmission.

Torque Capacity of Bands

The torque capacity of a transmission band must be great enough to hold the desired reaction member under all appropriate operating conditions. Torque capacity is determined by the friction characteristics of the band and drum (band friction lining and drum material); surface finish of the drum; drum and band diameter; servo piston diameter; hydraulic pressure to the servo piston; ratio of lever used to apply the band; and whether the servo is positioned to take advantage of the self-energizing effect of drum rotational direction when the band is applied. In the case of a low and reverse band, of course, the held member attempts to rotate counterclockwise in low and clockwise in reverse. It is therefore impossible to have self-energizing benefits in both directions. Higher hydraulic pressures are sometimes used in reverse to offset the difference there would otherwise be in torque capacity.

Higher hydraulic apply pressures are also used in reverse on transmissions equipped with a multiple-disc low and reverse holding clutch. This is required since the reaction force on the held member is greater in reverse than in low. In reverse, only the rear half of the planetary gearset provides the gear reduction while in low gear, reduction is achieved through both the front and rear section of the gear train. In the band and servo design, the self-energizing effect will work against hydraulic servo apply pressure, requiring even greater hydraulic pressure in reverse.

One-Way Clutches

Automatic transmissions use one or more one-way clutches to control planetary gear operation. A one-way clutch is also used in the torque converter stator.

One-way clutches can be used to allow rotation of a unit in one direction and prevent rotation of that unit in the other direction. The stator one-way clutch is a good example of this application. This unit prevents counterclockwise rotation of the stator during torque multiplication and allows clockwise rotation of the stator during the coupling phase of converter operation.

The one-way clutch can also be used to drive a unit in one direction and allow the unit to overrun the drive member under certain conditions. The turbine shaft driving the low-reverse sun gear through a one-way clutch in low gear in the Ford ATX transaxle is a good example of this application.

A common application of a one-way roller clutch in the planetary gear train is the use of the clutch as a reaction or holding device in drive-range low gear. While engine torque is being applied to the gearset, the one-way clutch prevents counterclockwise rotation of one member of the planetary gearset. The particular planetary gear member acting as the reaction member depends on transmission design. In the common Simpson three-speed automatic transmission gear train it is the rear carrier. In the Ravigneaux three-speed automatic arrangement it is the carrier.

Another application of the one-way roller clutch is its use in drive-range second gear on some transmissions, notably the THM 350. This allows freewheeling in drive-range second gear while still providing instant reaction through the intermediate roller clutch and sun gear when engine torque is applied to the gear train. In the manual second or L2 position, the intermediate band applies to render intermediate roller clutch overrun ineffective and provide engine braking by holding the sun gear.

In the THM 700–R4 four-speed overdrive transmission, a one-way roller clutch is used to drive the input sun gear. Other applications of the one-way clutch can be found by studying the power flow through the different makes and models of automatic transmissions.

Advantages of One-Way Clutches

One-way clutches have some distinct advantages that make them desirable for use in automatic transmissions. Automatic transmissions change gears by applying and releasing clutches and bands that control the drive and reaction in a planetary gearset. Precise timing of the application and release of these units is

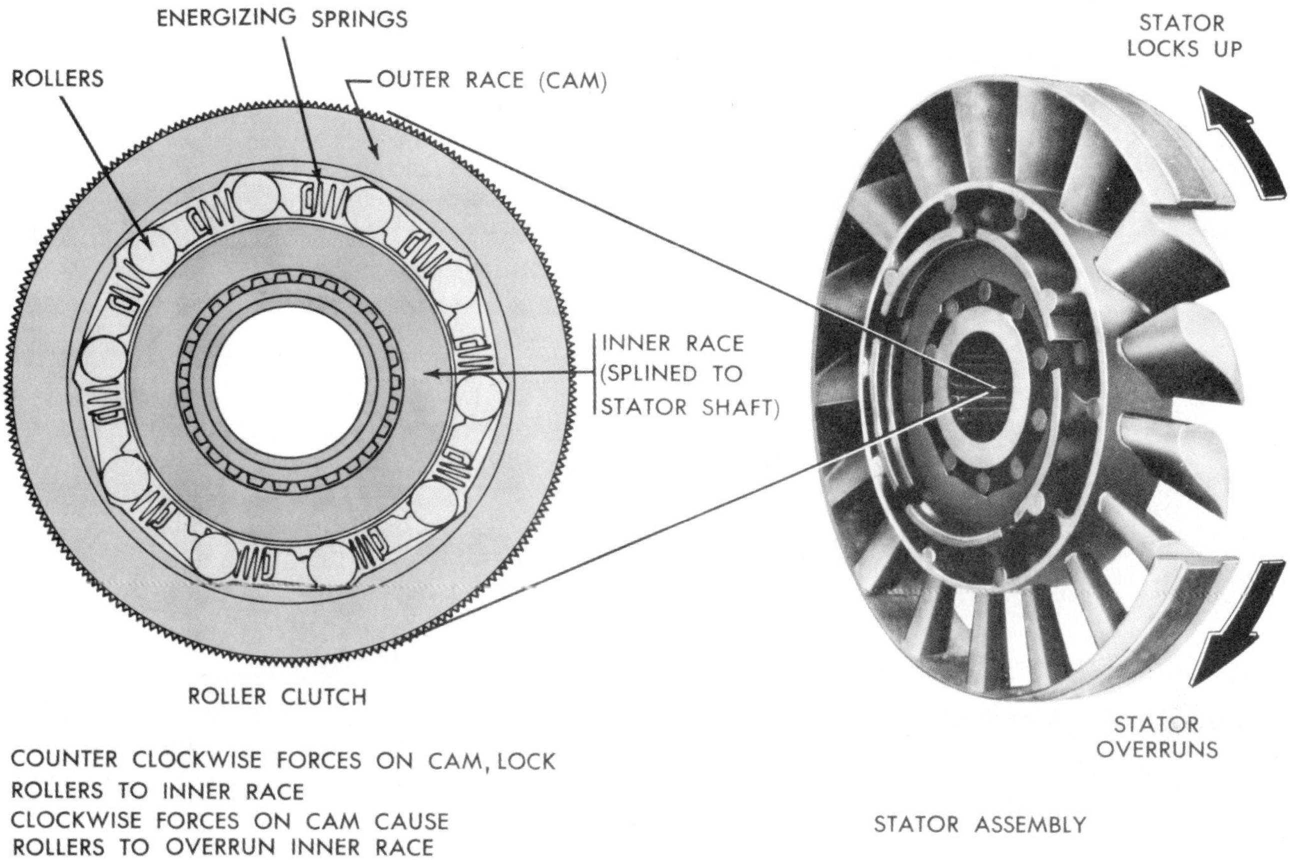

Figure 4-9. Roller-type one-way clutch. (*Courtesy of General Motors Corporation*)

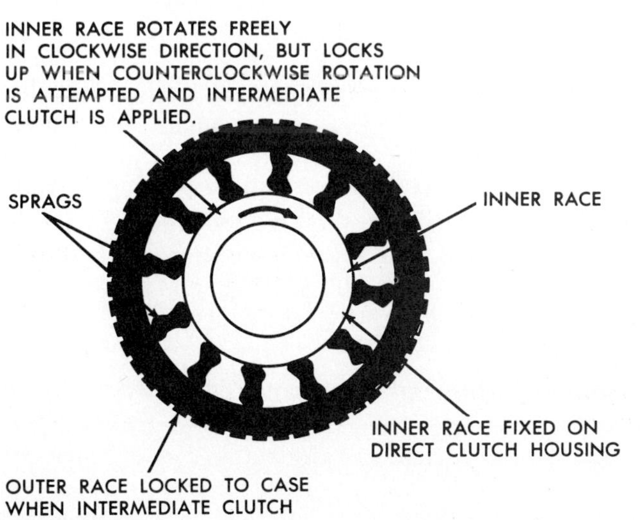

Figure 4-10. Sprag-type one-way clutch. (*Courtesy of General Motors Corporation*)

critical to smooth shifts. The one-way clutch acts and reacts instantly to application of torque and torque reversal. A good example of this is in the three-speed Simpson gear train. In drive low the one-way clutch provides instant reaction when engine torque is applied to the gear train, preventing the reaction carrier from turning backward (counterclockwise).

As soon as the transmission shifts from first to second, the reaction carrier must turn forward (clockwise). The one-way clutch provides instant capability for this action and allows the reverse to occur on a 2–1 downshift.

The one-way clutch is compact in design and requires no mechanical linkage or hydraulic devices for operation. This results in normally troublefree operation for the life of the transmission.

One-Way Clutch Construction and Operation

There are two types of one-way clutches: the sprag clutch and the roller clutch. Both rely on a wedging action for clutch lockup.

The sprag clutch consists of a sprag and cage

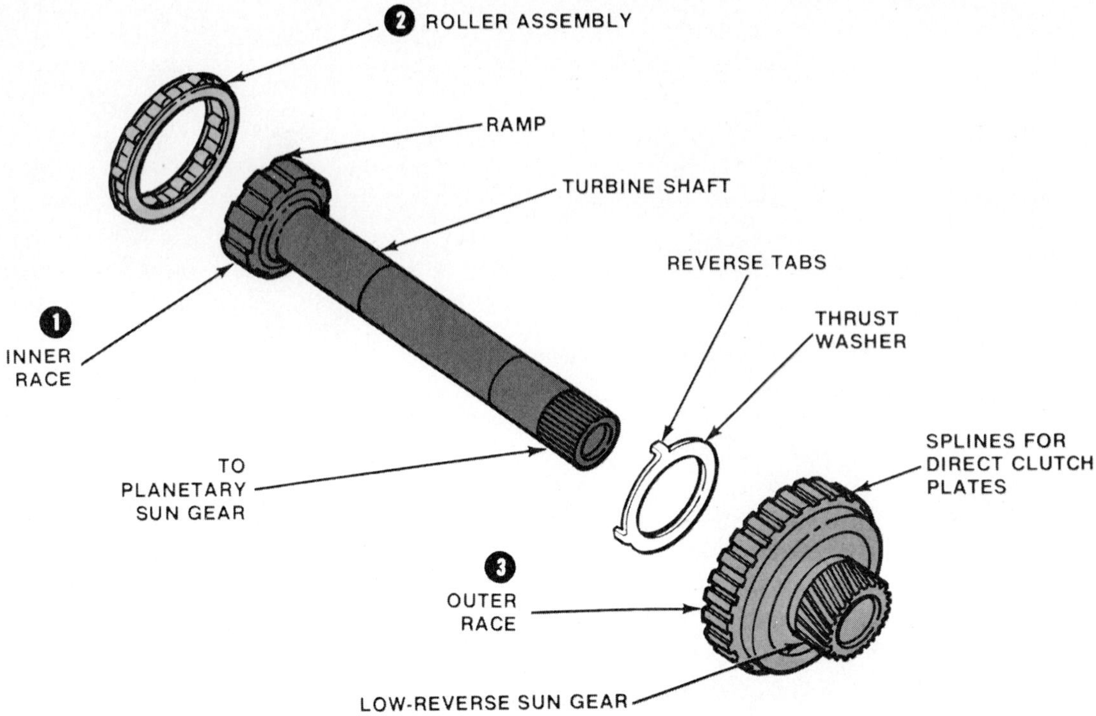

Figure 4-11. Planetary drive one-way clutch as used in Ford ATX transaxle to drive planetary gears. (*Courtesy of Ford Motor Company*)

assembly positioned between an inner and outer race. The roller clutch consists of an inner race, an outer race, and a series of spring-loaded rollers positioned between the two races.

The sprag clutch relies on wedging the larger diameter of the sprags between the inner and outer race for lockup and by tilting the sprags slightly on release, positioning the smaller diameter of the sprags between the two races. Spring action on the sprags and the direction of torque application to the race determine when the sprag clutch locks up or overruns.

The roller clutch relies on wedging rollers between the high part of the cams on the outer race and the inner race for lockup. Energizing springs tend to keep the rollers close to the high part of the cam. Torque application wedges the rollers firmly between the inner and outer race for lockup. Torque reversal causes the rollers to roll down the cam against spring action for freewheeling action.

In both of these designs one race is bolted, riveted, splined, or clutched to the transmission case when used as a reaction holding device. The other race is attached to the planetary gear member serving as the reaction member. When used as a driving device one race is clutched or attached directly to the drive shaft while the other race is connected to the planetary gear member to be driven.

It is critical to transmission operation that one-way clutches be properly assembled and installed. If installed in the opposite way to normal installation, the clutch will lock up when it should overrun and will overrun when it should lock up. This would result in either a no-drive condition or complete transmission lockup, depending on the type of transmission and the drive range selected.

Part 4–4 Worksheet

1. Examine the drive train from an automatic transmission. Learn to identify on sight the clutches, bands, and one-way clutches.
2. Determine which planetary gear member is affected by clutch application and clutch release.
3. Note whether servos are part of the transmission case or are separate bolted-on assemblies by examining various makes and models of automatic transmissions available for this purpose.
4. Using shop air pressure regulated at 25 psi (100 kPa) and a rubber-tipped nozzle on the air hose, direct the air pressure to the apply passage of a properly assembled clutch assembly provided by your instructor. Take note of the definite clunking noise indicating clutch apply.
5. Locate the check ball in an available clutch assembly.
6. Examine different types of clutch discs, steel, and friction, and note the differences in friction material and grooving.
7. Examine different types of bands such as single wrap, double wrap, rigid, and flexible, and take note of the different types of friction material, grooving, and strut- or lever-actuating designs.
8. Examine a disassembled transmission and determine whether the servos are positioned to take advantage of the self-energizing effect of band application.

PART 4-5 SELF-CHECK

1. A multiple-disc clutch is used in some transmissions to ____________ a planetary gear member or to ____________ a planetary gear member.
2. What is the minimum number of drive clutches required in the basic three-speed Simpson planetary gear transmission?
3. In a Simpson gear three-speed transmission the forward clutch drive plates are the composition plates. True or false?
4. Drive plates and driven plates in some clutches spin in opposite directions. True or false?
5. What is the purpose of the ball check valve in a clutch piston or drum?
6. What three types of springs are used in multiple-disc clutches?
7. How does a one-way clutch work?
8. What factors determine multiple-disc clutch torque capacity?
9. What causes band slippage?
10. What effect does low hydraulic pressure have on clutch and band operation?

PART 4-6 PERFORMANCE EVALUATION

After sufficient study of this chapter and the appropriate training models, are you able to and did you perform the following tasks as required by your instructor?

1. State the purpose of a driving device in an automatic transmission.
2. Describe the basic construction and operation of multiple-disc driving clutches, including apply and release.
3. Describe the basic construction and operation of multiple-disc holding clutches, including apply and release.
4. State the purpose of a holding device in an automatic transmission.
5. Describe the basic construction and operation of roller-type and sprag-type one-way holding clutches.
6. Describe the basic construction and operation of the band and servo in an automatic transmission, including apply and release.
7. List the different types of multiple-disc clutch plates and band types used in automatic transmissions and describe their operating characteristics.
8. Describe the effects of insufficient hydraulic apply pressure, wear, and excessive or insufficient clearance in multiple-disc driving and holding clutches.
9. Describe the effects of inoperative one-way driving or holding clutches.
10. Describe the effects of insufficient hydraulic apply pressure and excessive or insufficient clearances in band and servo operation.
11. Describe the factors that determine multiple-disc clutch torque capacity.

Chapter 5

Hydraulic System Principles

Performance Objectives

After sufficient study of this chapter and the appropriate training models, the student should be able to perform the following tasks as required by the instructor.

1. Define the terms *pressure* and *force* as they apply to hydraulic fluids.
2. Define Pascal's law.
3. Define force multiplication by hydraulic means and give an example of force multiplication by means of hydraulics.
4. Define the relationship between force and motion transmitted by hydraulic means, and give an example of this relationship.
5. Describe briefly the various types of automatic transmission hydraulic fluids and their properties.
6. State the purpose of the hydraulic pump in the automatic transmission.
7. Describe the basic construction and operation of the three types of hydraulic pumps discussed in this chapter, including the causes and effects of pump wear.
8. Describe the basic construction and operation of the following types of valves and flow control devices:
   (a) Ball check valve
   (b) Poppet valve
   (c) Spool valve
   (d) Pressure relief valve
   (e) Pressure regulating valve
   (f) Orifice
   (g) Accumulator
9. Describe the basic purpose, construction, and operation of automatic transmission gaskets, seals, and flow control seal rings.
10. Complete the worksheet in this chapter.
11. Complete the self-check in this chapter with at least 80 percent accuracy.
12. Evaluate your performance by using the performance evaluation section in this chapter.

PART 5-1 INTRODUCTION

This chapter deals with the basic principles of hydraulics as they apply to the automatic transmission hydraulic control system. Included in the discussion are basic definitions of hydraulic pressure, force, and Pascal's law.

How to multiply force hydraulically and how to transmit motion and force are also discussed as well as the relationship between force multiplication and transmitting motion by hydraulic means. A description of the various types of automatic transmission fluids and their properties is included in this chapter.

The purpose, construction, and operation of automatic transmission hydraulic pumps is covered, including the causes and effects of hydraulic pump wear. The various types of valves and flow control devices used by all automatic transmission hydraulic control systems and their construction and operation are discussed. The chapter concludes with a description of the types of gaskets, seals, and flow control seal rings used in automatic transmissions.

A good general understanding of hydraulic principles, transmission fluids, pumps, valves, and seals is essential to understanding the operation of the hydraulic control system of the automatic transmission. Since all automatic transmissions, domestic or imported, operate on the same hydraulic principles and use the same devices in their hydraulic control systems, a good grasp of these principles allows the serious student to apply this knowledge to any type of transmission.

PART 5-2 HYDRAULIC FORCE AND PRESSURE

A French scientist, Pascal, discovered the hydraulic lever principle in the early 17th century. By various experiments in his laboratory, he was able to prove that force and motion could be transmitted by means of a confined liquid. He also discovered that force multiplication or mechanical advantage could be achieved by the use of a hydraulic pressure system, proving that the relationships between force and distance are the same in a hydraulic system as they are in a simple mechanical lever.

From these laboratory experiments and the resulting data came Pascal's law. Pascal's law states that pressure on a confined fluid is transmitted equally in all directions and acts with equal force on all equal areas. A brief explanation of the terms *force, pressure,* and *force multiplication* will help to understand this law.

Force

Force can be defined as a push or pull exerted on an object. The force of gravity is one example. The force of gravity is the mass or weight of an object. A 50-pound (22.680-kg) block of concrete resting on the floor exerts a downward force of 50 pounds (22.680 kg) on the floor. To try to drag the block across the floor involves the force of friction between the block and the floor.

Another force, called spring force, results from stretching or compressing a spring. A compressed spring exerts a force at both ends of the spring against the restraining surfaces. A stretched spring exerts a pulling force at both ends of the spring on the spring anchors.

The unit used to describe the degree of force being applied in the U.S. customary system is the pound or the ounce. In the metric system the newton (N) is the unit used for force. One pound of force is equal to 4.448 newtons. Therefore, to convert pounds to newtons, simply multiply the number of pounds by 4.448. One thousand pounds, for example, would be 1000 x 4.448, or 4,448 newtons (4448N).

Pressure

Pressure is simply force per unit area (force divided by area). In the U.S. customary system the term *pounds per square inch* is commonly used. The abbreviation for this term is PSI or psi. To illustrate this, a 50-pound block of concrete with a surface area measuring 5 inches by 10 inches, or 50 square inches (50 in^2), resting on the floor exerts a pressure of 1 pound per square inch (1 psi). This is derived from

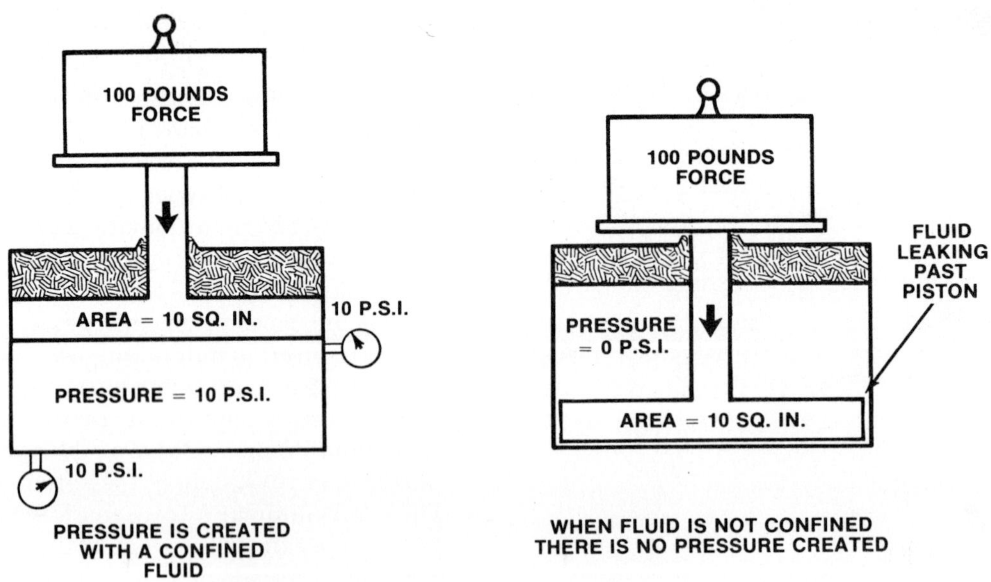

PRESSURE ON A CONFINED FLUID
Figure 5-1. *(Courtesy of Chrysler Corporation)*

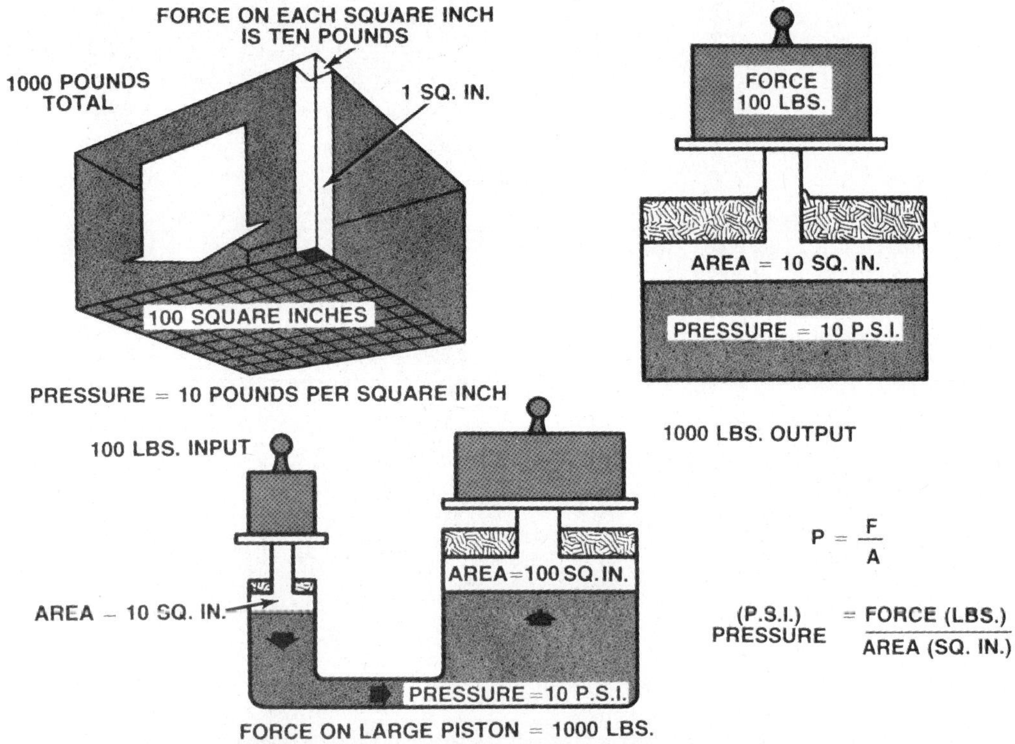

FORCE AND PRESSURE RELATIONSHIPS
Figure 5-2. *(Courtesy of Chrysler Corporation)*

dividing the force (50 pounds) by the area in contact ($50\ in^2$), $50\ lbs \div 50\ in^2 = 1$ psi.

In the metric system the unit for measuring pressure is the kilopascal (kPa). One pascal is the equivalent of 1 newton per square meter (N/m^2). One kilopascal is 1,000 pascals.

Manufacturers' service repair manuals may use the U.S. customary system, the metric system, or both when stating specifications for their products.

Pressure on a Confined Fluid

If a force is applied to a movable surface area in contact with a confined fluid, hydraulic pressure is created. If the fluid is not confined, no pressure will be created. In other words there must be a resistance to flow.

For example, if a cylinder filled with fluid and open at one end is fitted with a piston and a force is exerted on the piston, a pressure will be produced in the fluid. However, if the piston is not closely fitted, fluid will simply pass by the piston and no pressure will be produced in the fluid. It is extremely important in hydraulic systems that piston sealing prevent fluid leakage past the piston. Clutch and servo piston seals are good examples of this in an automatic transmission.

In a hydraulic system the effects on the confined fluid are the same whether the force is applied to the piston by gravity, by spring force, or by any other force. The principle remains the same regardless of the direction of the applied force on the fluid and regardless of the shape of the container in which the fluid is confined. The pressure on the fluid is the same everywhere in the container. Pressure on a confined fluid is transmitted undiminished in all directions.

Force Multiplication (Hydraulic Lever)

A simple hydraulic system can be used to multiply force. An example of this requires two cylinders, each fitted with pistons and filled with fluid. The cylinders must have a fluid passage connecting the bottom of the cylinders to each other. In this system a small weight on an input piston can be used to move a large weight on a large output piston.

Using the U.S. customary units in this example, a 100-pound input force applied to the input piston with a 10-square-inch area applying the force to the fluid results in a pressure of 10 pounds per square

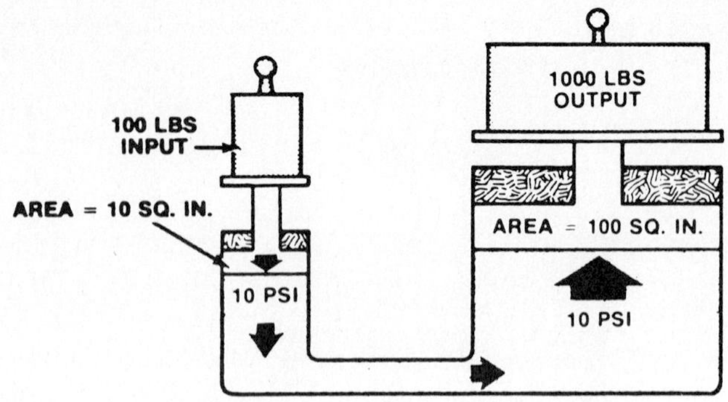

Figure 5-3. Hydraulic force multiplication. (*Courtesy of Chrysler Corporation*)

inch. The 10-psi pressure acting on 100 square inches of output piston area is able to create a 1,000-pound output force.

Using the formula

$$\text{pressure} = \text{force} \div \text{area}$$

the input force divided by the area of the input piston contacting the fluid, the pressure in the system is calculated like this:

$100 \text{ lbs.} \div 10 \text{ in}^2 = 10$ psi pressure in the system

The 10 psi pressure applied over the 100 square inches of output piston area results in an output force of 1,000 lbs. calculated in this way:

$$10 \text{ psi} \times 100 \text{ in}^2 = 1000 \text{ lbs of force}$$

This important concept is used in automatic transmission clutches, servos, and valves to create a difference in pressure in order to move an object. Another factor involved in this system must be remembered. Since it is not possible to create energy, it follows that a 10-square-inch piston with a 100-pound force applied to it cannot move a 100-square-inch piston an equal distance. This results from the fact that moving the input piston 1 inch displaces only enough fluid to move the 10 times larger output piston 1/10 the distance.

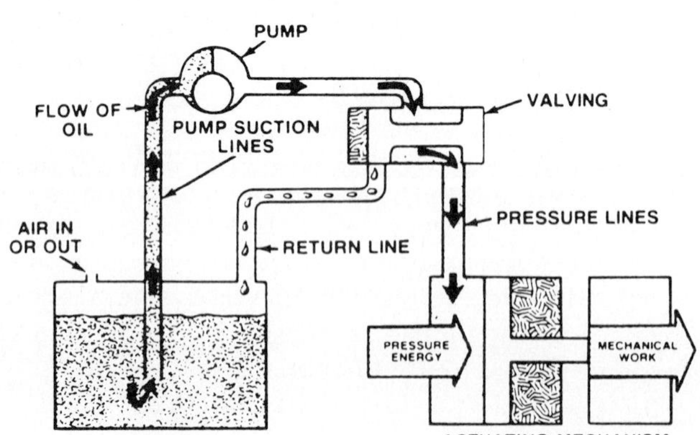

Figure 5-4. Basic hydraulic system. (*Courtesy of Chrysler Corporation*)

PART 5-3 AUTOMATIC TRANSMISSION FLUIDS

The entire hydraulic control system is totally dependent on the ability of the fluid to function in a consistent manner under all normal operating conditions. Consequently heavy demands are placed on the fluid in the system.

A hydraulic system consists basically of a source of fluid with a reserve supply in a reservoir or sump; a hydraulic pump to create hydraulic pressure and fluid flow; a number of pressure-regulating valves; a number of flow control valves; a number of hydraulic servos to operate the clutches and bands; some flow-delaying orifices or restrictions; some cushioning devices to prevent sudden and harsh clutch and band application; a lubrication system; a cooling system; and the necessary seals and gaskets to control fluid flow and leakage.

The fluid reservoir must contain a sufficient quantity of the correct type of fluid to meet all demands made on it by the hydraulic control system. The level of fluid in the reservoir rises and falls as a result of temperature changes in the fluid and as a result of varying quantities of fluid required to operate the clutches and bands in the different gear positions. For this reason the fluid reservoir is vented to the atmosphere. If it were not vented, a negative pressure (vacuum) would develop in the reservoir as the fluid level drops and a pressure above atmospheric pressure would develop as the fluid level rises.

The fluid in the automatic transmission is the means by which force and motion are transmitted to operate the various clutches, servos, and valves. The fluid also serves to help seal between moving parts and to lubricate and cool transmission parts. The characteristics of the fluid are also designed to provide the required coefficient of friction to produce the desired slip-lockup characteristics of clutches and bands.

The fluid is also the means of transferring engine torque through the torque converter to the transmission. The torque converter, clutches, and bands generate a tremendous amount of heat, which is absorbed by the fluid and must be dissipated to the atmosphere in order to prevent the fluid from being overheated and oxidized.

Approximate normal transmission fluid operating temperature is around 175°F (76°C). Operating conditions such as air conditioning, moving slowly in heavy urban traffic, pulling heavy trailers, or rocking a car from forward to reverse repeatedly cause fluid temperatures to rise dramatically.

For example, rocking a car back and forth in mud, snow, or sand for only one or two minutes can result in fluid temperatures of 300°F (150°C). If this practice is continued for any length of time, temperatures that will cause metal parts in the transmission to warp and twist and cause friction materials to separate from clutch discs and bands, may be reached. Clutch and servo piston seals may also become baked and hard so that they lose their ability to seal hydraulic pressures. Resulting slippage of clutches and bands soon cause complete transmission failure. Friction materials loosened from clutches and bands can also clog the filter, impede operation of valves, and find their way into the torque converter. Complete overhaul of the transmission and replacement of the torque converter as well as thorough flushing of the cooler and cooler lines are required to correct a condition of this kind.

Transmission Fluid Development

Automatic transmission fluids have undergone a number of developments and improvements over the years. Early transmissions simply used the same ordinary mineral oil that was used in the crankcase of the engine. Fluid capacity of the transmissions was generally high (12 to 14 quarts, approximately 14 to 17 liters) and demands on the transmission were not as great as they became later. Increased engine horsepower and the increased use of air conditioning and trailer towing placed much heavier demands on the transmission and the transmission fluid.

In 1949 General Motors developed the Type A fluid specification. Other vehicle manufacturers also adopted this fluid specification. Transmission fluids developed by oil companies and meeting the Type A specifications were approved to be labeled as Type A automatic transmission fluid.

In 1957 the Type A fluid was improved to reduce the buildup of varnish and was labeled Type A–Suffix A fluid. This fluid superseded and replaced the original Type A fluid for use in all automatic transmissions.

In the early 1960s the Ford Motor Company developed its own distinctive automatic transmission fluid for Ford transmissions. This fluid was designated Type F fluid and was to be used in all prior Ford transmissions as well.

In 1968 when high horsepower engines had almost reached their peak, improvements to transmission fluids were again realized. Ford's fluid was identified by a new Ford fluid designation number but continued to be known as Type F. General Motors' new improved fluid was designated as Dexron fluid. Chrysler and American Motors also specified Dexron fluid for their vehicles. The current Dexron II-D designation identifies the latest improved Dexron fluid.

In 1977 the Ford Motor Company introduced a new type of fluid designated CJ fluid for use in several Ford transmission designs, including their automatic overdrive transmission. Ford's C5 transmission with the centrifugal clutch lockup converter requires a special fluid designated Type H.

Although some original factory-fill transmission fluids may not be dyed red, most replacement fluids are red. The red dye aids in distinguishing oil leaks from the transmission from engine oil leaks.

Transmission Fluid Characteristics

Automatic transmission fluid is a mineral base oil with a number of additives to improve its operating characteristics. Among these are viscosity index improvers, oxidation inhibitors, foam inhibitors, and red dye for identification.

Viscosity Index Improver

Mineral oils thin out with increasing temperature and thicken when temperatures drop. A fluid that is too thin reduces the sealing ability of the fluid and decreases the ability of the hydraulic pump to maintain adequate operating pressure. A fluid that has thickened too much (as in severe winter weather) slows down the movement of shift valves, clutch, and servo pistons, which delays shifting.

Viscosity index improvers slow the rate of viscosity change in the fluid as temperatures change. This improves transmission operation.

Oxidation Inhibitors

Oxidation inhibitors reduce the fluids' tendency to oxidize. As mentioned, increasing fluid temperatures increase the fluid's tendency to oxidize.

Foam Inhibitors

Foam in transmission fluid is simply the result of rapid release of entrapped air. It is important that fluid in the automatic transmission be free of foam and air as it enters the hydraulic control system. Since air is compressible, the effectiveness of clutch and band operation and possible slippage result from air in the system. Foam inhibitors reduce the possibility of this problem. However, in spite of this it is possible for serious transmission damage to occur as a result of air in the hydraulic system simply from a fluid level that is either too high or too low.

The transmission fluid including the additives must be compatible with the various metals such as steel, bronze, aluminum, and the like used in the transmission. This also applies to the different seal materials and friction materials.

Transmission Fluid Recommendations

Since it is the nature of the automotive industry to continue development and improvement of its products, this can also be expected to continue with respect to automatic transmissions, transmission fluids, and fluid recommendations.

The best practice is to use only the fluid recommended by the vehicle manufacturer in the appropriate owner's handbook or service repair manual. The use of an incorrect fluid may affect both the quality of shifting of the transmission and its life expectancy.

The same should be said for the use of additives or conditioners. They cannot restore burned friction materials on clutch discs and bands nor can they correct cracked seals. If a transmission problem exists, such as a sticking valve due to varnish deposits, a commercial additive may correct the condition. However, the best policy is to follow the vehicle manufacturer's recommendations regarding additives.

PART 5-4 THE HYDRAULIC PUMP

The transmission hydraulic pump puts the fluid in motion and applies force to the fluid. Although the pump creates flow, there will be no pressure buildup unless there is a resistance to flow. Fluid flow is dead ended in some parts of the transmission hydraulic system but provides limited continuous flow for lubrication and cooling.

Pump Types

Transmission hydraulic pump types include the internal-external (IX) gear pump, the internal-external rotor pump, and the vane pump. The IX gear pump and the IX rotor pump are positive displacement pumps while the vane pump used with automatic transmissions is a variable-displacement or variable-capacity pump.

A positive displacement pump displaces a constant volume of fluid every revolution of the pump. Pump output varies only with pump speed. The vane-type variable-capacity pump is able to vary the volume of fluid displaced per pump revolution to meet the varying demands of the transmission hydraulic system.

Pump Construction and Operation

The gear pump consists of an externally toothed drive gear meshed with an internally toothed driven gear. The drive gear is small enough to allow the teeth to

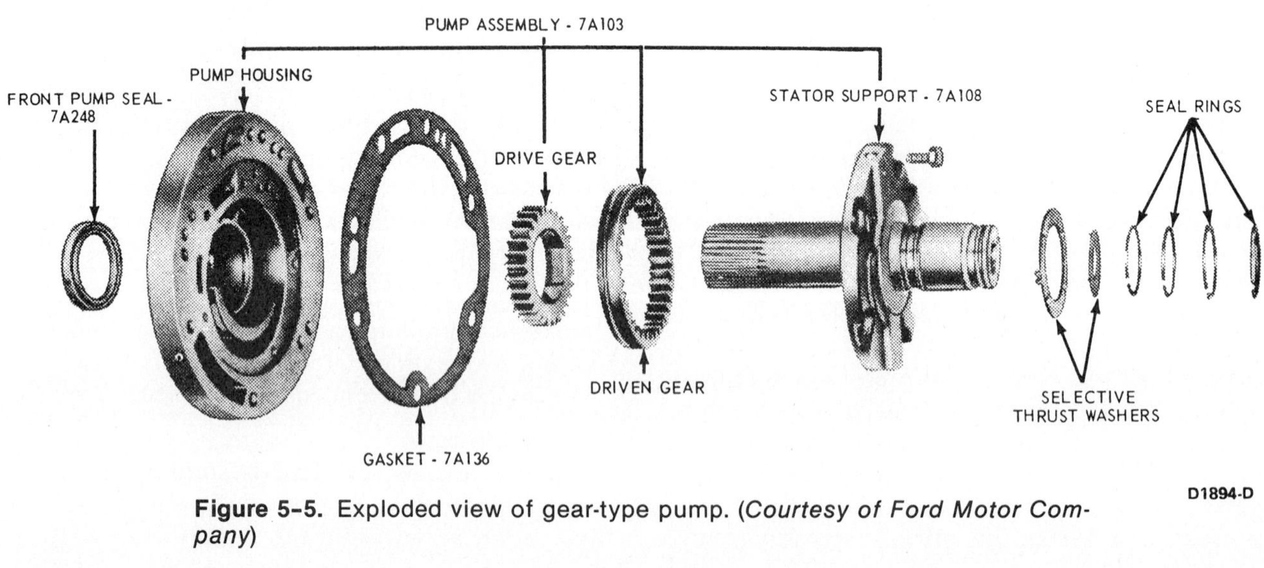

Figure 5-5. Exploded view of gear-type pump. (*Courtesy of Ford Motor Company*)

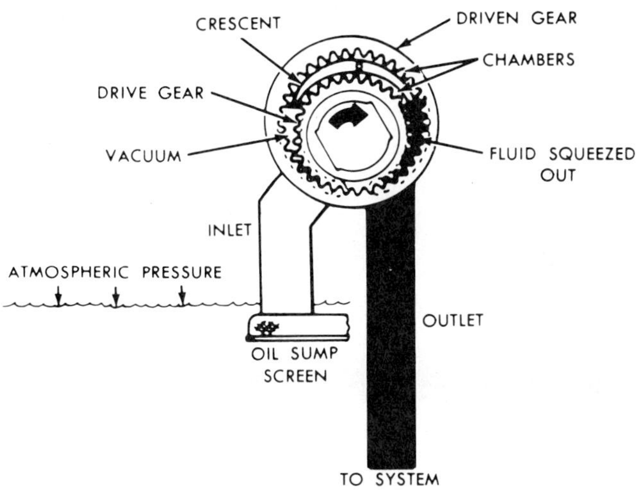

Figure 5-6. Gear pump operation. (*Courtesy of Ford Motor Company*)

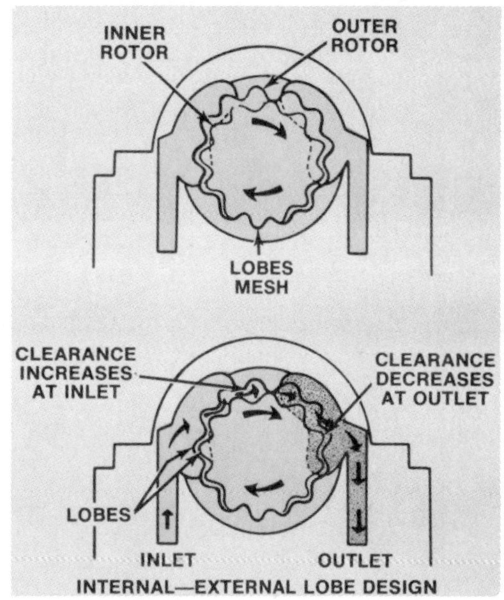

Figure 5-8. Rotor pump operation. (*Courtesy of Chrysler Corporation*)

mesh on one side only. A crescent-shaped section of the pump housing is positioned between the gears opposite to the side in mesh.

Rotation of the drive gear causes the gear teeth to separate on the inlet side of the pump, creating a low pressure area. Atmospheric pressure acting on the fluid in the reservoir causes fluid to fill the low-pressure inlet area of the pump. As the pump continues to turn, fluid is carried between the gear teeth past the crescent. As the gear teeth come into mesh again, the fluid is forced out from between the gears to the outlet passage with a resulting pressure rise. The limited clearance between the tips of the gear teeth and the crescent prevents fluid from returning to the inlet side of the pump.

The rotor pump has a drive rotor encircled by a driven rotor enclosed in the pump housing. The drive rotor is meshed with the driven rotor only on one side, similar to the gear pump design. However, no crescent is required to separate the inlet from the outlet. The inner and outer rotor size and design cause one drive rotor lobe and one driven rotor lobe to meet with minimal clearance between lobe tips at the midpoint between the inlet and outlet passages. This minimal clearance prevents output fluid from returning to the inlet passage.

The variable-capacity vane pump consists of a slotted rotor and vane ring assembly positioned in a circular slide. As the pump rotor turns, the vanes

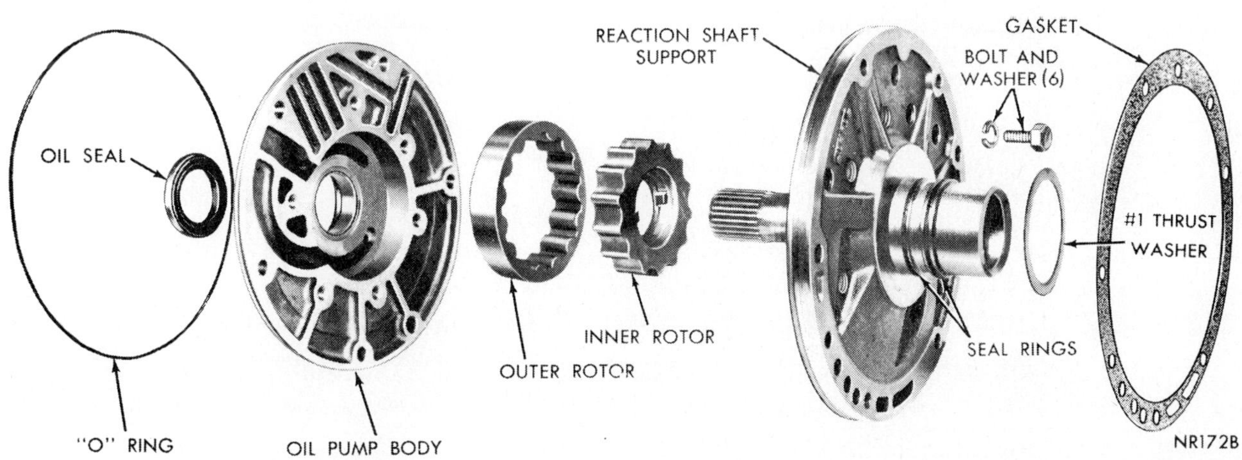

Figure 5-7. Exploded view of rotor-type hydraulic pump. (*Courtesy of Chrysler Corporation*)

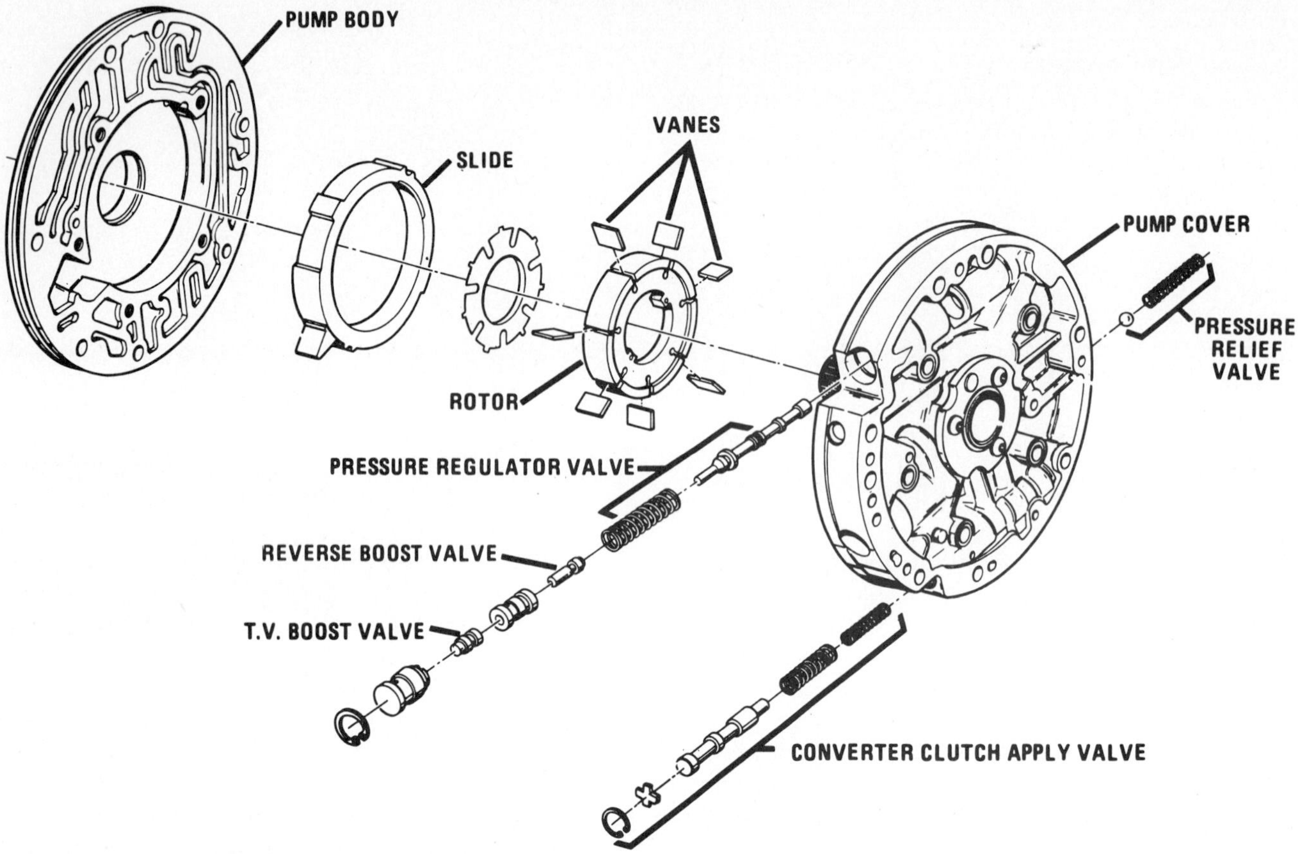

Figure 5-9. Variable-capacity vane pump components. (*Courtesy of General Motors Corporation*)

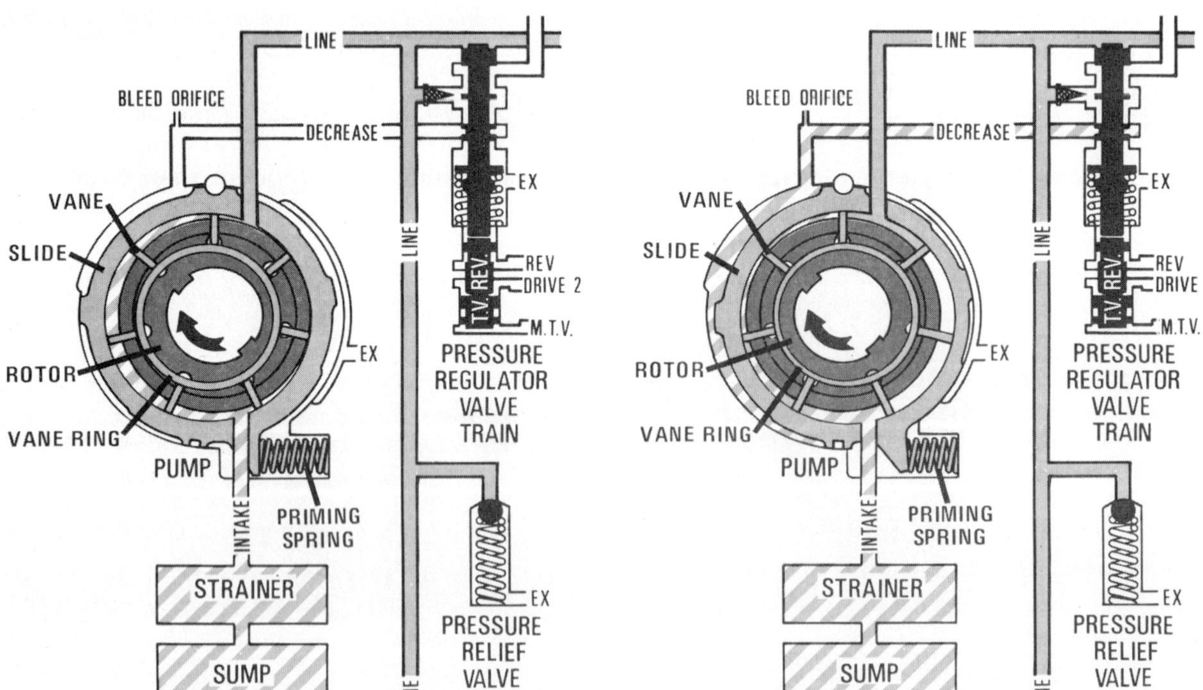

Figure 5-10. Operation of variable-capacity vane pump. Maximum output on left and minimum output on right. (*Courtesy of General Motors Corporation*)

slide around the inner surface of the slide. The slide is situated in the pump housing. A priming spring keeps the slide over to the maximum pumping position until transmission demands cause the pressure regulator valve to open a passage to an area between the slide and pump housing. Hydraulic pressure between the slide and pump housing pushes the slide back against spring pressure. This reduces the volume of the pump inlet area between the vanes and therefore reduces pump output to match transmission requirements. The advantage of the variable-capacity pump is that less engine power is required to drive the hydraulic pump any time the pump is not at maximum output position. This results in a saving in fuel consumption.

Pump Wear

The hydraulic pump is designed to last for the life of the transmission under normal operating conditions. Pump capacity is sufficient to allow for normal pump wear without pump capacity falling below transmission requirements.

As the drive and driven elements and the pump housing wear, more and more fluid is able to pass back from the outlet to the inlet side of the pump. Excessive wear reduces pump output below the requirements of the transmission. As a result the apply force for clutches and bands is reduced, causing clutch and band failure.

The major cause of excessive pump wear is contaminated or insufficient transmission fluid. If dirt is allowed to enter the transmission during the fluid level checking procedure, this dirt enters the hydraulic system. Abusing the transmission by towing excessive loads or rocking the car back and forth when stuck in mud, snow, or sand can cause overheating of fluid and friction materials, resulting in friction materials entering the hydraulic system. Any of these abrasives recirculating through the hydraulic pump will cause excessive and rapid pump wear.

PART 5–5 VALVES AND CUSHIONING DEVICES

Once the pump has started pumping the fluid, a system of valves is required to regulate pump pressure and to control the flow of fluid. These valves are designed to control the pressure and flow of fluid in the various transmission hydraulic circuits.

Two basic classifications of valves relate to their function in the system: those that regulate or vary the pressure in the system and those that simply open or close hydraulic passages without affecting pressure in the system. These latter valves may also be called relay valves or flow control valves. There are a number of each of these types in an automatic transmission.

This discussion deals with the basic principles of pressure regulation, flow control, and cushioning of hydraulic apply pressure or force. Also included is a basic discussion of the design and function of the various valves.

Regulating the Pressure

As stated earlier, the hydraulic pump produces hydraulic pressure and fluid flow. A pressure-regulating valve is required to regulate or control that pressure.

A very simple pressure regulator valve consists of a spring-loaded ball or plunger controlling a bleed-off passage from the pump outlet. When pressure builds up high enough to push the ball check valve or plunger valve back to open the bleed-off passage, some pump output is discharged through this opening. This fluid is routed back to the inlet side of the pump or is simply fed back to the sump. As pump speed and pump output increase, more fluid is bled off by the pressure-regulating valve.

In this simple pressure-regulating valve, spring force or spring tension is the only factor regulating pump output pressure. A heavier spring (not as easily compressed) would control output or line pressure at a higher level. A lighter, more easily compressed pressure regulator spring would maintain a line pressure of lower value or pressure. If an adjusting screw were used behind the pressure-regulating spring, it would be possible to vary line pressure by adjusting spring pressure against the pressure-regulating valve.

Spool-Type Pressure-Regulating Valve

The pressure-regulating valve in an automatic transmission is of the spool valve design. The cylindrically shaped spool valve has two or more "lands" with grooves between each pair of lands. The grooved area between the lands forms a chamber into which hydraulic pressure can be introduced. When present, hydraulic pressure acts on the flat surfaces of the valve ends or between the lands, depending on valve design and application. These surfaces are known as reaction areas.

The pressure regulator valve may be located in the valve body, the pump body, or the transmission case. A simple two-land spool valve can be used to regulate pump output pressure in the same way as the plunger- or ball-type regulator valve. However, the actual spool-type pressure regulator valve in an automatic transmission has a number of lands and grooves. This allows other hydraulic pressures to be

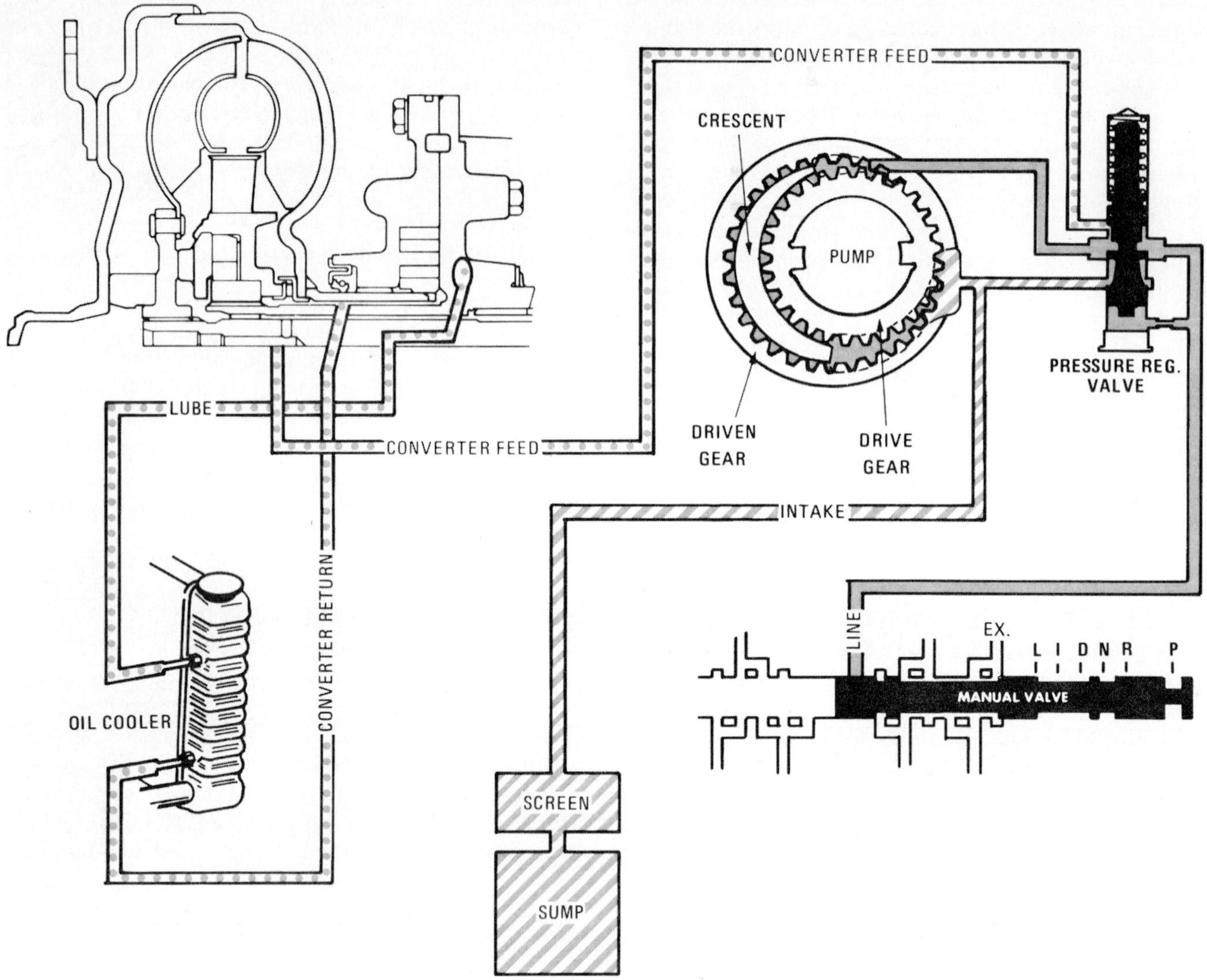

Figure 5-11. Pressure regulator valve operation in automatic transmission. (*Courtesy of General Motors Corporation*)

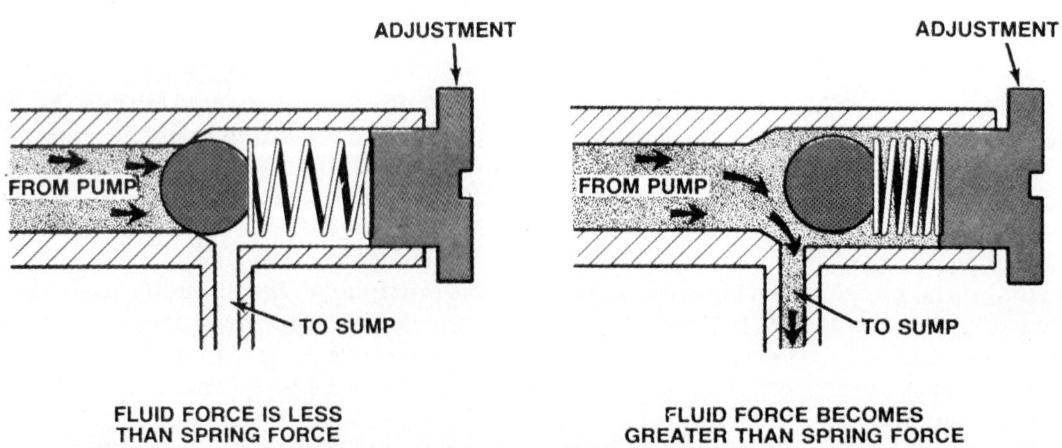

Figure 5-12. Ball-type pressure relief valve operation. (*Courtesy of Chrysler Corporation*)

P	R	N	D	2	1
4	4		1	1	1
5	5		2	2	4
	6		3	4	5

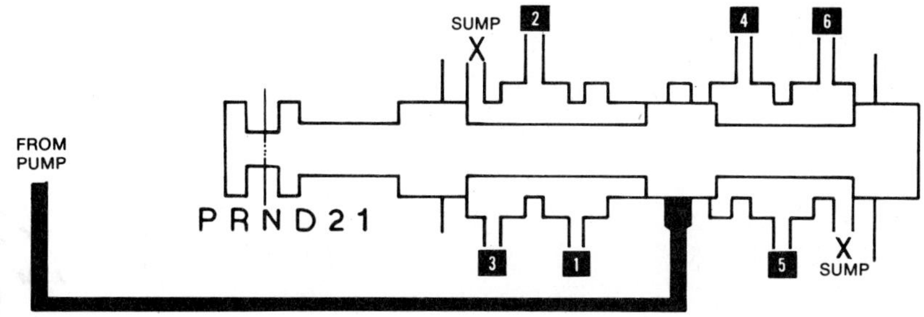

Figure 5-13. Typical manual valve operation. Chart shows passages charge at different valve positions. (*Courtesy of Ford Motor Company*)

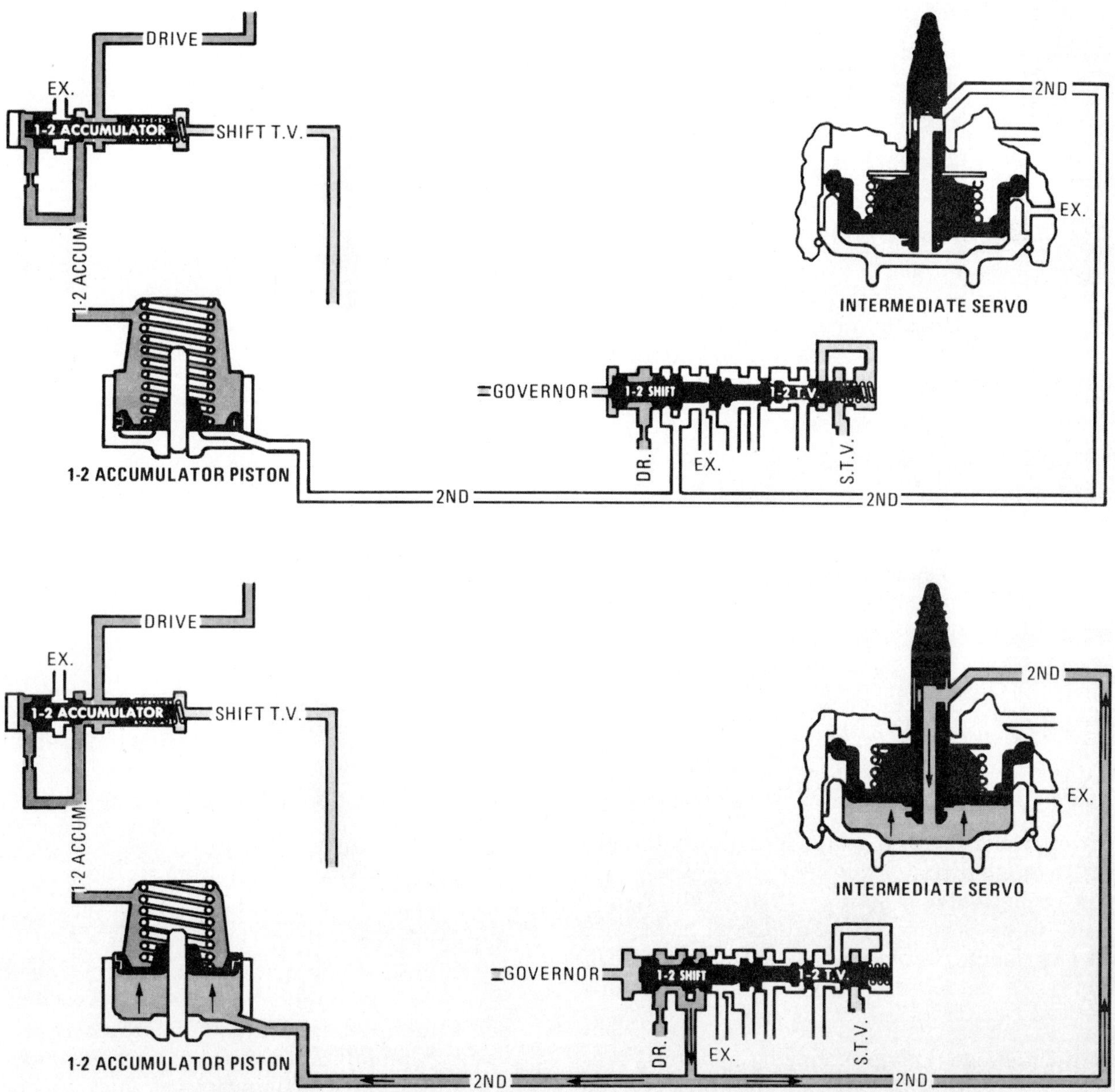

Figure 5-14. Typical 1-2 accumulator action in first gear (top) and second gear (bottom). The 1-2 accumulator valve provides a variable pressure to the accumulator (depending on the degree of throttle opening) to cushion intermediate band application. (*Courtesy of General Motors Corporation*)

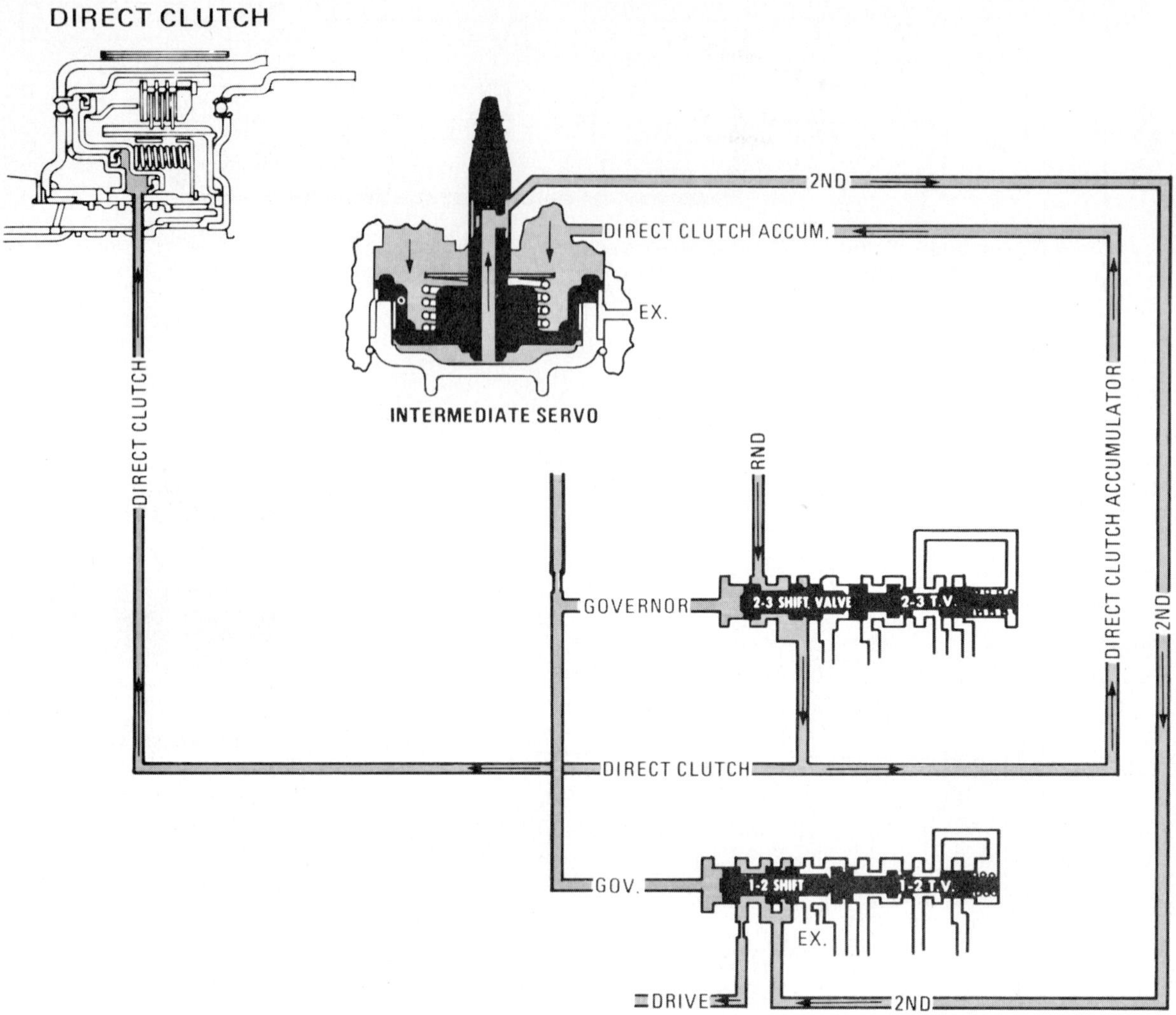

Figure 5-15. The release side of the intermediate servo acts as the accumulator to cushion direct clutch application during a 2-3 shift in this diagram. (*Courtesy of General Motors Corporation*)

introduced to the regulator valve to help control hydraulic system pressure to the exact needs of the transmission.

Pressure Relief Valve

A pressure relief valve is used in the hydraulic control system to limit maximum output or line pressure. The pressure relief valve is simply a ball or disc-type of blow-off valve, which opens to relieve system pressure when the maximum safe pressure is reached.

Relay Valve

A relay valve is an on-off valve used to open or close a hydraulic passage without changing hydraulic pressure. Transmission shift valves of the spool type described earlier are relay valves.

Orifice

An orifice is a restriction in a hydraulic passage designed to delay pressure buildup downstream from the orifice. An orifice may be used in a hydraulic clutch or band apply passage to delay slightly the full application or release of the clutch or band. Another effect of an orifice is to cushion or soften the application of a clutch or band.

Accumulator

A hydraulic accumulator is a cushioning device used to cushion the application of a clutch or band. This helps to smooth the shifts in the transmission.

Accumulators may be of the spring-loaded piston type or the valve or plunger type. They may be single acting or double acting.

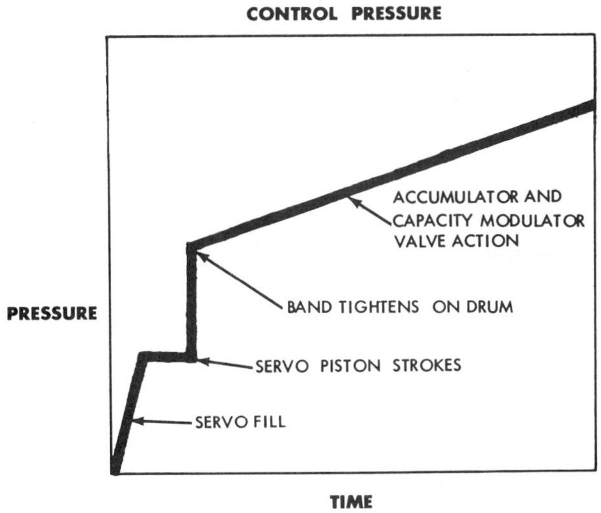

Figure 5-16. Intermediate band servo apply pressure rise and effect of accumulator action on apply pressure during a 1-2 upshift. (*Courtesy of Ford Motor Company*)

A typical accumulator in a clutch apply passage would "accumulate" some apply pressure during clutch application. In other words, the hydraulic pressure would cause the accumulator piston to be pushed back against spring pressure during clutch apply. When the accumulator piston bottoms in its bore, hydraulic pressure causes full clutch application.

A double-acting accumulator uses both sides of the accumulator piston to cushion application of clutches or bands. One side of the piston is subject to hydraulic apply pressure for one unit while the other side of the piston is subject to apply pressure from another hydraulic circuit.

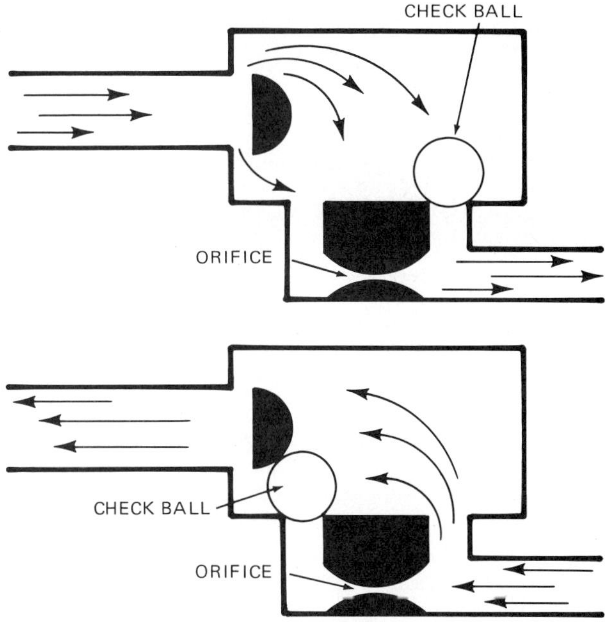

Figure 5-17. Action of an orifice control check ball when flow direction changes.

Ball Check Valves

In addition to the spring-loaded ball check valve type of pressure relief valve described earlier, a number of other ball check valves are used in automatic transmissions. These check valves are not spring-loaded but are controlled by fluid flow. Check balls are of either steel or some synthetic material construction. The ball operates in the hydraulic circuit as a result of hydraulic flow in one direction pushing the ball against its seat while hydraulic flow in the op-

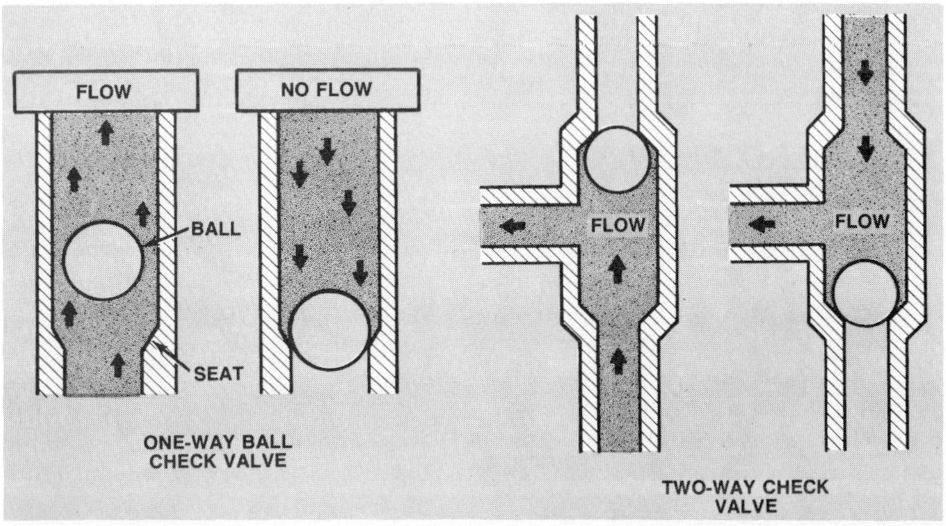

Figure 5-18. Two types of ball check valve operation. (*Courtesy of Chrysler Corporation*)

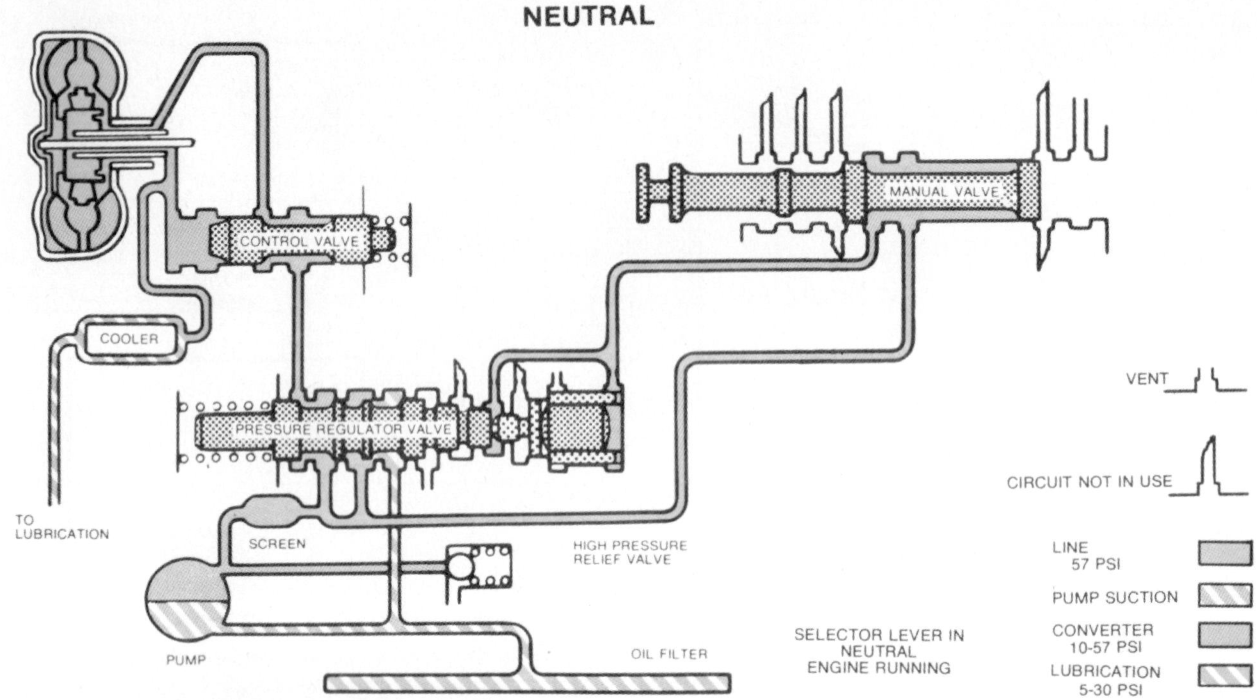

Figure 5-19. Typical application of a spring-loaded ball check valve is the high-pressure relief valve shown in this circuit. (*Courtesy of Chrysler Corporation*)

posite direction unseats the ball. Another design uses two seats to control flow in interconnecting passages.

A typical example of the two-seat ball check valve is the one used in the drive low and manual low hydraulic circuits in the three-speed Torqueflite transmission. A variation of the ball check valve is the clutch relief valve, which is seated by hydraulic pressure during clutch apply and unseated by centrifugal force when the clutch is released (apply pressure dumped).

A more detailed description of the application and operation of valves and other hydraulic control devices is given in Chapter 6.

PART 5-6 SEALS AND FLOW CONTROL SEAL RINGS

Sealing hydraulic pressures and fluids in an automatic transmission is a big job. Confining hydraulic pressures is critical to proper transmission operation.

The automatic transmission has several gaskets and many seals of different shapes and materials. Some are used between parts that do not move against each other while others are used to seal between a stationary part and a moving part.

This section deals with some of the more common types of seals and seal materials used in automatic transmissions.

Static Seals

A static seal is one that seals between two parts that do not move in relation to each other. Gaskets are perhaps the most common type of static seal. The automatic transmission uses gaskets in such places as between the hydraulic pump and the transmission case, between the oil pan and the case, between the extension housing and the case, and between some parts of the valve body.

Gasket materials include cork and paper compounds. Sometimes gaskets are used with a sealing compound. However, using a sealer on automatic transmissions may result in clogged hydraulic passages if not done properly. A gasket can do its job properly only if the right gasket is used, if it is installed as recommended, and if proper tightening (tightening sequence and correct torque) of fasteners is done.

Other static seals include O rings and lathe-cut or rectangular or square cross-sectional seal rings. These are usually made from synthetic rubber materials. Typical O-ring use is between the selector shaft and transmission case. Lathe-cut seals are often used

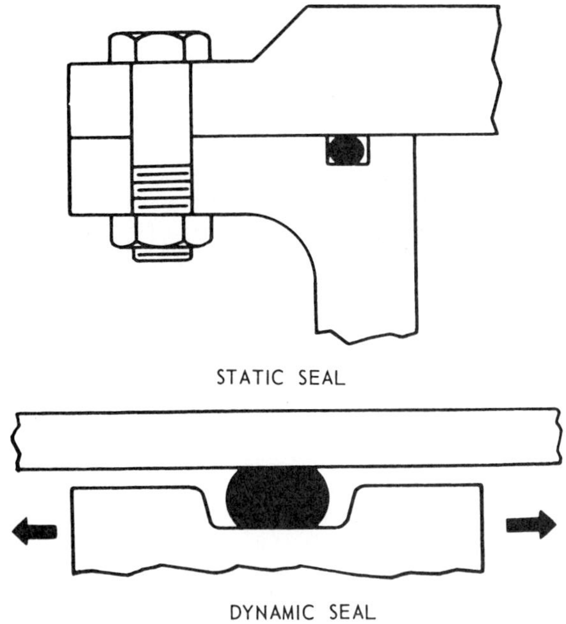

Figure 5-20. Typical application of hydraulic seals.

between the hydraulic pump and the transmission case.

Both O-ring and lathe-cut static seals must be of the correct cross-sectional dimensions as well as the correct total diameter for the particular application in order to do their job properly since they fit in

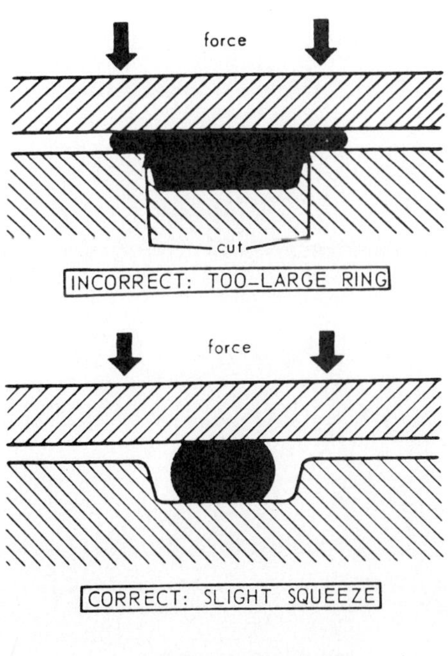

Figure 5-21. Effect of incorrect and correct cross-sectional size of static hydraulic seals.

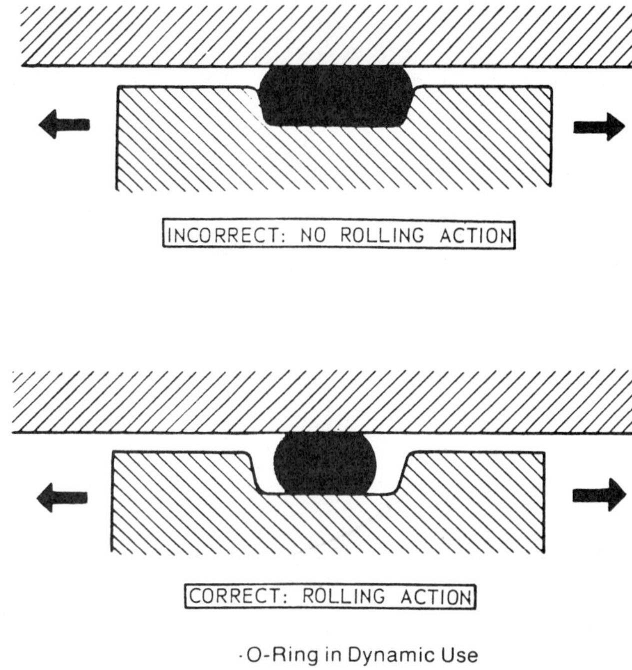

Figure 5-22. Effect of incorrect and correct cross-sectional size of dynamic seals.

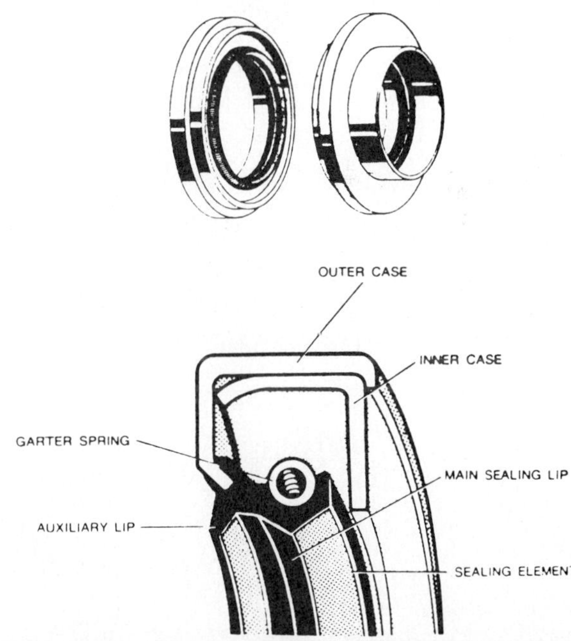

Figure 5-23. A typical dynamic seal and common seal terminology. Seals are used to keep in oils, fluids, and grease, to exclude dirt, or both. The main sealing element can be synthetic rubber, leather, or felt. Some seals have both an inner and outer case; others have only an outer case. Note garter spring, which provides added pressure of seal lip against sealing surface.

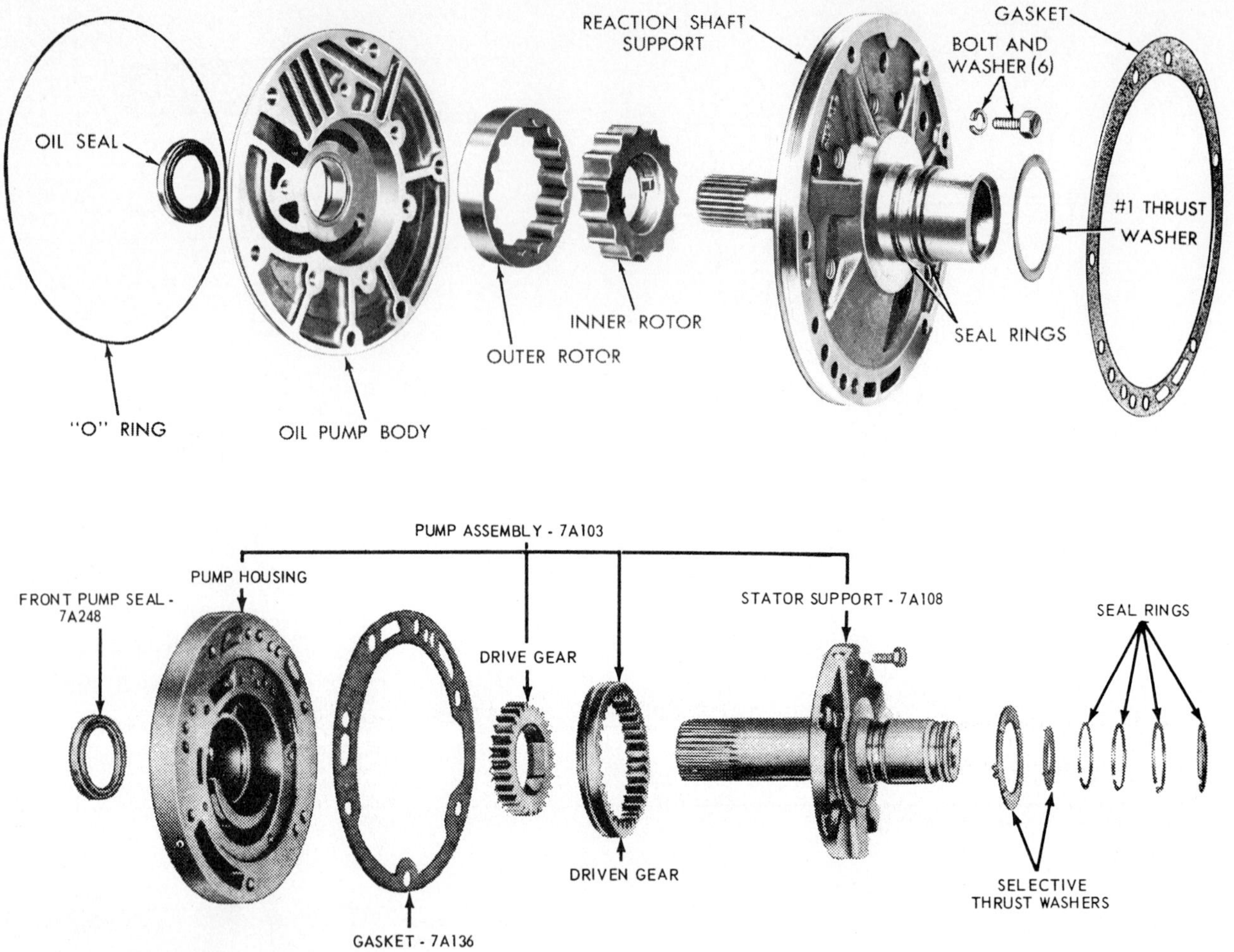

Figure 5-24. Top: Disassembled view of rotor type of automatic-transmission hydraulic pump. (*Courtesy of Chrysler Corporation*). Bottom: Gear type of transmission hydraulic pump. (*Courtesy of Ford Motor Co. of Canada Ltd.*)

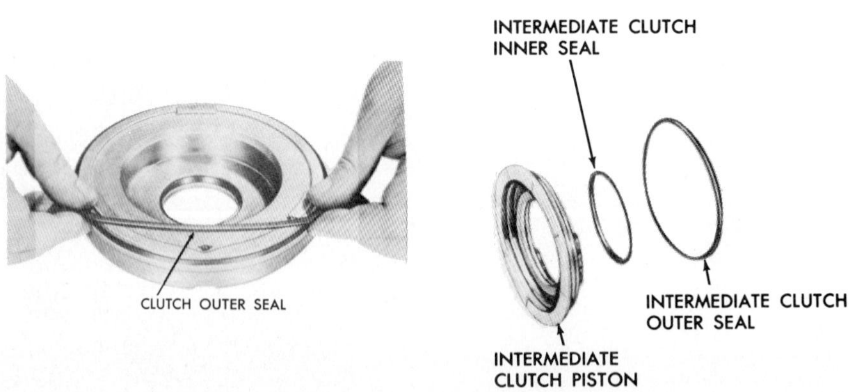

Figure 5-25. Hydraulic pistons (clutch and servo) use various types of piston seals, such as lip-type and lathe-cut (rectangular cross section) seals, for clutch pistons; and cast-iron, Teflon, or lip-type synthetic rubber seals on servo pistons. (*Courtesy of General Motors Corporation*)

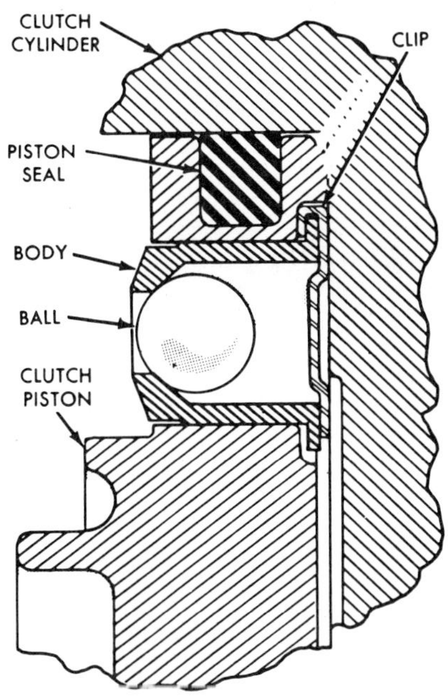

Figure 5-26. Cross-sectional view of lathe-cut clutch piston seal. (*Courtesy of Ford Motor Company*)

grooves in the parts that they seal. Seals of too small cross-sectional dimension do not provide for any compression. This results in lack of static pressure between parts to be sealed. Seals of too small a total diameter stretch, reducing cross-sectional dimensions, while too large a total diameter results in cutting and damage to the seal.

Transmission overhaul gasket and seal packages may contain seals and gaskets for several different

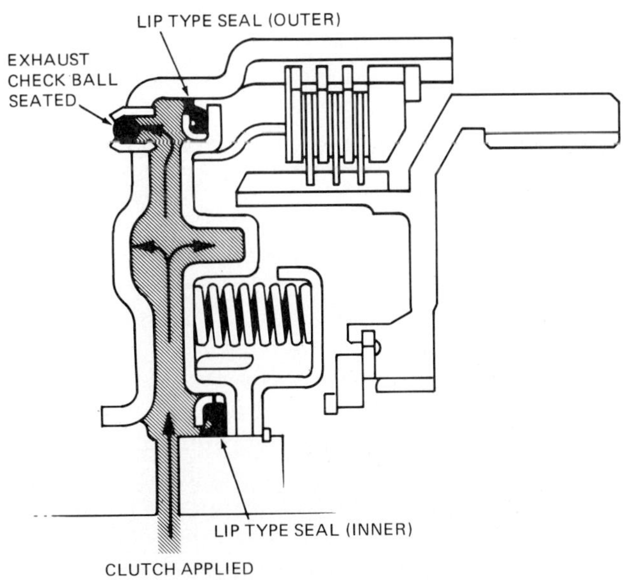

Figure 5-27. Cross-sectional view of lip-type clutch piston seal. (*Courtesy of General Motors Corporation*)

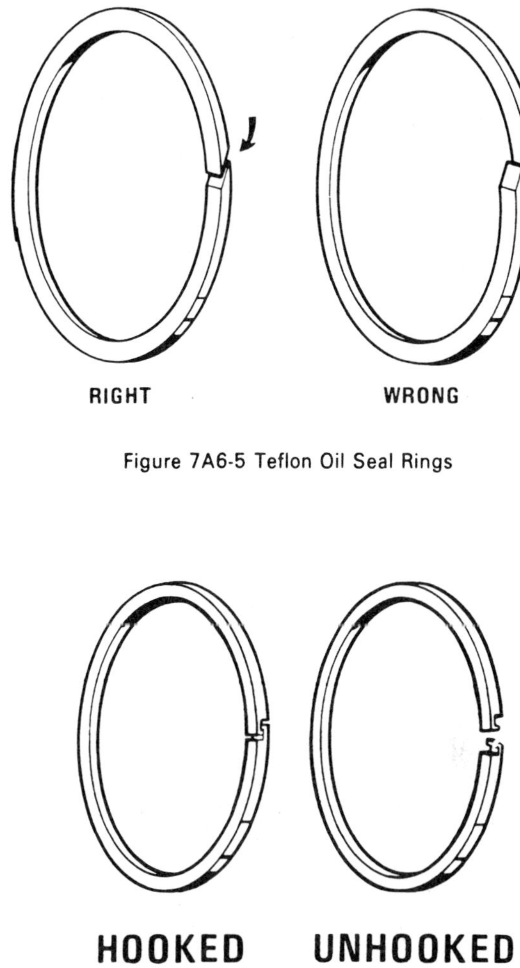

Figure 5-28. Typical oil transfer rings: Teflon type (above) with angled or scarf-cut joint and cast-iron type with hooked joint (below). (*Courtesy of General Motors Corporation*)

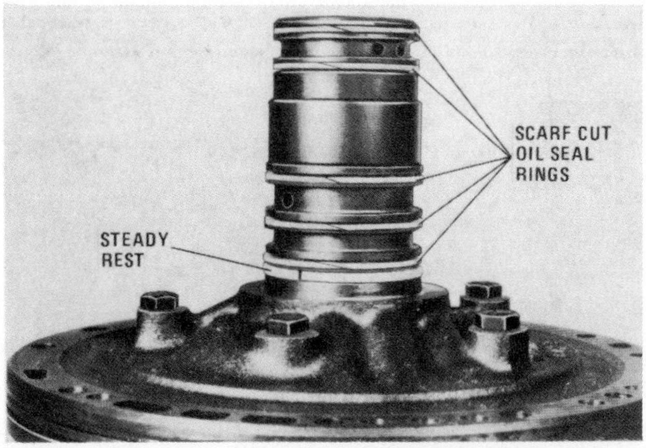

Figure 5-29. Scarf-cut Teflon oil transfer rings used on pump body. (*Courtesy of General Motors Corporation*)

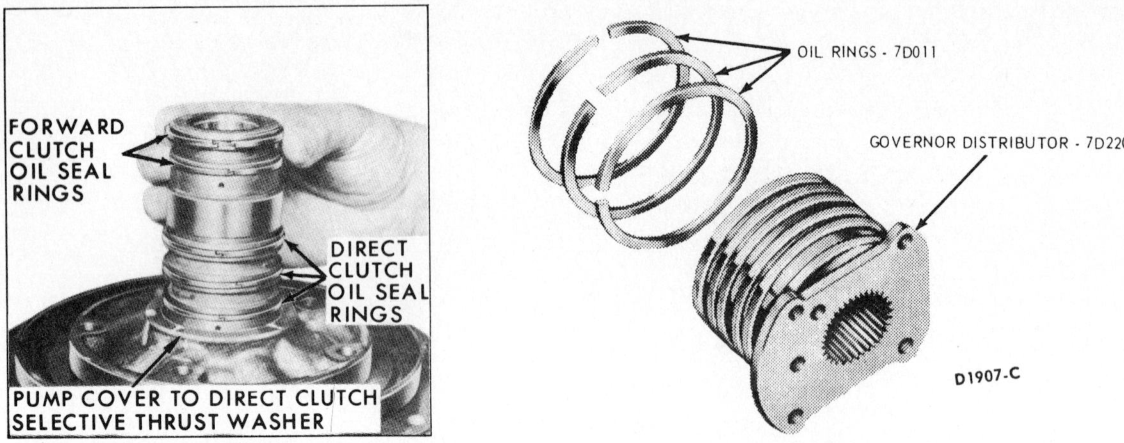

Figure 5-30. Cast-iron hooked-type seal rings (left) and butt-joint cast-iron rings (right). (*Courtesy of Ford Motor Company*)

models of the same transmission. There may be duplicate seals or gaskets with slight differences not noticeable unless closely examined. It is essential that the correct gaskets and seals be selected for the transmission being serviced. Selecting seals or gaskets from packages must be done carefully since an incorrect use of a seal or gasket could result in the transmission not functioning properly or in early failure.

Dynamic Seals

Dynamic seals are used to seal between a moving and a stationary part or between two moving parts. Dynamic seal designs include the O-ring type, the lathe-cut type, and the lip type of synthetic rubber construction. Other types include Teflon, nylon, and cast-iron, square or rectangular cross-section circular seal rings. These latter rings may be of the butt-end design, the tapered-end, or locking type.

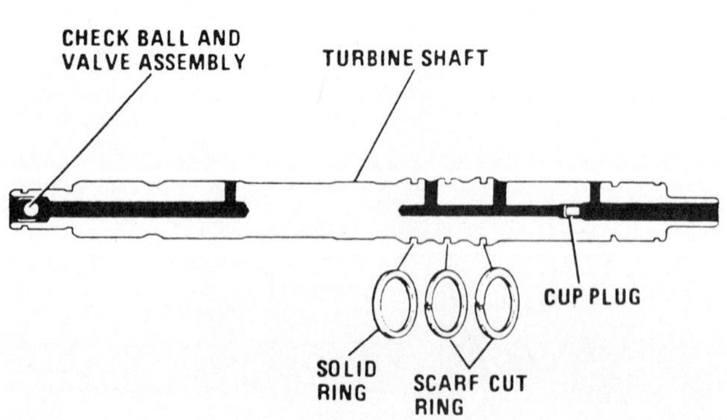

Figure 5-31. Typical application of oil transfer rings on turbine shaft. (*Courtesy of General Motors Corporation*)

Lip-type and lathe-cut seals are used to seal between clutch pistons and clutch cylinders as well as servo pistons and piston bores. Lip-type seals are always installed with the edge of the lip facing the fluid or pressure to be confined. Hydraulic pressure pushes the lip against the bore surface to increase the sealing ability of the seal.

Cast-iron seal rings are often used on accumulator pistons and on transmission shafts. Seal rings of synthetic materials such as Teflon are used in these areas by some transmission makers.

These seal rings are positioned in grooves on the parts to be sealed and must be of exact cross-sectional size as well as the proper diameter to function effectively. The use of an incorrect seal ring or its improper installation may result in early transmission failure or an inoperative transmission after overhaul.

Flow Control Rings (Oil Transfer Rings)

Flow control rings are dynamic seal rings. However, they are part of a hydraulic passage. These rings allow hydraulic flow and pressure to be transferred uninterrupted from a stationary part to a moving or rotating part and from a rotating part such as a shaft to a stationary part.

They fit into grooves in either the stationary part or the moving part. One example of the use of flow control rings is in directing hydraulic pressure from the pump housing to the forward clutch. Seal rings in the grooves of the stationary pump housing seal against the inner circumference of the rotating clutch drum. Fluid from a passage in the pump housing is directed between the two seal rings to a passage in the

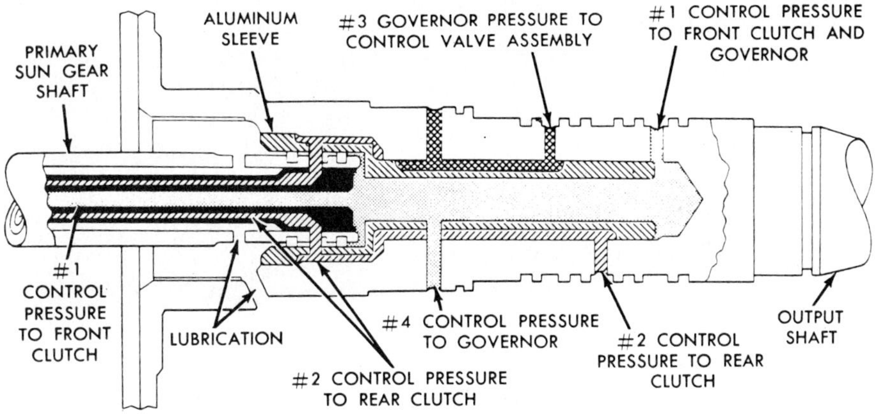

Figure 5-32. Oil transfer ring seals used to seal hydraulic pressures between the primary sun gear shaft and the output shaft. Shafts turn at different speeds in first and second gears and in opposite directions in reverse. (*Courtesy of Ford Motor Company*)

clutch drum leading to the area between the clutch piston and clutch drum.

Sealing between the outer circumference and the rotating drum is the result of mechanical pressure of the ring against the drum. Hydraulic pressure pushes the rings apart to seal against the sides of the grooves in the pump housing.

Materials used for flow control seal rings include cast iron and synthetic materials such as Teflon. Ring ends may be of the butt-end, tapered-end, or locking type.

Precise cross-sectional size and diameter are critical to the proper function of flow control seal rings. The use of an incorrect seal ring or its improper installation can result in faulty operation of the transmission and early transmission failure.

Part 5–7 Worksheet

1. Draw a sketch showing how force multiplication is achieved in a simple hydraulic system. Include figures to prove the degree of force multiplication achieved. State the relationship of distance moved by the pump piston as compared to the servo piston.
2. Examine training models of the three different types of automatic transmission hydraulic pumps if available. Find at least one application of each design in manufacturers' service manuals. List the make of vehicle, transmission model, and year for each of the three pump applications.
3. Examine the description and type designation given on Type F, Dexron II, and other automatic transmission fluid cans.
4. Examine an automatic transmission valve body if available. See if you can identify the valve types used, i.e., spool, ball, etc.
5. Examine any automatic transmission parts with seal rings and flow control rings available to you. Identify whether they are of the butt-end, tapered-end, or locking design. See if you can determine whether they are seal rings only or are flow control rings.

PART 5-8 SELF-CHECK

1. Define Pascal's law.
2. Explain how force multiplication can be achieved in a hydraulic system.
3. Explain why the distance moved by a servo piston is less than that of the pump piston in a simple force multiplication type of hydraulic system.
4. Why is it necessary to use only the fluid specified by the transmission manufacturers in their automatic transmissions?
5. Name the three types of hydraulic pumps used in automatic transmissions.
6. What is the advantage of a hydraulic pump capable of variable delivery capacity?
7. What is the purpose of a pressure relief valve?
8. What is the purpose of a pressure regulator valve?
9. What is the difference in purpose between a static seal and a dynamic seal?
10. What is the purpose of flow control seal rings?
11. List the three types of seal ring end designs.

PART 5-9 PERFORMANCE EVALUATION

After sufficient study of this chapter and the appropriate training models, did you perform the following tasks successfully at the request of your instructor?

1. Define the terms *pressure* and *force* as they apply to hydraulic fluids.
2. Define Pascal's law.
3. Define force multiplication by hydraulic means and give an example of force multiplication by means of hydraulics.
4. Define the relationship between force and motion transmitted by hydraulic means, and give an example of this relationship.
5. Describe briefly the various types of automatic transmission hydraulic fluids and their properties.
6. State the purpose of the hydraulic pump in the automatic transmission.
7. Describe the basic construction and operation of the three types of hydraulic pumps discussed in this chapter, including the causes and effects of pump wear.
8. Describe the basic construction and operation of the following types of valves and flow control devices:
   (a) Ball check valve
   (b) Poppet valve
   (c) Spool valve
   (d) Pressure relief valve
   (e) Pressure-regulating valve
   (f) Orifice
   (g) Accumulator
9. Describe the basic purpose, construction, and operation of automatic transmission gaskets, seals, and flow control seal rings.
10. Complete the worksheet in this chapter.
11. Complete the self-check in this chapter with at least 80 percent accuracy.
12. Evaluate performance based on this performance evaluation section.

Chapter 6

Transmission Hydraulic Control System

Performance Objectives

After sufficient study of this chapter and the appropriate training models, the student should be able to perform the following tasks at the request of the instructor.

1. State the purpose of and describe the basic construction of the transmission case.
2. State the purpose of the transmission cooling and lubrication systems and describe the basic construction and operation of each of these systems.
3. State the purpose of the pressure supply and regulation system and describe the basic construction and operation of each of the components in this system.
4. Describe the purpose, construction, and operation of the manual control system, including the selector lever, linkage, manual control valve, shift gate and detent, and the neutral and backup light switches.
5. Describe the purpose, construction, and operation of the 1–2, 2–3, and 3–4 shift valves.
6. Describe the purpose, construction, and operation of the mechanical and vacuum-operated throttle pressure control systems and the downshift system.
7. Describe the purpose, construction, and operation of the governor control system.
8. Describe the basic purpose, construction, and operation of the shift timing and shift quality control devices.
9. Complete the worksheet as required.
10. Complete the self-check in this chapter with at least 80 percent accuracy.
11. Evaluate your performance by using the performance evaluation section in this chapter.

PART 6-1 INTRODUCTION

This chapter deals with the purpose, construction, and operation of the automatic transmission hydraulic control system. The major emphasis is on the type of control system as it applies to a Ford C5 three-speed automatic transmission equipped with a Simpson compound planetary gear train. Other features such as the torque converter lockup control and overdrive control are added to complete the coverage. Included are both the mechanical-throttle-linkage–operated throttle pressure control system and the vacuum-modulator–controlled throttle pressure control system. A number of shift timing and shift quality control devices are also discussed.

Since basic hydraulic pump and pressure regulator construction and operation as well as basic principles of valving were discussed in Chapter 5, there is less emphasis on basic principles in this chapter. It is therefore necessary to have a good understanding of the principles covered in Chapter 5 before attempting to digest the material in this chapter.

It is also suggested that reference be made to hydraulic control system schematics provided to assist in gaining a more complete concept of hydraulic control system operation.

PART 6-2 TRANSMISSION CASE

The automatic transmission case houses the entire gear train, hydraulic control system, and torque converter. In front-wheel-drive vehicles the transaxle case may also house the differential and final drive components.

The transmission case for rear-wheel-drive–front-engine vehicles consists of the main transmission case, an extension housing bolted to the rear of the case, and an oil sump bolted to the bottom of the case. In some models the torque converter housing is integral with the case whereas in others it is a separate unit bolted to the transmission case. In this latter instance a precise fit exists between the converter housing and the transmission case since proper alignment of the torque converter with the transmission is essential. The converter housing is vented for air circulation on transmissions with air-cooled converters. A screened air inlet and outlet prevent entry of foreign objects.

Most automatic transmission cases are constructed of die-cast aluminum, the major exception being Ford's FMX transmission, which has a cast-iron case with an aluminum torque converter housing and extension housing.

The case is provided with a seal at the front (in the pump body) to seal between the torque converter hub, which drives the hydraulic pump, and the pump housing. This is usually a synthetic rubber lip-type seal in a metal case with a garter spring to provide sealing tension to the lip. The extension housing has a similar type of seal to seal between the slip yoke of the drive shaft and the extension housing. A bushing in the rear of the extension housing provides bearing support for the slip yoke and the rear of the splined output shaft. Other O-ring or lip-type seals are used at the speedometer drive adapter, the shift shaft, and the filler tube openings.

A vent is provided in the case to keep the case interior and the fluid in the sump at atmospheric pressure. A plugged vent could cause a pressure buildup in the case, resulting in fluid being forced out of the transmission. Since fluid expands at operating temperature by about 10 percent in volume, the transmission case must be able to "breathe" to prevent pressure buildup.

Since much of the transmission case is not subjected to hydraulic pressures, these areas may be repaired with any epoxy resin type of material if porous. However, repair must be restricted to the methods and areas prescribed in the manufacturer's service repair manual for the transmission concerned.

The transaxle transmission case is also constructed of die-cast aluminum and includes the same basic components as the transmission for rear-wheel-drive vehicles with the exception of the extension housing, which is not required in a transaxle. Similar types of seals are also used. Sealing is also required between the transmission section and the final drive section when different lubricants are used in these units.

The oil pan or sump must contain an adequate supply of the correct type of fluid at all times for the hydraulic control system to function properly. The oil pump driven by the torque converter pressurizes the fluid in the system and puts the fluid in motion. Oil pump design and operation were discussed in Chapter 5 as was basic pressure regulation and will therefore not be repeated in this chapter.

Fluid passages are either cast or drilled in a number of different transmission components. The majority of these passages are in the control valve body and the transmission case. Other fluid passages are located in the pump body, shafts, and clutch assemblies. Most of the valves are located in the valve body.

Automatic transmission operation requires a hydraulic control pressure or line pressure that varies to meet the demands of the transmission. For example, high engine torque output requires high control pressures for increased clutch and band capacity. On the other hand, lower hydraulic pressures are required to produce smoother shifts. The hydraulic control system is designed to provide varying control pressures suited to a variety of operating conditions.

The major components of the hydraulic control system in one transmission are very similar to those in other transmissions. Some of the differences (aside from differences in valve design) are that the transmission may have one or more of the following features:

Hydraulically operated or centrifugal lockup torque converter clutch

Fourth-gear overdrive

Vacuum-modulator–operated throttle pressure control system

Throttle-linkage–operated throttle pressure control system

Throttle-linkage–operated downshift valve

Electric-solenoid–operated downshift valve or converter clutch control valve

PART 6-3 BASIC OPERATION (FORD C5 TRANSMISSION)

Since the following discussion (Parts 6–3, 6–4, 6–5, and 6–6) is based on the Ford C5 automatic transmission (Courtesy of the Ford Motor Company), it should

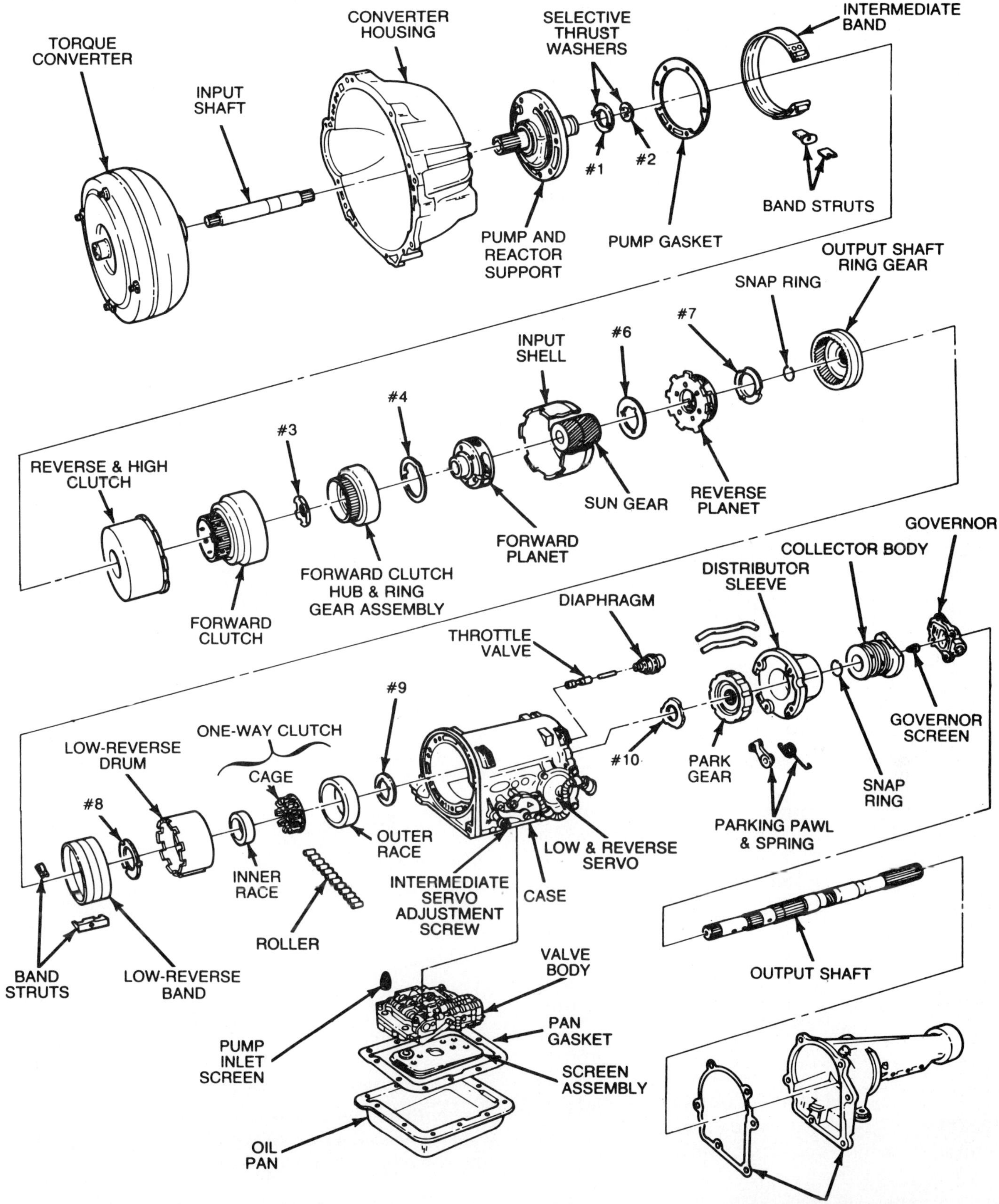

#3, #4 ETC. ARE GEAR TRAIN THRUST WASHERS

Figure 6–1. Ford C5 main components and assemblies. (*Courtesy of Ford Motor Company*)

GEAR	RATIO	HOLDING MEMBERS	FORWARD PLANETARY GEAR SET			REVERSE PLANETARY GEAR SET		
			Driven	Held	Output	Driven	Held	Output
Manual Low 1	2.46	Forward Clutch Low-and-Reverse Band	Ring Gear	*Carrier	Sun Gear	Sun Gear	Carrier	Ring Gear
D Low	2.46	Forward Clutch One-way Clutch	Ring Gear	*Carrier	Sun Gear	Sun Gear	Carrier	Ring Gear
*The carrier is actually turning with the output shaft, but at a slower speed than the input.								
D Second or 2	1.46	Forward Clutch Intermediate Band	Ring Gear	Sun Gear	Carrier	EFFECTIVELY IN NEUTRAL		
High	1.00	Forward Clutch Reverse and High Clutch	Sun Gear Ring Gear	None	Carrier	TURNS AS A UNIT		
Reverse	2.19	Reverse and High Clutch Low-and-Reverse Band	EFFECTIVELY IN NEUTRAL			Sun Gear	Carrier	Ring Gear

Figure 6-2. Gear train summary chart. (*Courtesy of Ford Motor Company*)

be noted that this is a three-speed transmission with a centrifugal lockup clutch torque converter and a Simpson planetary gear train. The selector lever positions and basic operation are as follows:

P—park. The transmission gear train is in neutral and the output shaft is locked to the transmission case by the park pawl engaged in the park gear, which is mounted on the output shaft. The engine can be started in this selector position. With column shift, this is the only position in which the steering column locks and the ignition key can be removed.

R—reverse. The transmission is in reverse at a reduced ratio.

N—neutral. The transmission is in neutral, but the output shaft is *not* locked to the case. The engine can be started, and the parking brake should be applied, since the rear wheels are free to turn.

D—drive. This is the normal driving range. The vehicle starts in low gear, with automatic upshifts to second (intermediate) and high (direct drive) as road speed increases. With the throttle closed, coasting down, or braking, the transmission downshifts from high to low at about 10 mph.

2—manual 2nd. If the gear selector is moved to the 2 (second) position, the transmission will shift to second gear immediately and remain in second gear regardless of speed or throttle position. This is particularly beneficial for maintaining engine braking when continuous second-gear operation is desirable. Intermediate range is useful for starting and driving on wet and icy roads.

1—manual low. The car starts in first (low) gear and stays in low gear, with no upshift. If the lever is moved to low while driving in high gear, the transmission will shift to intermediate and then to low as road speed decreases. Or it will shift directly to low below 25 mph. This range is provided mainly for engine braking when driving downhill.

Kickdown Shifts (Forced Downshifts)

Forced downshifts (as opposed to coasting downshifts) are possible at cruising speeds in the D range. It is not necessary to move the shift selector to "force" a downshift, only to press on the accelerator pedal.

Torque Demand Downshift

A torque demand downshift occurs because the torque needed is more than the engine can generate. It can happen in one of two ways or a combination of these.

1. The operator opens the throttle wider. The throttle valve causes an "increased load" signal to go to the valve body; the transmission downshifts. A "passing gear" kickdown shift can occur the same way as the driver calls for increased acceleration for passing.

2. Road speed drops off with high engine load so that the governor sends a decreased pressure signal to the valve body; a downshift occurs.

Torque demand downshifts occur at less than wide-open throttle (WOT) with only road speed or

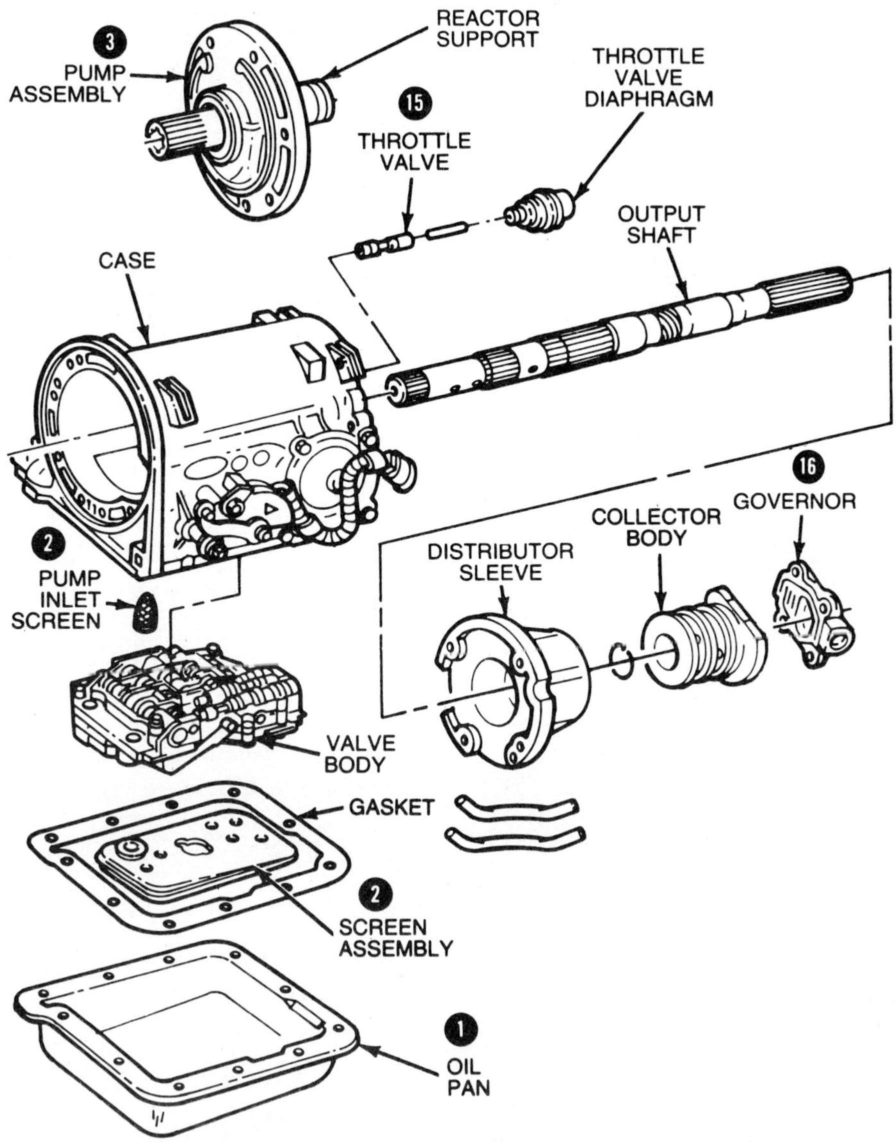

Figure 6-3. General arrangement of hydraulic system components. *(Courtesy of Ford Motor Company)*

engine load (throttle opening) changes. At wide-open throttle, for maximum acceleration, through-detent (kickdown) downshifts can be forced to lower gears at higher road speeds.

Through Detent Downshifts

Through detent means that the accelerator is pressed all the way to the floor. This causes the throttle lever to actuate a downshift valve in the valve body. Pressure passages are then opened to the shift valves. Kickdown shifts thus can happen at higher road speed than torque demand shifts.

The hydraulic system of the C5 transmission has several functions. It must:

- Supply fluid for torque converter operation,
- Direct fluid under pressure to apply band servos and clutches,
- Lubricate the working parts of the transmission, and
- Remove heat generated by vortex flow in the torque converter and by other parts of the transmission.

This section will first identify the hydraulic system components and functions. Then it will

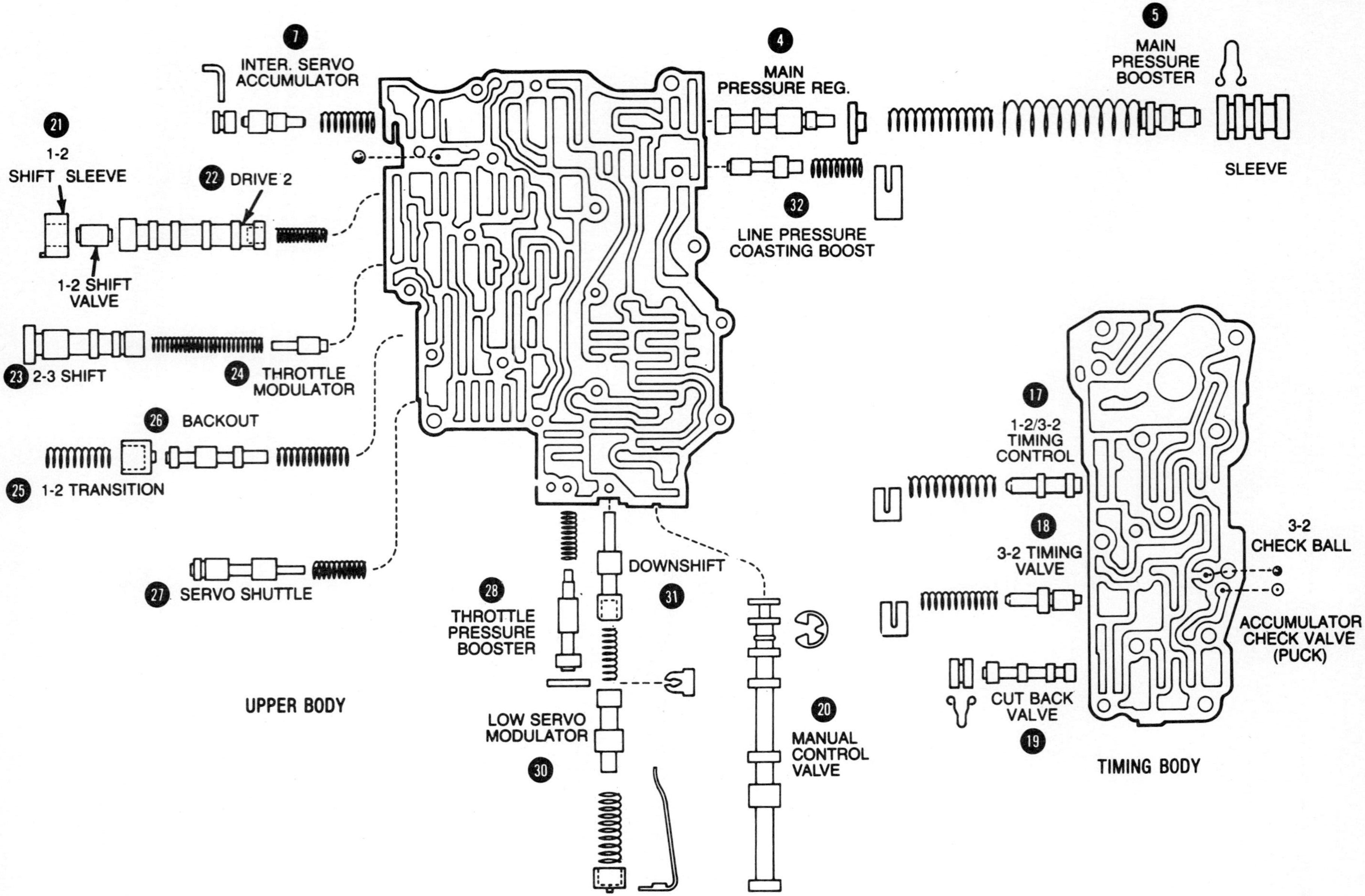

Figure 6-4. Valve body exploded view. (*Courtesy of Ford Motor Company*)

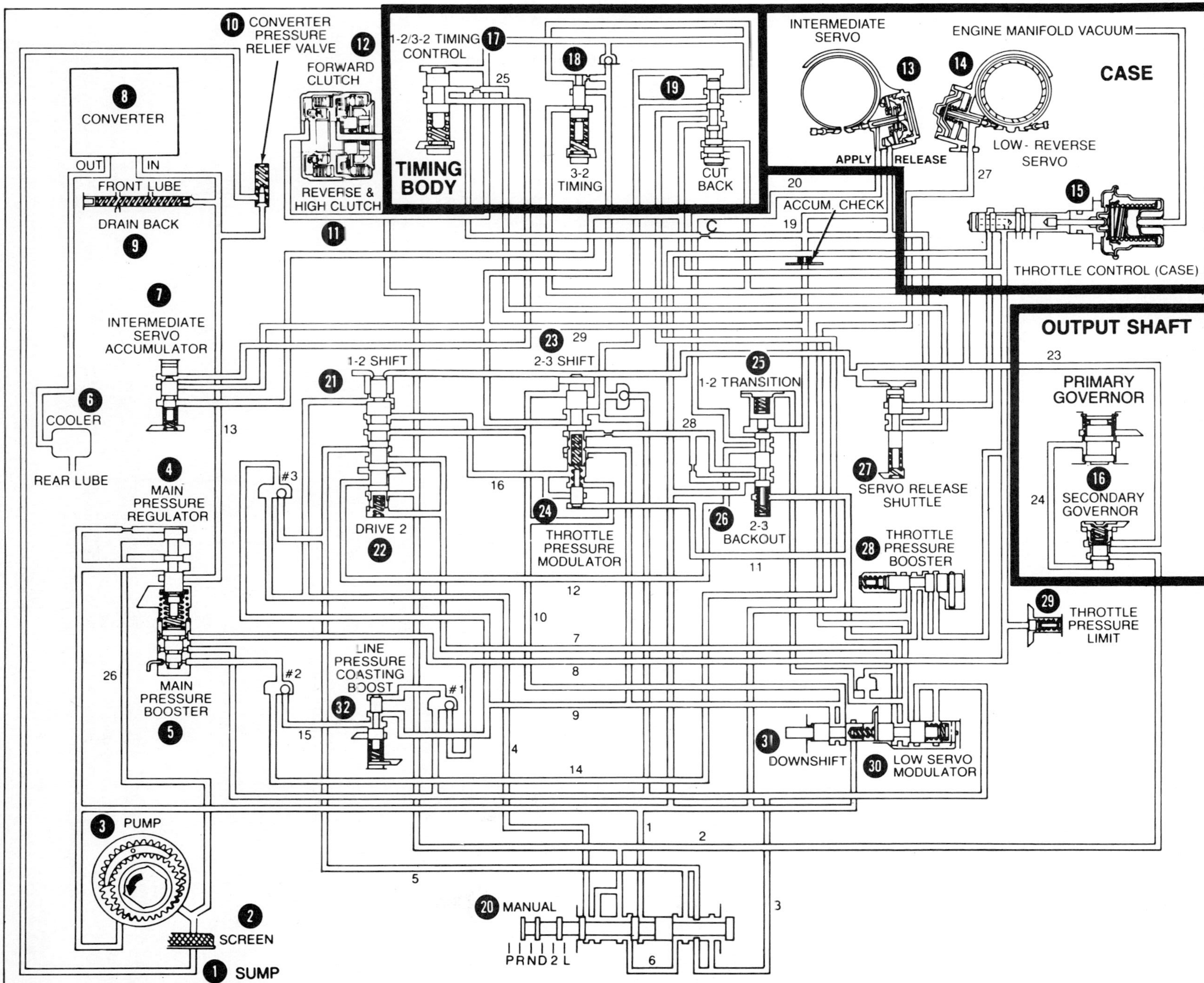

Figure 6-5. C5 hydraulic system schematic. (*Courtesy of Ford Motor Company*)

describe hydraulic system operation in various operating modes, explaining how the subsystems operate.

To explain the hydraulic system it will be necessary to represent it schematically. By comparing the valve body exploded view (Figure 6–4) and the system schematic (Figure 6–5), you can see that most of the components are contained in the valve body. The schematic is arranged in the same order that the valves are assembled in the main control valve body.

Most fluid flow passages are in the main control valve body and in the case (Figure 6–3). The pump reactor support contains passages to charge the reverse and high and forward clutches. It also feeds operating fluid to the torque converter and the cooler/lube system. At the rear of the case, the output shaft has lubrication fluid passages in its hollow center. The distributor sleeve and collector body connect the case fluid passages to the governor, which is driven by the output shaft.

PART 6-4 SYSTEM COMPONENTS (C5 FORD TRANSMISSION)

The main components of the C5 hydraulic system are listed, together with a brief description of their functions. All of the circled key numbers will be found on the schematic (Figure 6–5). They also are on Figures 6–3 and 6–4 whenever the component described appears in those illustrations.

Functions of Components and Valves (Figure 6-5)

1. Sump. The oil pan contains a supply of hydraulic oil (fluid) for the hydraulic system.

2. Screens. A pan screen and a pump inlet screen protect the system from dirt or other foreign material that may cling to the oil.

3. Pump. It pumps fluid to the system whenever the engine is running.

4. Main oil pressure regulator valve. This regulates line (main control) pressure in the system, also supplies converter, cooler, and lube systems.

5. Main pressure booster valve and sleeve. This causes line pressure to increase or decrease in relation to engine load. It also causes pressure to be boosted, depending on the range or gear.

6. Cooler. It removes heat generated in the torque converter and gear train. It is located in the radiator.

7. Intermediate servo accumulator. This cushions (smooths) 1–2 upshift.

8. Converter. It couples the engine to the gear train input shaft.

9. Drain back valve. This prevents lube oil from draining back to the sump when the engine is stopped.

10. Converter pressure relief valve. This

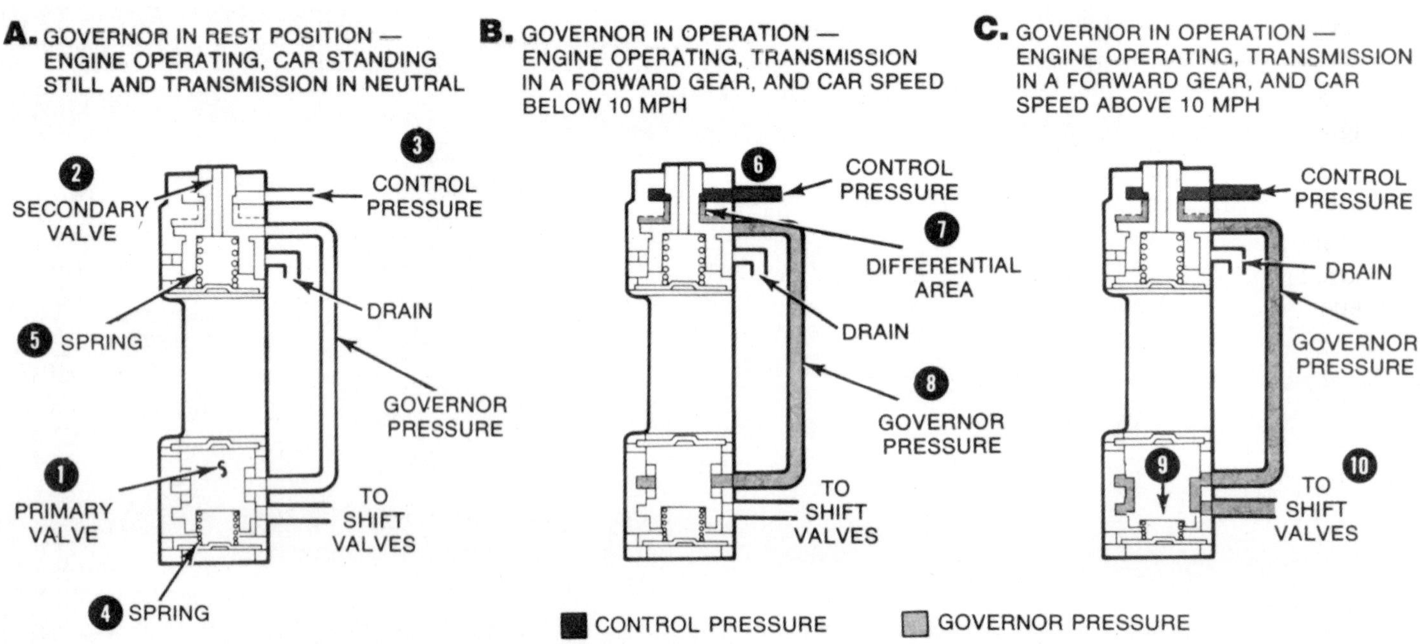

Figure 6-6. Governor operation. (*Courtesy of Ford Motor Company*)

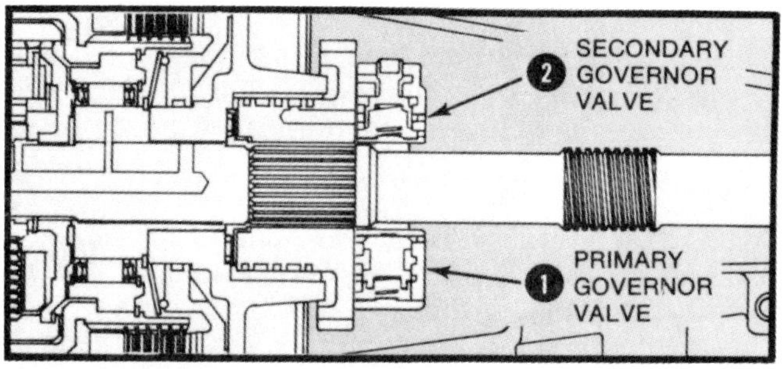

Figure 6-7. Governor installation. (*Courtesy of Ford Motor Company*)

prevents excess pressure buildup in the torque converter.

11. Reverse and high clutch. This is applied by pressure to couple the input shaft to the sun gear (see Chapter 3).

12. Forward clutch. This is applied by pressure to the couple input shaft to the forward ring gear (see Chapter 3).

13. Intermediate servo. This applies the intermediate band (see Chapter 3).

14. Low-reverse servo. This applies the low-reverse band (see Chapter 3).

15. Throttle valve. (TV) It regulates throttle pressure as an "engine load" signal to the hydraulic system and is located in the case.

16. Governor. It provides a "road speed" pressure signal to the system.

17. 1–2/3–2 timing control valve. This routes servo release oil to accumulator (7) on 1–2's and joins servo release pressure with the high clutch on 3–2's.

18. 3–2 timing valve. This regulates reverse and high clutch pressure for a smooth 3–2 shift.

19. Cutback valve. This causes a reduction or cutback in line pressure as road speed increases.

20. Manual valve. It directs line pressure to various passages to apply clutches and band servos, depending on the shift selector position.

21. 1–2 shift valve and sleeve. This is acted on by governor (road speed) pressure to control the drive 2 valve.

22. Drive 2 valve. This controls 1–2 upshift and 3–1 or 2–1 downshift.

23. 2–3 shift valve. It controls 3–2 downshift and 2–3 upshift.

24. Throttle pressure modulator. This modulates TV (or boosted TV) pressure to cause a lower TV signal to the drive 2 valve.

25. 1–2 transition valve. This prevents band-to-band tie-up on a manual 1–2 shift.

26. 2–3 backout valve. It prevents clutch-to-band tie-up on a closed throttle 2–3 shift.

27. Servo release shuttle valve. This selects servo release orifices on a 3–2 shift depending on road speed.

28. Throttle pressure booster valve. It provides a boosted TV pressure with high engine load.

29. Throttle pressure limit valve. This regulates maximum TV pressure.

30. Low servo modulator valve. This modulates reverse servo apply pressure in a manual shift to first gear.

31. Downshift valve. It causes through-detent kickdown shifts to occur at higher road speed than torque demand downshifts and also provides maximum delay on WOT upshifts.

32. Line pressure coasting boost valve. It causes a boost in line pressure in ranges 1 and 2.

PART 6-5 NEUTRAL PRESSURE REGULATION AT HIGH TV PRESSURE (FIGURE 6-9)

The following pages describe a full-system schematic for each gear and several operating modes. Not every operating condition is covered but those included are required to cover all the "subsystems" involved in controlling transmission operation.

In describing valve movement, *up, down, left,* and *right* refer to valve positions in the schematic. In the valve body, the valves actually move in or out in relation to the valve body.

Neutral at Full Throttle

Figure 6–9 shows the system in neutral or park, with the throttle open enough to cause maximum throttle pressure and to cause all the throttle pressure valves to regulate.

The operation of the line pressure regulating system, converter charge system, and the cooler and lube system are also shown. These remain the same in all conditions with the engine running.

Supply and Line Pressure Regulation

1. The pump inlet is "filled" by fluid from the sump. The pump gear rotation causes a partial vacuum at the inlet. Atmospheric pressure in the sump forces the fluid into the pump.

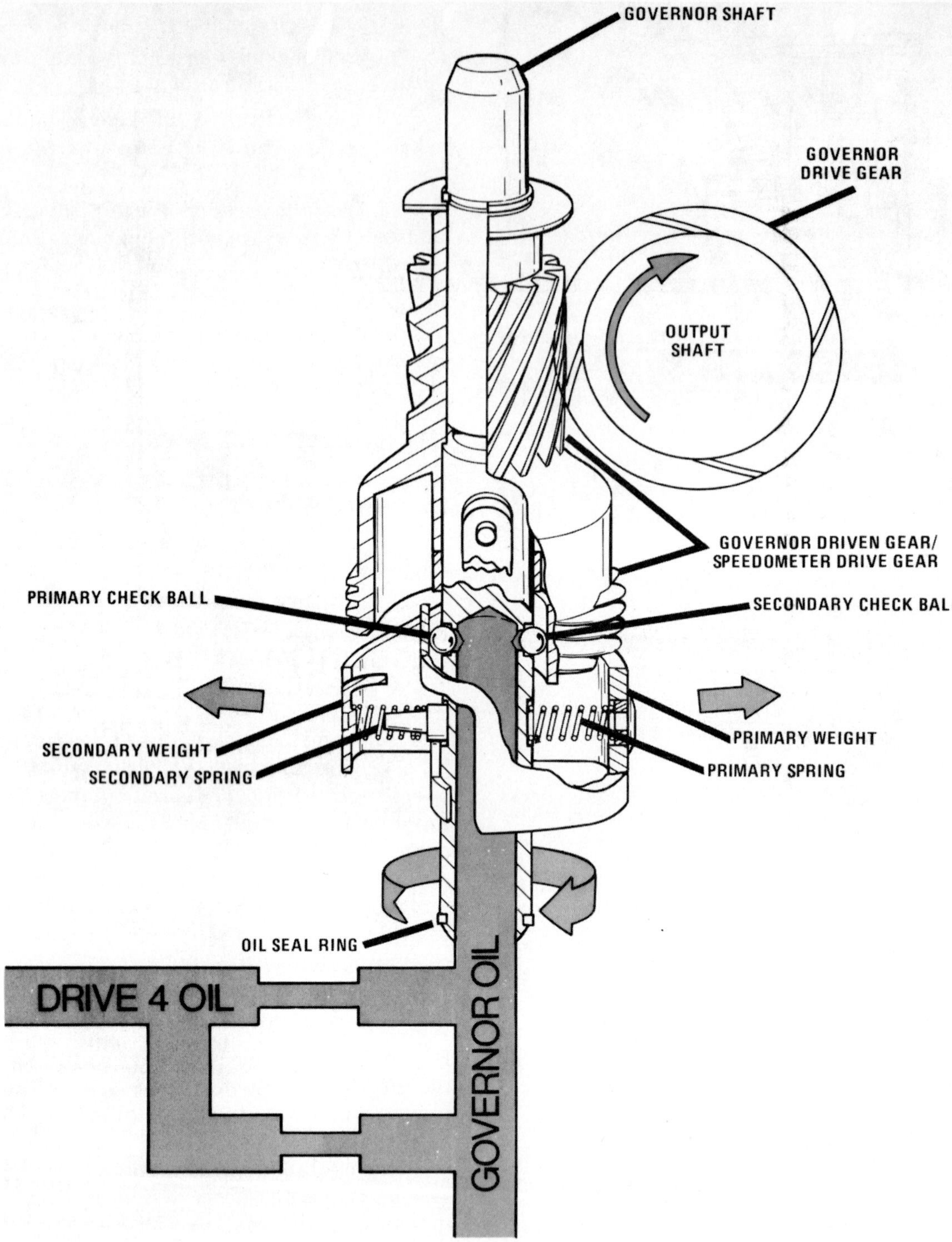

Figure 6-8. Typical General Motors governor. (*Courtesy of General Motors Corporation*)

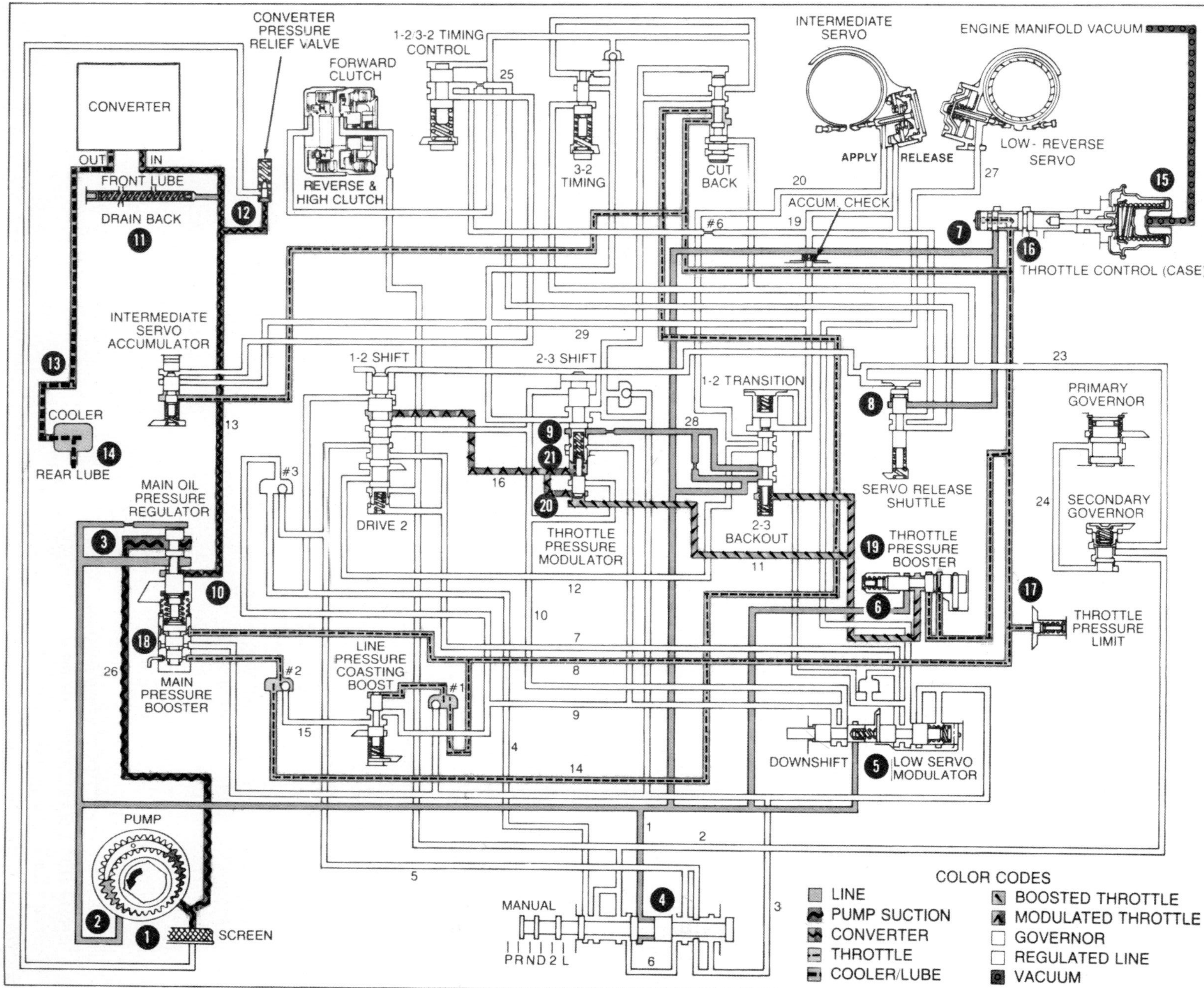

Figure 6-9. Main control pressure, converter, cooler, lube, and throttle pressure systems. (*Courtesy of Ford Motor Company*)

2. The "positive displacement" pump forces this fluid into the hydraulic system whenever the engine is running.

3. The main regulator valve operates by balancing line pressure against spring force and booster valve force. It constantly regulates line pressure in the system. Excess output from the pump is returned to the inlet through circuit 26.

4. Fluid at line pressure is always available to the manual valve through circuit 1.

Circuit 1 also supplies line pressure to the

5. Downshift valve,
6. Throttle pressure booster valve,
7. Throttle control valve (TV),
8. Servo release shuttle valve, and
9. 2–3 shift valve.

Converter, Cooler, and Lubrication

10. As it regulates, the main regulator valve opens a fluid supply to the torque converter and lube systems through circuit 13.

11. Front lube is fed from circuit 13 ahead of the converter and through the drainback valve.

12. The circuit is protected from too high pressure buildup by a relief valve.

13. Fluid leaving the converter goes to the cooler.

14. Return fluid from the cooler goes to the rear lube system and then drains into the sump.

Throttle Control Systems

15. Engine manifold vacuum acts on the throttle valve diaphragm. It effectively changes the spring force on the throttle valve (TV).

16. The throttle valve regulates by balancing throttle pressure at the inner (left) end of the valve against spring/vacuum force. The result is throttle pressure, or TV pressure.

17. TV pressure charges all the TV coded lines. The pressure limit valve prevents excessive pressure if the throttle valve should stick.

18. TV pressure acts on the main pressure booster at the end (circuit 14) and on a differential area (circuit 8). This "adds to" the spring force on the main regulator valve to increase line pressure.

19. At low TV pressure, the throttle pressure booster valve doesn't regulate. Circuit 11 also is charged with TV pressure. At about 68-psi throttle pressure (3 1/2-inches manifold vacuum), the valve goes into regulation. Circuit 11 then is charged with *boosted* (increased) throttle pressure.

20. Pressure in circuit 11 (TV or boosted TV) acts under the throttle pressure modulator valve. This increases the spring force against the 2–3 shift valve to delay the 2–3 upshift.

21. When the boosted throttle pressure reaches about 60 psi, the throttle pressure modulator valve goes into regulation and charges circuit 16. This pressure is applied to:

- The spring end of the 2–3 shift valve and
- A differential area of the drive 2 valve.

The effect is to delay the upshifts to their high road speed limits.

PART 6-6 OPERATION IN DIFFERENT GEAR RANGES

First Gear in D Range

In Figure 6–10 the hydraulic system is shown with the vehicle accelerating through 15 mph in first gear in the D range. This schematic will illustrate several additional systems: forward clutch apply, governor, control pressure cutback, and upshift control.

Basic Systems

1. Since the engine is running, the pump is delivering fluid. The following systems are charged as described earlier:

2. Main control pressure
3. Converter and front lube
4. Cooler and rear lube

Throttle Pressure Systems

5. Manifold vacuum to the throttle valve diaphragm is about 4 inches.

6. The TV system is producing a regulated TV pressure of about 65 psi. All the coded TV lines are charged with this throttle pressure. Circuit 8 is causing line pressure to be boosted.

7. The throttle pressure booster valve is *not* regulating. (TV pressure must be about 68 psi before it will regulate.)

8. Thus circuit 11 is charged with TV pressure, not boosted TV.

Forward Clutch and Governor

9. The manual valve is in the D position. Circuit 2, forward clutch and governor, is charged at the manual valve.

10. The forward clutch is applied from circuit 2.

11. Line pressure from circuit 2 is also supplied to the governor. (This happens whenever the forward clutch is applied.)

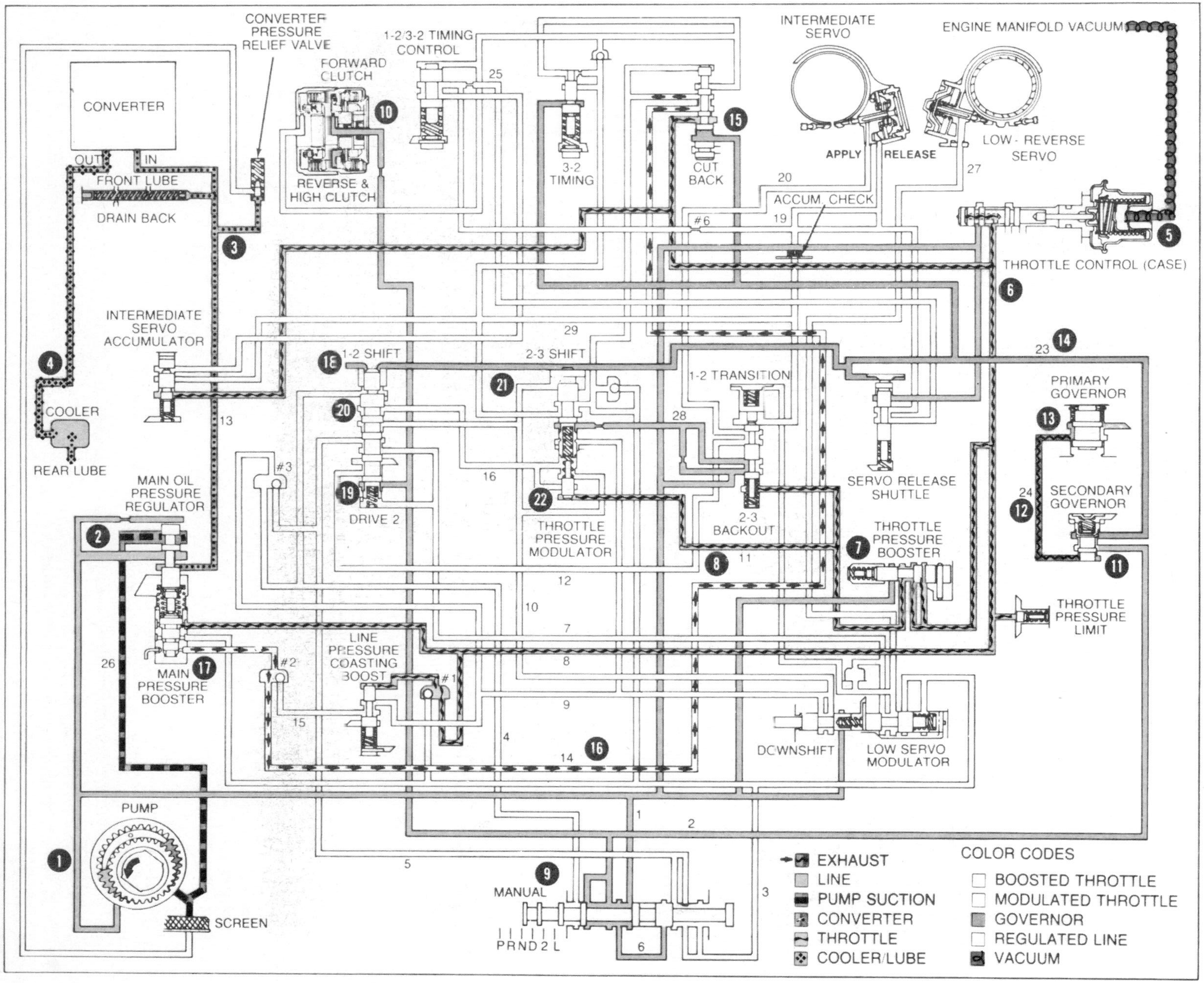

Figure 6-10. First-gear D range at 15 mph. (*Courtesy of Ford Motor Company*)

Governor Operation

12. Before the vehicle built up road speed, circuit 24 between the governor primary and secondary valves was also charged with line pressure. This prevented the governor from supplying a regulated pressure. Circuit 23 was at zero psi.

13. At about 10 mph, centrifugal force moved the primary valve outward (up). This exhausted circuit 24 to the sump.

14. The secondary governor valve then began to regulate a pressure to circuit 23, proportional to road speed. The regulation depends on a balance between centrifugal force aided by the spring, opposed by line pressure. Governor pressure is about 10 psi when regulation begins and can increase to about 80 psi at highway speed. It is effective in all the passages coded Governor.

Control Pressure Cutback

When the vehicle started, high line pressure was needed to hold the forward clutch applied against breakaway torque. This high pressure could cause a rough upshift if not reduced. Thus before the 1–2 upshift, a cutback occurs. (Governor pressure cuts in at about 10 psi. It takes 9 to 12 psi to shift the cutback valve.)

15. Governor pressure moves the cutback valve up.

16. Circuit 14 is cut off from TV pressure and exhausted.

17. There is now zero pressure at the end of the pressure booster valve. Line pressure is cut back from about 137 psi to 95 psi.

Upshift Control

In the D range, automatic upshifts will occur as road speed increases. The shifts are controlled by the shift valves.

18. Governor pressure above the 1–2 shift valve tries to move the 1–2 shift valve and drive 2 valve to cause the 1–2 shift.

19. Line pressure from circuit 2 acts on a differential area of the drive 2 valve to oppose the governor. Line pressure is high or low depending on TV and on the cutback valve. D2 spring force also opposes the governor.

20. If TV is very high, there may also be *modulated* throttle pressure at another differential area of the drive 2 valve. This also opposes the governor pressure.

21. Governor pressure on the end of the 2–3 shift valve will eventually cause the upshift to high gear.

22. This is opposed by TV or boosted TV pressure under the throttle modulator valve (Figure 6–10) or by modulated TV at the spring end of the 2–3 shift valve (Figure 6–9), depending on the throttle opening and the TV modulator spring load.

Second Gear in D Range (Figure 6-11)

Figure 6–11 shows the hydraulic system conditions in second gear in the D range at 20 mph. The forward clutch was already applied in first gear. For second gear, it was necessary to leave the clutch applied and also to apply the intermediate band. We will also use this illustration to describe how the accumulator valve cushions the 1–2 upshift.

Basic Systems

1. Since the engine is running, the pump is delivering fluid. The following systems are charged as described at the beginning of this section:

2. Main control pressure

3. Converter and front lube

4. Cooler and rear lube

Forward Clutch and Governor

5. In the D range, circuit 2 is charged at the manual valve. This supplies line pressure to the forward clutch and governor.

6. Circuit 23 is charged with governor pressure, which has increased to about 15 psi.

7. The cutback valve has shifted, and the control pressure cutback happened before the 1–2 upshift.

Throttle Valve System

8. Vacuum to the TV diaphragm is about 5 inches.

9. Regulated TV pressure is around 60 psi.

10. TV pressure in circuit 8 is causing a boost in line pressure.

Shift Valve Operation

11. Governor pressure forces the 1–2 shift valve and drive 2 valve down to cause the upshift.

12. The drive 2 valve has charged circuit 12 from circuit 2 line pressure.

13. At the backout valve, circuit 12 has charged circuit 20 with line pressure.

14. The intermediate servo is applied by pressure in circuit 20 to cause the 1–2 shift.

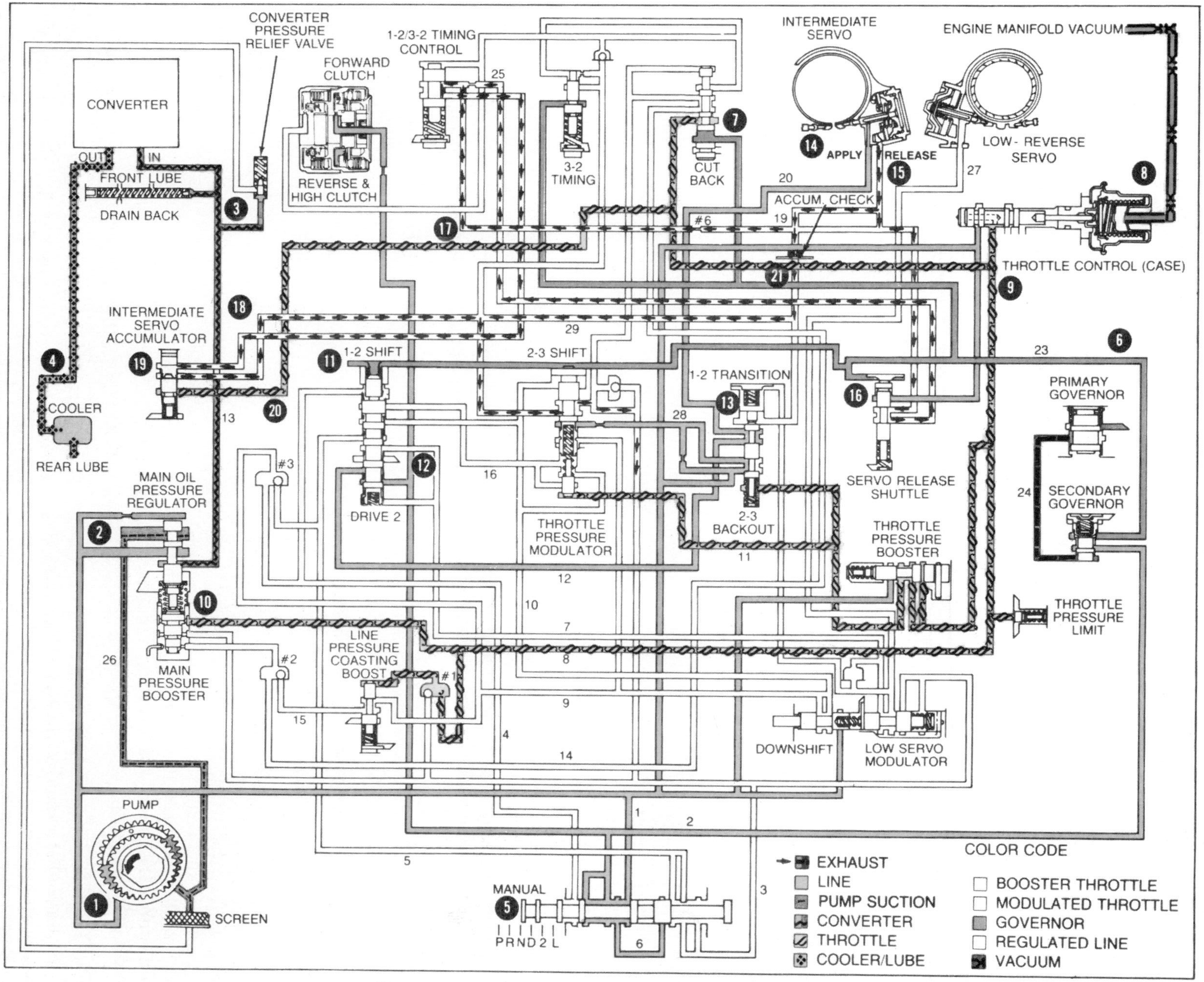

Figure 6-11. Second-gear D range at 20 mph. (*Courtesy of Ford Motor Company*)

Operation of Accumulator Valve on 1-2 Upshift

The purpose of the servo accumulator valve is to control the 1–2 shift feel by applying the band with just enough force to grab the drum firmly but not harshly. This is done by control of fluid exhausting from the servo "release" side.

15. The "release" side of the servo was not charged with pressure before the shift. But it was full of fluid (primed), and this fluid had to exhaust as the servo applied. There are two or three paths available, depending on road speed.

16. Below 30 mph, the servo release shuttle valve is held up by its spring. The exhaust fluid can flow around this valve and through circuit 25 to the 1–2/3–2 timing control valve.

17. At any road speed, release fluid flows through circuit 19 and the high speed downshift orifice to join the exhaust flow at the 1–2/3–2 timing control valve.

18. From there, the fluid flows between the valve lands and to the top of the accumulator valve.

19. When pressure builds up enough to overcome TV pressure, the accumulator moves down. The pressure balance causes a cushioning effect.

20. Now the fluid flows to the 2–3 shift valve. Here it is connected to the reverse circuit, which is exhausted in forward gears.

21. A third path for exhaust fluid is through the orifice in the accumulator check valve. The orifice creates some back pressure and a gradual bleed-down. Circuit 29 connects this orifice to the flow path from the accumulator.

Third (High) Gear in D Range (Figure 6-12)

To shift from second to high gear, it is necessary to apply the reverse and high clutch and to release the intermediate band servo. The timing of band release and clutch apply must be very precise to avoid "engine overspeeding" if the band comes off too quickly. In fact, the timing is controlled so that there is a slight "tie-up," causing the clutch to slip momentarily until the band is fully released. Orifices in the circuit control the clutch apply/band release pressure to make the shift smooth, as described next.

Basic Systems

1. Since the engine is running, the pump is delivering fluid. The following systems are charged as described at the beginning of this section:

2. Main control pressure

3. Connector and front lube

4. Cooler and rear lube

Forward Clutch and Governor

5. The forward clutch/governor system, circuit 2, is charged with line pressure at the manual valve. The clutch is applied.

6. The governor is regulating pressure in circuit 23.

7. The cutback valve has been forced up and control pressure cutback occurred before the 1–2 shift.

Throttle Valve System

8. Vacuum to the TV diaphragm is still around 5 inches.

9. Regulated TV is around 60 psi.

10. TV pressure in circuit 8 is causing a boost in line pressure.

11. TV pressure below the modulator valve, plus line pressure on the area differential on the 2–3 valve and spring, was not enough to resist the governor pressure on the 2–3 shift valve, so the upshift has occurred.

Shift Valve Operation

12. Governor pressure has shifted both shift valves.

13. The servo apply side is pressurized from circuit 12 at the 1–2 shift valve, through circuit 20 from the backout valve.

14. When the 2–3 shift valve moved down, it cut off the exhaust path for servo release, which was through the reverse circuit.

15. The valve also has opened circuit 28, *through the upshift orifice*, to the direct clutch apply circuit. The orifice controls the buildup of pressure to apply the clutch, until the servo is fully released.

16. Oil flow to the clutch is through the 3–2 check valve and bypasses the 3–2 timing valve.

17. A branch from the clutch apply circuit directs the same pressure to the release side of the servo through circuit 19. (This circuit branches into the three "exhaust" passages.)

18. Since the release side of the piston has a larger area than the apply side and is aided by the spring, the servo will stroke off and release the band as soon as the release pressure builds up to approach the apply pressure. As soon as the servo is fully released and the piston stops, pressure "backs up" through circuits 19 and 25. The entire servo release

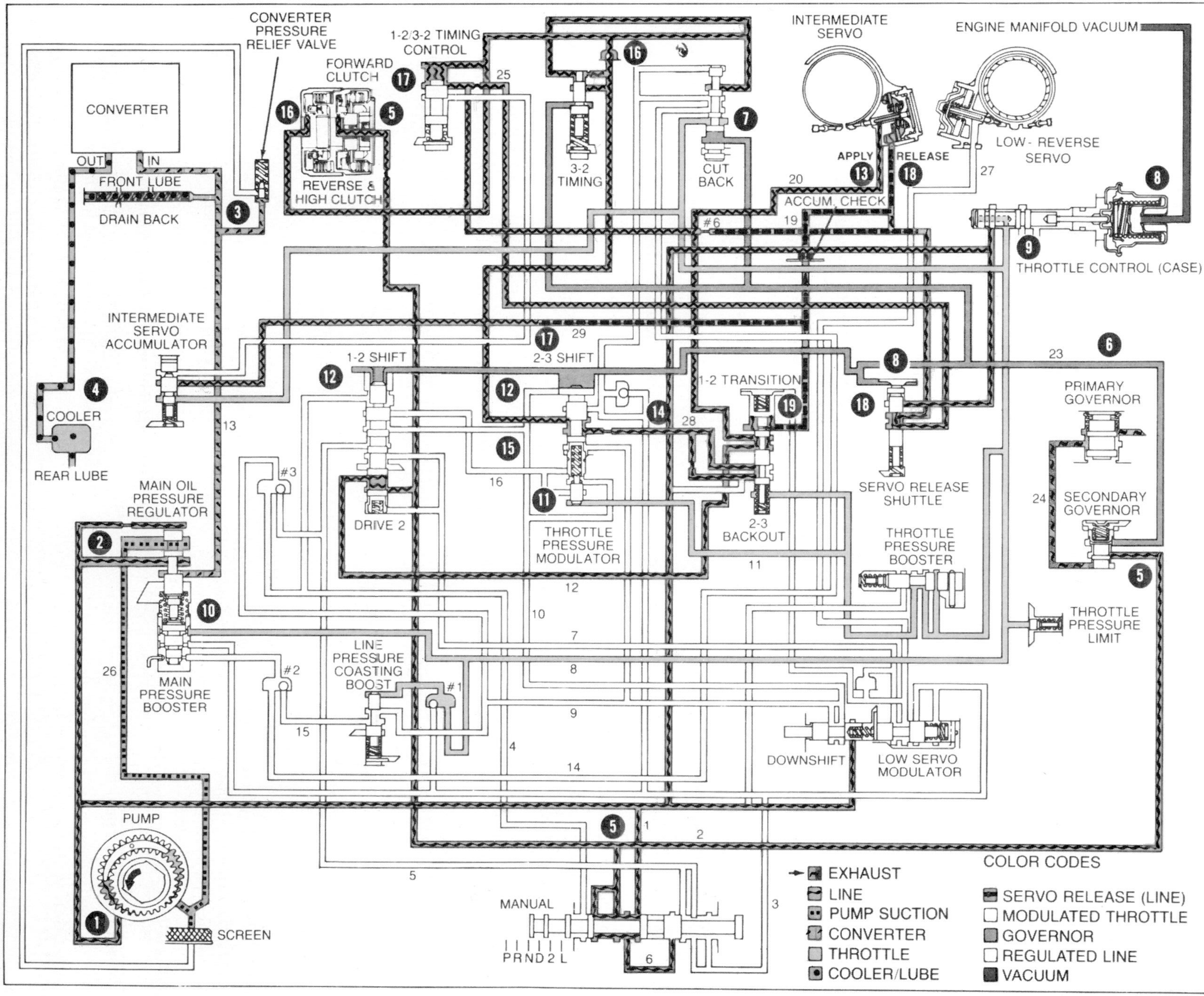

Figure 6-12. Third-gear D range at 30 mph. (*Courtesy of Ford Motor Company*)

and direct clutch apply system equalizes at line pressure (from circuit 28 ahead of the upshift orifice). The direct clutch no longer needs to slip slightly with the band completely off.

19. Servo release pressure is also applied to the top of the 2–3 backout valve. If the operator "backs out" on the accelerator during the shift, this valve prevents a clutch-band tie-up.

Backout 2-3 Upshift (Figure 6-13)

If the driver releases the accelerator before or during the 2–3 upshift, before the shift is completed, a clutch-band tie-up can happen. This is because TV pressure goes to zero and line pressure is accordingly reduced. The pressure may be enough to apply the clutch but not to release the band without excessive tie-up.

Basic Systems

1. Since the engine is running, the pump is delivering fluid. The following systems are charged as described at the beginning of this section:

2. Main control pressure

3. Converter and front lube

4. Cooler and rear lube

Forward Clutch and Governor

5. The forward clutch/governor system, circuit 2, is charged with line pressure at the manual valve. The clutch is applied.

6. The governor is regulating pressure in circuit 23.

7. The cutback valve has been forced up. The cutback occurred before the 1–2 upshift.

Throttle Valve System

8. Manifold vacuum went to maximum when the driver released the accelerator.

9. TV pressure dropped to zero or minimum.

10. There is no TV pressure to assist the spring of the 2–3 backout valve.

11. Circuit 8 is at zero pressure, so line pressure is regulating at minimum.

Shift Valves

12. The governor shifted the 1–2 shift and drive 2 valves for the 1–2 upshift.

13. The intermediate servo was applied through the 1–2 shift valve and backout valve.

14. The 2–3 shift valve has shifted because of governor pressure.

15. It has cut off the "release exhaust" path to the reverse circuit.

16. Clutch apply and servo release pressure are opened through the 2–3 shift valve and upshift orifice. The 2–3 upshift would occur as in Figure 6–12 except for the zero TV pressure.

2-3 Backout Valve Operation

As soon as TV drops to zero below the 2–3 backout valve 10:

17. Servo release pressure (circuit 19) forces the 2–3 backout valve down.

18. Now the servo release and servo apply pressures are *interconnected* in the bore between the 2–3 backout valve and the 1–2 transition valve. The two pressures immediately *equalize*.

19. Shifting the backout valve has cut off one passage of circuit 28. Thus, reverse and high clutch apply pressure is not regulated through the backout orifice.

20. With apply and release pressure equal, the servo immediately strokes off to release the band. This is because the servo release side has a larger area and release force is assisted by the spring. A clutch-band tie-up is thus prevented.

Again, as soon as the shift is complete, all the pressures involved (clutch apply, servo release, and servo apply) go to full-line pressure.

3-1 Coasting Downshift (Figure 6-14)

Coasting down, the C5 transmission stays in third gear until about 10 mph, when the governor exhausts. Then both shift valves move up to exhaust the direct clutch and the intermediate servo. The forward clutch stays applied, and the transmission is in free-wheeling first gear.

Basic Systems

1. The pump is delivering fluid. These basic systems remain unchanged:

2. Main control pressure

3. Converter and front lube

4. Cooler and rear lube

Forward Clutch and Governor

5. The forward clutch/governor circuit 2 is charged with line pressure at the manual valve.

6. The forward clutch is applied.

7. At about 10 mph road speed, the *primary* governor valve spring overcomes centrifugal force.

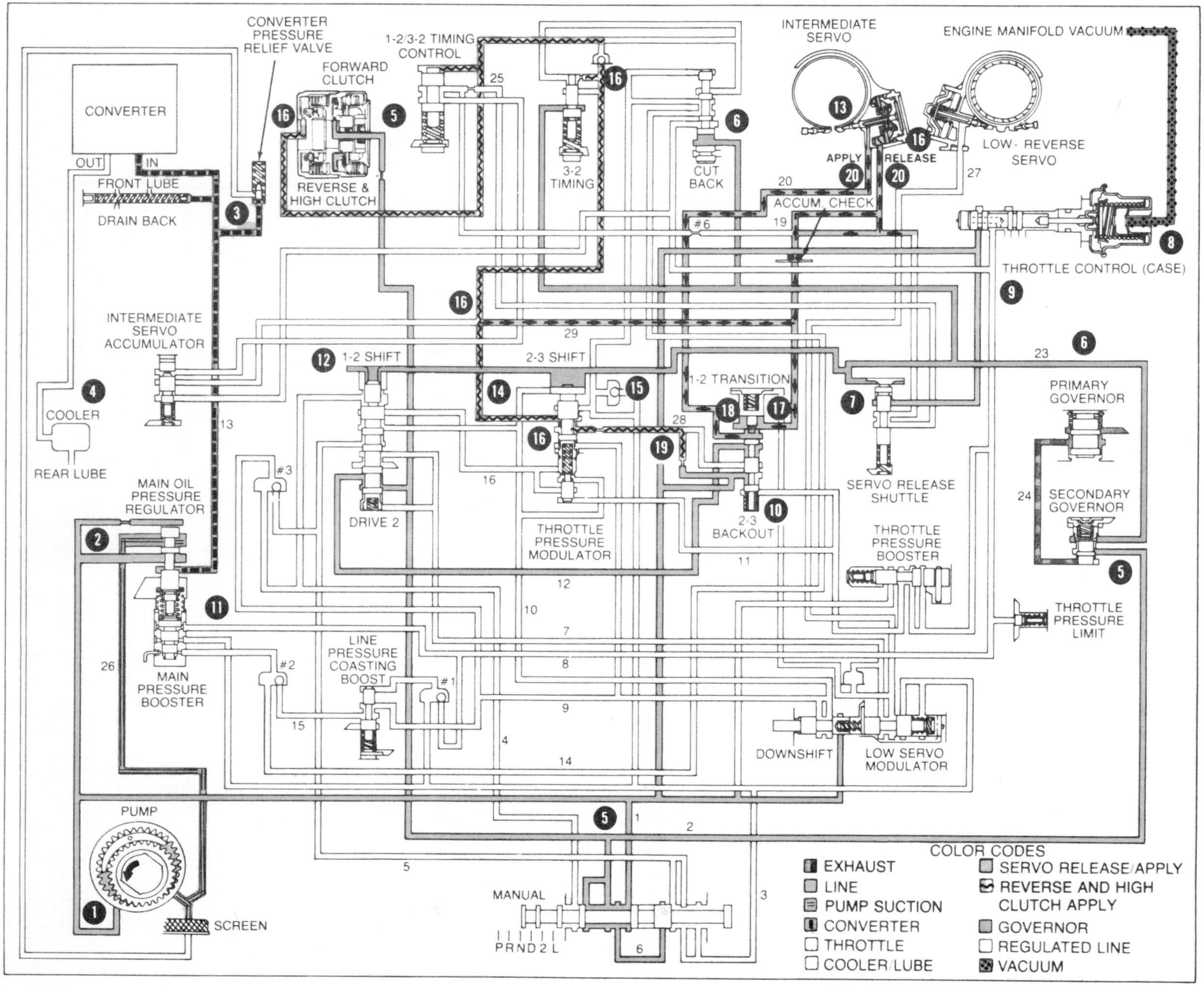

Figure 6-13. Closed-throttle 2-3 upshift. (*Courtesy of Ford Motor Company*)

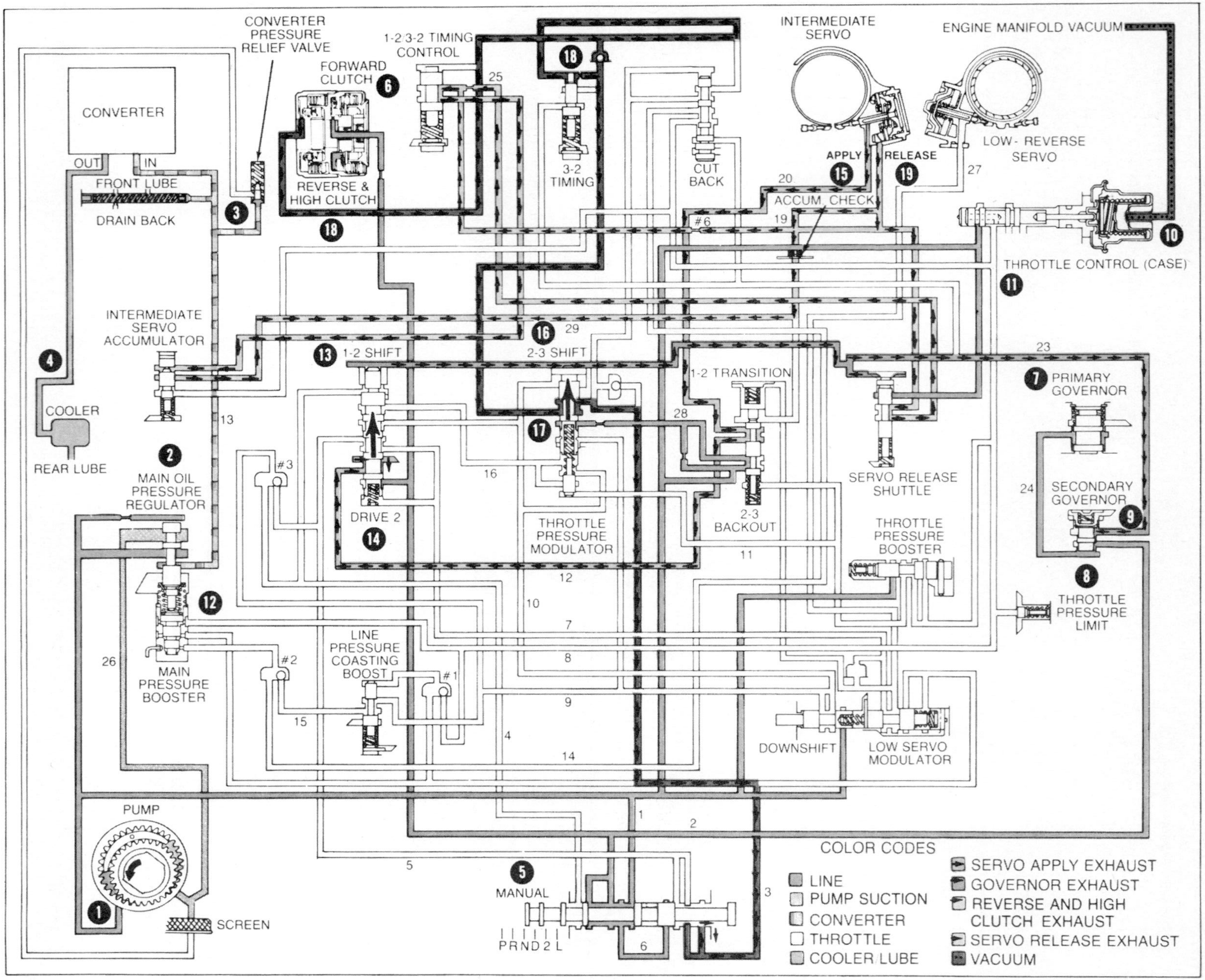

Figure 6-14. Closed-throttle 3-1 downshift at 10 mph. (*Courtesy of Ford Motor Company*)

The valve moves in and blocks the exhaust passage for circuit 24.

8. This causes full-line pressure to be applied to the outer end of the *secondary* governor valve. The valve is forced in (up).

9. Governor pressure in circuit 23 then is exhausted through the governor body. Circuit 23 goes to zero psi.

Throttle Valve System

10. With the throttle closed, manifold vacuum goes to maximum.

11. TV pressure goes to zero or minimum.

12. Circuit 8 is at zero psi, so the line pressure is regulated at minimum.

Downshift

13. With zero governor pressure on the 1–2 shift valve:

14. The drive 2 valve spring pushes both valves up. Circuit 12 is exhausted at the drive 2 valve.

15. Intermediate servo *apply* pressure exhausts through circuit 20, the backout valve, circuit 12, and the drive 2 valve. The intermediate band stays released.

16. Governor pressure is zero atop the 2–3 shift valve also.

17. The valve spring moves the valve up. This cuts off circuit 28 pressure to the direct clutch and servo release circuit and opens the circuit to exhaust through the reverse circuit 3.

18. Reverse and high clutch apply pressure exhausts through the 3–2 timing valve orifice, around the 2–3 shift valve, into circuit 3 and to the sump at the manual valve.

19. Servo release pressure exhausts through the accumulator check valve high- and low-speed orifices and 1–2/3–2 valve. Then it follows the same exhaust path as direct clutch exhaust. Since the servo is already released at the start of the shift, the pressure bleeds very quickly.

Pressure in circuits 19 and 25 and pressure at the intermediate servo accumulator also are exhausted with the 2–3 shift valve downshifted.

Torque Demand 3-2 Downshift (Figure 6-15)

A torque demand 3–2 shift requires careful timing to prevent any roughness. The vehicle is at cruising speed and the converter clutch is at high capacity. Thus there is very little slip in the converter to cushion the shift.

To downshift 3–2, it is necessary to release the reverse and high clutch and to apply the intermediate servo by exhausting the servo release pressure. A network of valves and orifices controls the exhaust of clutch and servo release oil to keep the shift smooth.

Basic Systems

1. Since the engine is running, the pump is delivering fluid. The following systems are charged as described at the beginning of this section:

2. Main control pressure

3. Converter and front lube

4. Cooler and rear lube

Clutch and Governor

5. The forward clutch/governor system, circuit 2, is charged with line pressure at the manual valve. The clutch is applied.

6. The governor is regulating pressure in circuit 23.

7. The cutback valve has been forced up and control pressure cutback occurred before the 1–2 shift.

Throttle Valve System

8. The driver has pushed the throttle to nearly full, so manifold vacuum is very low, perhaps 2 to 3 inches.

9. Throttle pressure is regulating near maximum.

10. TV pressure in circuit 8 is causing line pressure to be boosted.

11. The throttle pressure booster valve is regulating a boosted TV to the end of the throttle pressure modulator valve.

12. Modulated throttle pressure plus spring force has forced the 2–3 shift valve up. Line pressure in circuit 28 is cut off. The reverse and high clutch apply/servo release circuit is exhausted to the reverse circuit 3.

The 1–2 shift valve remains down, unaffected by TV, so there is still pressure on the apply side of the intermediate servo.

Downshift Timing

13. As soon as the reverse and high clutch apply circuit is exhausted, the 3–2 check valve closes. The reverse and high clutch is forced to exhaust through the 3–2 timing valve or the 3–2 orifice. The 3–2

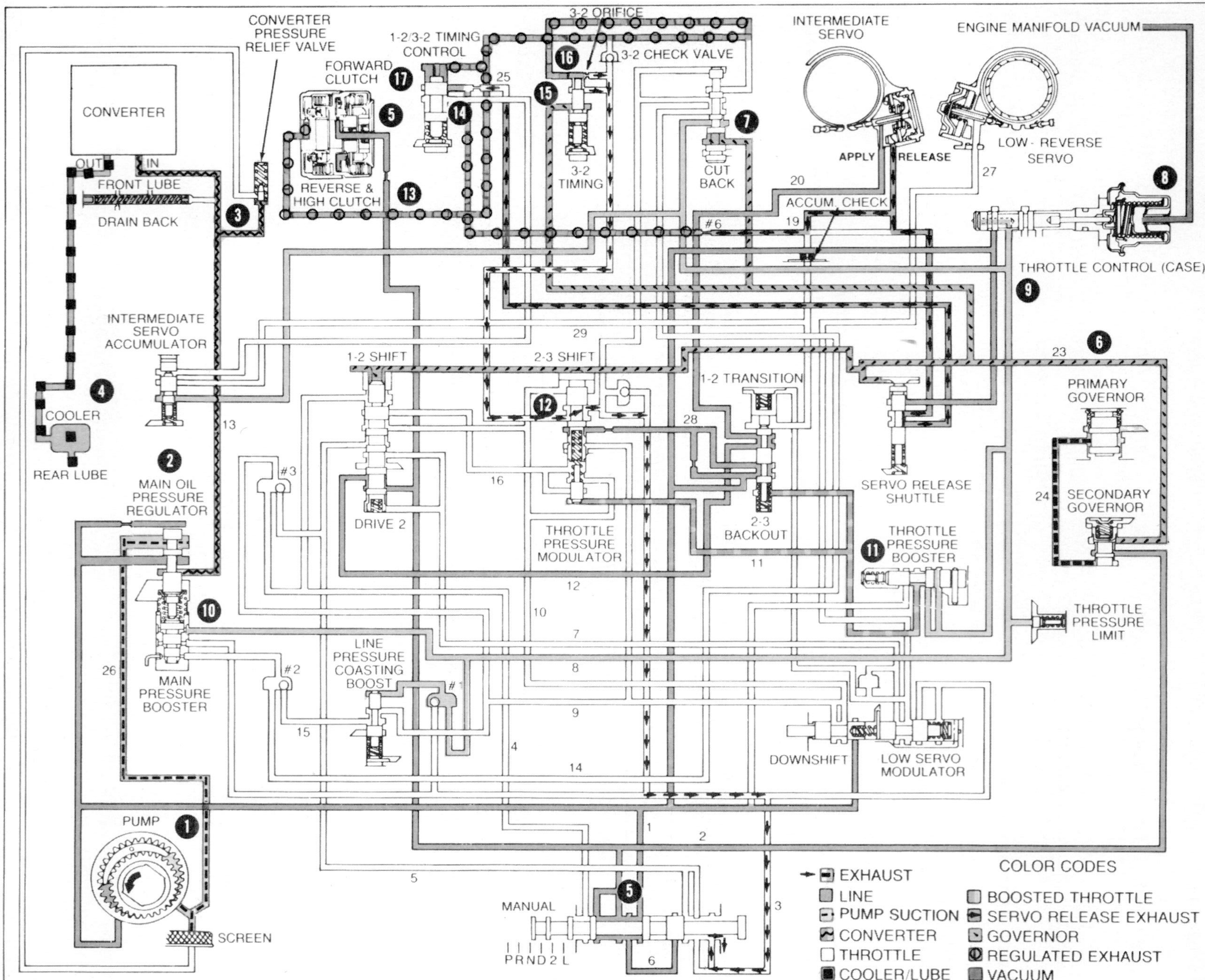

Figure 6-15. Part-throttle 3-2 shift (torque demand). (*Courtesy of Ford Motor Company*)

orifice is so small that the timing valve is forced down by the back-pressure in the clutch circuit.

14. At the same time, the 1–2/3–2 timing valve is being held down by reverse and high clutch pressure. This permits the servo release exhaust flow to join the reverse and high clutch flow. However, the servo release in the servo is not the same as clutch pressure because of the high- and low-speed orifices.

15. The 3–2 timing valve goes into regulation. It maintains a back-pressure in the clutch apply circuit and back to the low-speed and high-speed orifices in the servo release circuit until the servo piston has stroked to full apply. This pressure controls the clutch release, allowing it to slip until the servo is fully on. When the release exhaust flow stops, the 3–2 timing valve moves back up.

16. Clutch pressure then bleeds down through the 3–2 orifice.

17. When the clutch pressure drops below 13 psi, the 1–2/3–2 valve moves back up.

Through Detent 3-2 Downshift (Figure 6-16)

When the driver pushes the accelerator to the floor, the downshift rod is actuated by the throttle lever. This causes the downshift valve to move and charge the kickdown circuit 10 with line pressure. The 3–2 downshift then is forced at higher road speed than a torque demand shift. Figure 6–16 illustrates the hydraulic system operation during the through-detent 3–2 shift.

Basic Systems

1. Since the engine is running, the pump is delivering fluid. The following systems are charged as described at the beginning of this section:

2. Main control pressure

3. Converter and front lube

4. Cooler and rear lube

Forward Clutch and Governor

5. The forward clutch/governor system, circuit 2, is charged with line pressure at the manual valve. The clutch is applied.

6. The governor is regulating pressure in circuit 23.

7. The cutback valve has been forced up; control pressure cutback occurred before the 1–2 shift.

8. Governor pressure also may be high enough to shift the servo release shuttle valve, depending on road speed. If the valve is shifted, the servo release path through circuit 25 is cut off.

9. Governor pressure is also holding the 1–2 shift valve and drive 2 valve down, so the intermediate servo has pressure on the apply side of the piston.

Throttle Valve System

10. At wide-open throttle, the manifold vacuum is close to zero.

11. The throttle valve is regulating maximum TV pressure.

12. Circuit 8 pressure is causing a line pressure boost.

13. The throttle pressure booster valve is regulating a boosted TV pressure.

Kickdown Circuit

14. The downshift valve has moved in, opening line pressure in circuit 1 to the kickdown in circuit 10.

15. Circuit 10 line pressure is effective at the spring end of the 2–3 shift valve and on a differential area at the other end. The valve is shifted up.

16. The reverse and high clutch/intermediate servo release system is exhausted around the 2–3 shift valve to the reverse circuit 3.

As soon as the servo has stroked to apply and the clutch has stroked off, the transmission is in second gear.

3-1 or 3-2 Kickdown

17. Kickdown pressure is also effective on a differential area of the drive 2 valve. At low road speed, this pressure may force the drive 2 and 1–2 shift valves up. Thus the intermediate servo apply would exhaust. The transmission then would downshift to first gear.

Manual Second Gear (Figure 6-18)

In range 2, the transmission starts in second gear and stays in second. The forward clutch and intermediate servo are applied. Second-gear circuit 4 also provides a boost of line pressure and "locks out" a 2–3 or 2–1 shift by applying pressure to the shift valves.

Basic Systems

1. Since the engine is running, the pump is delivering fluid. The following systems are charged as described at the beginning of this section:

2. Main control pressure

3. Converter and front lube

4. Cooler and rear lube

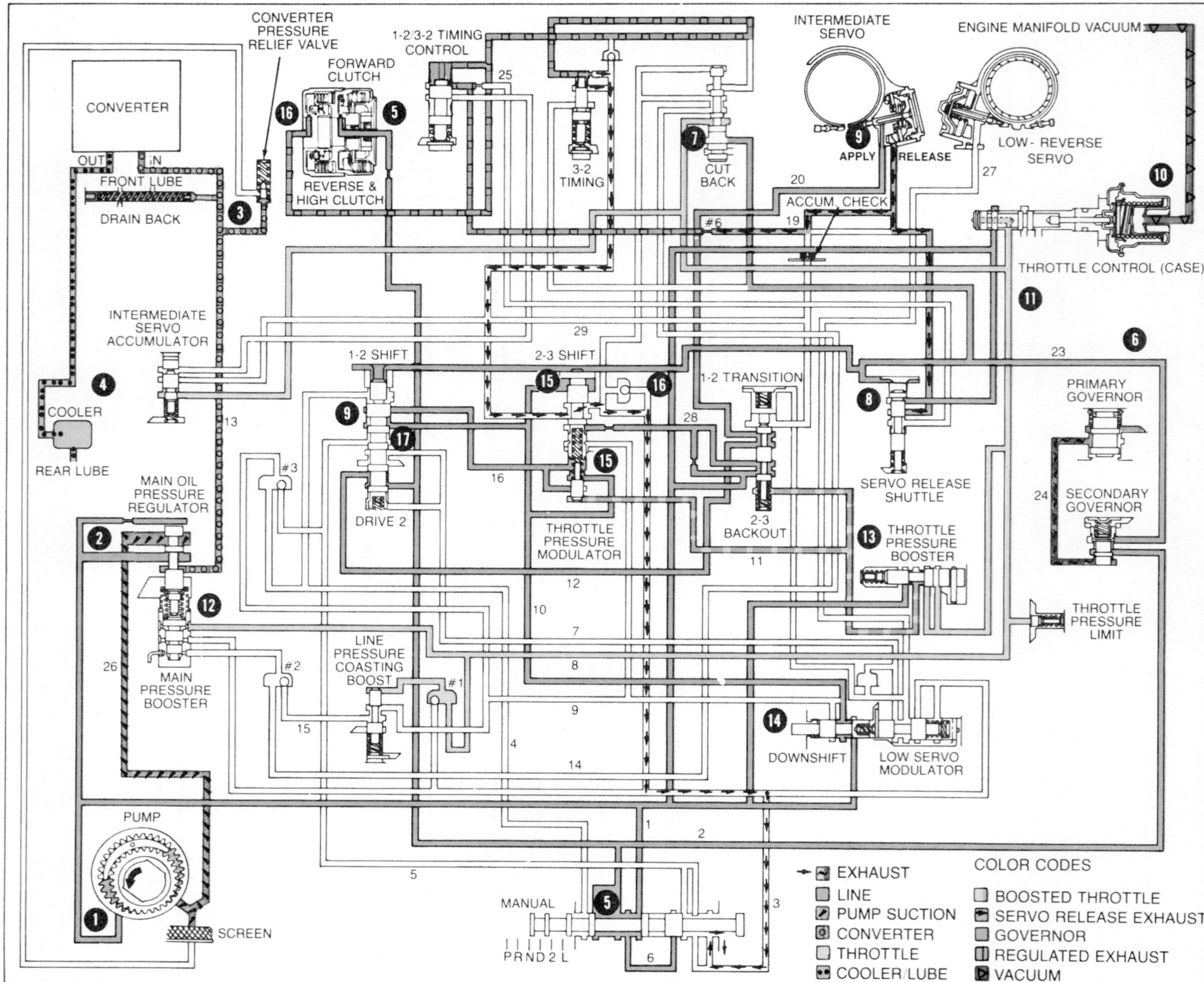

Figure 6-16. Wide-open throttle 3-2 shift (through detent). (*Courtesy of Ford Motor Company*)

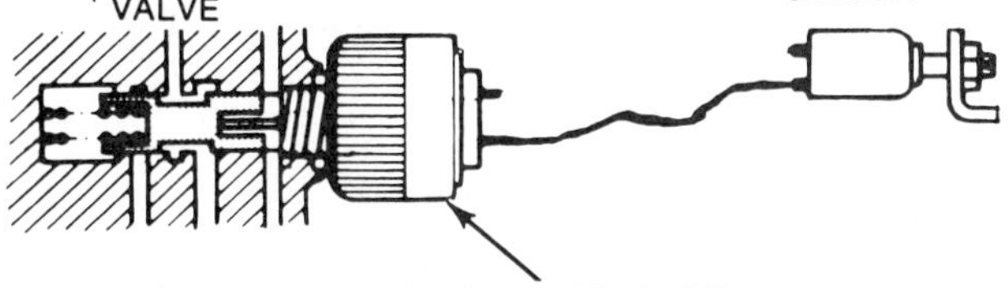

Figure 6-17. Kickdown switch operation. *(Courtesy of Ford Motor Company)*

Forward Clutch and Governor

5. The forward clutch/governor system, circuit 2, is charged with line pressure at the manual valve. The clutch is applied.

6. The governor is regulating pressure in circuit 23.

7. The cutback valve was forced up when the governor pressure cut in at about 10 mph.

Manual 2 Systems

8. Manual 2 circuit 4 also is charged with line pressure at the manual valve. This circuit feeds the lockout and pressure boost systems.

9. First-gear lockout. Pressure is applied between the 1–2 shift valve and the drive 2 valve. The drive 2 valve is forced down by this pressure and held down. This locks out a 2–1 downshift.

10. Pressure from circuit 2 charges circuit 12 to the intermediate servo.

11. The servo is applied.

12. Manual 2 pressure continues through the shuttle valve 3 into circuit 9 through the line coast boost valve and ball shuttle 2 to the line pressure boost valve.

13. Third-gear lockout. Circuit 9 charges circuit 10 around the downshift valve.

14. Circuit 10 applies pressure to the kickdown areas of the 2–3 shift valve. In all, three differential areas plus the spring end of the valve are subject to line pressure from circuits 9, 10, and 28. The valve cannot be upshifted no matter how high governor pressure gets. This locks out a 2–3 shift.

15. Line pressure boost. At the line coasting boost valve, the manual 2 line pressure is applied to a differential area between the valve lands. This causes a force opposing the valve spring and is assisted by TV pressure from circuit 8 (through shuttle valve 1). The valve goes into regulation at light throttle openings, producing a reduced pressure in circuit 15.

16. Regulated pressure is applied to the end of the main pressure booster valve. This provides a higher line pressure to the drive 2 valve and 2–3 shift valve. (This is required mainly for a 3–2 downshift if the selector is moved to 2 out of high gear.)

Throttle Valve System

17. At light throttle, the TV system is regulating a moderate TV pressure.

18. TV pressure in circuit 8 is causing an increase in line pressure.

19. At high TV pressure (more throttle opening), circuit 8 pressure at the top of the line pressure coasting boost valve begins to shut off pressure to circuit 15. When the TV pressure is enough to close off the regulated pressure, it is also high enough to provide as much line pressure as needed by acting on the booster valve differential area.

Manual Shift From High Gear

We have described manual second gear as it operates from a standstill. If the selector is moved to 2 from high gear in D, the same manual 2 systems are charged. However, the operation for downshifting is the same as for a 3–2 forced downshift.

Manual Low (First) Gear and Manual 1-2 Shift (Figure 6-19)

If the vehicle starts in manual low (range 1), the transmission engages in first gear and stays there. The forward clutch must be applied. Also, the low-reverse band is applied to cause engine braking. The governor system is active because its pressure supply is from the forward clutch circuit. The low-reverse circuit 5, besides applying the servo, activates second-gear lockout and high-gear lockout systems.

Basic Systems

1. Since the engine is running, the pump is delivering fluid. The following systems are charged as described at the beginning of this section:

2. Main control pressure

3. Converter and front lube

4. Cooler and rear lube

Forward Clutch and Governor

5. The forward clutch/governor system, circuit 2, is charged with line pressure at the manual valve. The clutch is applied.

6. The governor is regulating pressure in circuit 23.

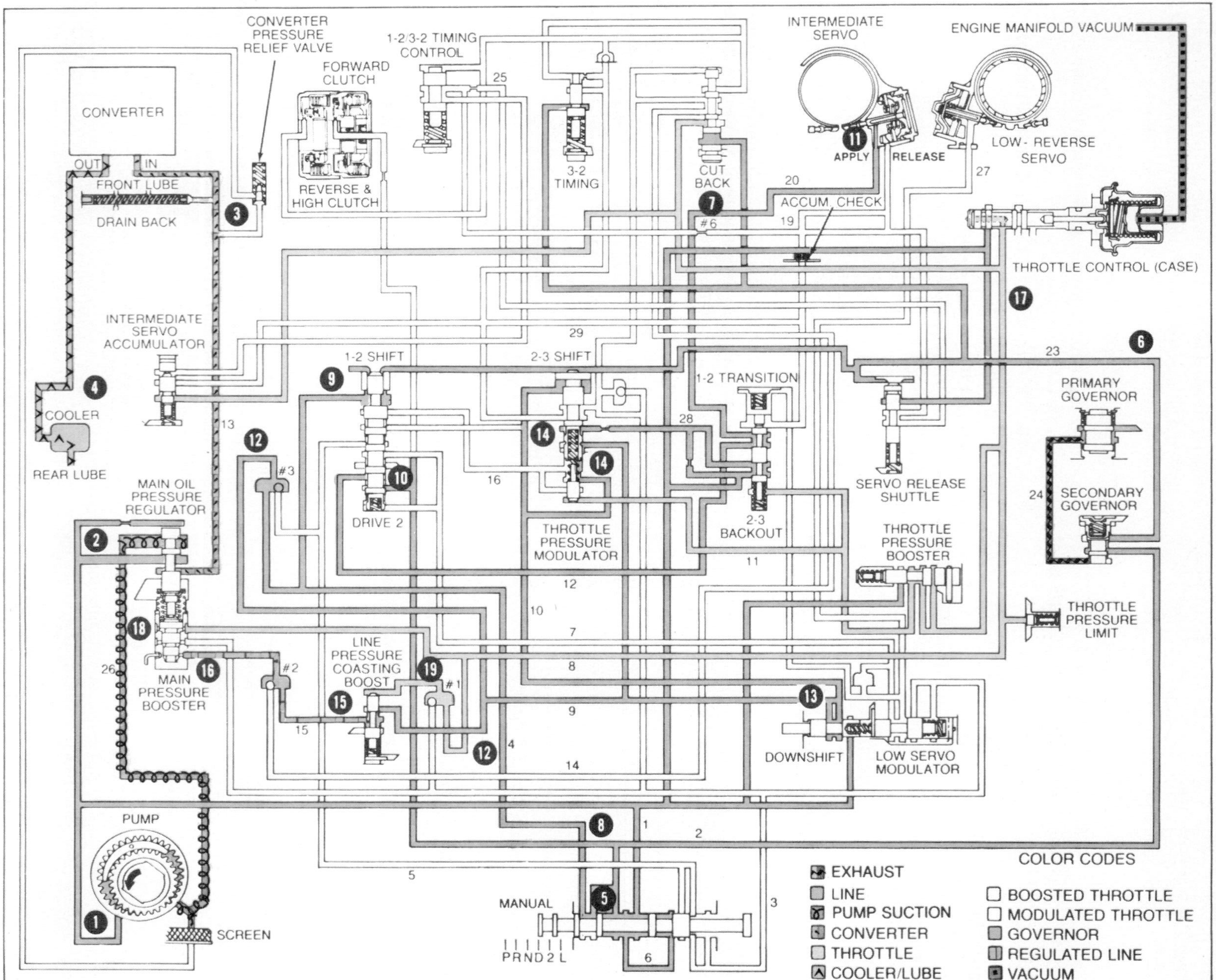

Figure 6-18. Manual second gear (range 2) at 15 mph. (*Courtesy of Ford Motor Company*)

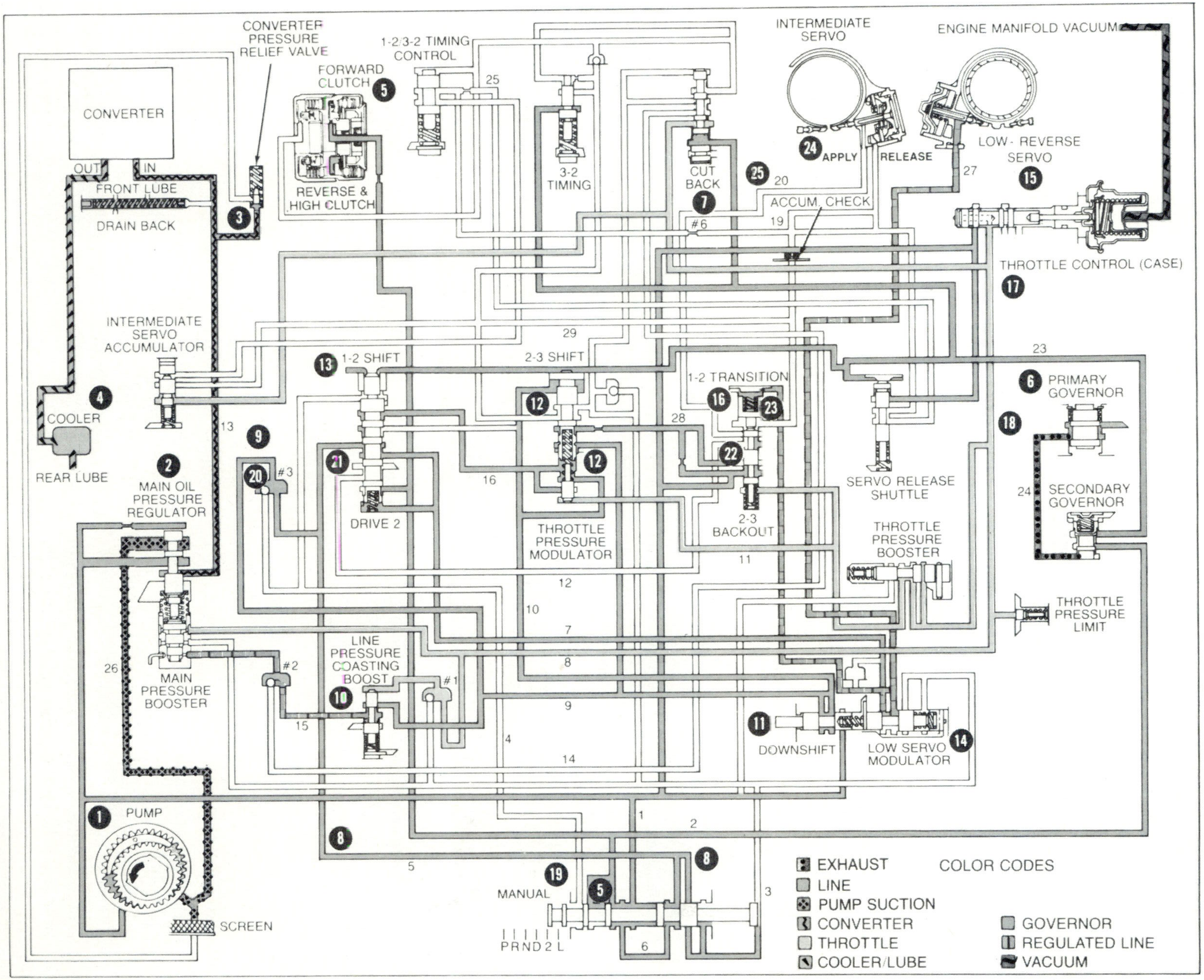

Figure 6-19. Manual low gear (range 1) at 15 mph. (*Courtesy of Ford Motor Company*)

7. The cutback valve has been forced up. Control pressure cutback occurred when the governor cut in at about 10 mph.

Low-Reverse Systems

8. The low-reverse circuit 5 is charged at the manual valve. This circuit feeds the band servo apply, the upshift lockouts, and a pressure boost system.

9. Line pressure boost. Circuit 5 pressure is applied, through shuttle valve 2, to the line pressure coasting boost valve and to lockout circuit 9.

10. The line pressure coasting boost valve supplies a regulated pressure to the end of the main pressure booster to increase line pressure.

11. Circuit 9 charges circuit 8 through the downshift valve.

12. High-gear lockout. Circuit 8 applies line pressure to the 2–3 shift valve, at the end and the large differential area. This is assisted by pressure to two more differential areas, supplied by circuits 9 and 28. Governor pressure cannot overcome these combined forces at any road speed.

13. Second-gear lockout. Circuit 5 also branches to charge circuit 7 at the drive 2 valve. Circuit 7 applies full line pressure to the spring end of the valve and to a differential area. This is assisted by pressure on two or more differential areas, applied by circuits 10 and 16. Governor pressure cannot overcome this combined force, so that a 1–2 upshift can't happen.

14. Circuit 7 continues to the low servo modulator valve and pressurizes passages to the low-reverse servo and 1–2 transition valve.

15. The low-reverse servo is applied with a pressure (25 psi) provided by regulation of the low servo modulator against its spring. The regulated pressure ensures a smooth band engagement on a 3–1 or 2–1 shift.

16. The 1–2 transition valve is forced down and mechanically forces the 2–3 backout valve down.

Throttle Valve System

17. At light throttle, the TV system is regulating a moderate pressure.

18. TV pressure in circuit 8 is causing an increase in line pressure.

19. If TV pressure increases (more throttle opening), TV pressure will shut off the regulation of the line pressure coasting boost valve. Circuit 15 will go to zero psi, and circuit 8 pressure will provide all the line pressure boost that is needed.

Manual 1-2 Shift

20. If the selector is moved from 1 to 2, circuit 5 will exhaust at the manual valve and circuit 4 (manual 2) will be charged. This will cause the low-reverse servo to stroke off and the intermediate servo on for the 1–2 upshift.

21. Shuttle valve 2 will shift so that circuit 4 can charge the pressure boost system.

22. The drive 2 will be forced down and charge circuit 12 from the forward clutch apply system.

23. Circuit 12 is momentarily blocked from circuit 20 at the 2–3 backout valve. If it were not, the intermediate servo might come on too fast and cause a band-to-band tie-up.

24. The 1–2 transition valve will move up and permit the 2–3 backout valve to move up as soon as the low-reverse servo pressure, which is exhausting, drops to about 5 psi.

25. Circuit 20 then will apply the intermediate servo.

Reverse (Figure 6-20)

For reverse gear, the reverse and high clutch and the low-reverse servo are applied. Also, line pressure is boosted to maximum to handle the high reverse band torque. There is no governor pressure, since the forward clutch circuit 2 is not charged. The shift valves are held downshifted by their springs and hydraulic pressures.

Basic Systems

1. Since the engine is running, the pump is delivering fluid. The following systems are charged as described at the beginning of this section:

2. Main control pressure

3. Converter and front lube

4. Cooler and rear lube

Throttle Valve System

5. With the throttle open moderately, the TV system is at about 40–50 psi.

6. Circuit 8 is causing a line pressure boost by pressurizing the upper differential area of the pressure booster valve.

7. Since there is no governor pressure, the cutback valve charges circuit 14 with TV pressure.

8. This is effective, through shuttle valve 2, to the end of the pressure booster valve. A further increase in line pressure results. Other TV system passages do not affect reverse operation.

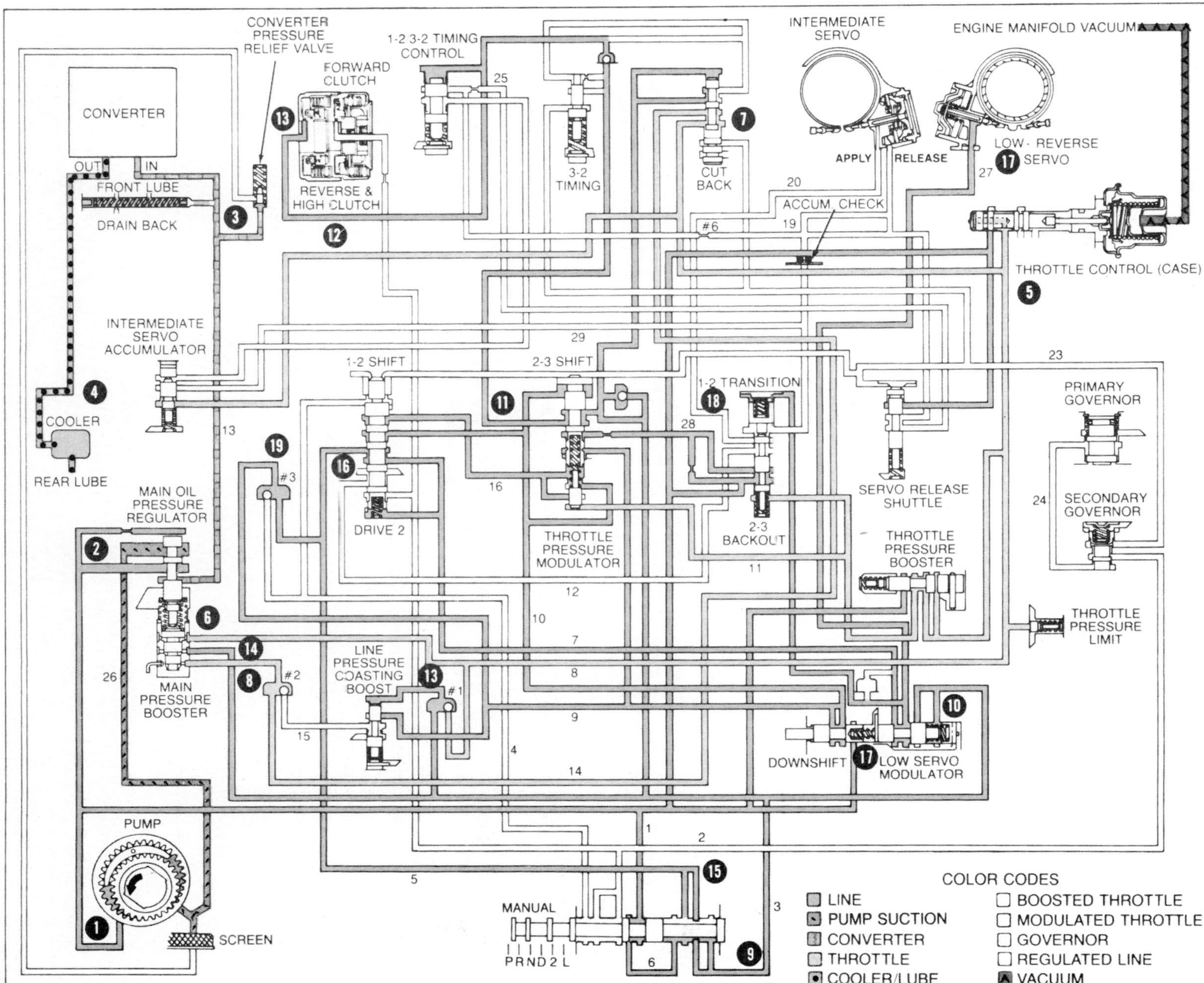

Figure 6-20. Reverse gear at light throttle. (*Courtesy of Ford Motor Company*)

Reverse Circuit Systems

9. Reverse circuit 3 is charged at the manual valve.

10. Pressure is applied to the end of the low servo modulator valve, so that it cannot "regulate" servo apply pressure.

11. The reverse and high clutch/intermediate servo release system is charged at the 2–3 shift valve.

12. The reverse and high clutch is applied.

The servo release systems also are charged but do not affect reverse operation.

13. Circuit 3 pressure is applied through shuttle valve 1 to the end of the line pressure coasting boost valve. This cuts off circuit 9 pressure, and the valve has no output to the main pressure booster.

14. Reverse pressure boost. Pressure at the lower differential area of the main pressure booster valve assists the two TV pressures. Line pressure is regulated at maximum.

Low-Reverse Systems

15. Low-reverse circuit 5 is also charged at the manual valve.

16. At the D2 shift valve, circuit 7 is charged.

17. At the low servo modulator valve, circuit 7 charges circuit 27 to apply the low-reverse servo with full line pressure.

18. Line pressure acts on the 1–2 transition valve and moves it down with the 2–3 backout valve. This is incidental to reverse operation.

19. Another branch of the circuit acts through shuttle valve 3 to charge the second-gear and high-gear lockout systems. This is also incidental to reverse operation, since there is no governor pressure to move the valves.

This concludes the coverage of the hydraulic control system as it applies to the Ford C5 automatic transmission.

Hydraulic control features other than those of the

Summary of Hydraulic Control Systems

System	Function
Main control pressure	Regulates line (control) pressure
Converter and cooler	Feeds converter, cools fluid, and lubricates transmission
Throttle pressure	Engine load input; delays upshift and forces downshift (torque demand)
Forward clutch	Engages transmission in forward gears
Governor	Road speed input; causes upshifts
Control pressure cutback	Reduces line pressure at increased road speeds
1–2 and 2–3 shift valves	Control upshifts and downshifts
1–2 accumulator	Controls 1–2 upshift feel
Reverse and high clutch/ servo release	Applies reverse and high clutch and releases intermediate servo together
2–3 backout	Prevents clutch-band tie-up on 2–3 shift
3–2 timing valve	Times 3–2 downshift to prevent tie-up or buzz-up
Kickdown	Causes downshift with accelerator through-detent or WOT upshift
Manual 2/first-gear lockout	Causes front servo application in range 2 and prevents downshift to first
Line pressure boost	Increases line pressure at low TV pressure in ranges 1 and 2
Third-gear lockout	Prevents 2–3 shift valve from moving to upshift position in ranges 1 and 2
Second-gear lockout	Prevents 1–2 shift valve and drive 2 valve from moving to upshift position in range 1
Low-reverse	Applies rear servo in manual low and R
1–2 transition	Prevents band-to-band tie-up on manual 1–2
Reverse pressure boost	Causes maximum line pressure in R

C5 include throttle-linkage–actuated throttle valve systems, hydraulic-controlled converter clutch lockup systems, fourth-gear overdrive control systems, and electric-solenoid–actuated downshift valves. These are shown in the hydraulic control system schematics in Chapter 7 as they apply to specific transmissions.

PART 6-7 TORQUE CONVERTER HYDRAULIC LOCKUP CONTROLS

General Motors Type

In this system the converter clutch pressure plate assembly is splined to the turbine assembly and is located between the turbine and the converter cover. Application of the pressure plate and clutch is controlled by hydraulic pressure on the turbine side, which exerts a force against the pressure plate. This seats the plate against the converter cover, which provides converter lockup for direct drive through the converter. To release the converter lockup clutch, hydraulic pressure is directed to the other side of the pressure plate, which moves the plate away from the converter cover. Clutch operation is dependent on selector lever position and a series of hydraulic controls.

Release Position

In park, neutral, reverse, and first gear, the converter clutch apply valve, located in the case, is held in the release position by the converter clutch apply valve spring. That is, the apply valve in this position takes converter feed oil, coming from the pressure regulator valve, and sends it into the release passage, through the turbine shaft, and into the cavity between the converter clutch pressure plate and the converter cover. This moves the clutch pressure plate away from the converter cover, releasing the converter clutch.

After the oil releases the converter clutch, it flows from behind the pressure plate and through the converter. From the converter, the oil flows backward through the apply passage to the apply valve. The apply valve then sends oil, returning from the converter, to the transmission cooler in the radiator. Oil, returning from the cooler, is then directed to the transmission lubrication system.

The apply valve is controlled by the converter clutch shift valve and will stay in the release position until it receives converter clutch signal oil from the converter clutch shift valve in the valve body.

Converter Clutch Apply

When the automobile has reached a sufficient speed where heavy acceleration is no longer needed, the torque multiplication of the converter is no longer needed. Because the converter is a fluid coupling, there is some slippage between the converter pump and the turbine. To eliminate this slippage and to improve fuel economy, the converter clutch can be applied in either second, third, or fourth gear, depending on the model of transmission.

The converter clutch apply is initiated by the converter clutch shift valve, located in the valve body. A spring acting on the converter clutch shift valve train tends to keep the valve closed, blocking second oil coming from the 1–2 shift valve, while governor pressure attempts to open the valve. When vehicle speed increases enough to develop sufficient governor pressure to open the converter clutch shift valve train, second oil enters the converter clutch signal passage. Converter clutch signal oil passes through an orifice to the converter clutch apply valve.

Converter Clutch Apply Valve

The converter clutch valve allows second oil to enter the converter clutch signal passage to the converter clutch apply valve. The converter clutch signal oil overcomes the converter clutch apply valve spring, shifting the valve and changing the direction of the converter feed oil. The converter clutch apply valve will now send converter feed oil from the pressure regulator into the apply passage. The apply oil will flow between the stator shaft and the converter hub to charge the converter with oil and force the converter clutch pressure plate against the converter cover, causing a mechanical link between the engine and the turbine shaft. At the same time the converter clutch apply valve will direct some converter feed oil through an orifice to the transmission cooler in the radiator. Oil returning from the cooler is then directed to the transmission lubrication system.

The converter clutch apply valve can shift to apply the converter only when the converter clutch solenoid is energized electrically. When the solenoid is not energized, the converter clutch signal oil is exhausted and the converter will stay in the release position. The converter clutch solenoid would be off under conditions such as high engine vacuum (idle), low engine vacuum (full throttle), braking, or cold engine operation.

In the hydraulic system the 1–2, 2–3, 3–4, and converter clutch shifts will always take place at the same vehicle speeds, that is, whenever the governor pressure overcomes the force of the springs on the

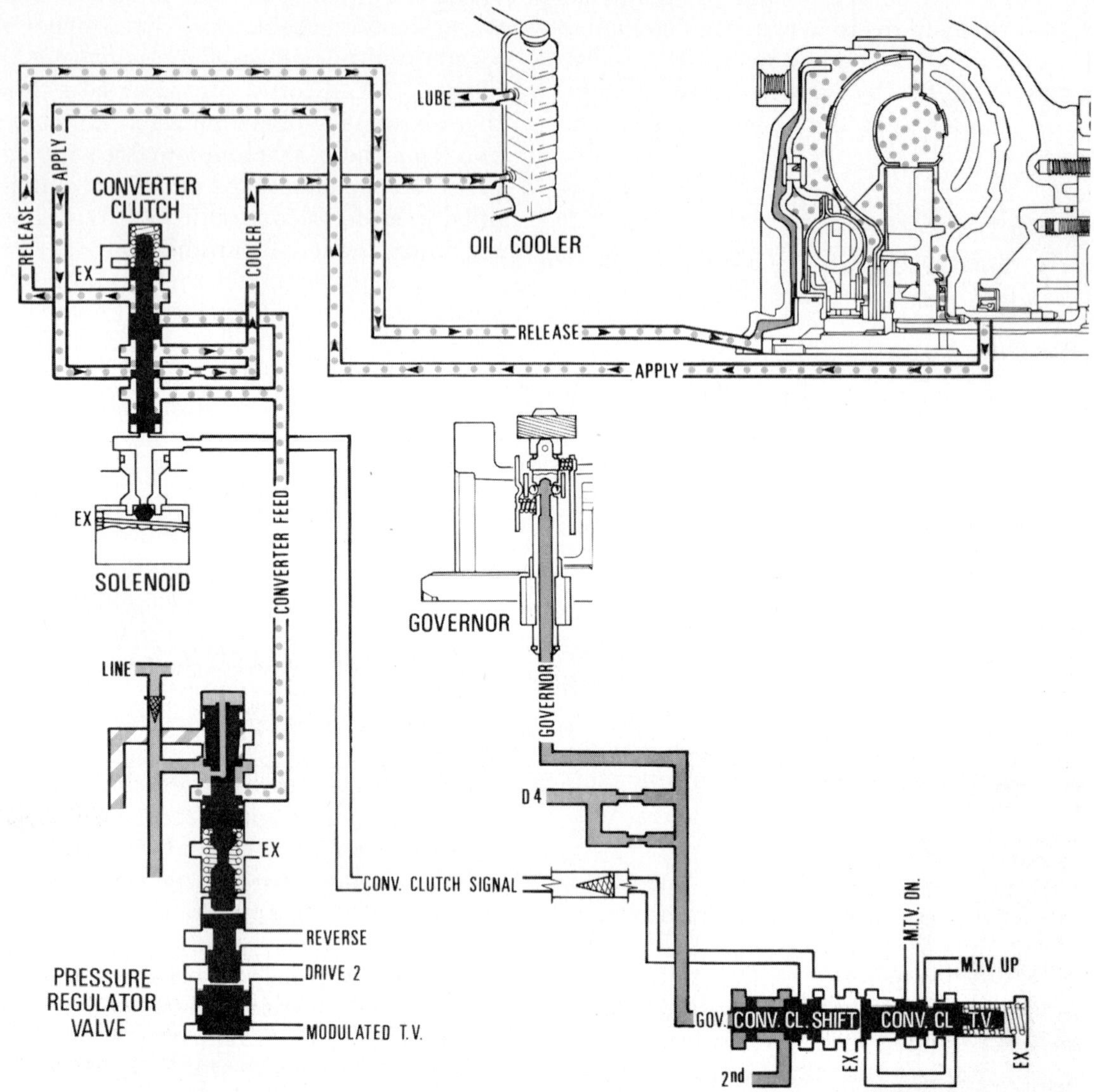

Figure 6-21. Converter clutch apply valve in release position. (*Courtesy of General Motors Corporation*)

shift valves. When accelerating under a heavy load or for maximum performance, the shifts occur at higher vehicle speeds.

Chrysler TorqueFlite Type

The lockup mode is activated only in direct drive. This means lockup only occurs after the 2–3 upshift. In order to prevent engine lugging due to lockup at too low an engine speed, the minimum lockup and disengagement points are approximately 24 to 34 mph, depending on engine size and axle ratio. At heavier throttle, the lockup shift will occur immediately after the 2–3 shift.

A switch valve provides both the on and off oil pressures, which move the lockup piston. The switch valve will not operate (to activate the lockup piston) until two other control valves are switched. These two valves are housed in a lockup control valve body.

The lockup valve automatically directs the switch valve to lock up if the vehicle is above a certain speed. The fail-safe valve will permit lockup in *direct gear only when the transmission front clutch has sufficient pressure.*

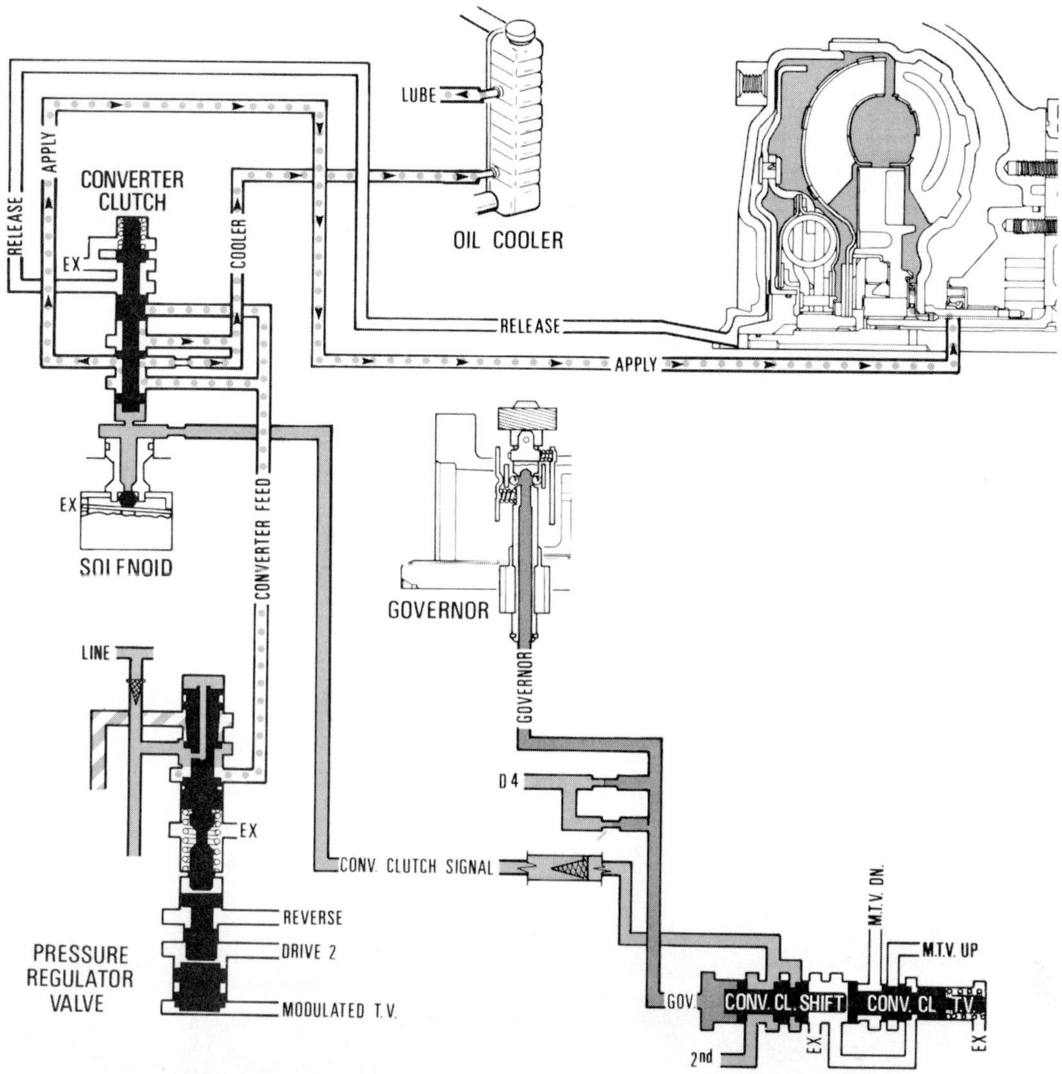

Figure 6-22. Converter clutch apply valve in apply position. (*Courtesy of General Motors Corporation*)

The fail-safe valve also provides fast lockup *release* during a kickdown. The smooth sensing actions of both the lockup valve and the fail-safe valve are necessary for good shift quality. In order to provide the oil pressures needed to move the lockup piston, there is an oil passage through the transmission input shaft.

In the *unlocked* mode, transmission converter pressure is applied through the shaft. The oil flows around the unlocked piston face, back through the converter, and out between the reaction shaft and the input shaft.

When the conditions are right for lockup, the switch valve vents the pressure on the *front* side of the piston. This causes the line pressure *behind* the piston to force it into lock.

The lockup friction material and an O-ring at the I.D. of the piston provide a pressure seal across the piston face to keep the converter in the locked-up mode. There is an additional lip seal located in the I.D. of the turbine hub to prevent internal leakage.

American Motors 904 and 998 transmissions equipped with lockup converters operate in a similar manner.

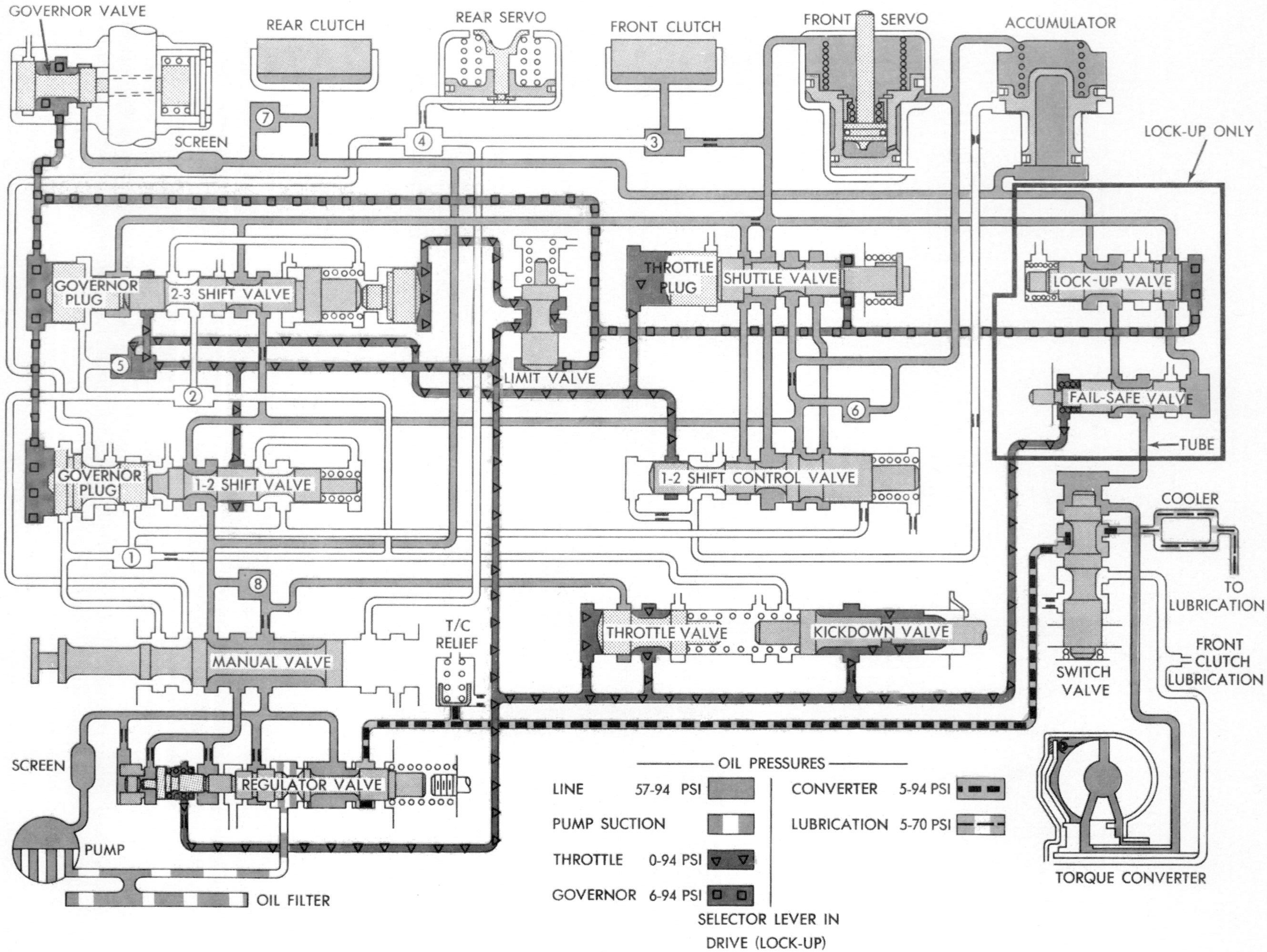

Figure 6–23. Drive-direct Lockup (eight-cylinder). (*Courtesy of Chrysler Corporation*)

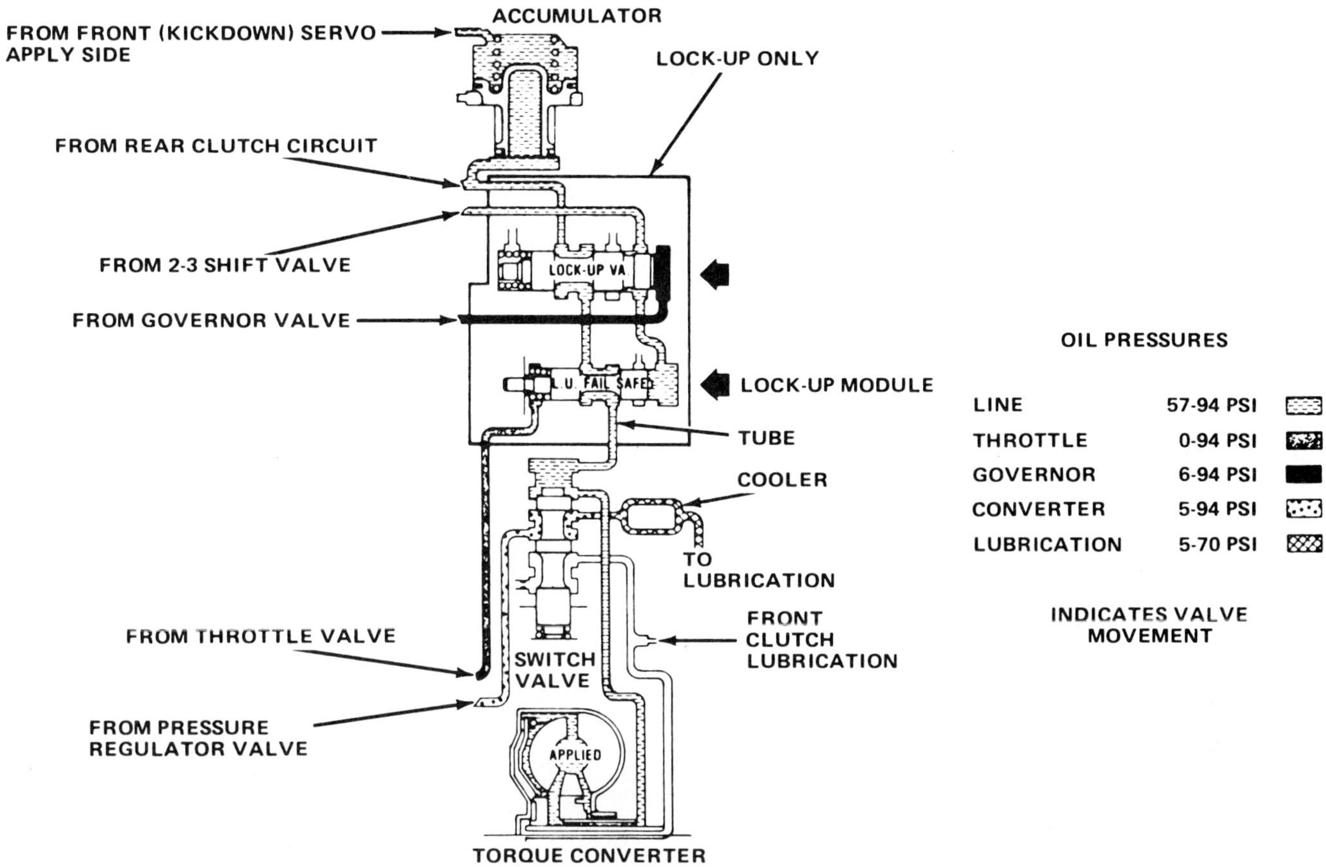

Figure 6-24. American Motors Torque Command torque converter clutch lockup controls. (*Courtesy of American Motors Corporation*)

PART 6-8 THROTTLE VALVE SYSTEMS

There are two different systems used by transmission manufacturers to operate the throttle valve in the hydraulic control system. One method is to use the mechanical accelerator linkage to actuate the valve. The other method is to use a vacuum modulator.

The mechanical linkage method causes the transmission throttle valve to move in direct proportion to accelerator pedal movement, which in turn represents carburetor throttle position. This throttle valve movement meters the amount of line pressure allowed into the throttle pressure passage. Throttle pressure is simply modified line pressure. Increased throttle opening results in increased throttle pressure, and of course throttle pressure acts in a manner that tends to delay the shift valves from moving to the upshifted position. Throttle pressure also acts on the control pressure regulator valve to increase line pressure.

The vacuum modulator system of throttle valve control uses a diaphragm and spring enclosed in a metal case. A push rod is used to transmit diaphragm movement to the throttle valve. (General Motors calls its modulator-operated valve a modulator valve. However, the resulting modulator pressure serves the same purpose in the hydraulic control system as throttle pressure does in other transmissions.)

Atmospheric pressure, spring pressure on one side of the diaphragm and engine intake manifold vacuum, and spring force on the other side control throttle valve or modulator valve movement. An evacuated metal bellows is used in some modulators to compensate for atmospheric pressure differences at high or low altitudes.

In operation, the vacuum modulator results in increasing throttle or modulator pressure as engine intake manifold vacuum decreases and decreasing throttle pressure as manifold vacuum increases. High vacuum causes low throttle pressure; low vacuum causes high throttle pressure. An engine under heavy load requires increased carburetor throttle opening, resulting in low manifold vacuum and higher throttle

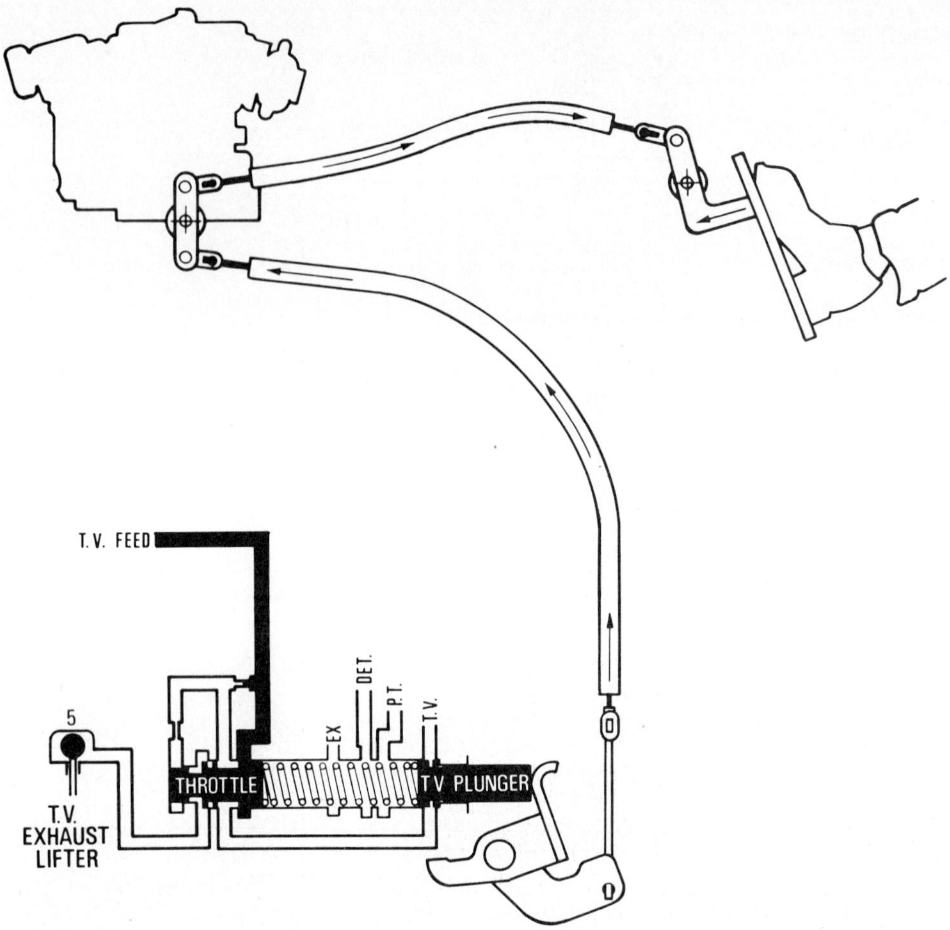

Figure 6-25. Accelerator-linkage-operated throttle valve system. (*Courtesy of General Motors Corporation*)

pressure to delay upshifts and provide increased apply pressure for clutches and bands.

Both mechanical and vacuum modulator systems have been used extensively by transmission manufacturers to provide an engine-torque–related hydraulic pressure to the transmission. Both systems have worked well. Some mechanical systems require accurate linkage adjustments to maintain proper shift timing and quality. A linkage adjustment causing too high a throttle pressure will delay upshifts and cause a harsher, more pronounced shift feel. Linkage adjustment causing too low a throttle pressure results in early upshifts and mushy shifts.

The vacuum modulator system is dependent on a leakfree vacuum supply and a leakfree vacuum modulator diaphragm. A vacuum leak anywhere in the system will result in abnormally high throttle pressure with the same results as in the mechanical system. A ruptured or leaking vacuum diaphragm can also result in transmission fluid being drawn into the engine intake system through the vacuum supply line.

The mechanical linkage throttle system is also used to actuate the downshift or kickdown valve. A torque demand or to-detent downshift does not result in enough downshift valve movement to open the kickdown passage and is dependent on throttle pressure increase only to cause a downshift. A through-detent or WOT (wide-open throttle) downshift moves the kickdown valve far enough to cause a downshift unless road speed and governor pressure are high enough to prevent this.

The vacuum modulator throttle system requires a mechanical linkage to operate the downshift valve. Some transmissions use a throttle-linkage–operated switch, which completes the circuit to an electric solenoid in the transmission. The solenoid plunger in turn operates the downshift valve.

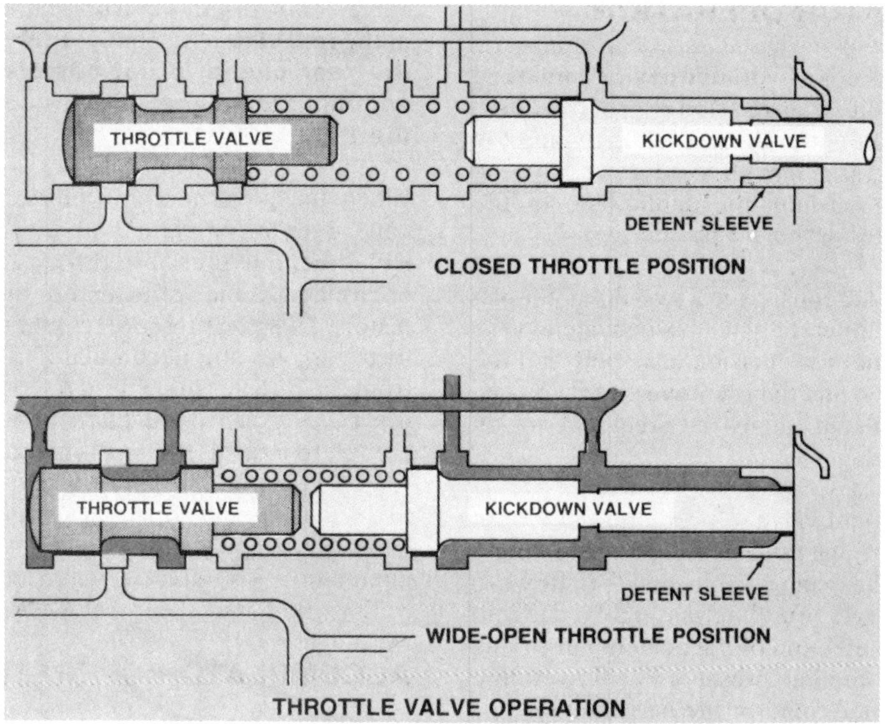

Figure 6-26. (*Courtesy of Chrysler Corporation*)

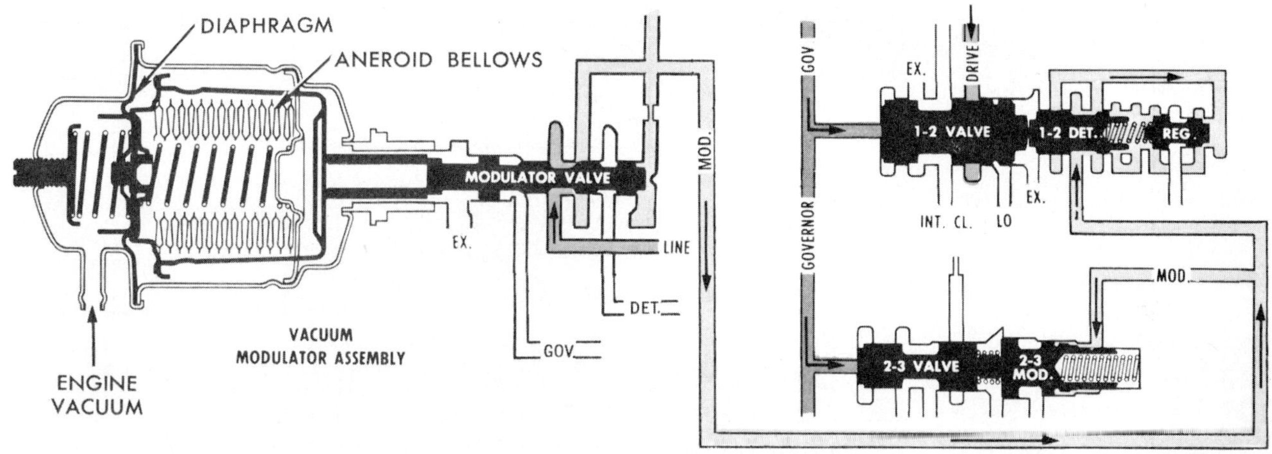

Figure 6-27. Vacuum-operated modulator valve system. (*Courtesy of General Motors Corporation*)

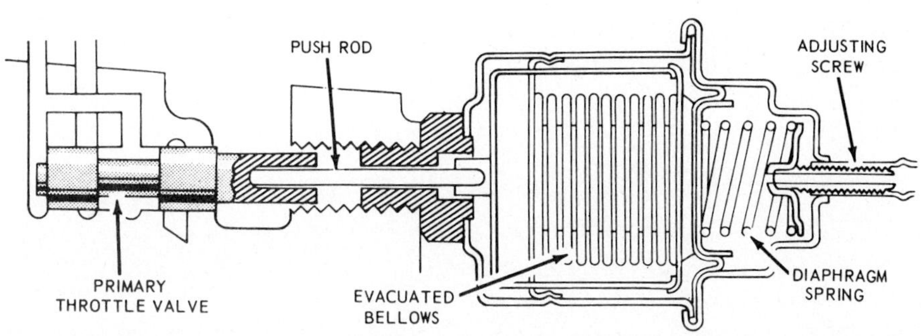

Figure 6-28. Altitude-compensating vacuum diaphragm and throttle valve arrangement. (*Courtesy of Ford Motor Company*)

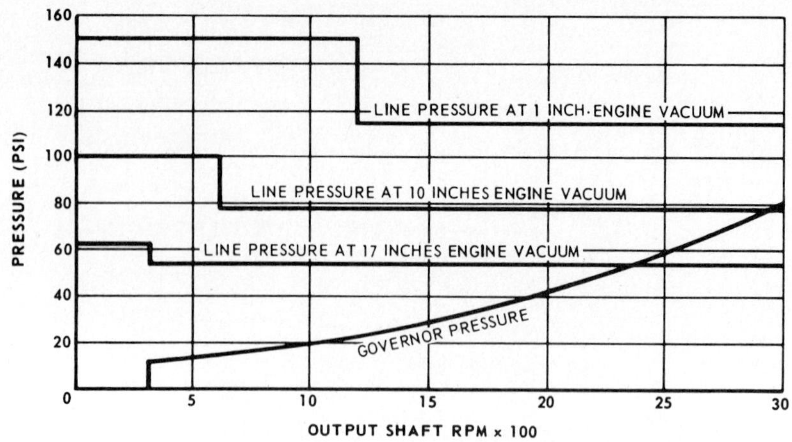

Figure 6-29. Control line pressure and governor pressure relationships to output shaft speed. (*Courtesy of Ford Motor Company*)

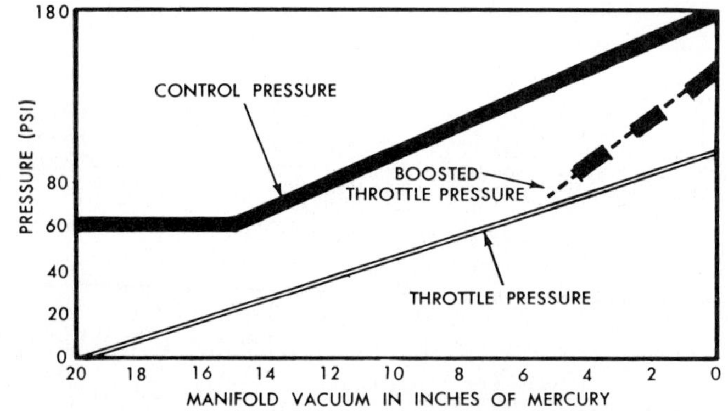

Figure 6-30. Relationship between engine manifold vacuum and transmission control pressure. (*Courtesy of Ford Motor Company*)

PART 6-9 FOURTH-GEAR OVERDRIVE CONTROL (3-4 SHIFT VALVE)

Four-speed overdrive automatic transmissions use a 3–4 shift valve system to shift the transmission from third-gear (direct) to fourth-gear overdrive. The 3–4 shift valve system is used to apply the appropriate bands or clutches (depending on transmission design) for an upshift into overdrive and to release them for a downshift.

The 3–4 shift system is sensitive to throttle pressure and governor pressure in the same way as the 1–2 and 2–3 shift systems. The 3–4 shift system controls line pressure to the clutches and bands to cause upshifts and downshifts.

Several different systems are used on different makes and models of transmissions. For detailed operations of these systems, see Chapter 7.

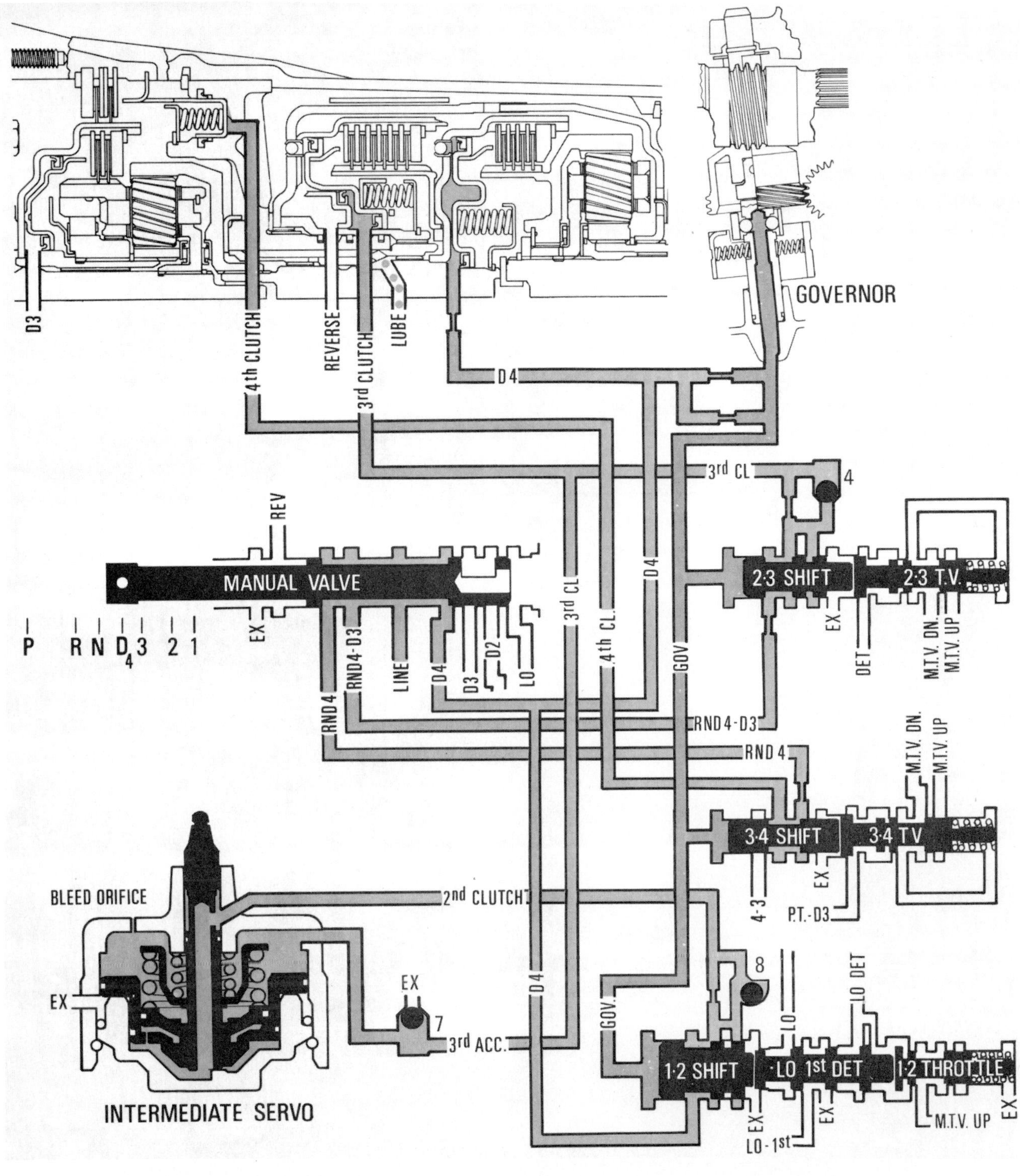

Figure 6-31. GM 200-4R third- to fourth-gear overdrive upshift. (*Courtesy of General Motors Corporation*)

Part 6–10 Worksheet

1. On available vehicles, examine the manual shift linkage systems, both column shift and console shift, and note the differences in range selector positions on the shift quadrant.
2. Using shop service manuals as references to prove your conclusions, identify one vehicle by make, year, and model for each of three different automatic transmission shift patterns and the gear ranges in each quadrant position.
3. Describe the importance of synchronizing shift lever position with manual valve detent position.
4. Study each of the hydraulic control system schematics in this chapter and trace the flow of hydraulic pressure on each chart.

PART 6–11 SELF-CHECK

1. What is the purpose of the transmission case?
2. Why must the case be vented?
3. Where are the majority of the hydraulic control system fluid passages located?
4. What is the purpose of manual low in a transmission?
5. Explain the difference between a torque demand downshift and a through-detent downshift.
6. What is the purpose of the main pressure regulator system?
7. Why must line pressure be boosted under certain conditions?
8. State the function of the following items:
   (a) Converter relief valve
   (b) Converter drain back valve
   (c) Manual valve
   (d) 1–2 shift valve
   (e) 2–3 shift valve
   (f) 3–4 shift valve
   (g) Throttle valve
   (h) Governor valve
   (i) Cutback valve
   (j) Downshift valve
   (k) Accumulator
9. Where is most of the heat generated in an automatic transmission?
10. What is the relationship between vehicle speed and governor pressure?
11. What three hydraulic pressures basically control the action of the shift valves?
12. What effect would delayed release of the direct clutch have on engine operation during a 3–2 downshift?
13. Manually selected second gear results in ______________ gear operation only in some transmissions and in ____________ and ____________ gears in others.
14. What is the effect of a vacuum leak in the modulator system on transmission operation?
15. What effect does excessive throttle pressure have on transmission operation?
16. What effect does poor engine condition have on transmission shifts?

PART 6–12 PERFORMANCE EVALUATION

After sufficient study of this chapter and the appropriate training models, did you perform the following tasks as required by your instructor?

1. State the purpose of and describe the basic construction of the transmission case.
2. State the purpose of the transmission cooling and lubrication systems, and describe the basic construction and operation of each of these systems.
3. State the purpose of the pressure supply and regulation system, and describe the basic construction and operation of each of the components in this system.
4. Describe the purpose, construction, and operation of the manual control system, including the selector lever, linkage, manual control valve, shift gate and detent, and the neutral and backup light switches.
5. Describe the purpose, construction, and operation of the 1–2, 2–3, and 3–4 shift valves.
6. Describe the purpose, construction, and operation of the mechanical and vacuum-operated throttle pressure control systems and the downshift system.
7. Describe the purpose, construction, and operation of the governor control system.
8. Describe the basic purpose, construction, and operation of the shift timing and shift quality control devices.
9. Complete the worksheet as required.
10. Complete the self-check in this chapter with at least 80 percent accuracy.

Chapter 7

Hydraulic System Schematics

Performance Objectives

After sufficient study of the preceding chapters as well as this chapter and the appropriate training models, the student should be able to use these schematics and others in the appropriate manufacturers' service repair manuals to perform these tasks:

1. Describe hydraulic system operation in any gear position.
2. Describe the operation of all components in the hydraulic system.
3. Diagnose possible malfunctions in the hydraulic control system.
4. Complete the self-check for this chapter with at least 80 percent accuracy.
5. Evaluate performance using the performance evaluation section at the end of this chapter.

PART 7-1 INTRODUCTION

This chapter is a collection of some of the more common automatic transmission hydraulic control system schematics. It is not intended to be nor could it be complete and exhaustive in a text of this scope. There is simply too much material for this to be attempted.

The intent of this chapter is to provide a number of common hydraulic system schematics, which should be studied circuit by circuit in order to be meaningful. Hydraulic system schematics are like city street maps in one sense since each street must be identified and followed individually in order to become completely familiar with them all. Only then does the entire map become meaningful and useful. In the same way each hydraulic circuit must be followed in each gear position in order to become acquainted with them all.

Presenting these schematics in this text allows an entire class to follow them together as a group and allows the individual student to study them independently without requiring a great number of manufacturers' service manuals for the entire class.

The hydraulic schematics presented in this chapter are for the following transmissions and transaxles.

PART 7-2 GENERAL MOTORS (COURTESY OF GENERAL MOTORS CORPORATION

- 125 and 125C transaxle (Figures 7–1 to 7–15)
- 200 — 4R overdrive transmission (Figures 7–16 to 7–29)
- 700 — R4 overdrive transmission (Figures 7–30 to 7–43)
- 325 — 4L transaxle (Figures 7–44 to 7–57)

PART 7-3 FORD (COURTESY OF FORD MOTOR COMPANY)

- ATX transaxle (Figures 7–62 to 7–71)
- AOD overdrive transmission (Figures 7–72 to 7–84)
- C5 transmission (schematics in Chapter 6) (pages 111 to 133)

PART 7-4 CHRYSLER (COURTESY OF CHRYSLER CORPORATION)

- A404, A413, and A470 transaxle (Figures 7–88 to 7–97)
- A904 and A727 TorqueFlite transmission (Figures 7–98 to 7–115)

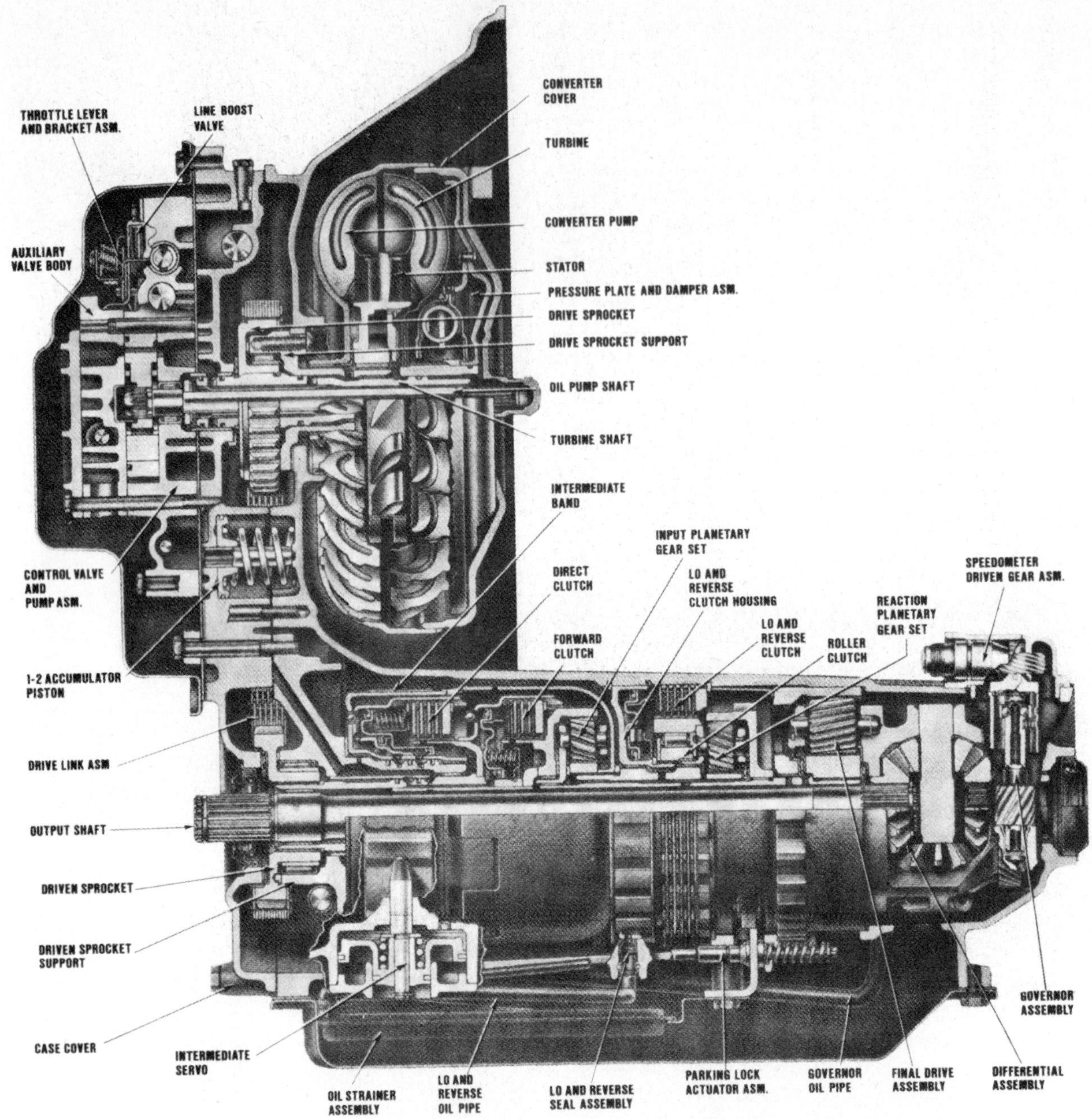

Figure 7-1. GM 125C automatic transaxle. (*Courtesy of General Motors Corporation*)

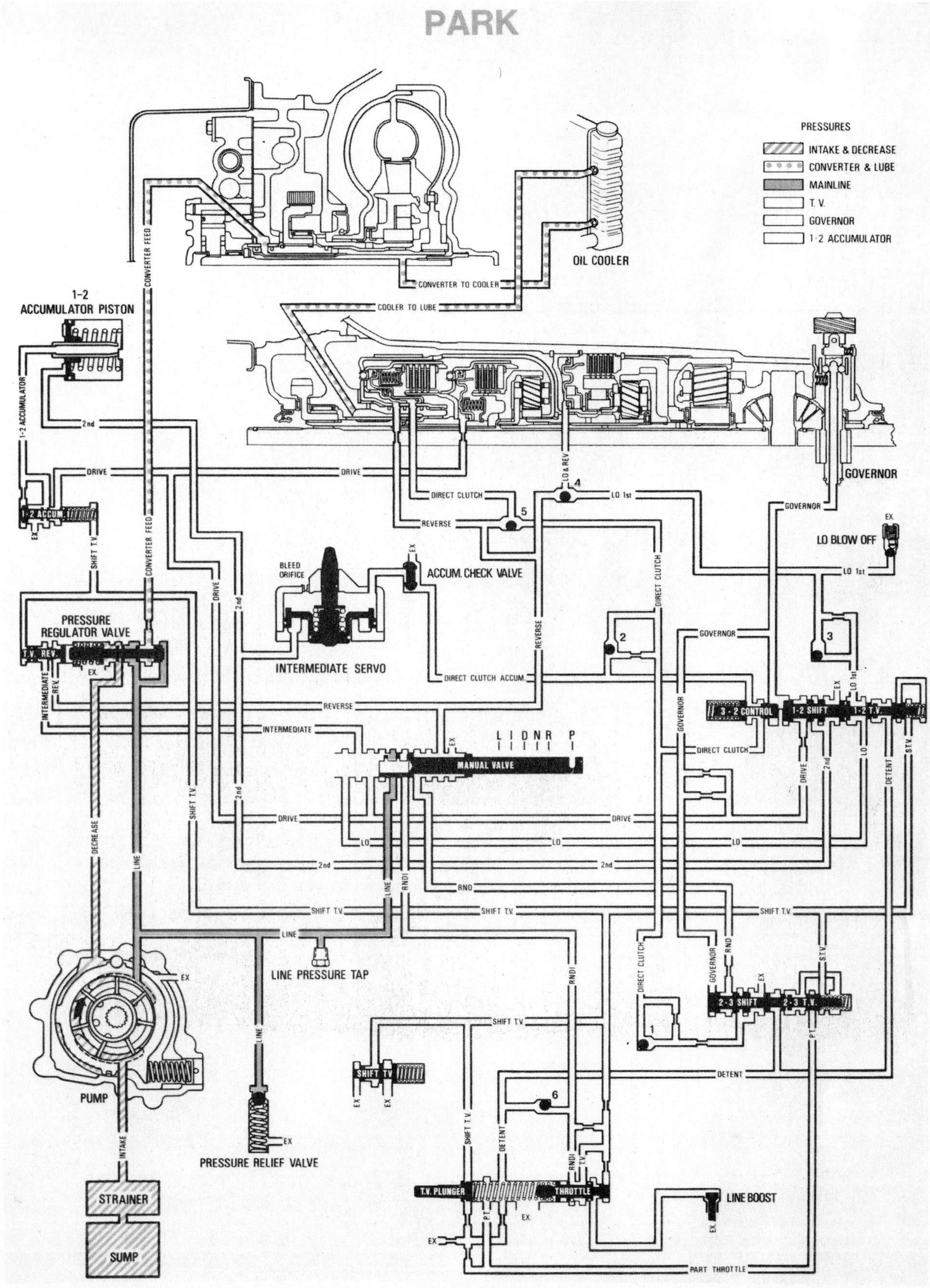
PARK
PRESSURES
INTAKE & DECREASE
CONVERTER & LUBE
MAINLINE
T.V.
GOVERNOR
1-2 ACCUMULATOR
OIL COOLER
CONVERTER TO COOLER
COOLER TO LUBE
CONVERTER FEED
1-2 ACCUMULATOR PISTON
1-2 ACCUMULATOR
2nd
DRIVE
1-2 ACCUM.
SHIFT T.V.
GOVERNOR
DIRECT CLUTCH
LO & REV
LO 1st
REVERSE
LO BLOW OFF
BLEED ORIFICE
ACCUM. CHECK VALVE
PRESSURE REGULATOR VALVE
T.V. REV.
INTERMEDIATE SERVO
DIRECT CLUTCH ACCUM.
INTERMEDIATE
3-2 CONTROL
1-2 SHIFT
1-2 T.V.
L I D N R P
MANUAL VALVE
DETENT
S.T.V.
LO
RND
RNDI
LINE
DECREASE
LINE PRESSURE TAP
PUMP
PRESSURE RELIEF VALVE
INTAKE
STRAINER
SUMP
2-3 SHIFT
2-3 T.V.
PT
T.V. PLUNGER
THROTTLE
LINE BOOST
PART THROTTLE
EX

Figure 7-2.

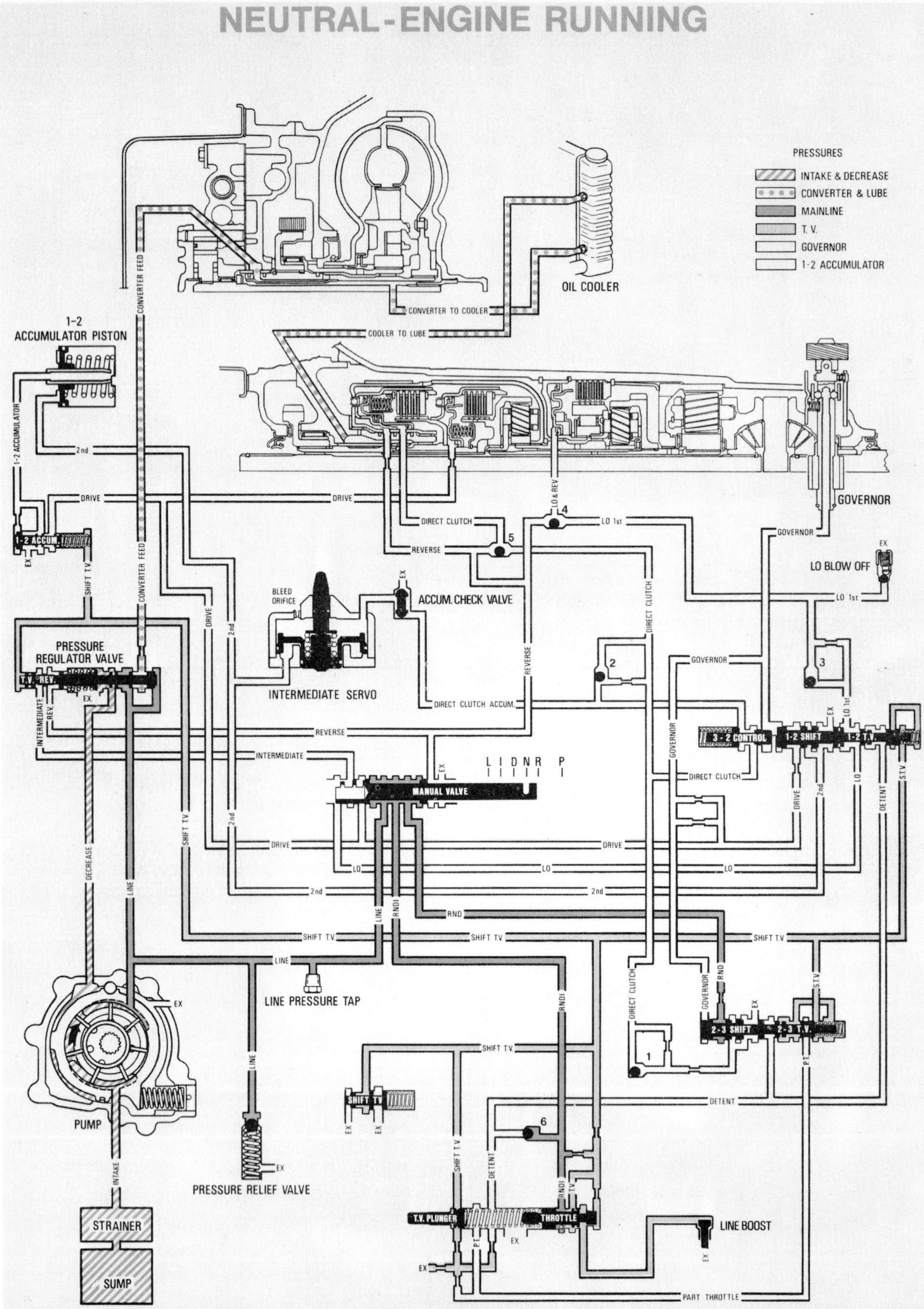
NEUTRAL-ENGINE RUNNING
PRESSURES
INTAKE & DECREASE
CONVERTER & LUBE
MAINLINE
T.V.
GOVERNOR
1-2 ACCUMULATOR
OIL COOLER
CONVERTER FEED
CONVERTER TO COOLER
COOLER TO LUBE
1-2
ACCUMULATOR PISTON
1-2 ACCUMULATOR
2nd
DRIVE
GOVERNOR
DIRECT CLUTCH
REVERSE
LO & REV
LO 1st
LO BLOW OFF
1-2 ACCUM.
SHIFT T.V.
BLEED
ORIFICE
ACCUM. CHECK VALVE
INTERMEDIATE SERVO
DIRECT CLUTCH ACCUM.
PRESSURE
REGULATOR VALVE
T.V.
REV.
INTERMEDIATE
EX
3-2 CONTROL
1-2 SHIFT
1-2 T.V.
DETENT
S.T.V.
L I D N R P
MANUAL VALVE
LO
RND
RNDI
LINE
DECREASE
LINE PRESSURE TAP
PUMP
INTAKE
STRAINER
SUMP
PRESSURE RELIEF VALVE
2-3 SHIFT
2-3 T.V.
SHIFT T.V.
PT
T.V. PLUNGER
THROTTLE
LINE BOOST
PART THROTTLE
1
2
3
4
5
6

Figure 7-3.

DRIVE RANGE-FIRST GEAR

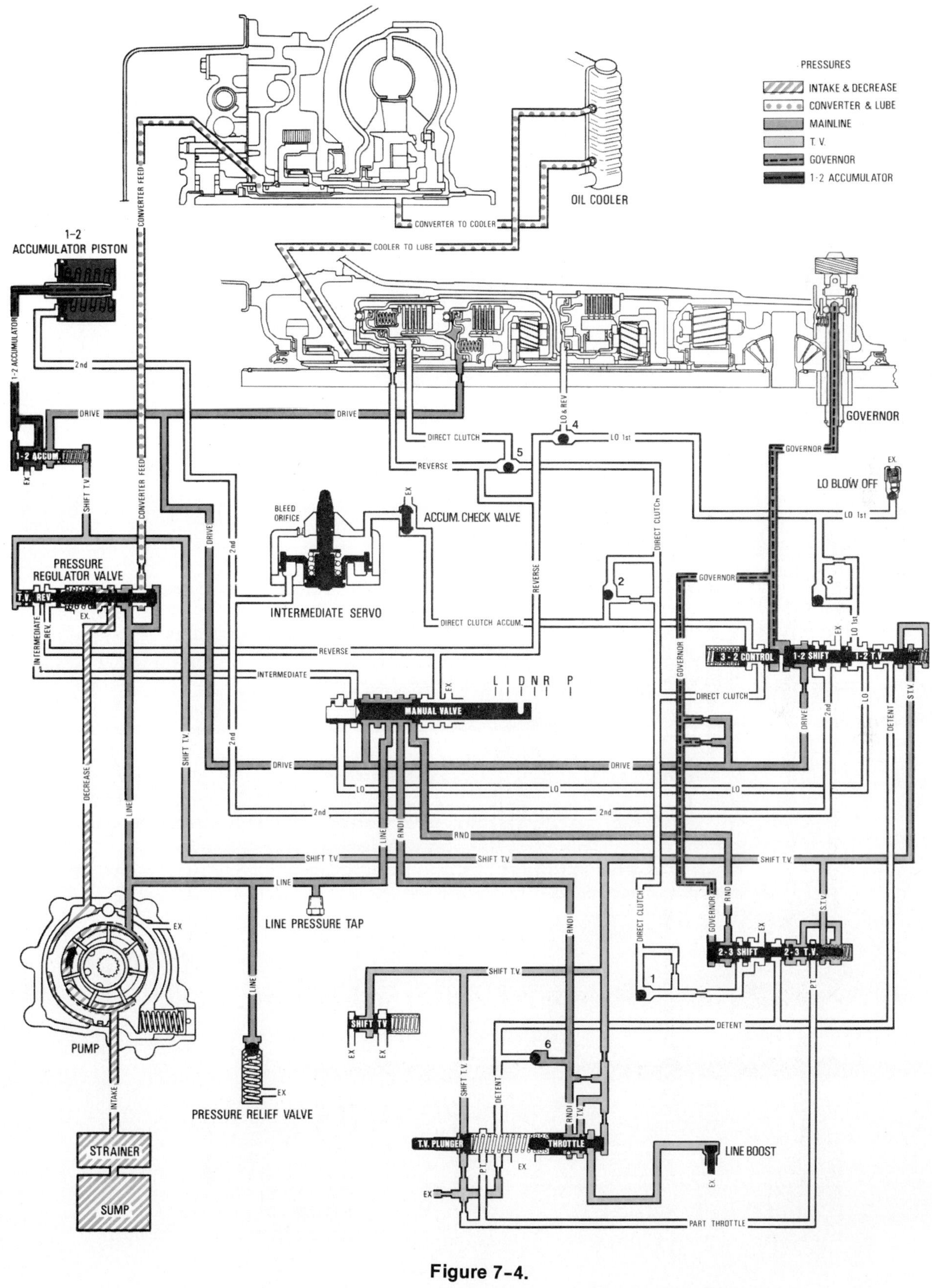

Figure 7-4.

DRIVE RANGE-SECOND GEAR

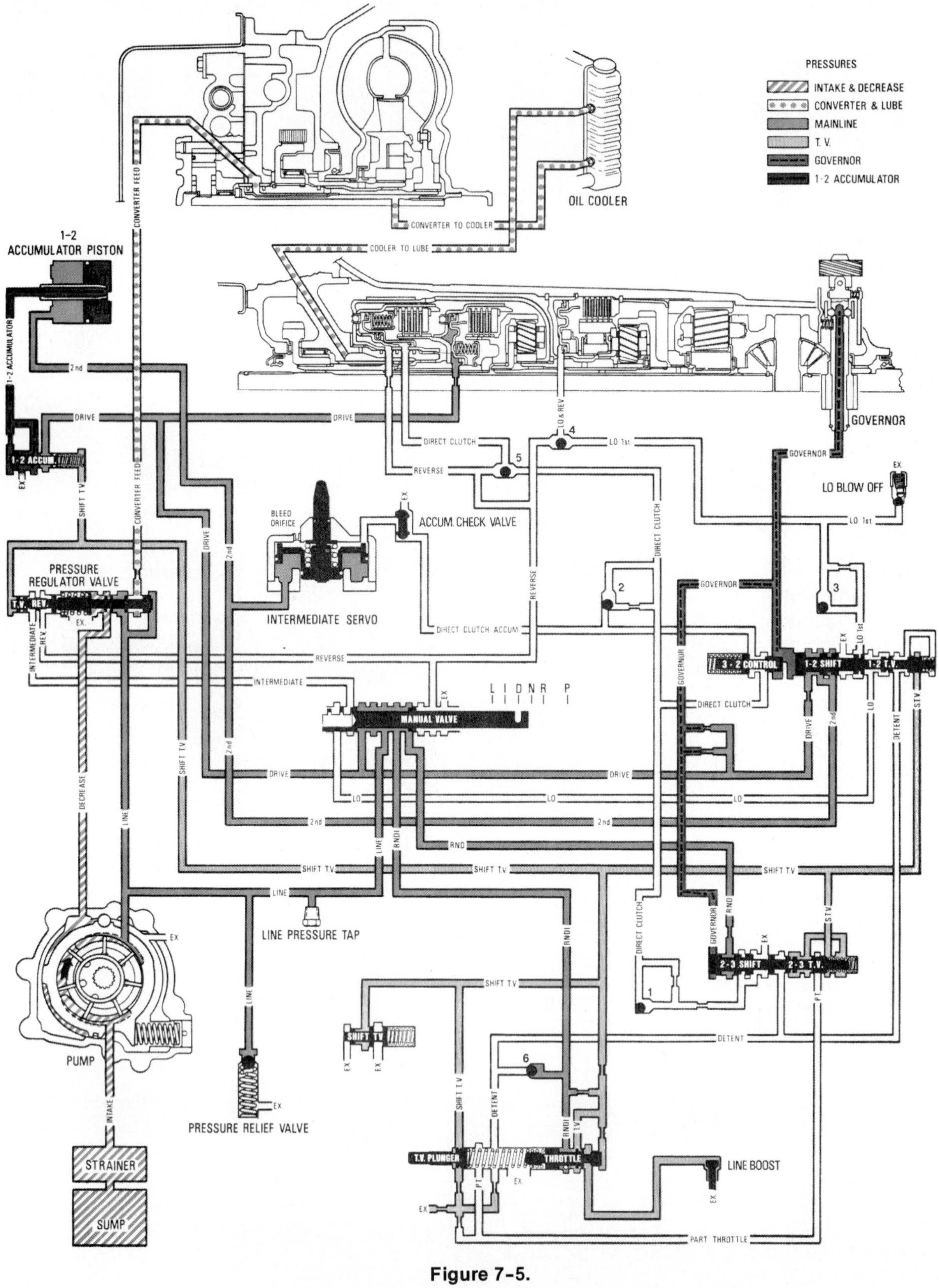

Figure 7-5.

DRIVE RANGE-THIRD GEAR

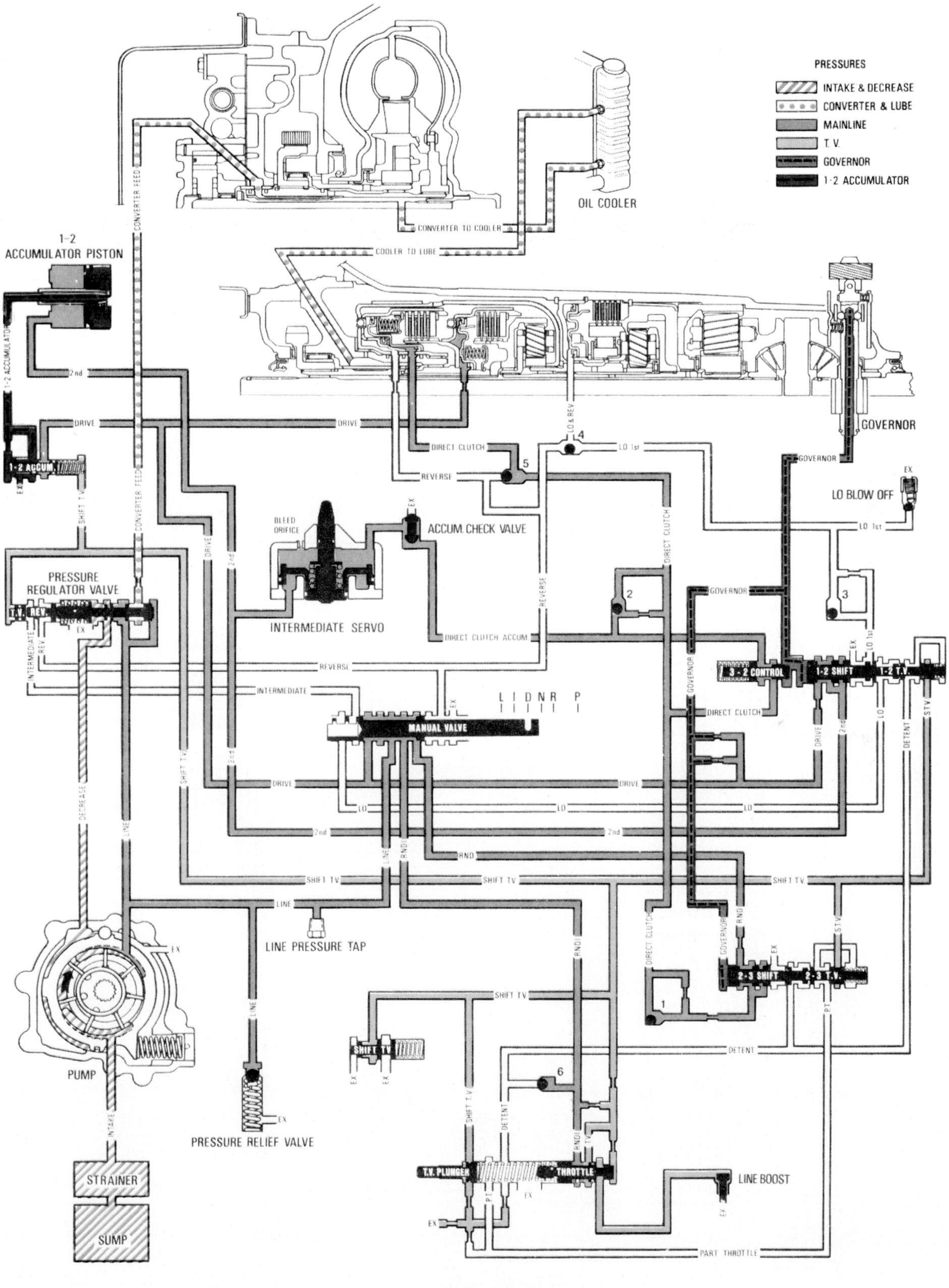

Figure 7-6.

DRIVE RANGE - THIRD GEAR

Converter Clutch Released

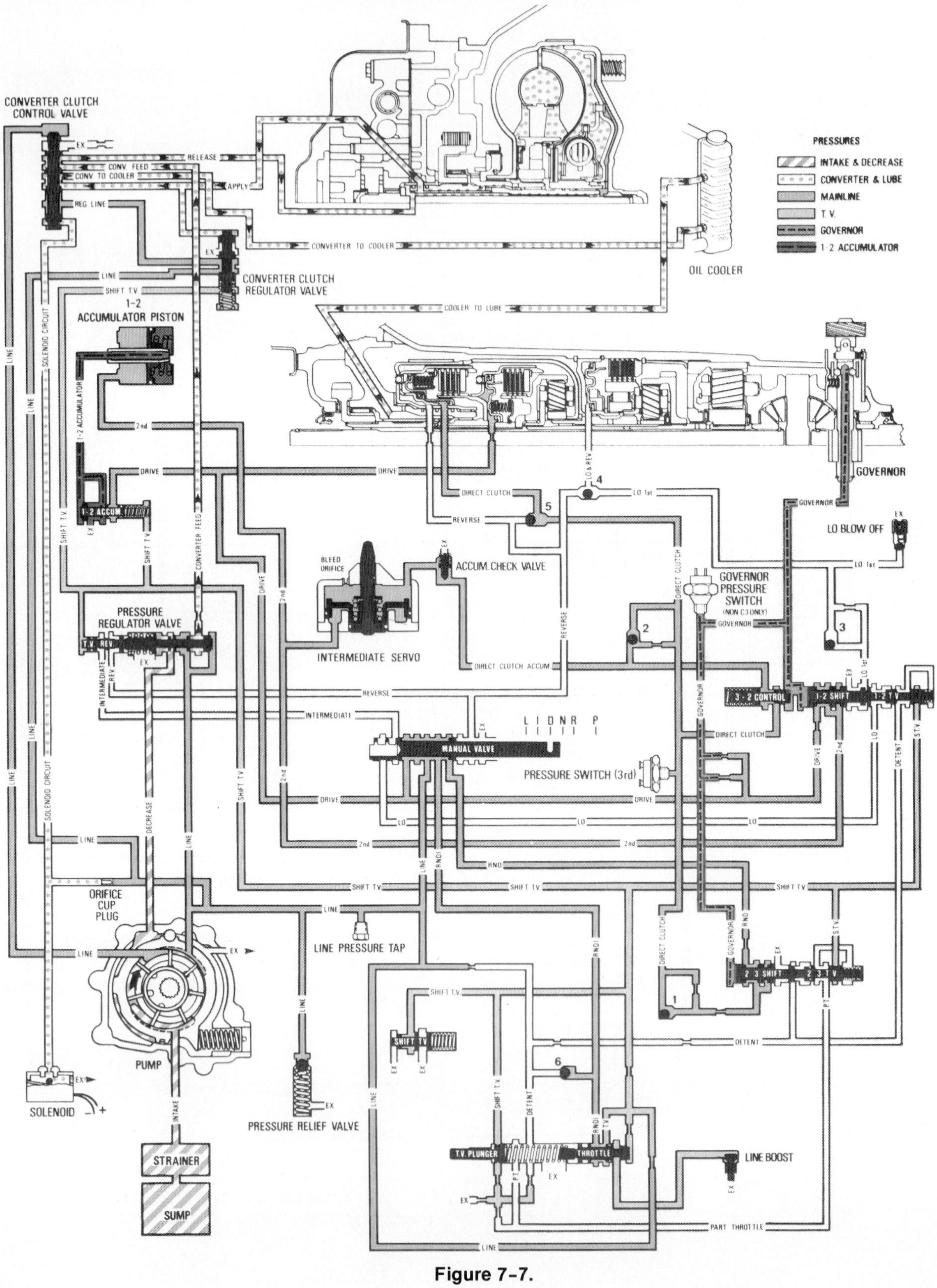

Figure 7-7.

DRIVE RANGE-THIRD GEAR

Converter Clutch Applied

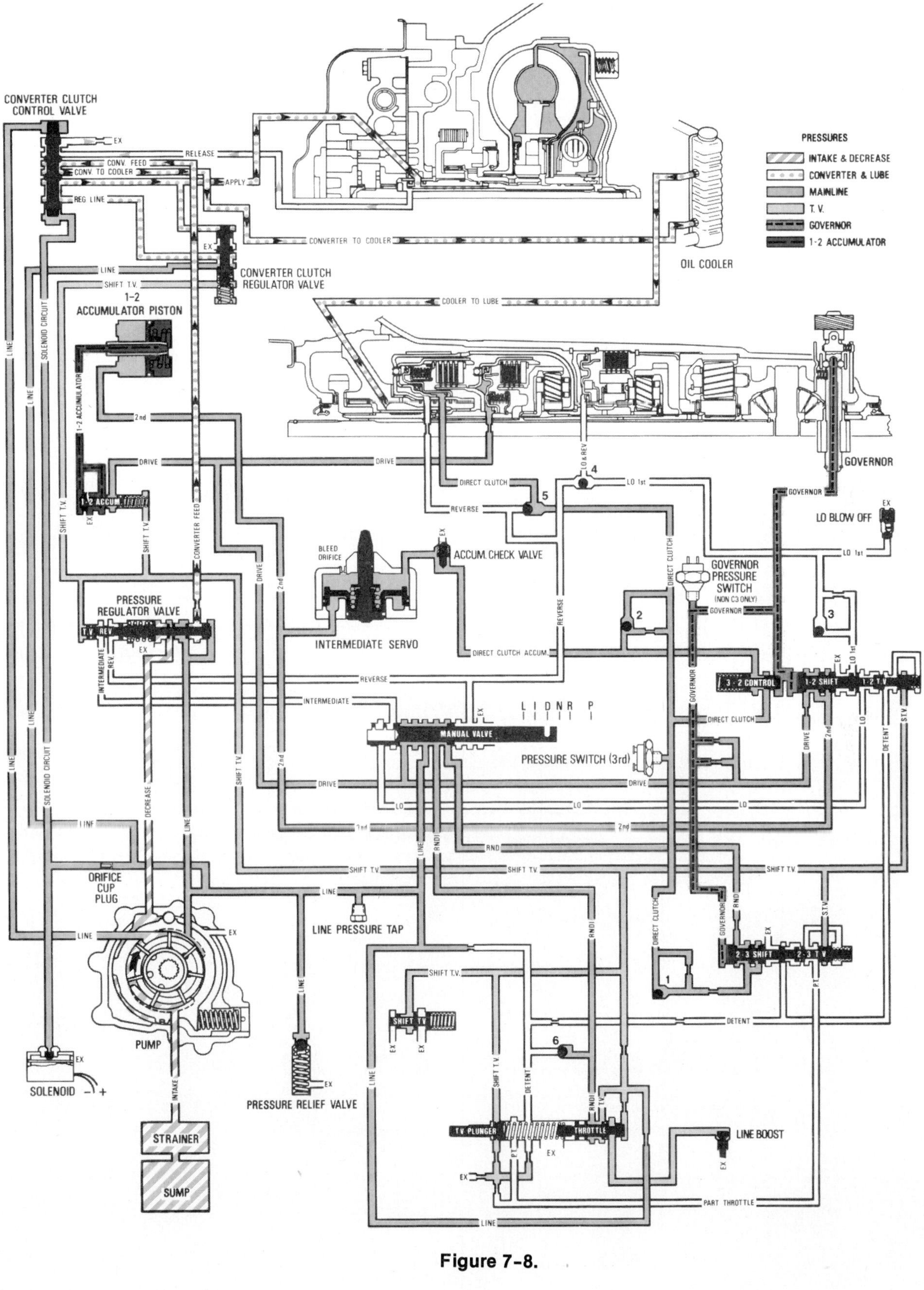

Figure 7-8.

PART THROTTLE 3-2 DOWNSHIFT

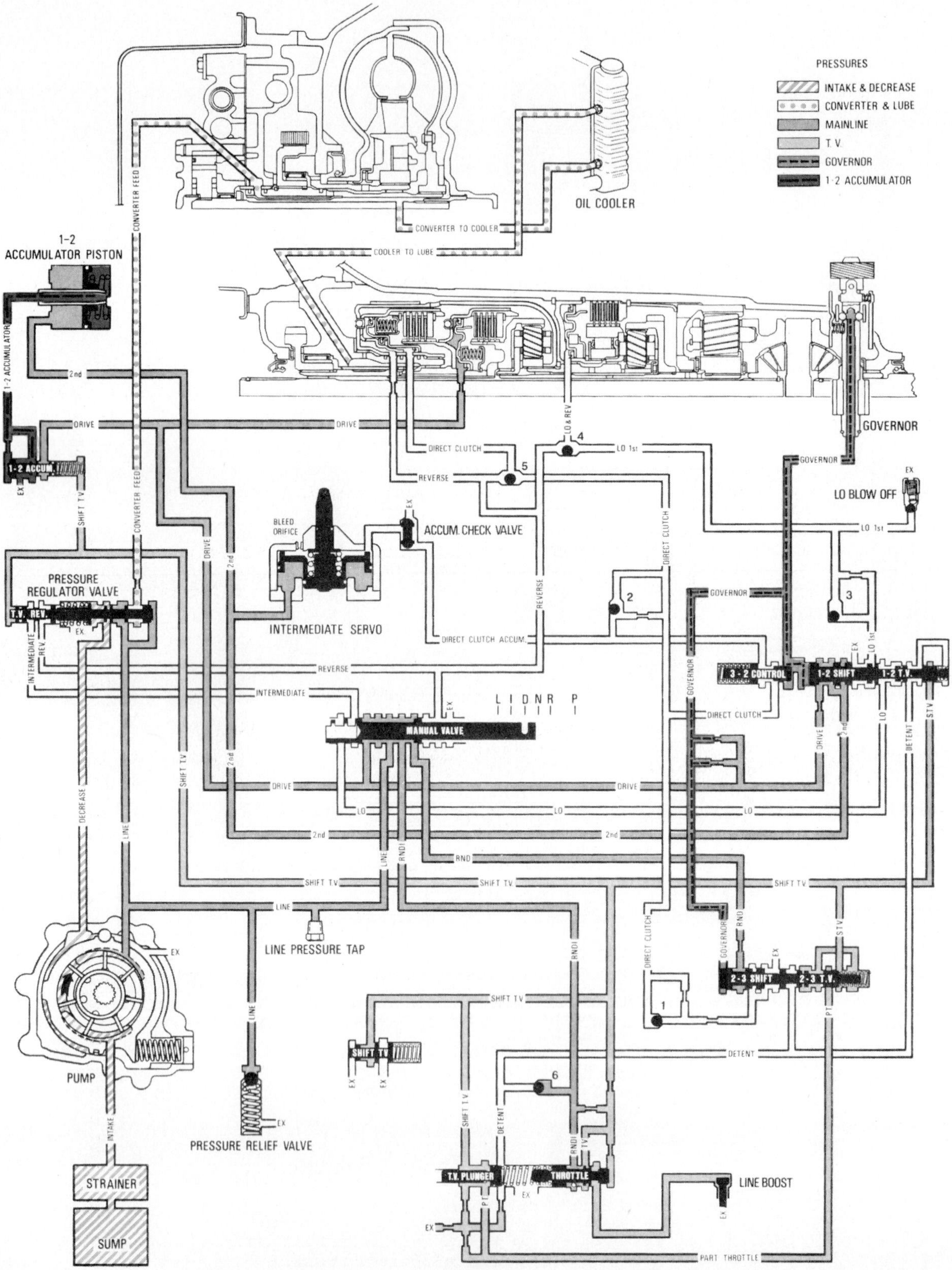

Figure 7-9.

DETENT DOWNSHIFT

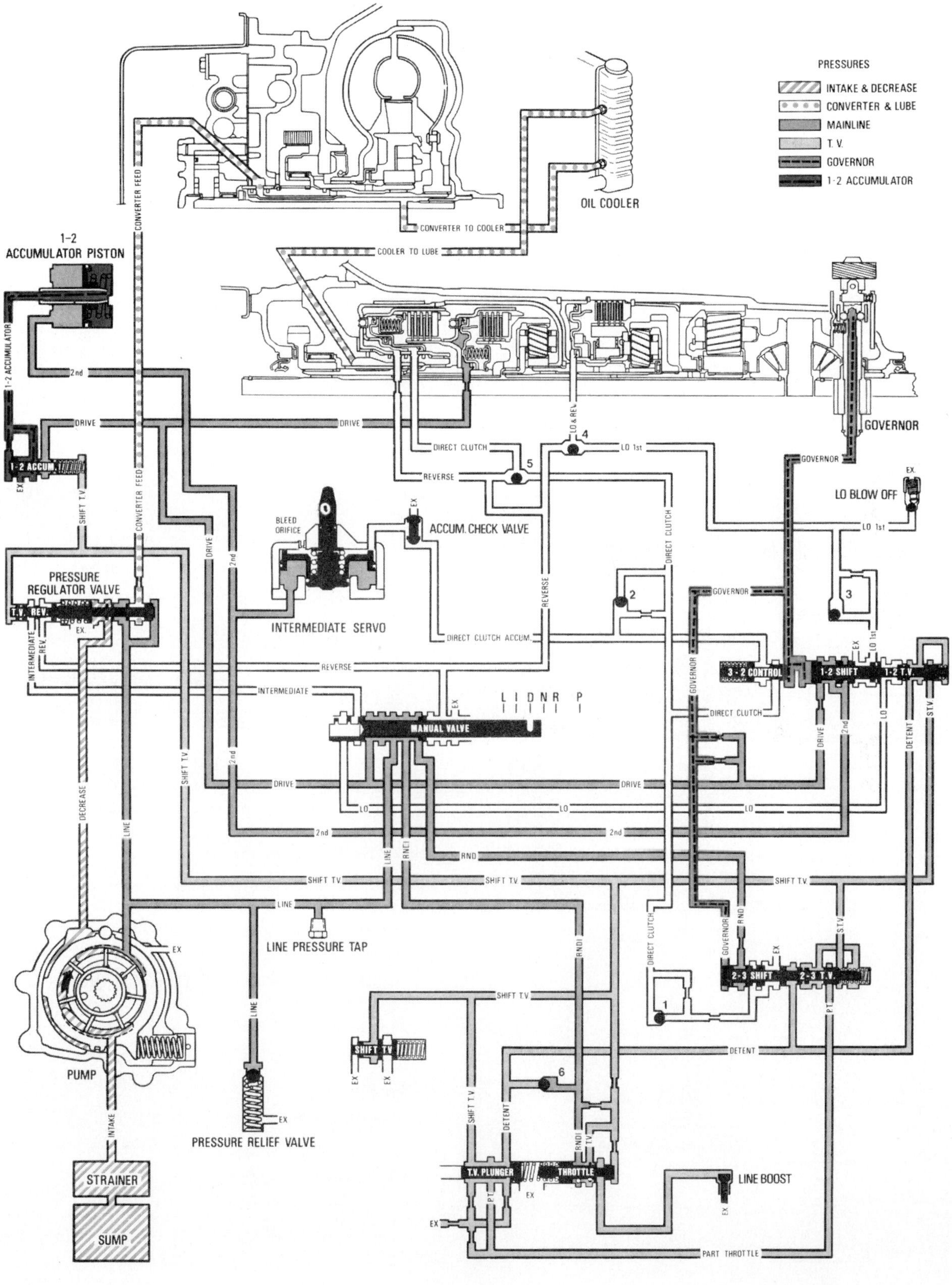

Figure 7-10.

INTERMEDIATE RANGE

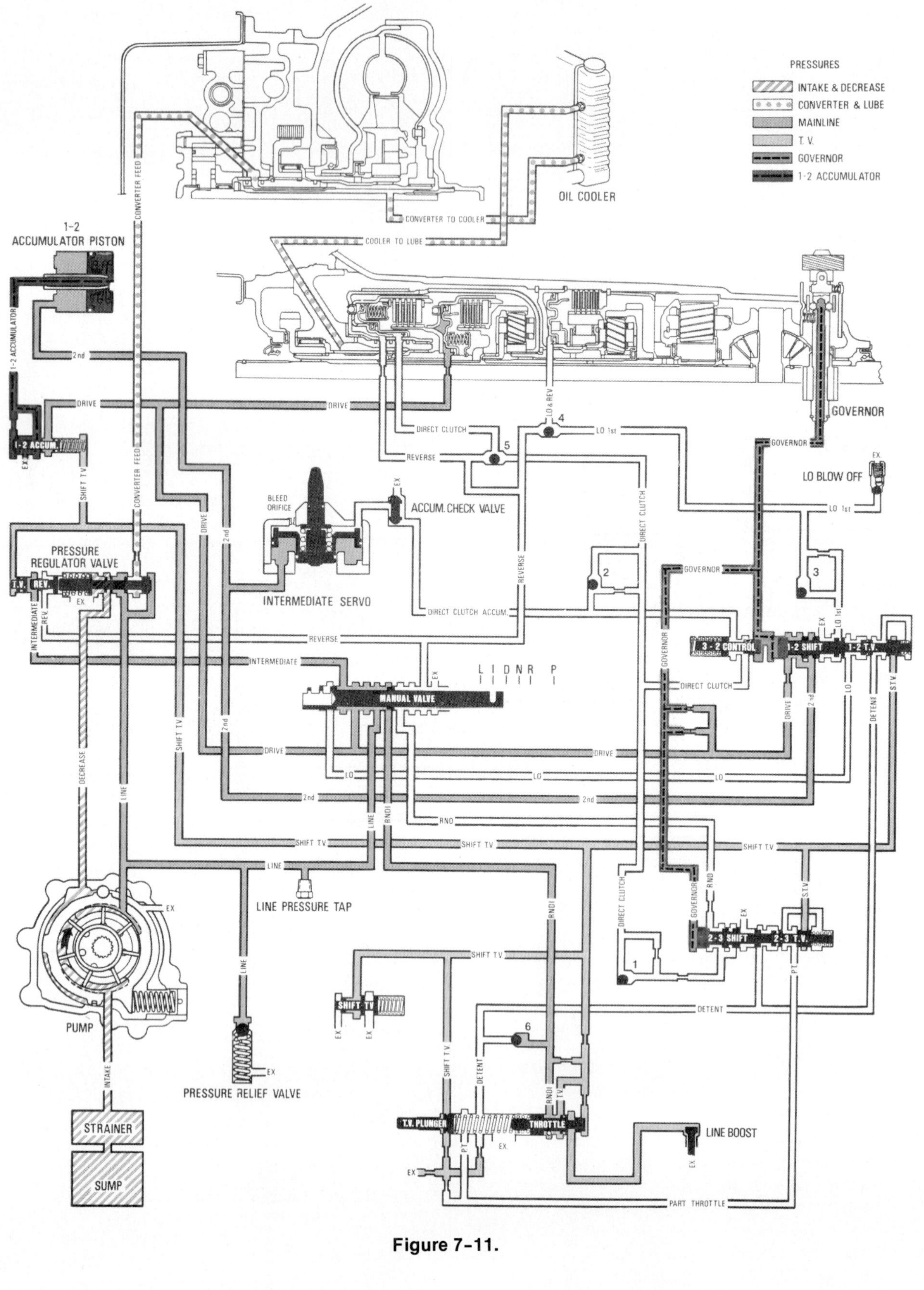

Figure 7-11.

LO RANGE

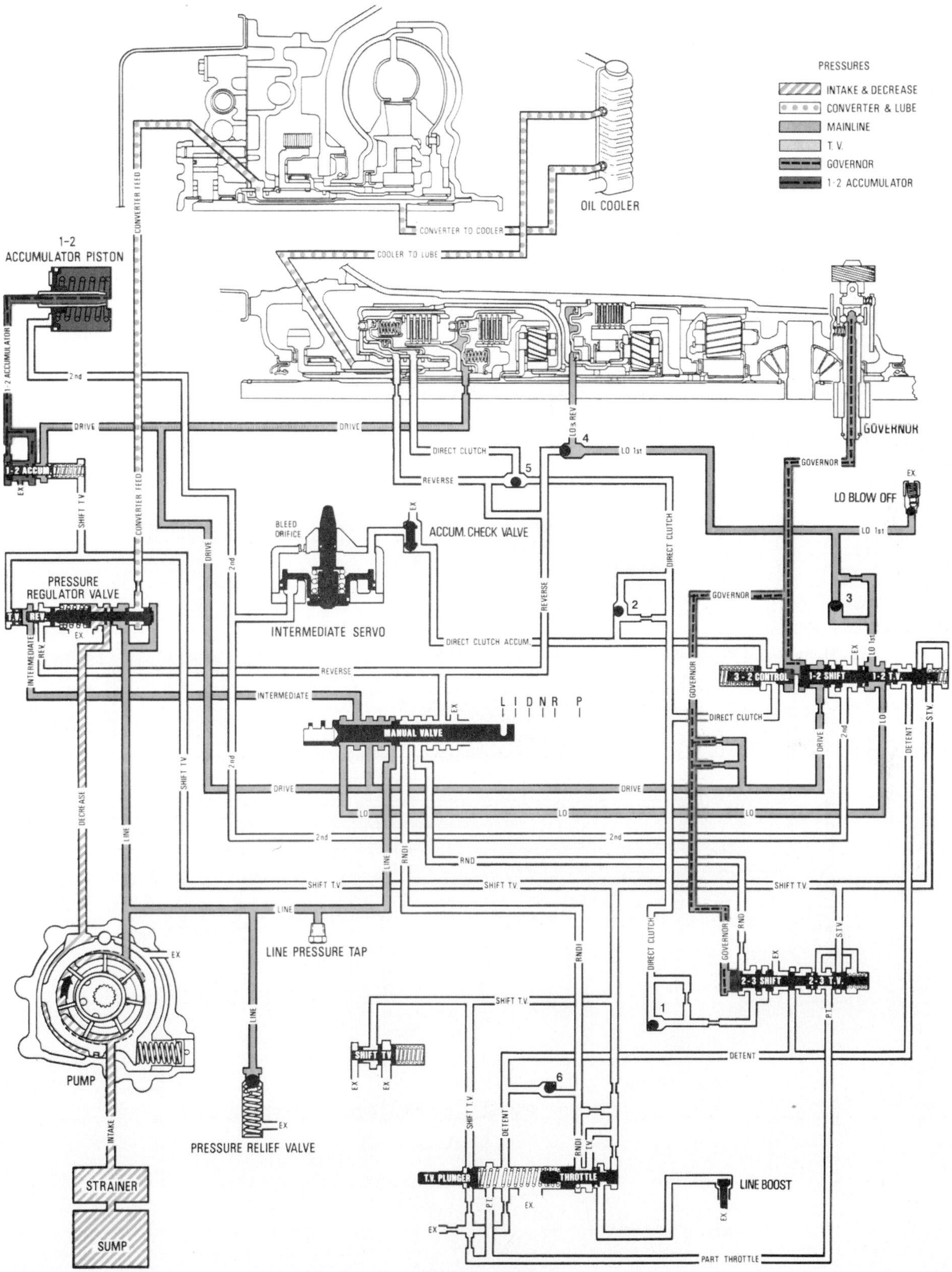

Figure 7-12.

REVERSE

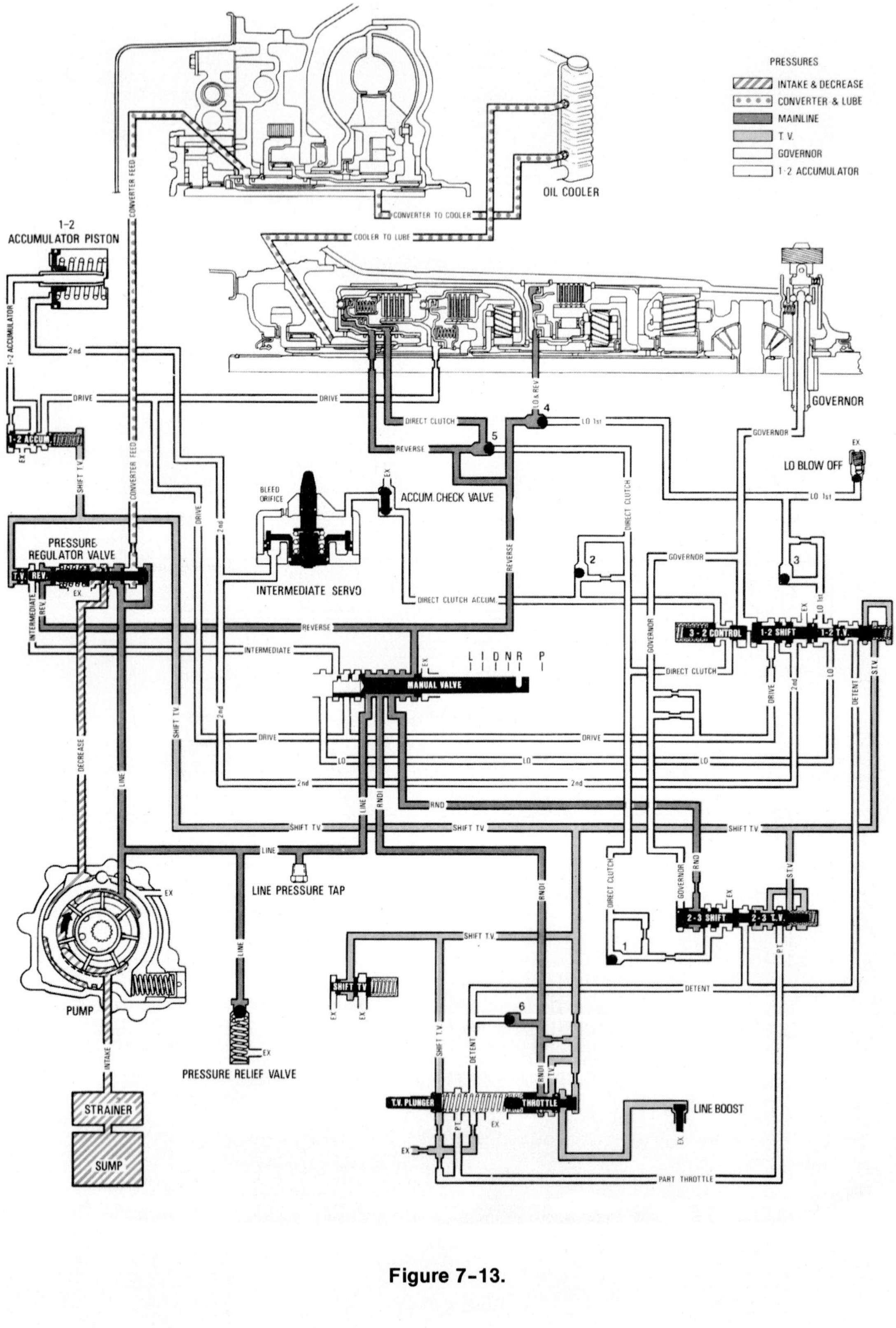

Figure 7-13.

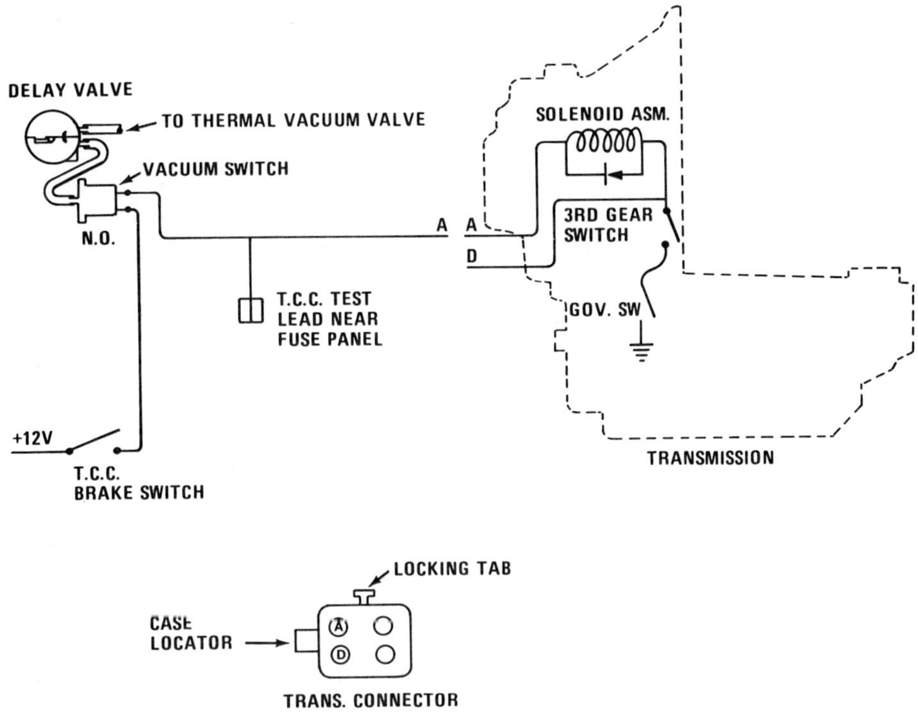

Figure 7-14. Wiring schematic—cars not equipped with Computer Command control (typical).

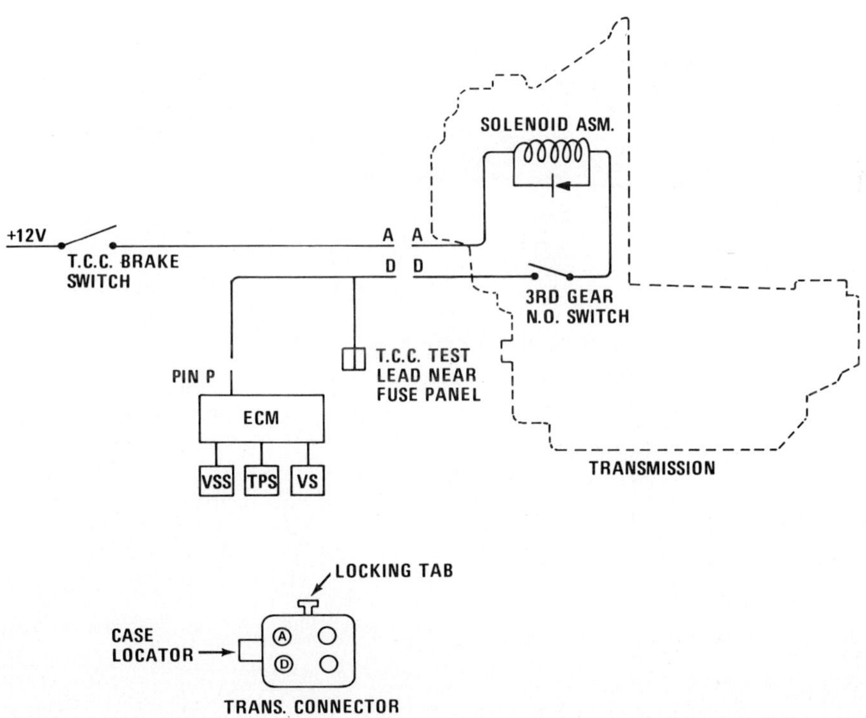

Figure 7-15. Wiring schematic—cars equipped with Computer Command Control.

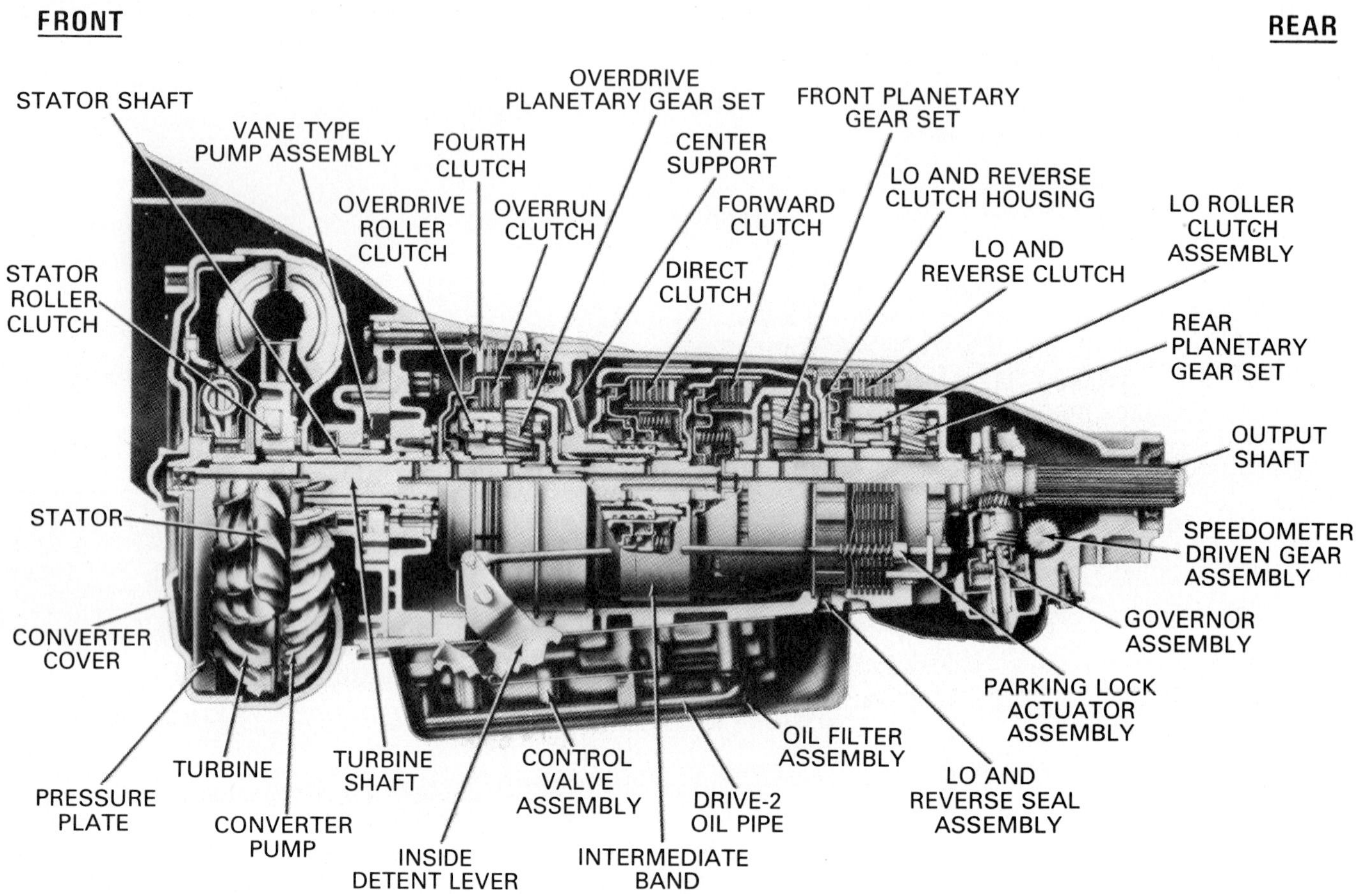

Figure 7-16. Cutaway view, THM 200-4R transmission.

PARK

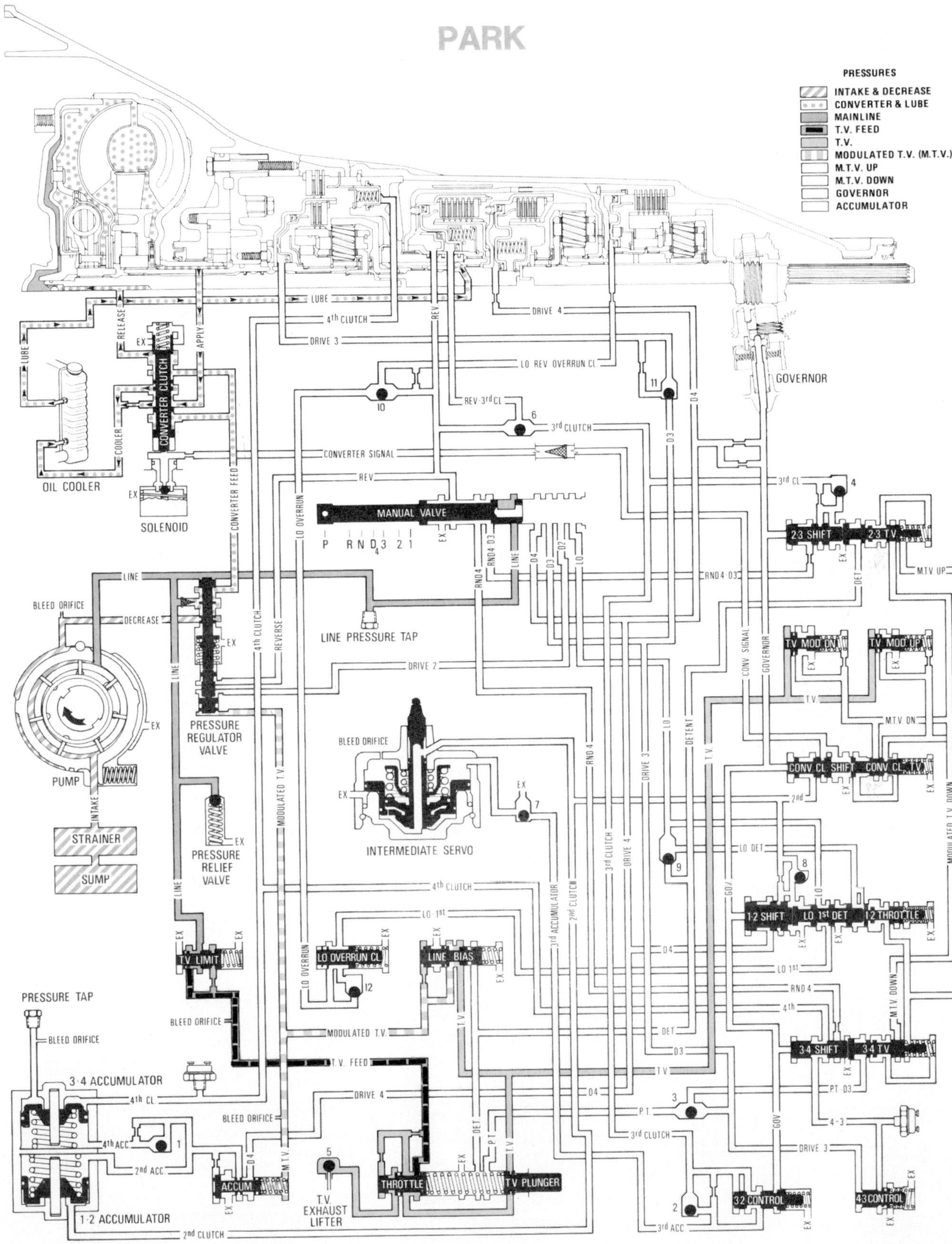

Figure 7-17.

NEUTRAL-ENGINE RUNNING

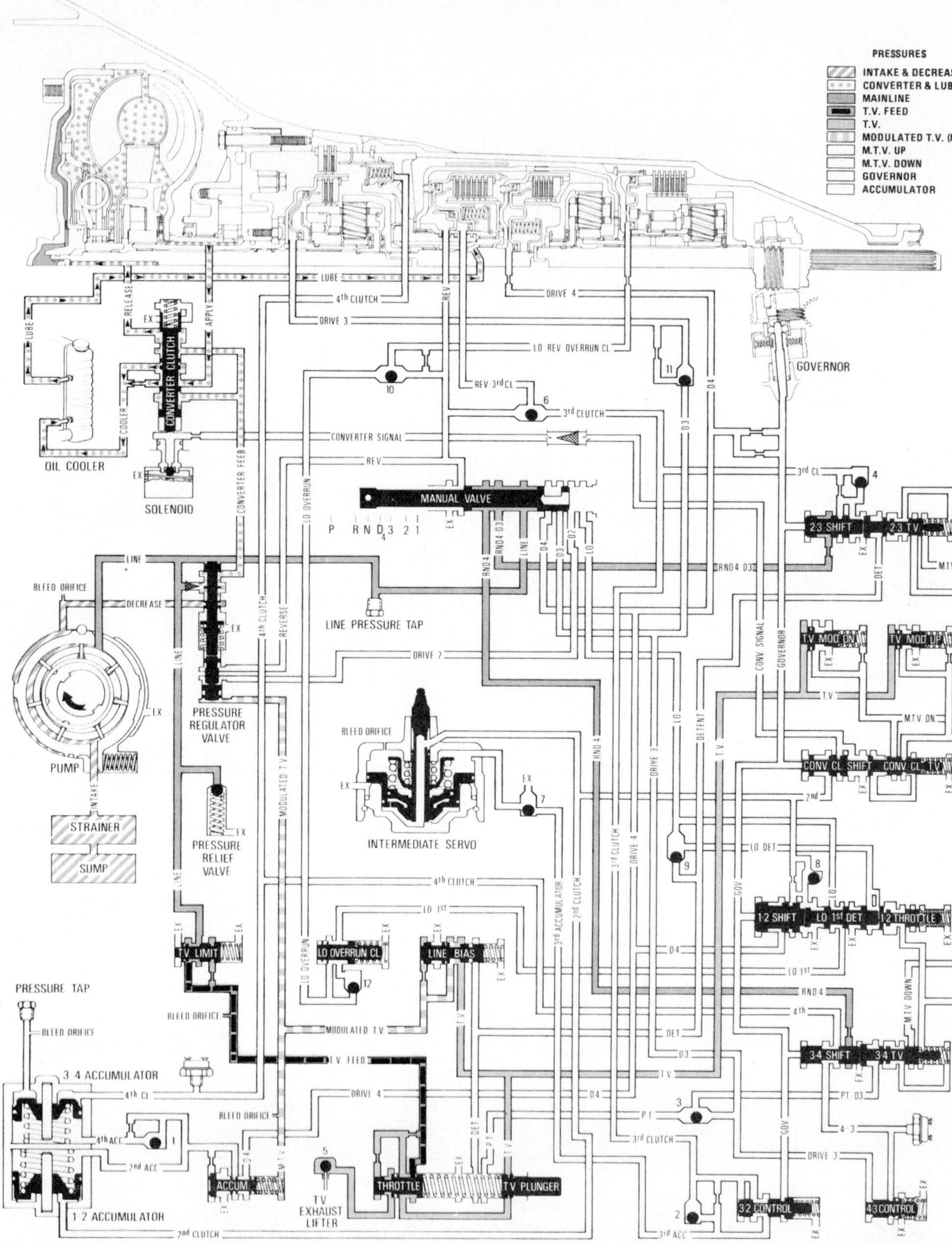

Figure 7-18.

DRIVE RANGE-FIRST GEAR

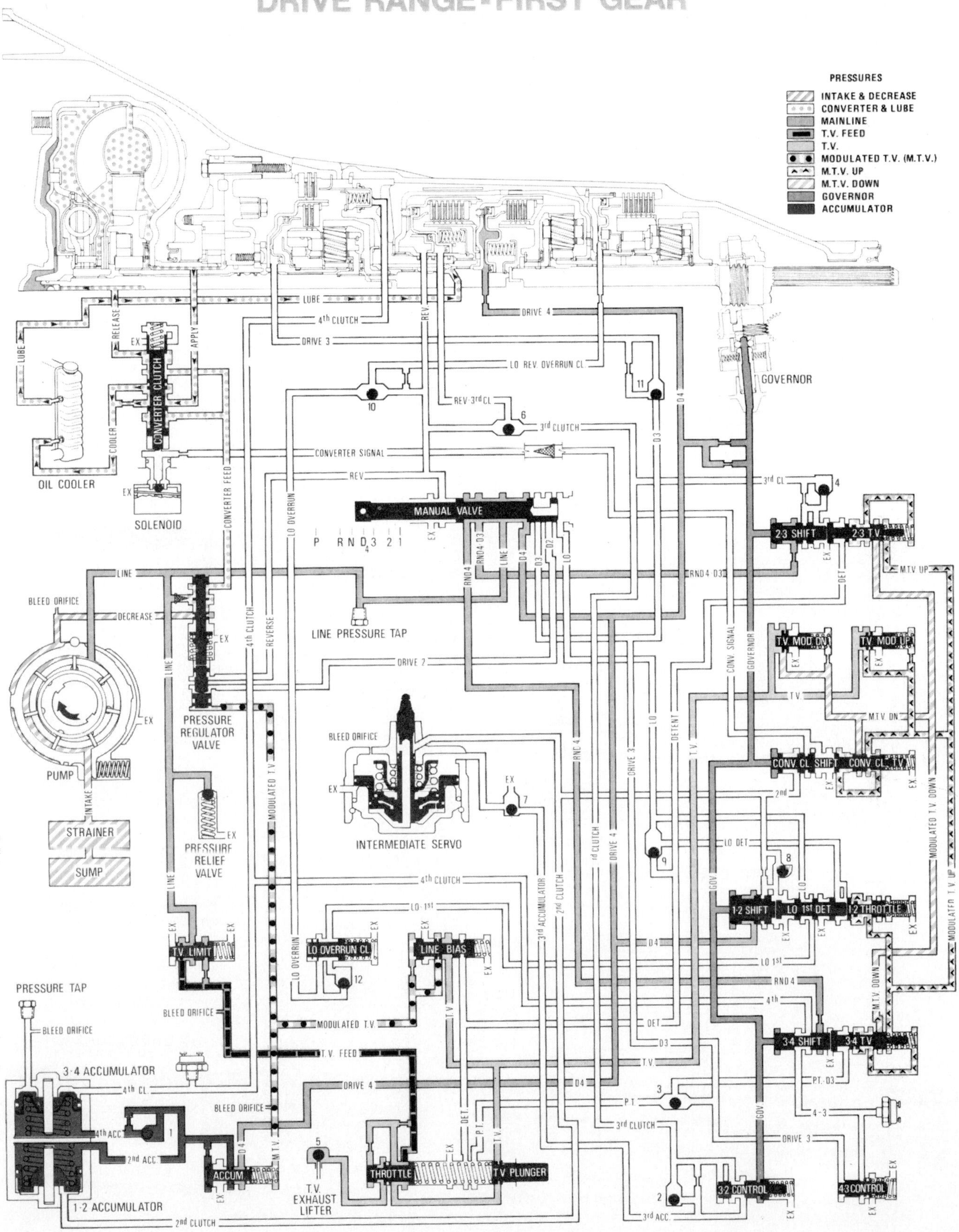

Figure 7-19.

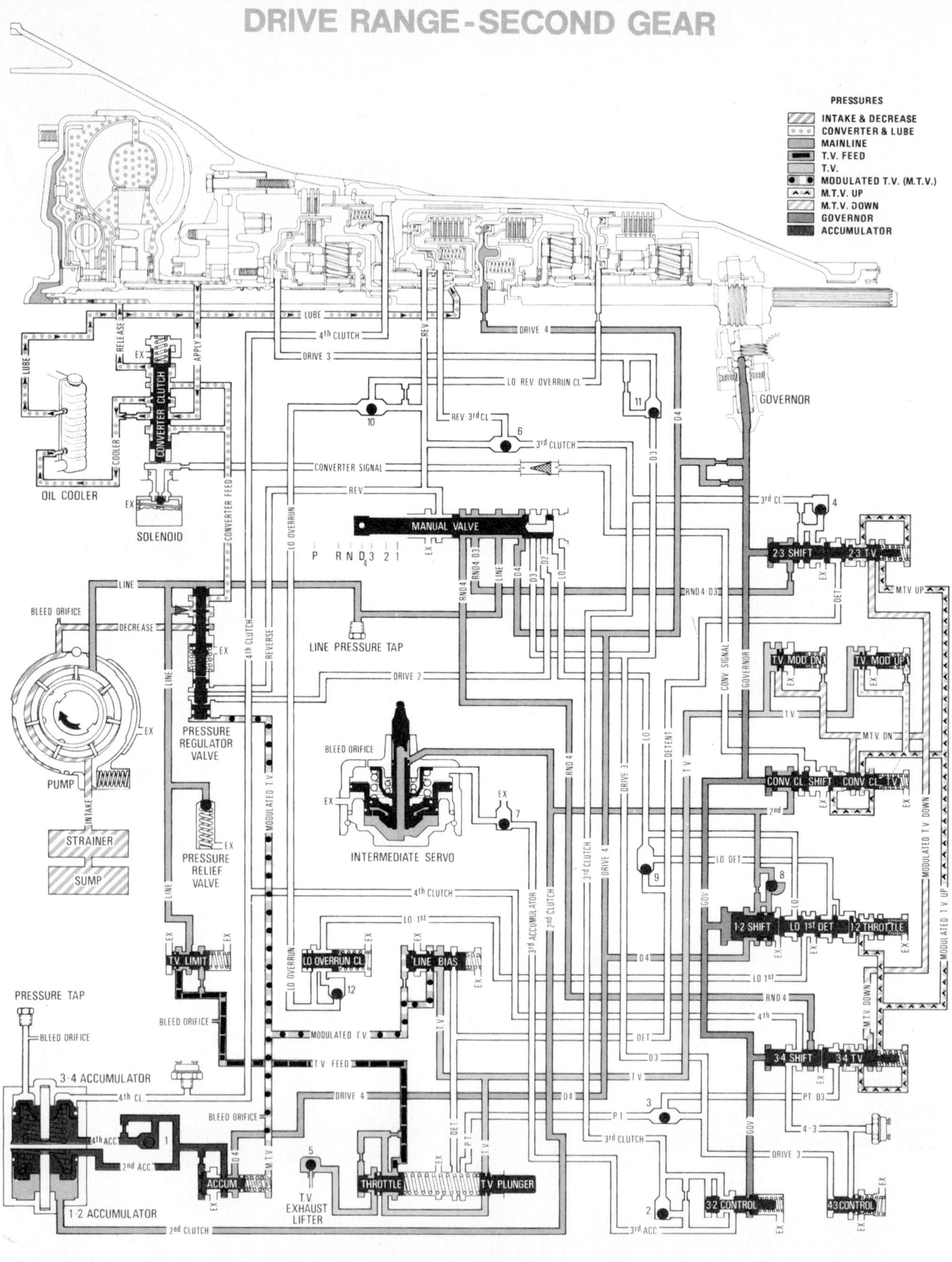
DRIVE RANGE-SECOND GEAR
PRESSURES
INTAKE & DECREASE
CONVERTER & LUBE
MAINLINE
T.V. FEED
T.V.
MODULATED T.V. (M.T.V.)
M.T.V. UP
M.T.V. DOWN
GOVERNOR
ACCUMULATOR
GOVERNOR
OIL COOLER
SOLENOID
CONVERTER CLUTCH
MANUAL VALVE
LINE PRESSURE TAP
PRESSURE REGULATOR VALVE
PUMP
STRAINER
SUMP
PRESSURE RELIEF VALVE
INTERMEDIATE SERVO
BLEED ORIFICE
PRESSURE TAP
3-4 ACCUMULATOR
1-2 ACCUMULATOR
T.V. LIMIT
LO OVERRUN CL
LINE BIAS
ACCUM
THROTTLE
T.V. PLUNGER
T.V. EXHAUST LIFTER
2-3 SHIFT
2-3 T.V.
T.V. MOD DN
T.V. MOD UP
CONV CL SHIFT
CONV CL T.V.
1-2 SHIFT
LO 1st DET
1-2 THROTTLE
3-4 SHIFT
3-4 T.V.
3-2 CONTROL
4-3 CONTROL

Figure 7-20.

DRIVE RANGE—CONVERTER CLUTCH APPLIED

Second Gear

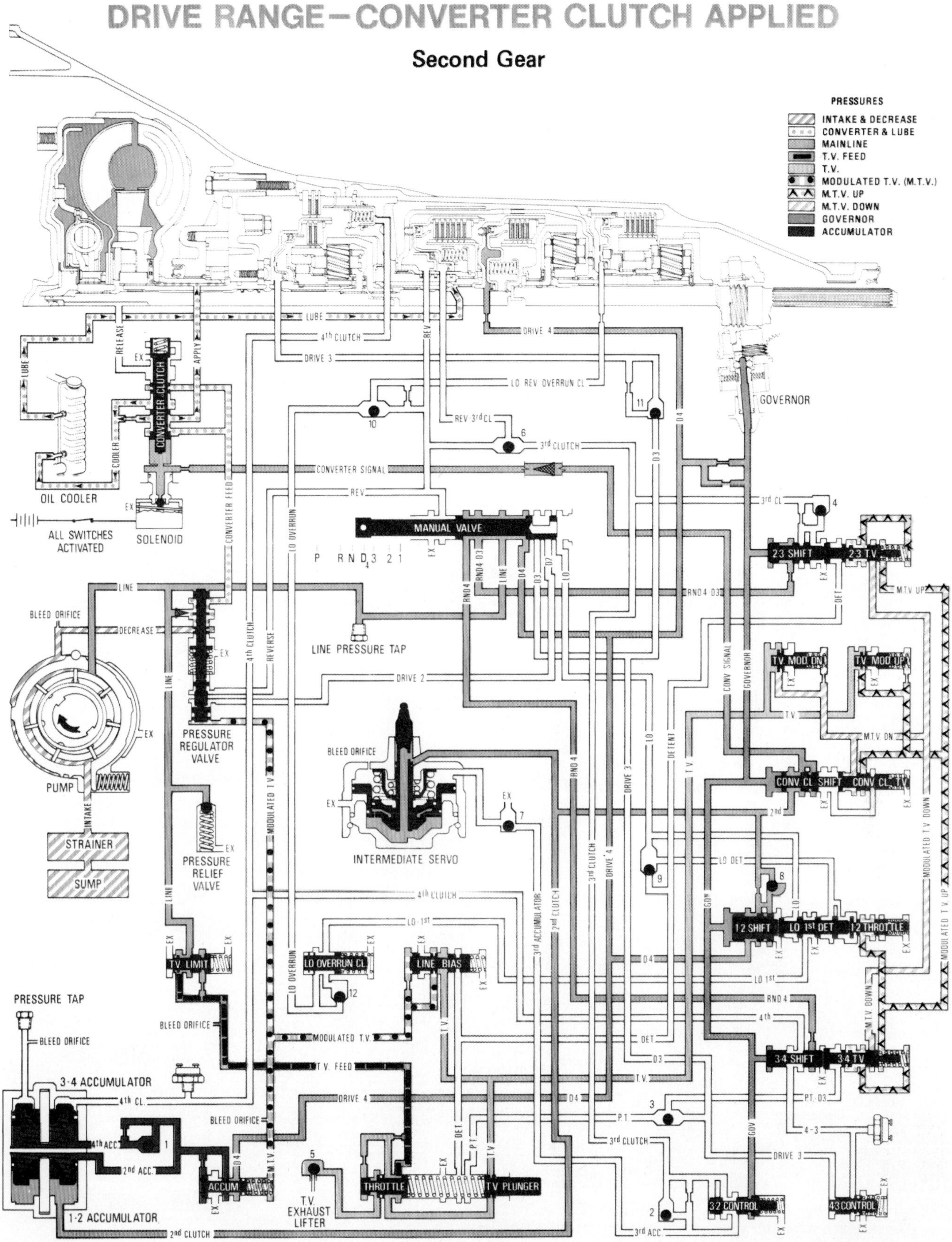

Figure 7-21.

DRIVE RANGE - THIRD GEAR

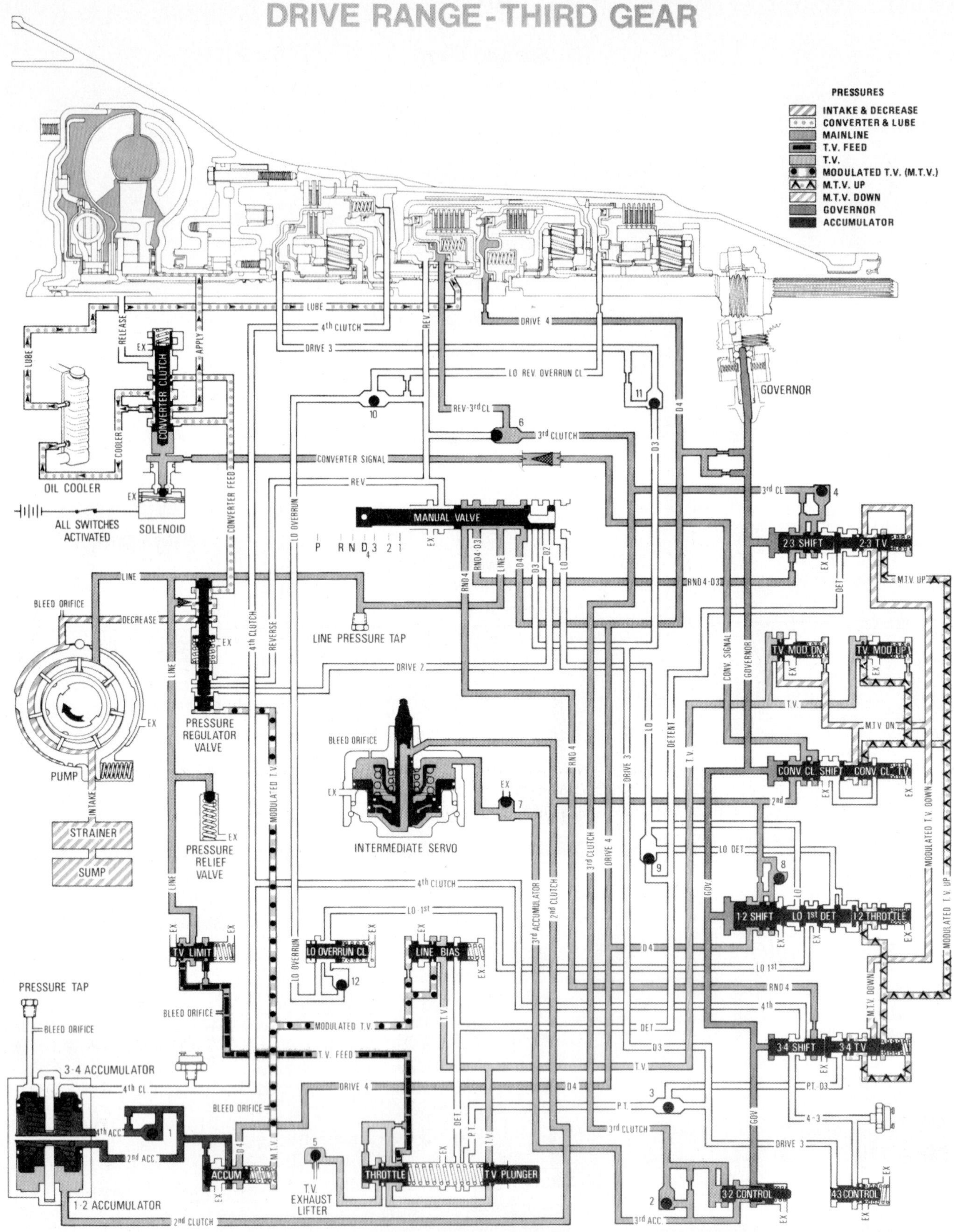

Figure 7-22.

DRIVE RANGE-OVERDRIVE

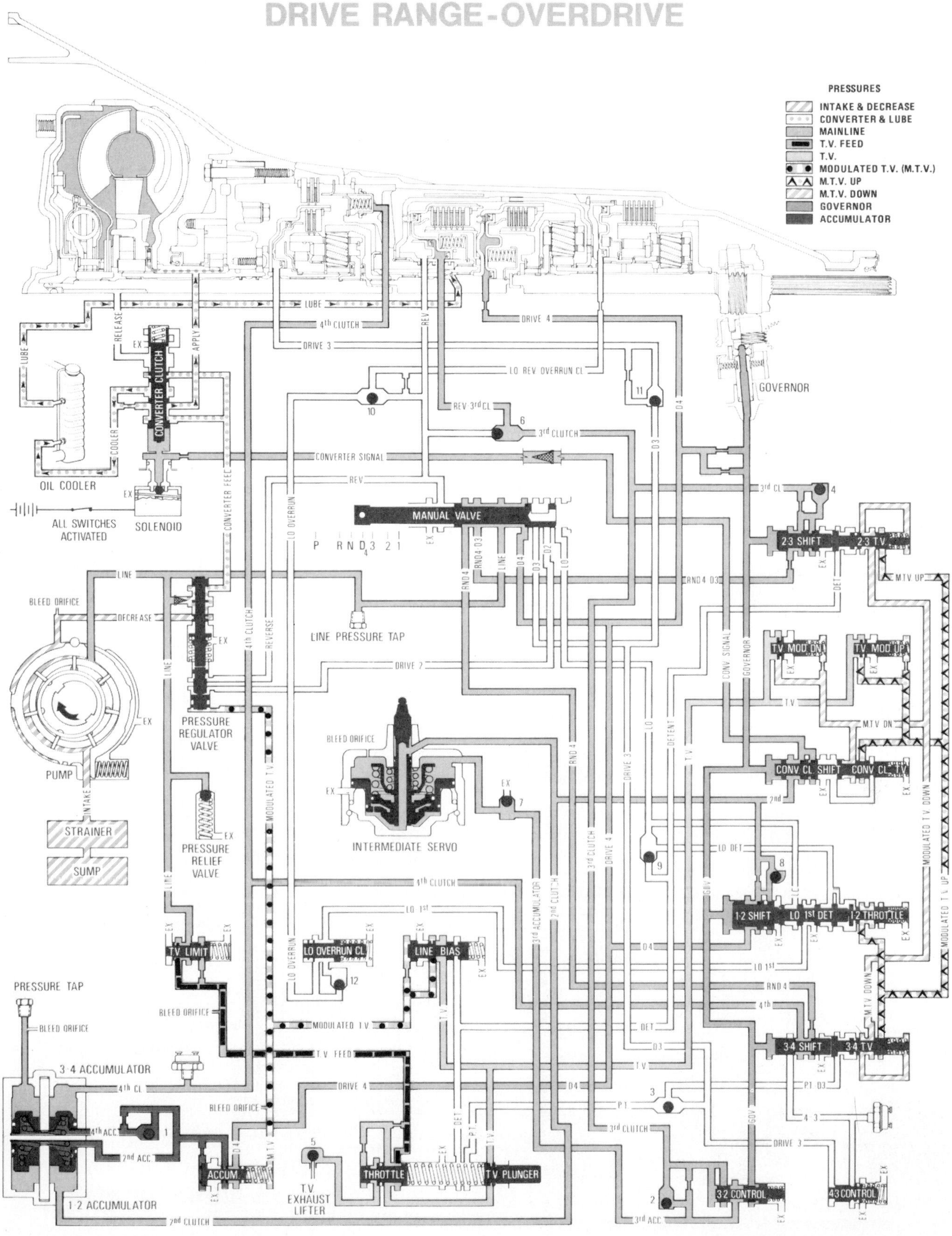

Figure 7-23.

PART THROTTLE 4-3 AND MODULATED DOWNSHIFTS

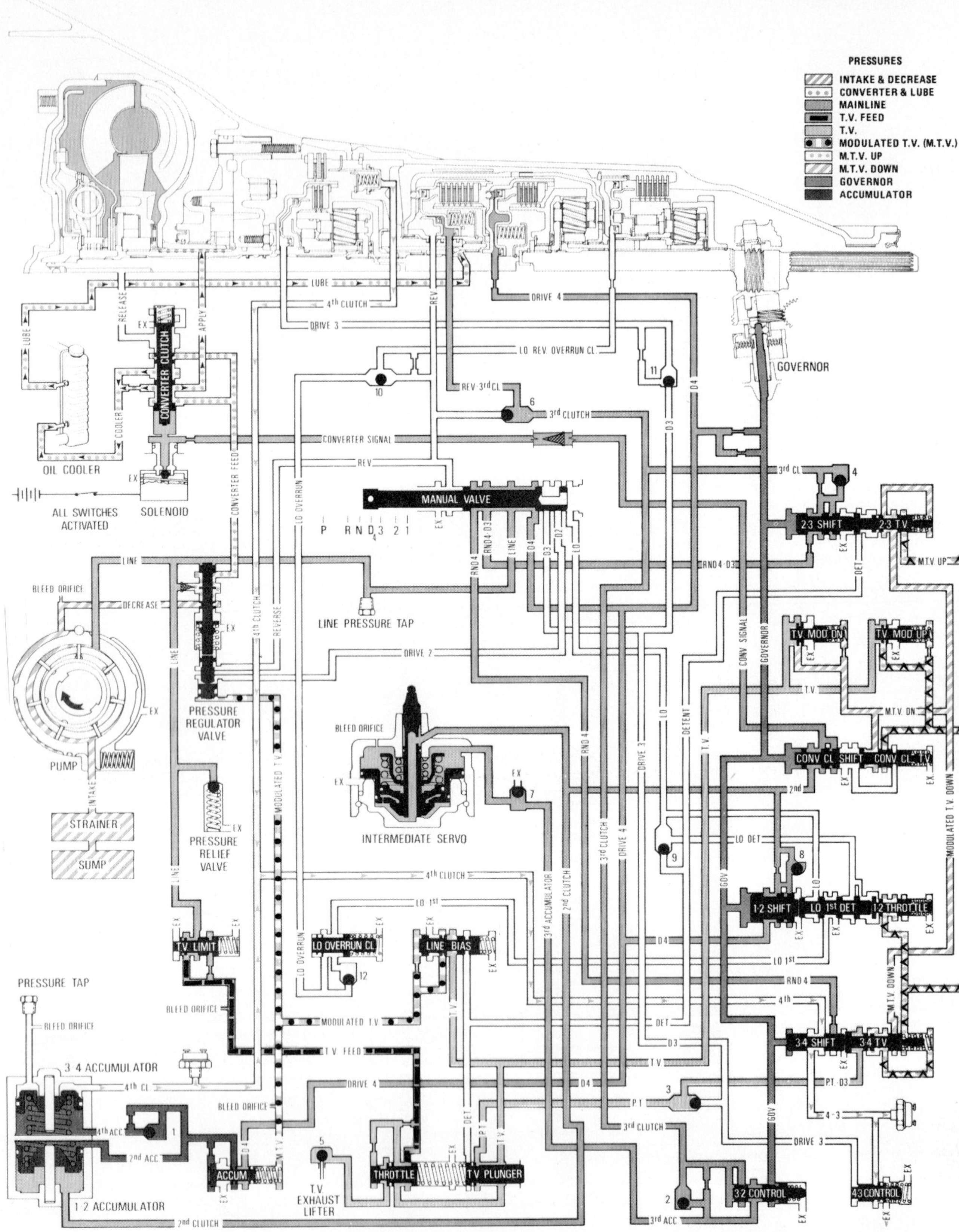

Figure 7-24.

DETENT DOWNSHIFTS

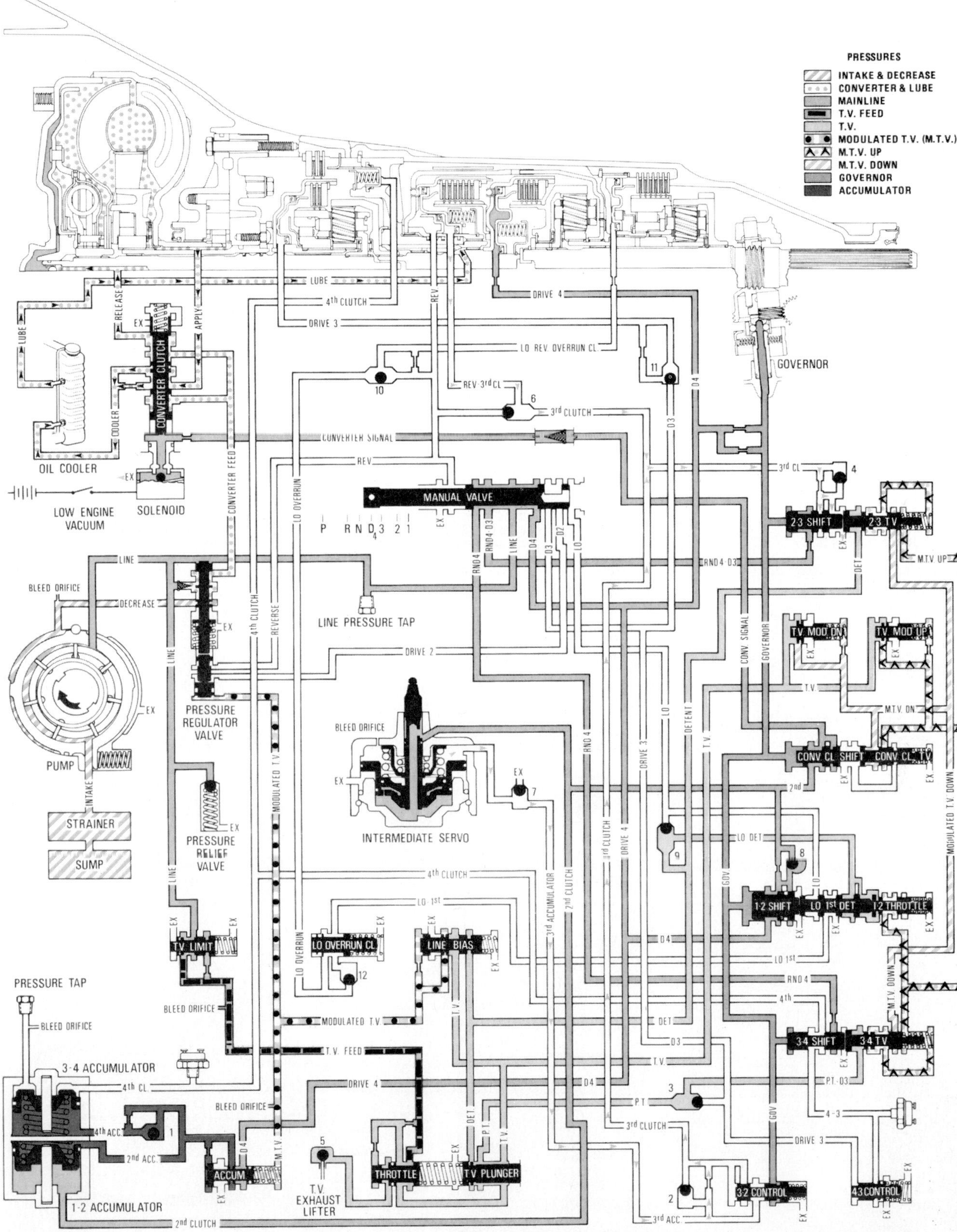

Figure 7-25.

MANUAL THIRD

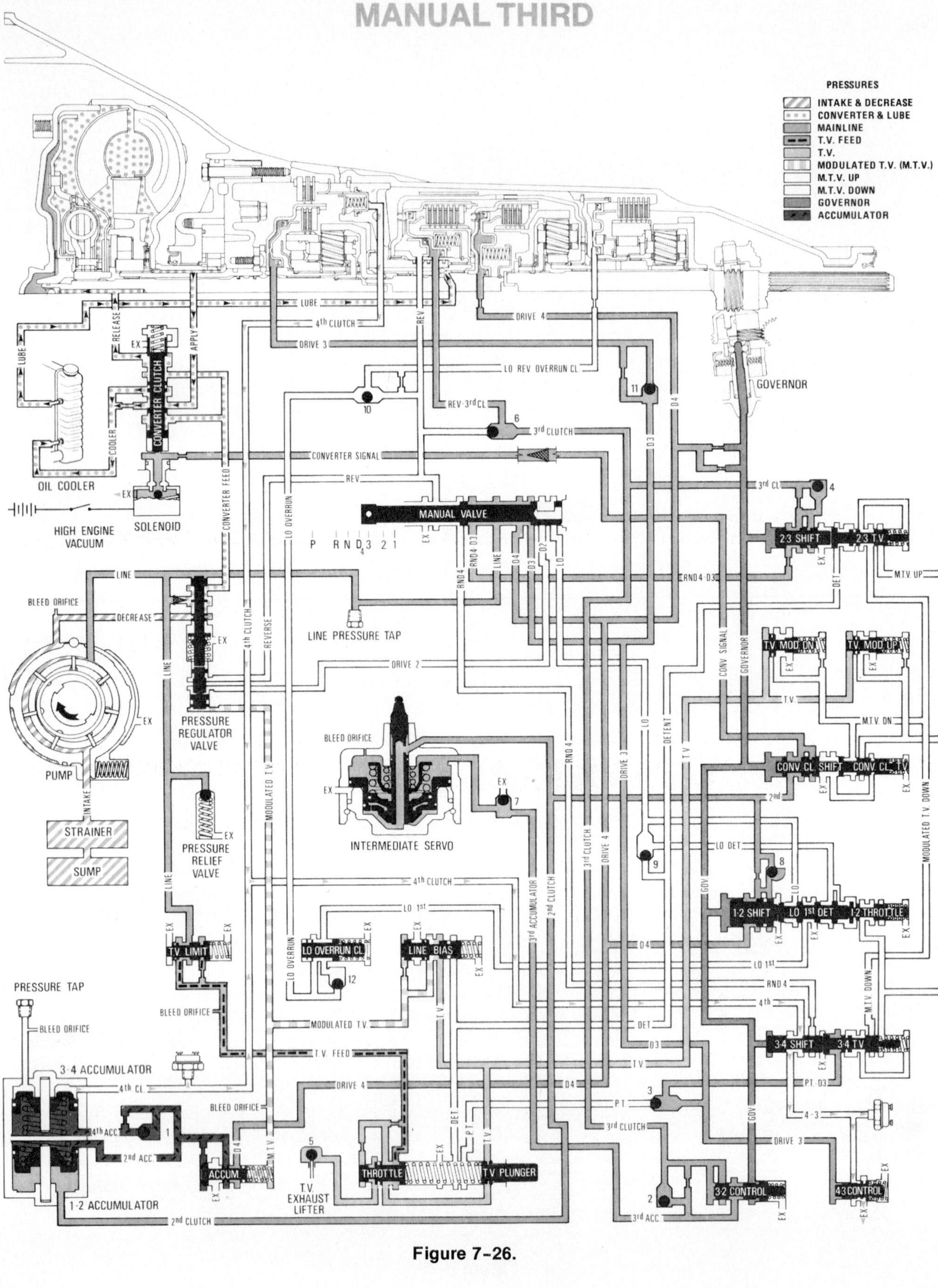

Figure 7-26.

MANUAL SECOND

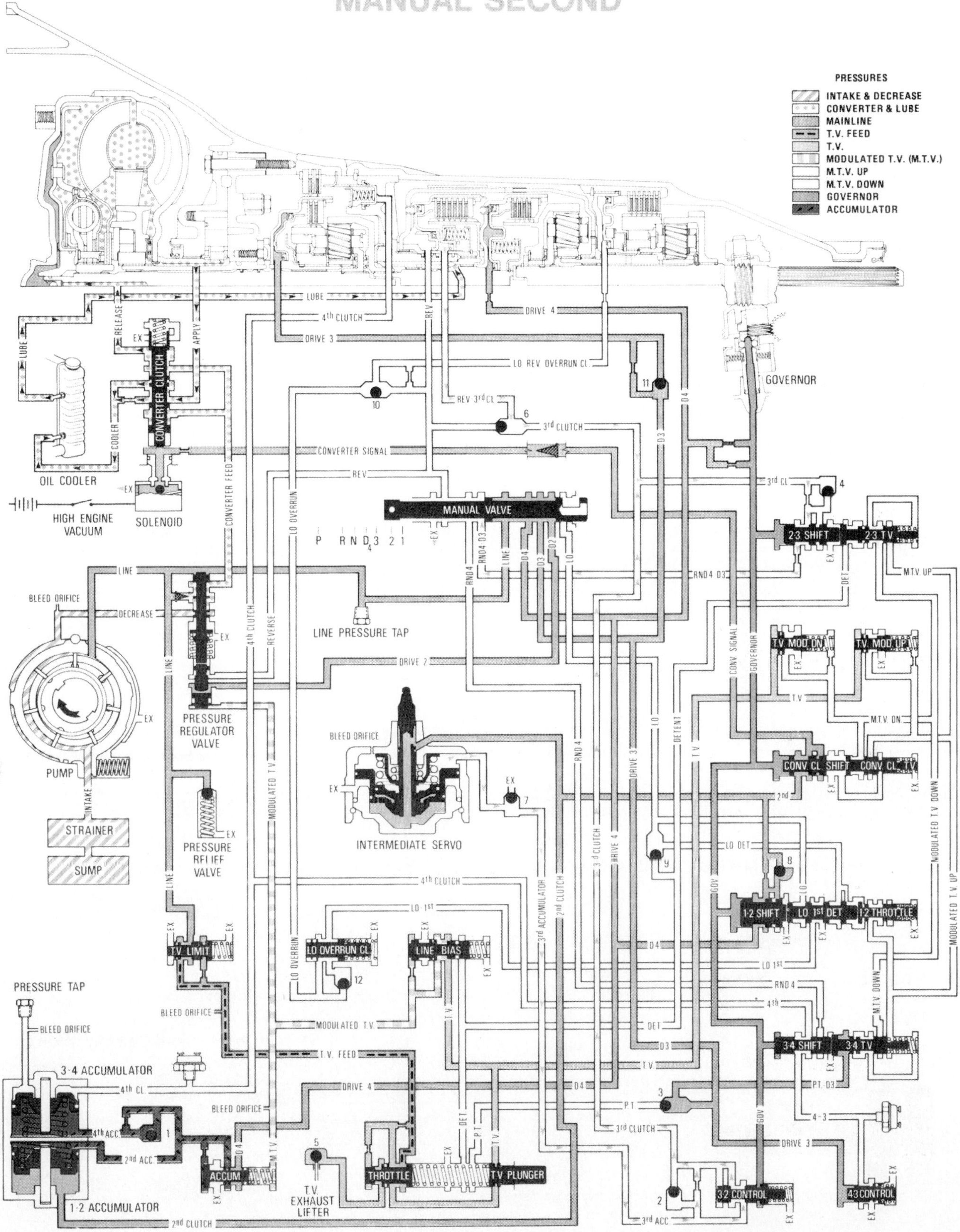

Figure 7-27.

MANUAL LO

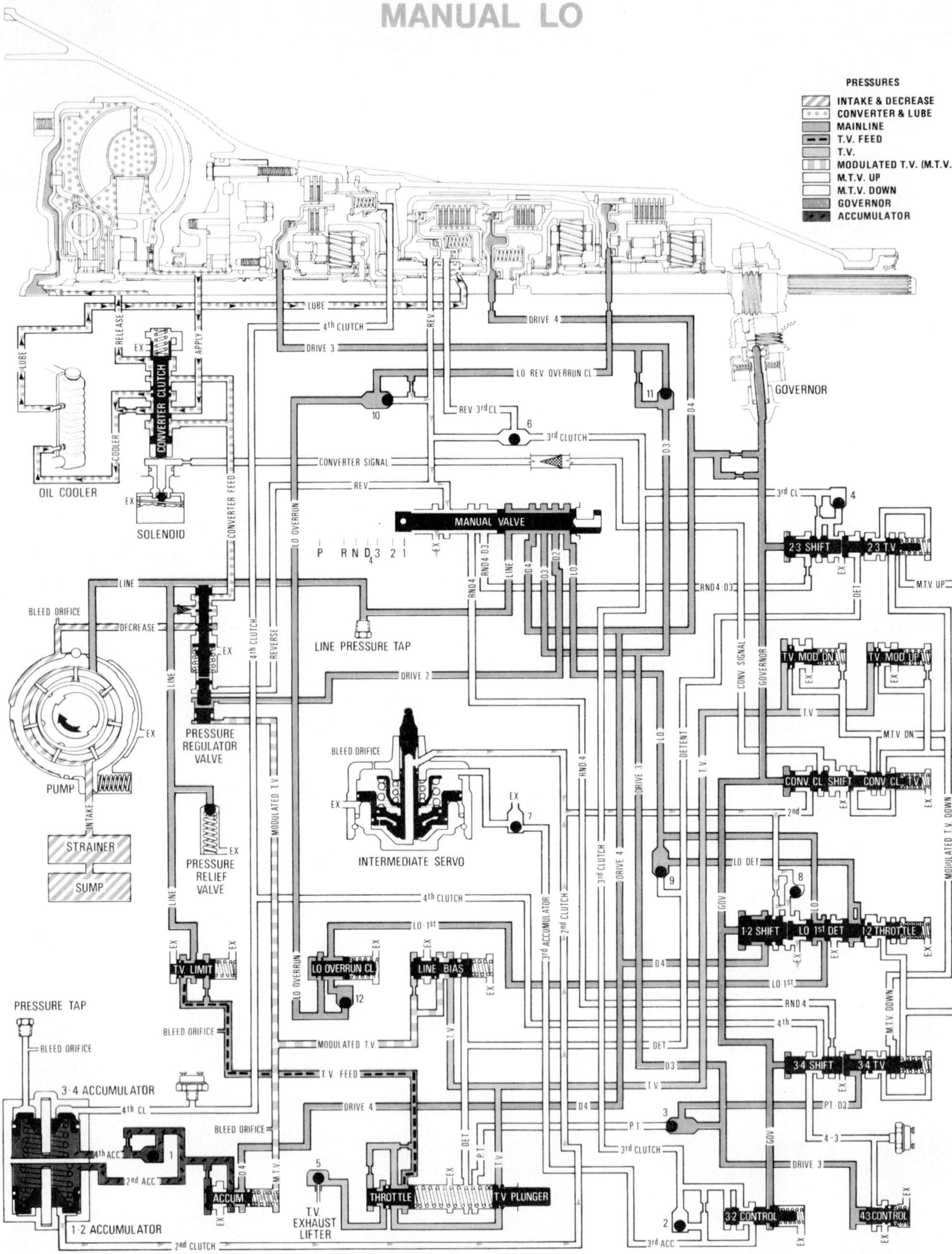

Figure 7-28.

REVERSE

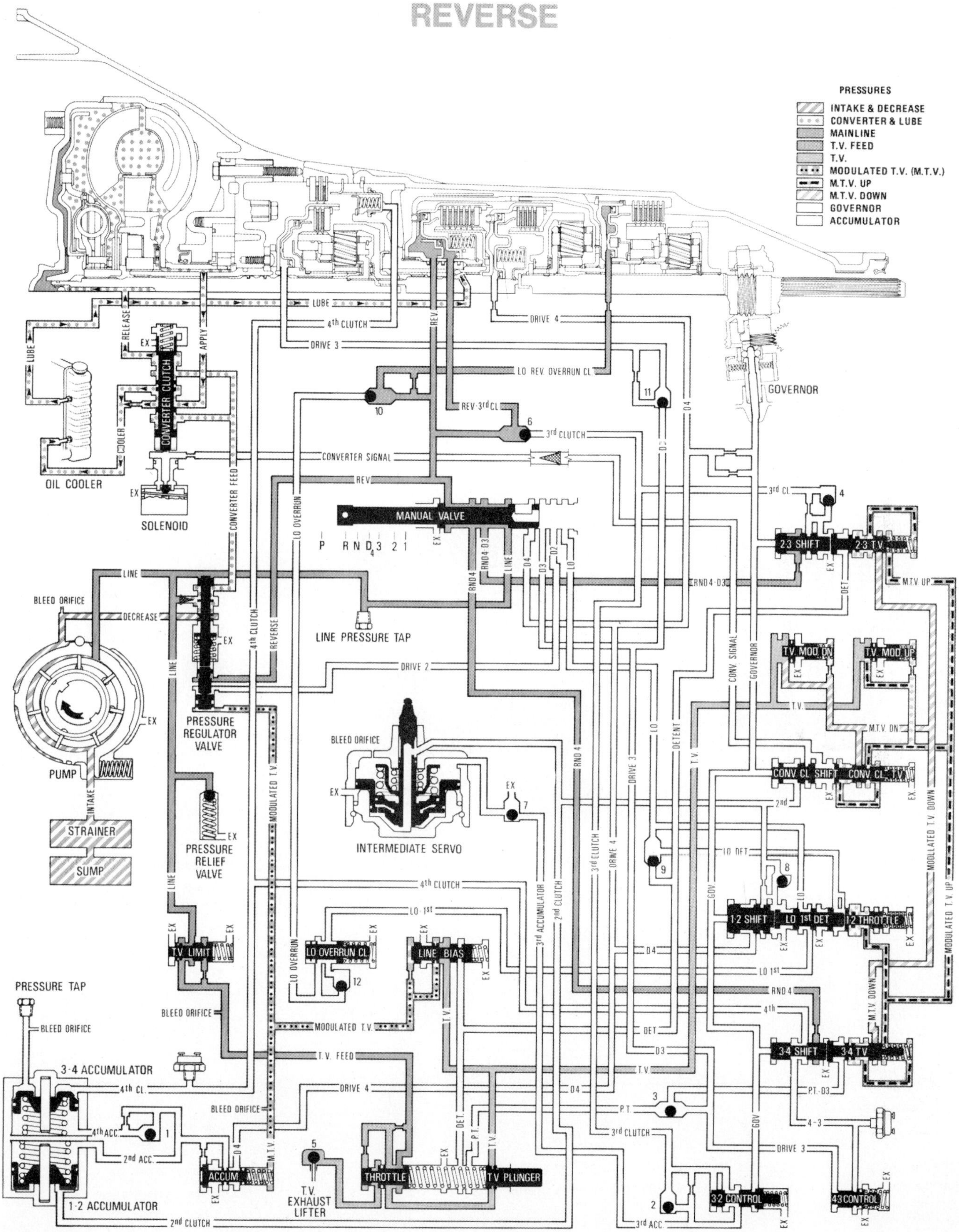

Figure 7-29.

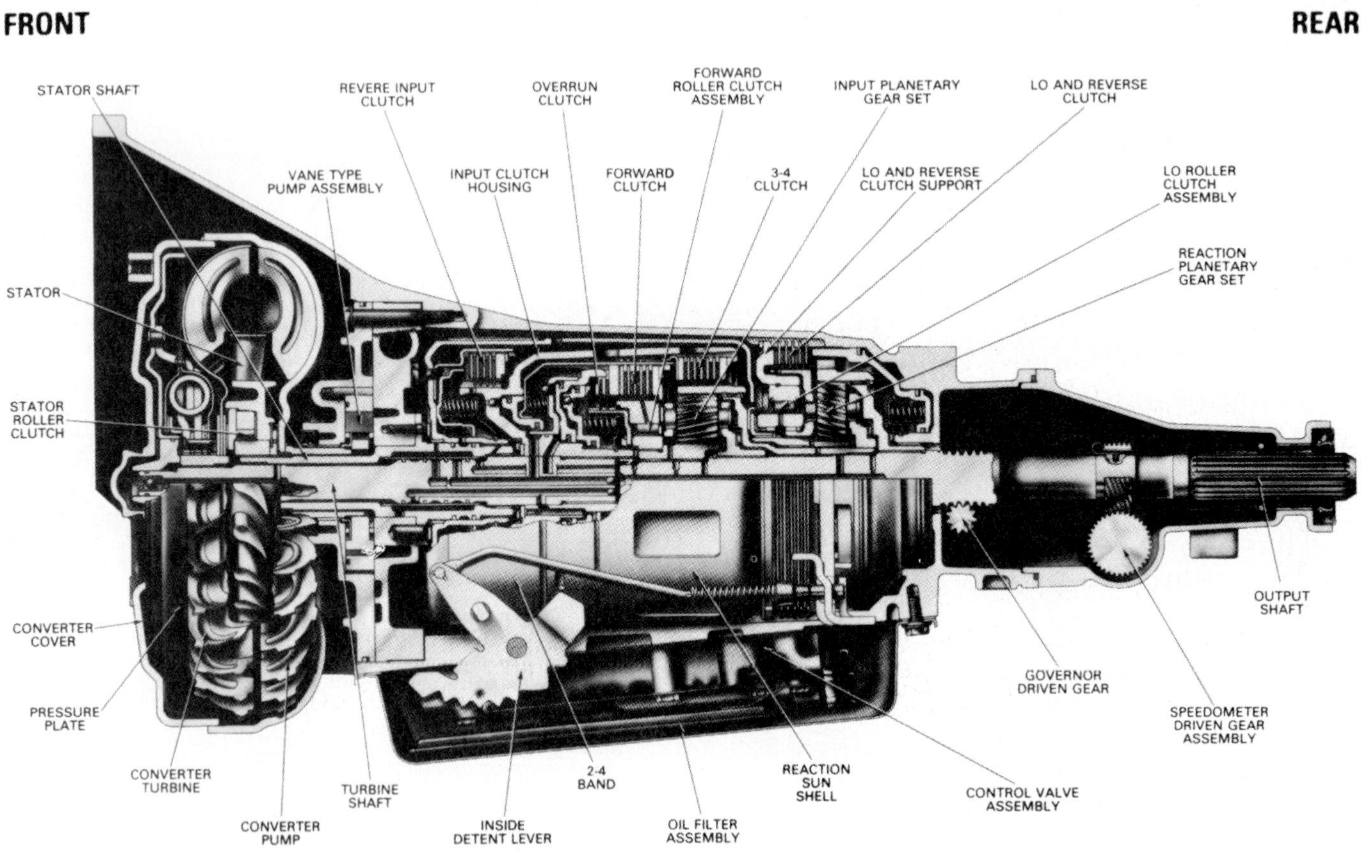

FRONT
REAR
STATOR SHAFT
REVERE INPUT CLUTCH
OVERRUN CLUTCH
FORWARD ROLLER CLUTCH ASSEMBLY
INPUT PLANETARY GEAR SET
LO AND REVERSE CLUTCH
VANE TYPE PUMP ASSEMBLY
INPUT CLUTCH HOUSING
FORWARD CLUTCH
3-4 CLUTCH
LO AND REVERSE CLUTCH SUPPORT
LO ROLLER CLUTCH ASSEMBLY
REACTION PLANETARY GEAR SET
STATOR
STATOR ROLLER CLUTCH
OUTPUT SHAFT
CONVERTER COVER
PRESSURE PLATE
GOVERNOR DRIVEN GEAR
SPEEDOMETER DRIVEN GEAR ASSEMBLY
CONVERTER TURBINE
TURBINE SHAFT
2-4 BAND
REACTION SUN SHELL
CONTROL VALVE ASSEMBLY
CONVERTER PUMP
INSIDE DETENT LEVER
OIL FILTER ASSEMBLY

PARK

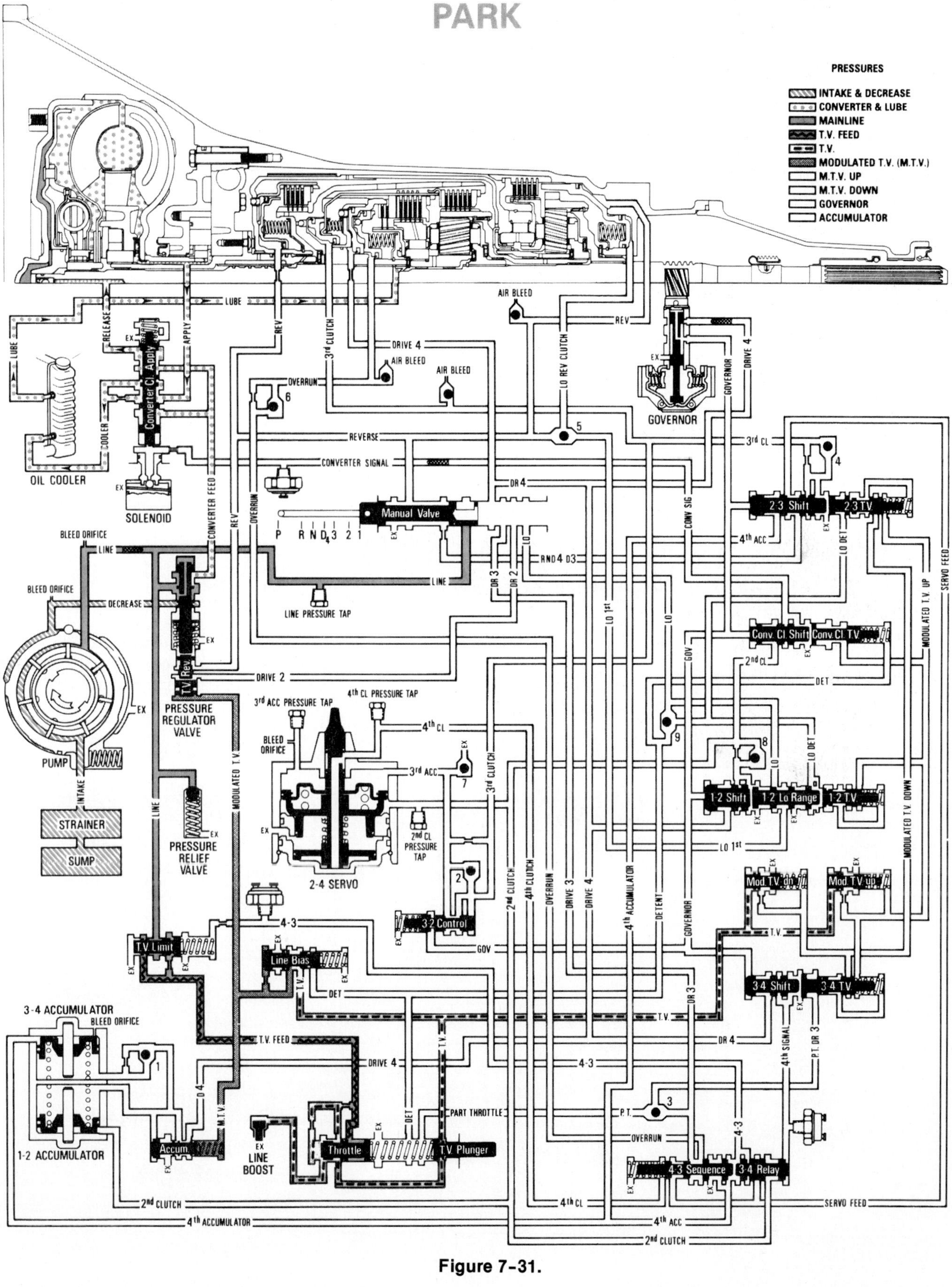

Figure 7-31.

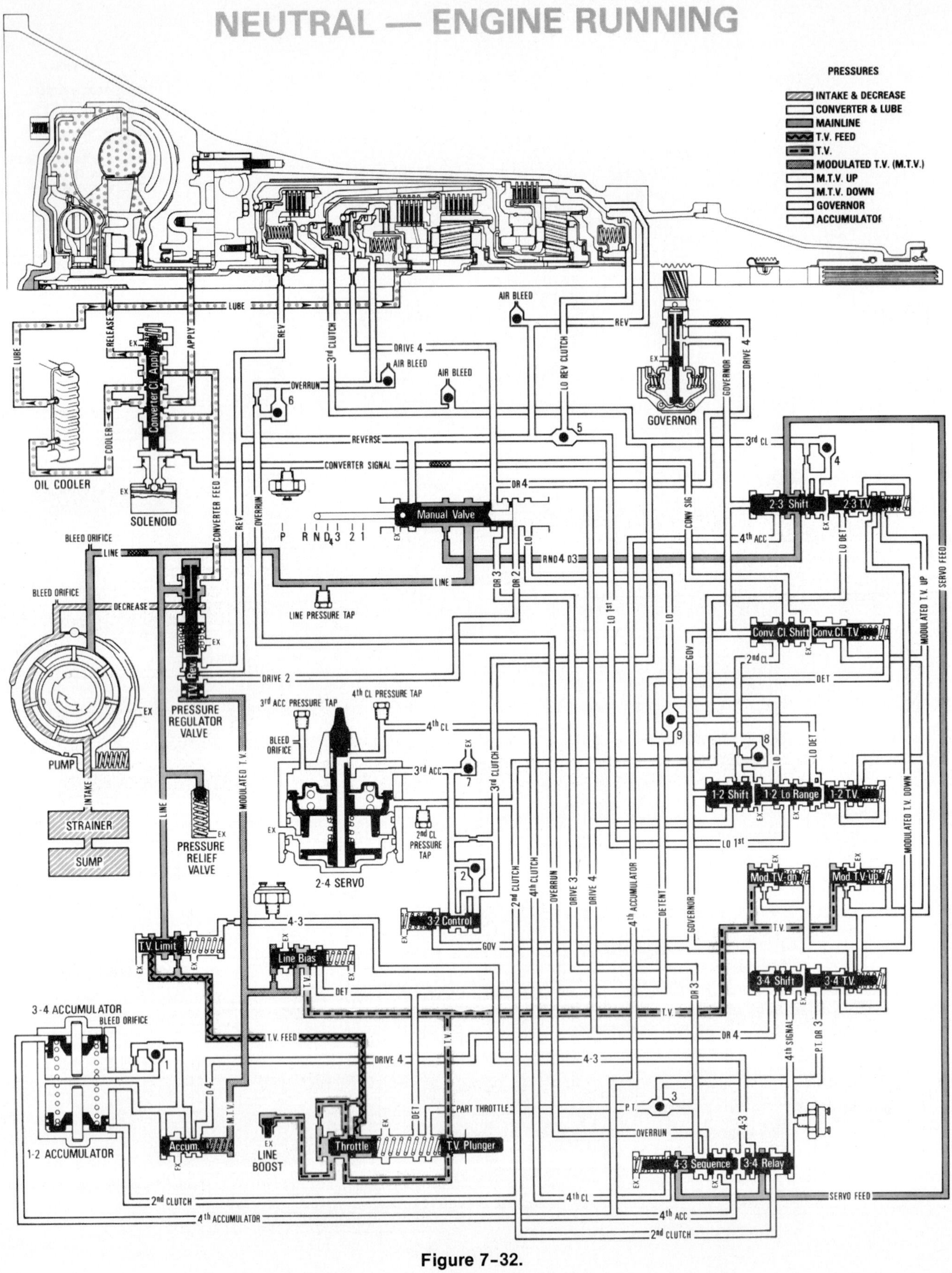
NEUTRAL — ENGINE RUNNING
PRESSURES
INTAKE & DECREASE
CONVERTER & LUBE
MAINLINE
T.V. FEED
T.V.
MODULATED T.V. (M.T.V.)
M.T.V. UP
M.T.V. DOWN
GOVERNOR
ACCUMULATOR
OIL COOLER
SOLENOID
Converter Cl. Apply
Manual Valve
PRESSURE REGULATOR VALVE
PUMP
STRAINER
SUMP
PRESSURE RELIEF VALVE
2-4 SERVO
GOVERNOR
2-3 Shift
2-3 T.V.
Conv. Cl. Shift
Conv. Cl. T.V.
1-2 Shift
1-2 Lo Range
1-2 T.V.
Mod. T.V. dn
Mod. T.V. up
3-4 Shift
3-4 T.V.
3-2 Control
T.V. Limit
Line Bias
3-4 ACCUMULATOR
1-2 ACCUMULATOR
Accum.
LINE BOOST
Throttle
T.V. Plunger
4-3 Sequence
3-4 Relay
LINE PRESSURE TAP
3rd ACC PRESSURE TAP
4th CL PRESSURE TAP
2nd CL PRESSURE TAP

Figure 7-32.

DRIVE RANGE — FIRST GEAR

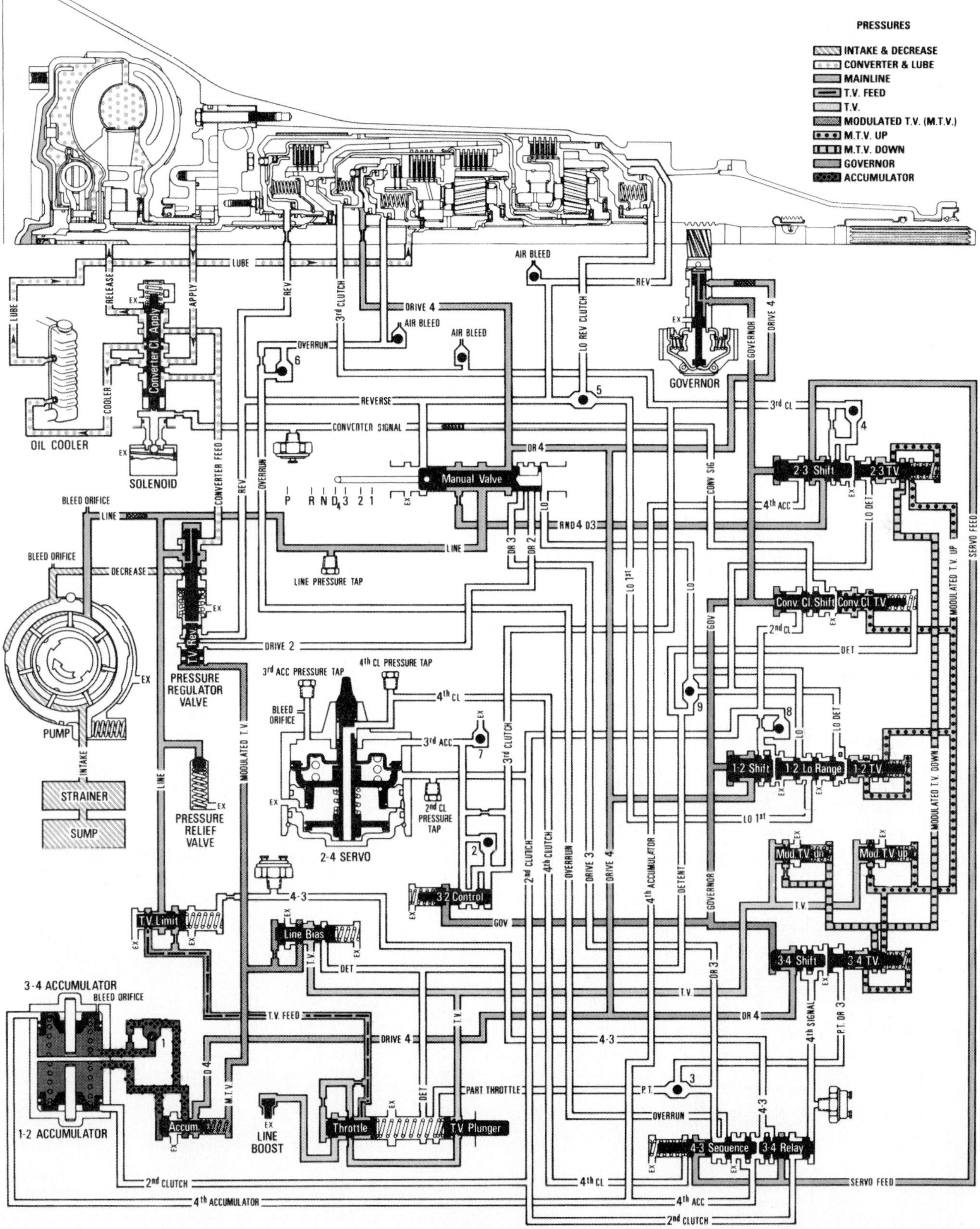

Figure 7-33.

DRIVE RANGE — SECOND GEAR

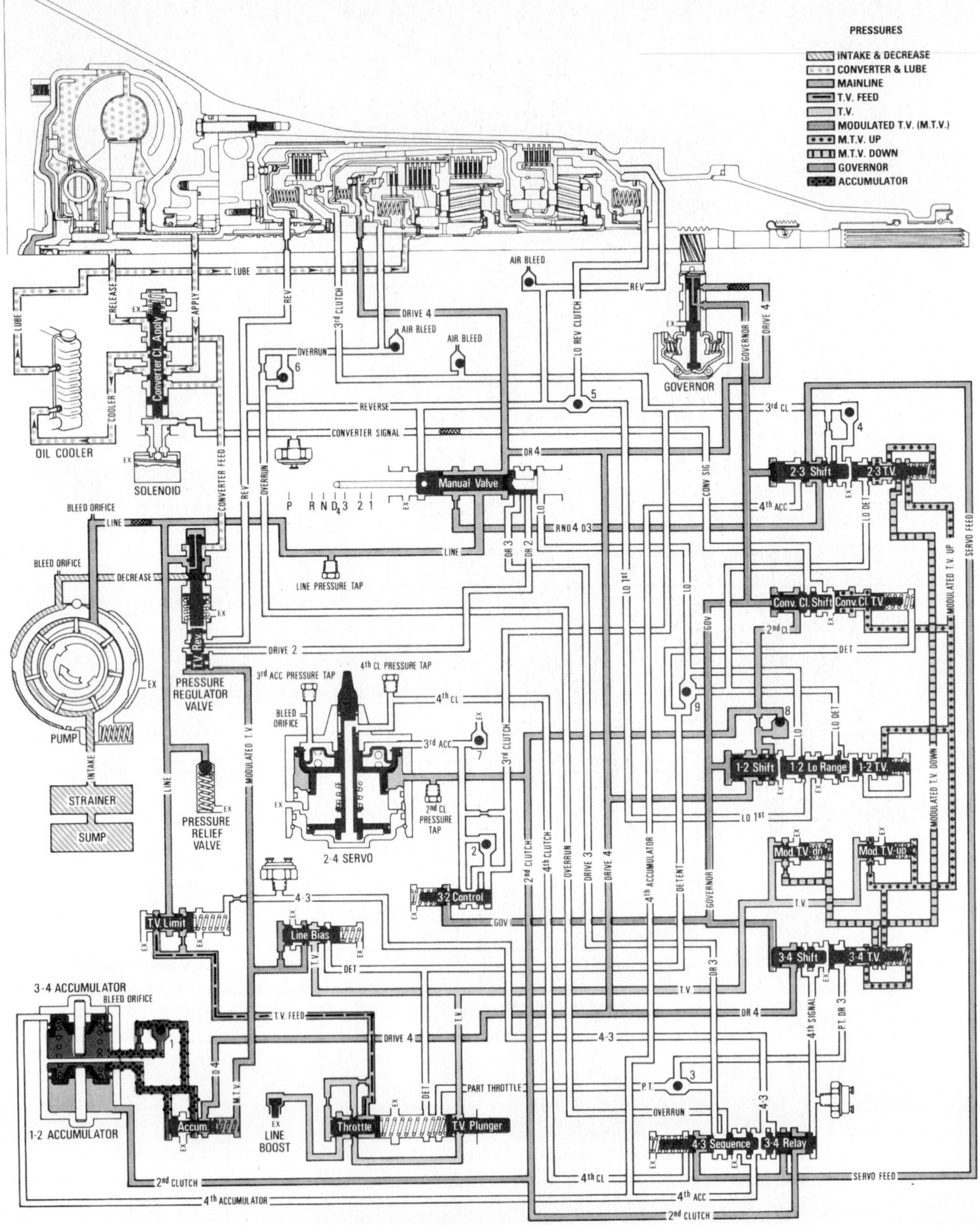

Figure 7-34.

DRIVE RANGE — CONVERTER CLUTCH APPLIED

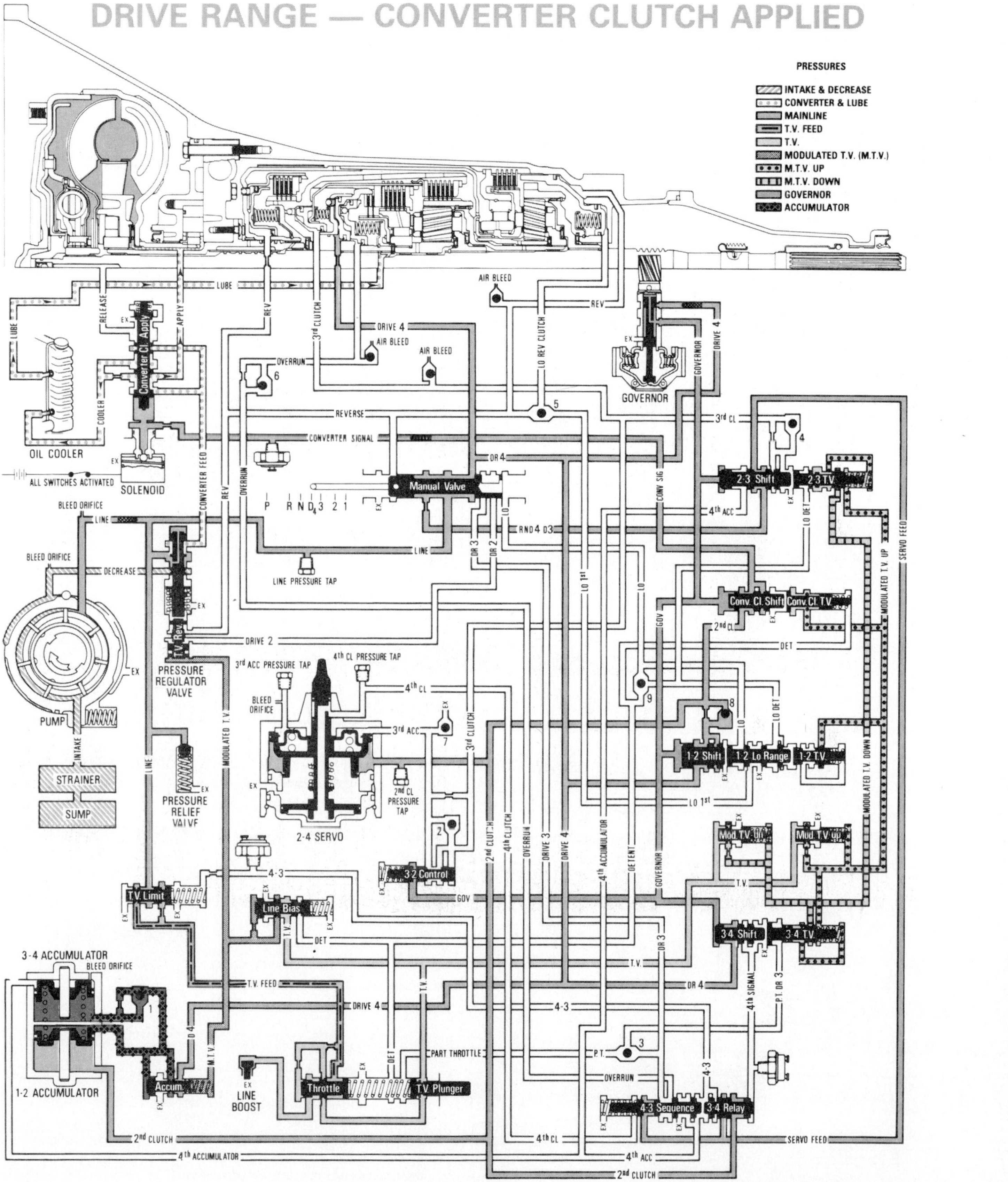

Figure 7-35.

DRIVE RANGE — THIRD GEAR

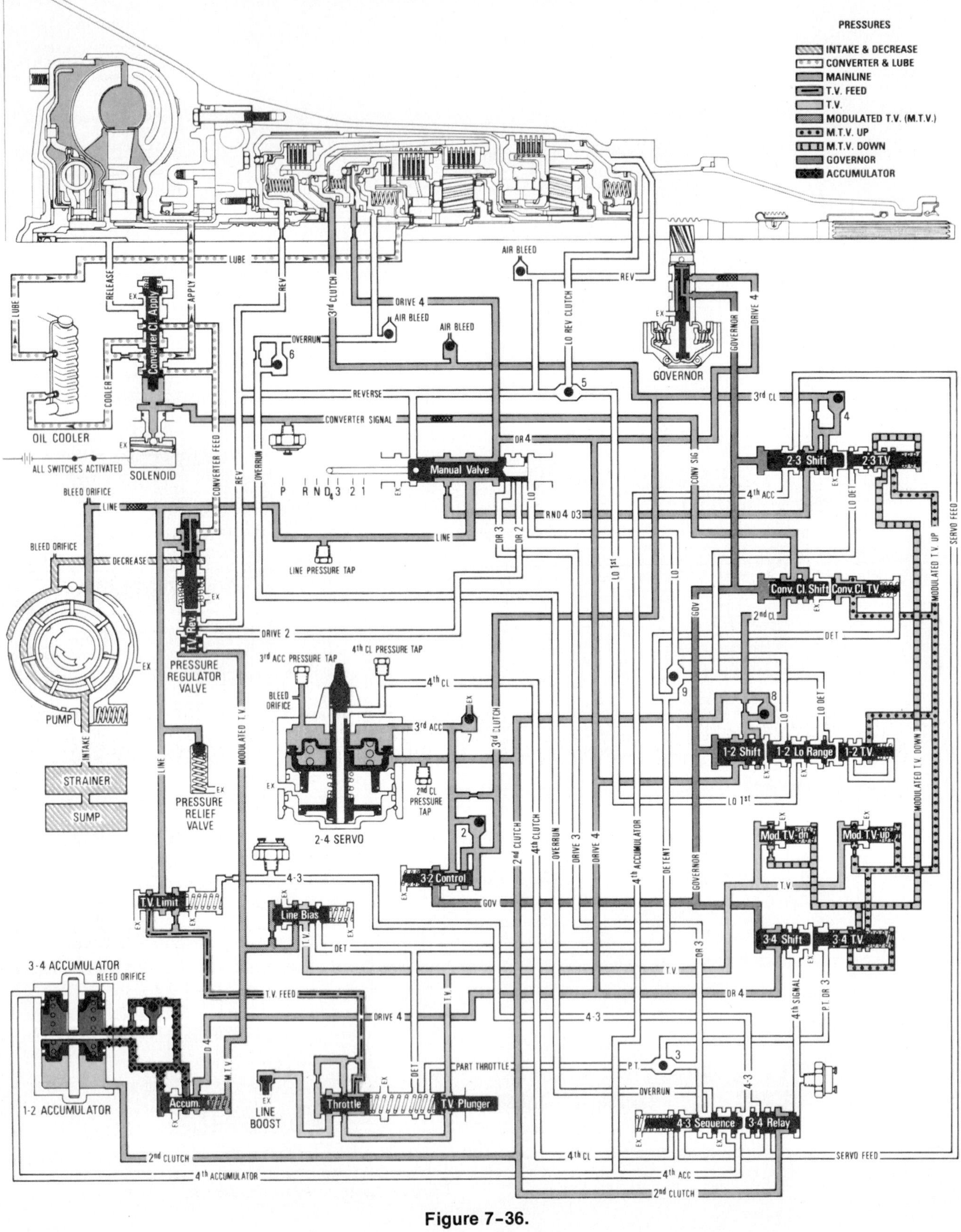

Figure 7-36.

DRIVE RANGE — OVERDRIVE

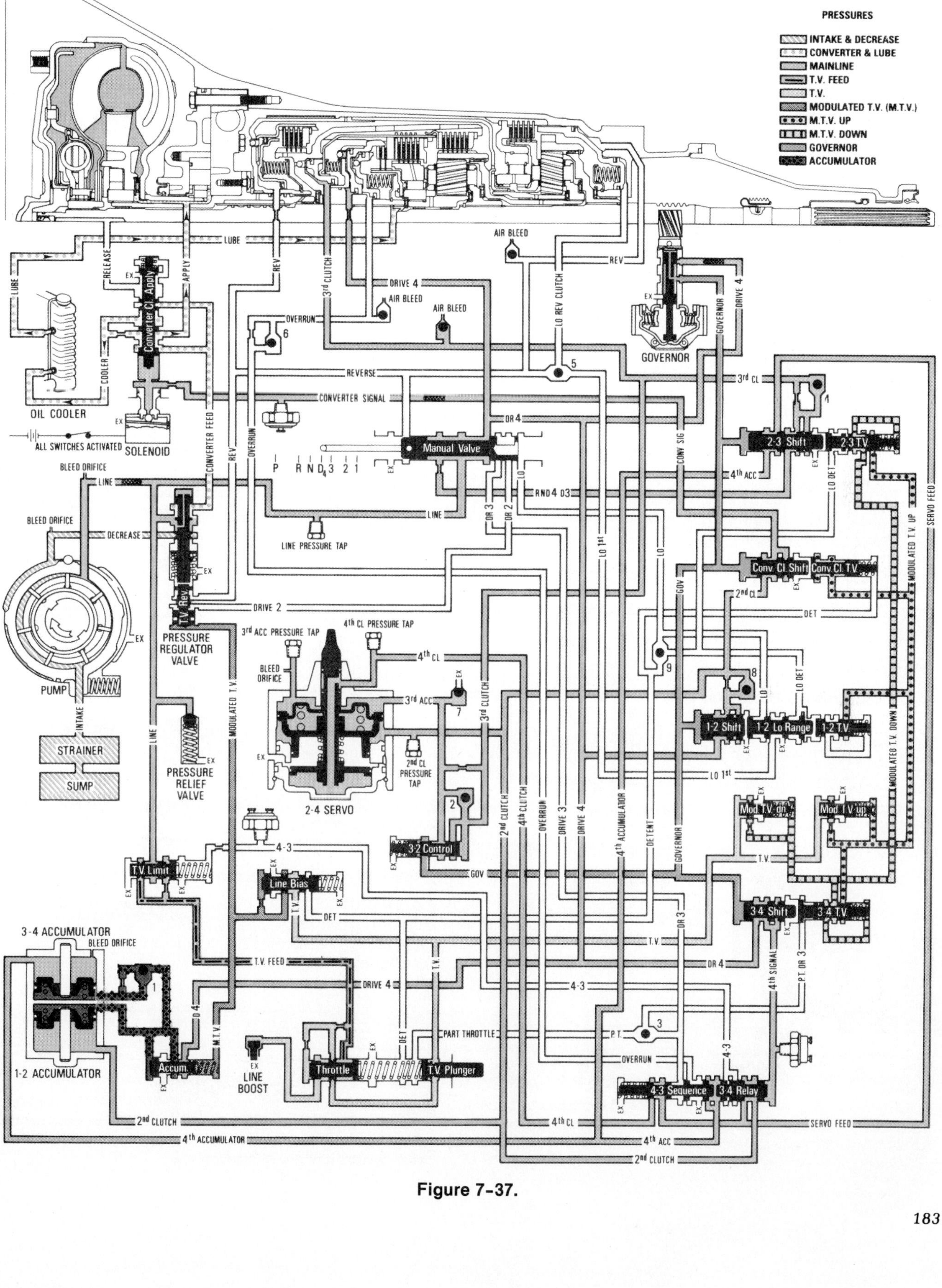

Figure 7-37.

PART THROTTLE 4-3 AND MODULATED DOWNSHIFTS

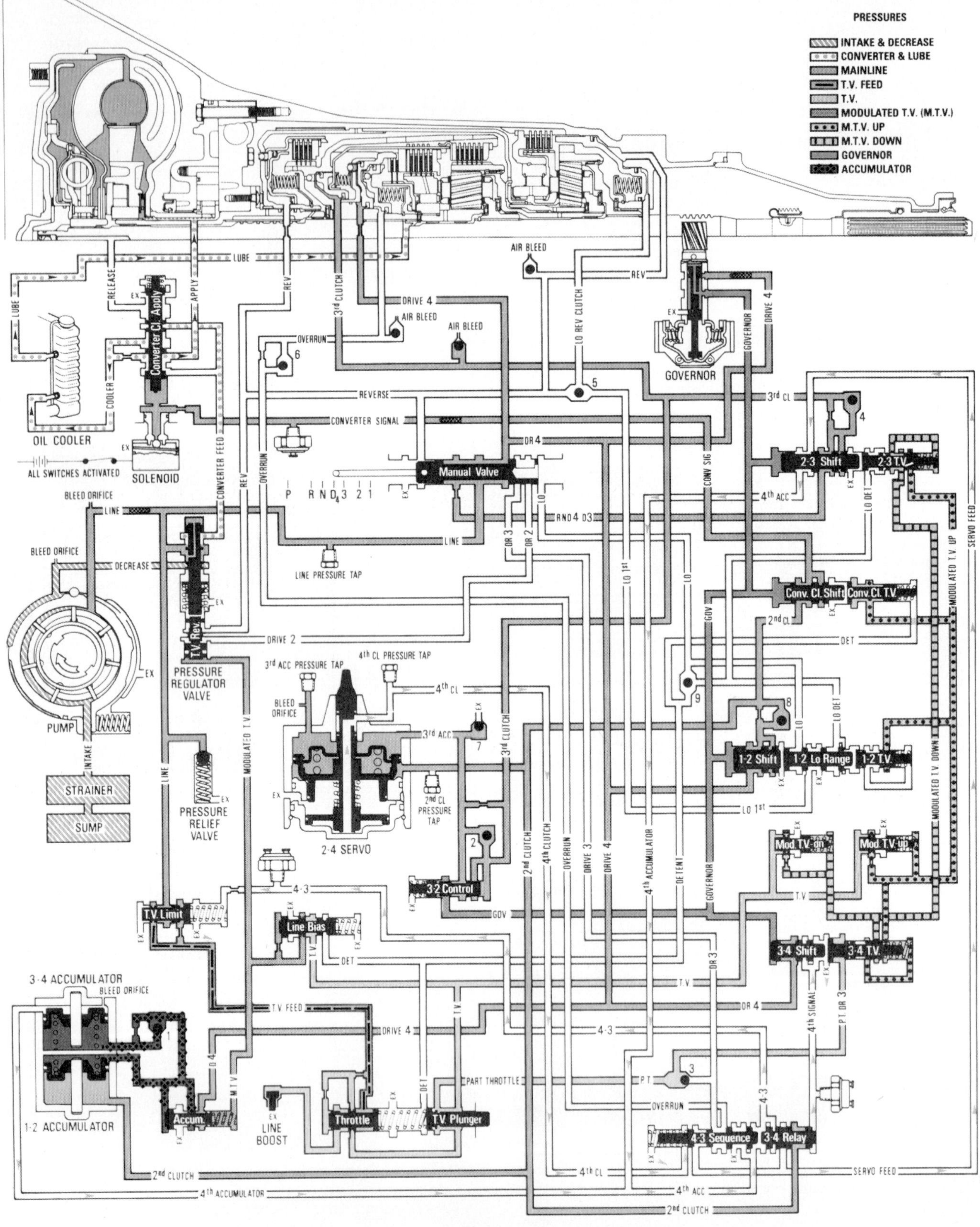

Figure 7-38.

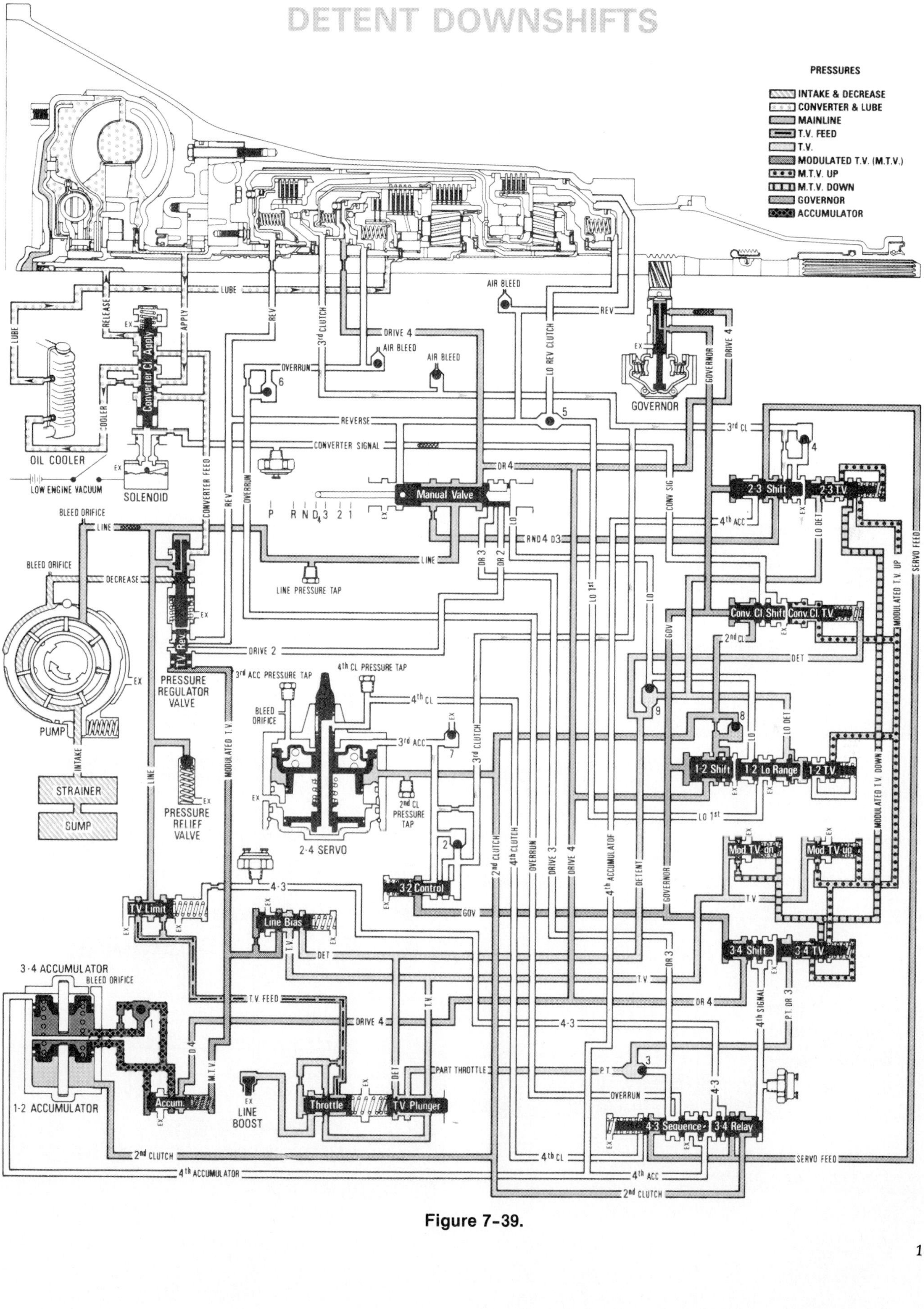
DETENT DOWNSHIFTS
PRESSURES
INTAKE & DECREASE
CONVERTER & LUBE
MAINLINE
T.V. FEED
T.V.
MODULATED T.V. (M.T.V.)
M.T.V. UP
M.T.V. DOWN
GOVERNOR
ACCUMULATOR
OIL COOLER
LOW ENGINE VACUUM
SOLENOID
Converter Cl. Apply
GOVERNOR
Manual Valve
P R N D 3 2 1
LINE PRESSURE TAP
2-3 Shift
2-3 TV
Conv. Cl. Shift
Conv. Cl. TV
PUMP
STRAINER
SUMP
PRESSURE REGULATOR VALVE
T.V. Rev
PRESSURE RELIEF VALVE
2-4 SERVO
3-2 Control
1-2 Shift
1-2 Lo Range
1-2 TV
Mod TV dn
Mod TV up
T.V. Limit
Line Bias
3-4 Shift
3-4 TV
3-4 ACCUMULATOR
1-2 ACCUMULATOR
Accum.
LINE BOOST
Throttle
TV Plunger
4-3 Sequence
3-4 Relay
2nd CLUTCH
4th ACCUMULATOR
SERVO FEED

Figure 7-39.

MANUAL THIRD

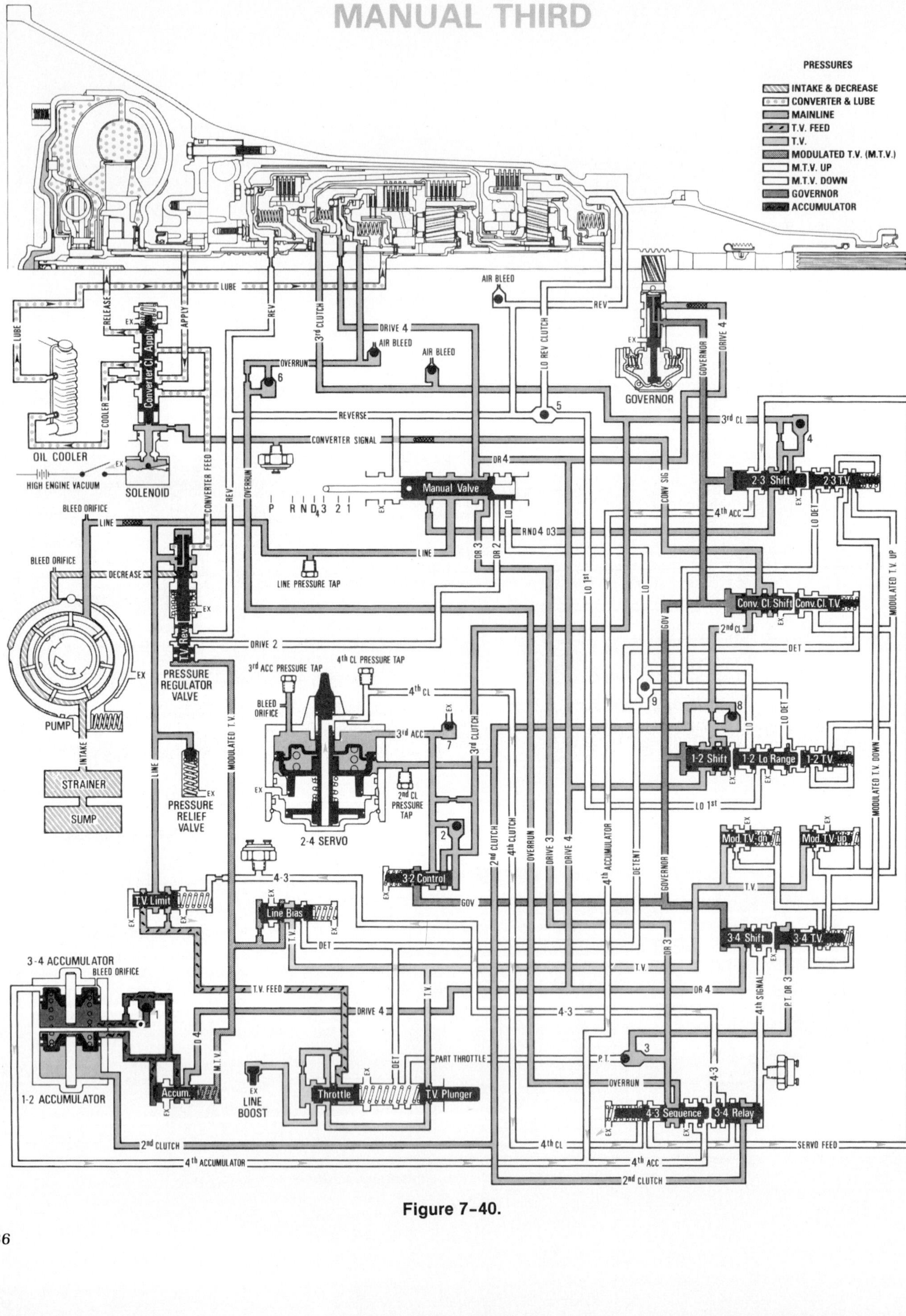

Figure 7-40.

MANUAL SECOND

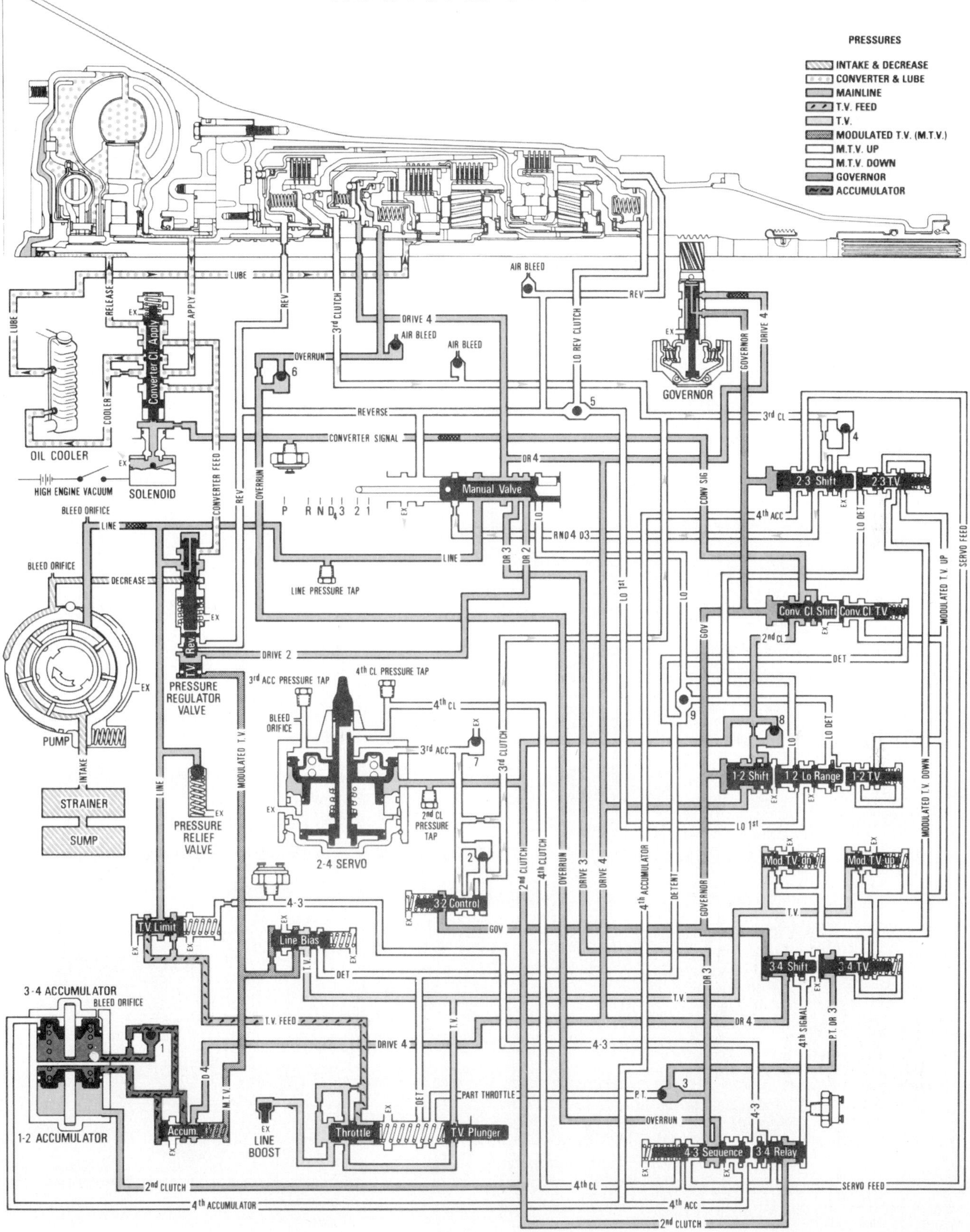

Figure 7-41.

MANUAL LO

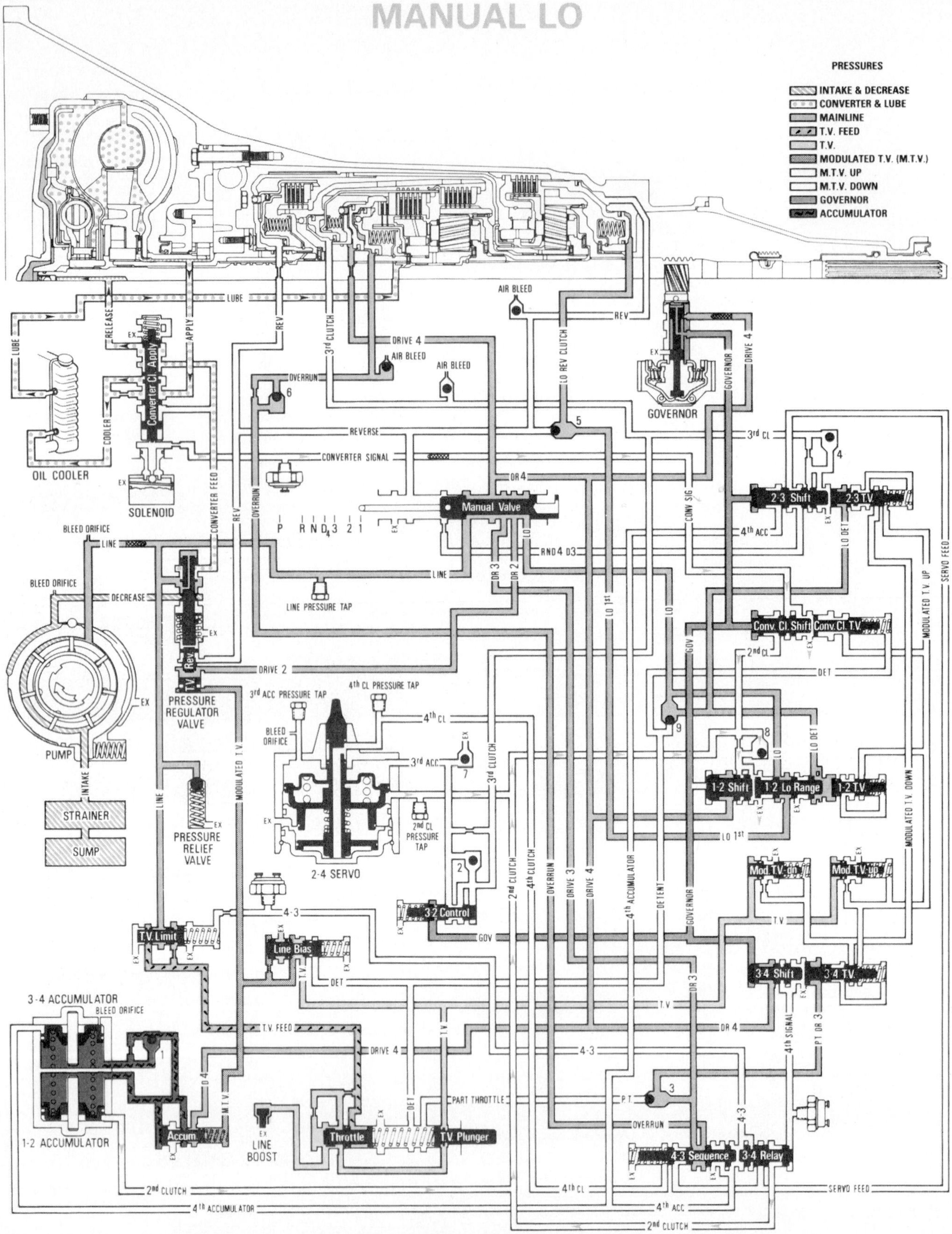

Figure 7-42.

REVERSE

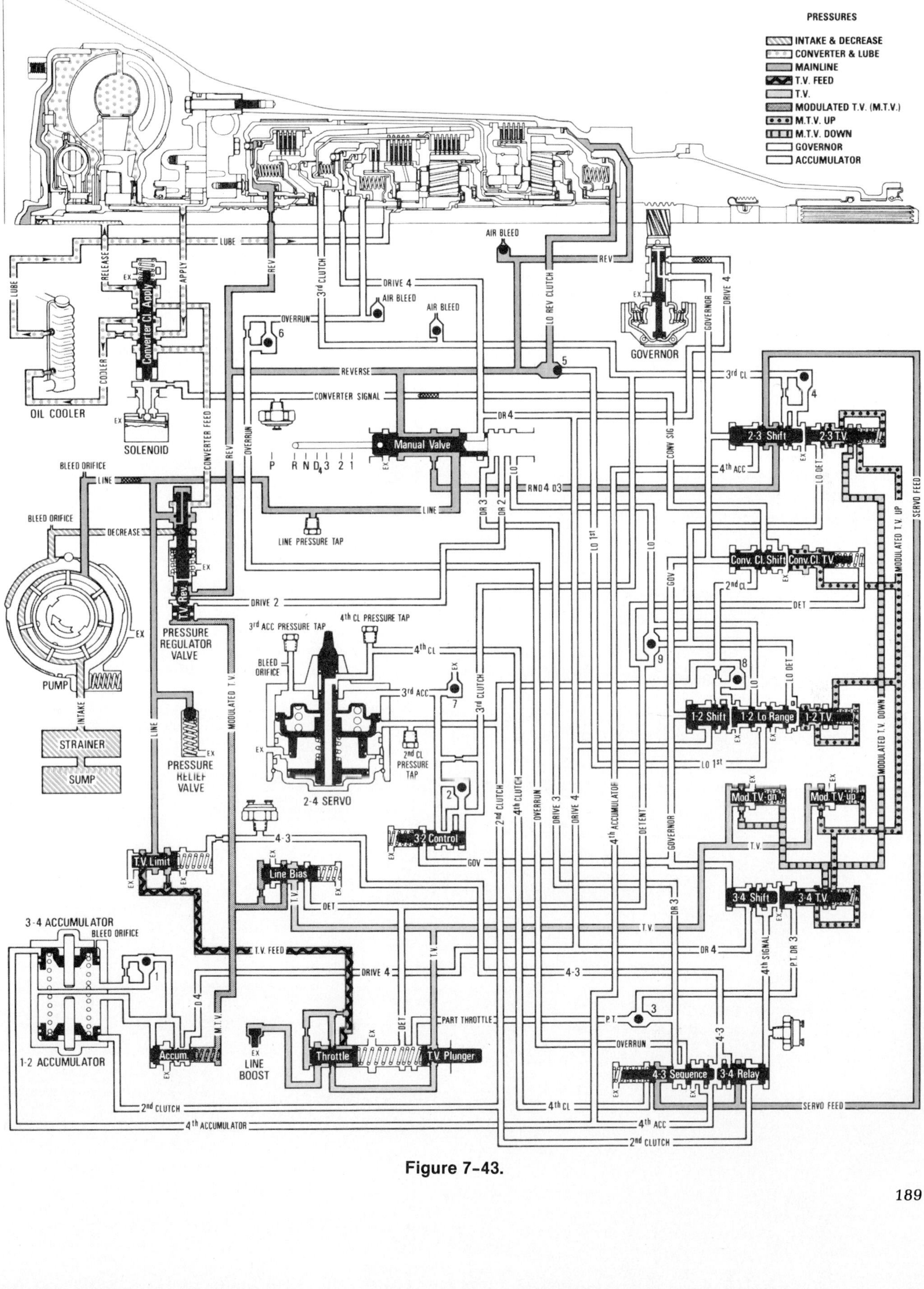

Figure 7-43.

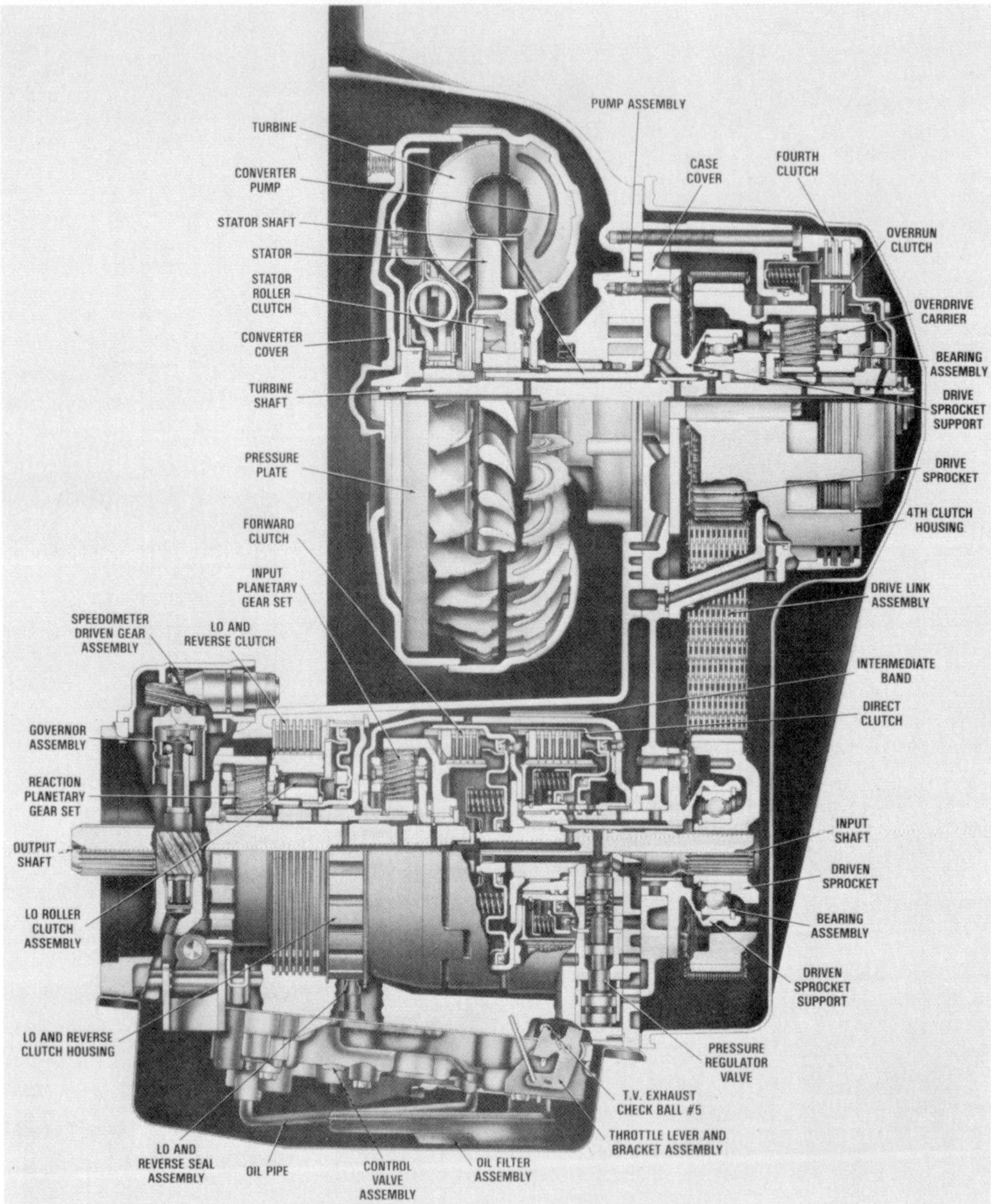

Figure 7-44. Cutaway view, THM 325-4L transmission.

PARK-ENGINE RUNNING

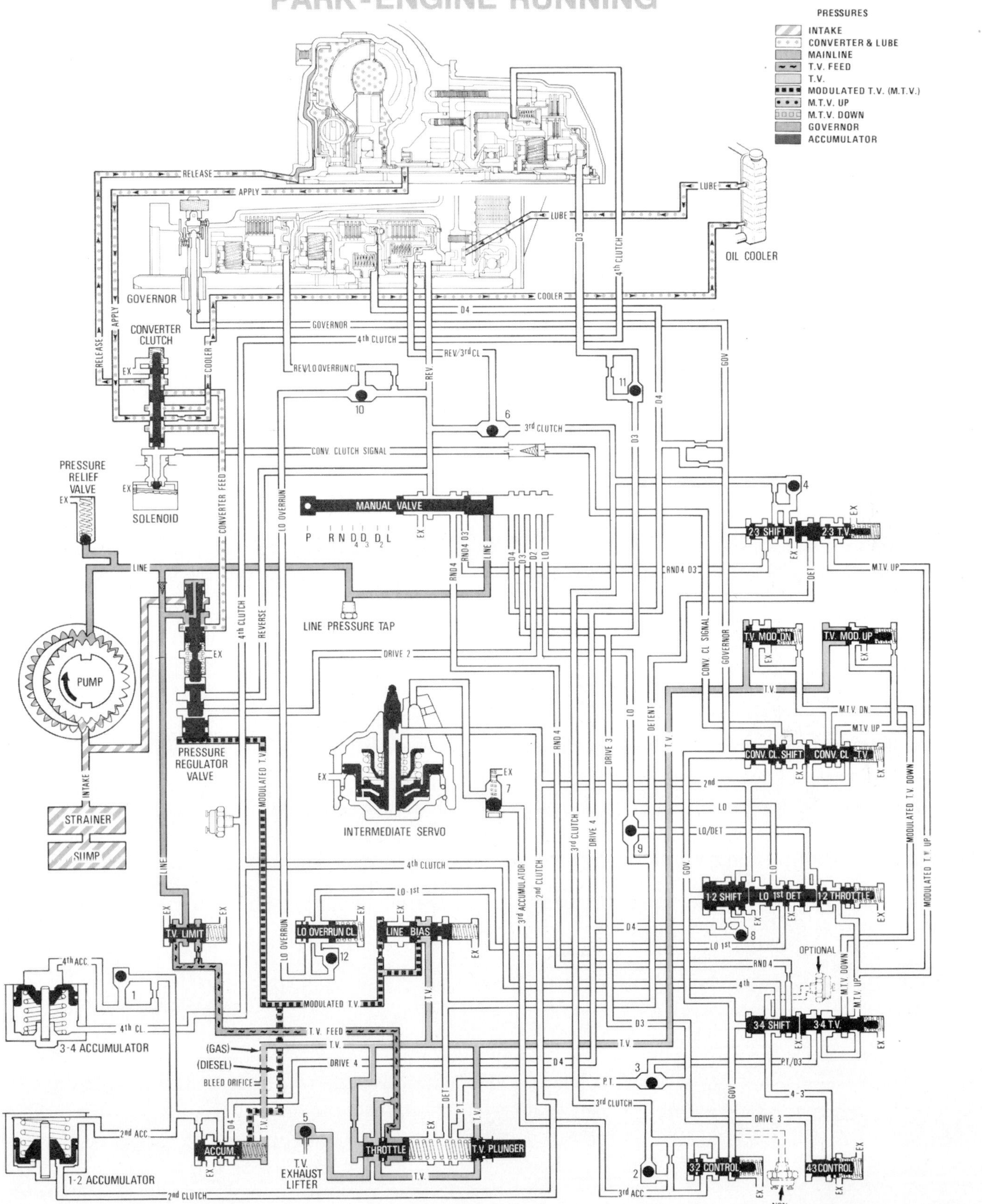

Figure 7–45.

NEUTRAL-ENGINE RUNNING

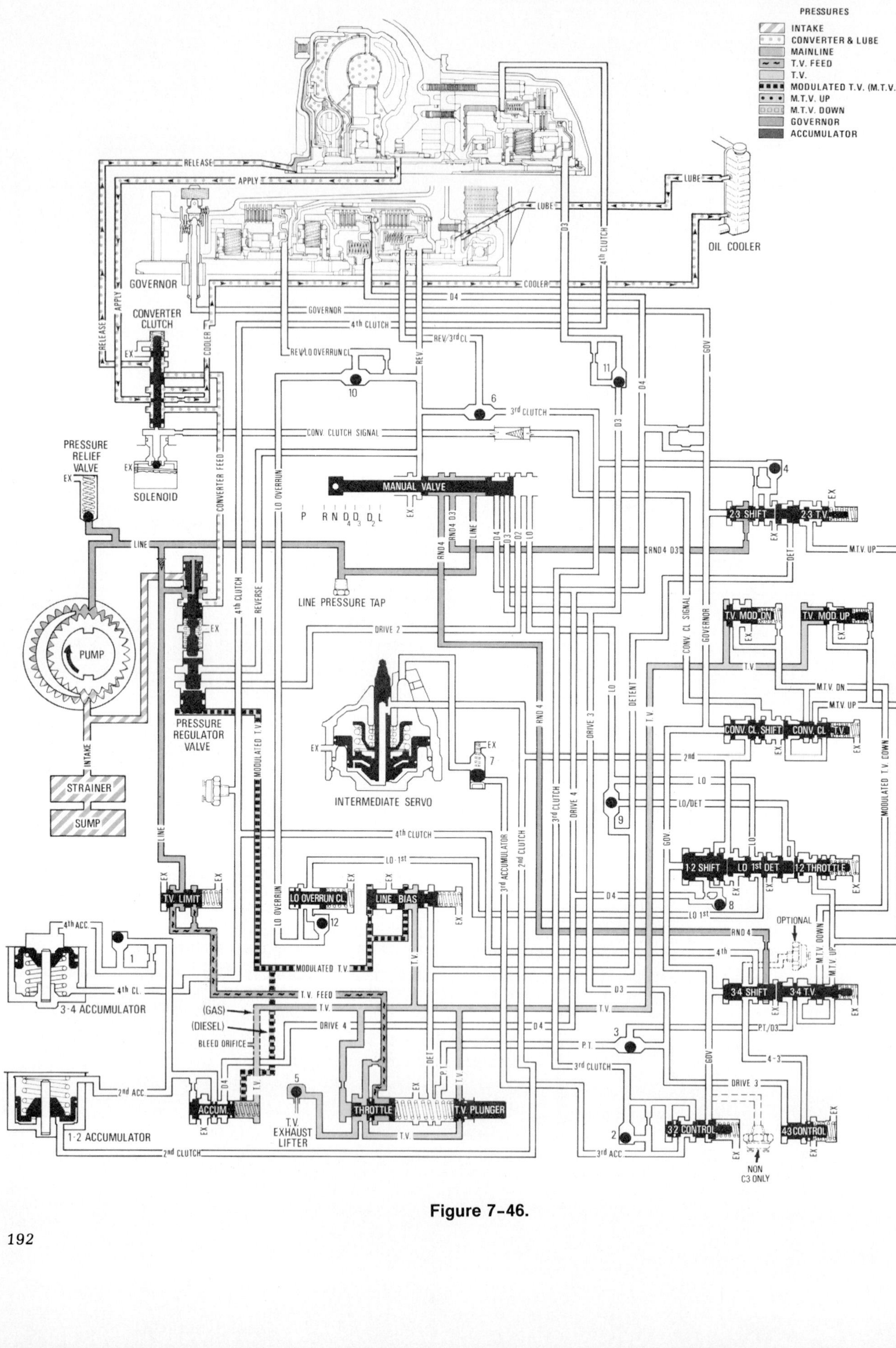

Figure 7-46.

DRIVE RANGE-FIRST GEAR

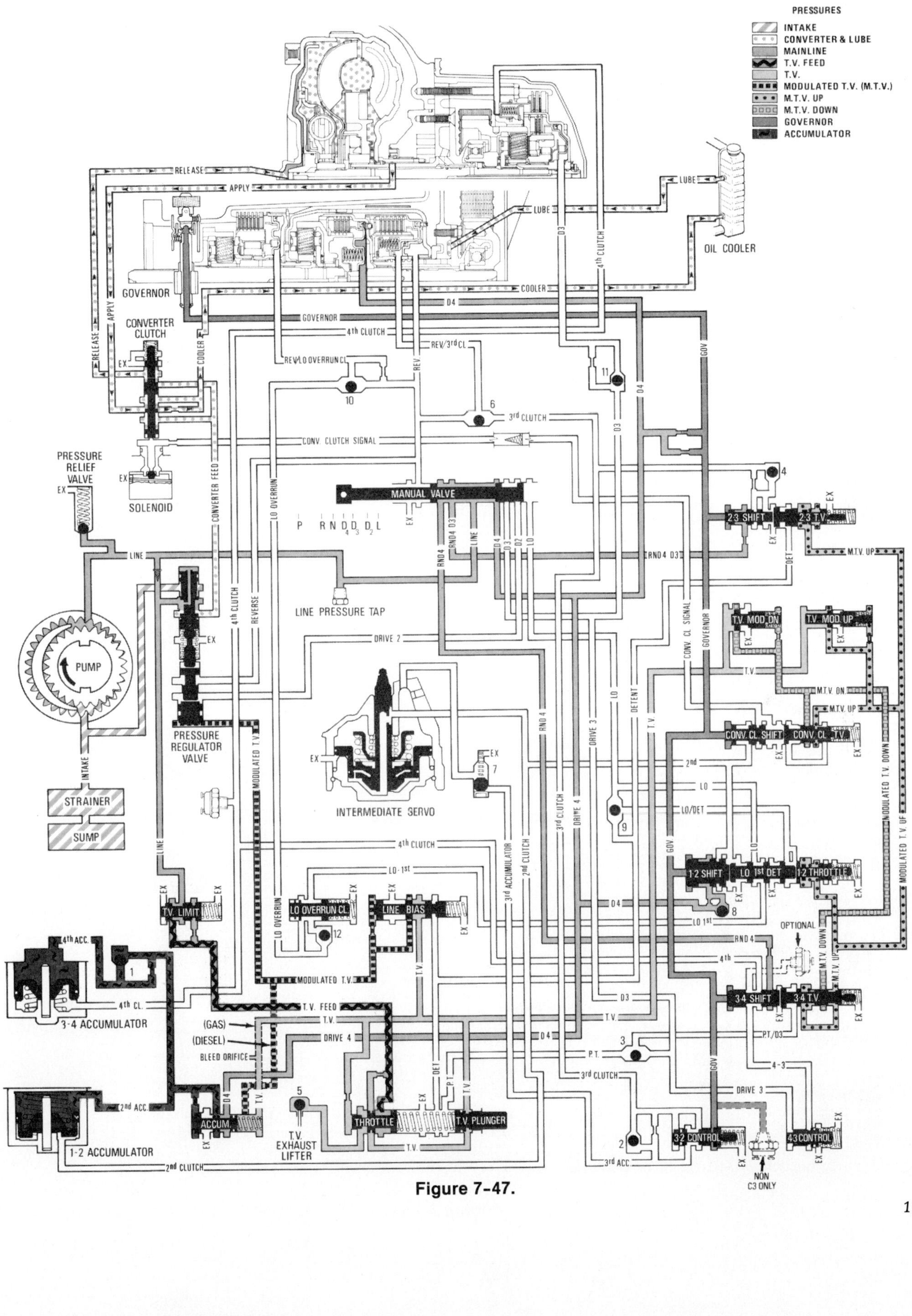

Figure 7-47.

DRIVE RANGE-SECOND GEAR

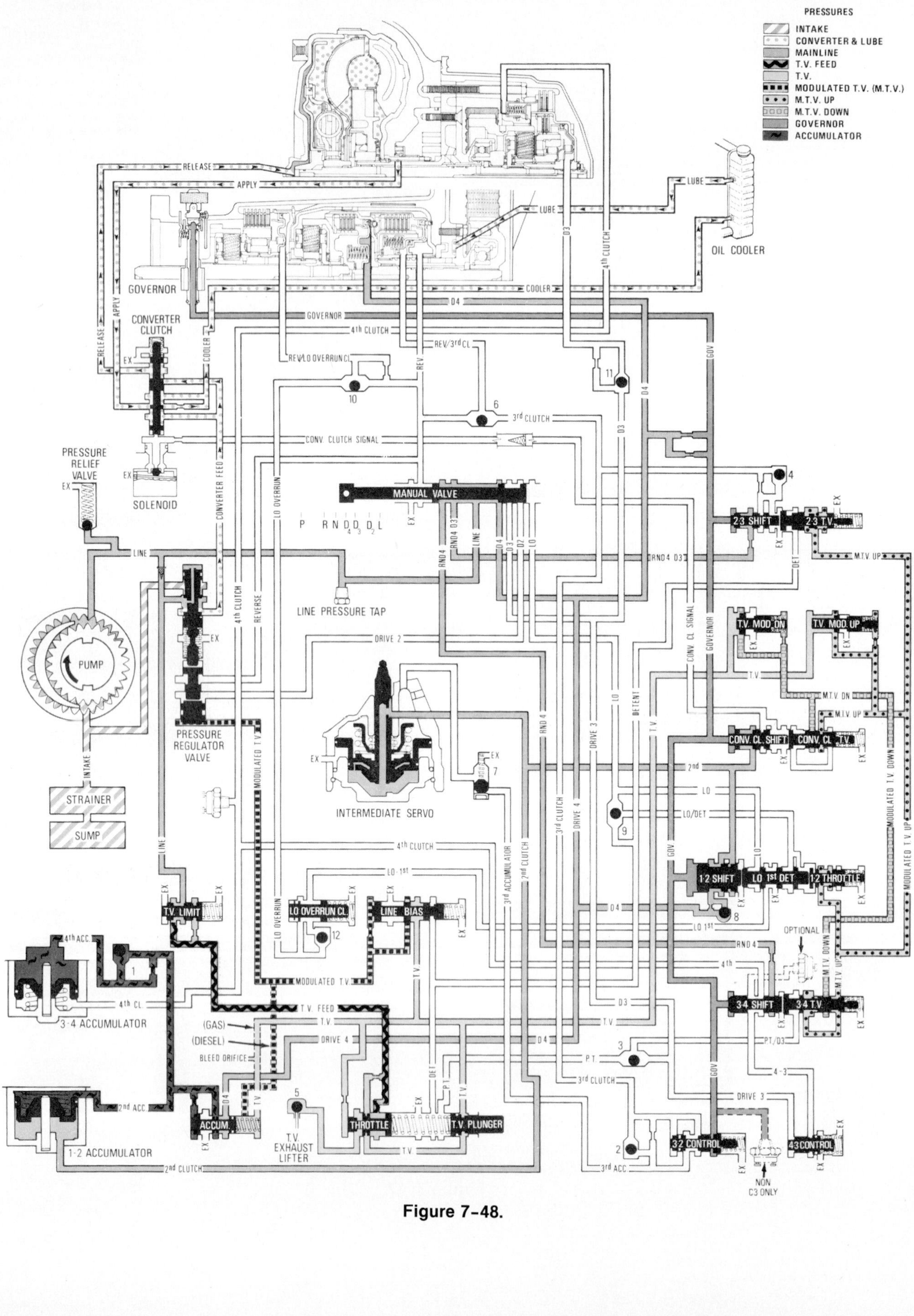

Figure 7-48.

DRIVE RANGE—CONVERTER CLUTCH APPLIED

Second Gear

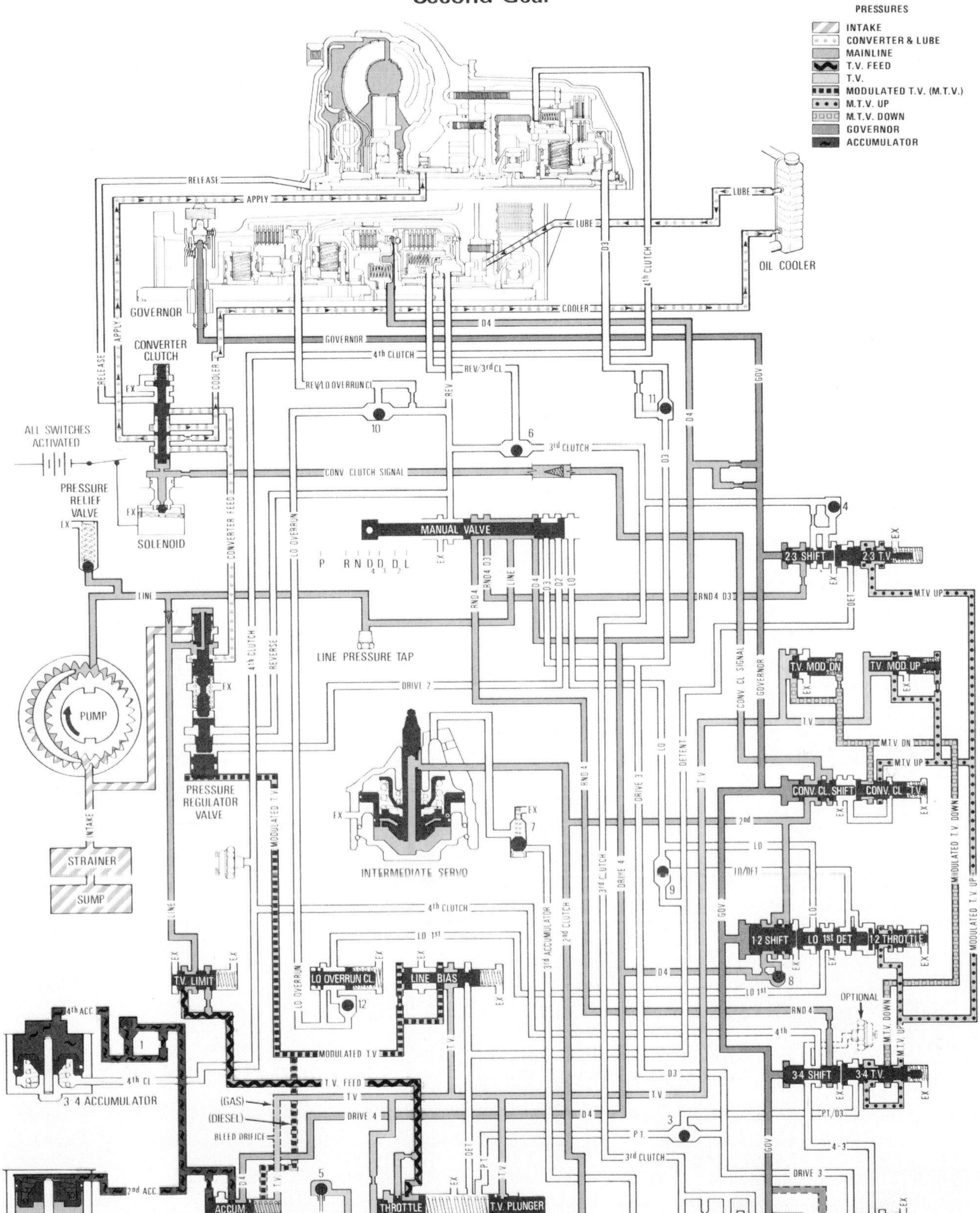

Figure 7-49.

DRIVE RANGE-THIRD GEAR

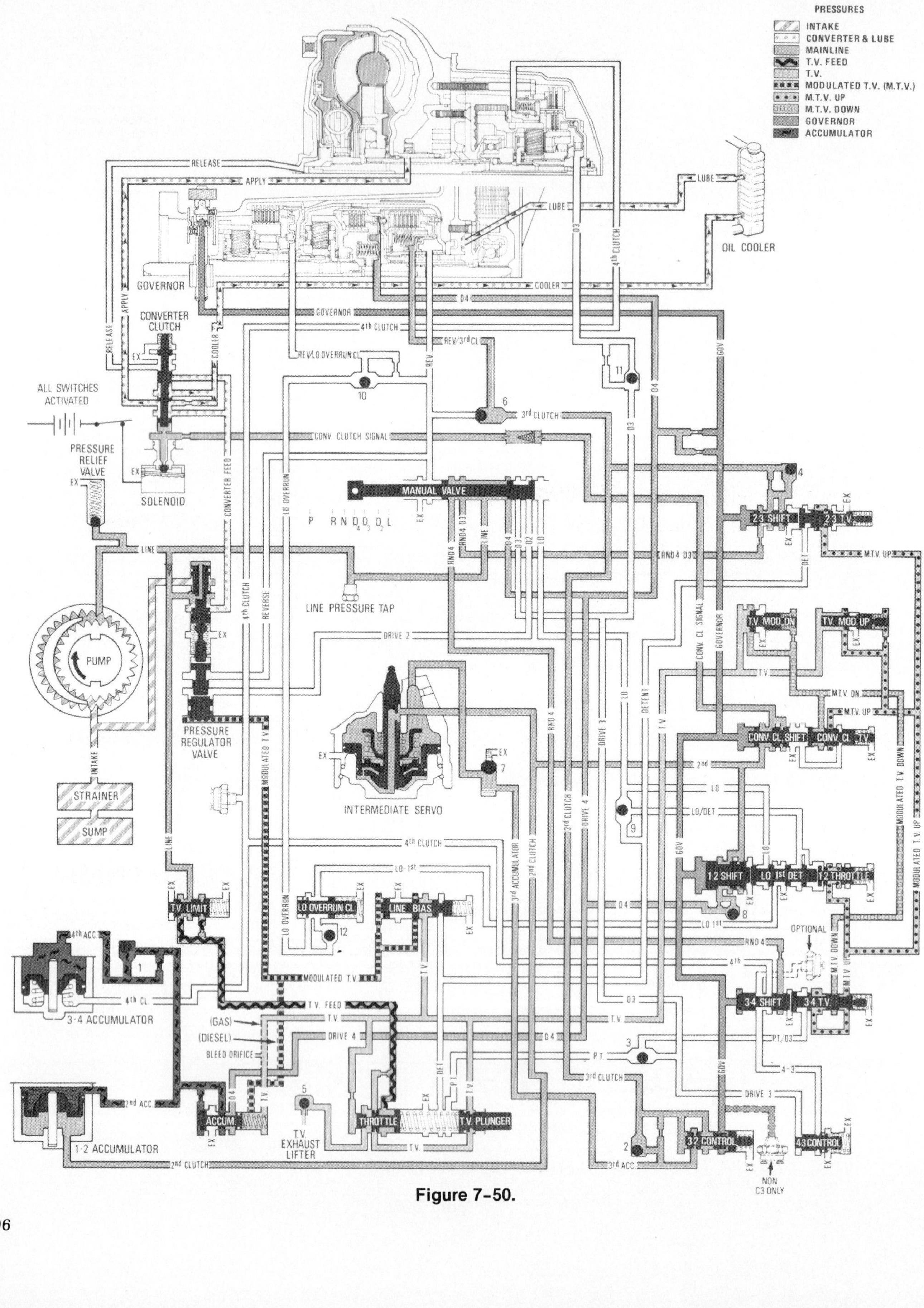

Figure 7-50.

DRIVE RANGE-OVERDRIVE

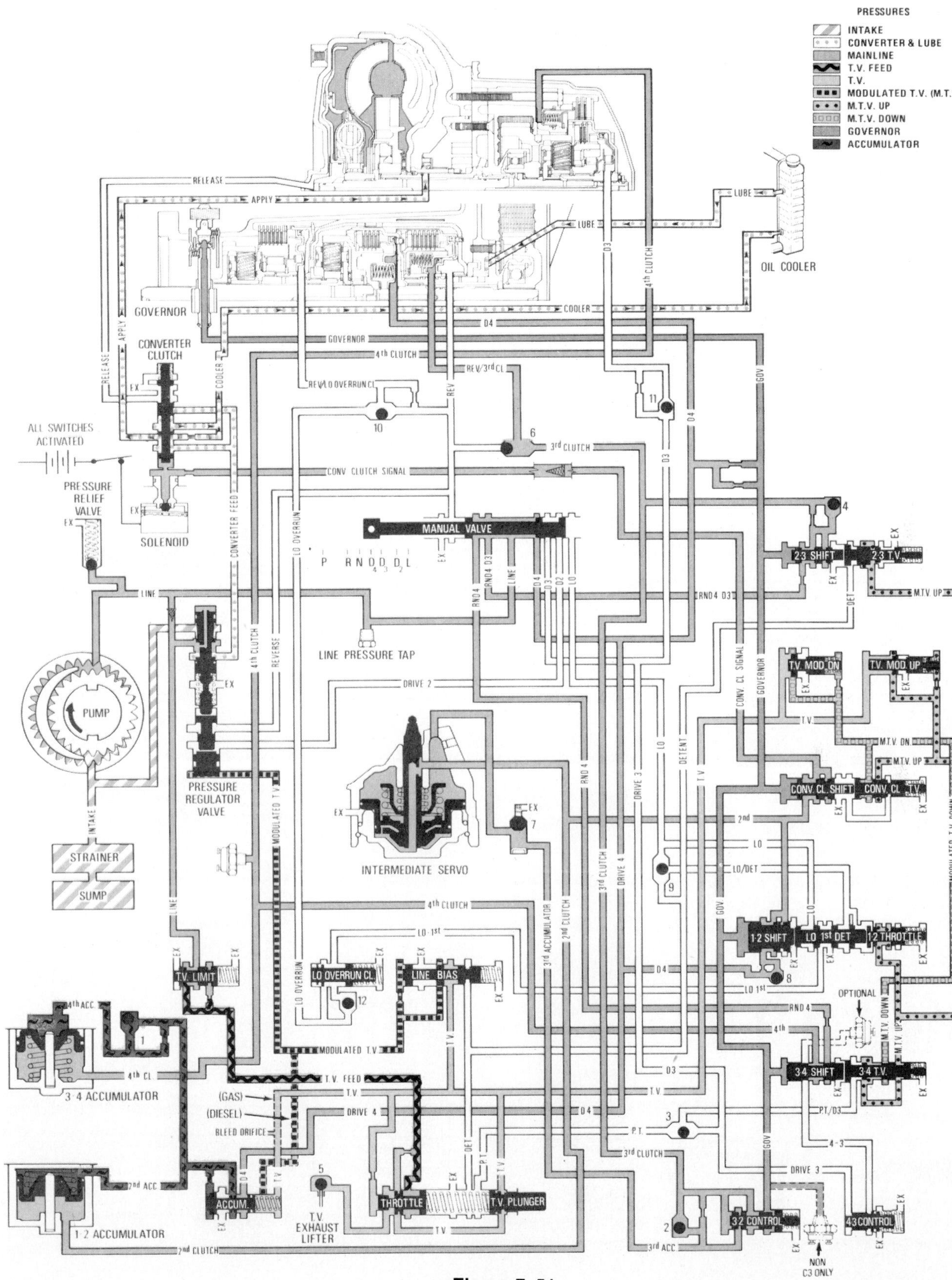

Figure 7-51.

PART THROTTLE 4-3 AND MODULATED DOWNSHIFTS

Valves In Third Gear Position

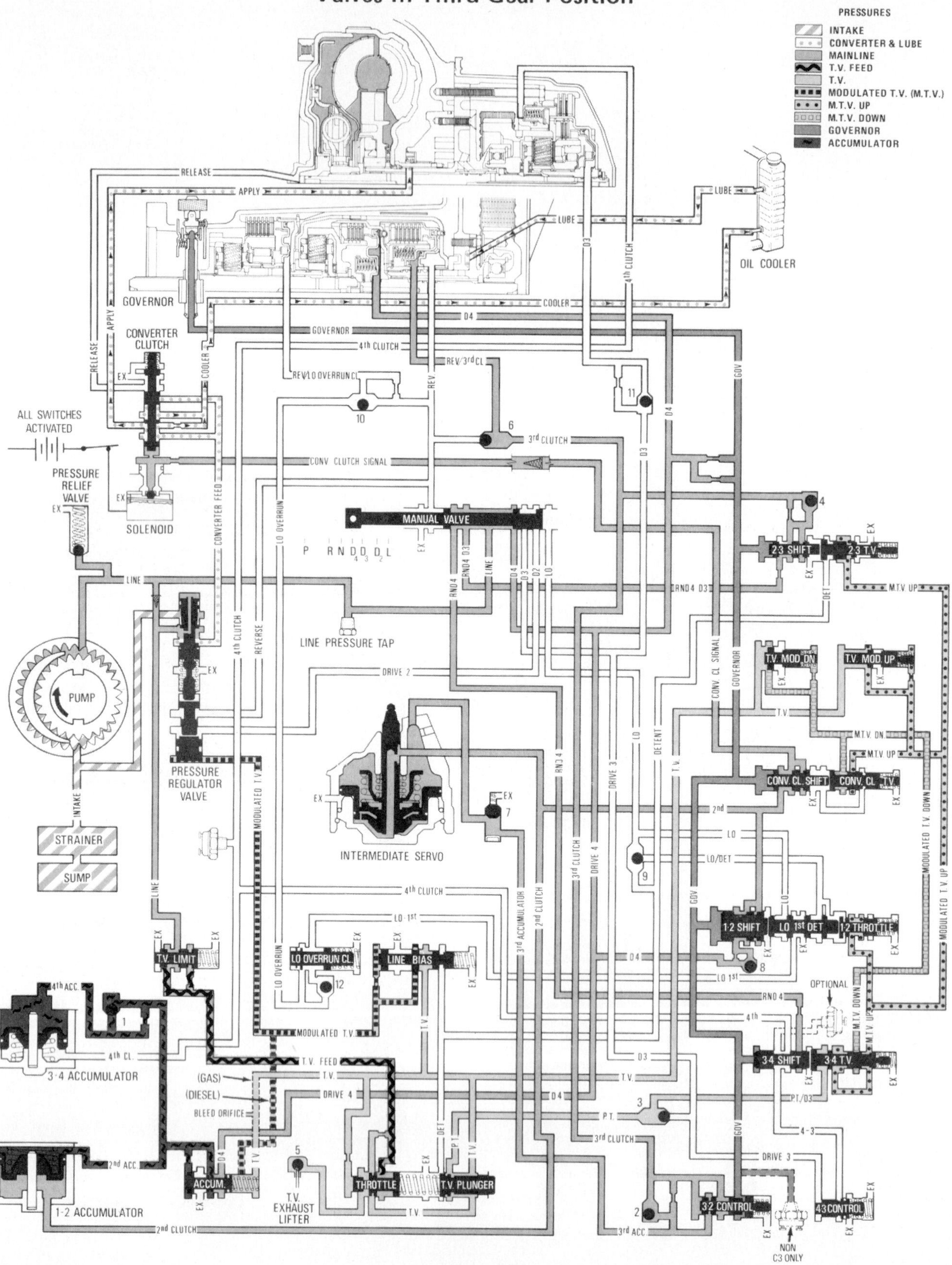

Figure 7-52.

DETENT DOWNSHIFTS

Valves In Second Gear Position

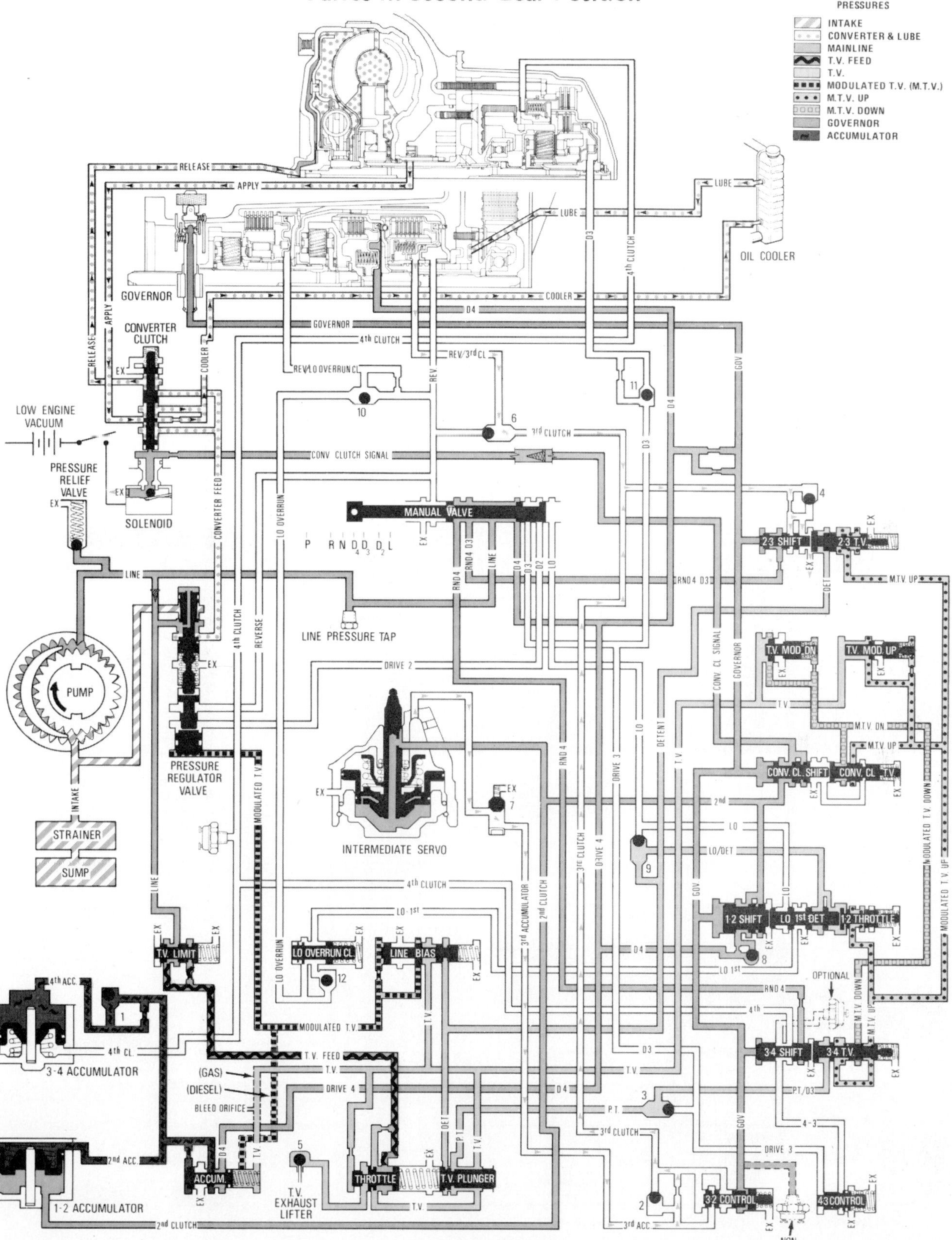

Figure 7-53.

MANUAL THIRD

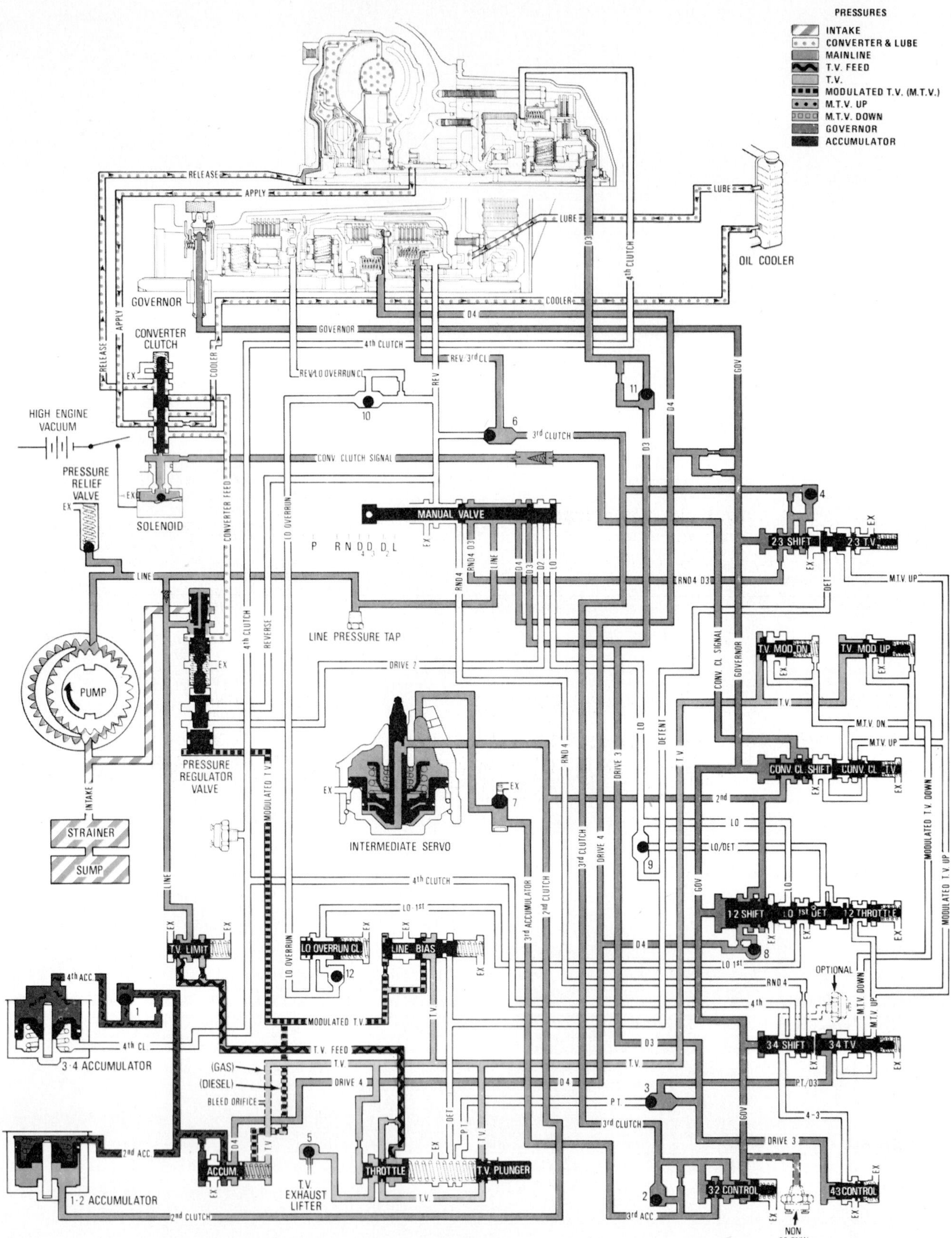

Figure 7-54.

MANUAL SECOND

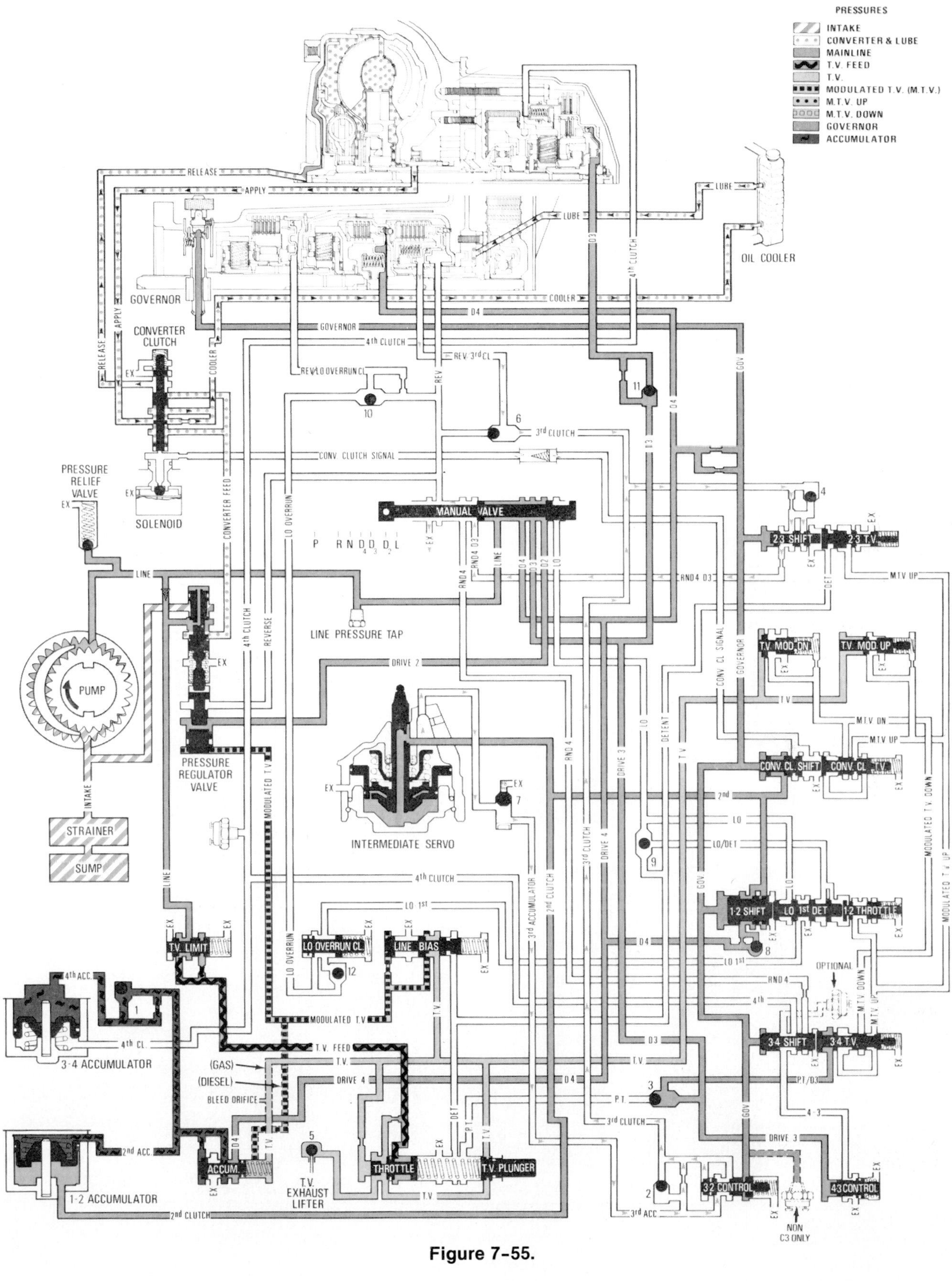

Figure 7-55.

MANUAL LO

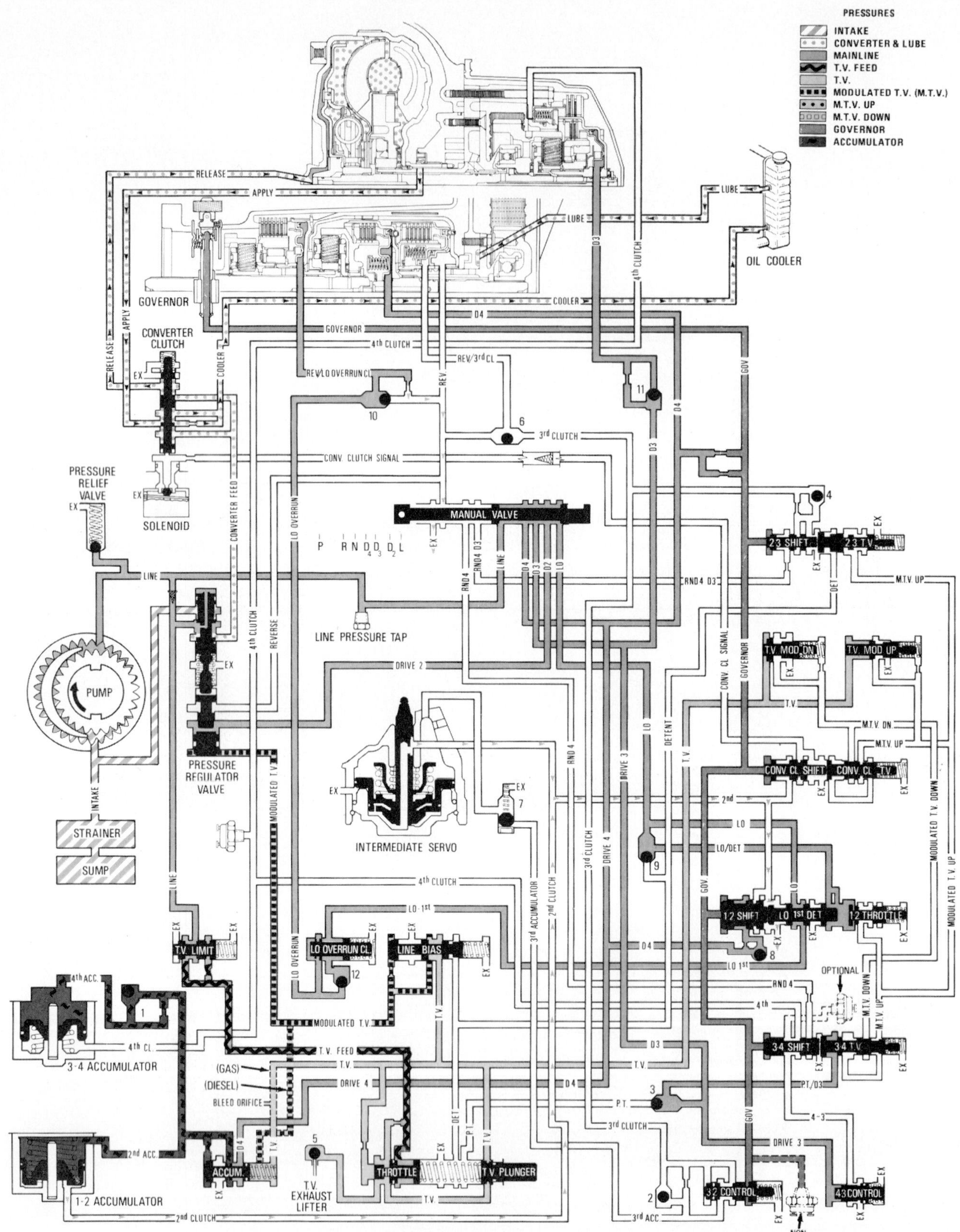

Figure 7-56.

REVERSE

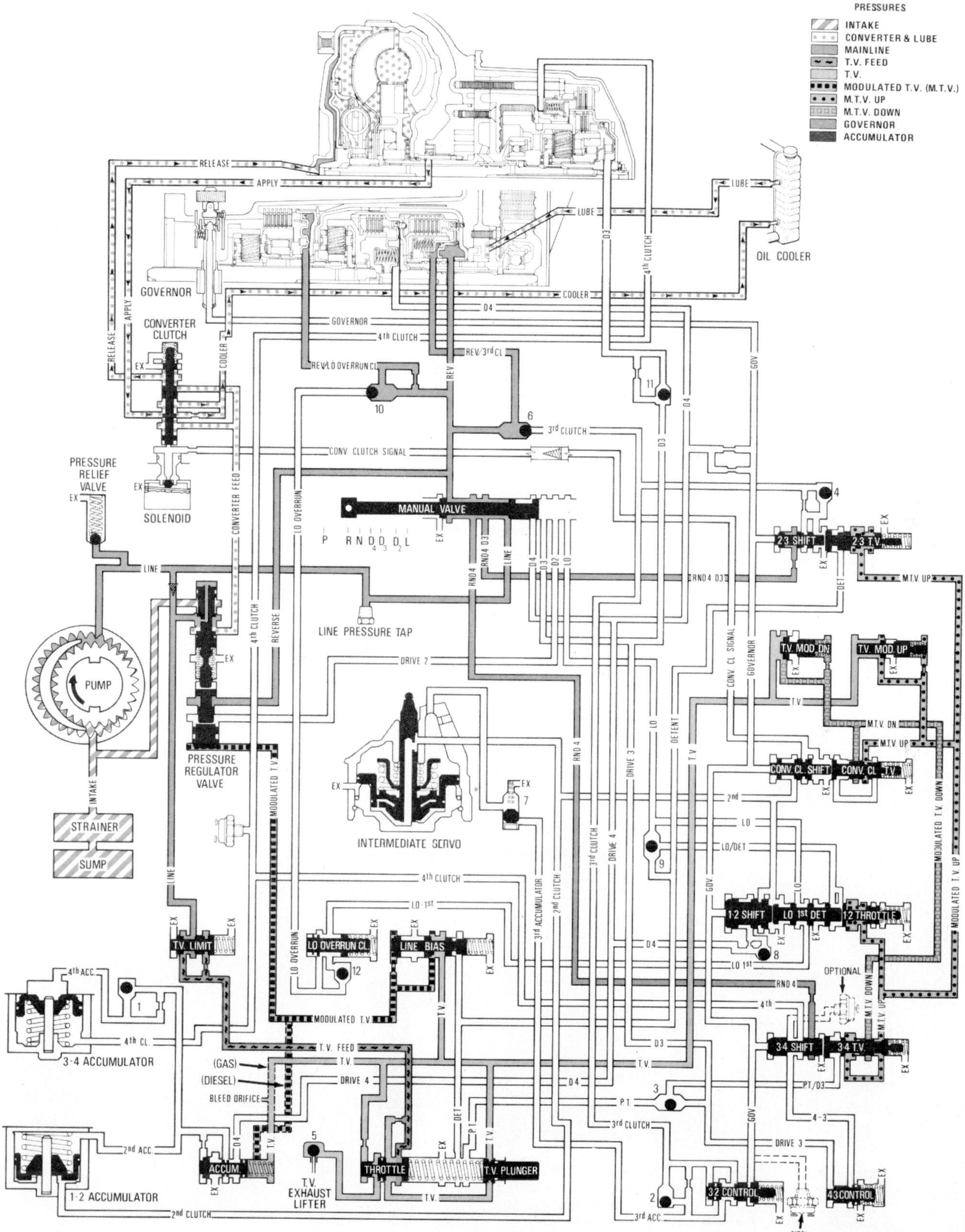

Figure 7-57.

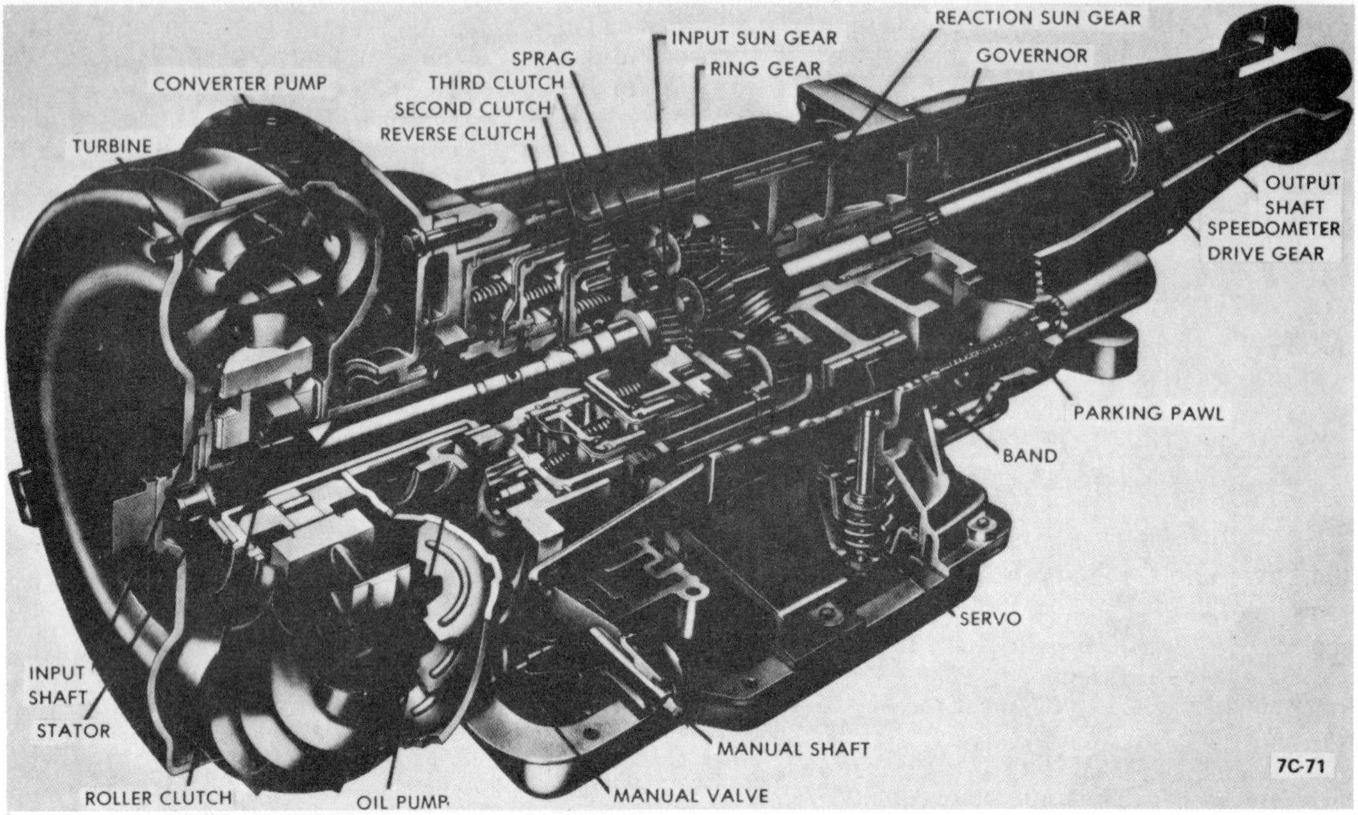

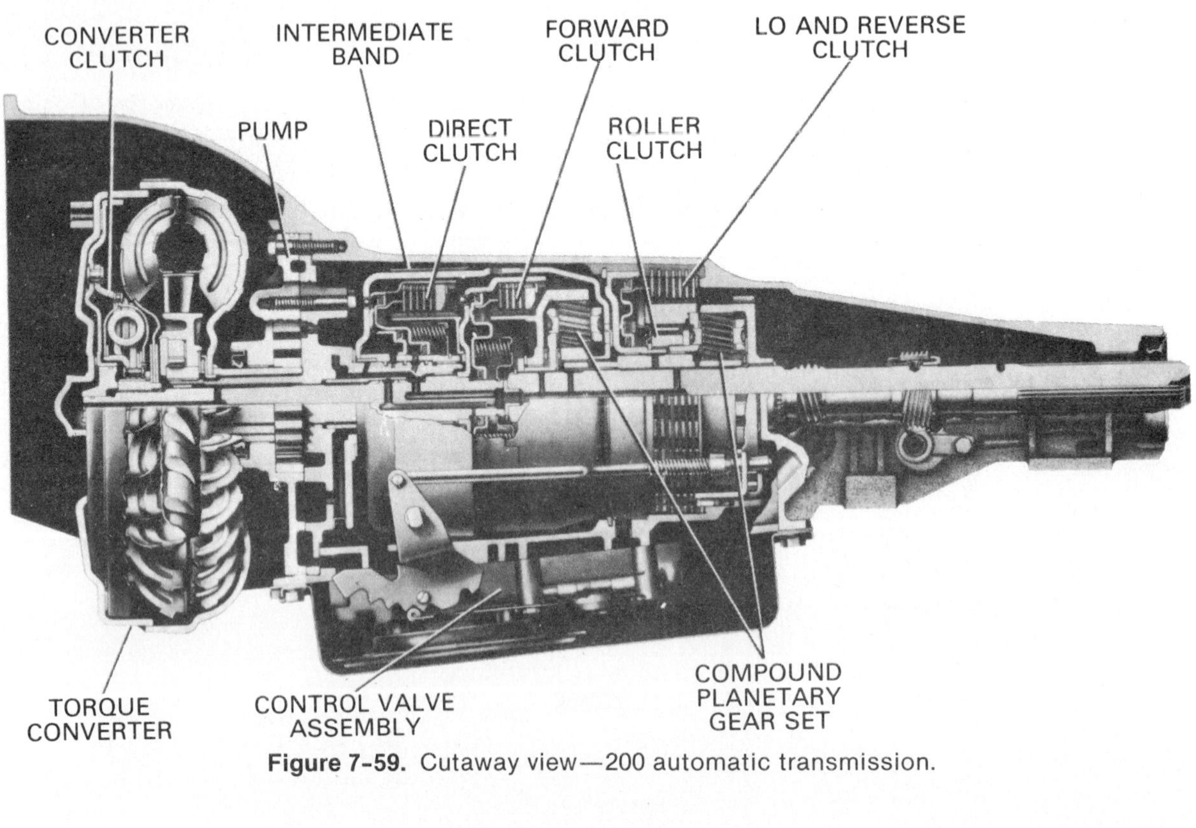

Figure 7-59. Cutaway view—200 automatic transmission.

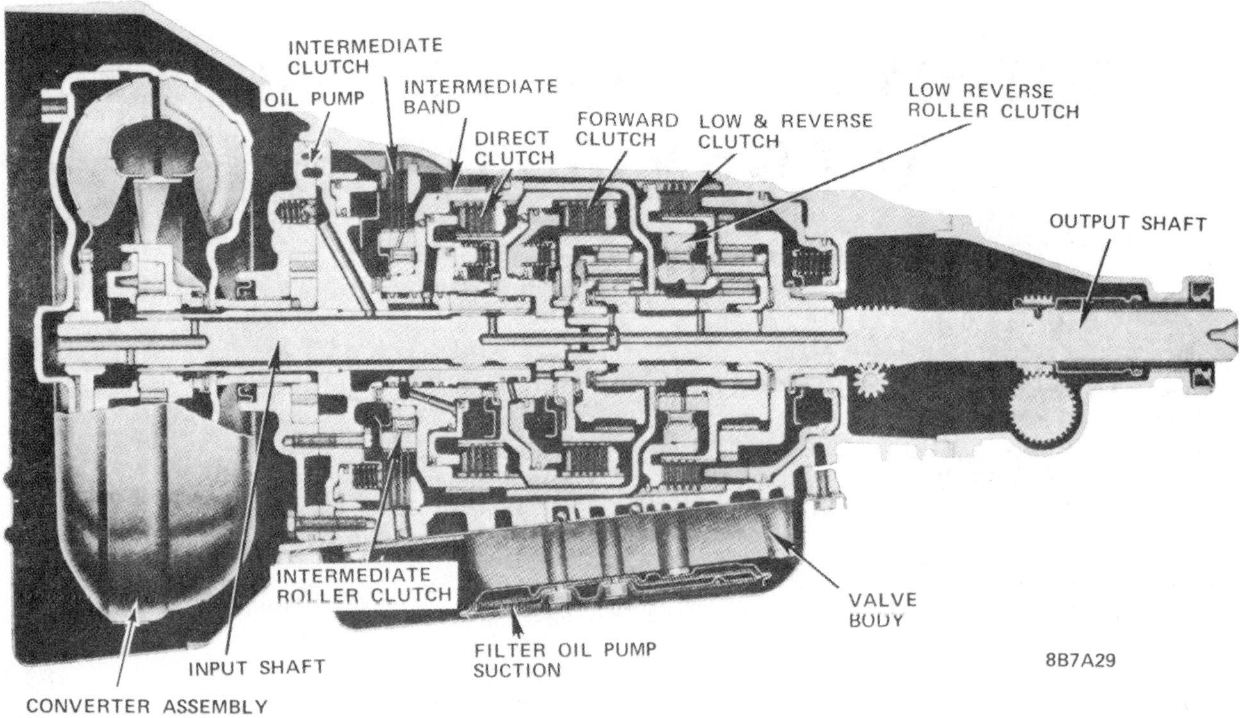

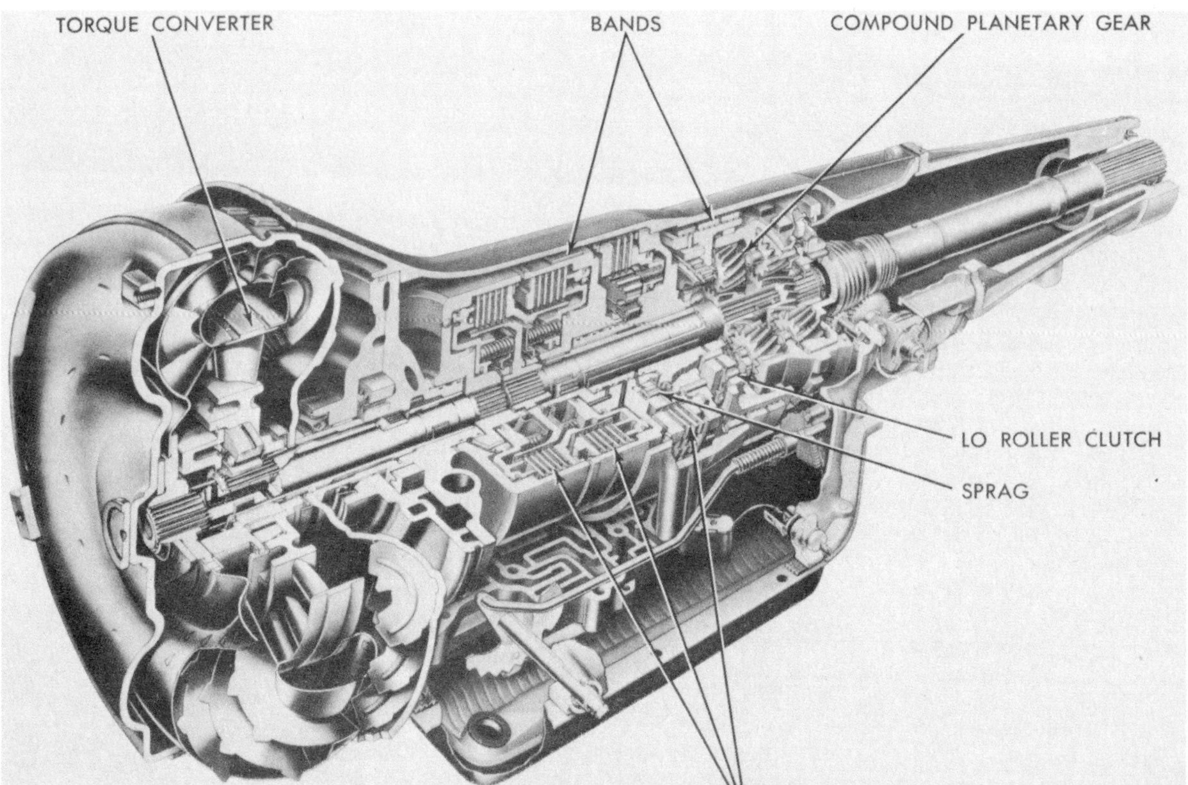

Figure 7-61. Model 400 Hydra-Matic.

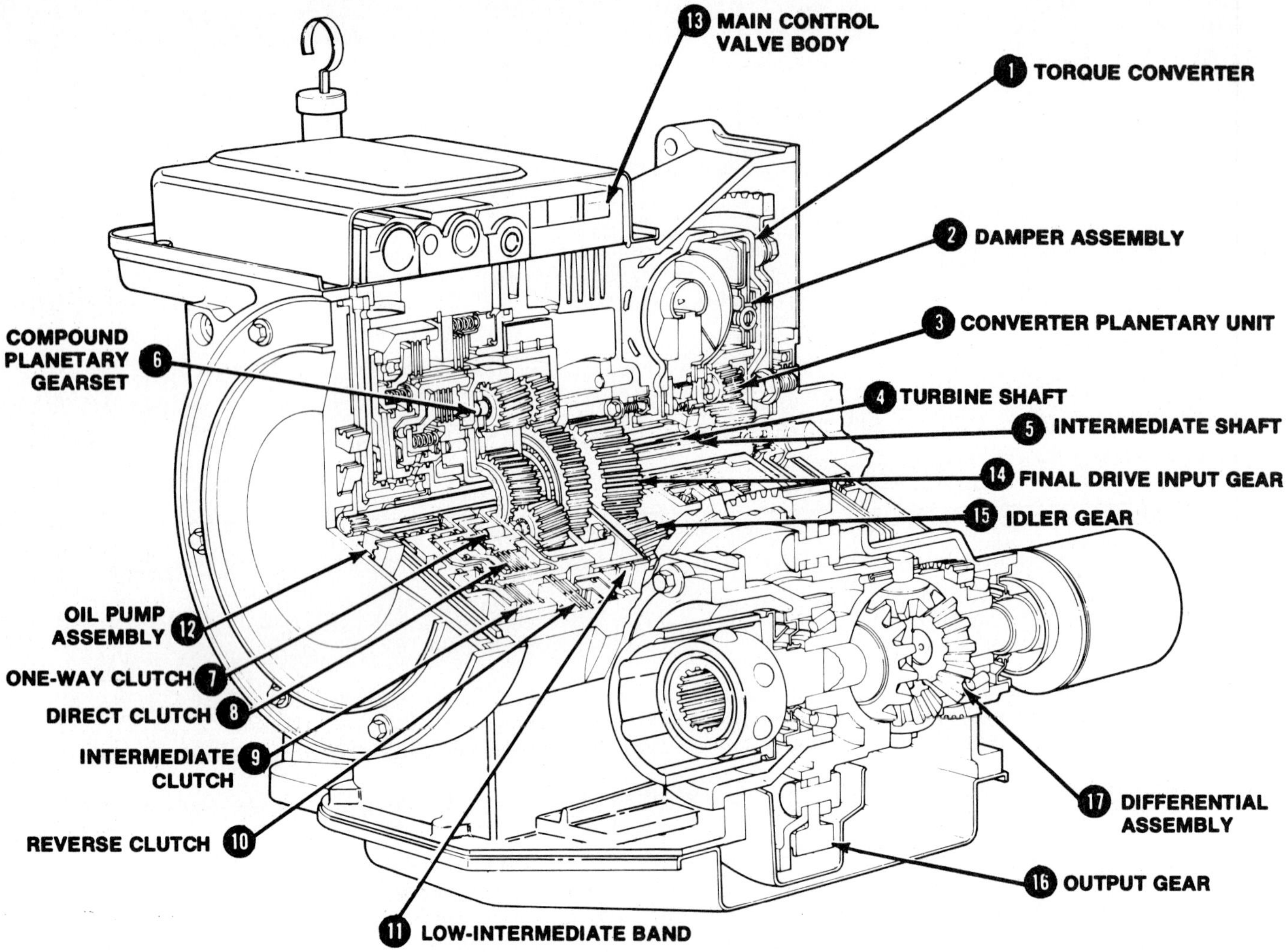

Figure 7-62. ATX automatic transaxle.

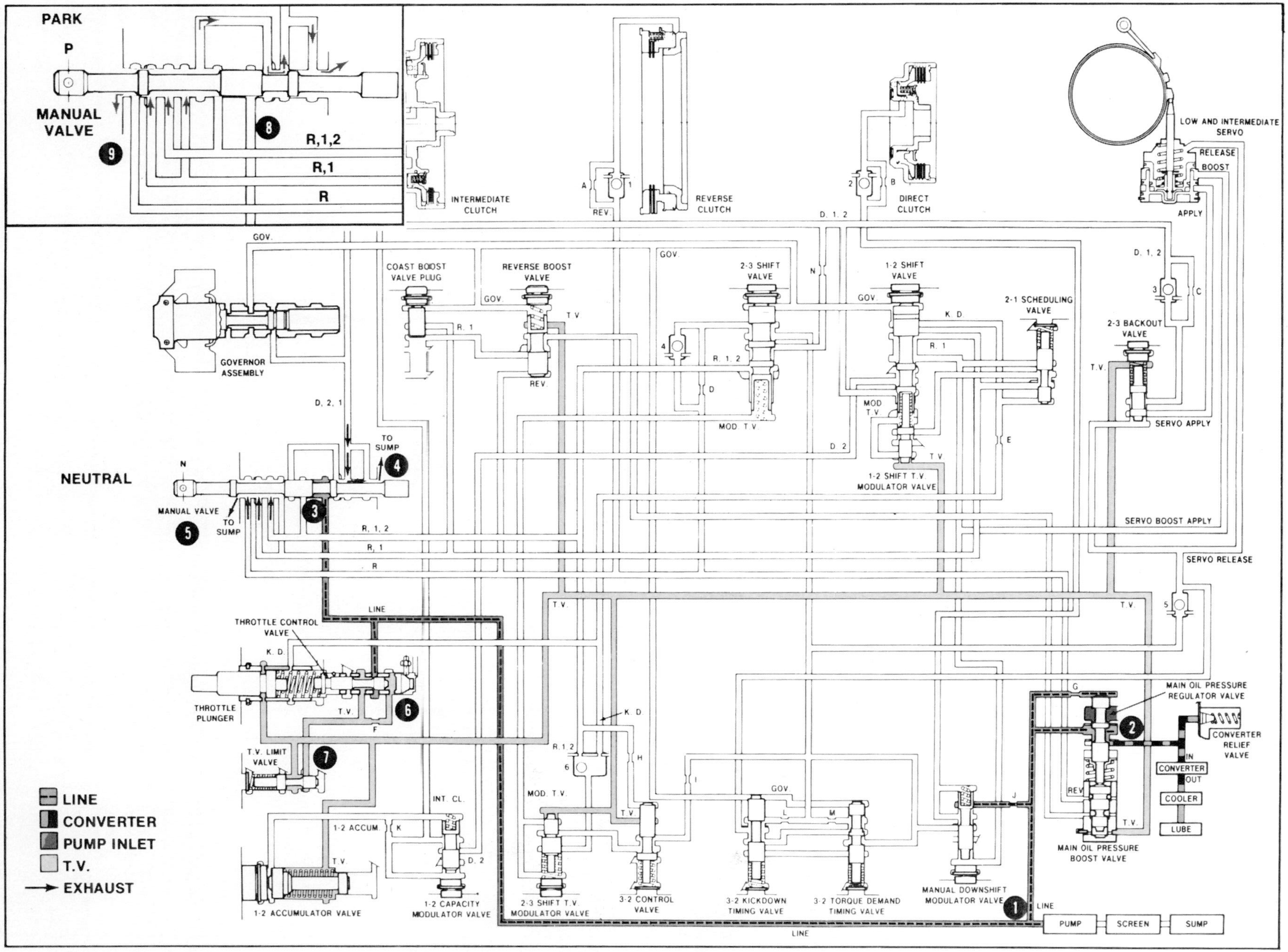

Figure 7-63. Hydraulic system in neutral or park at light throttle.

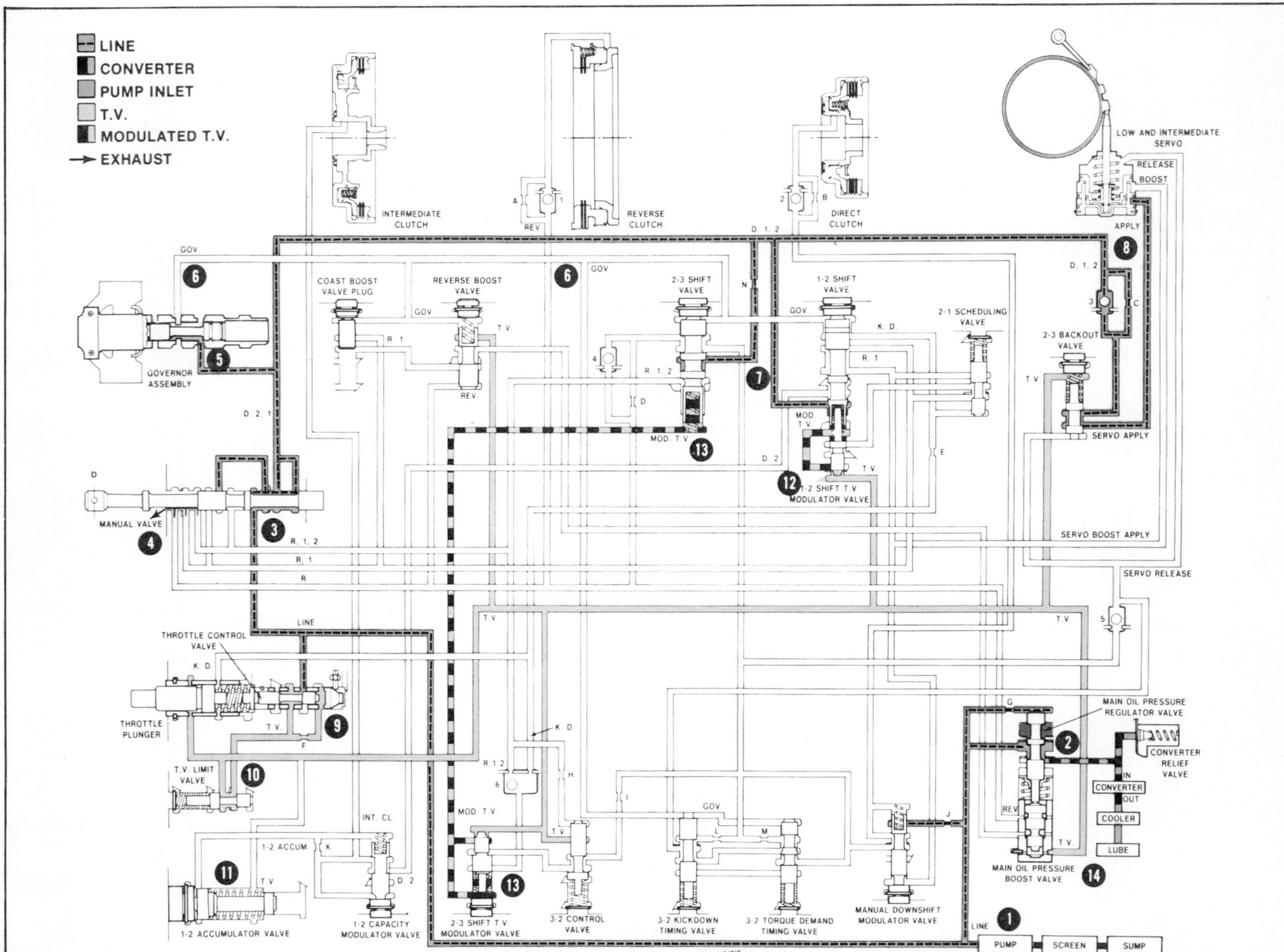

Figure 7-64. Hydraulic system in drive range, first gear.

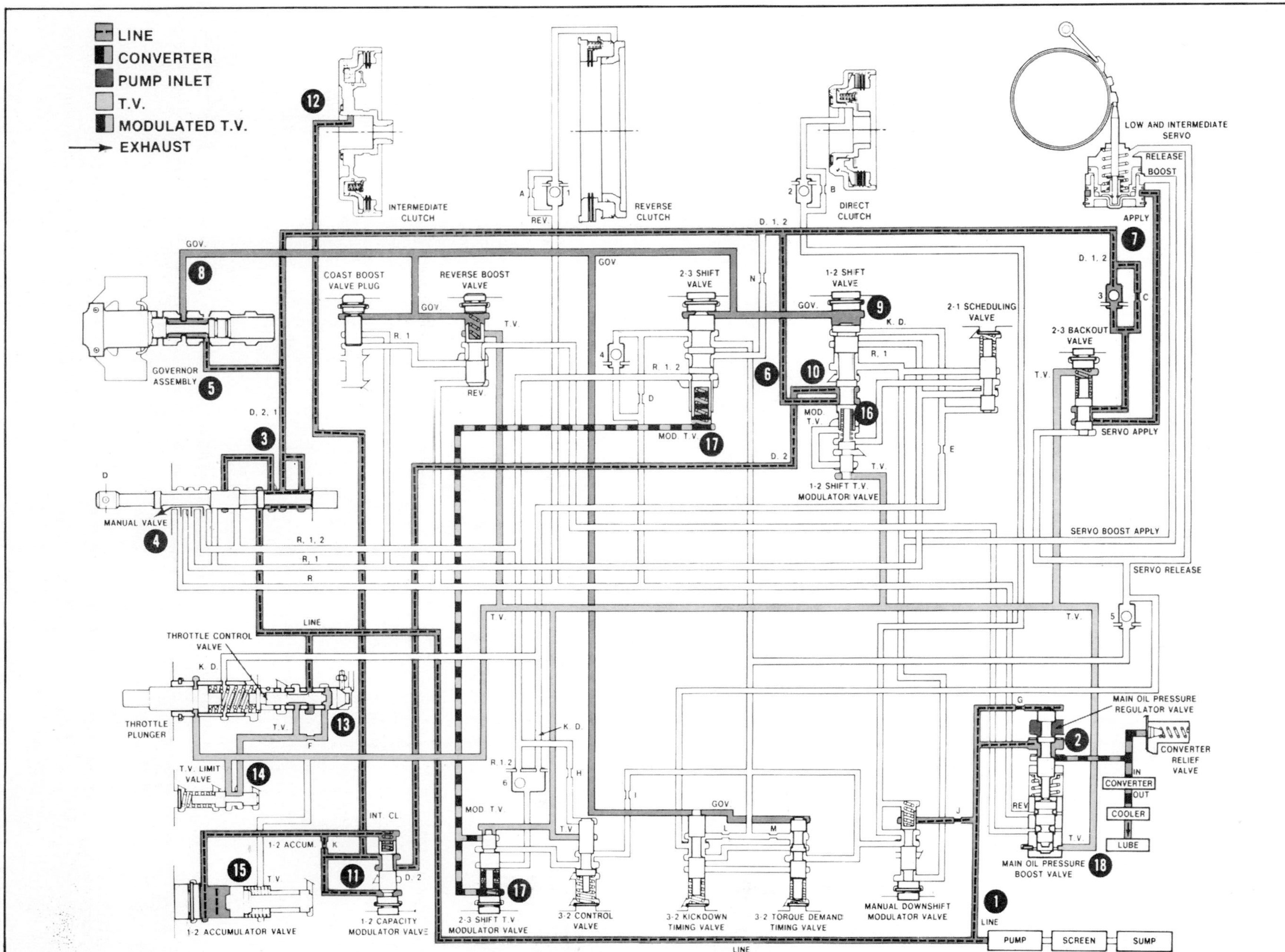

Figure 7-65. Hydraulic system in drive range, second gear.

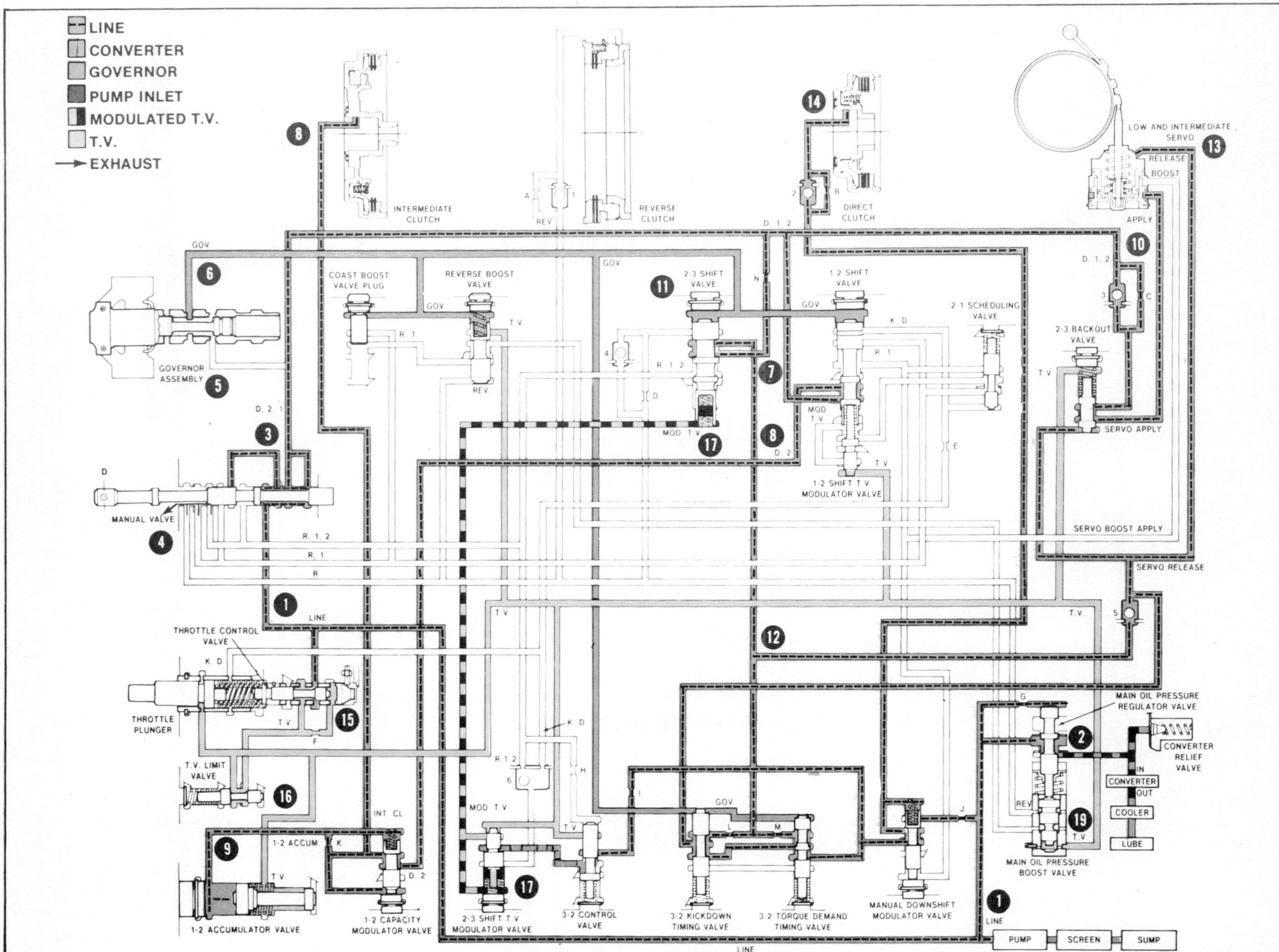

Figure 7-66. Hydraulic system in drive range, third gear.

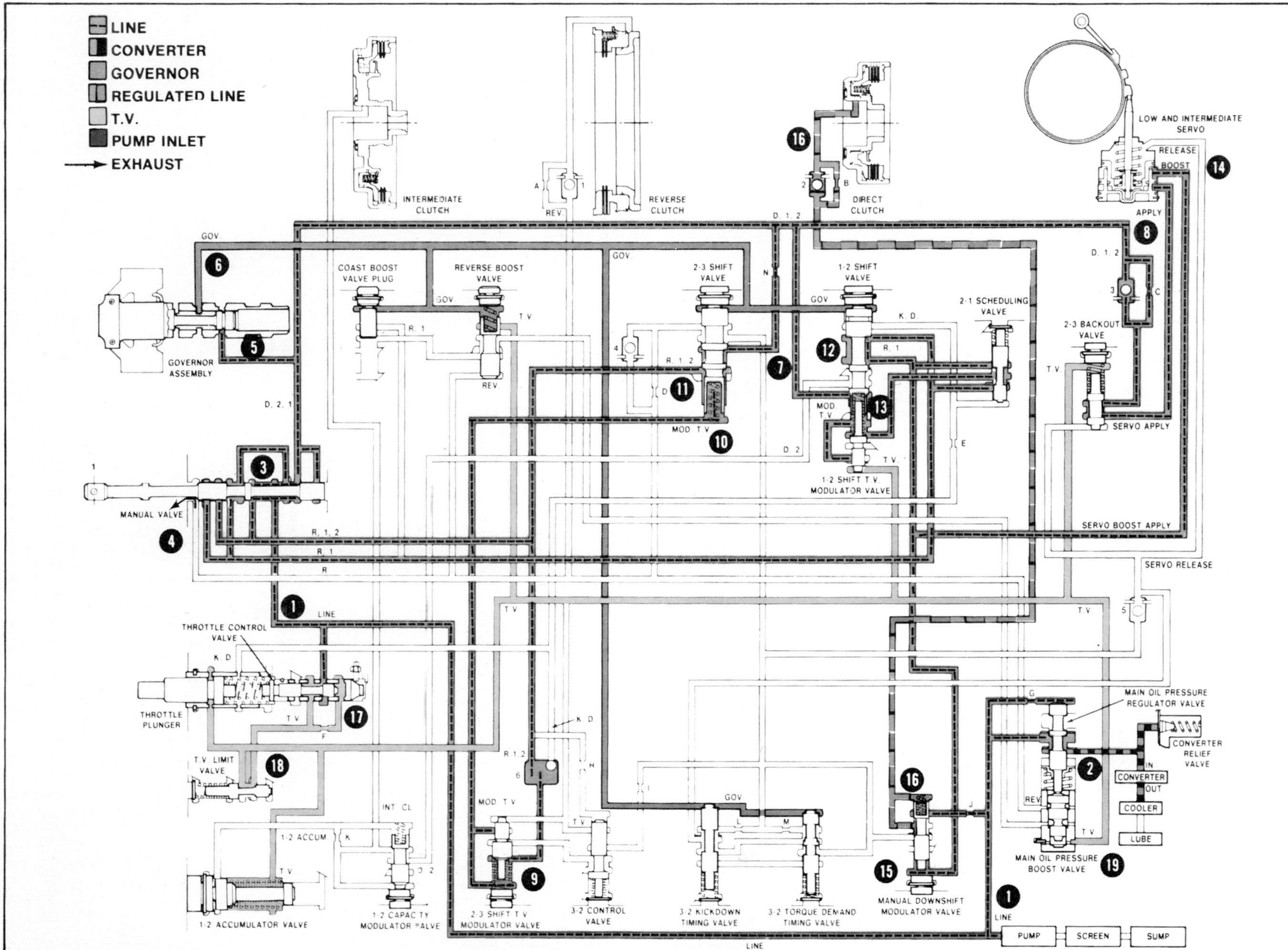

Figure 7-67. Hydraulic system in manual low gear.

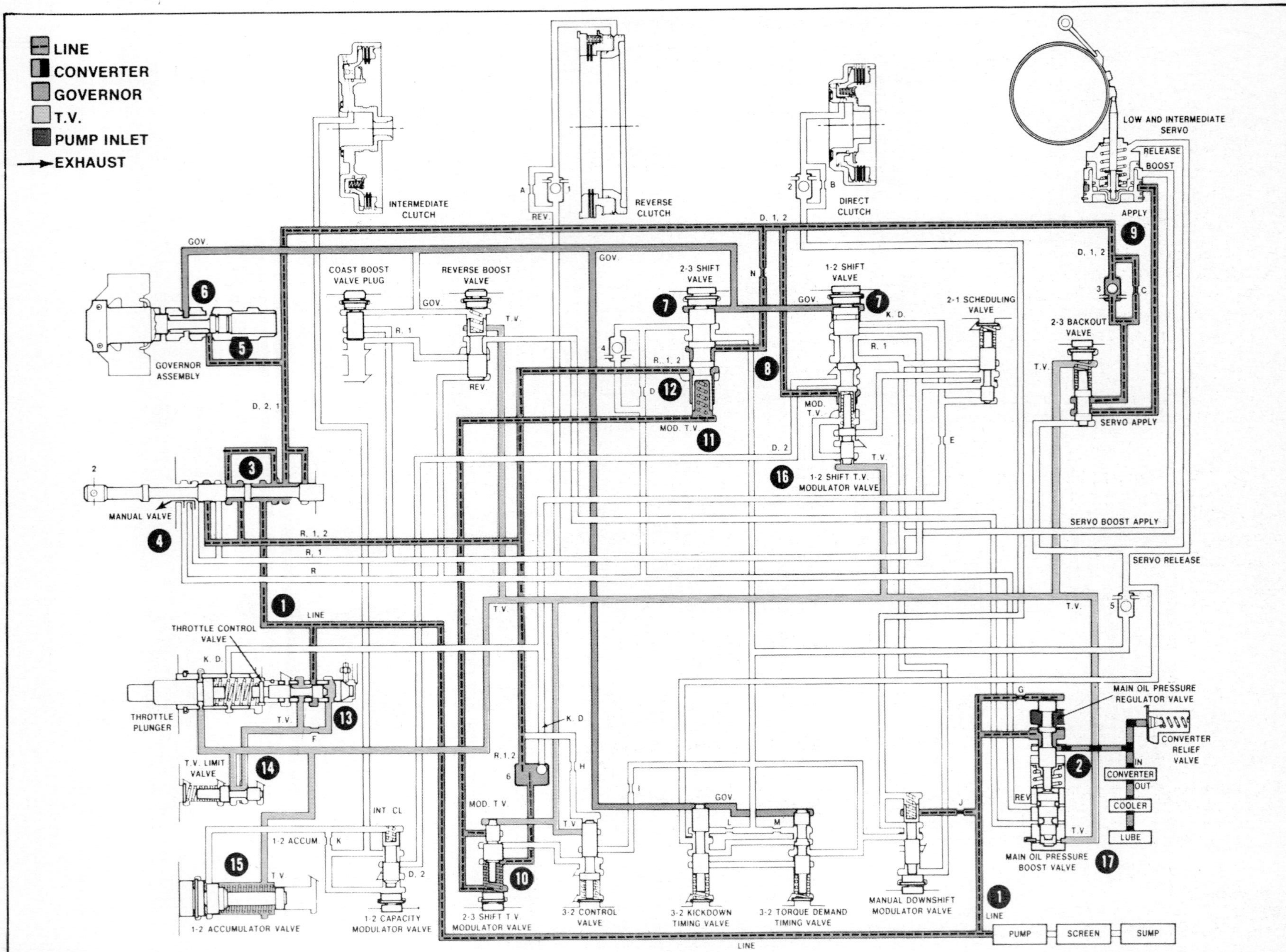

Figure 7-68. Hydraulic system in range 2, first gear.

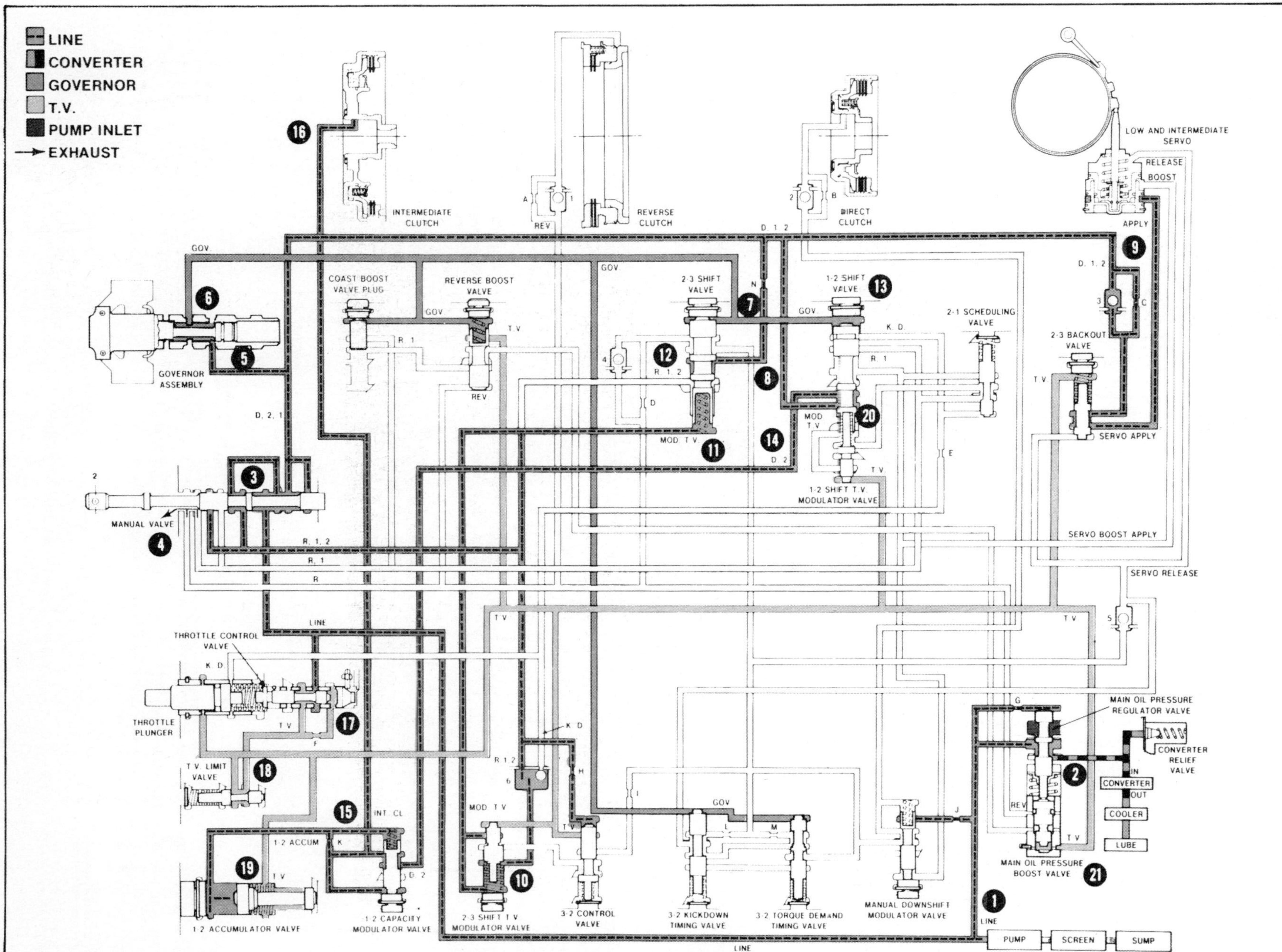

Figure 7-69. Hydraulic system in range 2, second gear.

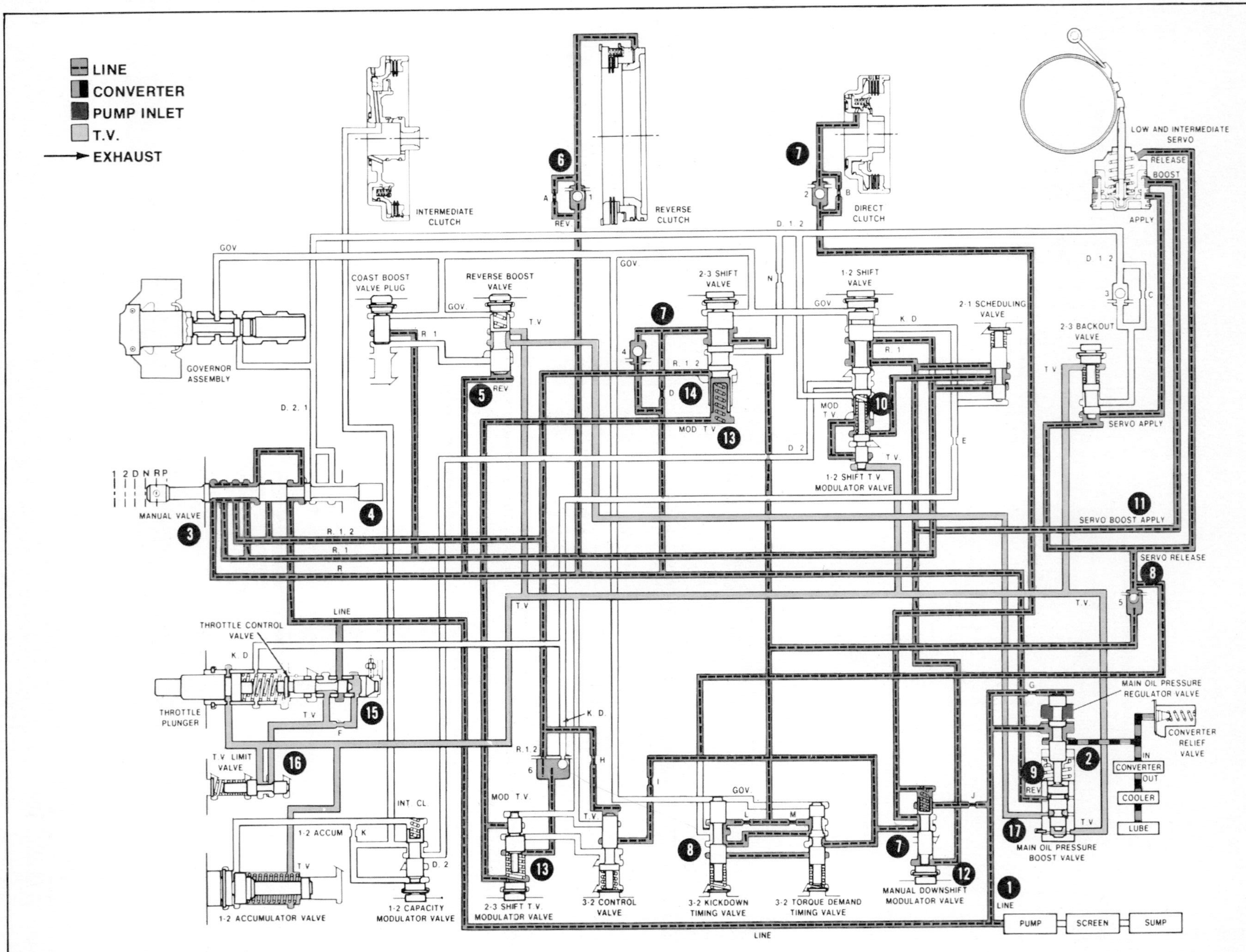

Figure 7-70. Hydraulic system in reverse.

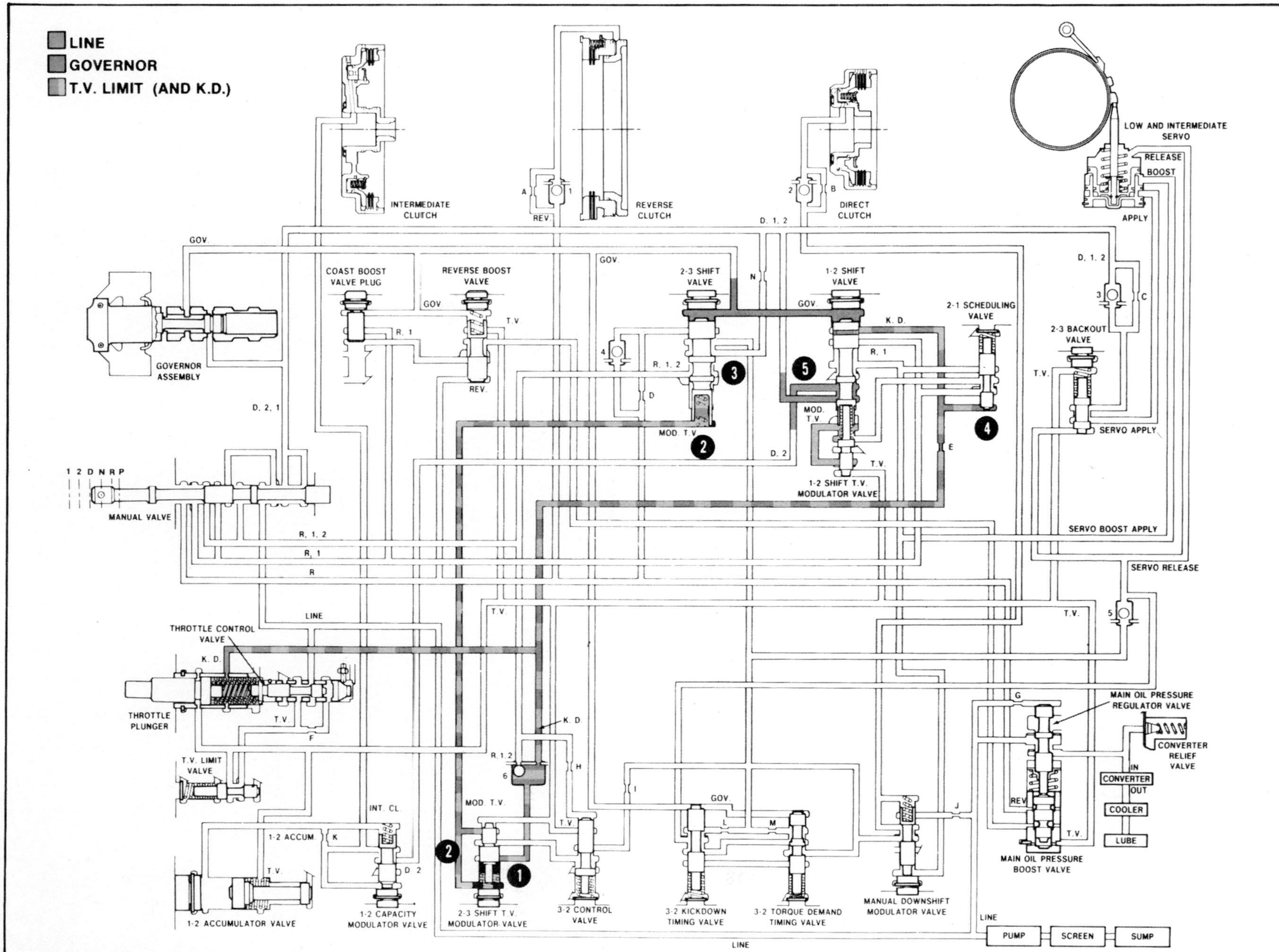

Figure 7-71. Kickdown system.

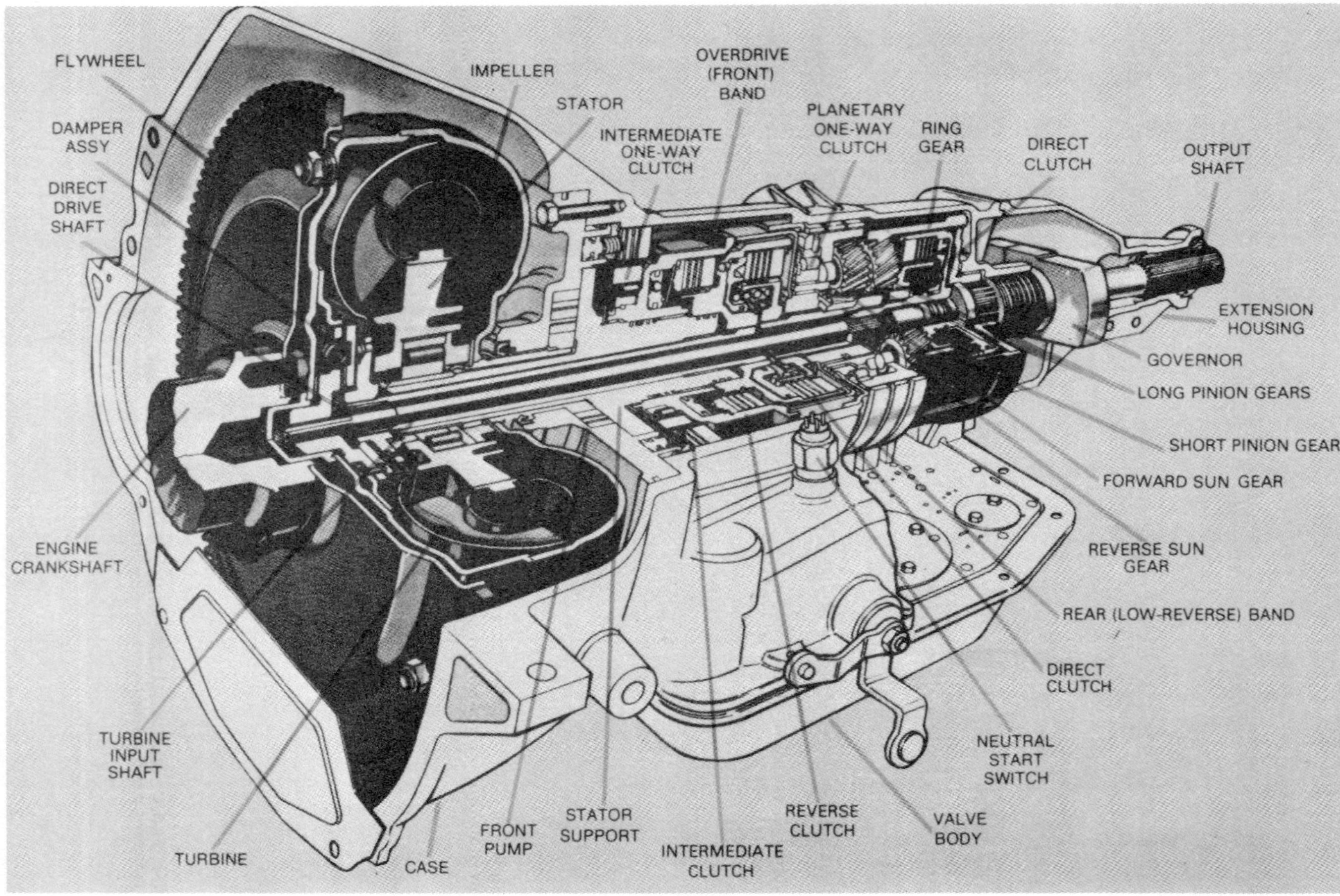

Figure 7-72. Automatic overdrive transmission.

NEUTRAL — CLOSED THROTTLE

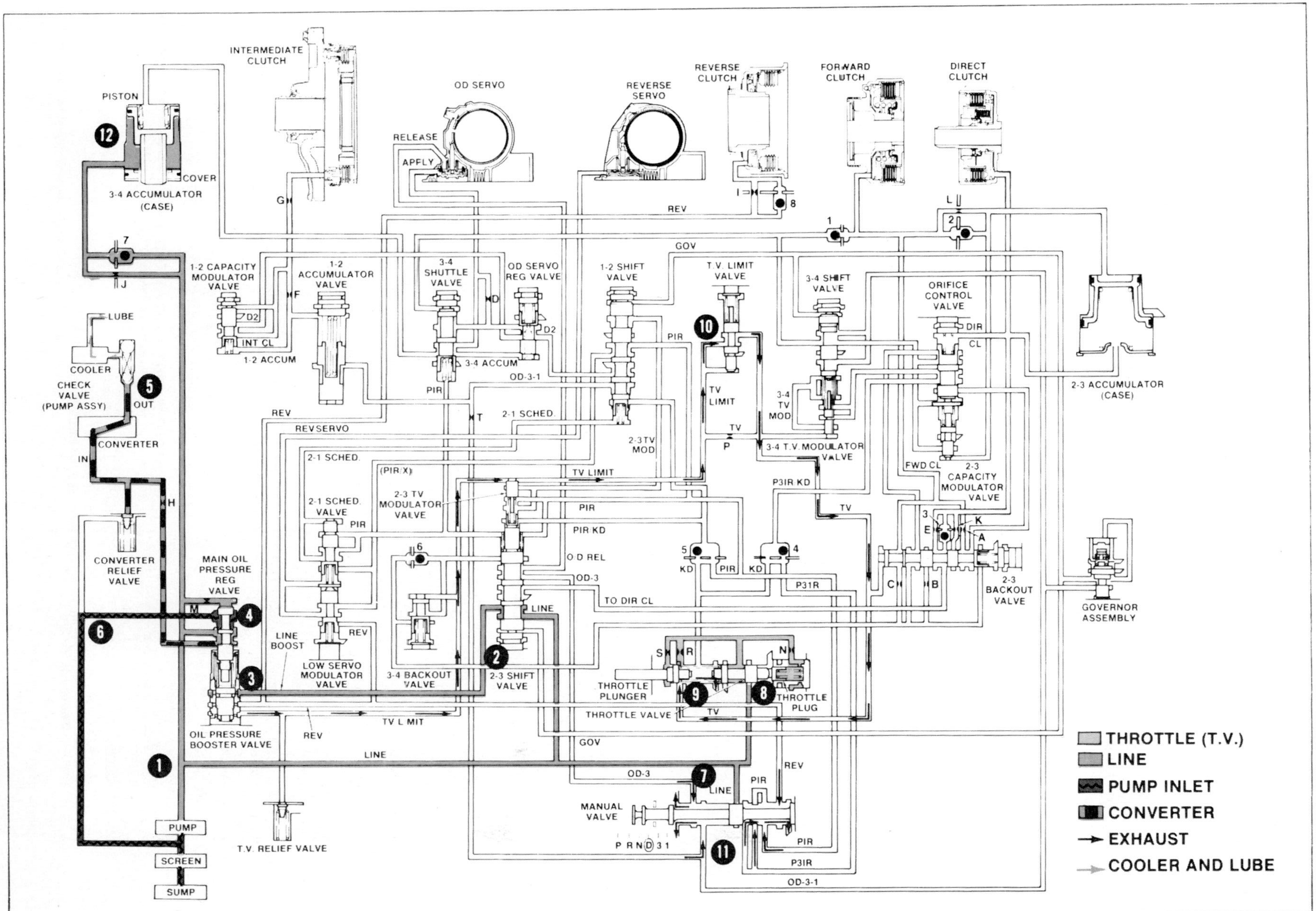

Figure 7-73.

PARK — CLOSED THROTTLE

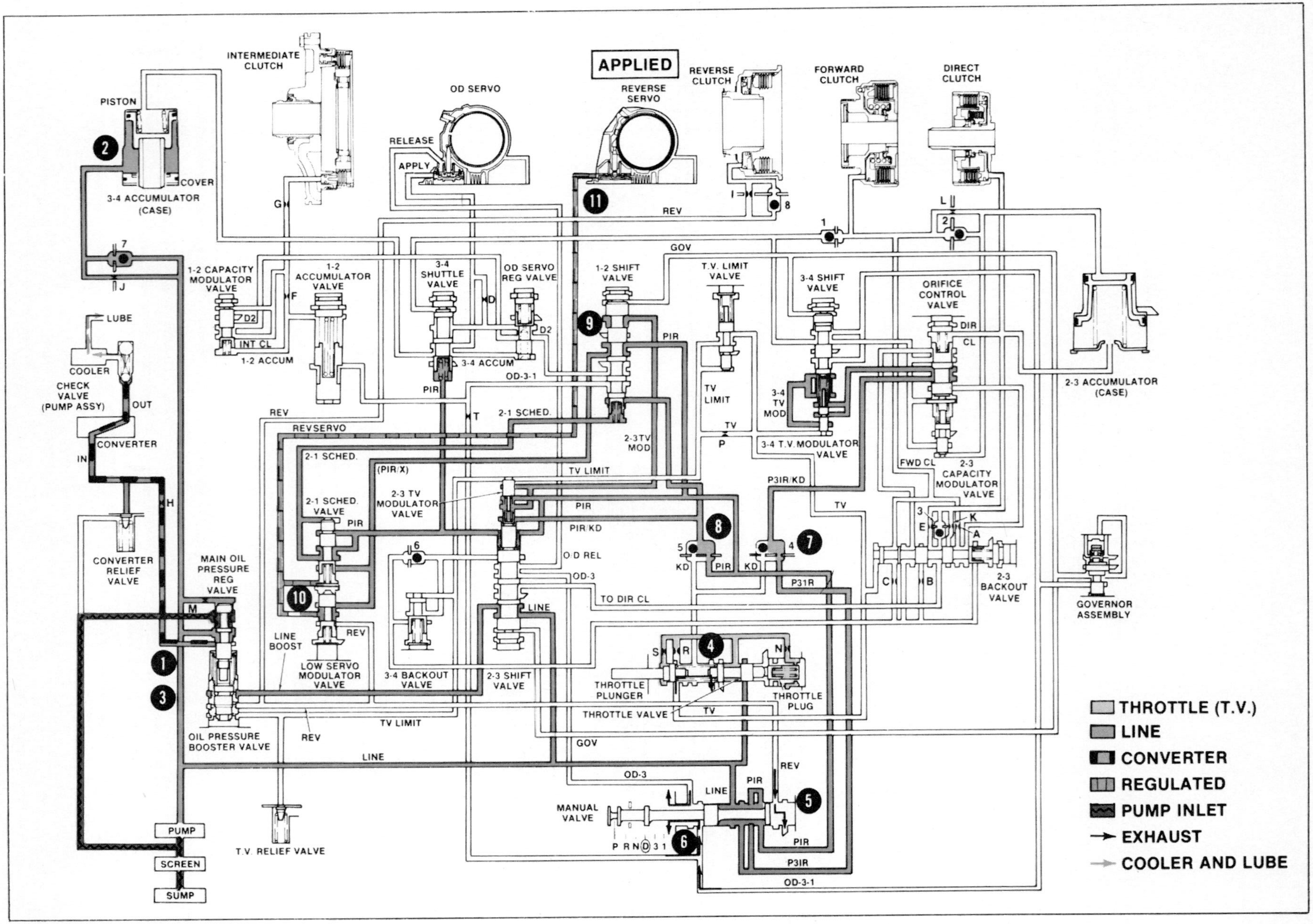

Figure 7-74.

1ST GEAR IN OVERDRIVE RANGE

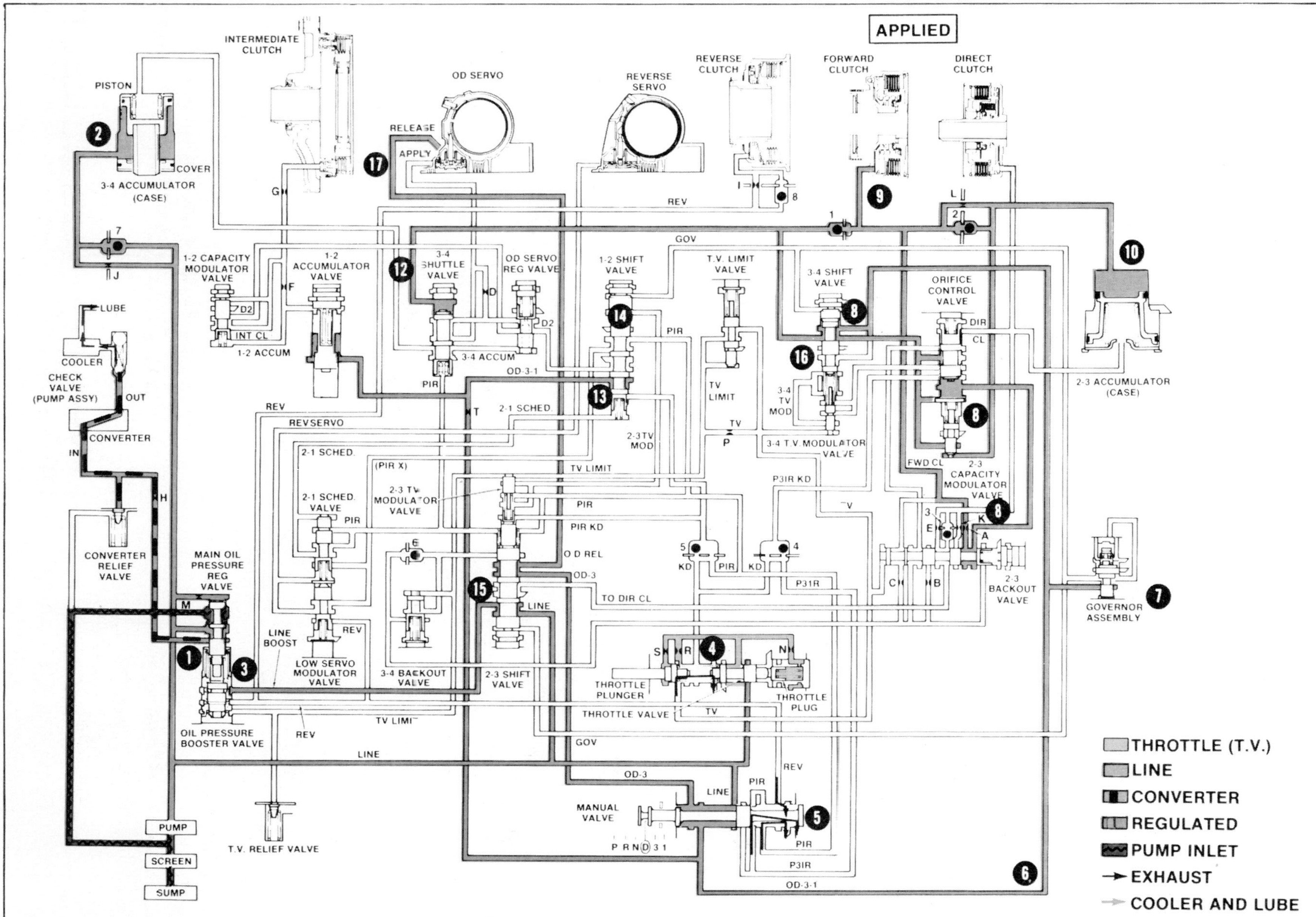

Figure 7-75.

2ND GEAR IN OVERDRIVE RANGE

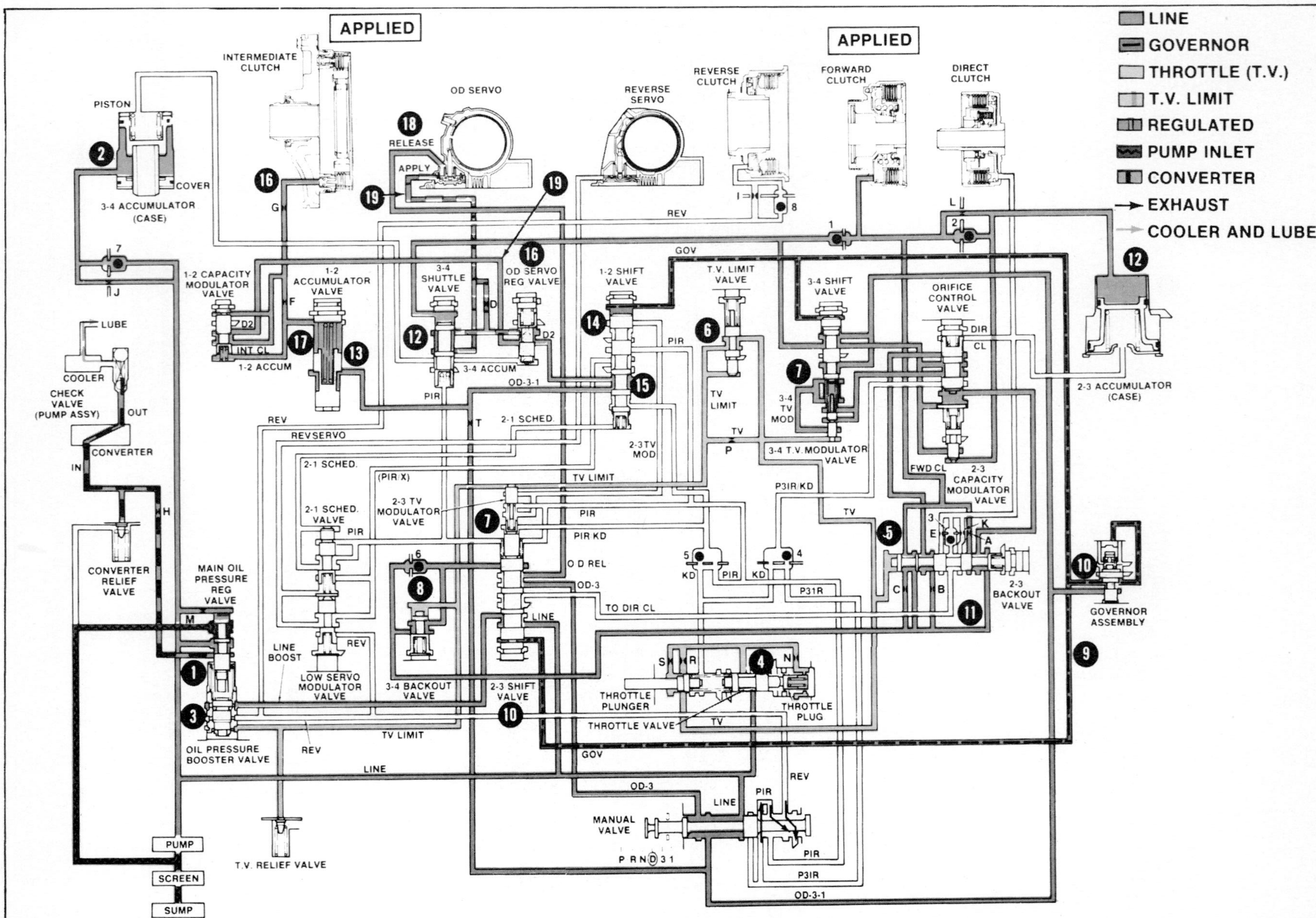

Figure 7-76.

3RD GEAR IN OVERDRIVE RANGE

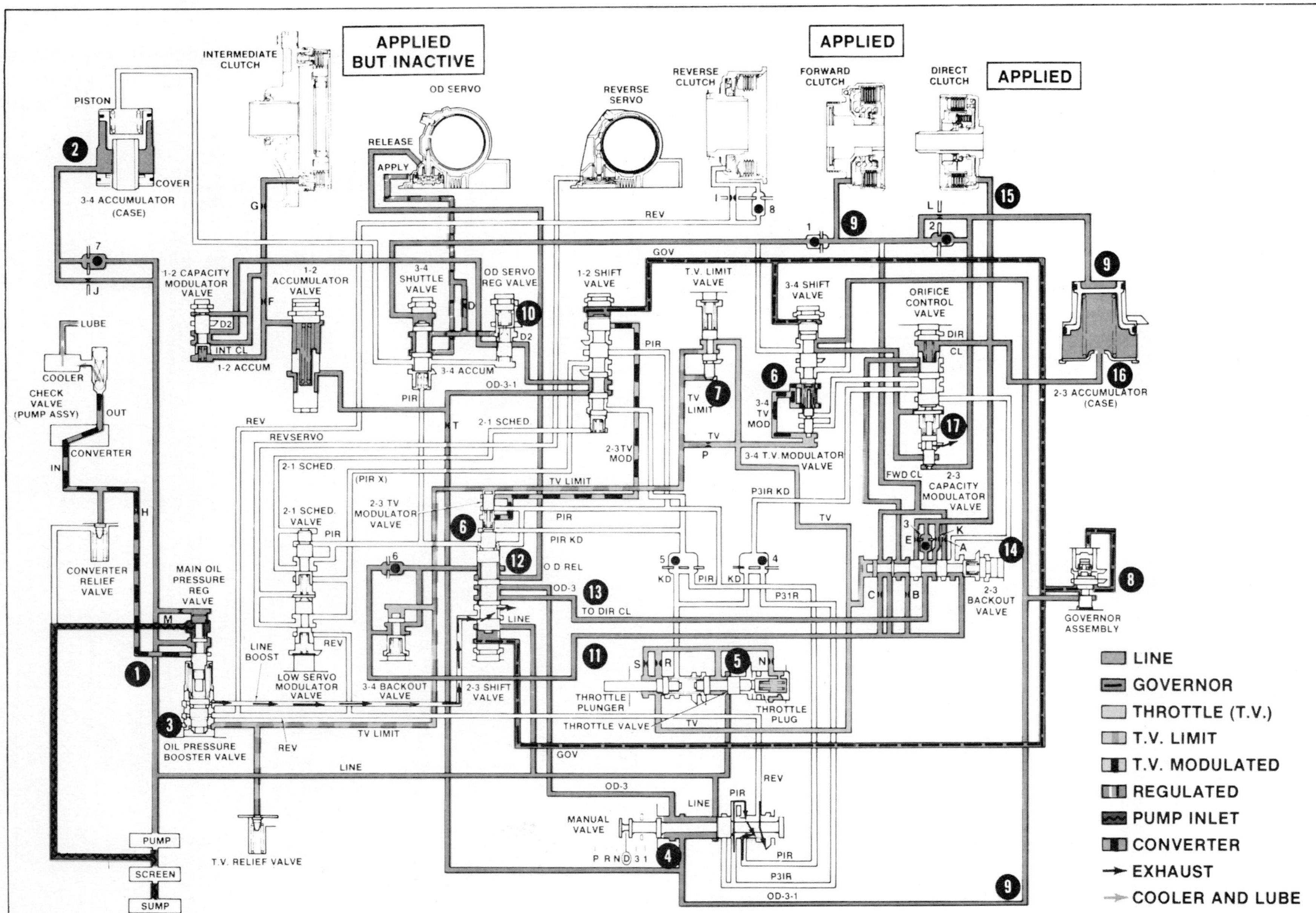

Figure 7-77.

3RD GEAR IN RANGE 3 (OVERDRIVE LOCKOUT)

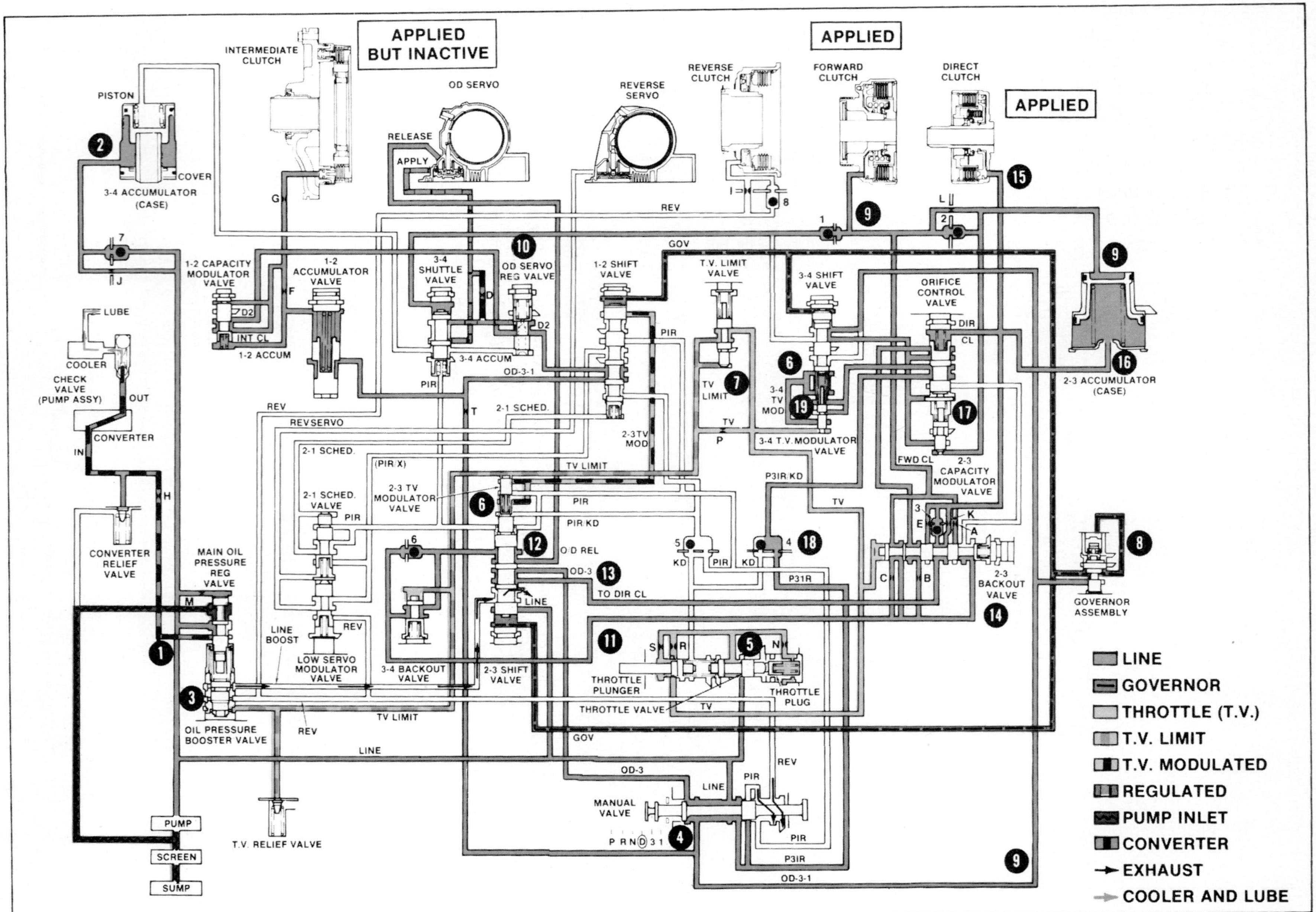

Figure 7-78.

OVERDRIVE (4TH GEAR)

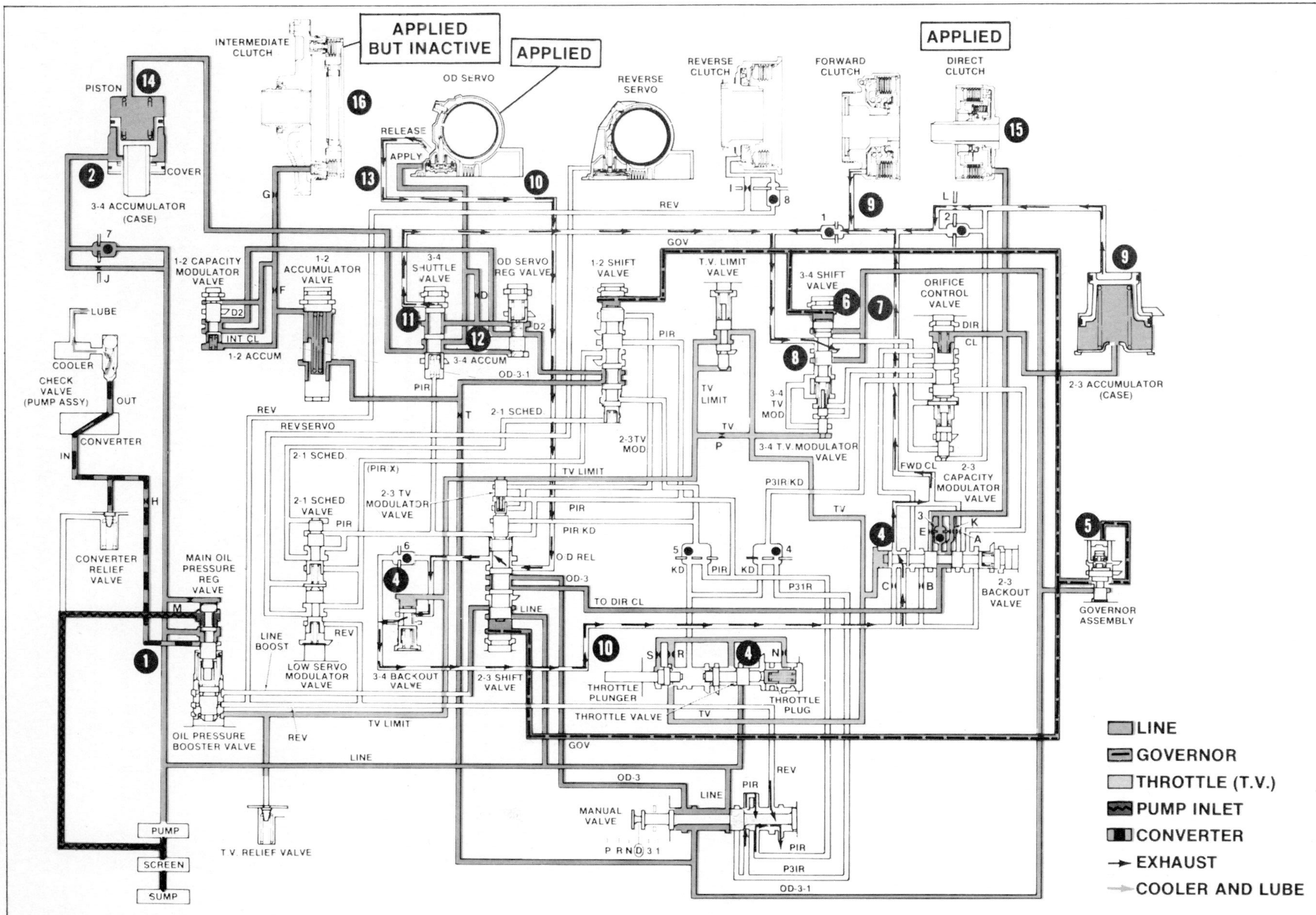

Figure 7-79.

KICKDOWN SYSTEM

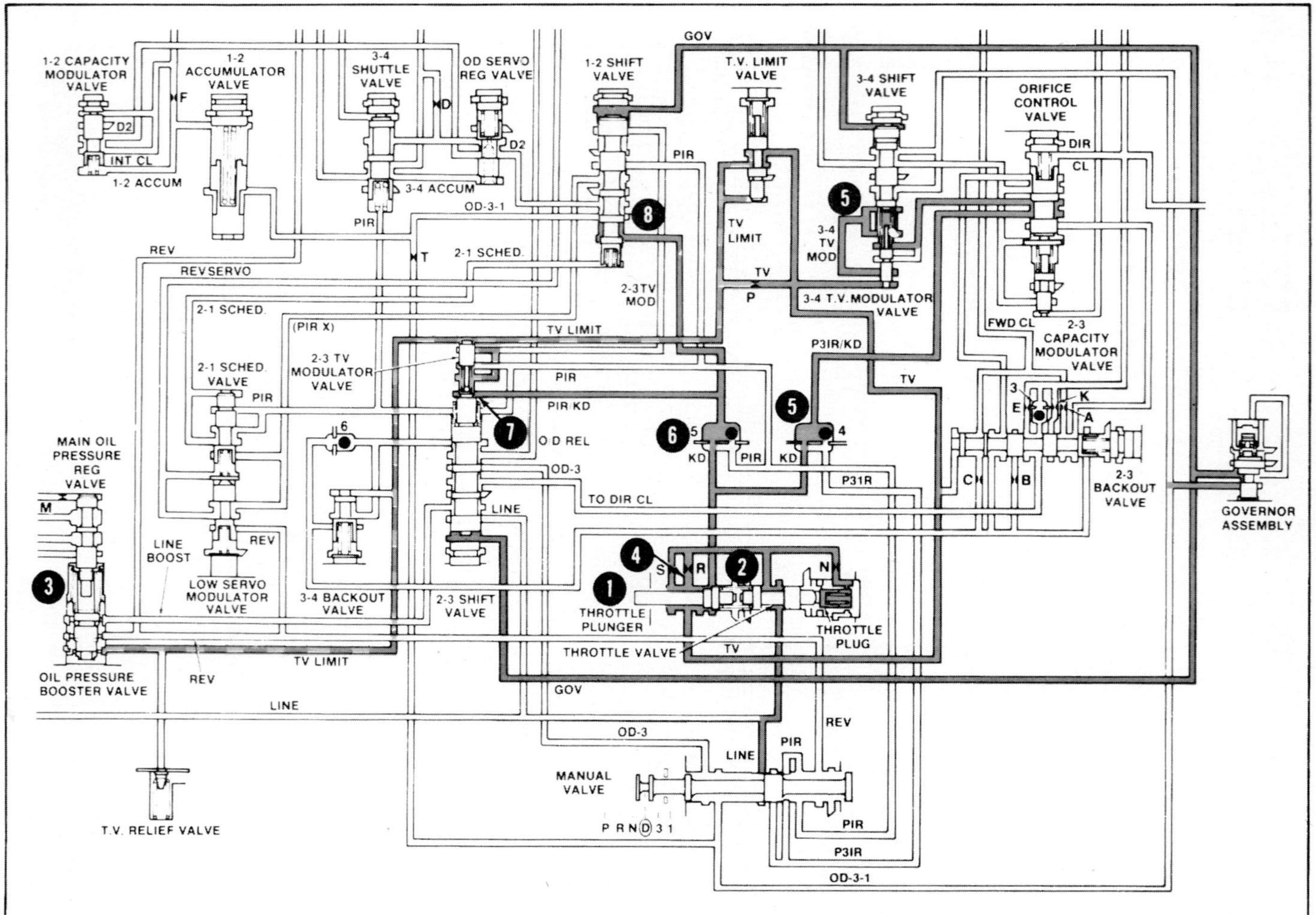

Figure 7-80.

MANUAL LOW (RANGE 1, 1ST GEAR)

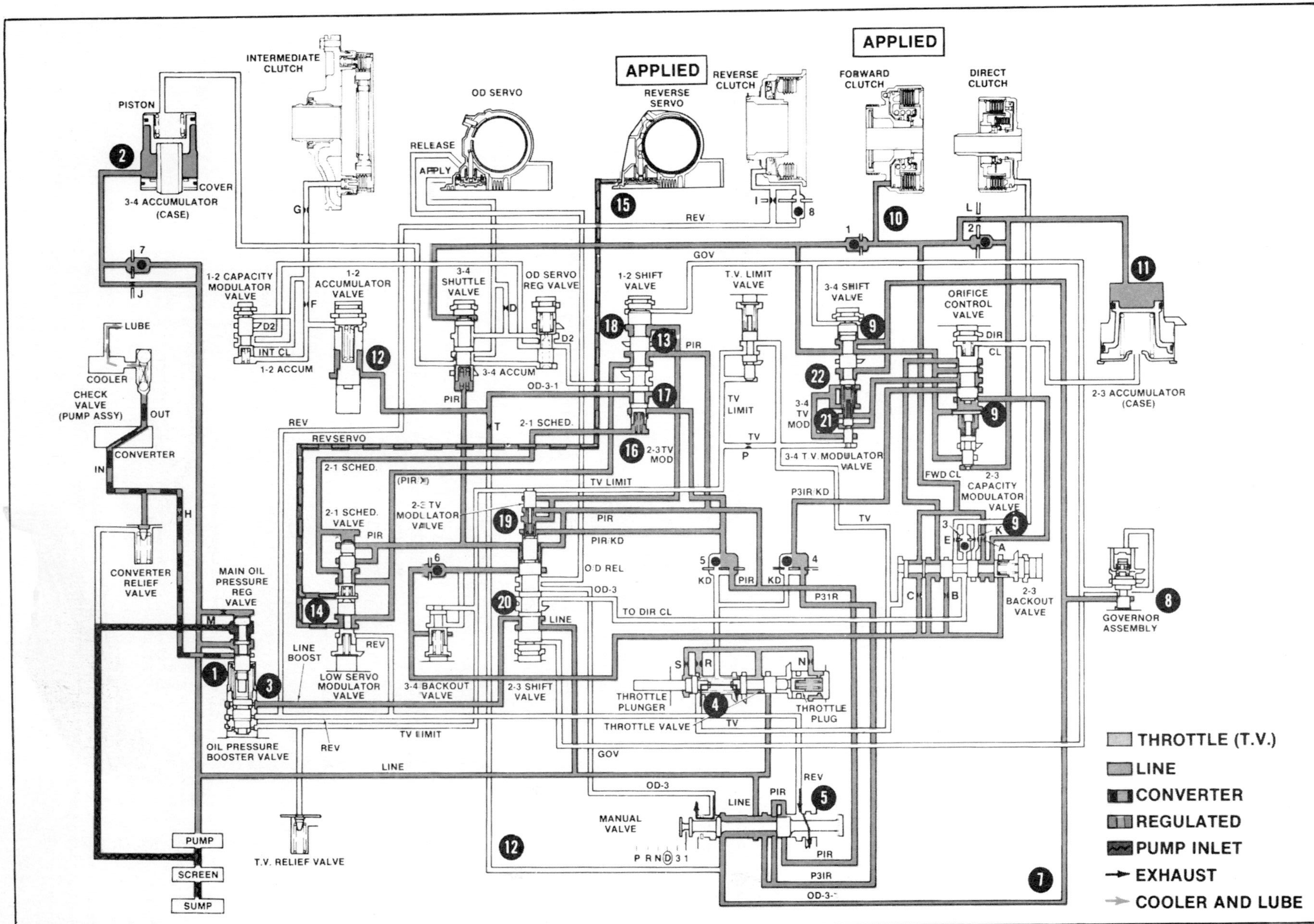

Figure 7-81.

2ND GEAR IN MANUAL LOW RANGE

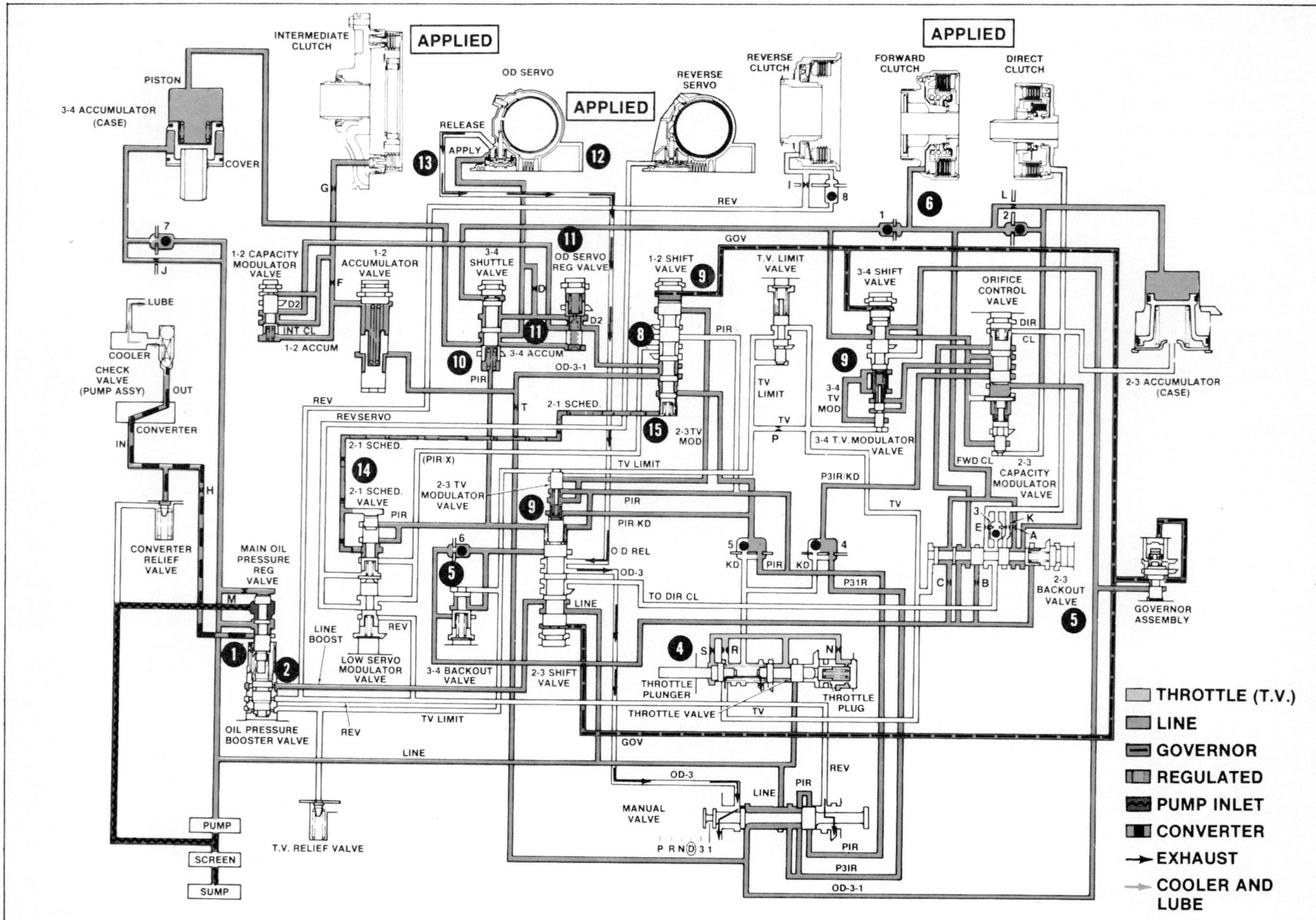

Figure 7-82.

REVERSE GEAR

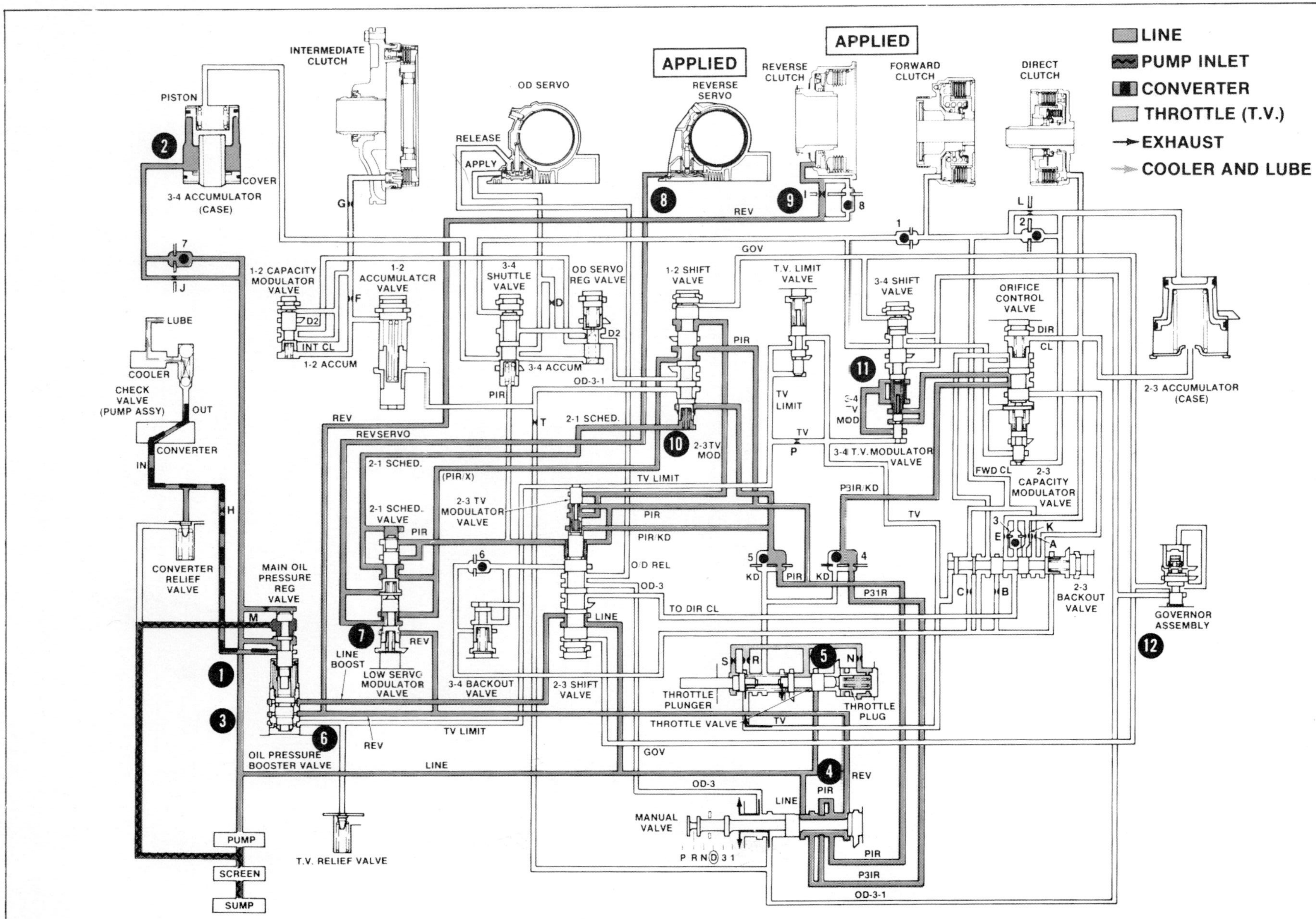

Figure 7-83.

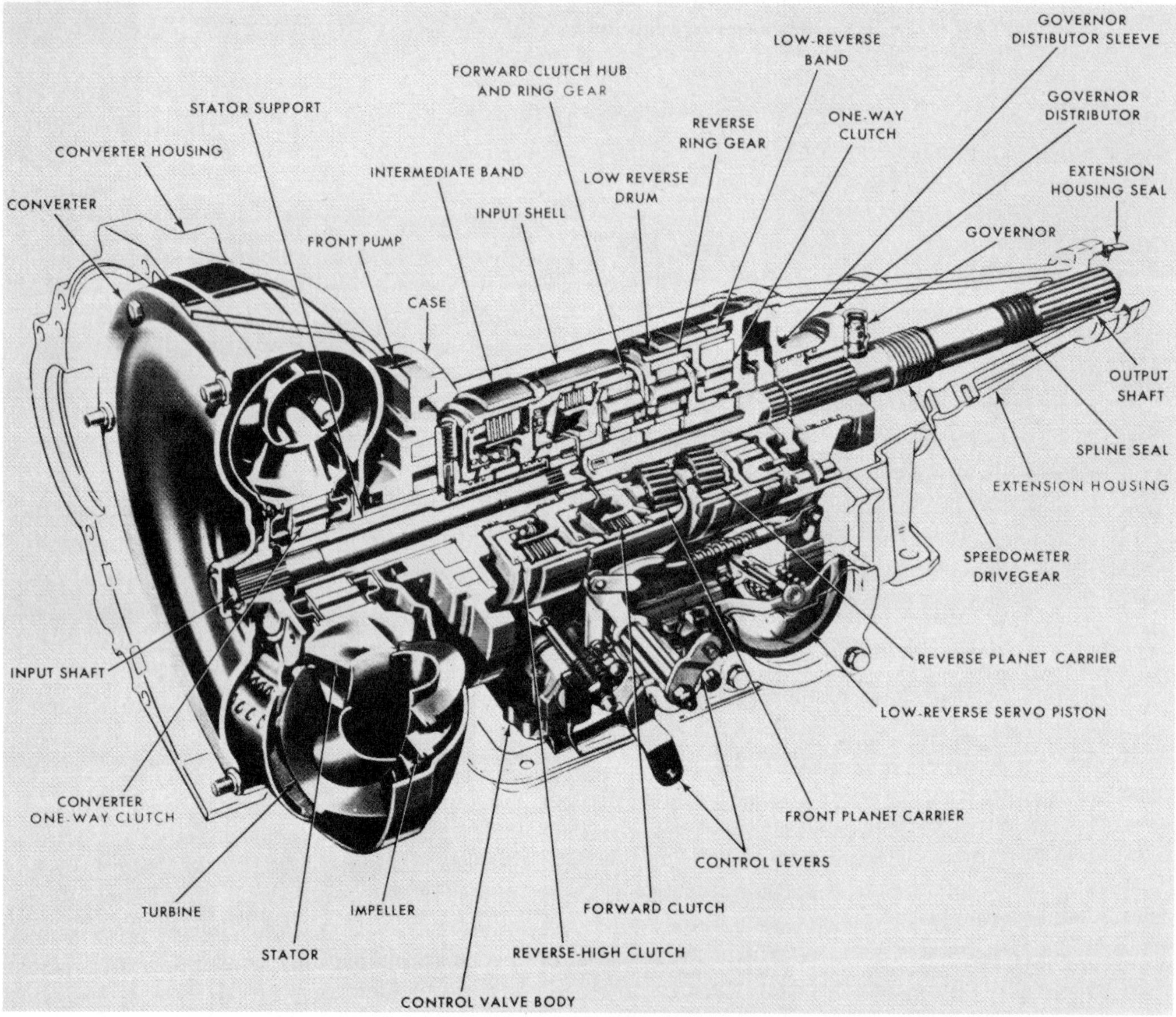

Figure 7-84. C4 transmission.

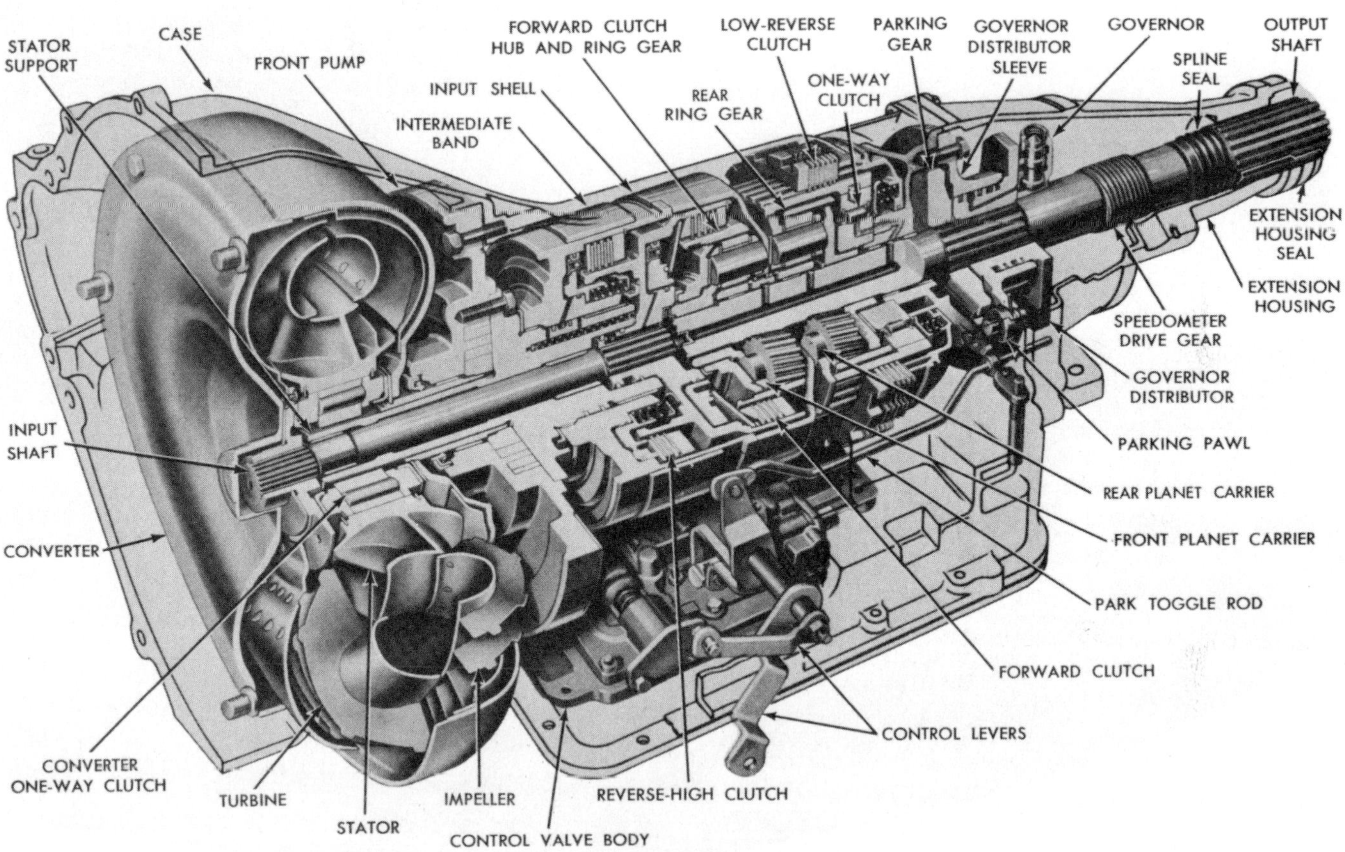

Figure 7-85. C6 transmission.

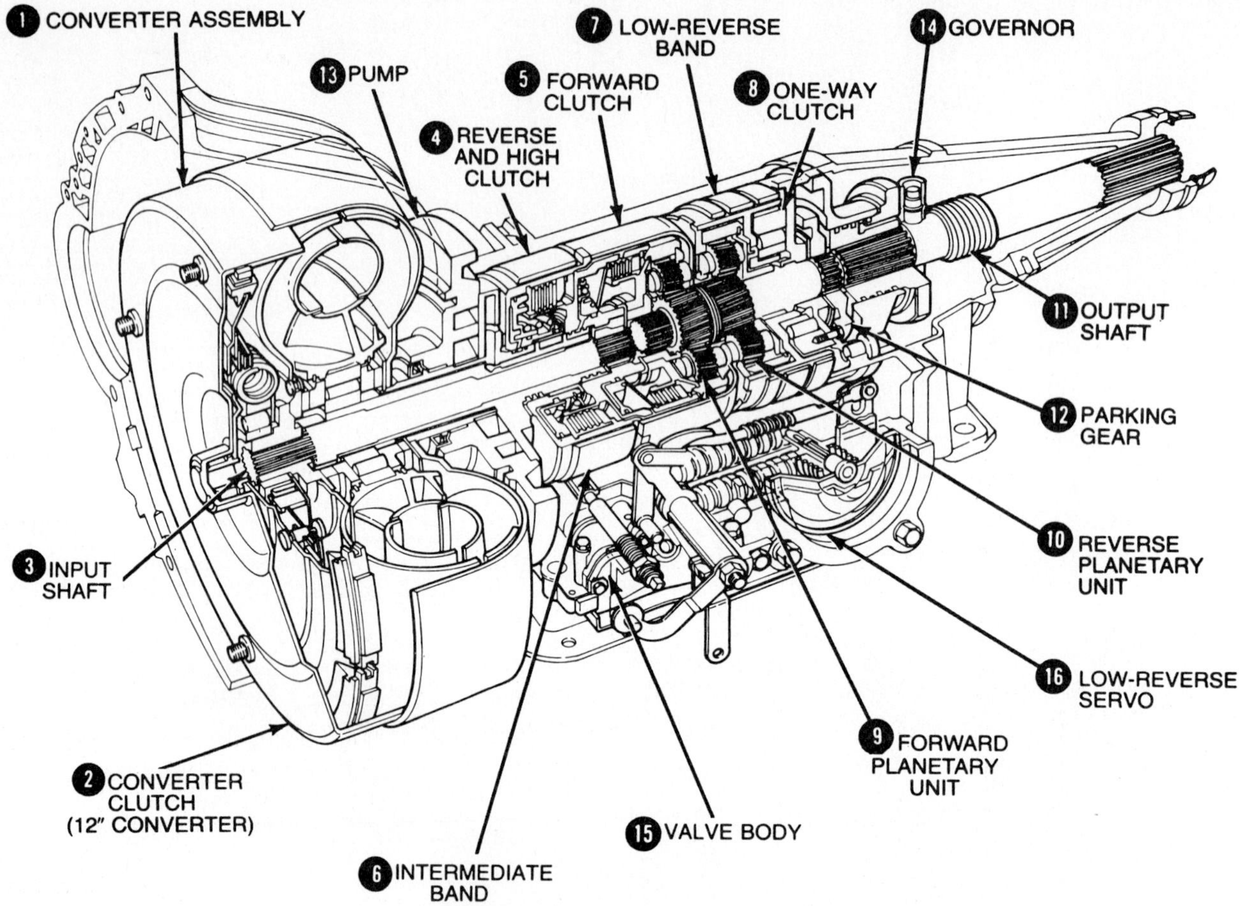

Figure 7-86. C5 transmission.

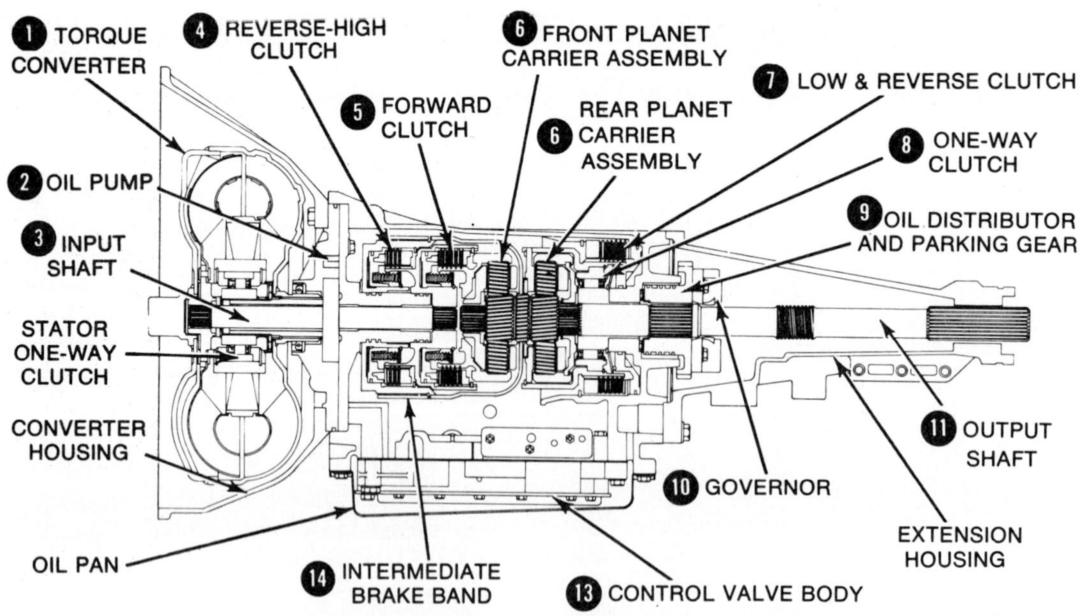

Figure 7-87. Jactco transmission.

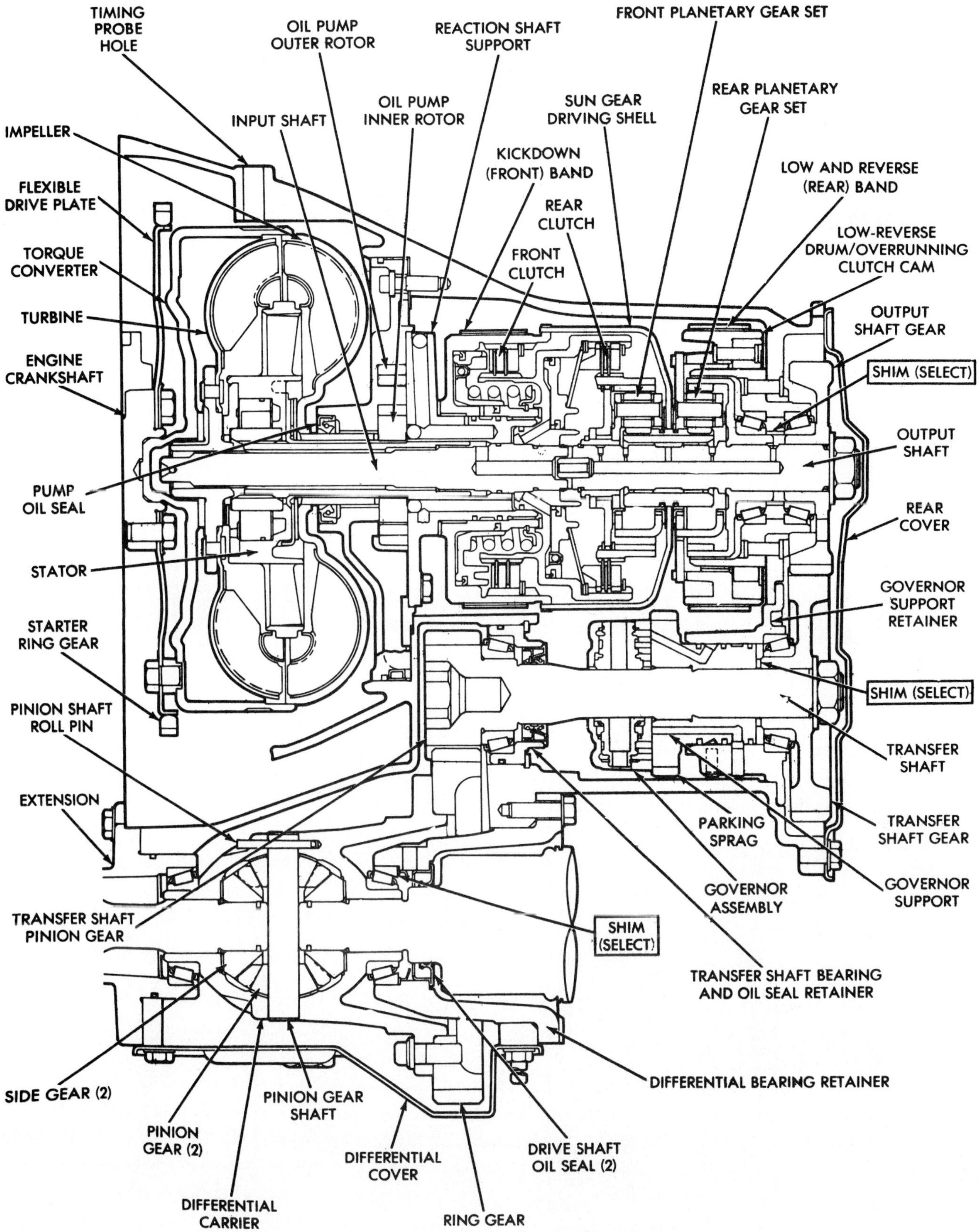

Figure 7-88. Typical TorqueFlite transaxle.

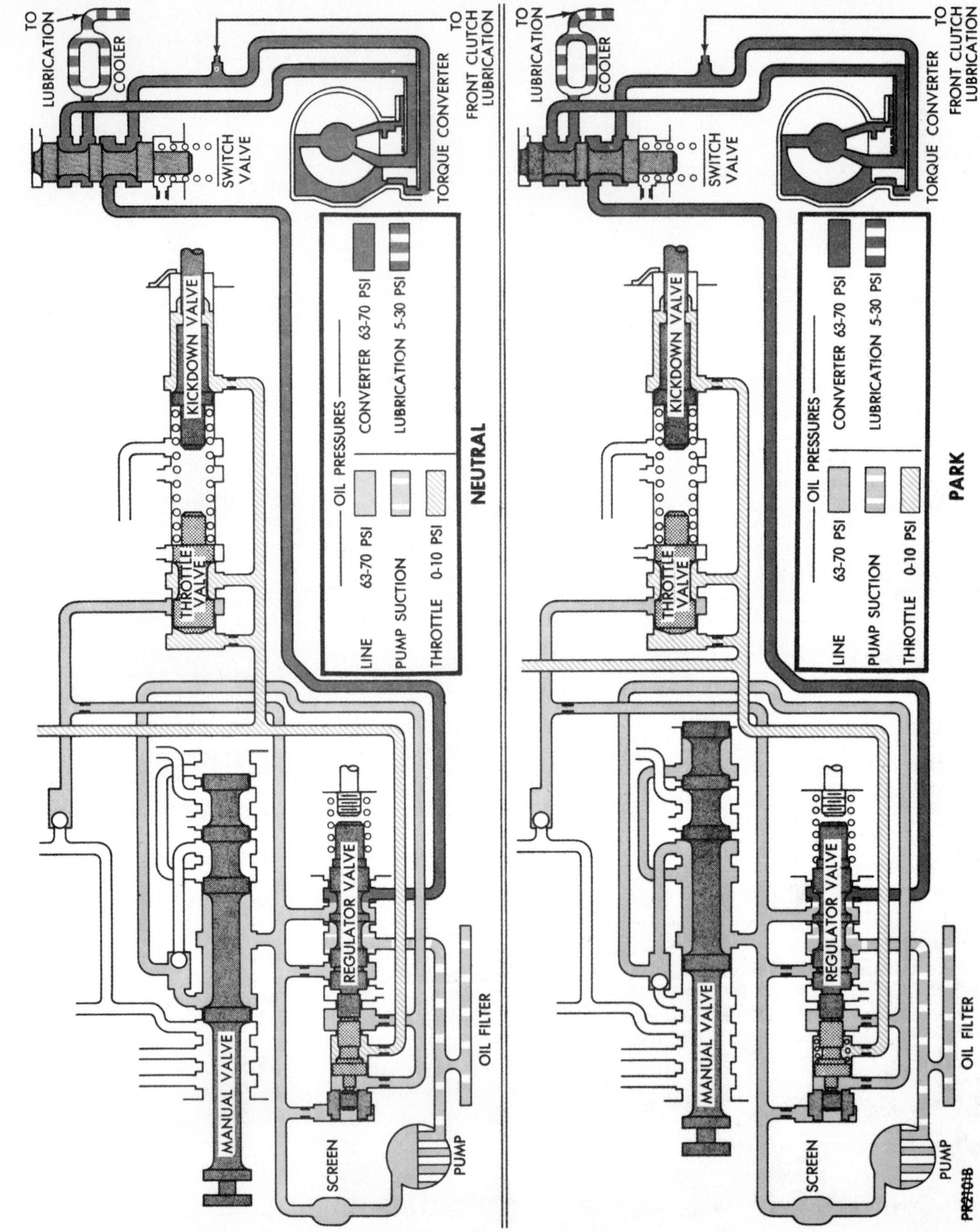
TO LUBRICATION
COOLER
TO FRONT CLUTCH LUBRICATION
SWITCH VALVE
TORQUE CONVERTER
KICKDOWN VALVE
THROTTLE VALVE
OIL PRESSURES
LINE 63-70 PSI
PUMP SUCTION
THROTTLE 0-10 PSI
CONVERTER 63-70 PSI
LUBRICATION 5-30 PSI
NEUTRAL
MANUAL VALVE
REGULATOR VALVE
SCREEN
PUMP
OIL FILTER
PR210-B
TO LUBRICATION
COOLER
TO FRONT CLUTCH LUBRICATION
SWITCH VALVE
TORQUE CONVERTER
KICKDOWN VALVE
THROTTLE VALVE
OIL PRESSURES
LINE 63-70 PSI
PUMP SUCTION
THROTTLE 0-10 PSI
CONVERTER 63-70 PSI
LUBRICATION 5-30 PSI
PARK
MANUAL VALVE
REGULATOR VALVE
SCREEN
PUMP
OIL FILTER

Figure 7–89.

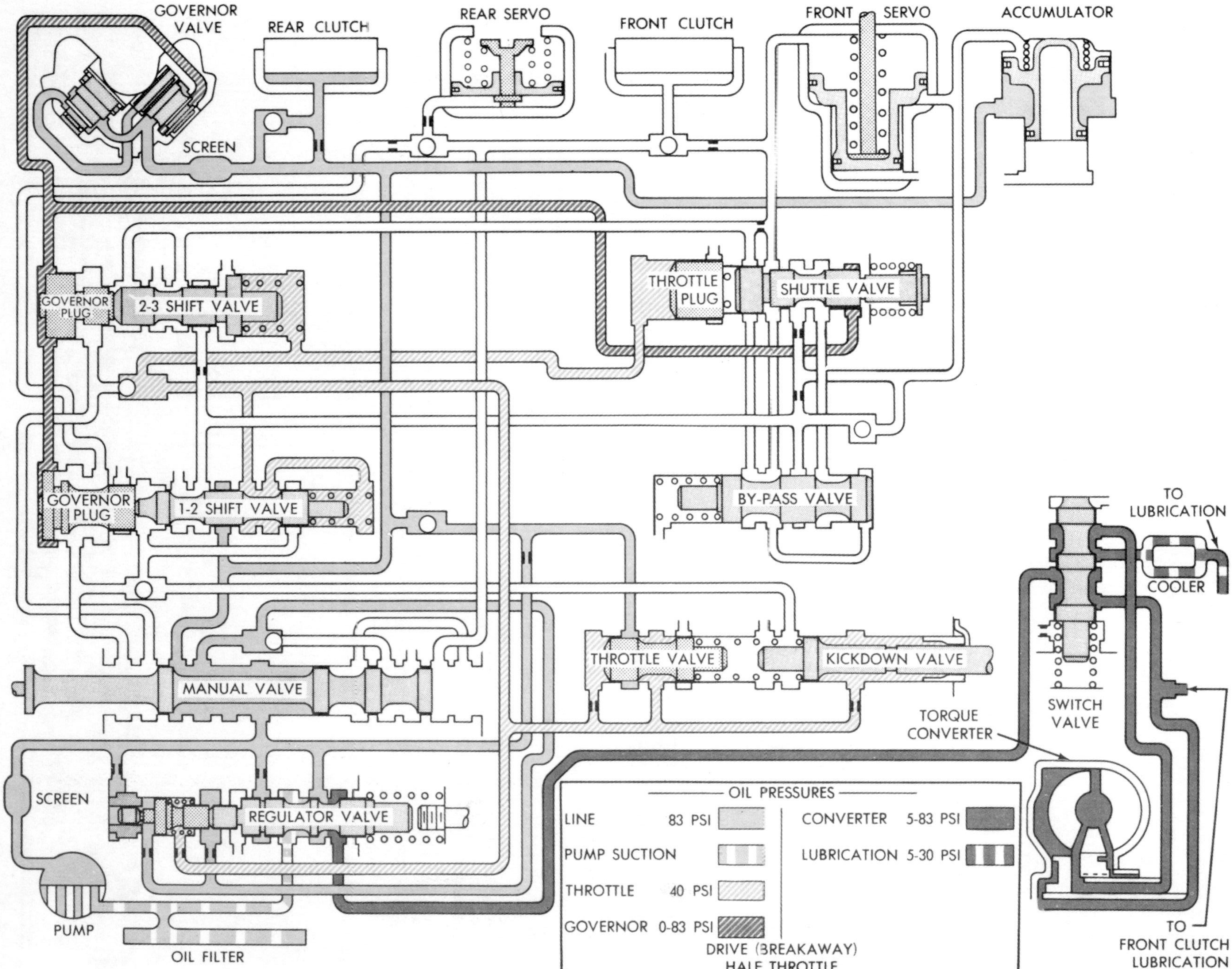
GOVERNOR VALVE
REAR CLUTCH
REAR SERVO
FRONT CLUTCH
FRONT SERVO
ACCUMULATOR
SCREEN
GOVERNOR PLUG
2-3 SHIFT VALVE
THROTTLE PLUG
SHUTTLE VALVE
GOVERNOR PLUG
1-2 SHIFT VALVE
BY-PASS VALVE
TO LUBRICATION
COOLER
THROTTLE VALVE
KICKDOWN VALVE
MANUAL VALVE
SWITCH VALVE
TORQUE CONVERTER
SCREEN
REGULATOR VALVE
PUMP
OIL FILTER
OIL PRESSURES
LINE 83 PSI
PUMP SUCTION
THROTTLE 40 PSI
GOVERNOR 0-83 PSI
CONVERTER 5-83 PSI
LUBRICATION 5-30 PSI
DRIVE (BREAKAWAY)
HALF THROTTLE
TO FRONT CLUTCH LUBRICATION

Figure 7–90.

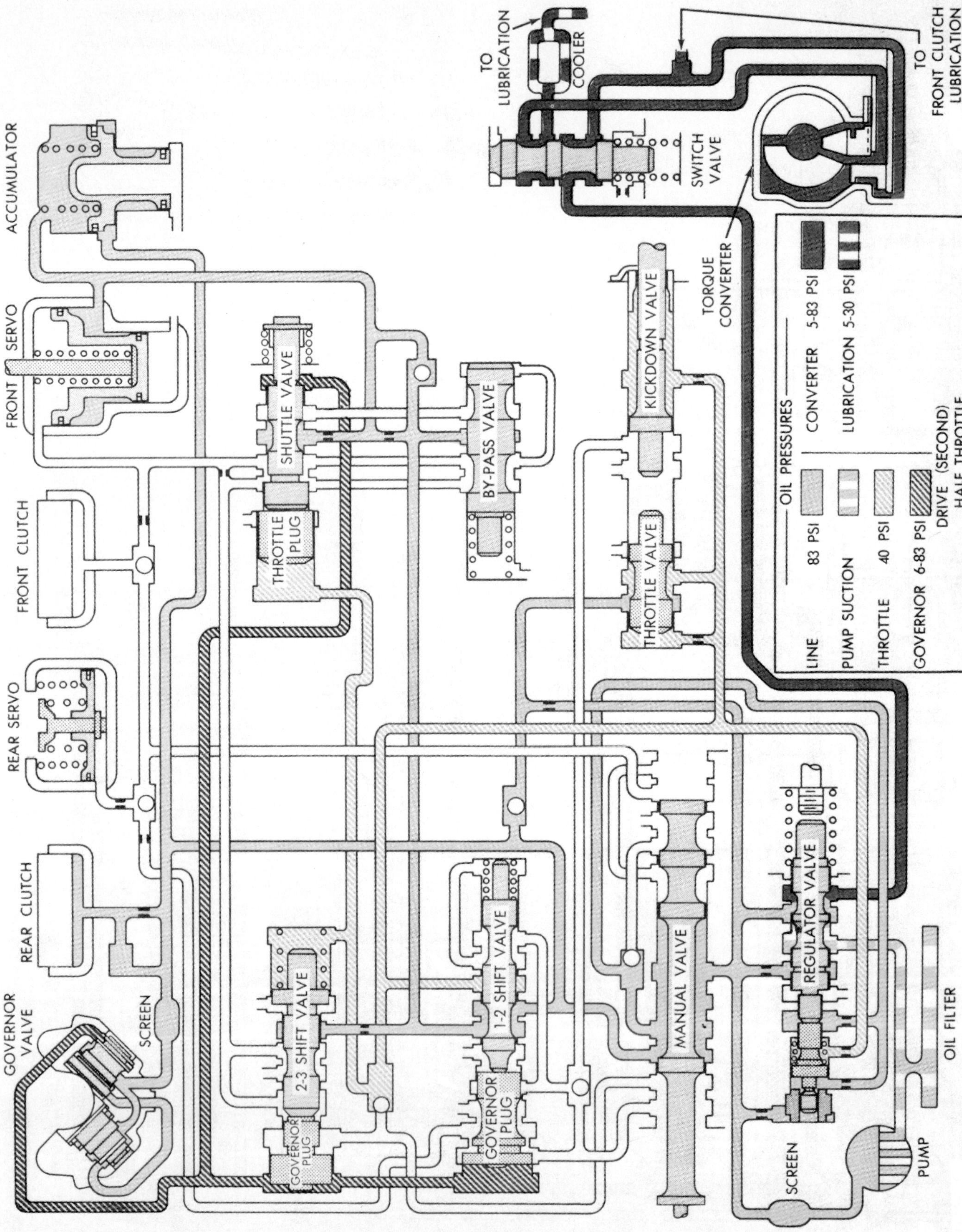
GOVERNOR VALVE
SCREEN
REAR CLUTCH
REAR SERVO
FRONT CLUTCH
FRONT SERVO
ACCUMULATOR
THROTTLE PLUG
SHUTTLE VALVE
BY-PASS VALVE
2-3 SHIFT VALVE
GOVERNOR PLUG
1-2 SHIFT VALVE
GOVERNOR PLUG
KICKDOWN VALVE
THROTTLE VALVE
MANUAL VALVE
REGULATOR VALVE
OIL FILTER
PUMP
SCREEN
TORQUE CONVERTER
SWITCH VALVE
TO LUBRICATION
COOLER
TO FRONT CLUTCH LUBRICATION
OIL PRESSURES
LINE 83 PSI
PUMP SUCTION
THROTTLE 40 PSI
GOVERNOR 6-83 PSI
CONVERTER 5-83 PSI
LUBRICATION 5-30 PSI
DRIVE (SECOND) HALF THROTTLE

Figure 7-91.

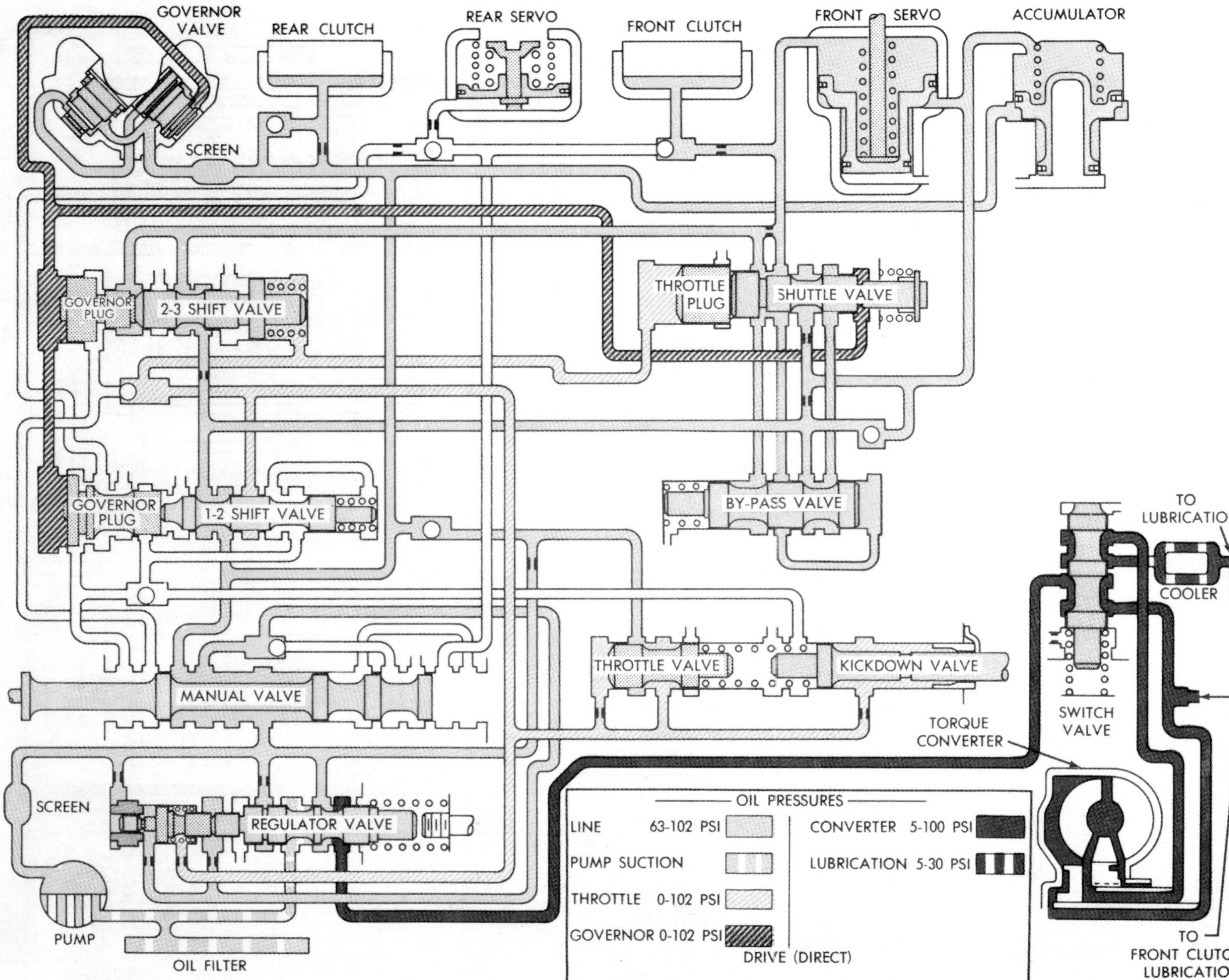
GOVERNOR VALVE
REAR CLUTCH
REAR SERVO
FRONT CLUTCH
FRONT SERVO
ACCUMULATOR
SCREEN
GOVERNOR PLUG
2-3 SHIFT VALVE
THROTTLE PLUG
SHUTTLE VALVE
GOVERNOR PLUG
1-2 SHIFT VALVE
BY-PASS VALVE
TO LUBRICATION
COOLER
THROTTLE VALVE
KICKDOWN VALVE
MANUAL VALVE
SWITCH VALVE
TORQUE CONVERTER
SCREEN
REGULATOR VALVE
PUMP
OIL FILTER
OIL PRESSURES
LINE 63-102 PSI
PUMP SUCTION
THROTTLE 0-102 PSI
GOVERNOR 0-102 PSI
CONVERTER 5-100 PSI
LUBRICATION 5-30 PSI
DRIVE (DIRECT)
TO FRONT CLUTCH LUBRICATION

Figure 7-92.

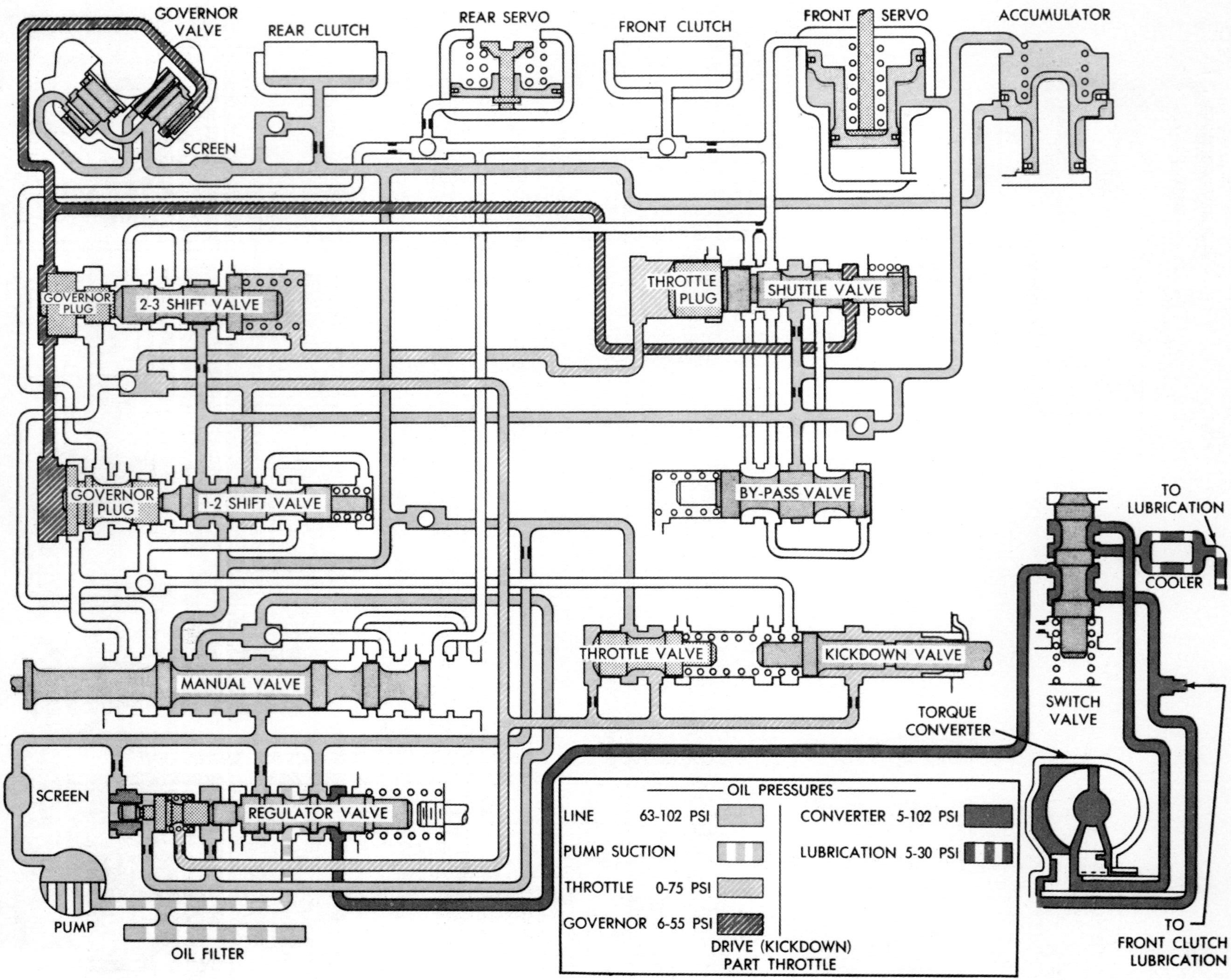
GOVERNOR VALVE
REAR CLUTCH
REAR SERVO
FRONT CLUTCH
FRONT SERVO
ACCUMULATOR
SCREEN
GOVERNOR PLUG
2-3 SHIFT VALVE
THROTTLE PLUG
SHUTTLE VALVE
GOVERNOR PLUG
1-2 SHIFT VALVE
BY-PASS VALVE
TO LUBRICATION
COOLER
THROTTLE VALVE
KICKDOWN VALVE
MANUAL VALVE
SWITCH VALVE
TORQUE CONVERTER
SCREEN
REGULATOR VALVE
PUMP
OIL FILTER
OIL PRESSURES
LINE 63-102 PSI
PUMP SUCTION
THROTTLE 0-75 PSI
GOVERNOR 6-55 PSI
CONVERTER 5-102 PSI
LUBRICATION 5-30 PSI
DRIVE (KICKDOWN)
PART THROTTLE
TO FRONT CLUTCH LUBRICATION

Figure 7-93.

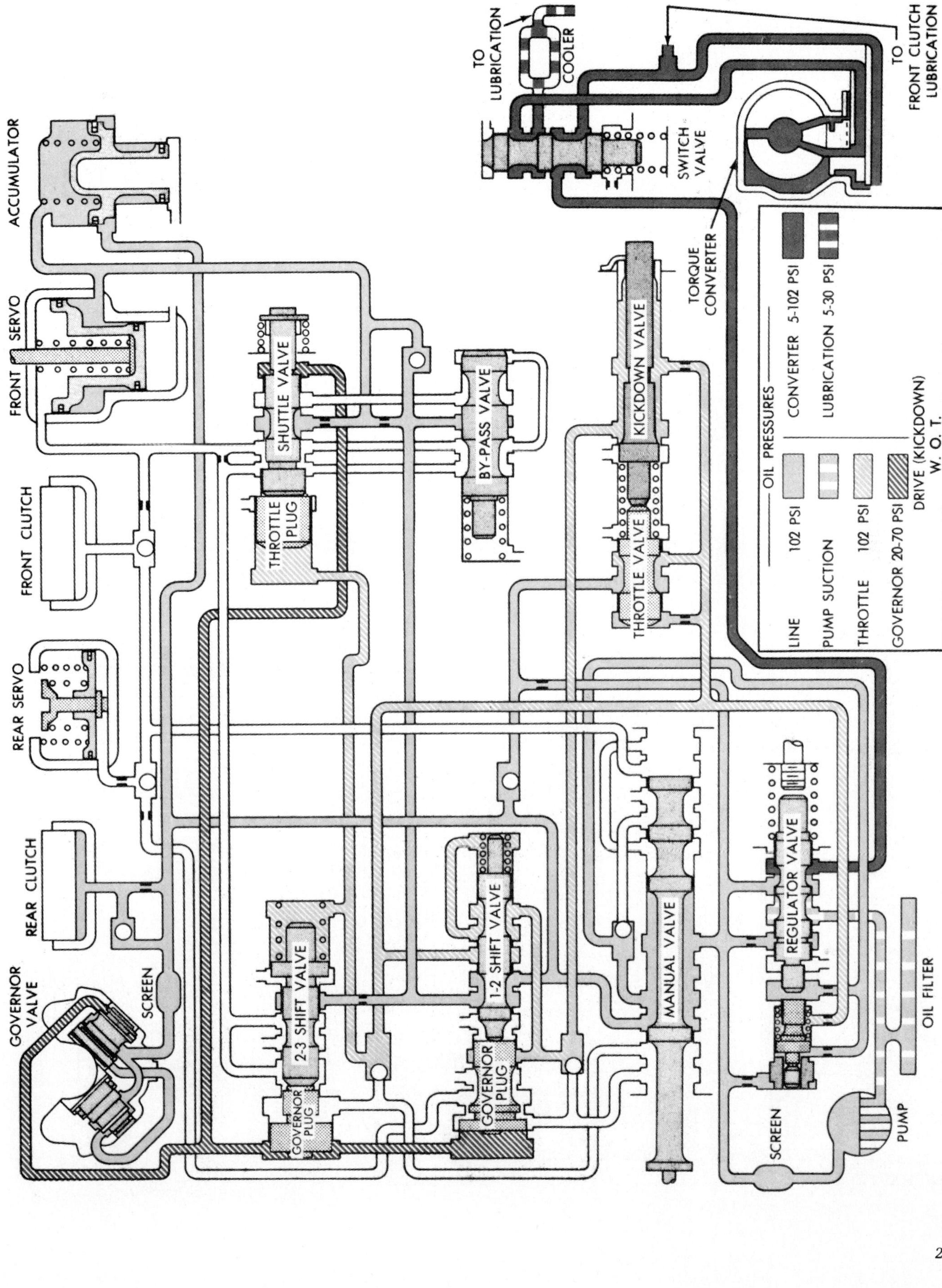

ACCUMULATOR
FRONT SERVO
FRONT CLUTCH
REAR SERVO
REAR CLUTCH
GOVERNOR VALVE
SCREEN
SHUTTLE VALVE
THROTTLE PLUG
BY-PASS VALVE
2-3 SHIFT VALVE
GOVERNOR PLUG
1-2 SHIFT VALVE
GOVERNOR PLUG
MANUAL VALVE
KICKDOWN VALVE
THROTTLE VALVE
REGULATOR VALVE
OIL FILTER
SCREEN
PUMP
TO LUBRICATION
COOLER
SWITCH VALVE
TORQUE CONVERTER
TO FRONT CLUTCH LUBRICATION
OIL PRESSURES
LINE 102 PSI
PUMP SUCTION
THROTTLE 102 PSI
GOVERNOR 20-70 PSI
DRIVE (KICKDOWN) W. O. T.
CONVERTER 5-102 PSI
LUBRICATION 5-30 PSI

Figure 7–94.

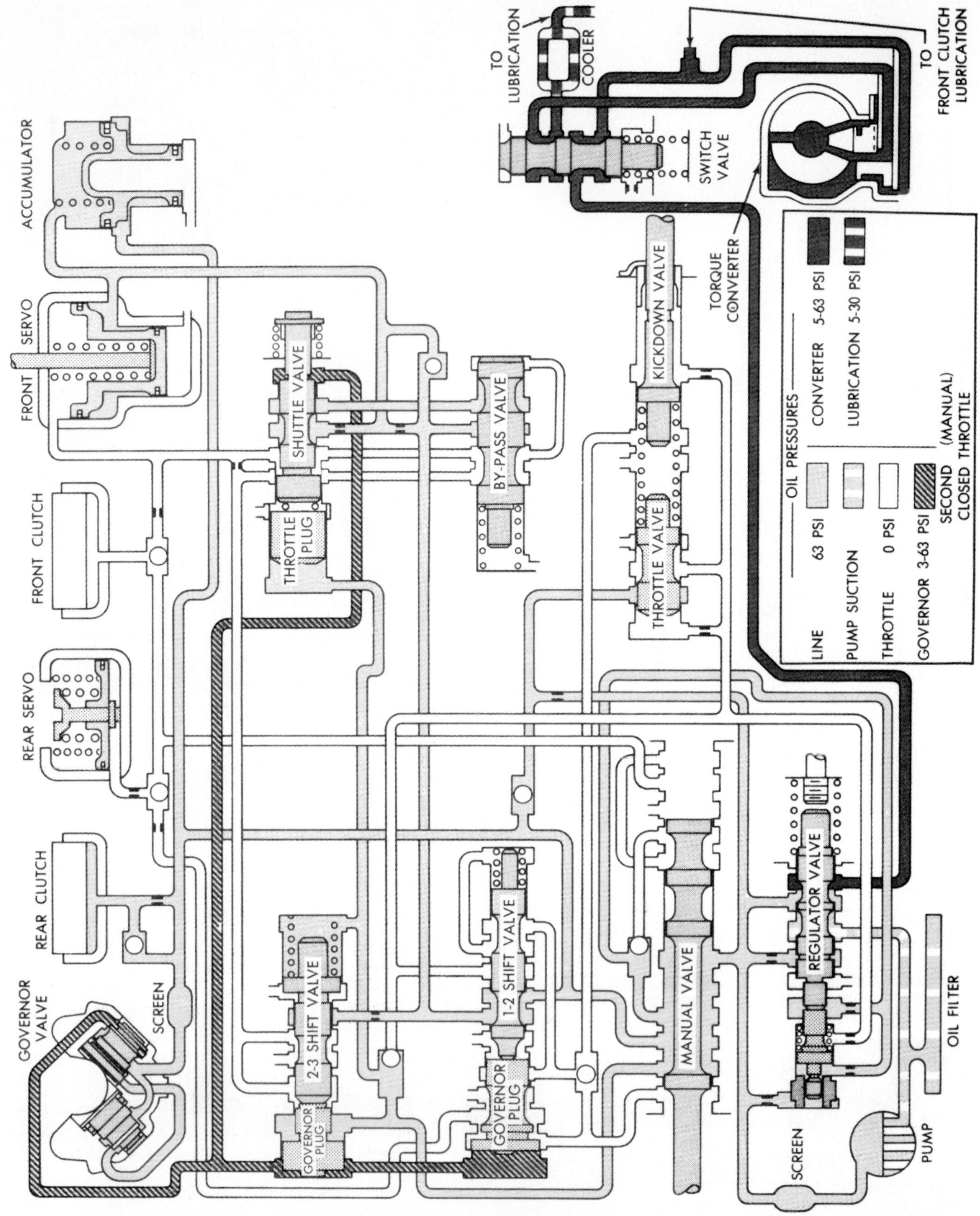
ACCUMULATOR
FRONT SERVO
FRONT CLUTCH
REAR SERVO
REAR CLUTCH
GOVERNOR VALVE
SCREEN
SHUTTLE VALVE
THROTTLE PLUG
BY-PASS VALVE
2-3 SHIFT VALVE
GOVERNOR PLUG
1-2 SHIFT VALVE
GOVERNOR PLUG
MANUAL VALVE
KICKDOWN VALVE
THROTTLE VALVE
REGULATOR VALVE
OIL FILTER
SCREEN
PUMP
TO LUBRICATION
COOLER
SWITCH VALVE
TO FRONT CLUTCH LUBRICATION
TORQUE CONVERTER
OIL PRESSURES
LINE 63 PSI
PUMP SUCTION
THROTTLE 0 PSI
GOVERNOR 3-63 PSI
CONVERTER 5-63 PSI
LUBRICATION 5-30 PSI
SECOND (MANUAL) CLOSED THROTTLE

Figure 7–95.

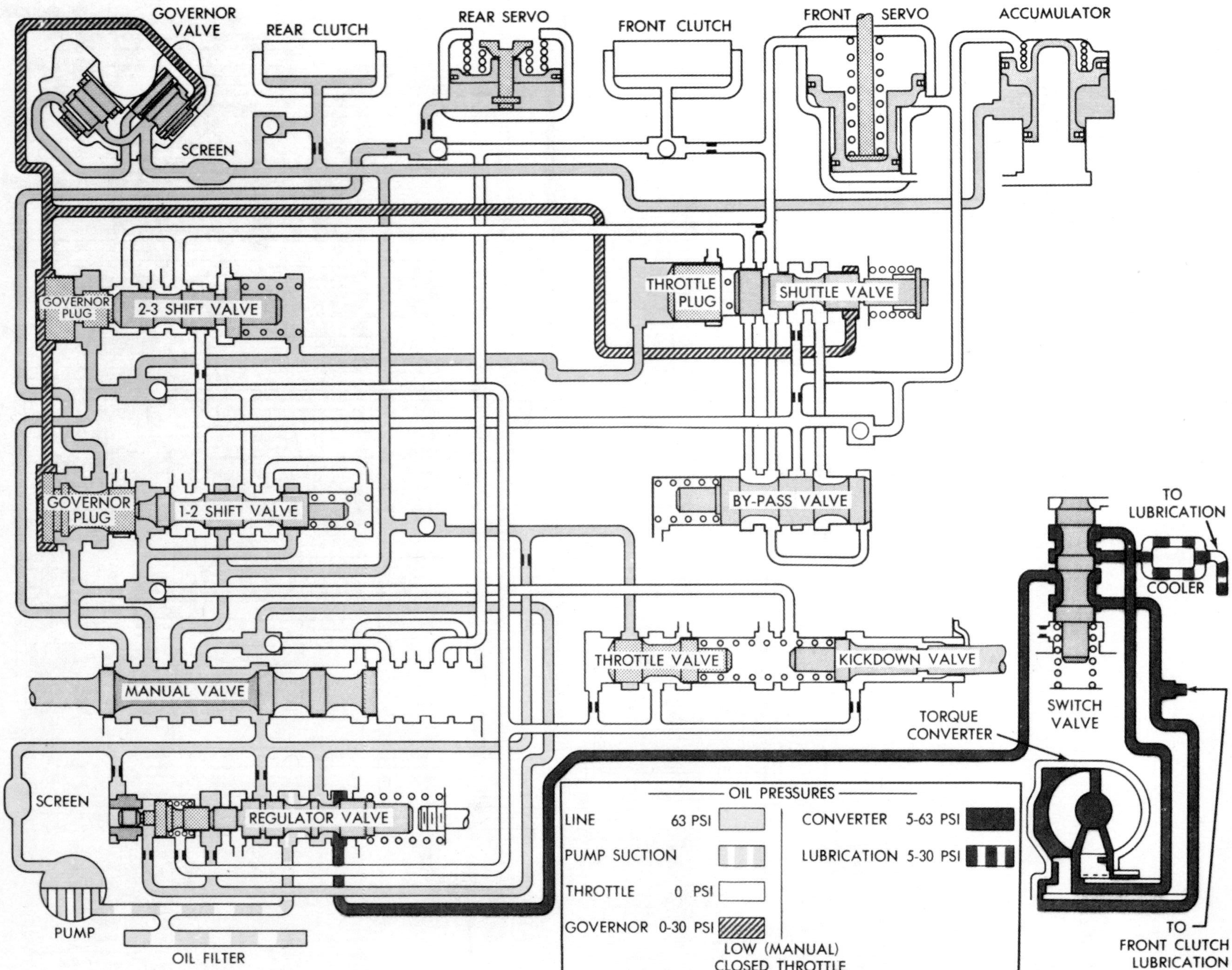
GOVERNOR VALVE
REAR CLUTCH
REAR SERVO
FRONT CLUTCH
FRONT SERVO
ACCUMULATOR
SCREEN
GOVERNOR PLUG
2-3 SHIFT VALVE
THROTTLE PLUG
SHUTTLE VALVE
GOVERNOR PLUG
1-2 SHIFT VALVE
BY-PASS VALVE
TO LUBRICATION
COOLER
THROTTLE VALVE
KICKDOWN VALVE
MANUAL VALVE
SWITCH VALVE
TORQUE CONVERTER
SCREEN
REGULATOR VALVE
PUMP
OIL FILTER
OIL PRESSURES
LINE 63 PSI
PUMP SUCTION
THROTTLE 0 PSI
GOVERNOR 0-30 PSI
LOW (MANUAL) CLOSED THROTTLE
CONVERTER 5-63 PSI
LUBRICATION 5-30 PSI
TO FRONT CLUTCH LUBRICATION

Figure 7-96.

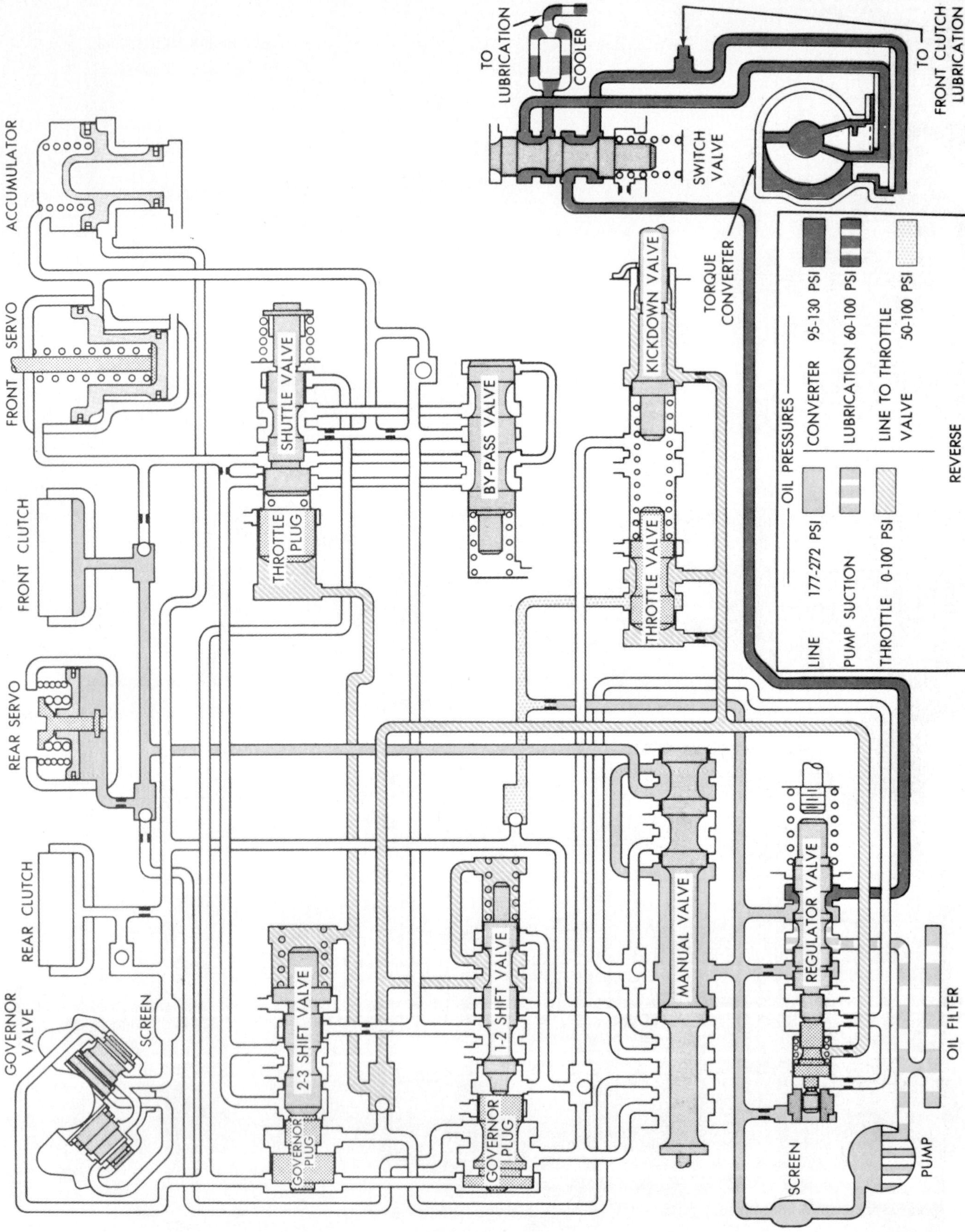

GOVERNOR VALVE
SCREEN
REAR CLUTCH
REAR SERVO
FRONT CLUTCH
FRONT SERVO
ACCUMULATOR
TO LUBRICATION
COOLER
TO FRONT CLUTCH LUBRICATION
SWITCH VALVE
TORQUE CONVERTER
SHUTTLE VALVE
THROTTLE PLUG
BY-PASS VALVE
KICKDOWN VALVE
THROTTLE VALVE
2-3 SHIFT VALVE
1-2 SHIFT VALVE
GOVERNOR PLUG
GOVERNOR PLUG
MANUAL VALVE
REGULATOR VALVE
OIL FILTER
SCREEN
PUMP
OIL PRESSURES
LINE 177-272 PSI
PUMP SUCTION
THROTTLE 0-100 PSI
CONVERTER 95-130 PSI
LUBRICATION 60-100 PSI
LINE TO THROTTLE VALVE 50-100 PSI
REVERSE

Figure 7–97.

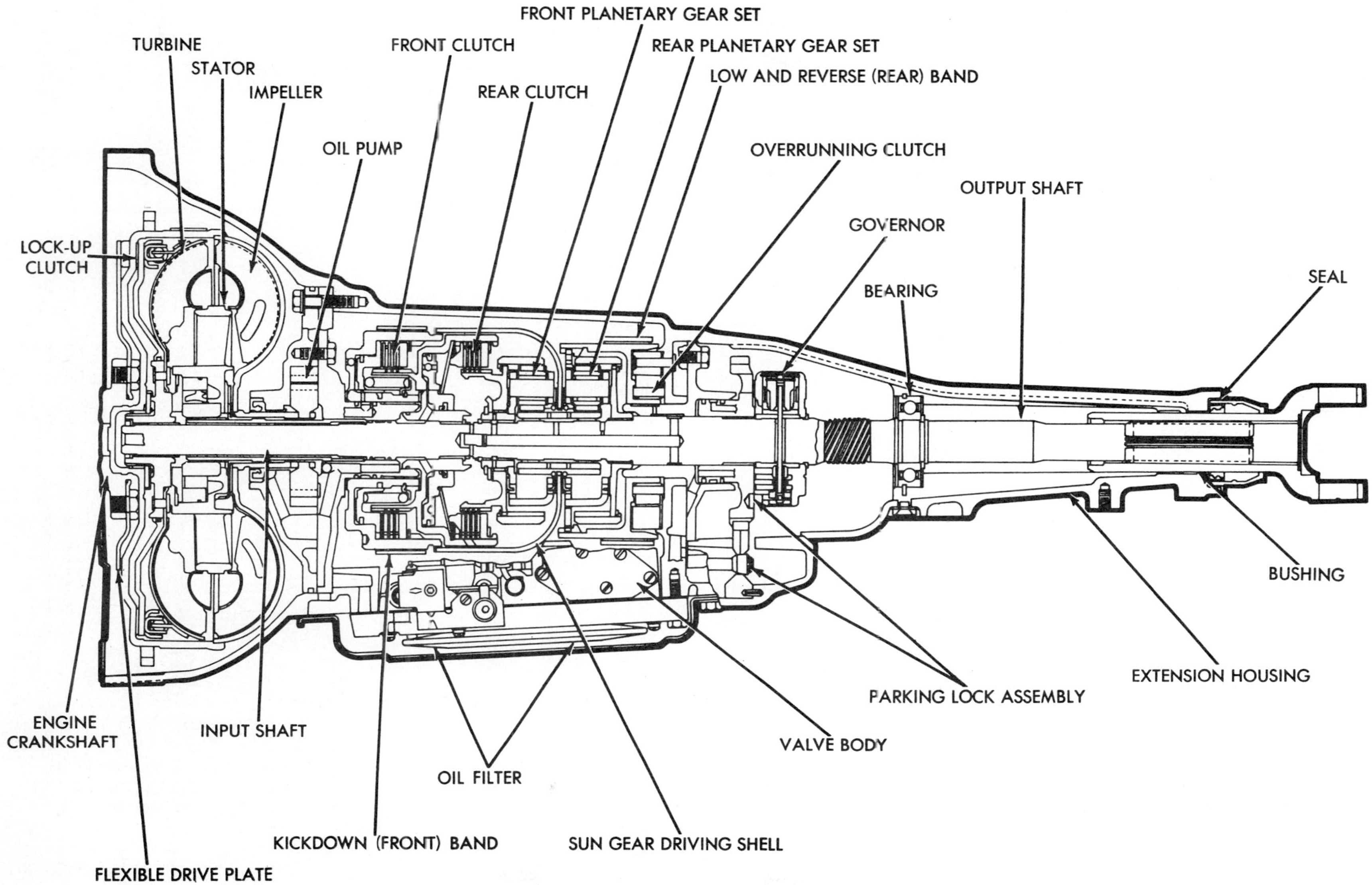

Figure 7-98. TorqueFlite transmission and torque converter (A-904).

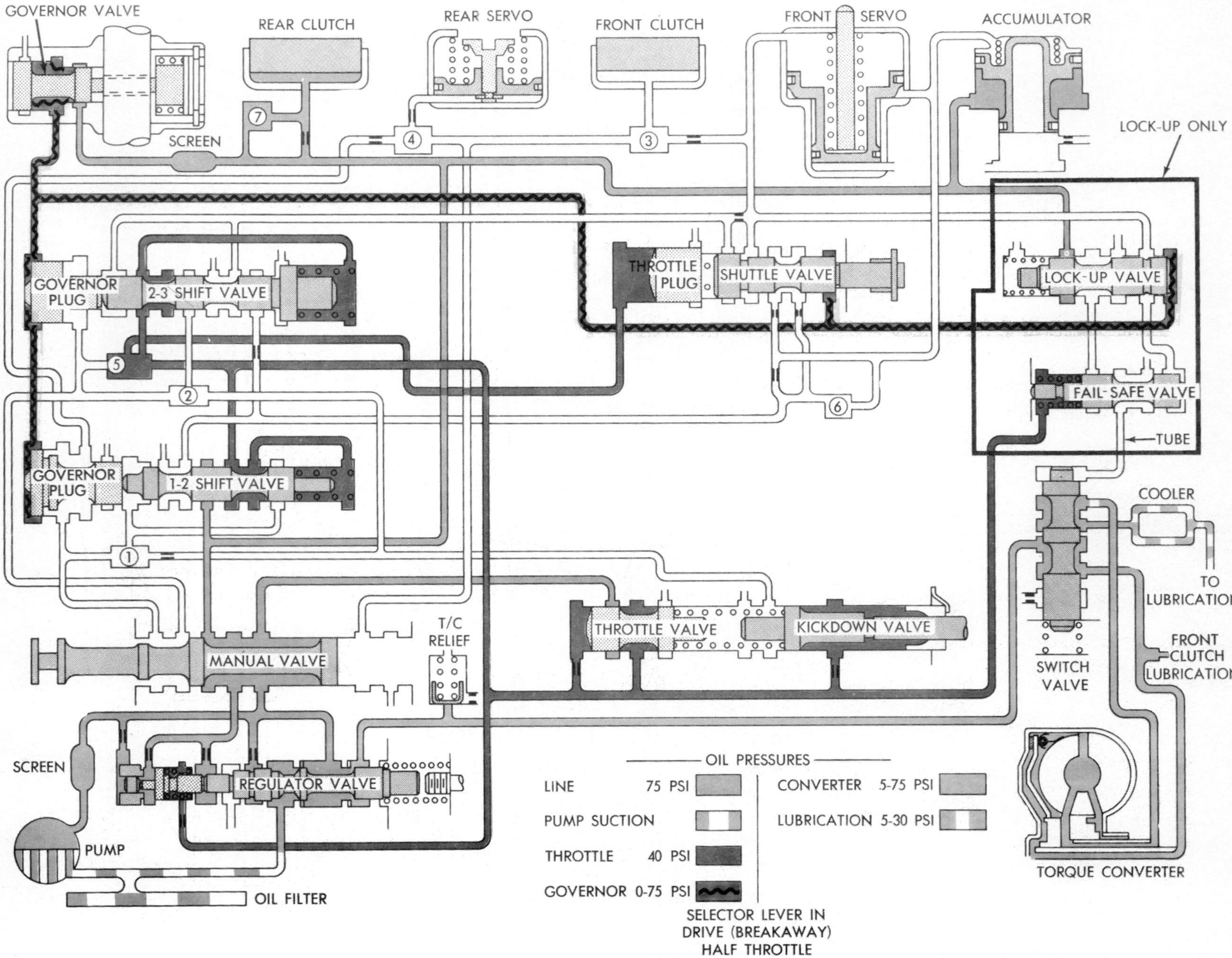
GOVERNOR VALVE
REAR CLUTCH
REAR SERVO
FRONT CLUTCH
FRONT SERVO
ACCUMULATOR
LOCK-UP ONLY
SCREEN
GOVERNOR PLUG
2-3 SHIFT VALVE
THROTTLE PLUG
SHUTTLE VALVE
LOCK-UP VALVE
FAIL-SAFE VALVE
TUBE
GOVERNOR PLUG
1-2 SHIFT VALVE
COOLER
TO LUBRICATION
T/C RELIEF
THROTTLE VALVE
KICKDOWN VALVE
MANUAL VALVE
FRONT CLUTCH LUBRICATION
SWITCH VALVE
SCREEN
REGULATOR VALVE
PUMP
OIL FILTER
TORQUE CONVERTER
OIL PRESSURES
LINE 75 PSI
PUMP SUCTION
THROTTLE 40 PSI
GOVERNOR 0-75 PSI
CONVERTER 5-75 PSI
LUBRICATION 5-30 PSI
SELECTOR LEVER IN DRIVE (BREAKAWAY) HALF THROTTLE

Figure 7-99.

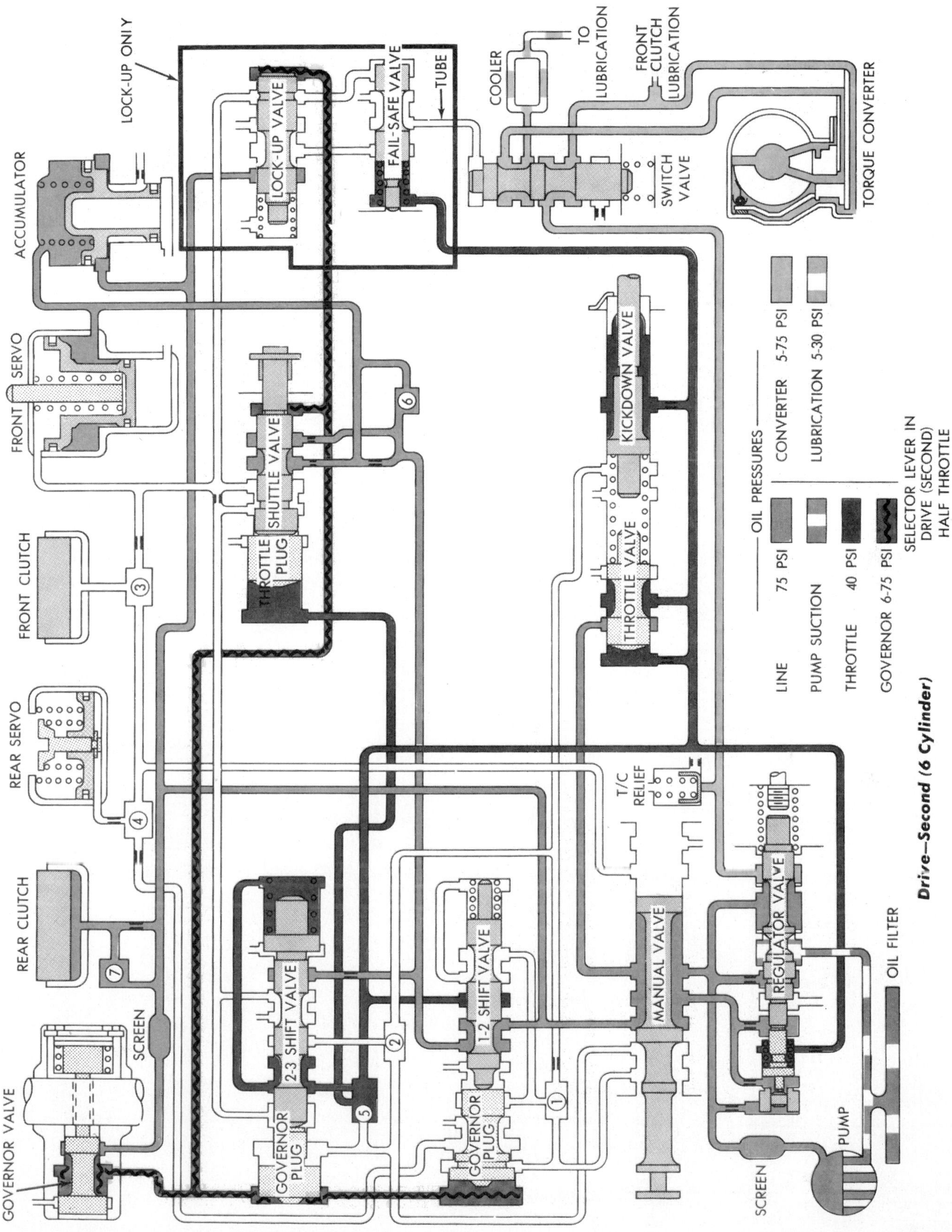

Drive—Second (6 Cylinder)

Figure 7–100.

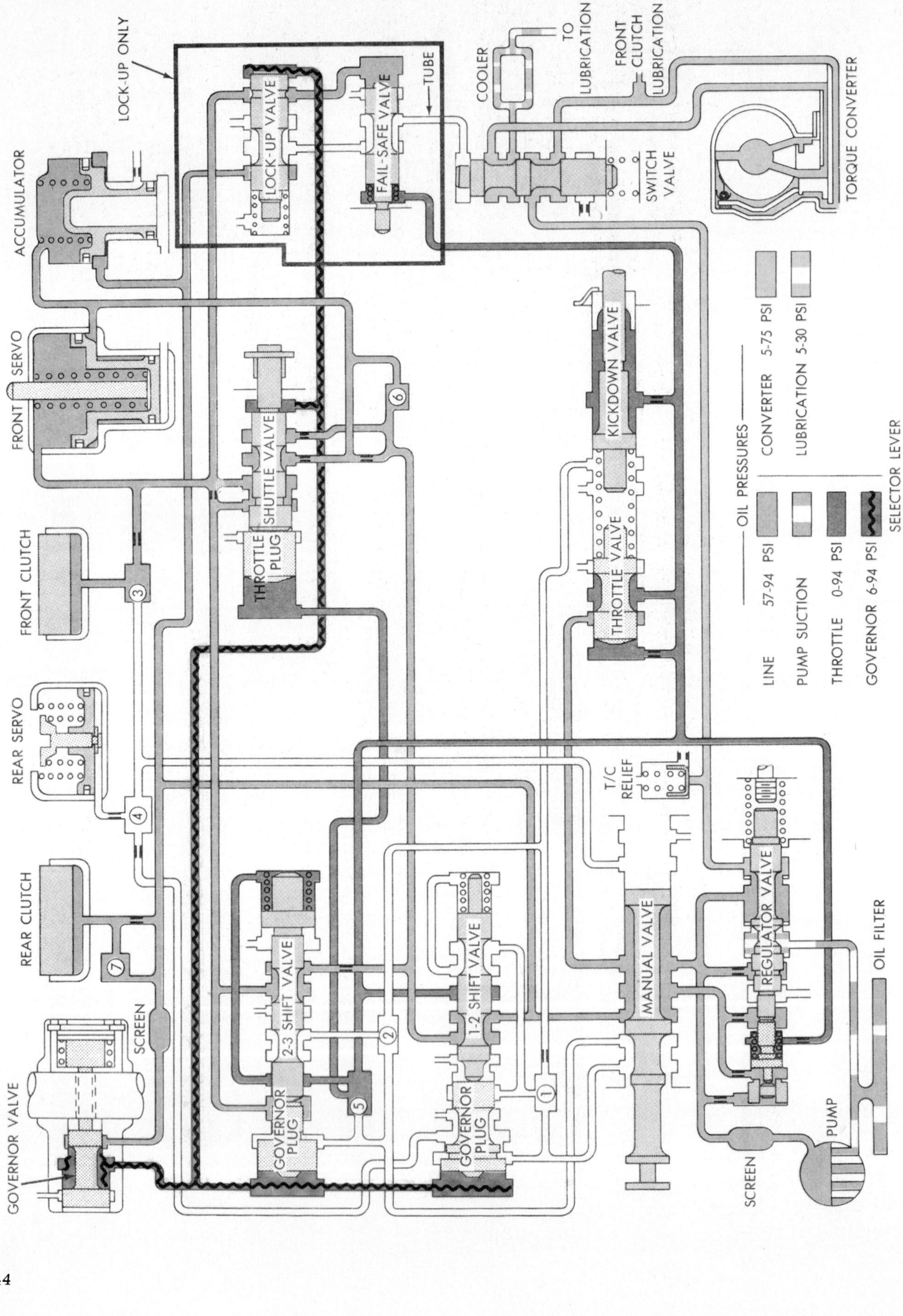
GOVERNOR VALVE
SCREEN
REAR CLUTCH
REAR SERVO
FRONT CLUTCH
FRONT SERVO
ACCUMULATOR
LOCK-UP ONLY
LOCK-UP VALVE
FAIL-SAFE VALVE
TUBE
COOLER
TO LUBRICATION
FRONT CLUTCH LUBRICATION
TORQUE CONVERTER
SWITCH VALVE
THROTTLE PLUG
SHUTTLE VALVE
KICKDOWN VALVE
THROTTLE VALVE
T/C RELIEF
MANUAL VALVE
REGULATOR VALVE
OIL FILTER
PUMP
SCREEN
2-3 SHIFT VALVE
1-2 SHIFT VALVE
GOVERNOR PLUG
GOVERNOR PLUG
1
2
3
4
5
6
7
OIL PRESSURES
LINE 57-94 PSI
PUMP SUCTION
THROTTLE 0-94 PSI
GOVERNOR 6-94 PSI
CONVERTER 5-75 PSI
LUBRICATION 5-30 PSI
SELECTOR LEVER IN DRIVE (DIRECT)

Figure 7-101.

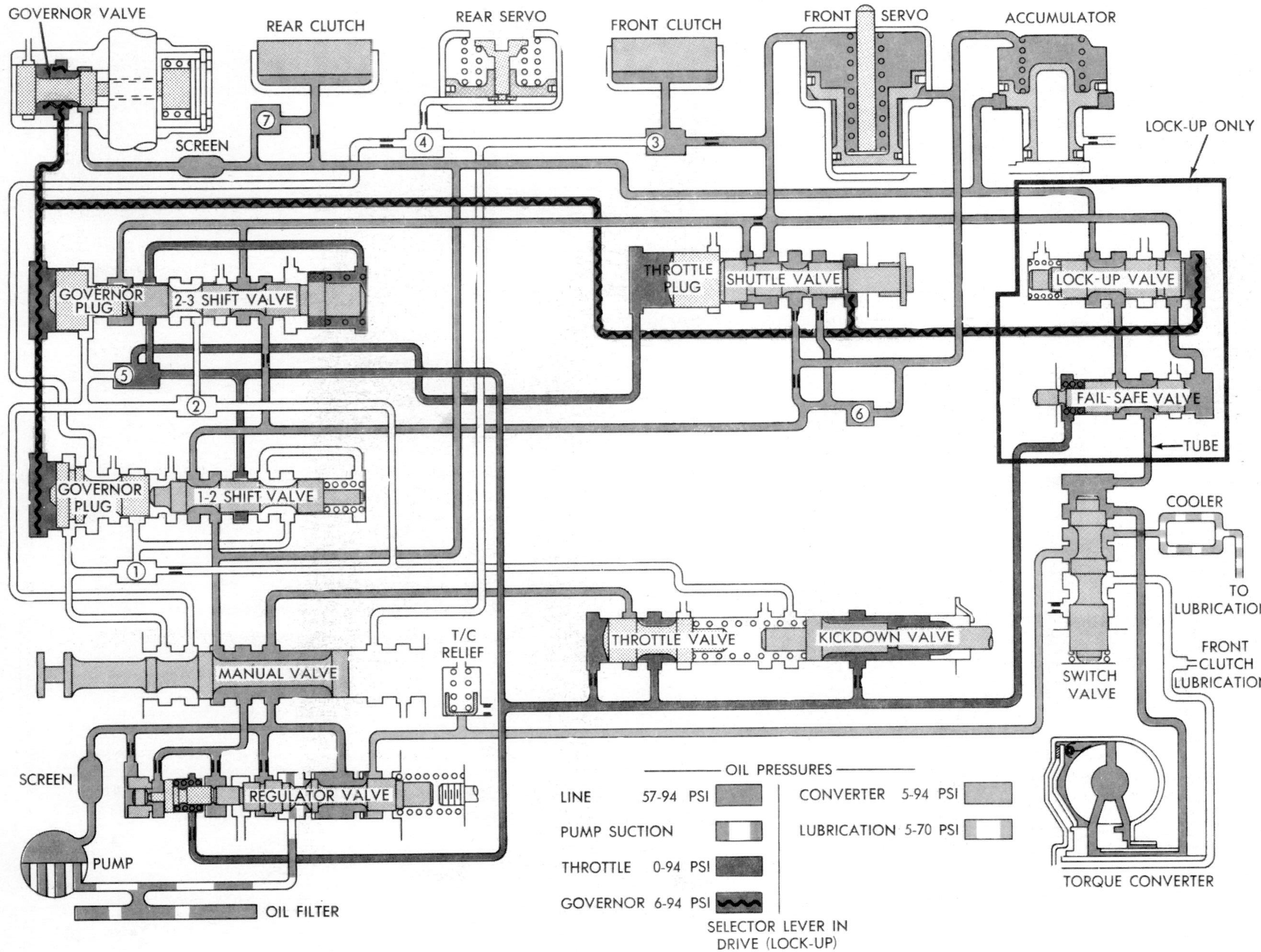

GOVERNOR VALVE
REAR CLUTCH
REAR SERVO
FRONT CLUTCH
FRONT SERVO
ACCUMULATOR
LOCK-UP ONLY
SCREEN
GOVERNOR PLUG
2-3 SHIFT VALVE
THROTTLE PLUG
SHUTTLE VALVE
LOCK-UP VALVE
FAIL-SAFE VALVE
TUBE
GOVERNOR PLUG
1-2 SHIFT VALVE
COOLER
TO LUBRICATION
THROTTLE VALVE
KICKDOWN VALVE
FRONT CLUTCH LUBRICATION
T/C RELIEF
MANUAL VALVE
SWITCH VALVE
SCREEN
REGULATOR VALVE
PUMP
OIL FILTER
TORQUE CONVERTER
OIL PRESSURES
LINE 57-94 PSI
PUMP SUCTION
THROTTLE 0-94 PSI
GOVERNOR 6-94 PSI
CONVERTER 5-94 PSI
LUBRICATION 5-70 PSI
SELECTOR LEVER IN DRIVE (LOCK-UP)

Figure 7-102.

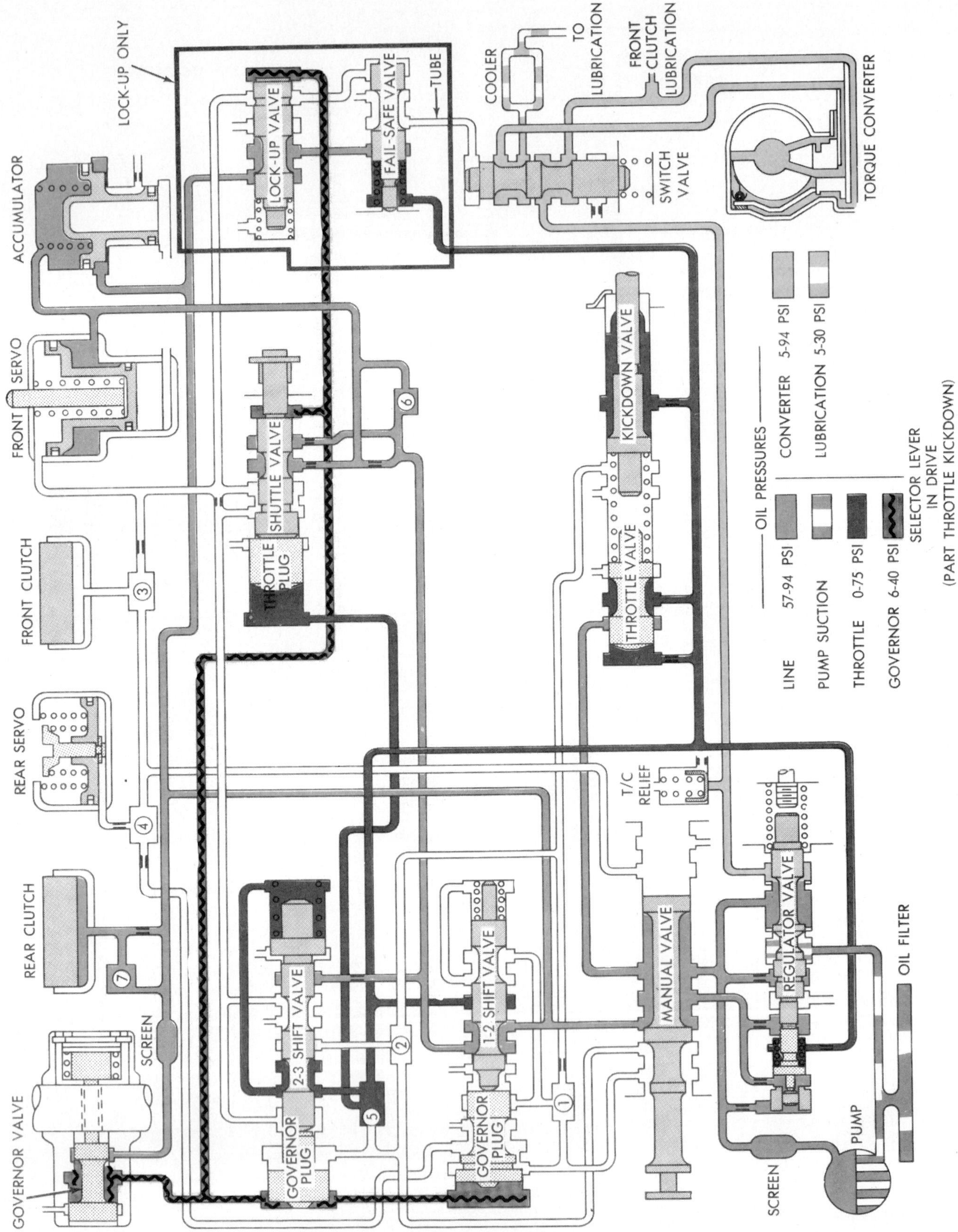
LOCK-UP ONLY
ACCUMULATOR
LOCK-UP VALVE
FAIL-SAFE VALVE
TUBE
COOLER
TO LUBRICATION
FRONT CLUTCH LUBRICATION
SWITCH VALVE
TORQUE CONVERTER
FRONT SERVO
SHUTTLE VALVE
THROTTLE PLUG
KICKDOWN VALVE
THROTTLE VALVE
FRONT CLUTCH
REAR SERVO
REAR CLUTCH
T/C RELIEF
2-3 SHIFT VALVE
1-2 SHIFT VALVE
MANUAL VALVE
REGULATOR VALVE
GOVERNOR VALVE
SCREEN
GOVERNOR PLUG
GOVERNOR PLUG
SCREEN
PUMP
OIL FILTER
1
2
3
4
5
6
7
OIL PRESSURES
LINE 57-94 PSI
PUMP SUCTION
THROTTLE 0-75 PSI
GOVERNOR 6-40 PSI
CONVERTER 5-94 PSI
LUBRICATION 5-30 PSI
SELECTOR LEVER IN DRIVE (PART THROTTLE KICKDOWN)

Figure 7-103.

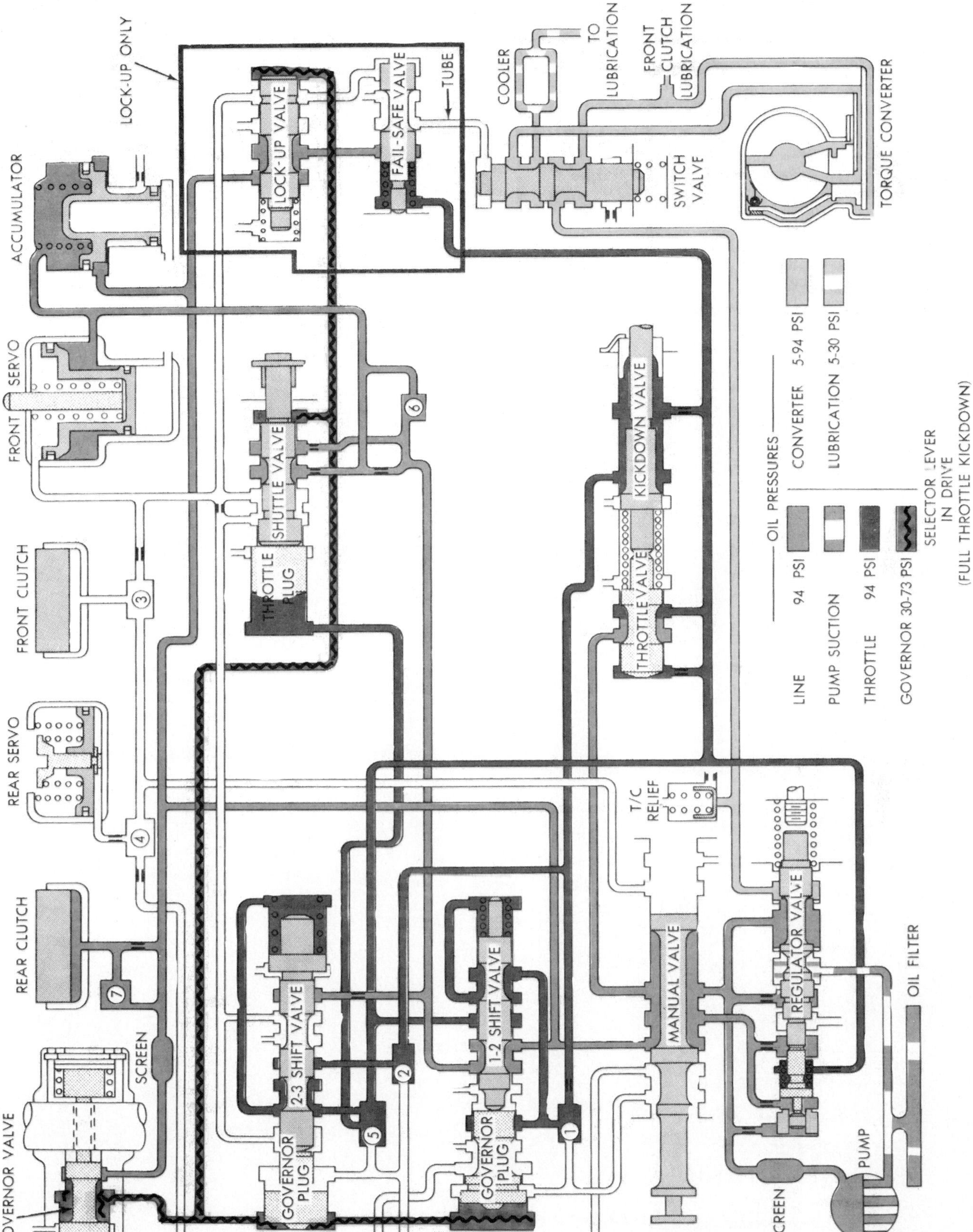
GOVERNOR VALVE
SCREEN
REAR CLUTCH
7
REAR SERVO
4
FRONT CLUTCH
3
FRONT SERVO
ACCUMULATOR
LOCK-UP ONLY
LOCK-UP VALVE
FAIL-SAFE VALVE
TUBE
COOLER
TO LUBRICATION
FRONT CLUTCH LUBRICATION
TORQUE CONVERTER
SWITCH VALVE
SHUTTLE VALVE
6
THROTTLE PLUG
KICKDOWN VALVE
THROTTLE VALVE
T/C RELIEF
2-3 SHIFT VALVE
1-2 SHIFT VALVE
GOVERNOR PLUG
GOVERNOR PLUG
5
2
1
MANUAL VALVE
REGULATOR VALVE
OIL FILTER
PUMP
SCREEN
OIL PRESSURES
LINE 94 PSI
PUMP SUCTION
THROTTLE 94 PSI
GOVERNOR 30-73 PSI
CONVERTER 5-94 PSI
LUBRICATION 5-30 PSI
SELECTOR LEVER IN DRIVE
(FULL THROTTLE KICKDOWN)

Figure 7–104.

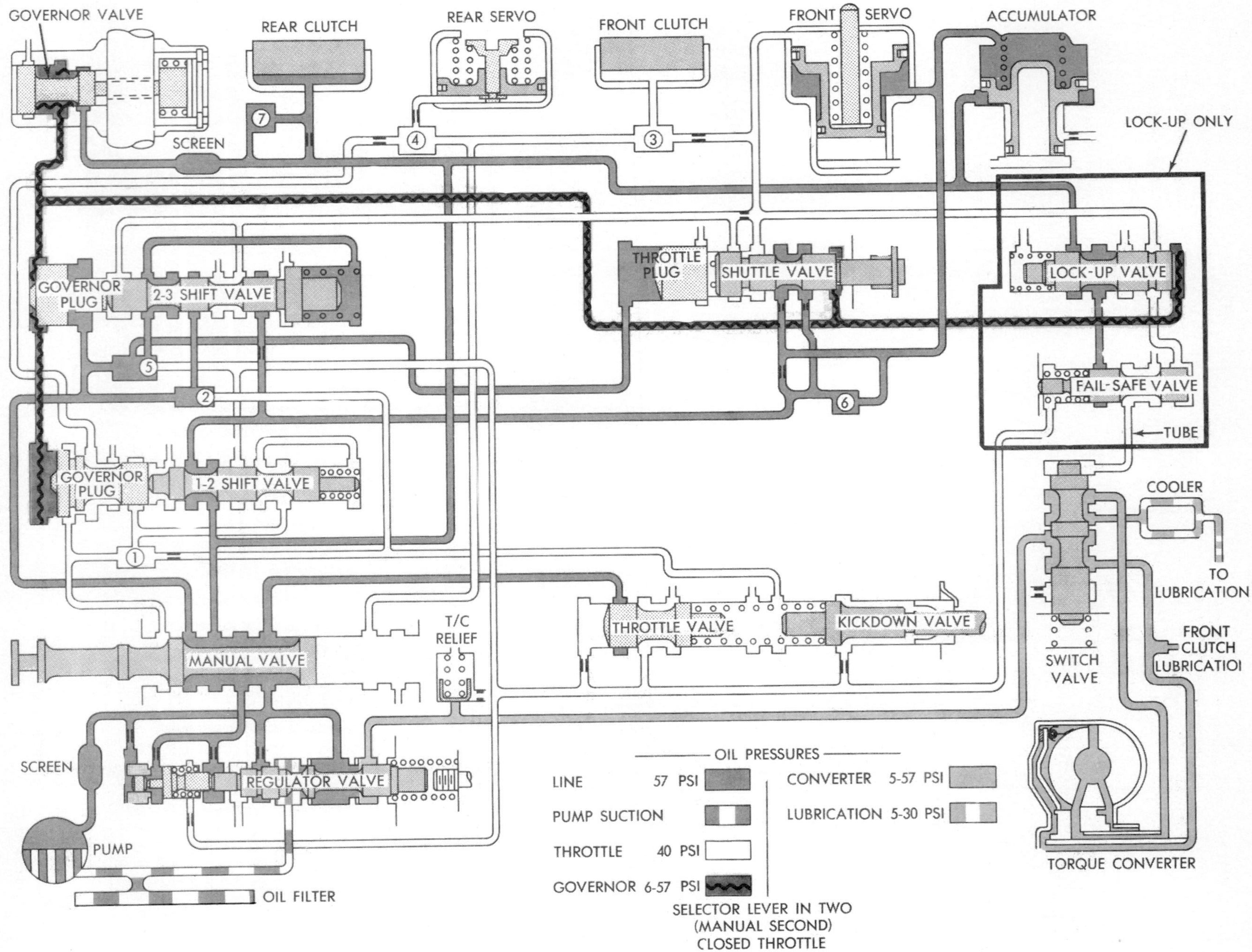
GOVERNOR VALVE
REAR CLUTCH
REAR SERVO
FRONT CLUTCH
FRONT SERVO
ACCUMULATOR
LOCK-UP ONLY
SCREEN
7
4
3
GOVERNOR PLUG
2-3 SHIFT VALVE
THROTTLE PLUG
SHUTTLE VALVE
LOCK-UP VALVE
5
2
6
FAIL-SAFE VALVE
TUBE
GOVERNOR PLUG
1-2 SHIFT VALVE
1
COOLER
TO LUBRICATION
T/C RELIEF
THROTTLE VALVE
KICKDOWN VALVE
FRONT CLUTCH LUBRICATIOI
SWITCH VALVE
MANUAL VALVE
SCREEN
REGULATOR VALVE
PUMP
OIL FILTER
OIL PRESSURES
LINE 57 PSI
PUMP SUCTION
THROTTLE 40 PSI
GOVERNOR 6-57 PSI
CONVERTER 5-57 PSI
LUBRICATION 5-30 PSI
SELECTOR LEVER IN TWO
(MANUAL SECOND)
CLOSED THROTTLE
TORQUE CONVERTER

Figure 7-105.

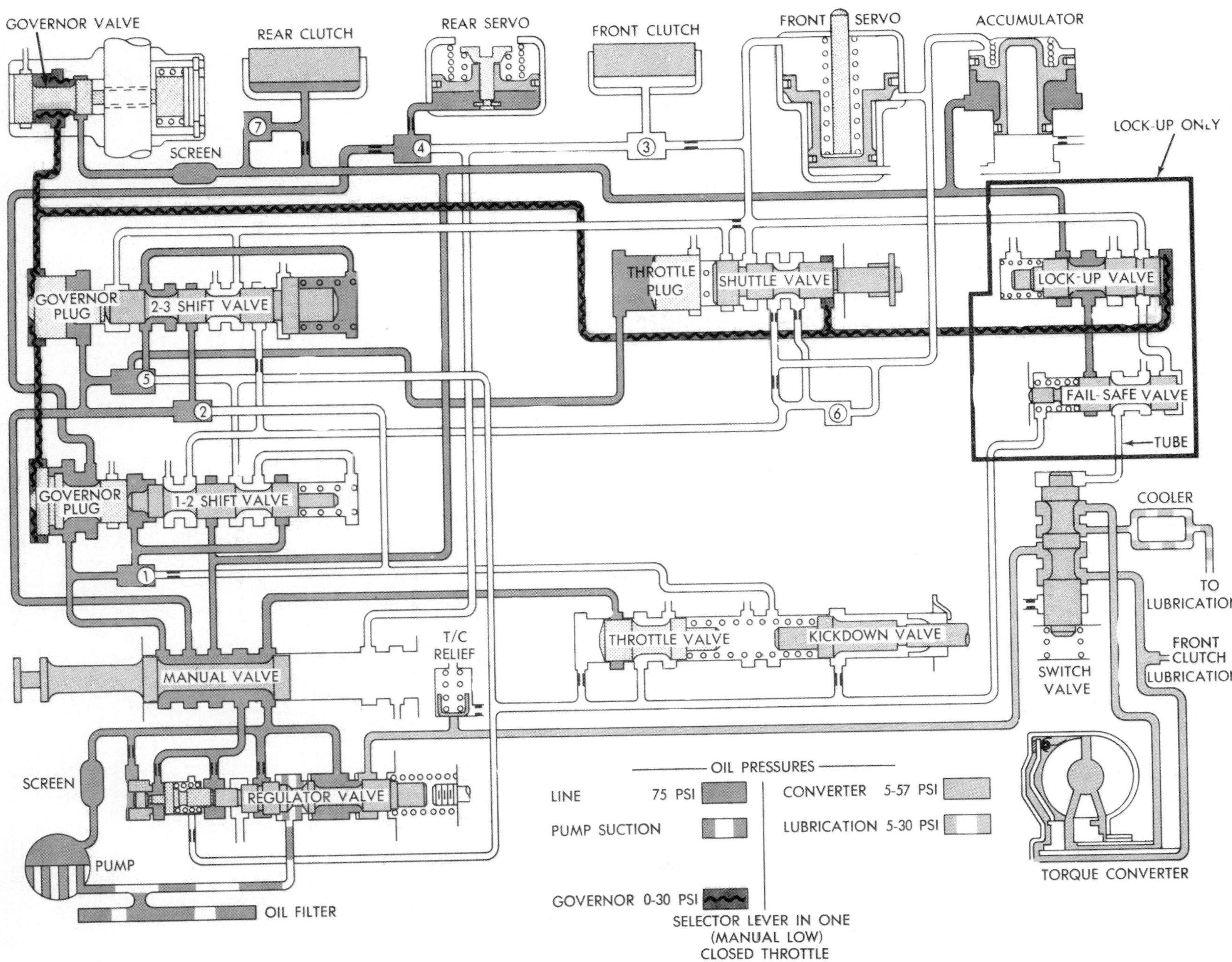
GOVERNOR VALVE
REAR CLUTCH
REAR SERVO
FRONT CLUTCH
FRONT SERVO
ACCUMULATOR
LOCK-UP ONLY
SCREEN
7
4
3
GOVERNOR PLUG
2-3 SHIFT VALVE
THROTTLE PLUG
SHUTTLE VALVE
LOCK-UP VALVE
5
2
FAIL-SAFE VALVE
6
TUBE
GOVERNOR PLUG
1-2 SHIFT VALVE
COOLER
1
TO LUBRICATION
T/C RELIEF
THROTTLE VALVE
KICKDOWN VALVE
FRONT CLUTCH LUBRICATION
MANUAL VALVE
SWITCH VALVE
SCREEN
REGULATOR VALVE
OIL PRESSURES
LINE 75 PSI
PUMP SUCTION
CONVERTER 5-57 PSI
LUBRICATION 5-30 PSI
PUMP
TORQUE CONVERTER
OIL FILTER
GOVERNOR 0-30 PSI
SELECTOR LEVER IN ONE
(MANUAL LOW)
CLOSED THROTTLE

Figure 7-106.

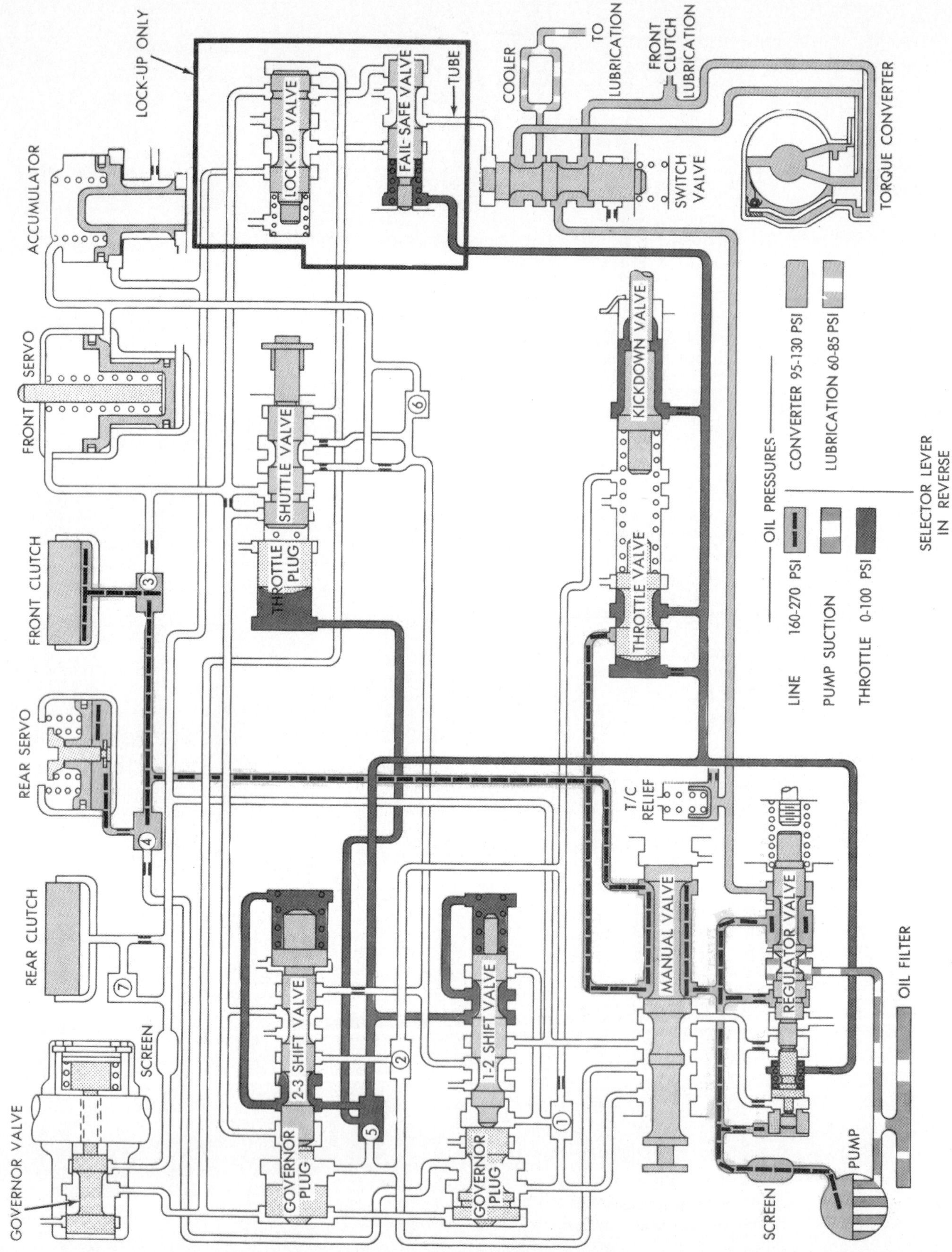

LOCK-UP ONLY
ACCUMULATOR
LOCK-UP VALVE
FAIL-SAFE VALVE
TUBE
COOLER
TO LUBRICATION
FRONT CLUTCH LUBRICATION
TORQUE CONVERTER
SWITCH VALVE
FRONT SERVO
SHUTTLE VALVE
THROTTLE PLUG
KICKDOWN VALVE
THROTTLE VALVE
FRONT CLUTCH
REAR SERVO
REAR CLUTCH
SCREEN
GOVERNOR VALVE
2-3 SHIFT VALVE
1-2 SHIFT VALVE
GOVERNOR PLUG
GOVERNOR PLUG
MANUAL VALVE
T/C RELIEF
REGULATOR VALVE
OIL FILTER
PUMP
SCREEN
1
2
3
4
5
6
7
OIL PRESSURES
LINE 160-270 PSI
PUMP SUCTION
THROTTLE 0-100 PSI
CONVERTER 95-130 PSI
LUBRICATION 60-85 PSI
SELECTOR LEVER IN REVERSE

Figure 7-107.

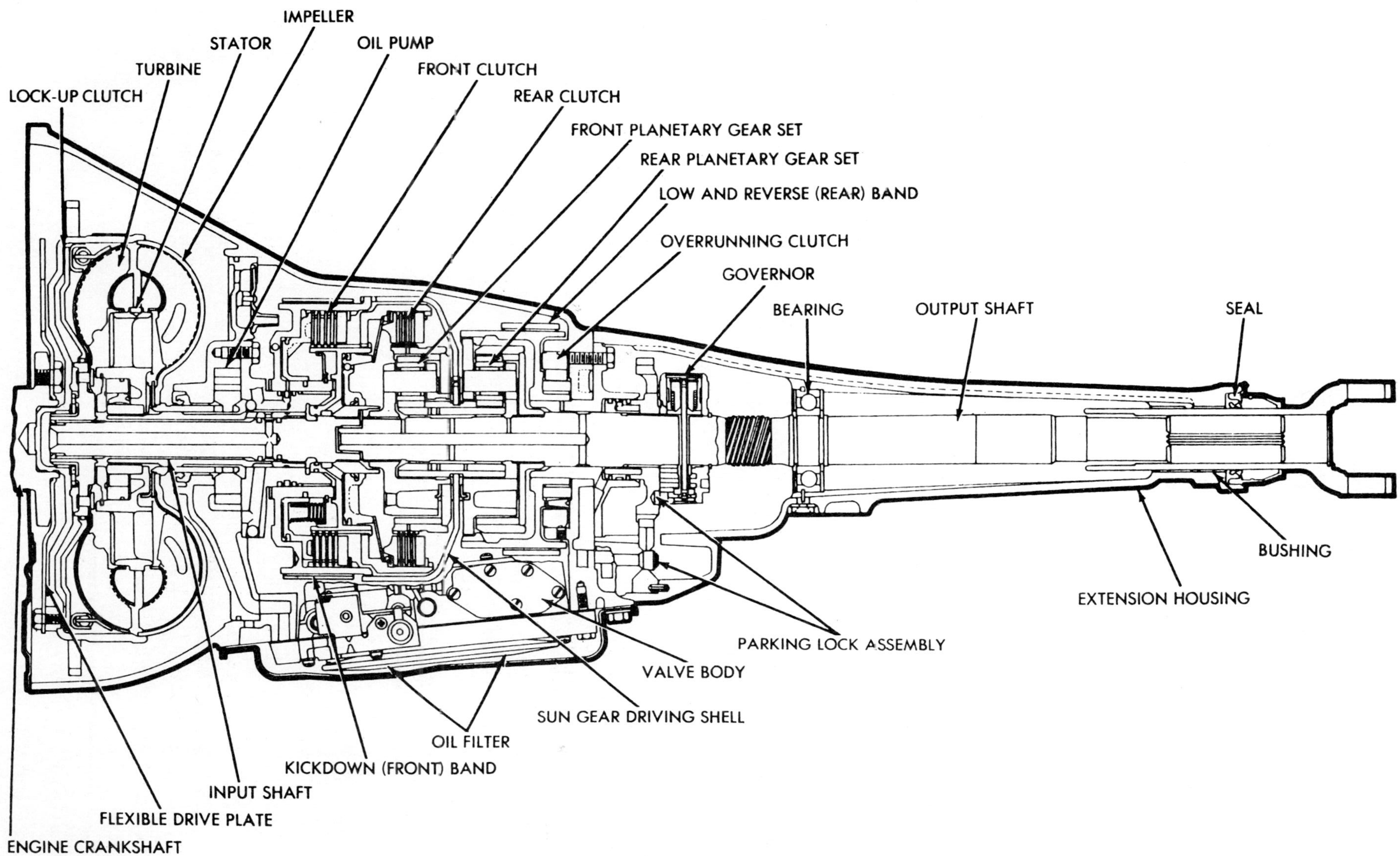

Figure 7-108. TorqueFlite transmission and torque converter (A-727).

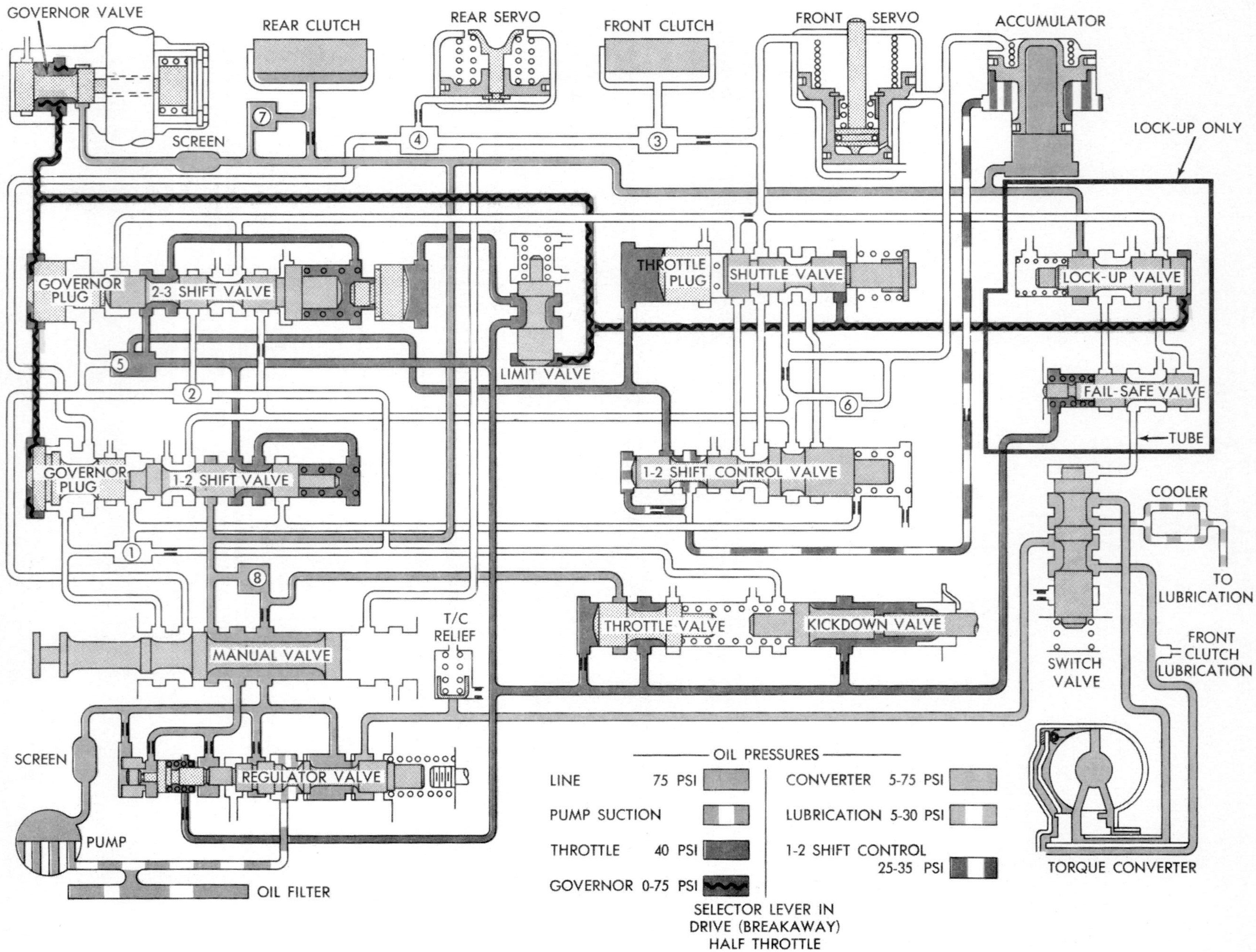
GOVERNOR VALVE
REAR CLUTCH
REAR SERVO
FRONT CLUTCH
FRONT SERVO
ACCUMULATOR
LOCK-UP ONLY
SCREEN
GOVERNOR PLUG
2-3 SHIFT VALVE
LIMIT VALVE
THROTTLE PLUG
SHUTTLE VALVE
LOCK-UP VALVE
FAIL-SAFE VALVE
TUBE
GOVERNOR PLUG
1-2 SHIFT VALVE
1-2 SHIFT CONTROL VALVE
COOLER
TO LUBRICATION
T/C RELIEF
MANUAL VALVE
THROTTLE VALVE
KICKDOWN VALVE
SWITCH VALVE
FRONT CLUTCH LUBRICATION
SCREEN
REGULATOR VALVE
PUMP
OIL FILTER
OIL PRESSURES
LINE 75 PSI
PUMP SUCTION
THROTTLE 40 PSI
GOVERNOR 0-75 PSI
CONVERTER 5-75 PSI
LUBRICATION 5-30 PSI
1-2 SHIFT CONTROL 25-35 PSI
TORQUE CONVERTER
SELECTOR LEVER IN DRIVE (BREAKAWAY) HALF THROTTLE

Figure 7-109.

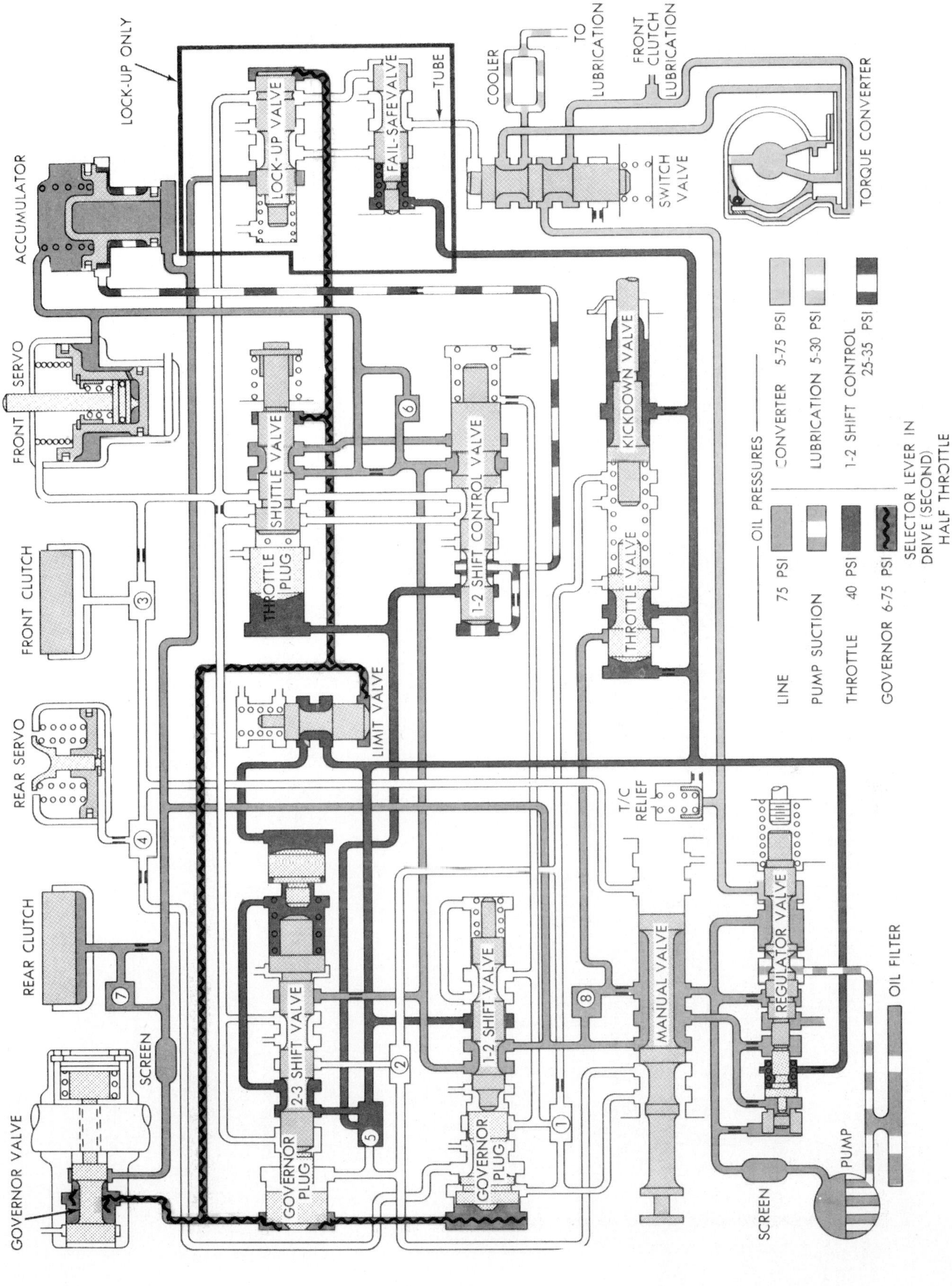

GOVERNOR VALVE
SCREEN
REAR CLUTCH
REAR SERVO
FRONT CLUTCH
FRONT SERVO
ACCUMULATOR
LOCK-UP ONLY
LOCK-UP VALVE
FAIL-SAFE VALVE
TUBE
COOLER
TO LUBRICATION
FRONT CLUTCH LUBRICATION
TORQUE CONVERTER
SWITCH VALVE
SHUTTLE VALVE
THROTTLE PLUG
1-2 SHIFT CONTROL VALVE
KICKDOWN VALVE
THROTTLE VALVE
LIMIT VALVE
T/C RELIEF
2-3 SHIFT VALVE
GOVERNOR PLUG
1-2 SHIFT VALVE
GOVERNOR PLUG
MANUAL VALVE
REGULATOR VALVE
OIL FILTER
PUMP
SCREEN
OIL PRESSURES
LINE 75 PSI
PUMP SUCTION
THROTTLE 40 PSI
GOVERNOR 6-75 PSI
CONVERTER 5-75 PSI
LUBRICATION 5-30 PSI
1-2 SHIFT CONTROL 25-35 PSI
SELECTOR LEVER IN DRIVE (SECOND) HALF THROTTLE

Figure 7-110.

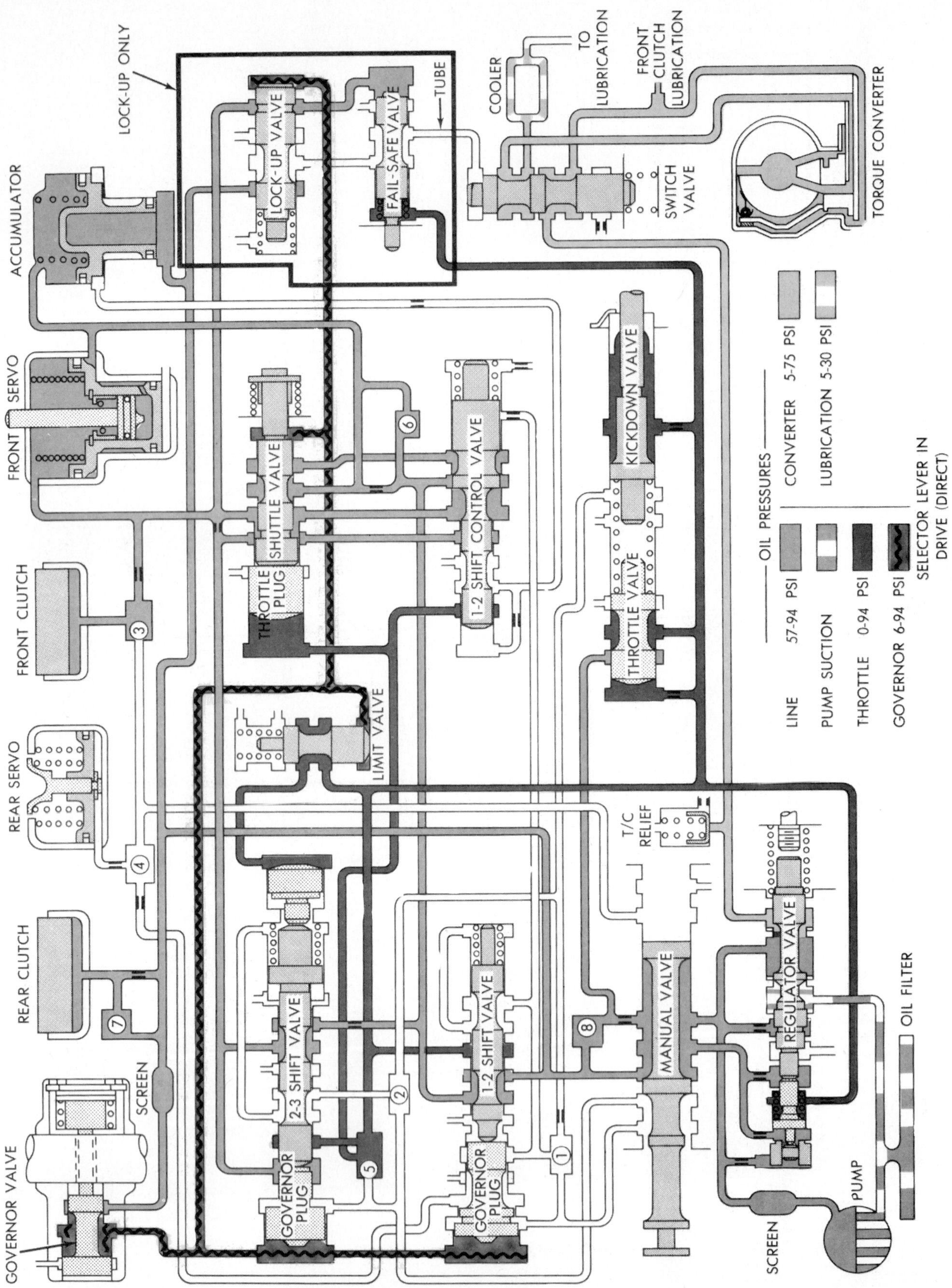
LOCK-UP ONLY
TO LUBRICATION
FRONT CLUTCH LUBRICATION
COOLER
TUBE
LOCK-UP VALVE
FAIL-SAFE VALVE
ACCUMULATOR
SWITCH VALVE
TORQUE CONVERTER
FRONT SERVO
SHUTTLE VALVE
THROTTLE PLUG
1-2 SHIFT CONTROL VALVE
KIDKDOWN VALVE
THROTTLE VALVE
FRONT CLUTCH
REAR SERVO
LIMIT VALVE
T/C RELIEF
REAR CLUTCH
2-3 SHIFT VALVE
1-2 SHIFT VALVE
MANUAL VALVE
REGULATOR VALVE
OIL FILTER
GOVERNOR VALVE
SCREEN
GOVERNOR PLUG
GOVERNOR PLUG
PUMP
SCREEN
1
2
3
4
5
6
7
8
OIL PRESSURES
LINE 57-94 PSI
PUMP SUCTION
THROTTLE 0-94 PSI
GOVERNOR 6-94 PSI
CONVERTER 5-75 PSI
LUBRICATION 5-30 PSI
SELECTOR LEVER IN DRIVE (DIRECT)

Figure 7-111.

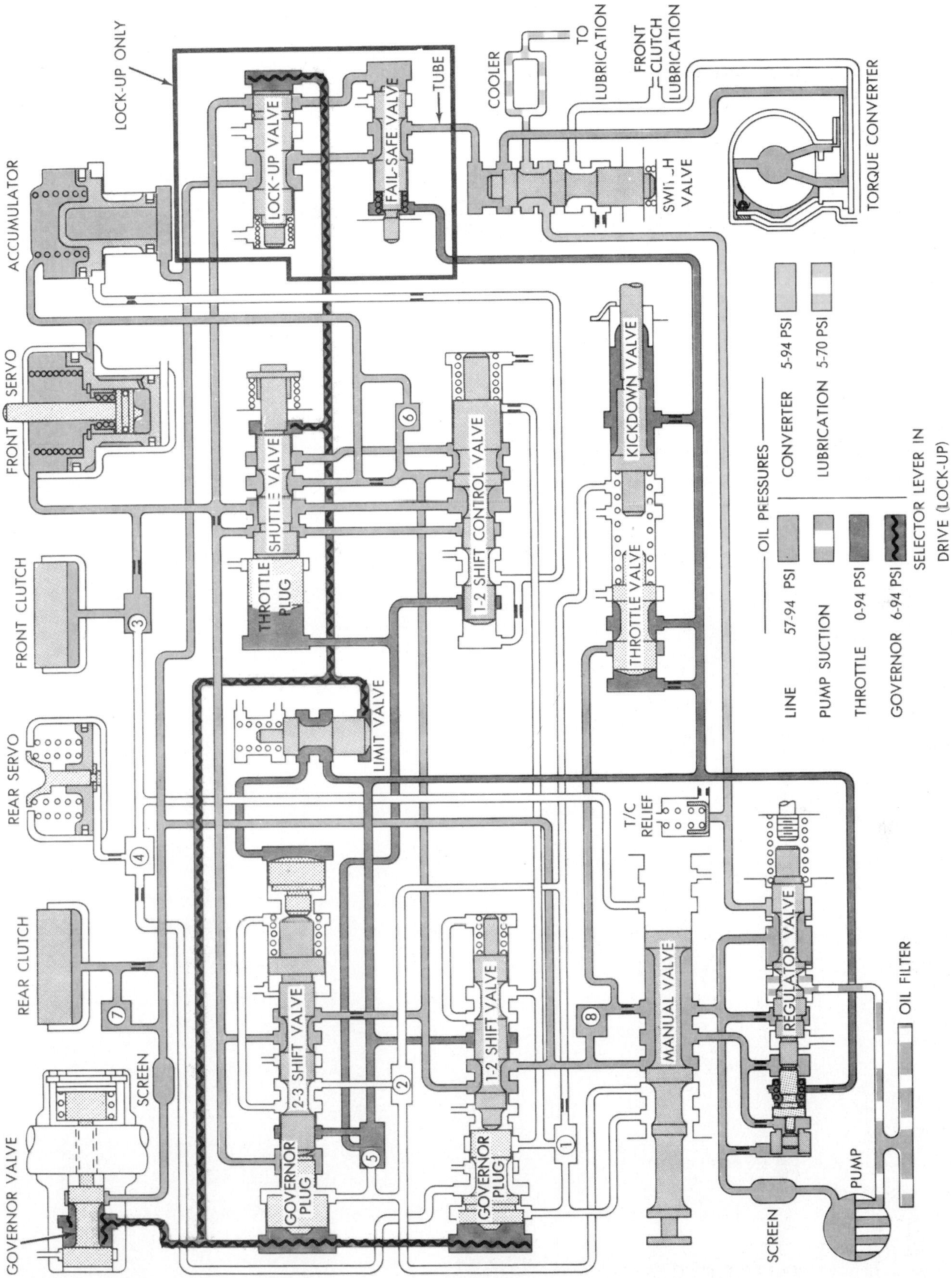
LOCK-UP ONLY
ACCUMULATOR
FRONT SERVO
FRONT CLUTCH
REAR SERVO
REAR CLUTCH
GOVERNOR VALVE
SCREEN
LOCK-UP VALVE
FAIL-SAFE VALVE
TUBE
COOLER
TO LUBRICATION
FRONT CLUTCH LUBRICATION
TORQUE CONVERTER
VALVE
SHUTTLE VALVE
THROTTLE PLUG
1-2 SHIFT CONTROL VALVE
KICKDOWN VALVE
THROTTLE VALVE
LIMIT VALVE
T/C RELIEF
2-3 SHIFT VALVE
1-2 SHIFT VALVE
GOVERNOR PLUG
GOVERNOR PLUG
MANUAL VALVE
REGULATOR VALVE
OIL FILTER
PUMP
SCREEN
OIL PRESSURES
LINE 57-94 PSI
PUMP SUCTION
THROTTLE 0-94 PSI
GOVERNOR 6-94 PSI
CONVERTER 5-94 PSI
LUBRICATION 5-70 PSI
SELECTOR LEVER IN DRIVE (LOCK-UP)

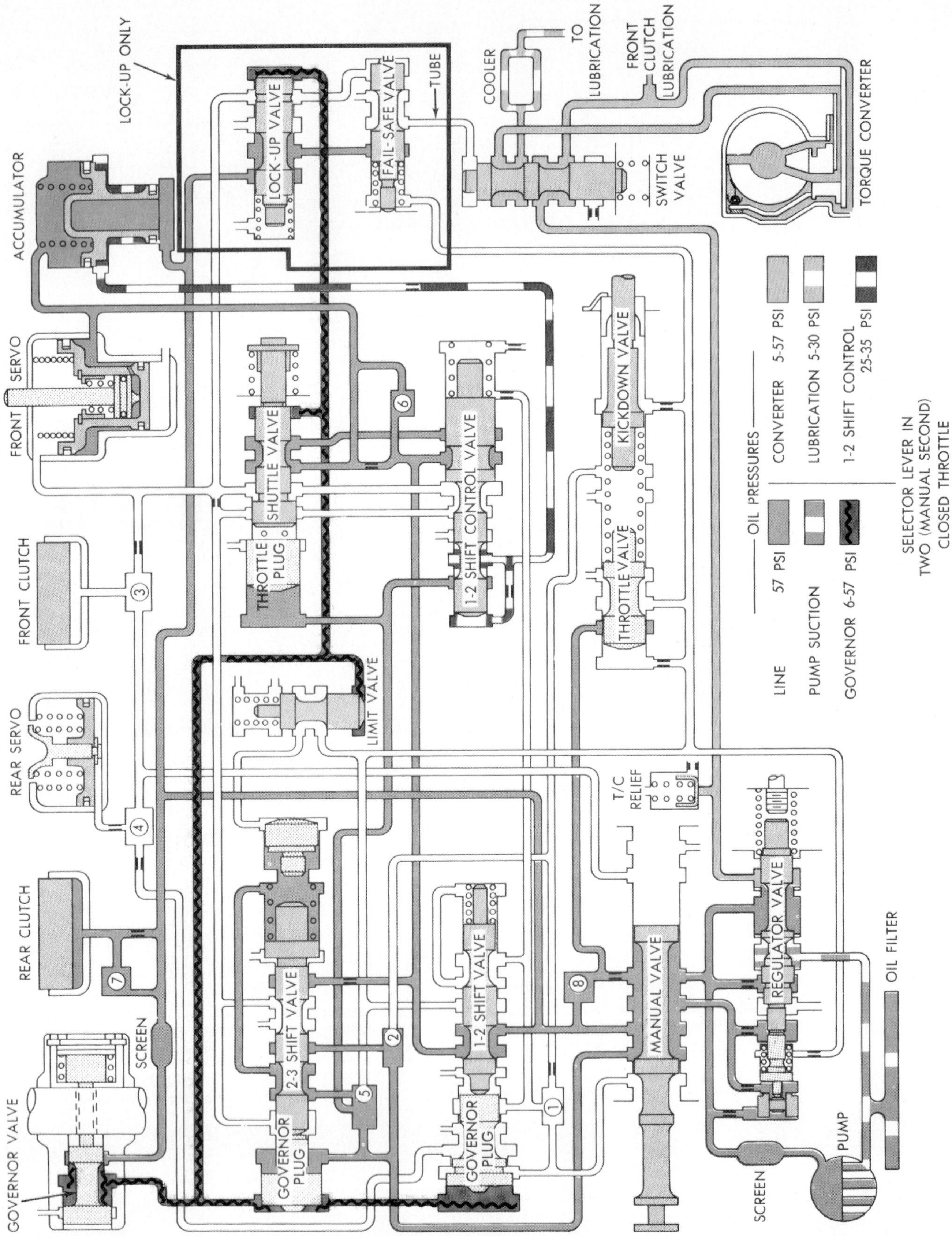
LOCK-UP ONLY
ACCUMULATOR
LOCK-UP VALVE
FAIL-SAFE VALVE
TUBE
COOLER
TO LUBRICATION
FRONT CLUTCH LUBRICATION
TORQUE CONVERTER
SWITCH VALVE
FRONT SERVO
FRONT CLUTCH
REAR SERVO
REAR CLUTCH
GOVERNOR VALVE
SCREEN
SHUTTLE VALVE
THROTTLE PLUG
1-2 SHIFT CONTROL VALVE
KICKDOWN VALVE
THROTTLE VALVE
LIMIT VALVE
T/C RELIEF
2-3 SHIFT VALVE
GOVERNOR PLUG
1-2 SHIFT VALVE
GOVERNOR PLUG
MANUAL VALVE
REGULATOR VALVE
OIL FILTER
PUMP
SCREEN
1
2
3
4
5
6
7
8
OIL PRESSURES
LINE 57 PSI
PUMP SUCTION
GOVERNOR 6-57 PSI
CONVERTER 5-57 PSI
LUBRICATION 5-30 PSI
1-2 SHIFT CONTROL 25-35 PSI
SELECTOR LEVER IN TWO (MANUAL SECOND) CLOSED THROTTLE

Figure 7-113.

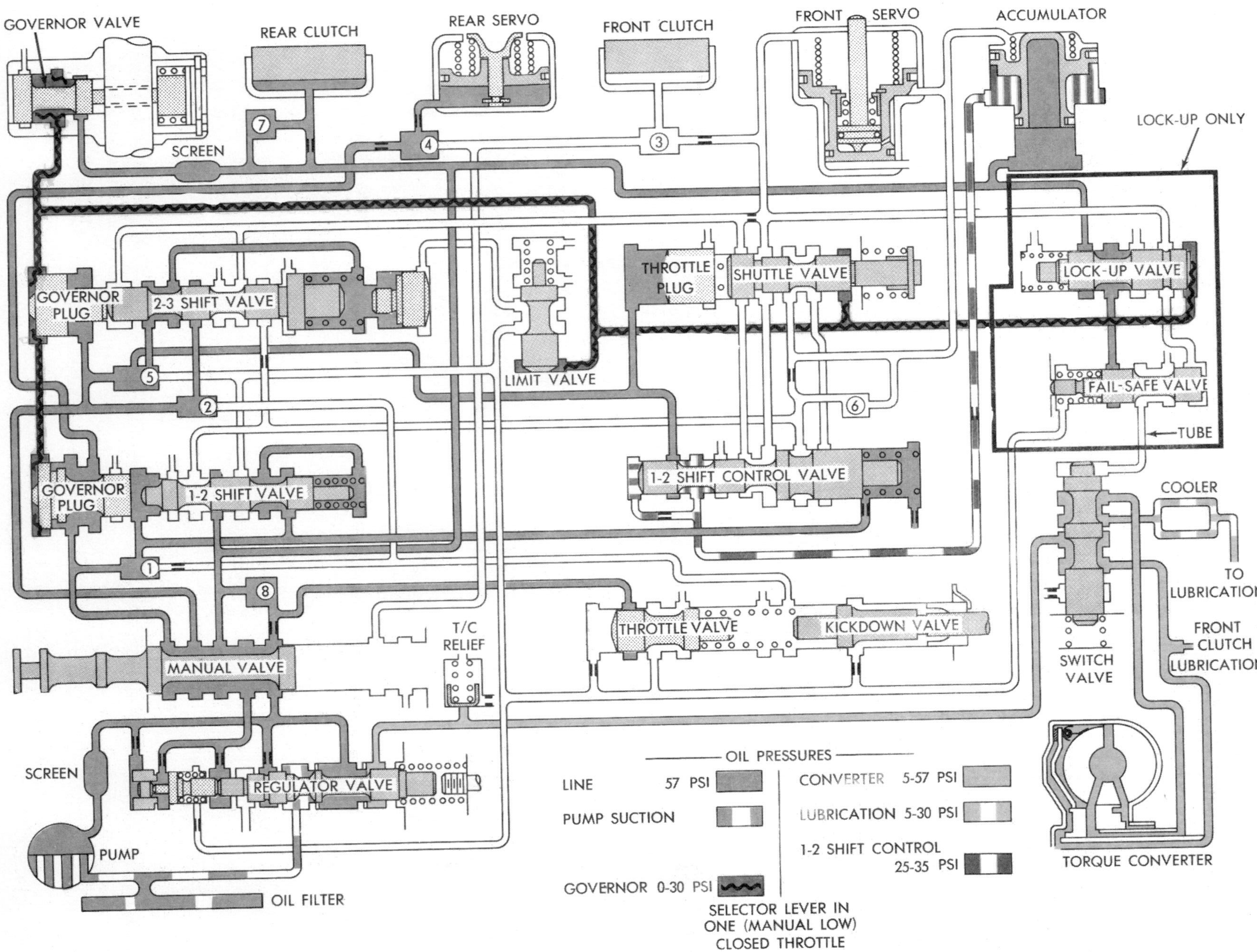
GOVERNOR VALVE
REAR CLUTCH
REAR SERVO
FRONT CLUTCH
FRONT SERVO
ACCUMULATOR
LOCK-UP ONLY
SCREEN
GOVERNOR PLUG
2-3 SHIFT VALVE
LIMIT VALVE
THROTTLE PLUG
SHUTTLE VALVE
LOCK-UP VALVE
FAIL-SAFE VALVE
TUBE
GOVERNOR PLUG
1-2 SHIFT VALVE
1-2 SHIFT CONTROL VALVE
COOLER
TO LUBRICATION
FRONT CLUTCH LUBRICATION
THROTTLE VALVE
KICKDOWN VALVE
T/C RELIEF
MANUAL VALVE
SWITCH VALVE
SCREEN
REGULATOR VALVE
PUMP
OIL FILTER
TORQUE CONVERTER
OIL PRESSURES
LINE 57 PSI
PUMP SUCTION
GOVERNOR 0-30 PSI
CONVERTER 5-57 PSI
LUBRICATION 5-30 PSI
1-2 SHIFT CONTROL 25-35 PSI
SELECTOR LEVER IN ONE (MANUAL LOW) CLOSED THROTTLE
1
2
3
4
5
6
7
8

Figure 7-114.

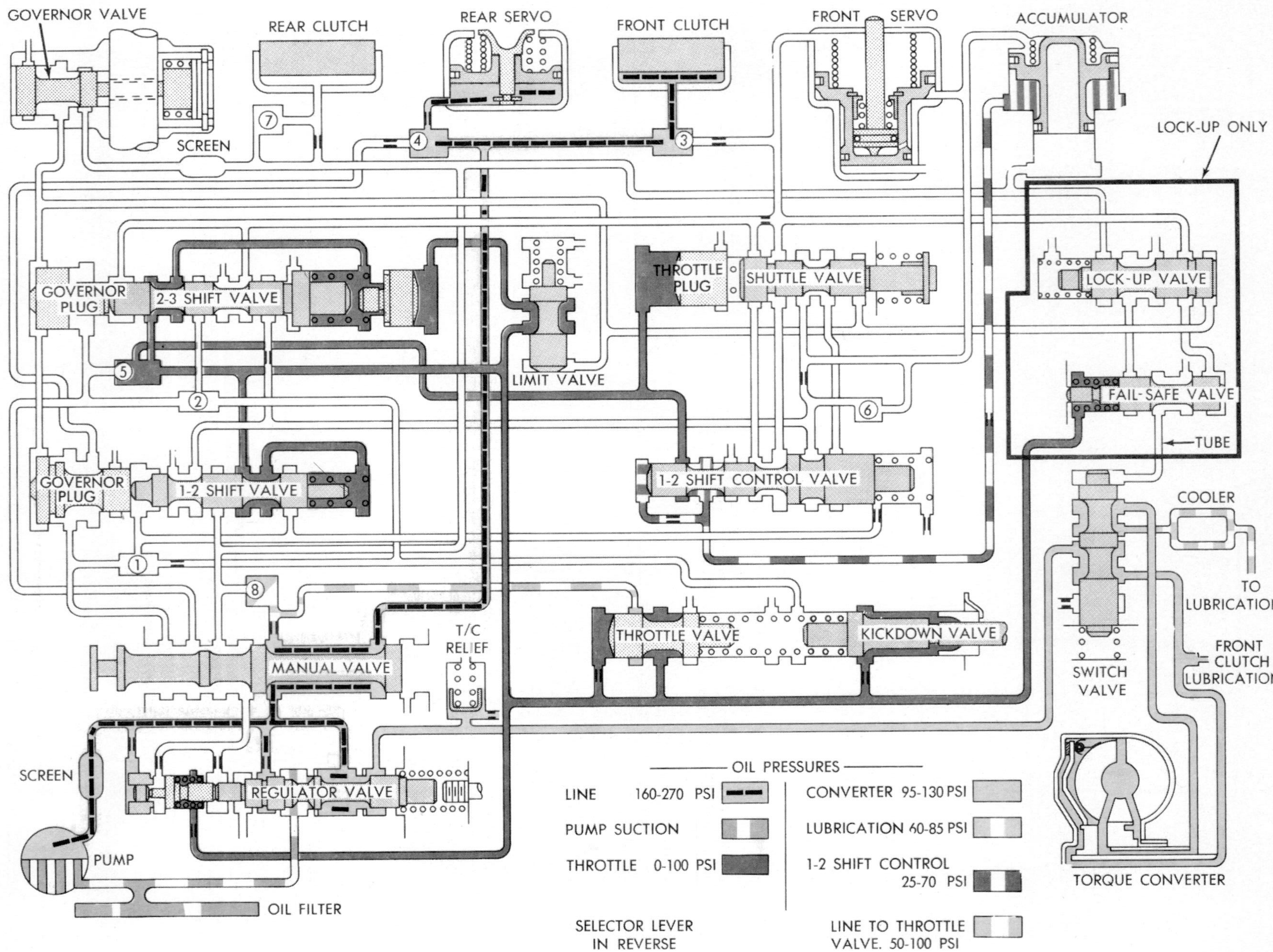
GOVERNOR VALVE
REAR CLUTCH
REAR SERVO
FRONT CLUTCH
FRONT SERVO
ACCUMULATOR
LOCK-UP ONLY
SCREEN
GOVERNOR PLUG
2-3 SHIFT VALVE
LIMIT VALVE
THROTTLE PLUG
SHUTTLE VALVE
LOCK-UP VALVE
FAIL-SAFE VALVE
TUBE
GOVERNOR PLUG
1-2 SHIFT VALVE
1-2 SHIFT CONTROL VALVE
COOLER
TO LUBRICATION
FRONT CLUTCH LUBRICATION
SWITCH VALVE
T/C RELIEF
MANUAL VALVE
THROTTLE VALVE
KICKDOWN VALVE
SCREEN
REGULATOR VALVE
PUMP
OIL FILTER
TORQUE CONVERTER
OIL PRESSURES
LINE 160-270 PSI
PUMP SUCTION
THROTTLE 0-100 PSI
CONVERTER 95-130 PSI
LUBRICATION 60-85 PSI
1-2 SHIFT CONTROL 25-70 PSI
LINE TO THROTTLE VALVE. 50-100 PSI
SELECTOR LEVER IN REVERSE

Figure 7-115.

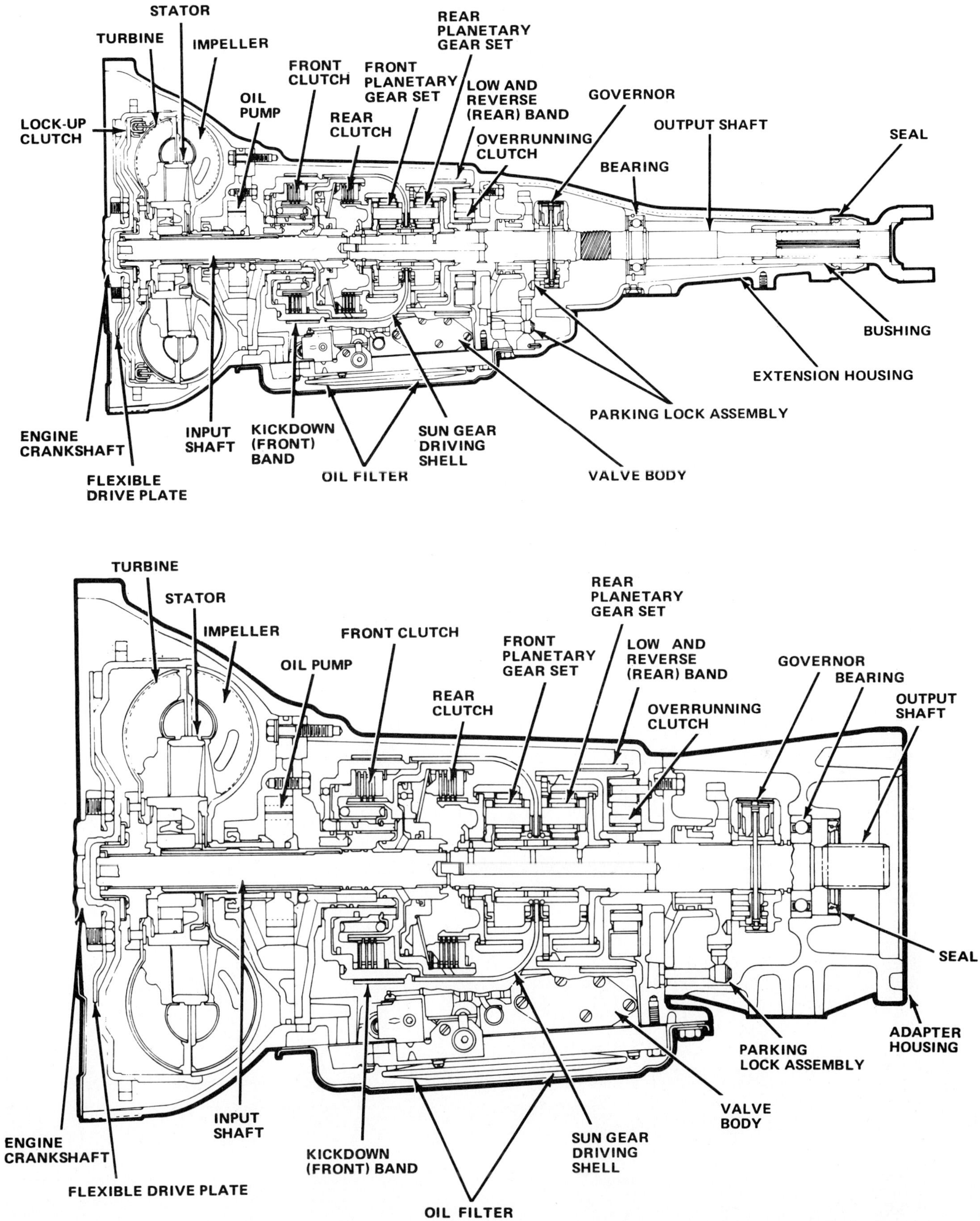

Figure 7-116. Automatic transmission—Models 904/998—AMC.

Part 7–5 Worksheet and Self-Check

Using the appropriate hydraulic system schematics, determine the following and be able to tell your instructor:

1. Which driving and holding units are applied in manual low in a transmission of your choice.
2. What prevents automatic upshifts in manual low.
3. How increased torque demand raises line pressure.
4. The difference between a torque demand or to-detent downshift and a full throttle kickdown downshift (through detent).
5. How line pressure cutback is achieved.
6. How line pressure is metered by the governor valve to produce governor pressure.
7. How line pressure is increased in reverse.

PART 7-6 PERFORMANCE EVALUATION

Did you accurately perform the following tasks?

1. Describe hydraulic system operation in any gear position.
2. Describe the operation of all components in the hydraulic system.
3. Diagnose hypothetical malfunctions in the hydraulic control system.
4. Complete the worksheet for this chapter.
5. Complete the self-check for this chapter with at least 80 percent accuracy.
6. Evaluate performance based on this performance evaluation section.

Chapter 8

Automatic Transmission Diagnosis and Service

Performance Objectives

After sufficient study of this chapter and with the appropriate practice on various automatic transmissions the student should be able to perform these tasks:

1. Follow the general precautions given in this chapter.
2. Diagnose correctly the cause of specific automatic transmission problems by following the accepted systematic approach to problem diagnosis given in this chapter, in the appropriate service repair manual, and by the instructor.
3. Perform the appropriate diagnostic tests, interpret the test results correctly, and make the necessary adjustments.
   - Check fluid level and condition.
   - Adjust manual shift linkage to specifications.
   - Adjust throttle linkage to specifications.
   - Check and adjust vacuum modulator system and vacuum lines to specifications.
   - Check and adjust engine idle speed to specifications.
   - Perform stall test; record and interpret results.
   - Perform hydraulic pressure tests; record and interpret results.
   - Perform road test and interpret results.
4. Perform the following in vehicle service checks and adjustments.
   - Change the oil and filter.
   - Remove valve body and perform air pressure tests.
   - Service hydraulic servos.
   - Adjust bands.
   - Service the governor.
   - Road test and interpret results.
5. Complete the worksheet in this chapter as required.
6. Complete the self-check with at least 80 percent accuracy.
7. Complete all service work with 100 percent accuracy.
8. Prepare the vehicle for customer acceptance.
9. Evaluate your performance according to the performance evaluation section in this chapter.

PART 8-1 INTRODUCTION

The proper diagnosis of automatic transmission problems requires a thorough understanding of transmission operation and it requires a systematic, step-by-step approach. A hit-or-miss, shot-in-the-dark approach is dangerous and open to creating more problems and wasting of valuable time.

This chapter presents a systematic, step-by-step approach to problem diagnosis that is generally applicable to all types of automatic transmissions. Diagnostic procedures and charts are included for some of the more common automatic transmissions. Also included are the more common external and internal in vehicle checks, adjustments, and service procedures.

The successful automatic transmission technician is one who has learned to apply knowledge of the transmission and its operating characteristics to a quick and accurate diagnosis of transmission problems. When a transmission problem is presented, the competent technician immediately considers all the possible causes and pursues the diagnostic procedure with these in mind.

The diagnostic procedures then are used to verify one or more of the possible causes suspected, and the appropriate corrections are then made. The road test is used to validate the corrections and assure proper transmission operation.

PART 8-2 GENERAL PRECAUTIONS

All the usual personal safety factors, such as proper clothing, work habits, and shop practice, should be observed when working on automatic transmissions. Proper tools and equipment should be used in the correct manner as outlined by the equipment manufacturer.

Particular attention should be given to proper hoisting and jacking methods. No one should be under a vehicle that is being raised or lowered, and no one should be under a vehicle supported only by a jack. Safety stands properly positioned must be used when working under a vehicle raised by a jack.

A clean, uncluttered work area is also essential to safe shop practice. Jack handles, electrical cords and hoses, or tools and equipment left lying around on the shop floor are open invitations to accidents and injury. Dirt and oil on shop floors are similarly hazardous.

Some additional points to remember follow.

- Disconnect the battery when not actually required for test procedures.
- Be careful of hot engine, exhaust, and transmission parts.
- Be careful of hot transmission fluid. Fluid temperatures may be hot enough to cause severe burns.
- Transmissions are heavy. Use only the proper equipment to remove and install.
- Handle all parts carefully. Many transmission parts are machined to very close tolerances and are easily damaged, requiring replacement.
- Never drop transmission parts. Place all parts so that they will not roll off workbenches.
- Absolute cleanliness of all parts for assembly is essential. Even a small particle of dirt or lint can cause transmission problems.
- Make sure that whatever you are doing will not result in consequences that can cause personal injury or damage to parts or vehicles.

PART 8-3 DIAGNOSTIC PROCEDURES, TESTS, AND ADJUSTMENTS

Successful accurate diagnosis of automatic transmission problems is not difficult or complicated. However, it can only be achieved if transmission operation is thoroughly understood and if a logical, step-by-step procedure is followed. This approach soon tells the technician what is working properly in the transmission and what is not working properly.

Preliminary Diagnosis Procedure

1. Vehicle not drivable. Check for broken or disconnected throttle linkage to transmission. Check for disconnected shift linkage. Check the fluid level and condition. Check and correct leaks and fluid level. Raise the vehicle on a hoist, start the engine, and place the transmission in gear. If the drive shaft turns, the problem is in the differential.

If propeller shafts do not turn and the transmission is noisy, stop the engine, remove the oil pan, and check for debris. If debris is not found, remove the transmission and check for a broken drive plate or drive-plate-to-converter bolts, broken converter hub, broken input or output shaft, broken oil pump, or stripped torque converter turbine hub splines. If propeller shafts do not turn and the transmission is not noisy, perform the hydraulic pressure test to determine if the problem is a malfunction of a hydraulic or mechanical component.

2. Vehicle drivable. Proceed with checks and adjustments outlined in the following groups 1 and 2.

It is important that the checks and procedures in group 1 be performed first, then those in group 2, and then those in group 3. Unless this is done, some of the tests in group 2 may not be valid. Of course, it would not make much sense to remove the transmission to effect a repair or adjustment that can be made in group 1 or group 2 without transmission removal.

The following procedures are general and should be used in conjunction with service manuals for specific transmissions.

Group 1. Those that can be done without any disassembly of the transmission.

Group 2. Those that can be done by partial transmission disassembly but without transmission removal.

Group 3. Those that require transmission removal (covered in Chapter 9).

Group 1 Diagnostic Procedures

Step 1 Verify the Complaint

If the shop where you are working allows direct contact with the customer, discuss the problem with the owner of the vehicle. This can help in identifying the problem later during a road test. If customer contact is not allowed, discuss the complaint with those responsible for customer contact in your shop. Deter-

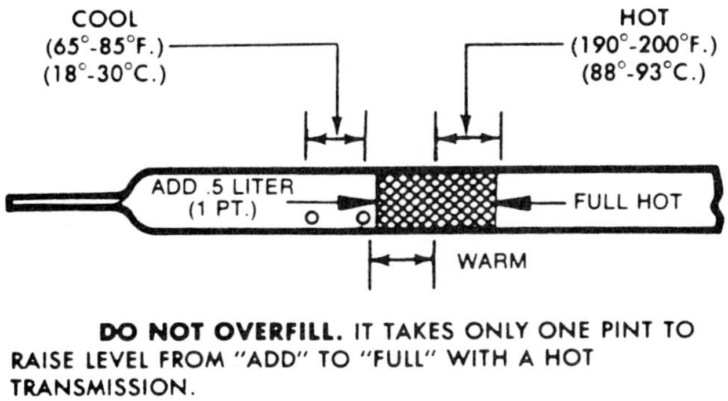

Figure 8-1. Typical General Motors transmission dipstick. (*Courtesy of General Motors Corporation*)

mine whether any recent transmission service or fluid change have been performed.

Step 2 Check Fluid Level and Condition

The following conditions should apply when checking the fluid level.

1. Transmission fluid is at operating temperature 185°F (85°C).
2. The vehicle is in a level position.
3. The parking brake is fully applied.
4. With the engine running at idle, move the selector lever through all gear positions.
5. Leave the engine running at idle (most vehicles) and place the selector lever in the position specified by the vehicle manufacturer (usually park or neutral). Many vehicles show a different fluid level in park from that in neutral; therefore it is important to place the selector lever in the specified position.
6. Use a clean, lintfree, preferably white cloth. Pull the dipstick and wipe the blade clean. Insert the dipstick, making sure that it is fully seated. Remove the dipstick and check the fluid level and condition (color, smell, contamination). The car should not be road tested while the fluid level is too low or too high since further damage to the transmission is possible. Fluid level must be corrected first.

A fluid level that is too low can result in air in the hydraulic system, which in turn can cause insufficient apply pressure to friction elements. Clutch and band slippage can cause overheating and serious transmission damage. Varnish forms on internal transmission parts; it can cause valve malfunctions. Delayed engagement and mushy shifts can also be caused by low fluid levels.

The cause for the low fluid level should be determined and the problem corrected (see the section on leak diagnosis in this chapter). A fluid level that is too high can cause similar damage. The fluid is churned up by the rotating gear train, causing fluid aeration and foam. Foaming allows air to get into the hydraulic system. This causes clutch and band slippage, wear, overheating, and varnish deposits. Foaming can also result in fluid being forced out of the transmission vent or filler tube, which may be mistaken for a fluid leak.

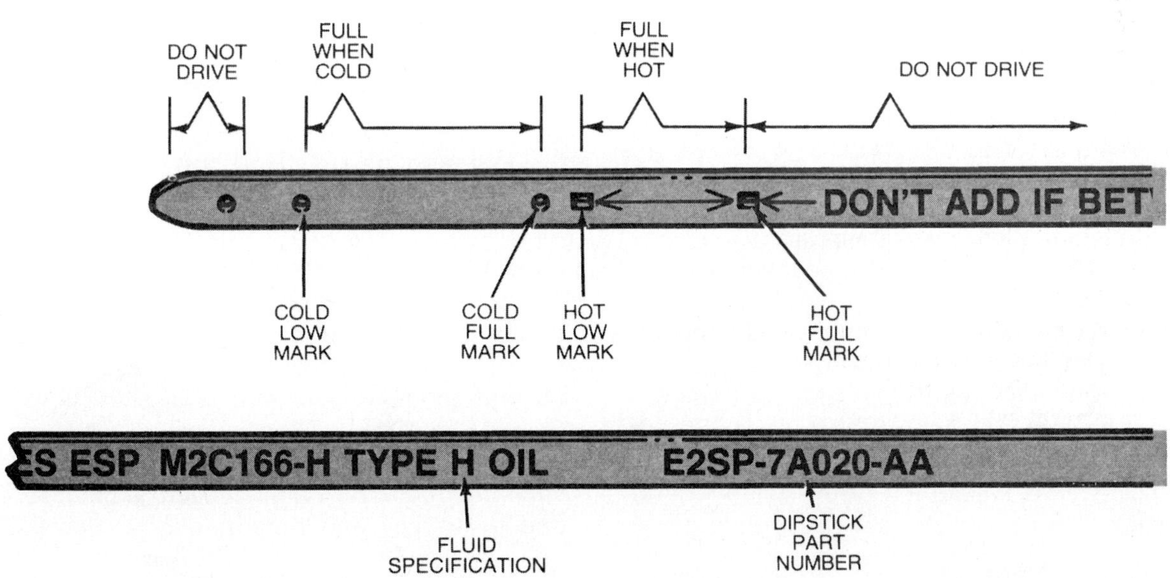

Figure 8-2. Typical Ford C5 transmission dipstick. (*Courtesy of Ford Motor Company*)

Fluid Condition. Transmission fluid colors range from an almost clear fluid to a deep red. Some fluids tend to discolor and darken somewhat after use. This should be taken into account when checking fluid condition.

The transmission oil pan should be removed for further diagnosis of fluid and deposits if the following conditions are present in the fluid:

- Very dark colored fluid, which has a pungent, burned smell
- Presence of metallic particles and particles of friction material from bands or clutches
- Milky colored fluid, which indicates a coolant leak into the transmission fluid via the fluid cooler in the radiator
- Varnish deposits on the dipstick and fluid that is dark brown in color

Depositing several drops of fluid from the dipstick onto a clean white cloth helps determine fluid condition and the presence of metal or friction particles. The fluid should feel smooth and slippery when rubbed between the finger and thumb. If there is a gritty feel to the fluid, further diagnosis is required. Make sure that hands are clean for this test.

If further verification of fluid and transmission condition is required, the transmission oil pan should be removed and the fluid and deposits examined.

Oil Pan Removal and Inspection. Remember that the pan and the fluid in it may be hot enough to cause severe burns.

To remove the pan, use a large drain pan to catch the transmission fluid when the pan is removed. First, loosen all the oil pan bolts a turn or two. Remove all the bolts from the three sides of the pan, which will allow the pan to drop slightly to drain the fluid.

Loosen the remaining bolts an additional few turns to allow the other side of the pan to drop some more. When all the fluid has drained, remove the remaining bolts and the pan.

Carefully pour the remaining fluid out of the oil pan. Before disturbing the deposits in the pan, examine them for type and quantity. Metallic particles are shiny. Friction material deposits are very dark or black. A small amount of deposit is normal. Any larger deposits indicate that a transmission overhaul is required.

Fluid contaminated by engine coolant (water or antifreeze) appears milky and requires repair of the leak, reverse flushing of the cooler and lines, a complete draining of the transmission and torque converter, replacement of the filter, and filling with new fluid.

Step 3 External Checks and Adjustments

The number and types of possible external checks and adjustments vary slightly among the various makes and models of automatic transmissions. Those included here cover the majority of makes and models.

1. Manual shift linkage. It is essential to proper transmission operation that the selector lever position and the manual valve position in the transmission be synchronized. In other words, when the driver selects reverse, for example, it is essential that the manual valve in the valve body also be in the reverse position.

The position of the selector lever in each range is determined by gates in the shift mechanism. The position of the manual shift valve in each range is held in place by detents. Synchronization of selector lever position and manual valve position is determined by adjustment of the linkage between the two.

A typical linkage adjustment involves loosening or disconnecting the linkage at the transmission shift lever bracket. The selector lever is then placed in park against the gate, and the shift valve lever is placed in the park detent position. With these in position, the clamp on the sliding adjustment is tightened to maintain this relationship. On the threaded type of linkage adjustment, the connection is lengthened or shortened as required to make the connection.

A further check is required by selecting all gear positions and making sure that the engine will start in the park and neutral positions but not in any other positions. Some vehicles may require a neutral start switch adjustment to be made after the linkage adjustment has been corrected.

Incorrect manual shift linkage adjustment can cause delayed engagement, slippage, overheating, and shift problems. Manual shift linkage types include those using rods and those with a cable

2. Throttle linkage. There are two types of throttle linkage: the type that controls only the downshift valve (these transmissions are equipped with a vacuum modulator to control the throttle valve) and the type that controls both the downshift valve and the throttle valve (these transmissions do not have a vacuum modulator). Linkage design includes the rod and lever type and the cable type. Some of these linkage systems are not adjustable.

The importance of throttle valve linkage adjustment stems from the fact that throttle pressure rises proportionately with increased carburetor throttle

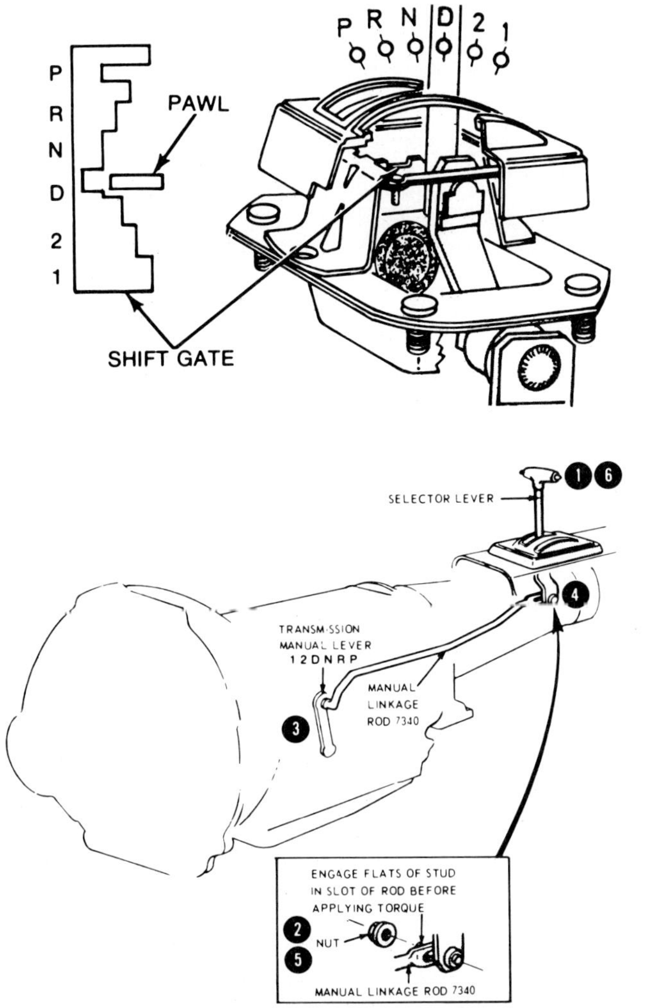

Figure 8-3. Manual linkage adjustment requires synchronization of shift lever, pawl, gate, and detent. (*Courtesy of Ford Motor Company*)

opening. If TV (throttle valve) linkage adjustment is incorrect, it can result in:

(a) Throttle pressure that is too high in relation to carburetor throttle opening. This results in harsh upshifts and delayed upshifts. Part-throttle (to detent) and WOT (through detent) downshifts will occur earlier or sooner than normal. Part-throttle downshifts can occur at a relatively small increase in throttle opening. This can be annoying in city traffic and on the highway.

(b) Throttle pressure that is too low in relation to carburetor throttle opening. This results in upshifts occurring too early and "mushy" upshifts and may result in slippage of bands and clutches. Greater than normal carburetor throttle opening will be required to effect forced downshifts.

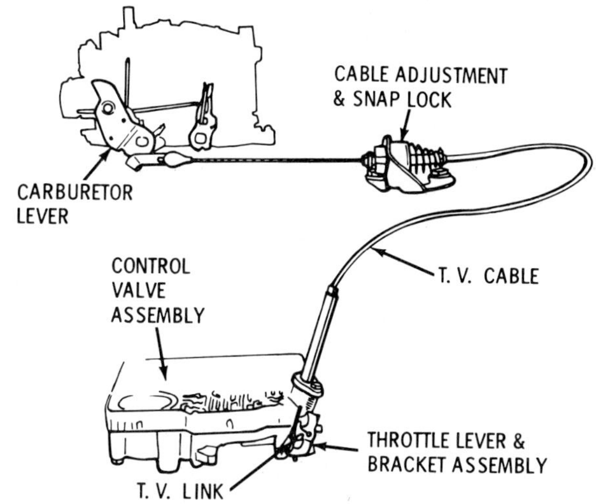

Figure 8-4. Throttle valve cable and linkage. (*Courtesy of General Motors Corporation*)

Procedures for adjusting throttle valve and downshift linkage vary considerably. Since transmission operating characteristics and transmission life are affected by TV and downshift linkage, it is important that procedures and specifications given in the appropriate service repair manual are followed for every unit being serviced. Differences in specifications may exist, for example, for the same model and year of transmission on a car equipped with a six-cylinder engine as compared to the same unit on a car

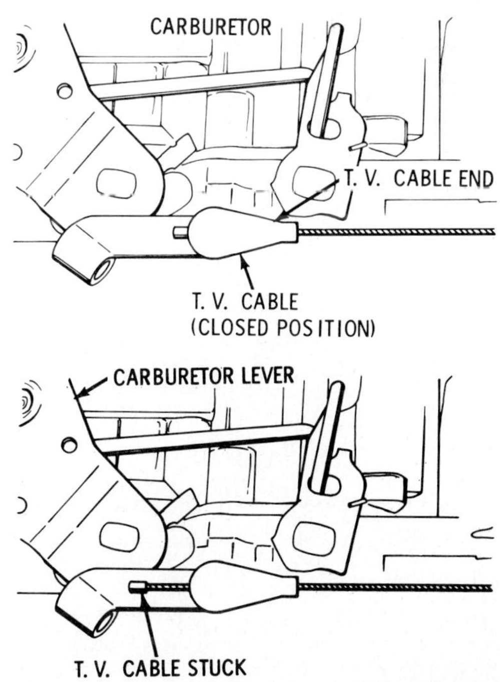

Figure 8-5. (*Courtesy of General Motors Corporation*)

ADJUSTING SELF-ADJUSTING TYPE T.V. CABLE GAS ENGINE MODEL ONLY

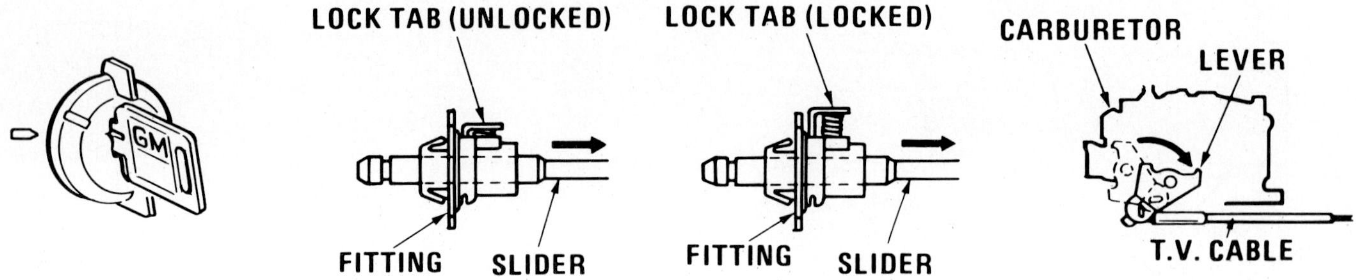

Stop engine.

Depress lock tab and hold. Move slider back through fitting in direction away from throttle body or pump lever until slider stops against fitting.

Release lock tab.

Open carburetor lever to "full throttle stop" position to automatically adjust T.V. cable. Release Carburetor lever.

Figure 8-6. (*Courtesy of General Motors Corporation*)

equipped with an eight-cylinder engine. Some examples of linkage adjustment are shown in the illustrations.

3. Vacuum modulator system. The vacuum modulator system has the same effect on the transmission hydraulic control system as the throttle valve linkage system. Engine intake manifold vacuum varies with engine speed and load and is related to carburetor throttle opening. Any leakage in the vacuum system will have the effect of reducing

ADJUSTING MANUAL TYPE T.V. CABLE GAS ENGINE MODEL ONLY

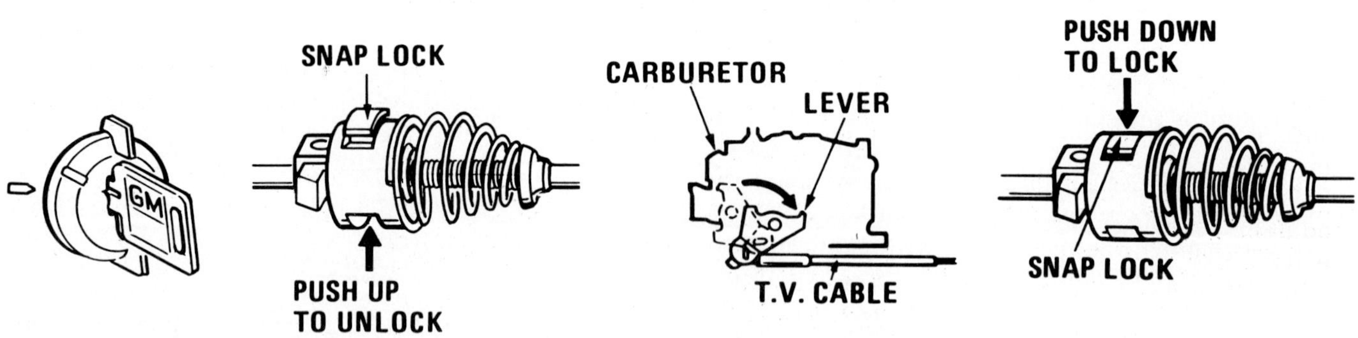

Stop engine.

Unlock T.V. cable "snap-lock" button.

Rotate carburetor lever by hand to wide open throttle and hold open.

Engage T.V. cable "snap-lock" button.

Figure 8-7. (*Courtesy of General Motors Corporation*)

ADJUSTING MANUAL TYPE T.V. CABLE DIESEL ENGINE MODEL ONLY

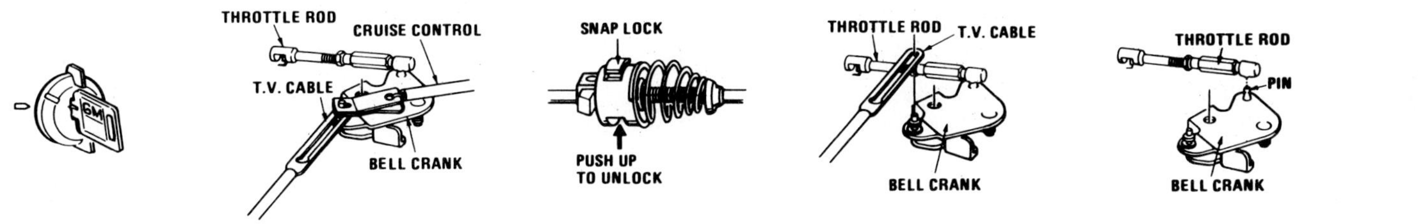

1. Stop engine.
2. Remove cruise control rod (if so equipped).
3. Unlock T.V. cable "snap-lock" button.
4. Remove T.V. cable from bell crank.
5. Remove throttle rod from bell crank.

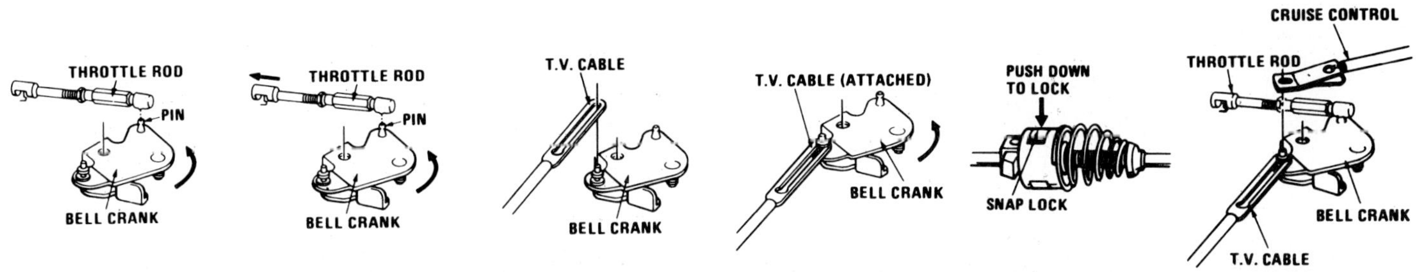

6. Rotate bell crank to full throttle stop and hold open.
7. Hold throttle rod & pump lever at full throttle stop; adjust throttle rod to meet bell crank pin at full throttle stop. (Do Not Connect.) Release bell crank.
8. Reconnect T.V. cable to bell crank.
9. Rotate bell crank to full throttle stop and hold open.
10. Engage T.V. cable "snap-lock" button. Release bell crank.
11. Connect throttle rod and cruise control.

NOTICE

If bell crank full throttle stop is not obtained when accelerator pedal is completely depressed, all full throttle adjustments must be made by completely depressing the accelerator pedal instead of rotating the bell crank by hand.

Figure 8–8. (*Courtesy of General Motors Corporation*)

vacuum. This results in higher than normal throttle (or modulator, GM) pressure with the same effect on transmision operation as described in the preceding item 2, throttle linkage, section a.

The general procedure for checking the vacuum modulator system requires checking the vacuum at the line removed from the modulator. If vacuum readings taken at this point are normal, there are no vacuum leaks in the line and the engine condition is assumed to be satisfactory. If any transmission fluid is noticeable when the vacuum line is disconnected at the modulator, the vacuum diaphragm in the modulator is leaking and the engine is consuming some transmission fluid through the vacuum line to the intake manifold. The vacuum modulator must be replaced. If the vacuum source, vacuum lines, and vacuum modulator are in good condition but shift characteristics indicated a vacuum modulator system problem, the vacuum modulator may have to be adjusted. This requires removing the modulator from the transmission. Special modulator wrenches are required to remove some screw-in–type units since there is very little room between the transmission case and the modulator body in some cases. The push-in–type modulator has a clamp and cap screw holding it in place. The cap screw and clamp must be removed and the modulator can then be pulled out. In some cases it may have to be pried out. Whenever a vacuum modulator is removed from a transmission, have a pan ready to catch any fluid that may drain from the transmission (about 1 quart or 1 liter).

The modulator should be removed carefully in

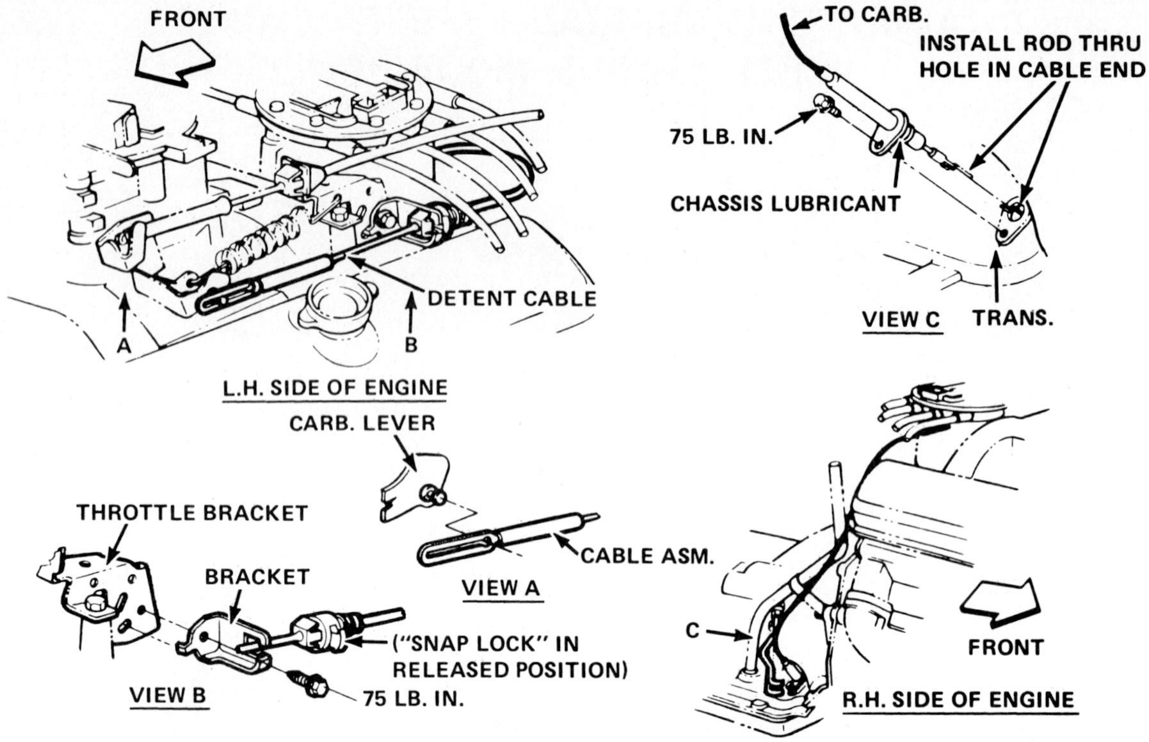

DETENT CONTROL CABLE ADJUSTMENT PROCEDURE

1. AFTER INSTALLATION TO TRANSMISSION, AND CLIP IF SPECIFIED, INSTALL DETENT CONTROL CABLE IN CONTROL CABLE BRACKET WITH "SNAP LOCK" DISENGAGED (DETENT CABLE SHOULD BE FREE TO SLIDE THROUGH "SNAP LOCK" TO ALLOW FOR ADJUSTMENT).
2. INSTALL DETENT CONTROL CABLE TO CARBURETOR LEVER.
3. OPEN CARBURETOR LEVER TO "FULL THROTTLE STOP" AND HOLD IN THIS POSITION.
4. CARBURETOR LEVER MUST BE TIGHT AGAINST FULL THROTTLE STOP, THEN PUSH "SNAP LOCK" UNTIL IT IS FLUSH WITH THE REST OF THE DETENT CABLE FITTING.
5. RELEASE CARBURETOR LEVER.

Figure 8-9. (*Courtesy of General Motors Corporation*)

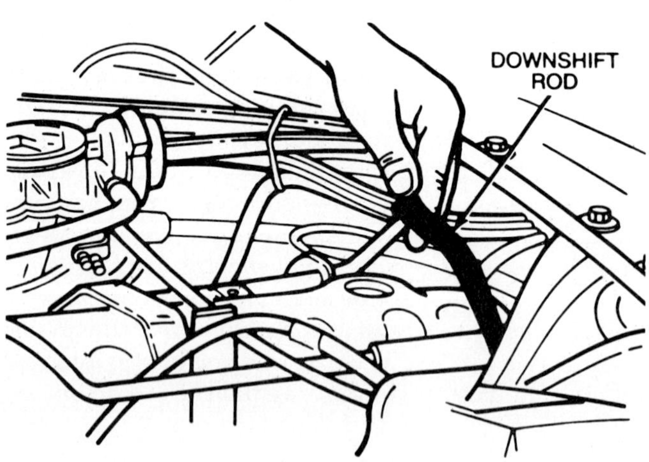

Figure 8-10. Checking downshift rod for binding. (*Courtesy of Ford Motor Company*)

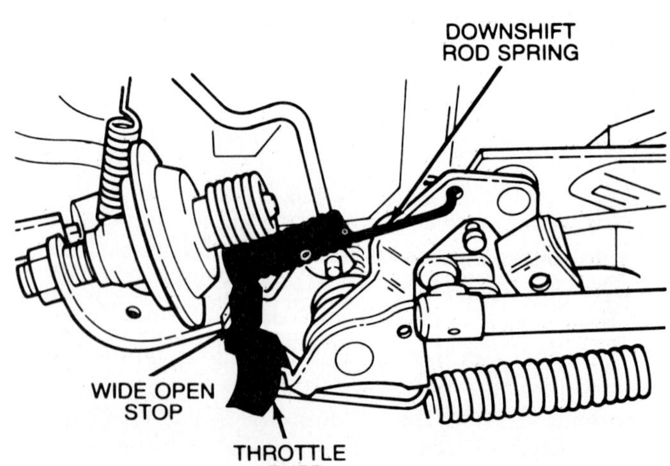

Figure 8-11. Wide-open throttle check. (*Courtesy of Ford Motor Company*)

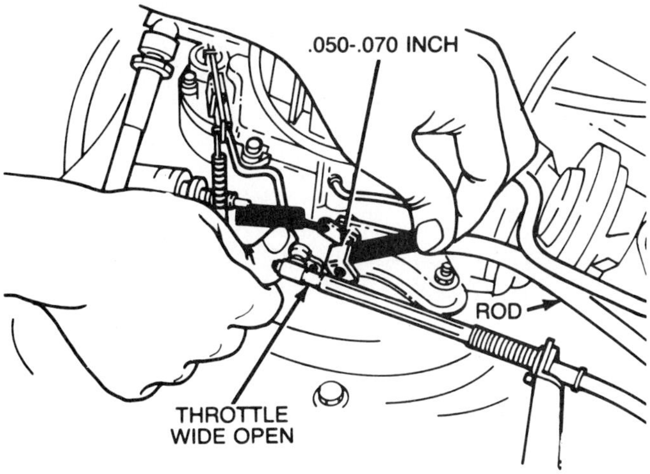

Figure 8-12. Downshift rod adjustment. (*Courtesy of Ford Motor Company*)

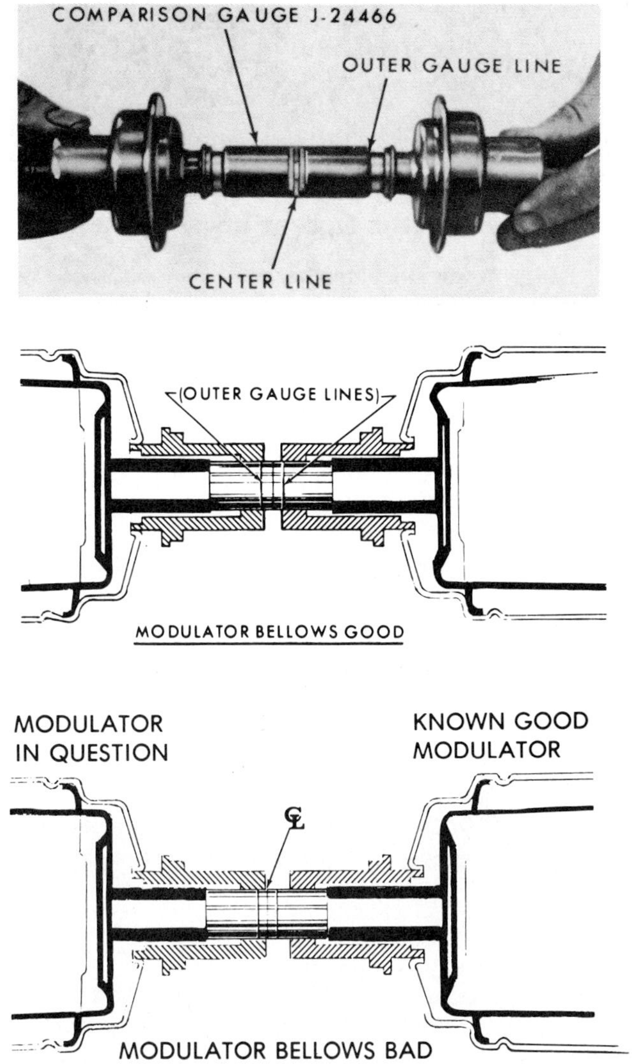

Figure 8-14. Checking vacuum modulator with special gauge. (*Courtesy of General Motors Corporation*)

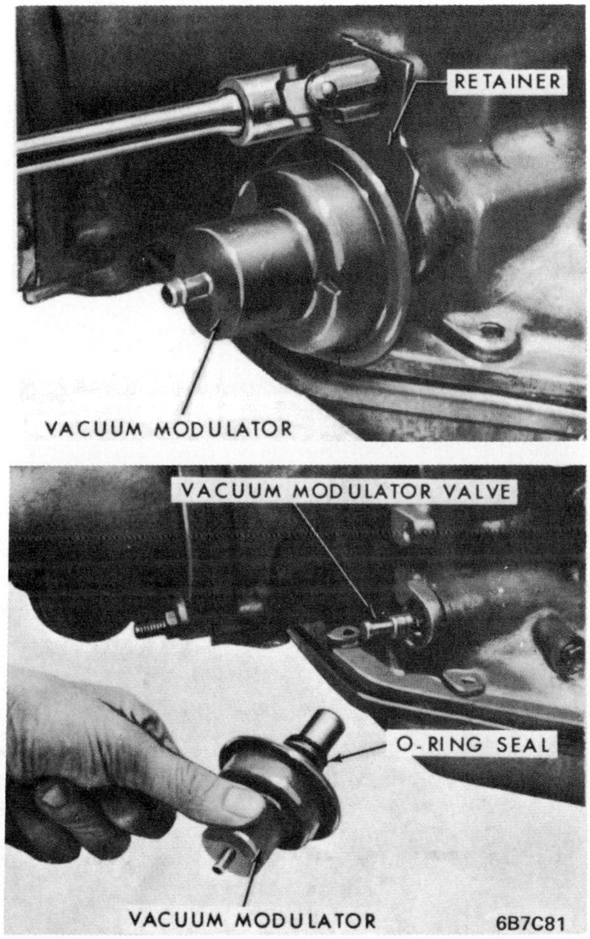

Figure 8-13. Removing vacuum modulator—push-in type. (*Courtesy of General Motors Corporation*)

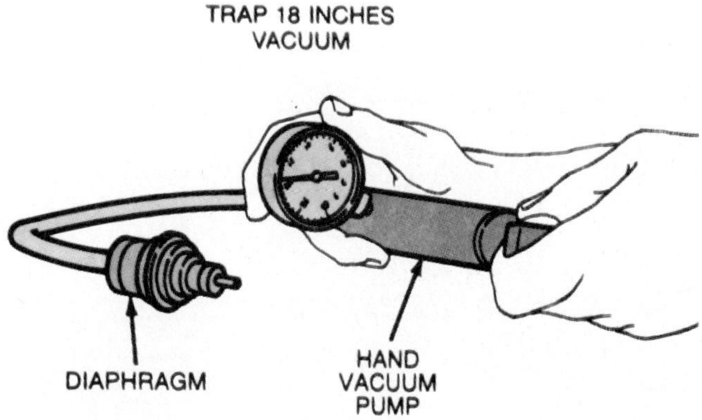

Figure 8-15. Diaphragm leak test. (*Courtesy of Ford Motor Company*)

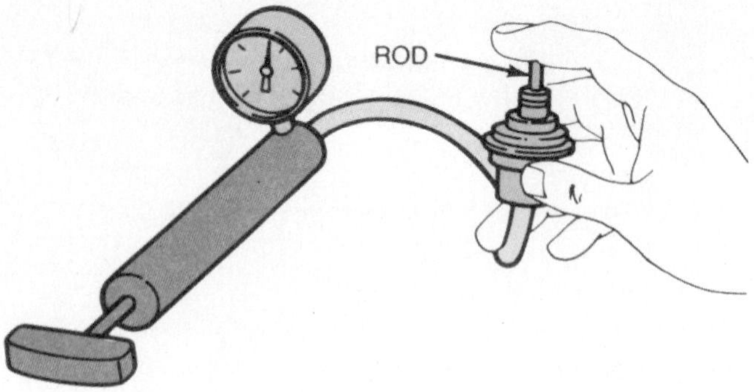

Figure 8-16. Checking diaphragm rod for binding. (*Courtesy of Ford Motor Company*)

order not to lose the actuating pin, which may drop during modulator removal. After removal, the modulator can be bench tested with the use of a vacuum pump and gauge and with the use of special gauge pins in some cases. If adjustment is required, this should only be done as specified in the appropriate service manual.

Some examples of testing and adjusting procedures for vacuum modulators are shown in the accompanying illustrations. A new O-ring seal should be used.

4. EGR (exhaust gas recirculation) system. Since EGR system operation affects engine intake manifold vacuum and manifold vacuum affects transmission operation, this system should be checked. If not operating properly, the necessary corrections should be made.

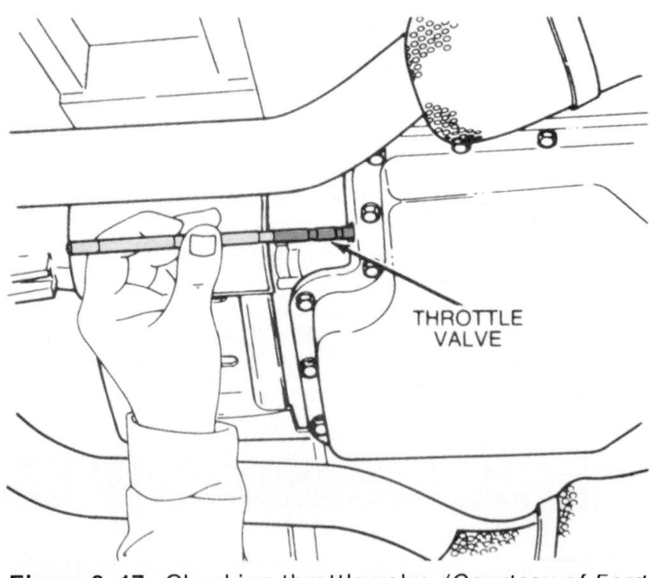

Figure 8-17. Checking throttle valve. (*Courtesy of Ford Motor Company*)

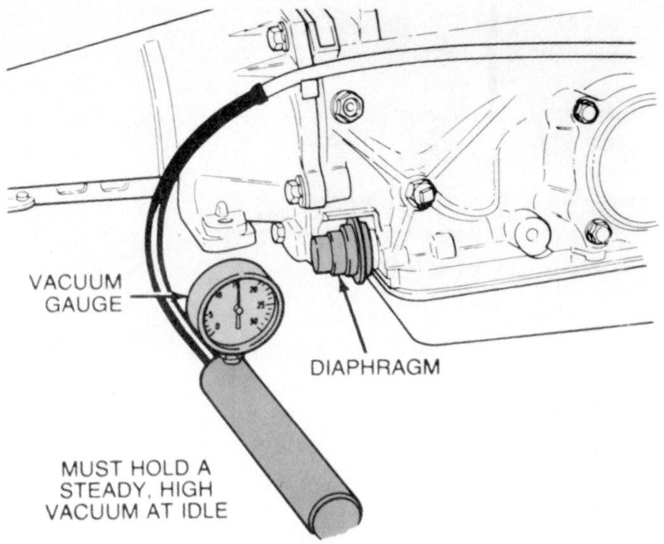

Figure 8-18. Manifold vacuum supply test. (*Courtesy of Ford Motor Company*)

A quick check can be made as follows:

- Check the engine at operating temperature and at idle speed.
- Locate and observe movement of the EGR valve stem as engine speed is increased to at least 1,700 rpm.

If no movement is evident, check the EGR vacuum source. If OK, the EGR valve is stuck and requires service as specified in the service manual.

5. Engine idle speed. The engine fast-idle and hot-idle speed should be adjusted to specifications. Too high an engine idle speed results in harsh engagement of clutches in both forward and reverse. This is perceived by the customer as a clunk during engagement. A harsh closed throttle (coasting) downshift may also be caused by too high an engine idle speed.

Use a tachometer and adjust both the fast idle and hot idle to the manufacturer's specifications. These are usually stated on the underhood emission sticker.

6. WOT check. With the engine off, check to make sure that pressing the accelerator pedal to the floor does in fact position the carburetor throttle valves to the wide-open position. If not, adjust the linkage to achieve this condition. Through detent (WOT), downshift operation is affected by incorrect adjustment.

Step 4 Stall Testing

1. *Caution:* Perform a stall test only on vehicles when the manufacturer recommends a stall test.

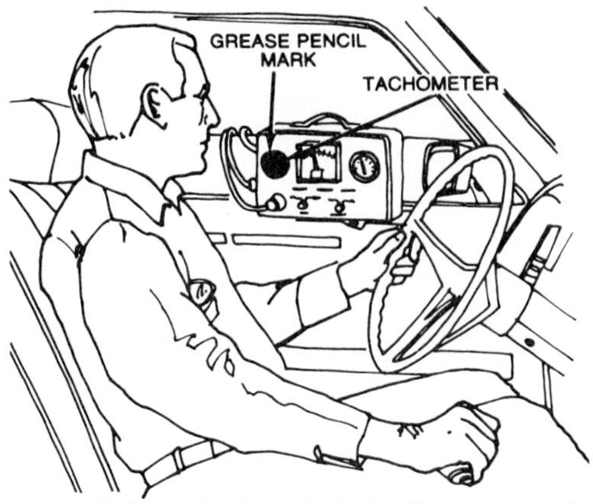

RANGE	SPECIFIED RPM	HIGH, LOW OR OK
D		
2		
1		
R		

Figure 8-19. Stall testing transmission. (*Courtesy of Ford Motor Company*)

Follow all safety precautions required during a stall test.

Depending on the make and model of the transmission being tested, the following tests may be made during a stall test:

D position—forward clutch, one-way clutch in transmission, stator one-way clutch, engine power output

2 position—forward clutch, one-way clutch in transmission, stator one-way clutch, intermediate band and servo (some models only), engine power output

1 position—forward clutch, one-way clutch in converter, engine power output

R position—reverse and high clutch, low and reverse band and servo or low and reverse holding clutch, stator one-way clutch

2. Procedure and safety precautions. Since a stall test is performed at wide-open throttle, the following procedure and safety precautions should be rigidly observed during a stall test. Perform a stall test only if recommended by the vehicle manufacturer.

(a) Connect the tachometer to the engine. You must know your power flow for this test to be meaningful. This will tell you which clutch, band, or servo is being tested in each case.

(b) The engine cooling system should be in good condition and engine and transmission fluid levels correct.

(c) The engine and transmission should be at normal operating temperature. This requires approximately 15 minutes of operation.

(d) Block vehicle wheels for safety front and back.

(e) Parking brake and service brakes must be fully applied during the stall test.

SELECTOR POSITIONS	STALL SPEED(S) HIGH (SLIP)	STALL SPEEDS LOW
D only	Low (Planetary) One-Way Clutch	1. Does engine misfire or bog down under load? Check Engine for Tune-Up. If OK . . . 2. Remove torque converter and bench test for reactor one-way clutch slip.
D, 2 and 1	Forward clutch	
All Driving Ranges	Perform Control Pressure Test	
R Only	Reverse and High Clutch or Low-Reverse Band or Servo	

Figure 8-20. Stall test results diagnosis. (*Courtesy of Ford Motor Company*)

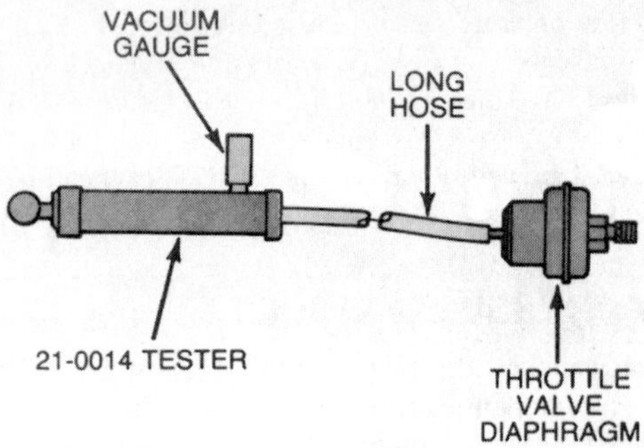

Figure 8-21. Vacuum supply for governor test. *(Courtesy of Ford Motor Company)*

(f) No one should be in front of the vehicle or behind the vehicle during stall tests.

(g) Never hold the throttle at the wide-open position for more than 5 seconds at a time.

(h) After each 5-second test, place the selector lever in neutral and run the engine at 1,000 to 1,500 rpm for at least 15 seconds to cool the converter before making the next test.

(i) If specified engine speed is exceeded during any specific test, release the accelerator pedal immediately since clutch or band slippage is indicated. Continued testing in this position would result in further transmission damage.

(j) Place the selector lever in each specified position in turn and make the test. Record engine rpm in each case.

3. Stall test results. All stall test results should be compared with those given in the appropriate service

Figure 8-22. Raising vehicle for governor test.

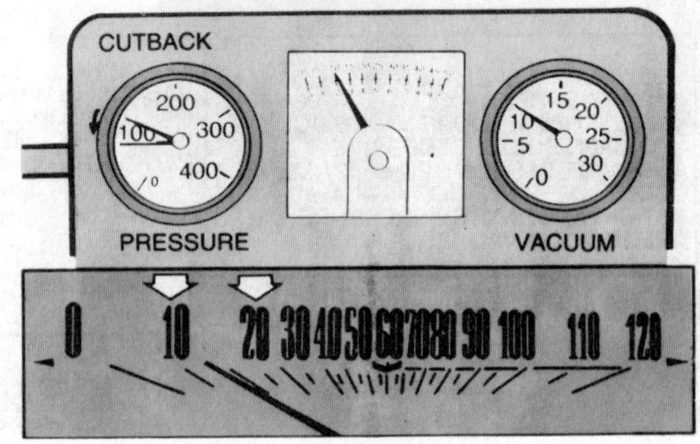

Figure 8-23. Governor light load test indicated by line pressure cutback at specified vacuum and road speed. *(Courtesy of Ford Motor Company)*

repair manual. Some variation in readings should be allowed for possible reduced engine power output and for higher altitudes.

- Stall speed above normal = clutch or band slippage.
- Stall speed slightly below normal = engine power output below normal; engine requires tuneup or mechanical repair.
- Stall speed well below (about 2/3) normal = stator one-way clutch in converter not holding. Fluid leaving the turbine is acting against impeller and engine rotation, which reduces engine speed. Poor low-speed acceleration is evident since the converter is acting as a simple fluid coupling.

Step 5 Road Testing

A road test is performed to obtain the following information about transmission operation: *what is normal* about the transmission and *what is abnormal* about transmission operation.

Caution: Transmission fluid level, engine oil level, and coolant level should be corrected before road testing. The engine and transmission must be at operating temperature for a road test to be valid.

Caution: Traffic conditions and consideration for others on the road are prime safety considerations during road testing procedures. Unexpected maneuvers can cause accidents. Select a section of road where speed limits and traffic conditions will allow safe road testing.

Look for the following points during a road test:

1. Proper engagement as the selector lever is

RANGE	CHECK FOR	CONDITION (OK OR NOT OK)
1	Engagement	
	Should be **no** 1-2 upshift	
	Engine braking in low gear	
	Shifts 3-2 and then 2-1 coming out of D at cruise	
	No slipping	
2	Engagement/Starts in second gear	
	Should be **no** 2-3 upshift	
	Shifts 3-2 coming out of D at cruise	
	No slipping	
D	Engagement/Starts in first gear	
	Upshifts not mushy or harsh	
	Upshifts and downshifts at specified speeds	
	• Minimum Throttle 1-2	
	• Minimum Throttle 2-3	
	• Minimum Throttle 3-2	
	• Minimum Throttle 2-1	
	• To-detent (heavy throttle) 1-2	
	• To-detent (heavy throttle) 2-3	
	• To-detent (heavy throttle) 3-2	
	• Through-detent (W.O.T.) 1-2	
	• Through-detent (W.O.T.) 2-3	
	• Through-detent (W.O.T.) 3-2	
	• Through-detent (W.O.T.) 3-1 or 2-1	
R	Engagement	
	Back up without slip	

Figure 8–24. Road test check chart. (*Courtesy of Ford Motor Company*)

moved to each gear position including park. Things to look for are delayed or harsh engagement.

2. Proper transmission operation in all forward ranges. This includes:

(a) The 1–2, 2–3, 3–4 upshifts and converter lockup during light throttle operation. Shifts should be smooth, yet noticeable, with no engine flare-up or run-up during shifts. Upshifts should occur at proper speeds.

(b) The same shifts during medium and heavy throttle operation. Shifts should be firmer and have a more positive feel with no engine flare-up or run-up during upshifts. Compare to specifications.

3. To detent (part-throttle or torque demand) downshift operation. Downshifts and downshift pattern (3–2, 3–1, etc.) should correspond to those given for the particular vehicle make, model, engine, and differential ratio combination. Tire size will also have a bearing on shift speed patterns.

4. Through detent (WOT downshift or kickdown) operation. Compare to specifications in item 3.

5. Coasting downshift (closed-throttle downshift) operation. Brakes can be used to retard vehicle for this check. Note downshift speed and quality and compare with specifications.

6. Noisy operation. Note any abnormal noise (this may require a window to be open) and note the gear position in which noise occurs. If disassembly is required later, this will help pinpoint the problem if you know the power flow patterns.

7. Manually selected downshifts. While driving in third or fourth gear at approximately 35 mph (56 km/h), manual selection of second- or first-gear position can help determine intermediate band (or clutch) operation, both apply and release.

Some typical conclusions that can be drawn from abnormal transmission operation during a road test include the following:

- Forward clutch slippage, causing delayed engagement in all forward gears, poor acceleration.

- Direct clutch (reverse and high clutch) slippage, causing delayed engagement in reverse, delayed 2–3 upshift, engine speedup (neutral feel) during 2–3 upshift, and poor acceleration in third or direct.
- Intermediate band or clutch slippage, causing delayed 1–2 upshift, engine speedup (neutral feel) during 1–2 upshift, and poor acceleration in second gear.
- Leaking vacuum modulator or vacuum supply line, causing delayed or no upshift.
- Governor sticking, possibly causing delayed or no upshift or early upshift regardless of throttle opening.
- Worn hydraulic pump, causing delayed engagement and slippage in all gear positions if severe wear present.
- Stator one-way clutch not holding, causing very poor low-speed acceleration.

For more detailed and more specific problems, refer to the diagnostic charts and procedures included in this chapter for some of the more common automatic transmissions.

Step 6 Hydraulic Pressure Testing

Although road testing provides the means to identify which area in the transmission is causing a problem, hydraulic pressure testing may be required to verify the diagnosis and isolate the problem further. If road testing yields conclusive results, pressure testing is not required.

The number of hydraulic circuits that can be tested varies between different makes and models of transmissions. Generally, all transmissions can be tested for proper line (or control) pressures, including proper line pressure increase and decrease in the various modes of transmission operation. From this it can easily be seen that pressure testing may be required to solve a shift timing or shift quality problem since line pressure changes affect both.

Pressure testing will indicate whether the following systems are operating properly:

- The pressure supply system—pump and pressure regulator valve

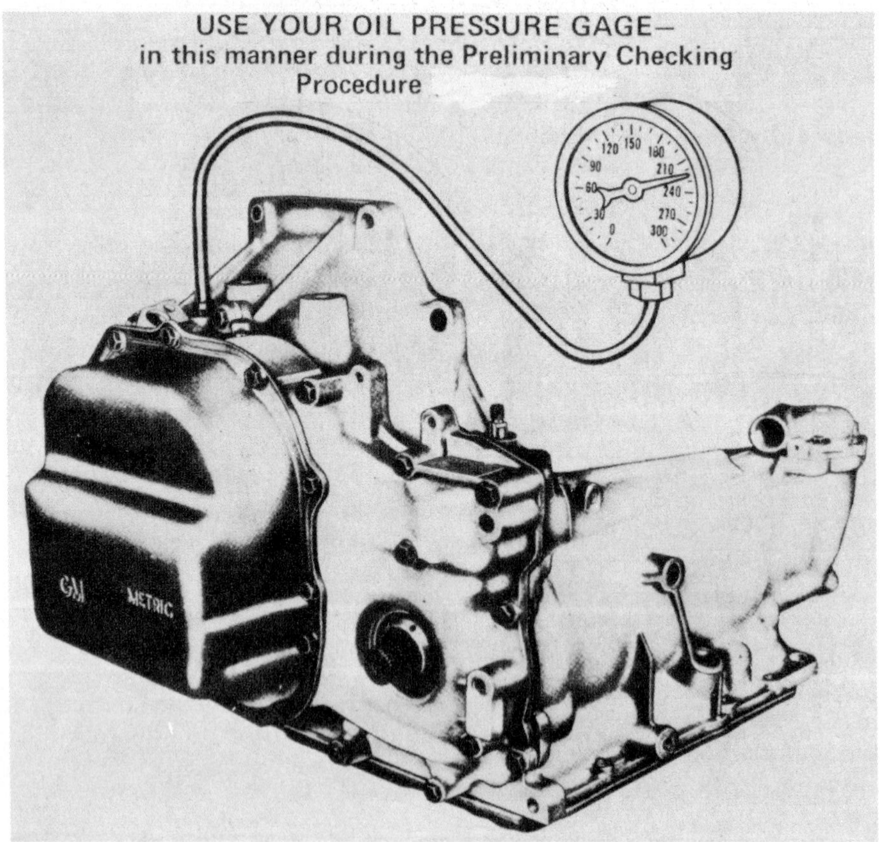

Figure 8-25. GM-125C pressure test hookup. (*Courtesy of General Motors Corporation*)

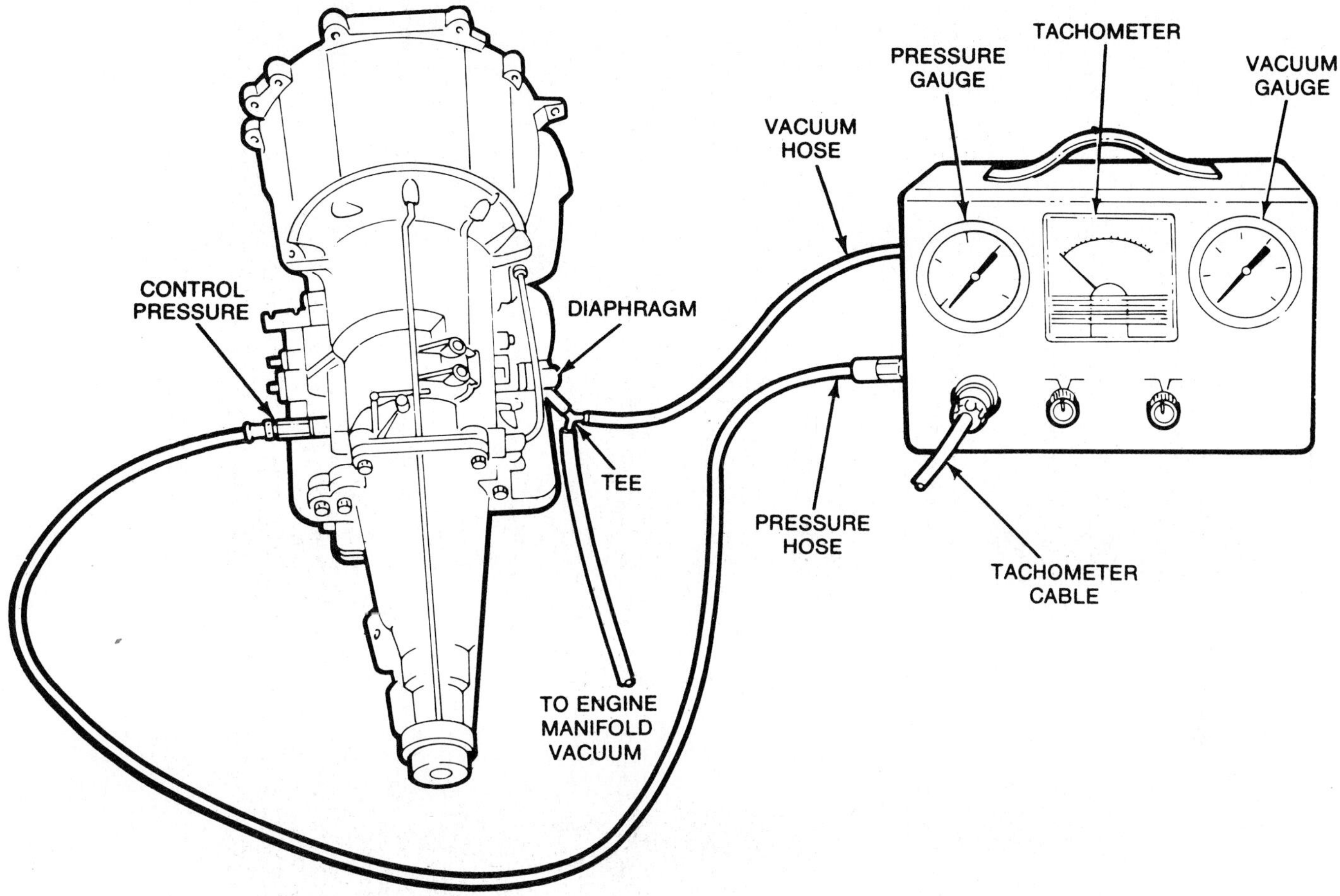

Figure 8-26. Pressure test connections for Ford C3 transmission (*Courtesy of Ford Motor Company*)

- The throttle pressure system—throttle valve and vacuum modulator system (if so equipped)
- The governor pressure system—the governor valve, weights, and springs

The following equipment is required to perform a hydraulic pressure test:

- A tachometer (capable of being read from the driver's position).
- One or two hydraulic pressure gauges (depending on the transmission being tested) with fittings and hoses long enough to hang gauges inside the vehicle for the driver to observe. Gauges should be capable of at least 400 psi (2,800 kPa).
- A vacuum gauge (for those transmissions equipped with a vacuum modulator) with a long hose capable of reaching the driver's position for easy reading.
- A vacuum supply pump, possibly needed to provide adequate vacuum in some cases to verify accurately and distinguish between transmission-caused problems and engine-caused transmission problems.

Many automatic transmission testers provide all the required equipment in one portable unit, which makes it easier to use than individual gauges.

Some transmissions have more than one plugged pressure tap. The specific problem encountered in the transmission and the step-by-step diagnostic procedure in the manual will decide by process of elimination how many pressure taps must be used. The tap plug is removed and the pressure gauge fitting screwed into the threaded opening. Make sure that you have the correct fitting with the correct size and type of thread. The fitting should screw easily into the hole by hand at least several turns; then tighten it with a wrench.

The vacuum gauge is connected to the vacuum

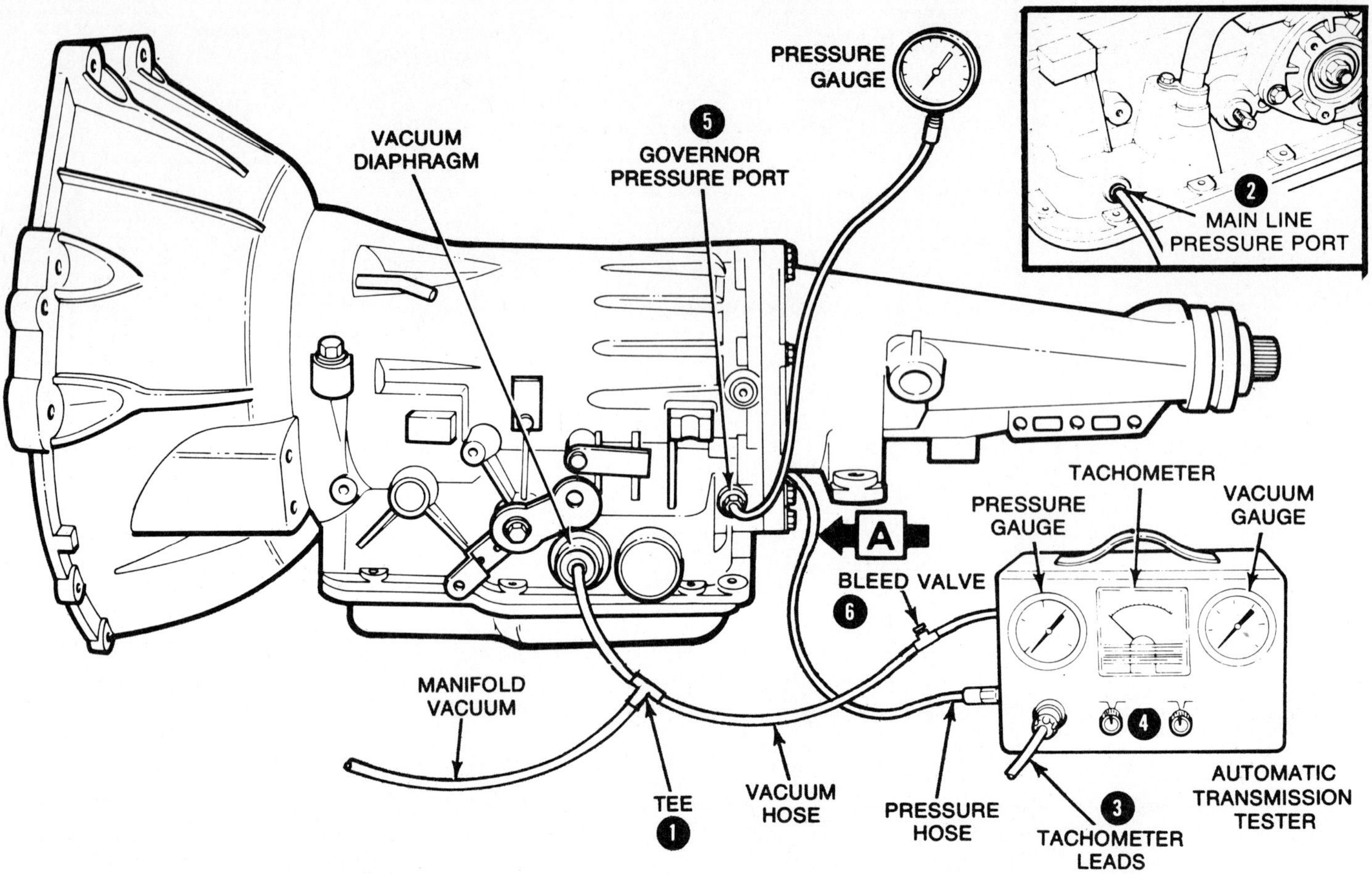

Figure 8-27. Pressure test connections for Ford Jatco transmission. (*Courtesy of Ford Motor Company*)

line at the modulator with a T-fitting. The vacuum gauge is necessary to determine whether specified hydraulic pressures are achieved at specified vacuum values.

All hydraulic pressure test results, both vacuum and hydraulic pressures, should be recorded for tests in order to be valid and for accurate comparison and analysis with the manufacturer's specifications. Use the appropriate chart from the service manual for comparison and analysis.

PART 8-4 DIAGNOSTIC CHARTS AND PROCEDURES

The diagnostic charts and procedures in this section provide more specific information and detail on some of the more popular automatic transmissions. Included are the General Motors 125C transaxle, 200, 200–4R, 250C, 325, 325–4L, and 350 automatic transmissions; the Ford ATX transaxle, C3, C4, C5, and AOD automatic transmissions; the Chrysler A404, A413, and A470 transaxles and the A904 and A727 automatic transmissions, and the American Motors 904 and 998 automatic transmissions.

Note: Although the charts and procedures included in this section refer to specific makes and models of transmissions, they should only be used in conjunction with the appropriate manufacturer's service manuals and any service bulletins that may have been issued by the manufacturer that apply to a specific transmission.

GM 125C Transaxle Diagnosis Procedure (Courtesy of General Motors Corporation)

Before diagnosis of any transmission complaint is attempted, there must be an understanding of the oil checking procedure and what appearance the oil should have. Many times a transmission malfunction can be traced to low oil level or improper reading of dipstick. Due to the transmission fluid now being used, it may appear to be darker and have a stronger odor. This is normal and not a positive sign of required maintenance or transmission failure.

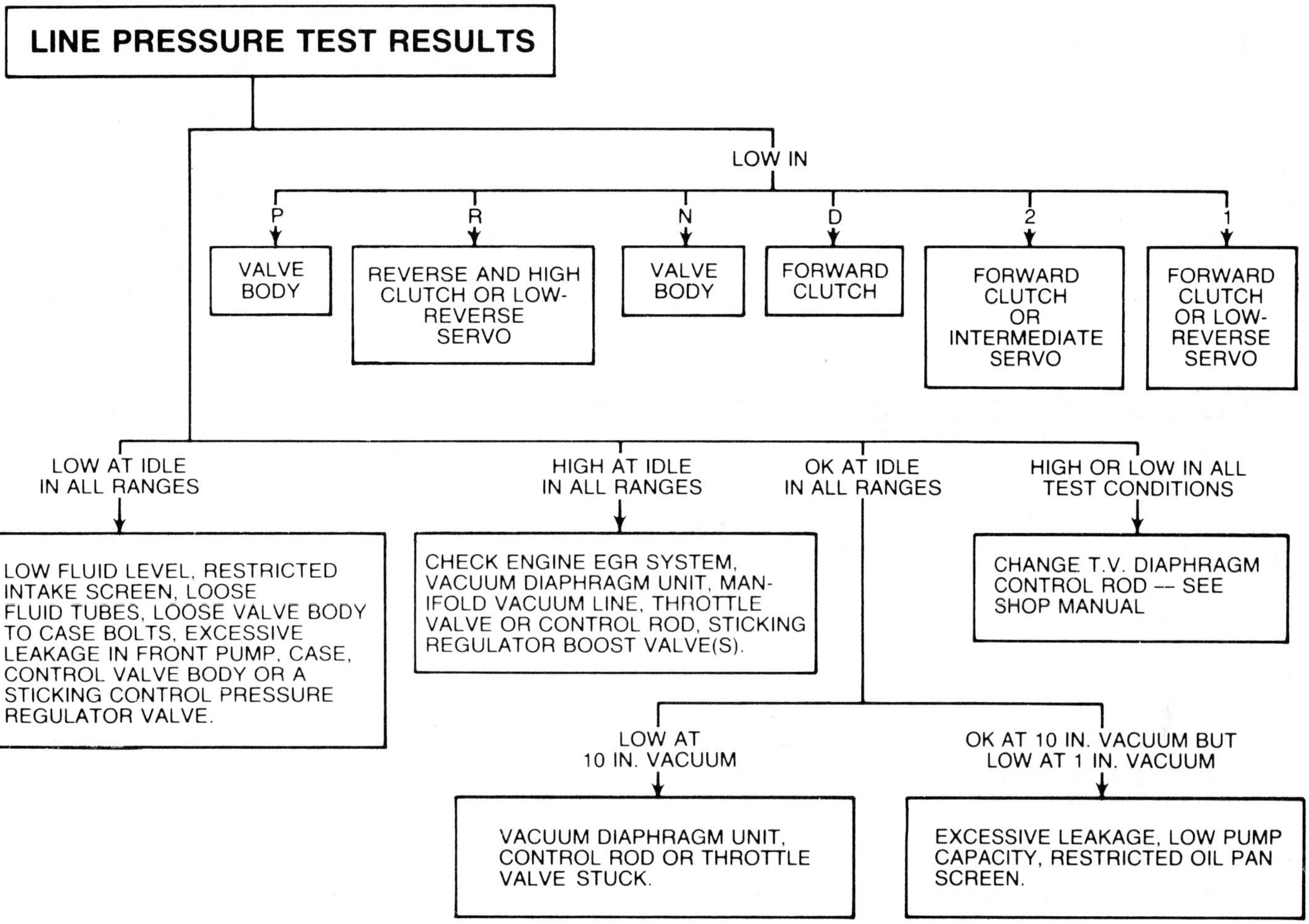

Figure 8-28. Interpretation of line pressure test results. (*Courtesy of Ford Motor Company*)

Also, when the dipstick is removed, it should be noted whether the oil is devoid of air bubbles. Oil with air bubbles gives an indication of an air leak in the suction lines, which can cause erratic operation and slippage. Water in the oil imparts a milky, pink cast to the oil and can cause spewing.

Preliminary Check Procedure

1. Check transmission oil level.
2. Check and adjust TV cable.
3. Check outside manual linkage and correct.
4. Check engine tune.
5. Install oil pressure gauge.
6. Connect tachometer to engine.

Check oil pressures in the following manner.

Minimum TV Line Pressure Check. Set the TV cable to specification; with the brakes applied, take the line pressure readings in the ranges and at the engine rpm's indicated in the following chart.

Full TV Line Pressure Check. Full TV line pressure readings are obtained by tying or holding the TV cable to the full extent of its travel; with the brakes applied, take the line pressure readings in the ranges and at the engine rpm's indicated in the following chart.

Line pressure is basically controlled by pump output and the pressure regulator valve. In addition, line pressure is boosted in reverse, intermediate, and low by the reverse boost valve.

Also, in the neutral, drive, and reverse positions of the selector lever, the line pressure should increase with throttle opening because of the TV system. The TV system is controlled by the TV cable, the throttle lever and bracket assembly, and the TV link, as well as the control valve pump assembly.

Note: Total running time for this combination should not exceed two minutes.

Caution: Brakes must be applied at all times.

Model	Range	Normal Oil Pressure at Minimum TV		Normal Oil Pressure at Full TV	
		kPa	psi	kPa	psi
PO CJ, HC, CI, HY	Park at 1,000 rpm	410-550 345-485	60-80 50-70	No TV pressure in park. Line pressure is the same as park at minimum TV.	
PO CJ, HC, CI, HY	Reverse at 2,000 rpm	725-860 655-795	105-125 95-115	1450-1585 1380-1515	210-230 200-220
PO CJ, HC, CI, HY	Neutral at 1,000 rpm	410-550 345-485	60-80 50-70	825-965 760-895	120-140 110-130
PO CJ, HC, CI, HY	Drive at 1,000 rpm	Same as neutral.		Same as neutral.	
PO CJ, HC, CI, HY	Intermediate at 1,000 rpm	720-855 650-790	103-123 93-113	720-855 650-790	103-123 93-113
PO CJ, HC, CI, HY	Low at 1,000 rpm	Same as intermediate.		No TV pressure in low range. Line pressure is the same as intermediate at minimum TV.	

Checking Transaxle Oil Level

Approximately 3.8 liters (4 quarts) of fluid are required to refill the transaxle after the pan has been drained. The fluid capacity of the 125C transaxle and converter assembly is approximately 8.5 liters (9 quarts), but the correct level is determined by the mark on the dipstick rather than by amount added. Use only Dexron® II automatic transmission fluid or its equivalent.

If the oil changes to a darker color at an early stage from the usual red color or to a strong odor, this is normal even though it is usually associated with overheated transmission fluid. It is not a positive sign of required maintenance or transaxle failure.

Due to the shape of the filler tube, oil level readings may be misleading. Look carefully for a full oil ring on both sides of dipstick. Recheck if any question of oil level occurs.

Check the oil level as follows:

1. Place the transaxle in park and leave it in park. Apply the parking brake and block the wheels; start the engine and let it idle (carburetor off fast idle step). *Do not race the engine. Move the lever through the ranges.*

2. The vehicle should be on a level surface.

3. Check the oil level on the dipstick.

The oil level at room temperature of approximately 21°C (70°F) should be 13mm (1/2 inch) above the full mark or between the two dimples on the dipstick, Figure 8–1. The oil level should be between the add and full marks at normal operating temperature, 200°F (90°C). This temperature is obtained after at least 15 miles (24 km) of expressway driving or equivalent city driving. The transaxle oil is stored in the valve body cover. Oil level will decrease on the dipstick when warm. Also, at normal operating temperature, the oil will heat the gauge end of the dipstick to a degree where the average person cannot grasp it firmly with a bare hand without discomfort.

Maintain the oil level between the add and full marks at normal operating temperature.

Do not overfill the transaxle, as this may cause foaming and loss of oil through the vent.

4. If oil is low, check for possible causes.

Drain Intervals

The transaxle operating temperature resulting from the type of driving conditions under which the car is used is the main consideration in establishing the proper frequency of transaxle fluid changes.

Change the transaxle fluid and replace the strainer every 15,000 miles if the car is usually driven under one or more of the following severe transaxle service conditions:

1. In heavy city traffic

2. Where the outside temperature regularly reaches 90°F (32°C)

3. In very hilly or mountainous areas

4. Frequent trailer pulling

5. Commercial use, such as taxi, police car, or delivery service

If you do not use your car under any of these conditions, change the fluid and replace the strainer every 100,000 miles.

Note: Do not overfill. It takes only one pint to raise level from add to full with a hot transaxle. Overfilling can cause damage to the transaxle.

Fluid Capacities

Pan removal—3.8 liters (4 quarts)
Overhaul—5.7 liters (6 quarts)

Fluid and Filter Change

1. Raise the car.

2. With the drain pan placed under the transaxle oil pan, remove the oil pan attaching bolts from the front and side of the pan.

3. Loosen the rear pan attaching bolts approximately four turns.

4. Carefully pry the transaxle oil pan loose with a screwdriver, allowing fluid to drain.

5. Remove remaining bolts and remove the oil pan and gasket. Discard the gasket.

6. Drain fluid from the oil pan. Clean the pan with solvent and dry thoroughly with clean compressed air.

7. Remove the strainer and O-ring seal.

8. Install the strainer and new O-ring seal, locating the strainer against the dipstick stop.

9. Install a new gasket on the oil pan and install the oil pan. Tighten the attaching bolt to 16 N.m (12 foot-pounds).

10. Lower the car; add approximately 3.8 liters (4 quarts) of Dexron® II automatic transmission fluid or its equivalent through the filler tube.

11. With the selector lever in the park position, apply the parking brake and block the wheels, start the engine, and let it idle (carburetor off fast idle step). *Do not race the engine. Move the lever through the ranges.*

12. With the selector lever in the park range, check for proper fluid level.

Service Procedures

Engine Coolant in Transmission

If the transmission oil cooler, located in the radiator assembly, has developed a leak allowing engine coolant to enter the transmission, use the following procedure.

1. Remove the transmission from the car.

2. Disassemble the transmission and replace all rubber-type seals. (The coolant will attack the seal material, causing leakage.)

3. Replace the composition-faced clutch plate assemblies. (The facing material may become separated from the steel center portion.)

4. Replace the nylon washers, speedometer gears, and governor gear. (The nylon can swell and become damaged.)

5. Flush the converter.

6. Thoroughly clean and rebuild the transmission, using new gaskets and oil filter.

7. Flush the cooler lines after the transmission cooler has been properly repaired or replaced.

Case Porosity Repair

External leaks caused by case porosity have successfully been repaired with the transmission in the car by using the following recommended procedures.

1. Road test and bring the transmission to operating temperature, approximately 93°C (200°F).

2. Raise the car on a hoist or jack stand, with the engine running, and locate the source of oil leak. Check for leakage in all operating positions. A mirror may be helpful in finding leaks.

3. Shut the engine off, thoroughly clean the area to be repaired with a cleaning solvent, and air dry.

4. Using instructions of the manufacturer, mix a sufficient amount of epoxy, No. 1360016 or equivalent, to make the repair. Observe the cautions of the manufacturer in handling.

5. While the transmission case is still *hot,* apply the epoxy to the area to be repaired. A clean, dry soldering acid brush can be used to clean the area and also to apply the epoxy cement. Make certain that the area to be repaired is fully covered.

6. Allow the cement to cure for three hours before starting engine.

7. Road test and check for leaks.

Oil Cooler Pipes

If replacement of transmission steel tubing cooler lines is required, use only double-wrapped and brazed steel tubing meeting GM specification 123M or the equivalent. Under no condition use copper or aluminum tubing to replace steel tubing. Those materials do not have satisfactory fatigue durability to withstand normal car vibrations.

Steel tubing should be flared using the double-flare method.

Cooler Line Flushing

In a major transmission failure, where particles of metal have been carried with the oil throughout the units of the transmission, it will be necessary to flush out the oil cooler and connecting lines. To flush the oil cooler and lines, use the following procedure.

1. Disconnect both cooler lines from the transmission.

2. Place a hose over the end of the cooler inlet line (from the bottom of the cooler) and insert the hose into an empty container.

3. Flush clean oleum solvent or the equivalent through the return line (from the top of the cooler), using an oil suction gun until clean solvent comes out of the hose. This will back flush the cooler.

4. Remove the hose from the inlet cooler line and place it on the return line.

5. Flush clean oleum solvent or the equivalent through the inlet line until clean solvent comes out the return line. Remove remaining solvent from the cooler with compressed air applied to the return line and flush with transmission fluid.

6. Reconnect oil cooler lines and torque nuts to 17 N.m (12 foot-pounds).

Checking Transmission Mount

Raise the car. Push up and pull down on the transmission tailshaft while observing the transmission mount. If rubber separates from the metal plate of the mount or if the tailshaft moves up but not down (mount bottomed out), replace the mount. If there is relative movement between a metal plate of the mount and its attaching point, tighten screws or nuts attaching the mount to the transmission or crossmember.

Adjustments

Manual Linkage

The transaxle manual linkage must be adjusted so that the indicator quadrant and stops correspond with the transaxle detents. If the linkage is not adjusted properly, an internal leak could occur, which could cause a clutch or band to slip.

Caution: If a manual linkage adjustment is made with the selector lever in the park position, the parking pawl should freely engage the reaction internal gear to prevent the car from rolling. Transmission, vehicle, or personal injury may occur if not properly adjusted.

Checking TV Cable in Vehicle

To check the TV cable for freeness, pull out on the upper end of the cable. The cable should travel a short distance with light spring resistance. This light resistance is due to the small coiled return spring on the TV lever and bracket that returns the lever to the zero TV or closed-throttle position. Pulling the cable farther out moves the lever to contact the TV plunger, compressing the TV spring, which has more resistance. By releasing the upper end of the TV cable, it should return to the zero TV position. This checks the cable in the housing, the TV lever and bracket, and the TV plunger in its bushing for freeness.

Checking TV Cable Adjustment

1. Install the line pressure gauge.

2. Check line pressure with the selector lever in park and the engine at 1,000 rpm.

3. Check line pressure with the selector lever in neutral at 1,000 rpm. Pressure should be the same, to no more than 34 kPa (10 psi) higher than in park.

4. Increase engine rpm to 1,400; line pressure should increase.

General Service Precautions

1. When servicing the transaxle, it is recommended that upon disassembly of a unit, all parts be cleaned and inspected.

2. The unit should be reassembled before disassembly of other units to avoid confusion and interchanging of parts.

3. Before disassembly of the unit, thoroughly clean the exterior.

4. Disassembly and reassembly of the unit and the subassemblies must be made on a clean workbench. As in repairing any hydraulically operated unit, cleanliness is of the utmost importance; therefore, the bench tools and parts must be kept clean at all times.

5. Before installing cap screws into aluminum parts, *always dip screws into transmission oil* to prevent cap screws from galling the aluminum threads and also to prevent the screws from seizing.

6. Always use a torque wrench when installing cap screws into aluminum parts to prevent the possibility of stripping the threads.

7. If tapped threads in aluminum parts are stripped or damaged, the part can be made serviceable by the use of Heli-coils or the equivalent.

8. Seal-protecting tools must be used when assembling the units in order to prevent damage to the seals. The slightest flaw in the sealing surface of the seal can cause an oil leak.

9. The aluminum castings and the valve parts are very susceptible to nicks, burrs, etc., and care should be exercised when handling them.

10. The internal snap rings should be expanded and the external snap rings compressed if they are to be reused. This will ensure proper seating when installed.

11. Replace all O-rings, gaskets, and oil seals that are removed. Teflon oil seal rings should not be removed unless damaged.

12. During assembly of each unit, all internal parts must be lubricated with oil.

Oil Leaks

The suspected area should be wiped clean of all oil before inspecting for the source of the leak. Red dye is used in the transaxle oil and will indicate if the oil leak is from the transaxle.

When an oil leak is suspected, the following procedure will assist in locating the leak.

1. Clean the area with solvent to remove all traces of oil.

2. Remove the converter shield if a leak is suspected at the engine transaxle connection.

3. Spray suspected leak areas with a pressurized foot spray powder.

4. Start the engine and run at high idle.

5. Repair any leaks that are detected and recheck as necessary. Reinstall removed parts.

The use of a black light to identify the oil at the source of the leak is also helpful. Comparing the oil from the leak to that on the engine or transaxle dipstick (when viewed by black light) will determine the source of the leak.

Oil leaks around the engine and transaxle are generally carried toward the rear of the car by the air stream. In determining the source of an oil leak, it is most helpful to keep the engine running.

6. Degrease the underside of the transaxle.

7. Road test to get the unit at operating temperature, 88°C (190°F).

8. Inspect for a leak with the engine running.

9. With the engine off, check for oil leaks due to the raised oil level caused by drainback.

Possible Points of Oil Leak

1. Transaxle pan or valve body cover

(a) Attaching bolts not correctly torqued

(b) Improperly installed or damaged gasket

(c) Oil pan or valve body cover mounting face not flat

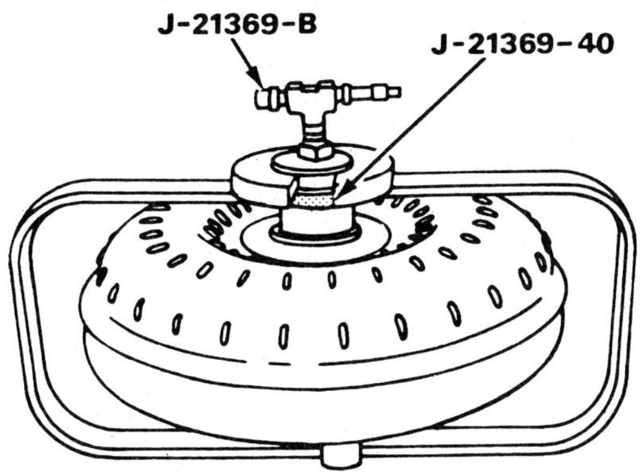

Figure 8-29. Tools installed to check for converter leaks. (*Courtesy of General Motors Corporation*)

2. Case Leak

(a) Filler pipe multilip seal damaged or missing; misposition of filler pipe bracket to engine "loading" one side of multilip seal

(b) TV cable multilip seal missing, damaged, or improperly installed

(c) Governor cover and O-rings damaged or missing

(d) Speedometer-driven gear O-ring damaged

(e) Manual valve bore plug

(f) Oil cooling connector fittings

(g) Axle oil seals

(h) Parking pawl shaft cup plug

(i) Governor pressure pickup plug

(j) Line pressure pickup pipe plug

(k) Case to case cover gasket

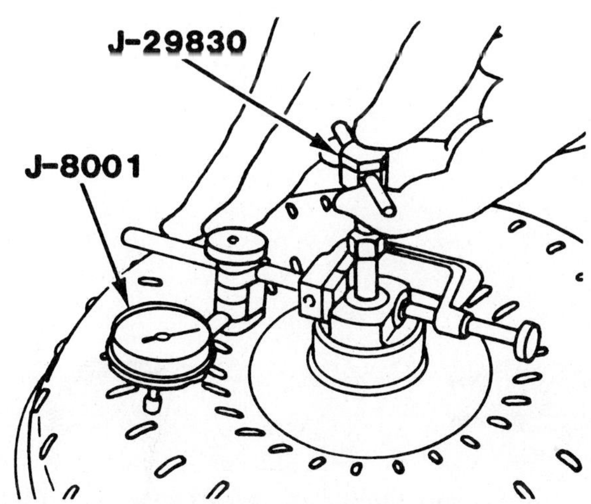

Figure 8-30. Checking converter turbine end play. (*Courtesy of General Motors Corporation*)

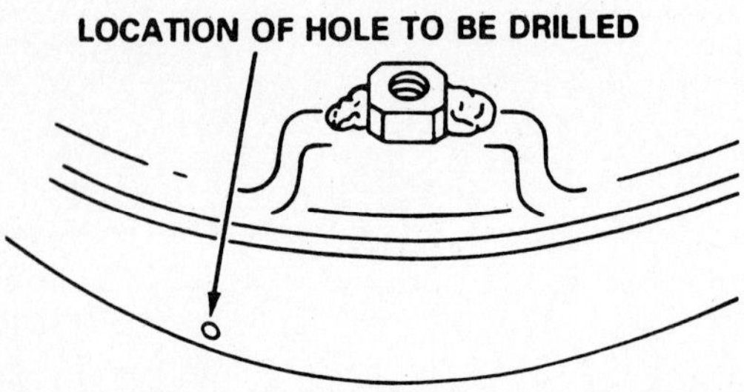

Figure 8-31. One-eighth-inch hole drilled in GM converter for flushing purposes. (*Courtesy of General Motors Corporation*)

3. Leak at the converter end of the transaxle
   (a) Converter seal damage
      (1) Seal lip cut (check the converter hub for nicks, etc.)
      (2) Bushing moved forward and damaged
      (3) Garter spring missing from seal
   (b) Converter leak in weld area
   (c) Porous casting (case or drive sprocket support)
   (d) Turbine shaft oil seal
4. Oil coming out of the vent pipe
   (a) Transaxle overfilled
   (b) Water in oil
   (c) Case porous
   (d) Incorrect dipstick

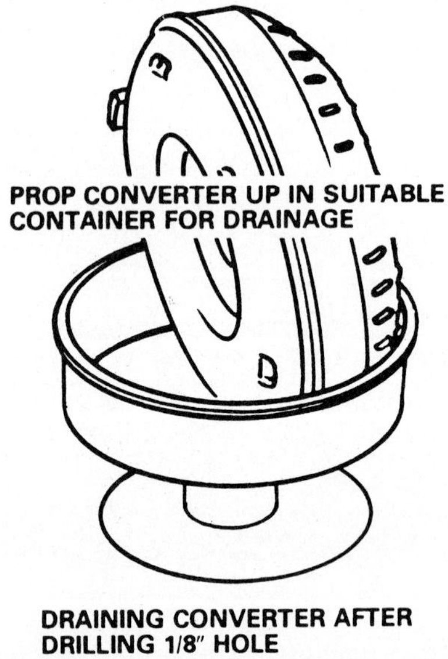

Figure 8-32. (*Courtesy of General Motors Corporation*)

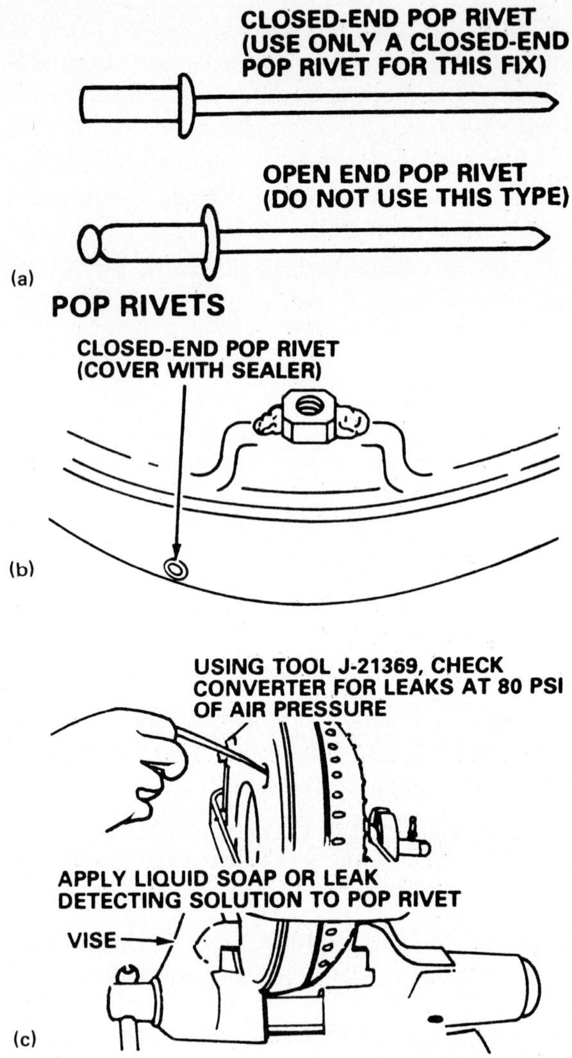

Figure 8-33. Converter hole must be closed with pop rivet and suitable sealer after flushing and pressure-tested for leaks. (*Courtesy of General Motors Corporation*)

(e) Pump shy of stock on mounting faces, porous casting

(f) Drainback holes plugged

On-Car Serviceable Components

The following parts can be serviced with the transaxle in the car. For part removal and installation procedures not listed in this section, refer to the disassembly and reassembly sections in the service manual.

1. Throttle valve control cable or sleeve seal
2. Transaxle filler pipe or sleeve seal
3. Transaxle governor assembly and speedometer gear assembly

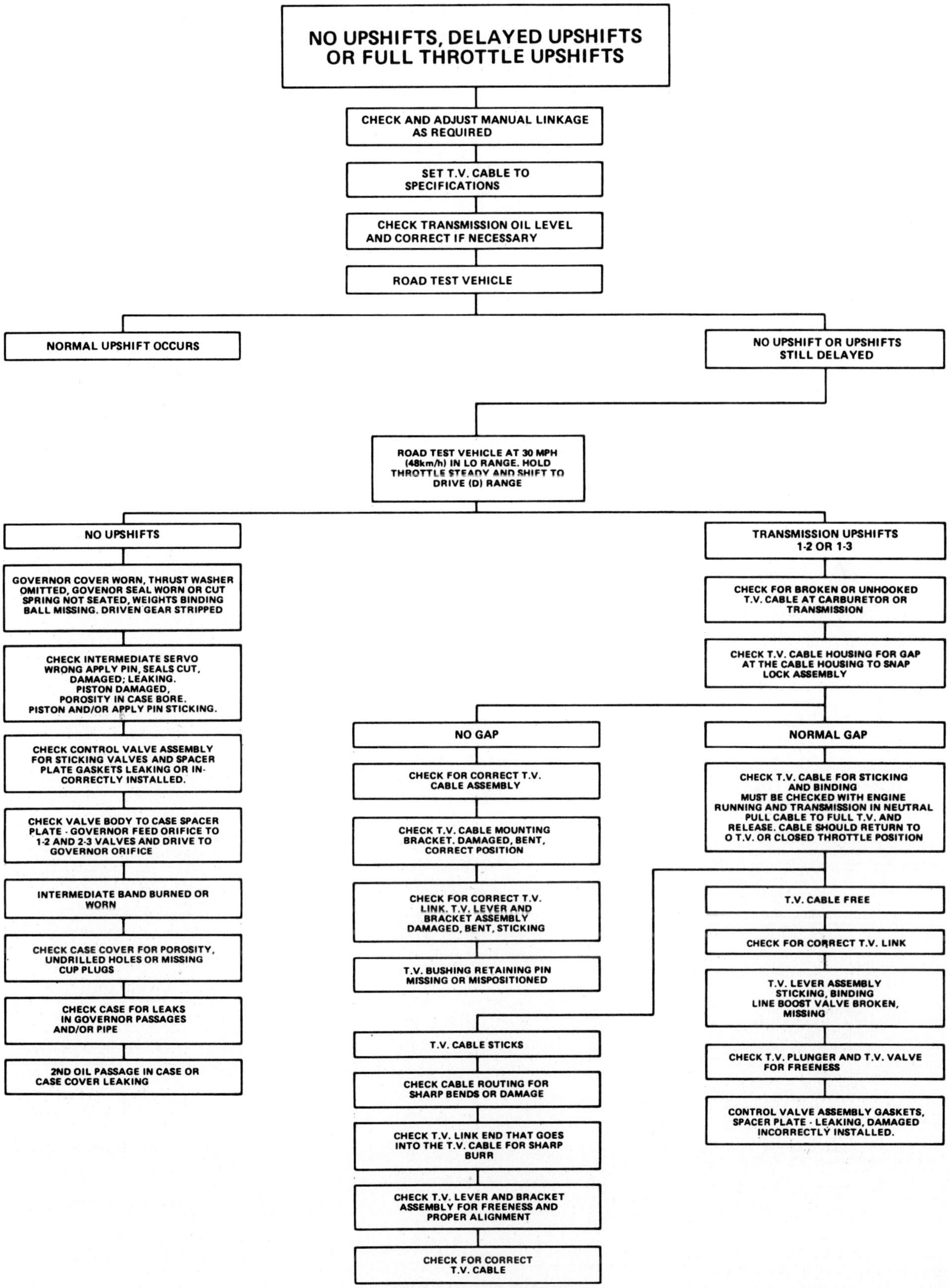

Figure 8-34. GM 125C.

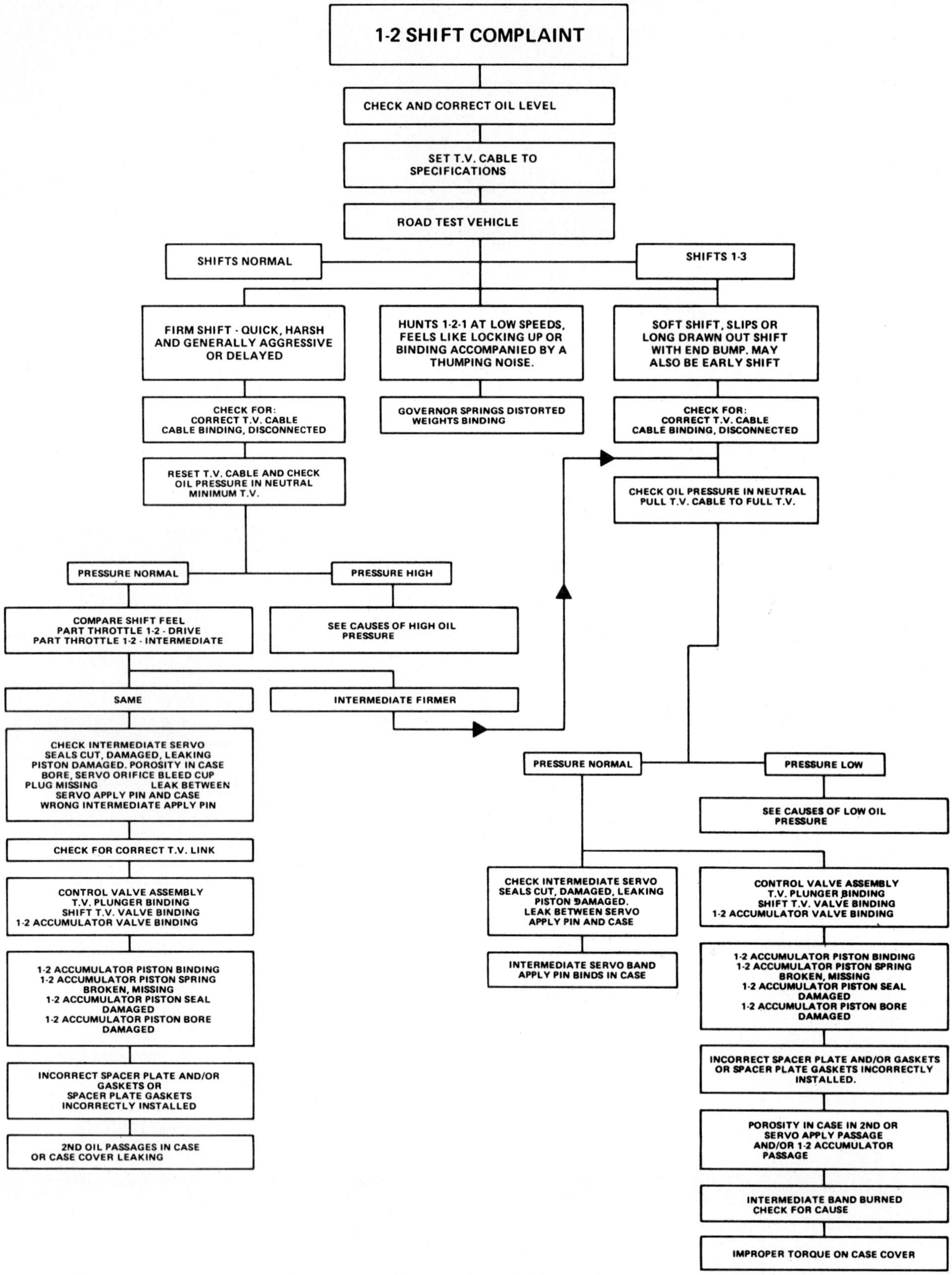

Figure 8-35. GM 125C.

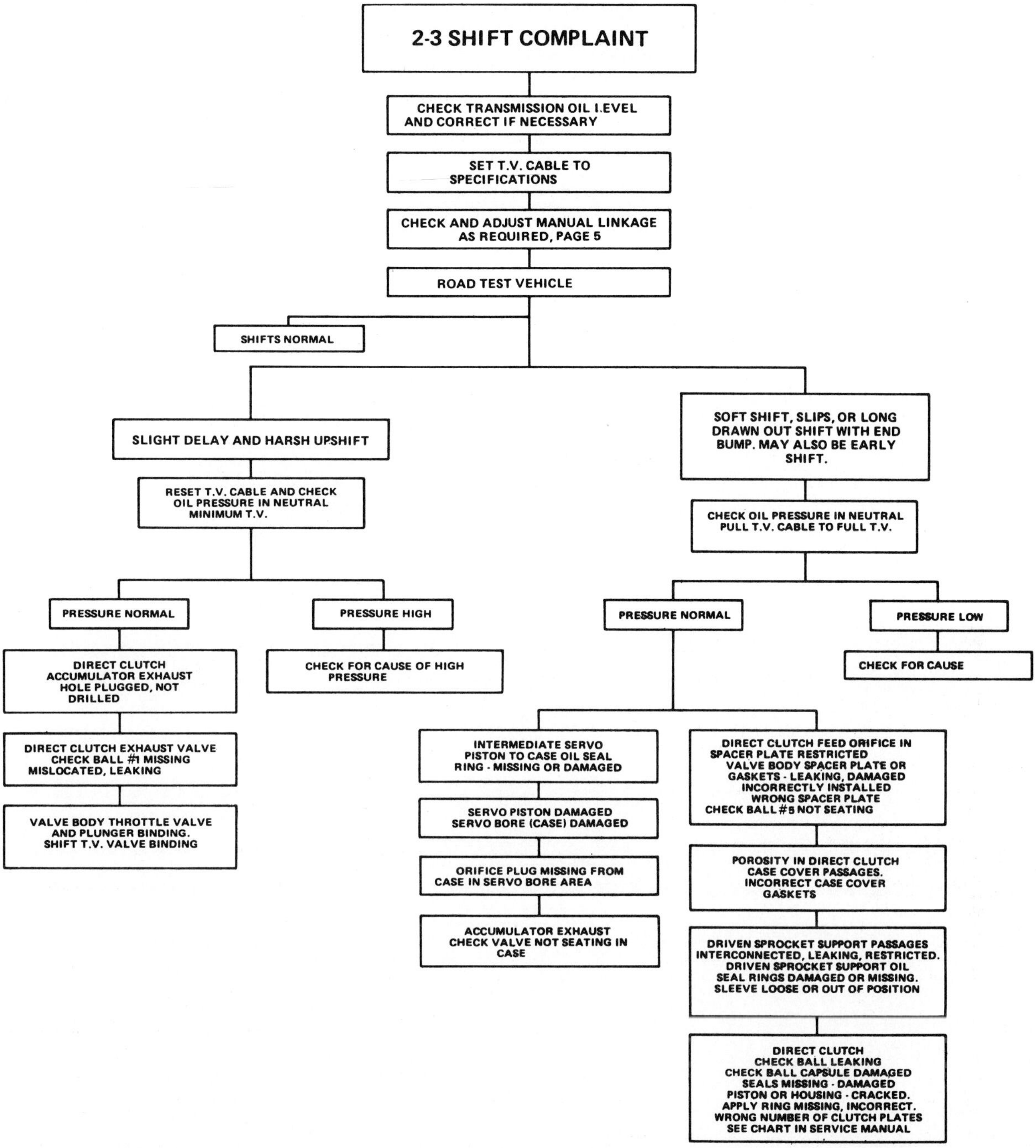

Figure 8-36. GM 125C.

4. Transaxle intermediate servo assembly and direct clutch accumulator check valve

5. Transaxle valve body assembly, spacer plate, gaskets, throttle lever and bracket assembly, pump shaft, valve body cover, and gasket

6. Transaxle converter to flexplate bolts

7. Transaxle oil pan and gasket, strainer assembly, and O-ring

8. Low and reverse pipe, O-ring seal, and oil seal

9. Dipstick stop bracket, parking pawl, and return spring

10. Output shaft, axle joint retaining ring, snap ring (shaft), and axle oil seals

11. Case cover, cooler fittings, manual valve, electrical connector, 1–2 accumulator piston, spring

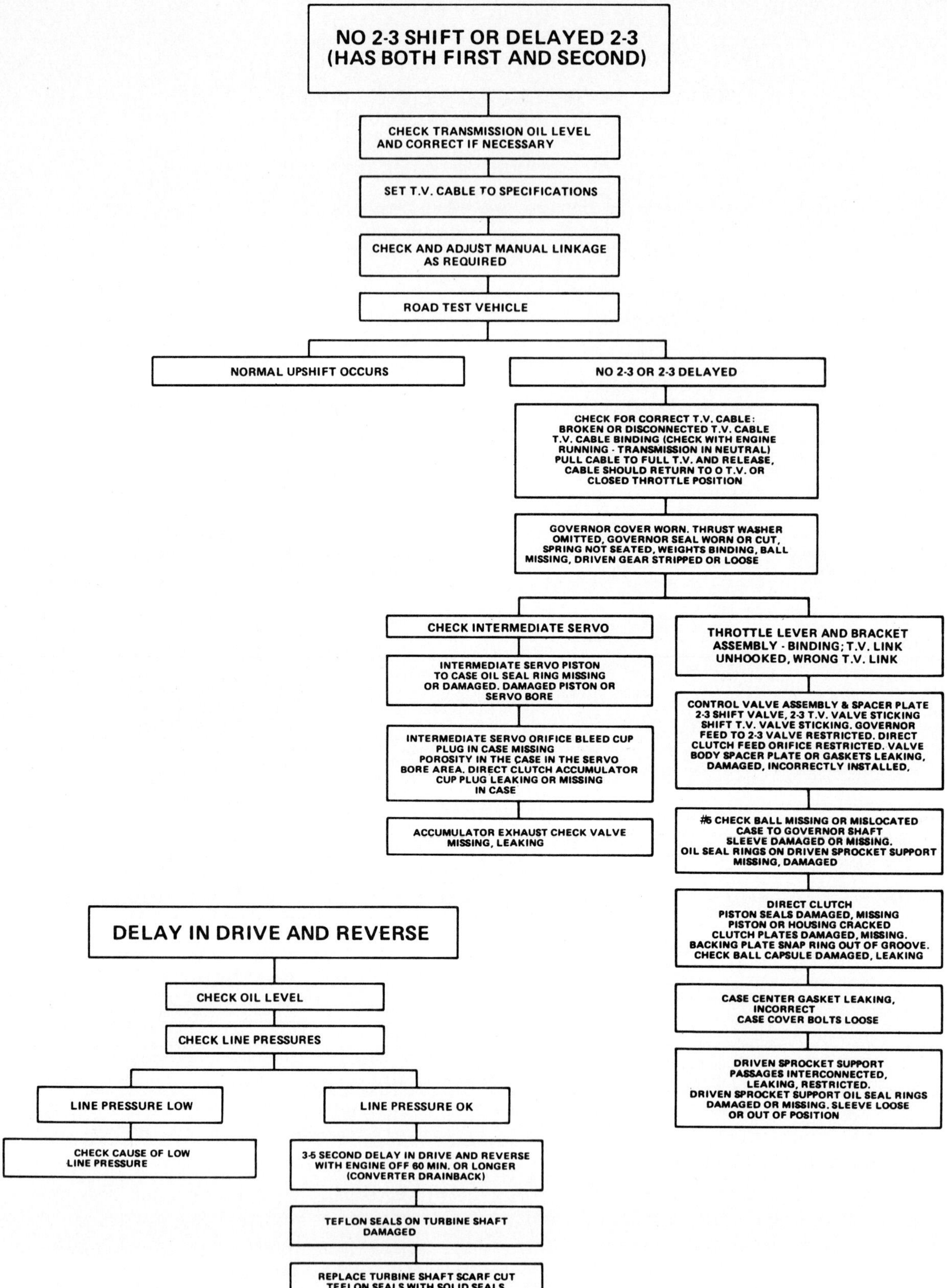

Figure 8-37. GM 125C.

NO DRIVE

A "NO DRIVE" COMPLAINT CAN BE REPORTED UNDER SEVERAL CONDITIONS OR IN DIFFERENT OPERATING RANGES. SELECT FROM THE FOLLOWING CONDITIONS THE ONE THAT BEST REPRESENTS THE PROBLEM.

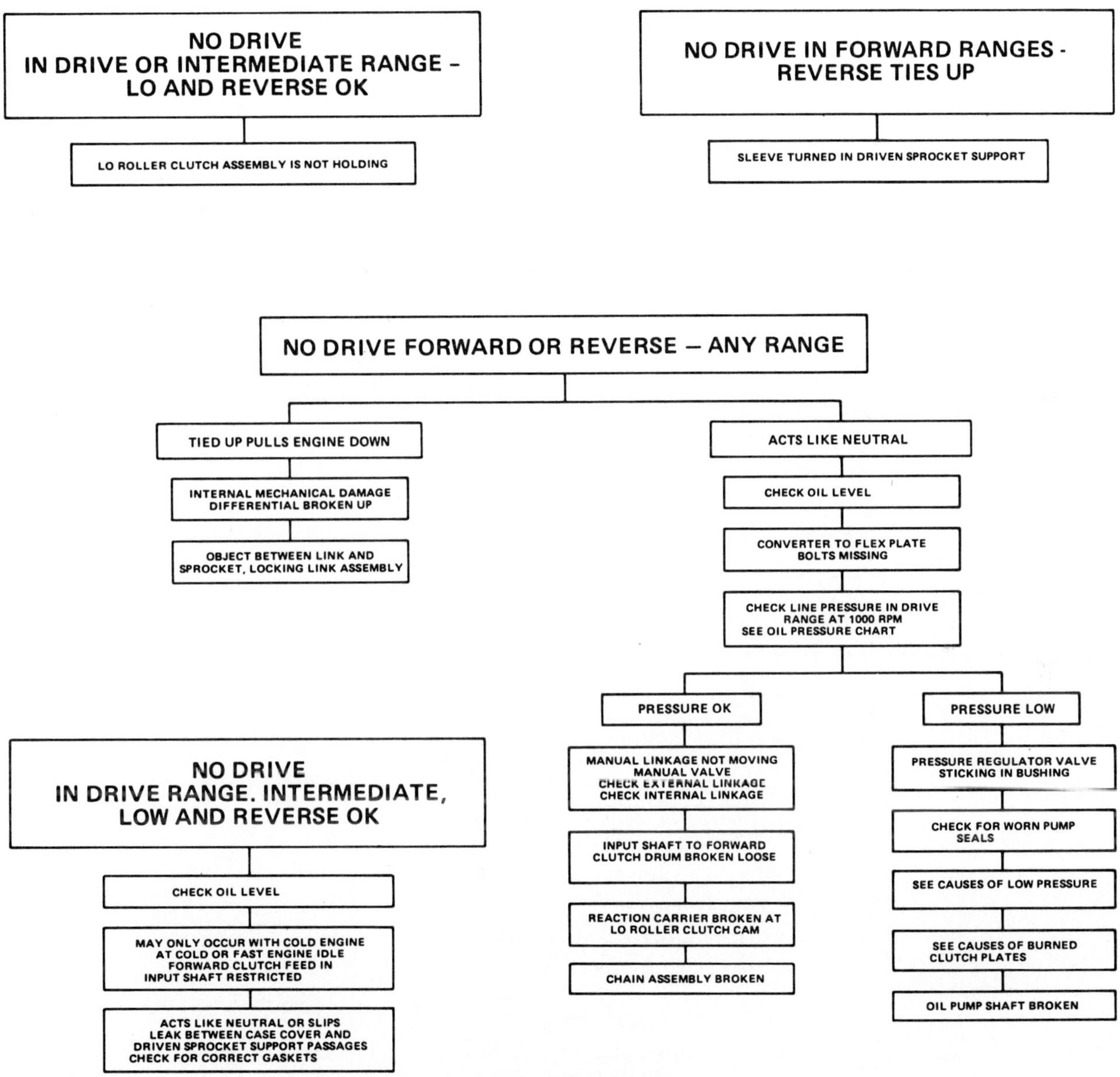

Figure 8-38. GM 125C.

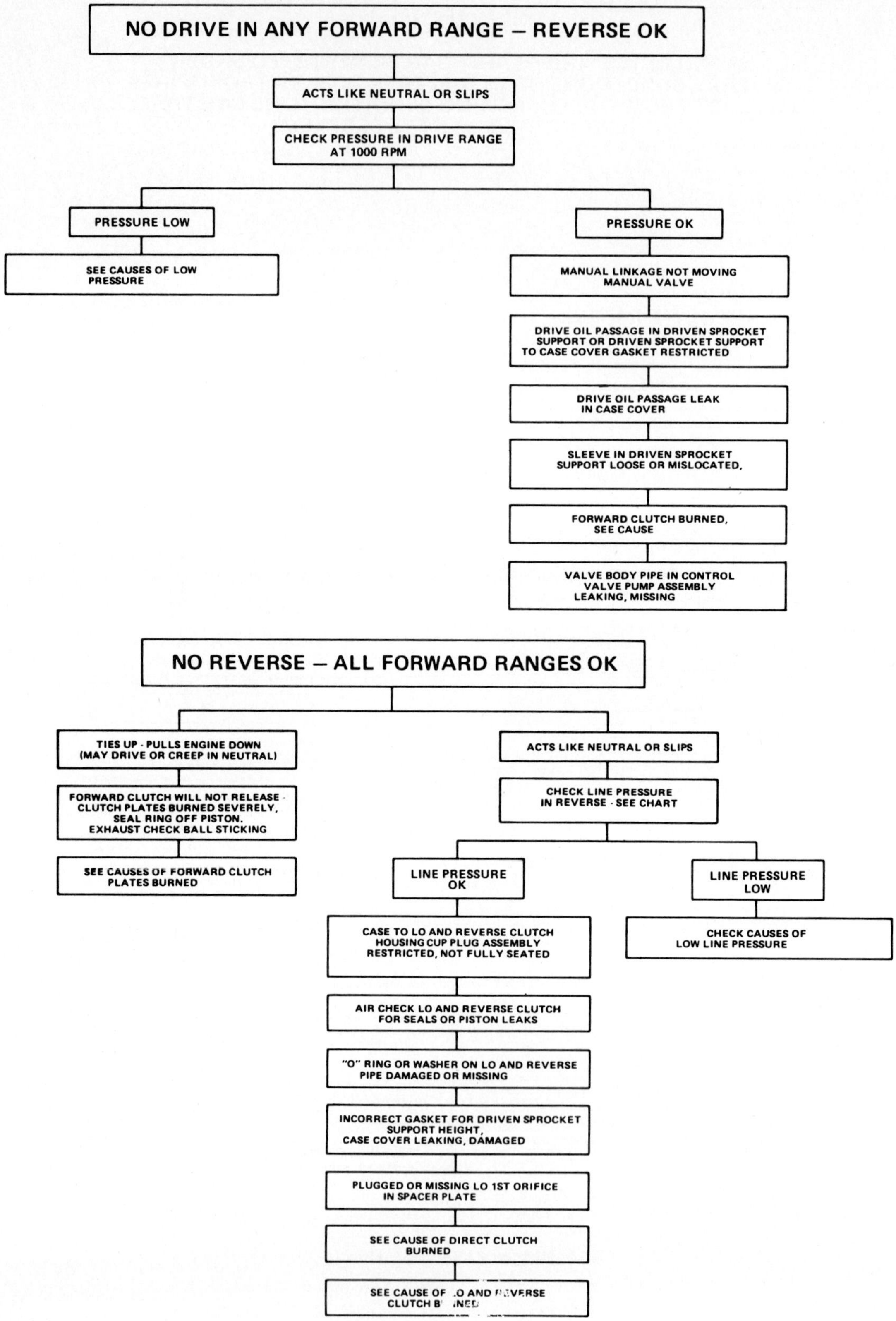

Figure 8-39. GM 125C.

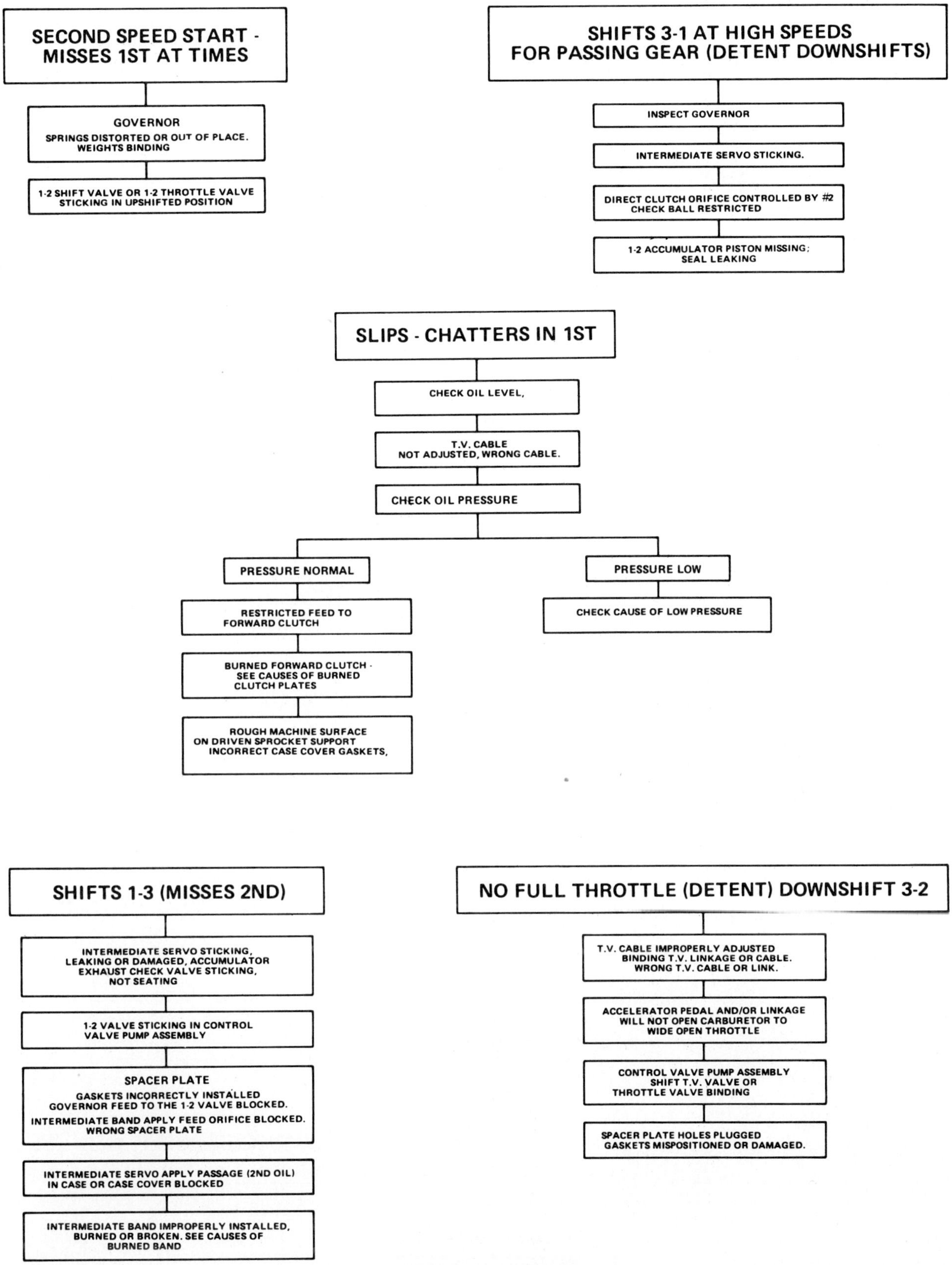

Figure 8-40. GM 125C.

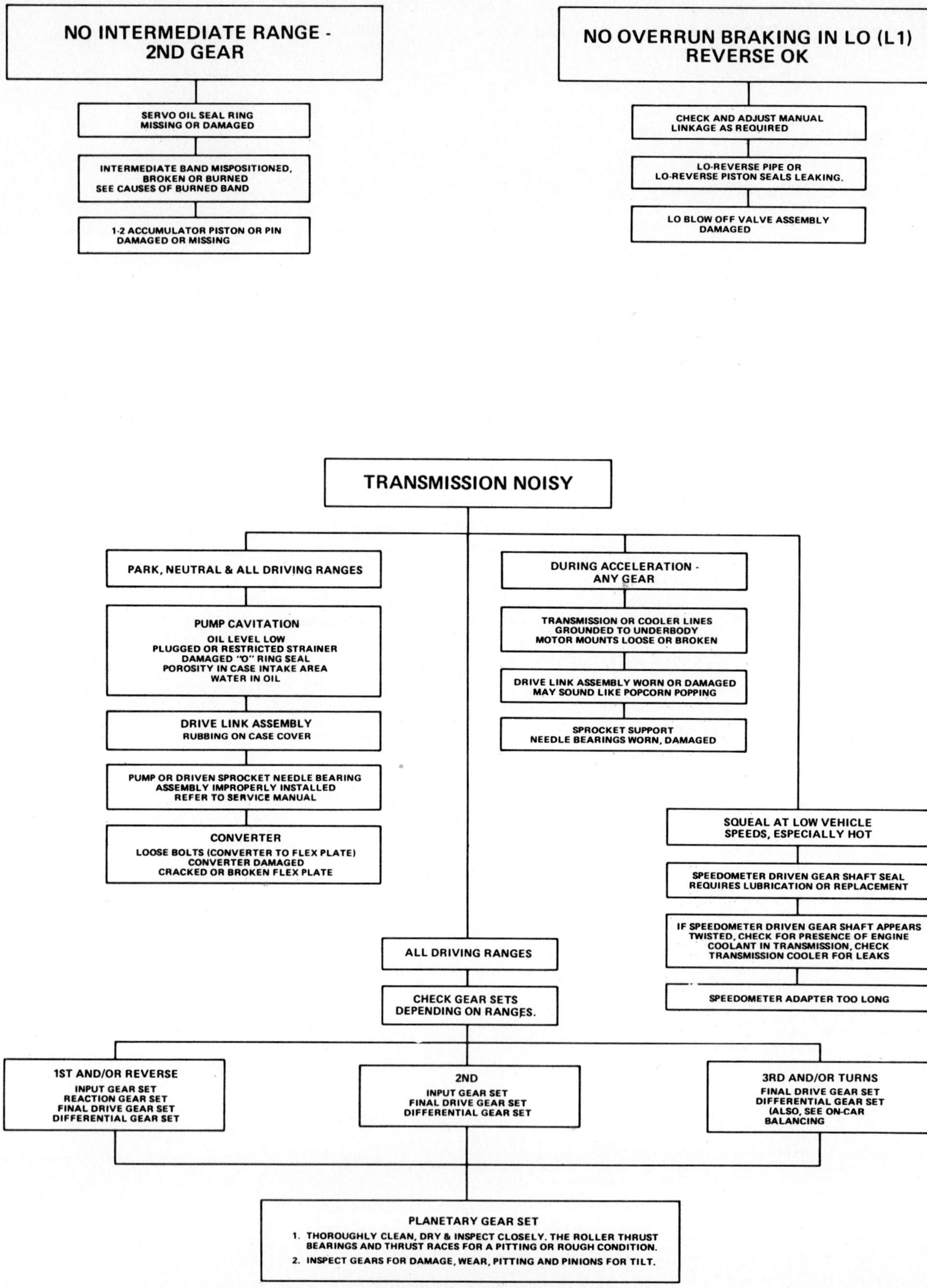

Figure 8-41. GM 125C.

TRANSMISSION CONVERTER CLUTCH (TCC)
ELECTRICAL DIAGNOSIS

ALL EQUIPPED WITH COMPUTER COMMAND CONTROL

MECHANICAL CHECKS SUCH AS LINKAGE, OIL LEVEL, ETC. SHOULD BE PERFORMED PRIOR TO USING THIS CHART.

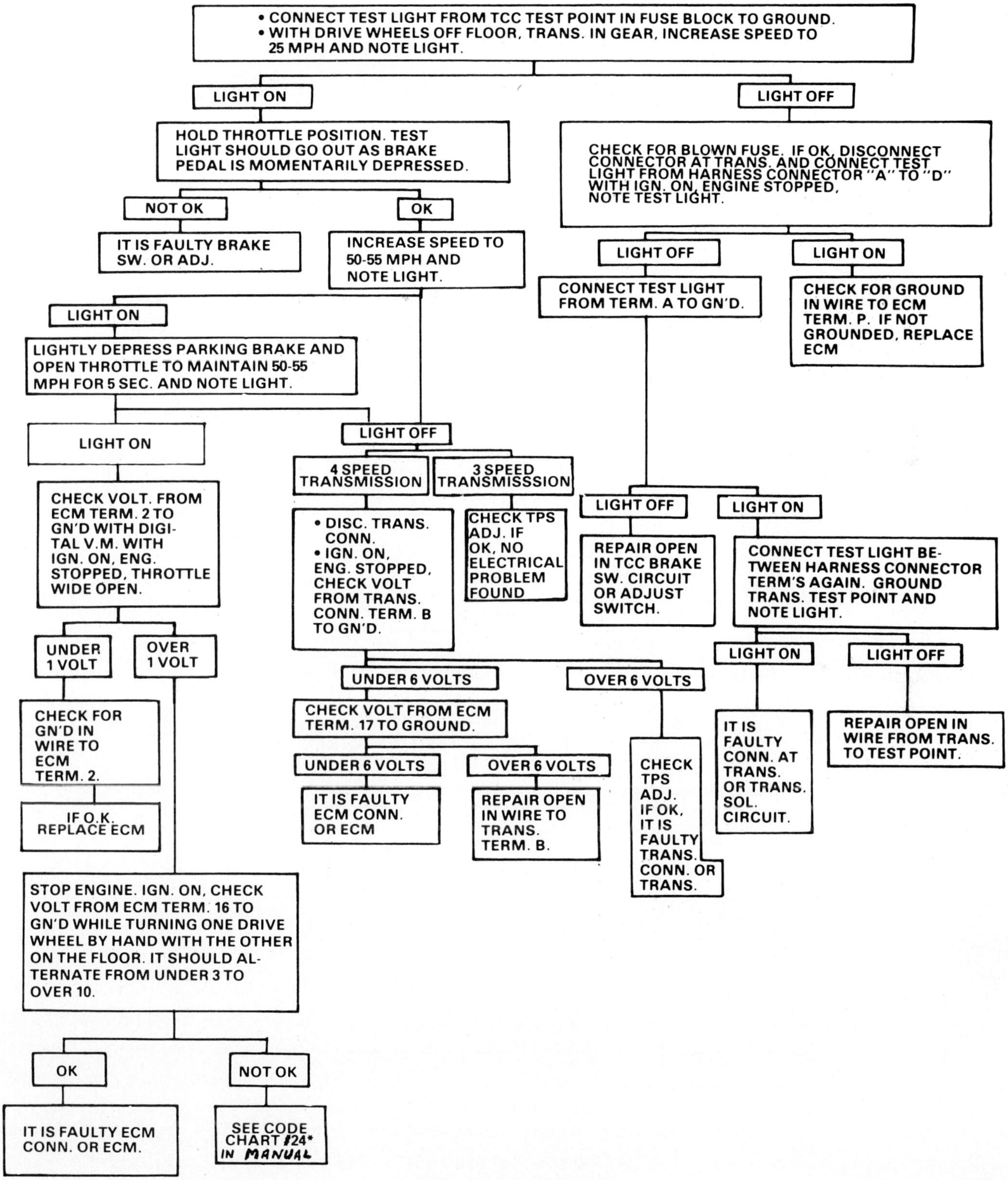

Figure 8-42. GM 125C.

TRANSMISSION CONVERTER CLUTCH (TCC) ELECTRICAL DIAGNOSIS

LESS COMPUTER COMMAND CONTROL EQUIPPED

MECHANICAL CHECKS, SUCH AS LINKAGE, OIL LEVEL, ETC., SHOULD BE PERFORMED PRIOR TO USING THIS CHART.

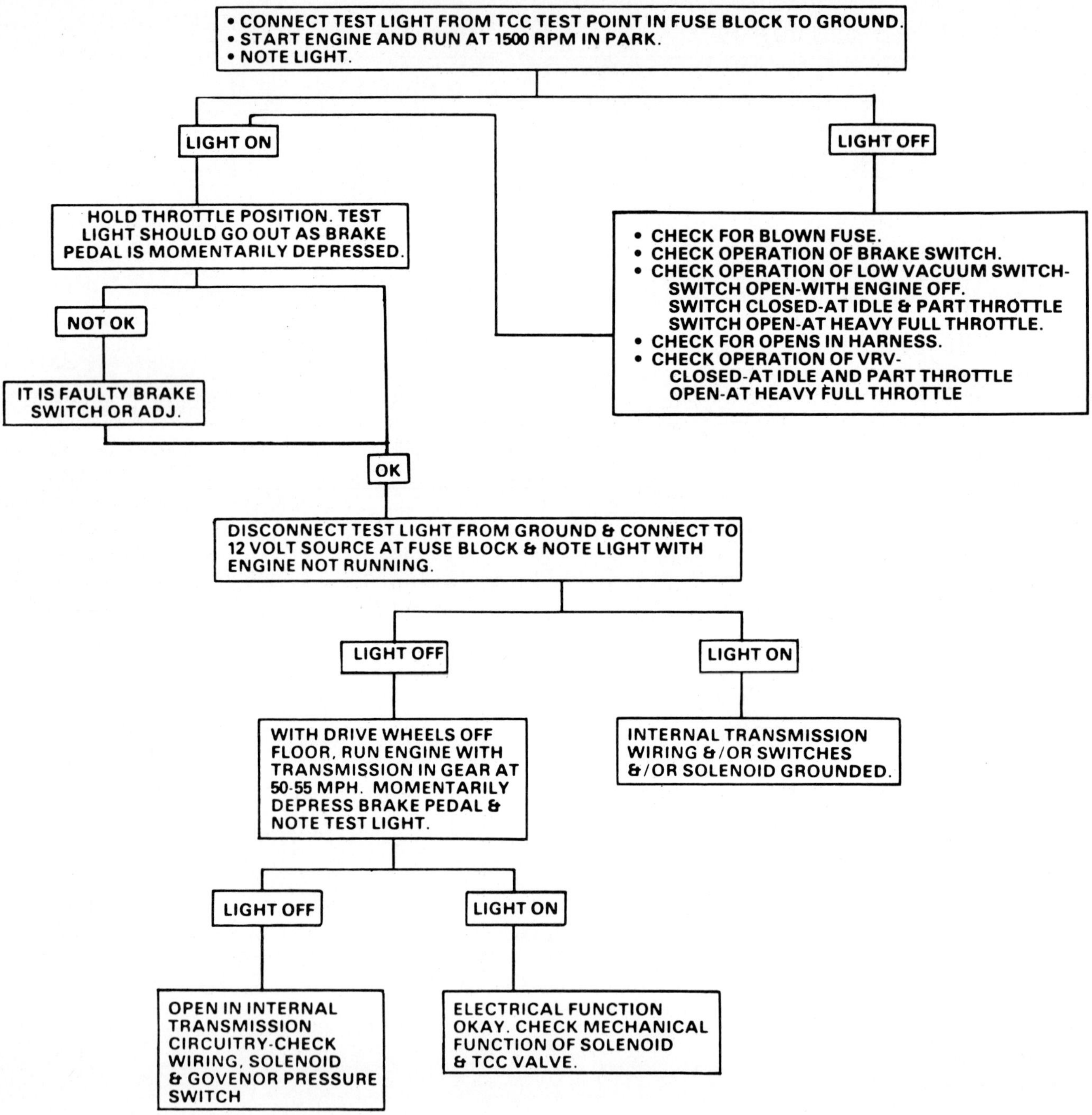

T.C.C. Diagnosis - Less Computer Command Control

Figure 8-43. GM 125C.

and seal, thermostatic element assembly, manual detent spring and roller assembly, sprockets, drive link, and thrust washers

12. Third-clutch pressure switch, governor pressure switch, solenoid, auxiliary valve body, cover, and gasket

GM 200 Diagnosis Procedure (Courtesy of General Motors Corporation)

Fluid Capacity

Approximately 2.8 liters (3 quarts) of fluid are required to refill the transmission after the oil pan has been drained. The fluid capacity of the 200 automatic transmission and converter assembly is approximately 6.2 liters (6.5 quarts), but the correct level is determined by the mark on the dipstick rather than by the amount added. Use only Dexron® II-D number automatic transmission fluid or its equivalent.

Note: If the oil changes to a darker color at an early stage, from the usual red color or to a strong odor, this is normal even though it is usually associated with overheated transmission fluid. It is not a positive sign of required maintenance or transmission failure.

Sequence for Diagnosis

1. Check and correct the oil level.
2. Check and adjust the TV cable.
3. Check and correct the manual linkage.
4. Install oil pressure gauge.
5. Road test the car.

(a) Road test using all selective ranges, noting when differences in operation or oil pressure occur.

(b) Attempt to isolate the unit or circuit involved in the malfunction.

(c) If engine performance indicates an engine tuneup is required, this should be performed before road testing is completed or transmission correction attempted. Poor engine performance can result in rough shifting or other malfunctions.

Maintenance

The fluid level should be checked at each oil change. When adding or changing fluid, use the Dexron® II-D number or equivalent automatic transmission fluid. Under normal driving conditions, change transmission fluid every 60,000 miles (100,000 km).

Checking and Adding Fluid

Transmission at Operating Temperature

The automatic transmission is designed to operate at the full hot mark on the dipstick at the normal operating temperature of 88° to 93°C (190° to 200°F) and should be checked under these conditions. The normal operating temperature is obtained only after at least 15 miles of highway type driving or the equivalent of city driving.

To determine proper level, proceed as follows:

1. Apply the parking brake and block the car wheels.

2. With the selector lever in the park position, start the engine. *Do not race the engine. Move the selector lever through each range.*

3. Immediately check fluid with the selector lever in park, the engine running at slow idle, and the car on a level surface. The fluid level on the dipstick should be at the full hot mark.

4. If additional fluid is required, add sufficient fluid to bring to the full hot mark on the dipstick.

Transmission at Room Temperature

Automatic transmissions are frequently overfilled because the fluid level is checked when the fluid is cold and the dipstick indicates fluid should be added. However, the low reading is normal since the level will rise as the fluid temperature increases. A level change of over 19 mm (3/4 inch) will occur as fluid temperature rises from 16° to 82°C (60° to 180°F).

Overfilling can cause foaming and loss of fluid through the vent. Slippage and transmission failure can result. Fluid level that is too low can cause slipping, particularly when the transmission is cold or the car is on a hill.

Check the transmission fluid level with the engine running, the shift lever in park, and the car level.

Note: If the car has recently been operated for an extended period at high speed or in city traffic in hot weather or the car is being used to pull a trailer, an accurate fluid level cannot be determined until the fluid has cooled down (usually about 30 minutes after the car has been parked).

Remove the dipstick and touch the transmission end of the dipstick cautiously to find out if the fluid is cool, warm, or hot. Wipe it clean and reinsert until the cap seats. Remove the dipstick and note the reading.

1. If the fluid feels cool, about room temperature—18° to 29°C (65° to 85°F)—the level should be between the two dimples below the add mark.

2. If it feels warm, the level should be close to the add mark (either above or below).

3. If it feels hot (cannot be held comfortably), the level should be between the add and full marks.

Fluid Drain Intervals

The transmission operating temperature resulting from the type of driving conditions under which the car is used is the main consideration in establishing the proper frequency of transmission fluid changes.

Change the transmission fluid and service screen every 15,000 miles (24,000 km) if the car is usually driven under one or more of the following conditions, which are considered severe transmission service:

1. In heavy city traffic
2. When the outside temperature regularly reaches 32°C (90°F)
3. In very hilly or mountainous areas
4. Frequency trailer pulling
5. Commercial use, such as taxi, police car, or delivery service

If you do not use your car under any of these conditions, change the fluid and service screen every 60,000 (100,000 km).

Note: Do not overfill. It takes only 1 pint to raise the level from add to full with a hot transmission.

Checking and Adding Fluid (After Transmission Service)

If work has been performed on the transmission, add the following quantity of fluid, then check fluid as in the preceding and following outline.

Changing Fluid

Pan removal—2.8 liters (3 quarts)
Unit repair—4.2 liters (4.5 quarts)

1. Raise the car.
2. With the drain pan placed under the transmission oil pan, remove oil pan attaching bolts from the front and side of the pan.
3. Loosen rear pan attaching bolts approximately four turns.
4. Carefully pry the transmission oil pan loose with a screwdriver, allowing fluid to drain.
5. Remove remaining bolts and remove the oil pan and gasket. Discard the gasket.
6. Drain fluid from the oil pan. Clean the pan with solvent and dry thoroughly with clean compressed air.
7. Remove two screen-to-valve body bolts, screen, and gasket. Discard the gasket.
8. Thoroughly clean the screen assembly in solvent and dry throughly with clean compressed air.
9. Install the new gasket on the screen and two bolts. Tighten the bolt to 15 N.m (11 pound-feet).
10. Install the new gasket on the oil pan and install the oil pan. Tighten attaching bolt and washer assemblies to 16 N.m (12 pound-feet).
11. Lower the car, add approximately 2.8 liters (3 quarts) of Dexron® II-D number automatic transmission fluid or its equivalent through the filler tube.
12. With the selector lever in the park position, apply the hand brake, start the engine and let it idle (carburetor off fast idle stop). *Do not race the engine.*
13. Measure the selector lever through each range and, with the selector lever in the park range, check the fluid level.
14. Add additional fluid to bring the level between the dimples on the dipstick.

Adding Fluid to Fill Dry Transmission and Converter

In cases of transmission overhaul, when a complete fill is required, including the converter, proceed as follows:

1. Add 4.2 liters (4.5 quarts) of transmission fluid through the filler tube.
2. With the manual control lever in the park (P) position, start the engine and place it on cold idle cam. *Do not race the engine.* Move the manual control lever through each range.
3. Immediately check the fluid level with the selector lever in park (P), with the engine running and vehicle on a level surface, and add additional fluid to bring the level to a point between the two dimples on the dipstick. Do not overfill.

Fluid Level and Capacity

The transmission dipstick is located in a filler tube at the right rear of the engine. To bring the fluid level from the add mark to the full mark requires 1 pint of fluid. The fluid level should be checked at every engine oil change.

The fluid level should be to the full mark with transmission fluid at normal operating temperature, 93° C (200°F). With warm fluid at room temperature, 21°C (70°F), the level will be between the two dimples on the dipstick. The normal operating temperature is obtained only after at least 15 miles of highway type driving or equivalent of city driving.

Oil Leaks

The suspected area should be wiped clean of all oil before inspecting for the source of the leak. Red dye is used in the transmission oil at the assembly plant and will indicate if the oil leak is from the transmission.

When an oil leak is found, the following procedure will assist in locating the leak.

1. Clean the area with solvent to remove all traces of oil.

2. Remove the flywheel lower cover, if the leak is suspected at the rear of engine or the front of the transmission.

3. Spray suspected leak areas with a pressurized foot powder.

4. Start the engine and run at high idle.

5. Repair any leaks that are detected and recheck as necessary. Reinstall removed parts.

The use of a black light to identify the oil at the source of the leak is also helpful. Comparing the oil from the leak to that on the engine or transmission dipstick (when viewed by black light) will determine the source of the leak.

Oil leaks around the engine and transmission are generally carried toward the rear of the car by the air stream. For example, a transmission oil-filler-tube-to-case leak will sometimes appear as a leak at the rear of the transmission. In determining the source of an oil leak, it is most helpful to keep the engine running.

A black light testing unit such as J 6640 may be obtained from service tool suppliers.

6. Degrease the underside of the transmission.

7. Road test to get the unit at operating temperature, 82°C (180°F).

8. Inspect for the leak with the engine running.

9. With the engine off, check for oil leaks due to the raised oil level caused by drainback.

Possible Points of Oil Leak

1. Transmission oil pan leak

(a) Attaching bolts not correctly torqued

(b) Improperly installed or damaged pan gasket

(c) Oil pan gasket mounting face not flat

2. Case leak

(a) Filler pipe O-ring seal damaged or missing; misposition of filler pipe bracket to engine "loading" one side of O-ring

(b) TV cable O-ring seal missing, damaged, or improperly installed

(c) Rear seal assembly damaged or improperly installed

(d) Governor cover and O-rings damaged or missing

(e) Speedometer-driven gear O-ring damaged

(f) Manual shaft lip seal damaged or improperly installed

(g) Line pressure tap plug

(h) Vent pipe

(i) Porous casting

(j) Intermediate servo O-ring damaged

3. Leak at the front of the transmission

(a) Front pump seal leaks

- Seal lip cut (Check the converter hub for nicks, etc)
- Bushing moved forward and damaged
- Garter spring missing from seal

(b) Front pump attaching bolts loose or bolt seals damaged or missing

(c) Front pump housing O-ring damaged or cut

(d) Converter leak in the weld area

(e) Porous casting (pump or case)

4. Oil coming out the vent pipe

(a) Transmission overfilled

(b) Water in oil

(c) Foreign matter between the pump and case or between the pump cover and body

(d) Case porous, front pump cover mounting face shy of stock near breather

(e) Pump-to-case gasket mispositioned

(f) Incorrect dipstick

(g) Pump shy of stock on mounting faces, porous casting, breather hole plugged in the pump cover

Case Porosity Repair

External leaks caused by case porosity have successfully been repaired with the transmission in the car by using the following recommended procedures.

1. Road test and bring the transmission to operating temperature, approximately 90°C (200°F).

2. Raise the car on a hoist or jack stand, with the engine running, and locate the source of the oil leak. Check for the leak in all operating ranges.

3. Shut off the engine, thoroughly clean the area to be repaired with a cleaning solvent, and air dry.

4. Using instructions of the manufacturer, mix a sufficient amount of epoxy, No. 1360016 or the equivalent, to make the repair. Observe cautions of the manufacturer in handling.

5. While the transmission case is still *hot*, apply the epoxy to the area to be repaired. A clean, dry soldering acid brush can be used to clean the area and also to apply the epoxy cement. Make certain that the area to be repaired is fully covered.

6. Allow cement to cure for three hours before starting the engine.

7. Road test and check for leaks.

Converter Stator Operation Diagnosis

The torque converter stator assembly and its related roller clutch can have one of two different type malfunctions.

1. The stator assembly freewheels in both directions.

2. The stator assembly remains locked up at all times.

Stator Assembly Freewheels

If the stator roller clutch becomes ineffective, the stator assembly freewheels at all times in both directions. With this condition, the car will tend to have poor acceleration from a standstill. At speeds above 48 to 56 km (30 to 35 mph), the car may act normal. If poor acceleration problems are noted, it should first be determined that the exhaust system is not blocked, the engine is in good tune, and the transmission is in first gear when starting.

If the engine will freely accelerate to high rpm in neutral (N), it can be assumed that the engine and exhaust system are normal.

Stator Assembly Remains Locked Up

If the stator assembly remains locked up at all times, the engine rpm and car speed will tend to be limited or restricted at high speeds. The car performance when accelerating from a standstill will be normal. Engine overheating may be noted. Visual examination of the converter may reveal a blue color from the overheating that will result.

Under these conditions, if the converter has been removed from the transmission, the stator roller clutch can be checked by inserting a finger into the splined inner race of the roller clutch and trying to turn the race in both directions. The inner race should turn freely in the clockwise direction but not turn or be very difficult to turn in the counterclockwise direction.

Note: Do not use such items as the pump cover or stator shaft to turn the race as the results may be misleading.

Checking the Converter

1. Check the converter for leaks as follows.

(a) Install J 21369-2 and J 21369-9 and tighten the hex nut on J 21369-2.

(b) Fill the converter with air, 551 kPa (80 psi).

(c) Submerge in water and check for leaks.

(d) Bleed pressurized air from J 21369-2 before removing tools from the converter, then remove tools.

2. Check converter hub surfaces for signs of scoring or wear.

3. Check the converter end clearance as follows.

(a) Fully release the collet end of tool J 25020.

(b) Install the collet end of tool J 25020 into the converter hub until it bottoms; then tighten the cap nut to 6.7 N.m (5 lb. ft.).

(c) Install tool J 21371-3 and tighten the hex nut to 4.0 N.m (3 lb. ft.).

(d) Install the dial indicator and set it at zero while its plunger rests on the cap nut of tool J 25020.

(e) Loosen the hex nut while holding the cap nut stationary. With the hex nut loose and tool J 21371-3 held firmly against the converter hub, the reading obtained on the dial indicator will be the converter end clearance. The end clearance should be less than 0–1.27 mm (0–0.050 inch). If the end clearance is greater, the converter must be replaced.

4. With the transmission in a cradle or portable jack, install the converter assembly into the pump assembly, making sure that the converter hub drive slots are fully engaged with the pump drive gear tangs and the converter installed fully toward the rear of the transmission.

Note: The converter will be properly installed if the distance is 25.4 mm (1.00 inch) minimum between the engine mounting face of the case and the front face of the converter cover drive lugs.

5. Retain the converter with J 21366.

Oil Pressure Checks

1. Connect an oil pressure gauge to the line pressure tap.

2. Connect a tachometer to the engine.

The TV cable used with the 200 automatic transmission should not be thought of as a 350 downshift cable. The TV cable used on the 200 controls line pressure, shift points, shift feel, part-throttle downshifts, and detent downshifts. The function of the cable is similar to the combined functions of the vacuum modulator and the downshift cable used in vehicles that use the 350 transmission. The cable operates the throttle lever and bracket assembly.

The throttle lever and bracket assembly serves two basic functions.

1. The primary function of this assembly is to transfer the carburetor throttle plate movement to the TV plunger in the control valve assembly as related by the TV cable and linkage. This causes TV pressure and line pressure to increase according to throttle opening and also controls part-throttle and detent downshifts. The proper adjustment of the TV cable is based on the TV plunger being fully depressed to flush with the TV bushing at wide-open throttle.

2. The second function of the assembly involves the TV exhaust valve lifter rod, spring, and TV exhaust ball. The function of this system is to prevent the transmission from operating at low (idle position) pressures, if the TV cable should become broken or disconnected. If the cable is connected, not broken or stretched, the TV lifter rod will not move from its normal, spring-loaded, up position, which holds the TV exhaust check ball off its seat. The TV lifter rod will drop to allow the TV exhaust ball to seat only if the cable is broken, disconnected, or extremely out of adjustment. With the transmission pan removed, it should be possible to pull down on the TV exhaust valve lifter rod and the spring should return the rod to its normal up position. If the throttle lever and bracket assembly or lifter rod binds or sticks so that the TV lifter rod cannot lift the exhaust ball off its seat, high line pressures and delayed upshifts will result. The right angle leg must not be bent to any other angle or it will not function properly.

If the TV cable is broken, sticky, misadjusted or the incorrect part for the model, the vehicle may exhibit various malfunctions.

Sticking or binding TV linkage can result in delayed or full-throttle shifts. The TV cable must be free to travel to the full-throttle position and return to the closed throttle position without binding or sticking. Some binding or sticking of the TV cable and associated parts may occur only with the engine running and will not be noted or obtained with the engine off.

Inspection of the TV linkage for sticking or binding should be made with the engine running at idle speed, the transmission selector in neutral, and the vehicle brakes set. Pull the TV cable full travel beyond the carburetor lever pin that it attaches to and then release the cable; it should return to the closed throttle position against the carburetor lever pin. If the TV cable sticks and remains ahead of the carburetor lever pin, it may be caused by one or more of the following:

1. Sharp bends or damaged TV cable housing. Correct by rerouting the cable or replace it if required.

2. Sharp end or burr on the TV link, dragging in the TV cable housing. Correct by making the end smooth, using a file or stone.

3. Bent TV link. Correct by straightening or replace as required.

4. Misalignment of the throttle lever and bracket assembly on the coiled pin in the control valve assembly.

5. Damaged or binding throttle lever and bracket assembly. Correct by straightening or replace as required.

6. Throttle lever spring unhooked or damaged. Correct by assembling the spring properly or replace the throttle lever and bracket assembly as required.

If the TV cable is adjusted too long, it may result in one of the two following conditions:

1. Early and slipping shifts or no detent downshifts.

2. Delayed or full-throttle shifts by causing the transmission to operate in high mode. Line pressures checked under the minimum TV conditions in neutral and drive will be in the range of the full TV pressures if the transmission is in the high pressure mode. The complaint could be described as no upshifts or delayed or sharp upshifts; the closed throttle 3–2 shifts may automatically occur as high as 72 km/h (45 mph).

If the TV cable is adjusted too short or not adjusted at all, it will result in raising the line pressure and shift points. It may also limit the carburetor opening to prevent full-throttle operation.

After performing the recommended TV cable adjustment procedure, some units may still have a slightly delayed minimum throttle shift pattern and drag the 1–2 out to 23–27 km/h (14–17 mph). Tailoring the cable adjustment longer by up to 3.18 mm (1/8 inch) generally will correct the complaint. Care must be taken after this tailoring to be sure that (1) the detent downshifts are still obtainable and

(2) the transmission has not been placed into the high pressure mode.

Diagnosis Procedure

1. Check the transmission oil level and correct as required.

2. Be sure that the engine is operating properly and brakes are not dragging.

3. Check for correct TV cable, according to the parts catalog.

4. Check that the TV cable is connected at both ends.

5. Adjust the TV cable as follows. (Engine off.)

(a) Unlock the snap lock on the TV cable.

(b) Rotate the carburetor lever to the wide-open position.

(c) While holding the carburetor lever at wide-open throttle, push the snap lock flush in order to secure the cable housing.

(d) Return the carburetor lever to the closed position.

6. Check the TV cable for sticking and binding.

7. When the TV cable adjustment is made and locked, the cable housing will extend through the cable snap lock assembly about 1.57–7.95 mm (1/16–5/16 inch).

8. Road test; if OK, return to the owner. If delayed or only full-throttle shifts still occur, proceed with step 9.

9. Install an oil pressure gauge.

(a) The engine is running and at hot idle speed.

Note: Neutral pressure should be equal to or no more than 69 kPa (10 psi) above park pressure.

(b) If neutral pressure is 69 kPa (10 psi) or more than park pressure, proceed to step 10.

10. Remove the oil pan and inspect the throttle lever and bracket assembly on the control valve assembly. Check that the TV exhaust check ball moves up and down as the lifter does. Also be sure the lifter spring holds the lifter rod up against the bottom of the control valve assembly. Make sure that TV plunger is not stuck. Inspect the transmission for correct throttle lever to cable link.

GM 200 Diagnosis Procedure (Courtesy of General Motors Corporation)

Problem	Possible Cause	Possible Condition
No drive in drive range (Check oil pressure.)	A. Low oil level	1. Incorrect level. 2. External leaks.
	B. Manual linkage	1. Maladjusted.
	C. Low or no oil pressure	1. Pump—pressure regulator valve stuck with chips or burrs. 2. Pump drive or driven gear damaged or broken. 3. Cup plugs missing from pump cover. 4. Plugged or restricted oil screen. 5. Oil screen gasket off location. 6. Case-porosity in intake bore.
	D. Control valve assembly	1. Manual valve disconnected from inside detent lever pin. 2. Control valve assembly gasket mispositioned.
	E. Forward clutch	1. Forward clutch does not apply—piston cracked, seals missing, damaged; clutch plates burned; snap ring out of groove. 2. Forward clutch oil seal rings missing or damaged on turbine shaft; leak in feed circuits; pump to case gasket mispositioned or damaged.

GM 200 Diagnosis Procedure (Courtesy of General Motors Corporation) (*cont.*)

Problem	Possible Cause	Possible Condition
		3. Clutch housing check ball stuck or missing. 4. Cup plug leaking or missing in the rear of the turbine shaft in the clutch apply passage. 5. Wrong forward clutch apply ring or wrong number of clutch plates. 6. Feed orifice plugged in turbine shaft. 7. Turbine shaft not welded to housing. 8. Feed passage in turbine shaft intersecting lube passage.
	F. Roller clutch assembly	1. Springs missing in the roller clutch. 2. Rollers galled or missing.
	G. Converter	1. Turbine hub broken loose from turbine.
High or low oil pressures	A. Throttle valve assembly	1. Misadjusted, binding, unhooked, or broken. 2. Wrong throttle lever to cable link. 3. Wrong TV cable.
	B. Manual valve	1. Not connected to inside detent lever pin.
	C. Throttle lever and bracket assembly	1. Binding, unhooked, or mispositioned.
	D. TV exhaust valve lifter rod (causing high line pressure)	1. Bent or binding.
	E. Pressure regulator valve and spring	1. Valve binding. 2. Wrong spring. (Check pressure regulator bore plug leaking.)
	F. Control valve assembly	1. Throttle valve or throttle valve plunger binding. 2. Shift TV valve binding. 3. Intermediate boost valve binding—pressures will be incorrect in intermediate and low ranges only. 4. Reverse boost valve binding —pressures will be incorrect in reverse range only.
	G. Valve body spacer plate	1. Intermediate boost orifice plugged in spacer plate. 2. Reverse boost orifice plugged in the spacer plate.
	H. Pump assembly	1. Low oil pressure. 2. Damaged gears and damaged pump gear pocket. 3. Cup plugs missing. 4. Mismatched pump cover/ pump body.

(*continued*)

GM 200 Diagnosis Procedure (Courtesy of General Motors Corporation) (*cont.*)

Problem	Possible Cause	Possible Condition
		5. Oil pressure control orifice in pump cover plugged, causing high line pressure.
	I. Forward clutch assembly	1. Cup plug missing from turbine shaft.
1-2 and 2-3 shifts—full throttle only	A. Throttle valve cable	1. Binding, unhooked, or broken. 2. Misadjusted. 3. Wrong TV cable.
	B. Throttle lever and bracket assembly	1. Binding or unhooked. 2. Wrong throttle lever to cable link.
	C. TV exhaust valve lifter rod (high-line pressure)	1. Binding, mispositioned, or unhooked. *Note:* Allowing ball 5 to seat causes full TV pressure regardless of throttle valve position.
	D. Control valve assembly	1. Control valve assembly gaskets or spacer plate leaking, damaged, or incorrectly installed. 2. Throttle valve and throttle valve plunger binding.
	E. Case assembly	1. Porosity.
First speed only, no 1-2 upshift	A. Throttle valve cable	1. Misadjusted. 2. Binding, disconnected, or broken. 3. Wrong TV cable.
	B. Governor and governor cover	1. Governor assembly missing. 2. Governor balls missing in governor assembly. 3. Inner governor cover rubber O-ring seal missing or leaking. (*Note:* If the outer governor cover O-ring seal leaks, an external leak will be present along with no upshifts.) 4. Governor shaft seal worn, missing, or damaged. 5. Governor-driven gear stripped. 6. Governor weights binding on pin. 7. Distorted or misaligned governor springs. 8. Governor cover shaft worn or damaged. 9. Governor weight tangs must seat check balls when weights are extended completely outward.
	C. Intermediate servo assembly	1. Servo-to-cover oil seal ring missing or damaged. 2. Porosity in servo cover or piston.

GM 200 Diagnosis Procedure (Courtesy of General Motors Corporation) (*cont.*)

Problem	Possible Cause	Possible Condition
		3. Wrong intermediate band apply pin. (Refer to chassis service manual for proper pin selection.) 4. Incorrect usage of cover and piston.
	D. Throttle lever and bracket assembly	1. Binding or unhooked. 2. Wrong throttle lever to cable link.
	E. Control valve assembly	1. 1–2 shift valve or 1–2 throttle valve stuck in downshift position. 2. Wrong spacer plate gaskets or in wrong position.
	F. Valve body spacer	1. Plugged governor-to-valve-body orifice in spacer plate—feeds governor oil to shift valves. 2. Plugged drive-to-governor orifice in spacer plate.
	G. Case	1. Porosity in case channels or undrilled 2nd speed feed holes. 2. Excessive leakage between case bore and intermediate band apply pin. 3. Intermediate band anchor pin missing or unhooked from band. 4. Broken or missing band.
First and second speeds only, no 2–3 shift	A. Throttle valve cable	1. Misadjusted. 2. Binding, disconnected, or broken. 3. Wrong TV cable.
	B. Governor assembly and governor cover	1. Governor shaft oil seal missing, damaged, or worn. 2. Check balls missing. 3. Governor cover seal damaged.
	C. Intermediate servo assembly (direct clutch accumulator oil passages)	1. Servo-to-case oil seal ring broken or missing on intermediate servo piston. 2. Exhaust hole in case between servo piston seal rings, plugged, or undrilled. 3. Orifice plug missing from case in the intermediate servo bore area.
	D. Throttle lever and bracket assembly	1. Binding or unhooked. 2. Wrong throttle lever to cable link.
	E. Control valve assembly and spacer plate	1. 2–3 shift valve or 2–3 throttle valve stuck in the downshift position.

(*continued*)

GM 200 Diagnosis Procedure (Courtesy of General Motors Corporation) (*cont.*)

Problem	Possible Cause	Possible Condition
		2. Direct clutch feed orifice in spacer plate plugged. 3. Valve body spacer plate or gaskets leaking, damaged, or incorrectly installed. 4. Wrong spacer-plate-to-valve body gaskets.
	F. Case	1. Porosity in case channels. 2. Orifice cup plug in intermediate servo bore area in case missing.
	G. Pump assembly	1. Channels in pump plugged or leaking. 2. Check ball missing or mislocated. 3. Pump-to-case gasket off location. 4. Rear oil seal ring on pump cover leaking or missing. 5. Cup plug missing.
	H. Direct clutch	1. Oil seals missing or damaged on piston. 2. Direct clutch piston or housing cracked. 3. Direct clutch plates damaged or missing. 4. Direct clutch backing plate snap ring out of groove. 5. Wrong piston apply ring. 6. Check ball in housing or piston missing or damaged. 7. Direct clutch release spring guide mislocated, preventing check ball in the piston from seating, causing plates to burn.
Drive in neutral	A. Manual linkage	1. Misadjusted or disconnected.
	B. Forward clutch	1. Clutch does not release. Exhaust check ball sticking. Plates burned together.
	C. Pump	1. Cross leakage in pump passages.
	D. Case	1. Cross leakage to forward clutch passages.
No drive in reverse or slips in reverse (Refer to oil pressure checks.)	A. Throtte valve cable	1. Binding or misadjusted.
	B. Manual linkage	1. Maladjusted.
	C. Pump assembly	1. Pressure regulator valve binding or stuck with chips. 2. Check ball missing or mislocated.

GM 200 Diagnosis Procedure (Courtesy of General Motors Corporation) (*cont.*)

Problem	Possible Cause	Possible Condition
		3. Pump to case gasket mislocated or damaged. 4. Direct clutch feed holes leaking, cross feeding or not drilled. 5. Cup plugs missing. 6. Oil rings damaged or missing.
	D. Throttle valve assembly	1. Throttle valve binding. 2. Shift TV valve binding. 3. Reverse boost valve binding. 4. Low overrun clutch valve binding. (Line pressure readings will be normal.)
	E. Valve body spacer plate and gaskets	1. Orifices plugged. 2. Control valve assembly gaskets in reverse order.
	F. Case assembly	1. Low and reverse clutch housing-to-case seal or cup plug missing, damaged or no feed hole in cup plug. Also, seal may not be seated against the housing. 2. Porosity in case passages—direct clutch feed.
	G. Direct clutch assembly	1. Piston or housing cracked. 2. Piston seals cut or missing. 3. Check ball in housing or piston stuck, leaking, or missing. 4. Plates burned. 5. Incorrect piston. 6. Clutch release spring guide mislocated, preventing check ball in the piston from seating, causing plates to burn.
	H. Low and reverse clutch assembly	1. Piston—cracked, broken, or missing seals. Clutch plates burned. 2. Wrong selective spacer ring.
Slips on 1-2 shift	A. Low oil level	1. Oil aerating. Correct oil level.
	B. Throttle valve cable	1. Not adjusted properly.
	C. Intermediate servo piston and cover assembly	1. Incorrect usage of piston and cover. 2. Wrong band apply pin. (Refer to chassis service manual for selection procedure.) 3. Porosity in piston. 4. Seals damaged or missing. 5. Excessive leakage between band apply pin and case.
	D. Control valve assembly	1. Throttle valve binding. 2. Shift TV valve binding.

(*continued*)

GM 200 Diagnosis Procedure (Courtesy of General Motors Corporation) (*cont.*)

Problem	Possible Cause	Possible Condition
		3. 1-2 accumulator valve sticking, causing low 1-2 accumulator pressure.
	E. 1-2 accumulator piston and spring	1. Piston seal cut or missing. 2. Leak between piston and pin in control valve assembly. 3. Spring broken or missing.
	F. Case	1. Porosity in 2nd clutch passages.
	G. Pump assembly	1. Forward clutch feed hole in the pump cover sleeve not lined up with drive oil in the pump cover casting.
	H. Intermediate band	1. Worn or burned.
	I. Forward clutch assembly	1. Forward clutch feed passage intersecting with lube passage in the turbine shaft.
Rough 1-2 shift	A. Throttle valve cable	1. Not adjusted properly. 2. Binding. 3. Wrong or damaged cable.
	B. Intermediate servo assembly	1. Wrong band apply pin. (See chassis service manual for correct selection.) 2. Servo piston to case oil seal ring damaged or missing.
	C. Control valve assembly	1. TV plunger binding. 2. Throttle valve binding. 3. Shift TV valve binding. 4. 1-2 accumulator valve binding.
	D. 1-2 accumulator piston and spring	1. Oil ring damaged. 2. Piston stuck. 3. Broken or missing spring. 4. Bore damaged.
Slips 2-3 shift	A. Oil level low	1. Correct oil level.
	B. Throttle valve cable	1. Not adjusted properly. 2. Binding.
	C. Intermediate servo assembly	1. Servo-to-case oil seal ring damaged.
	D. Control valve assembly	1. Throttle valve binding.
	E. Spacer plate and gaskets	1. Direct clutch orifice partially blocked in spacer plate. 2. Gaskets mispositioned or damaged.
	F. Case assembly	1. Porosity in direct clutch feed channels in case. 2. Orifice cup plug missing from case in the intermediate servo bore area.
	G. Pump assembly	1. Pump-to-case gasket mispositioned or damaged.

GM 200 Diagnosis Procedure (Courtesy of General Motors Corporation) (*cont.*)

Problem	Possible Cause	Possible Condition
		2. Pump channels cross feeding, leaking, or restricted. 3. Pump cover oil seal rings damaged or missing. 4. Pump cover sleeve loose.
	H. Direct clutch assembly	1. Direct clutch piston or housing cracked. 2. Piston seals cut or missing. 3. Direct clutch plates burned. 4. Check ball in housing and/or piston damaged or missing.
Rough 2-3 shift	A. Throttle valve cable	1. Not adjusted properly. 2. Binding. 3. Wrong or damaged cable.
	B. Intermediate servo assembly	1. Exhaust hole undrilled or plugged between intermediate servo piston seals, not allowing intermediate servo piston to complete its stroke.
	C. Control valve assembly	1. TV plunger binding. 2. Throttle valve binding. 3. Shift TV valve binding. 4. Direct clutch check ball 4 missing or mislocated.
No engine braking in intermediate range, 2nd gear (Check oil pressures.)	A. Intermediate servo assembly	1. Small oil seal on piston cut or missing.
	B. Control valve assembly	1. Intermediate—reverse check ball 3 mislocating or missing. 2. Shift TV check ball 1 mislocated or missing. 3. Intermediate boost valve binding.
	C. Intermediate band	1. Not located on band anchor pin. 2. Broken or burned.
No engine braking in low range, 1st gear (Check oil pressures.)	A. Control valve assembly	1. Low overrun clutch valve binding.
	B. Low and reverse clutch assembly	*Note:* No reverse should also be a complaint with any of the 3 following conditions.) 1. Piston seals broken or missing. 2. Porosity in piston or housing. 3. Clutch housing snap ring out of case.
	C. Case	1. Low and reverse clutch housing-to-case seal or cup plug missing, damaged, or no feed hole in cup plug. Also, seal may not be seated against the housing.

(*continued*)

GM 200 Diagnosis Procedure (Courtesy of General Motors Corporation) (*cont.*)

Problem	Possible Cause	Possible Condition
No part throttle or detent downshifts (Check oil pressures.)	A. Throttle valve cable	1. Improperly adjusted. 2. Wrong or damaged cable.
	B. Throttle lever and bracket assembly	1. Wrong throttle lever to cable link.
	C. Control valve assembly	1. Passages not open in 2–3 TV bushing. 2. Throttle valve bushing passages not open. 3. Shift TV valve binding. 4. Throttle valve binding.
	D. Control valve assembly	1. Mispositioned or damaged.
	E. Valve body space plate	1. Hole plugged or undrilled.
Low or high shift points (Check oil pressures.)	A. Throttle valve cable	1. Binding, disconnected, or misadjusted. 2. Wrong or damaged cable.
	B. Throttle lever and bracket assembly	1. Binding, unhooked, or loose at mounting valve body bolt. 2. Not positioned at the throttle valve plunger bushing pin. 3. Wrong throttle lever to cable link.
	C. Governor shaft to cover seal ring	1. Broken or missing.
	D. Governor cover O-ring	1. Broken or missing. *Note:* The outer ring will leak externally and the inner ring internally.
	E. Pump assembly	1. Pressure regulator valve binding or stuck.
	F. Control valve assembly	1. TV exhaust valve lifter binding, unhooked, or missing. 2. 1–2 or 2–3 valves sticky. 3. Throttle valve plunger binding. 4. Throttle valve binding. 5. Shift TV valve binding. 6. TV shift check ball 1 missing or mislocated.
	G. Valve body spacer plate or gaskets	1. Damaged or wrong gaskets.
	H. Case	1. Porosity
Won't hold in park	A. Manual linkage	1. Maladjusted.
	B. Manual detent roller and spring assembly	1. Loose bolt that holds roller assembly to control valve assembly. 2. Pin or roller damaged or mispositioned or missing.
	C. Inside detent lever and pin assembly	1. Nut loose 2. Hole in lever worn or damaged.
	D. Parking lock components	1. Park pawl binding in case. 2. Actuator rod or plunger damaged.

GM 200 Diagnosis Procedure (Courtesy of General Motors Corporation) (*cont.*)

Problem	Possible Cause	Possible Condition
		3. Parking pawl broken. 4. Parking bracket, loose, or damaged.
Transmission noisy	A. Pump noise	1. Oil level setting low. 2. Cavitation due to plugged screen, porosity in intake circuit, water in oil. 3. Pump gears—damaged.
	B. Gear noise	1. Transmission grounded to body. 2. Roller bearings worn or damaged.
Thumping or clunking sound at 1-5 mph—drive or intermediate range	A. Governor assembly	1. Governor springs damaged, mispositioned, or tilted.

GM 200-4R Diagnosis Procedure (Courtesy of General Motors Corporation)

Preliminary Check Procedure

1. Check transmission oil level.
2. Check and adjust TV cable.
3. Check outside manual linkage and correct.
4. Check engine tune.
5. Install the oil pressure gauge.
6. Connect the tachometer to the engine.

Check oil pressures in the following manner.

Minimum TV Line Pressure Check. Set the TV cable to specification; with the brakes applied, take the line pressure readings in the ranges and at the engine rpm's indicated in the service manual.

Full TV Line Pressure Check. Full TV line pressure readings are obtained by tying or holding the TV cable to the full extent of its travel; with the brakes applied, take the line pressure readings in the ranges and at the engine rpm's indicated in the following chart.

Notice: Total running time for this combination not to exceed two minutes.

Caution: Brakes must be applied at all times.

Line pressure is basically controlled by pump output and the pressure regulator valve. In addition, line pressure is boosted in reverse, second, and low by the reverse boost valve.

Also, in the neutral, drive, intermediate, and reverse positions of the selector lever, the line pressure should increase with throttle opening because of the TV system. The TV system is controlled by the TV cable, the throttle lever and bracket assembly, and the TV link, as well as the control valve assembly.

Possible Points of Oil Leak

1. Transmission oil pan leak
   (a) Attaching bolts not correctly torqued
   (b) Improperly installed or damaged pan gasket
   (c) Oil pan gasket mounting face not flat
2. Case leak
   (a) Filler pipe multilip seal damaged or missing; misposition of filler pipe bracket to engine "loading" one side of seal
   (b) TV cable multilip seal missing, damaged, or improperly installed
   (c) Rear seal assembly damaged or improperly installed
   (d) Speedometer driven gear O-ring damaged
   (e) Manual shaft lip seal damaged or improperly installed
   (f) Line pressure tap plug
   (g) Fourth-clutch pressure tap plug
   (h) Vent pipe
   (i) Porous casting
   (j) Intermediate servo O-rings damaged
3. Leak at the front of the transmission
   (a) Front pump seal leaks

- Seal lip cut (check the converter hub for nicks, etc.)
- Bushing moved forward and damaged
- Garter spring missing from the seal

(b) Front pump attaching bolts loose or bolt seal damaged or missing

(c) Front pump housing O-ring damaged or cut

(d) Converter leak in the weld area

(e) Porous casting (pump or case)

4. Oil coming out the vent pipe

(a) Transmission overfilled

(b) Water in oil

(c) Foreign matter between the pump and case or between the pump cover and body

(d) Case porous; front pump cover mounting face shy of stock near breather

(e) Pump-to-case gasket mispositioned

(f) Incorrect dipstick

(g) Pump shy of stock on mounting faces, porous casting, breather hole plugged in pump cover

No Drive in Drive Range (Install Pressure Gauge)

1. Low oil level

(a) Incorrect level

(b) External leaks

2. Manual linkage

(a) Misadjusted

3. Low oil pressure

(a) Plugged or restricted oil filter

(b) Cut or missing oil filter O-ring seals

(c) Pump assembly–pressure regulator stuck

(d) Pump rotor–tangs damaged by converter

(e) Porosity in oil filter to pump intake bore

4. Overdrive unit

(a) Springs missing in the roller clutch

(b) Rollers galled or missing

5. Forward clutch

(a) Forward clutch not applying—piston cracked, seals missing or damaged; clutch plates burned; snap ring out of groove

(b) Forward clutch oil seal rings missing or damaged on turbine shaft; leak in feed circuits; pump-to-case gasket mispositioned or damaged

(c) Clutch housing check ball stuck or missing

(d) Cup plug leaking or missing in the rear of the forward clutch shaft in the clutch apply passage

6. Low and reverse roller clutch

(a) Springs missing in the roller clutch

(b) Rollers galled or missing

High or Low Oil Pressure (Refer to Oil Pressure Checks)

1. Throttle valve cable

(a) Misadjusted, binding, unhooked, broken, or wrong link

2. Throttle valve assembly damaged or leaking

(a) Throttle lever and bracket assembly binding, unhooked, or mispositioned

(b) Throttle valve or plunger valve binding

3. Pressure regulator valve

(a) Valve binding

4. TV boost valve

(a) Valve binding

(b) Wrong valve, causing low oil pressure only

5. Reverse boost valve

(a) Valve binding

6. Manual valve

(a) Unhooked manual valve

(b) Mispositioned manual valve

7. Pressure relief valve

(a) Ball missing

(b) Spring damaged

8. Pump

(a) Slide stuck

(b) Pump slide seal damaged or missing

(c) Decrease air bleed orifice missing or damaged, causing high oil pressure

(d) Decrease air bleed orifice plugged, causing low oil pressure

9. TV limit valve

(a) Valve binding

10. Line bias valve

(a) Valve binding in open position, causing high oil pressure

(b) Valve binding in closed position, causing low oil pressure

The control valve assembly spacer plate and case

should be closely inspected for correct orifices and passages.

1-2 Shift—Full Throttle Only

1. Throttle valve cable
   (a) Binding, unhooked, or broken
   (b) Misadjusted
2. Throttle lever and bracket assembly
   (a) Binding or unhooked
3. TV exhaust ball lifter or ball 5
   (a) Binding, mispositioned, or unhooked

Allowing ball 5 to seal causes full TV pressure regardless of throttle valve position.

4. Throttle valve and plunger
   (a) Binding
5. Control valve assembly
   (a) Valve body gaskets leaking, damaged, or incorrectly installed
6. Case assembly
   (a) Porosity

First Speed Only, No 1-2 Shift

1. Governor and governor feed passsages
   (a) Plugged governor oil-feed orifice in spacer plate
   (b) Governor ball or balls missing in governor assembly
   (c) Inner governor cover rubber O-ring seal missing or leaking
   (d) Governor shaft seal missing or damaged
   (e) Governor-driven gear stripped
   (f) Governor weights binding on pin
   (g) Governor-driven gear not engaged with governor shaft
2. Control valve assembly
   (a) 1–2 shift, low 1st/detent, or 1–2 throttle valve stuck in downshift position
   (b) Spacer plate gaskets in wrong position
3. Case
   (a) Porosity in case channels or undrilled second oil-feed hole
   (b) Excessive leakage between case bore and intermediate band apply rings
   (c) Intermediate band anchor pin missing or unhooked from band
   (d) Broken or missing band
4. Intermediate servo assembly
   (a) Servo cover oil seal ring missing
   (b) Porosity in servo; cover, inner piston, or outer piston
   (c) Wrong intermediate band apply pin
   (d) Incorrect usage of cover and piston
5. 1–2 accumulator
   (a) 1–2 accumulator housing bolts loose
   (b) 1–2 accumulator housing face damaged
   (c) Missing or damaged accumulator plate

First and Second Speeds Only, No 2-3 Shift

1. Control valve assembly and spacer plate
   (a) 2–3 shift valve or 2–3 throttle valve stuck in the downshift position
   (b) Valve body gaskets leaking, damaged, or incorrectly installed
   (c) Reverse/third check ball not seating, damaged, or missing
2. Case
   (a) Porosity in case channels
3. Center support
   (a) Direct clutch feed passage in the center support plugged or not drilled through
   (b) Steel oil seal rings on center support damaged
4. Direct clutch
   (a) Inner oil seal ring missing or damaged on piston
   (b) Center oil seal ring missing or damaged on direct clutch hub
   (c) Check ball or retainer damaged or missing from direct clutch piston
   (d) Direct clutch piston or housing damaged or missing
   (e) Direct clutch plates damaged or missing
   (f) Direct clutch backing plate snap ring out of groove
   (g) Release spring guide mislocated, preventing piston check ball from seating in retainer
5. Intermediate servo assembly (third-clutch accumulator oil passages)
   (a) Servo-to-case oil seal ring broken or missing on intermediate servo piston
   (b) Intermediate servo or capsule missing or damaged

(c) Exhaust hole in case between servo piston seal rings plugged or undrilled

(d) Bleed orifice cup plug missing from intermediate servo pocket in case

No Drive in Reverse or Slips in Reverse (Install Pressure Gauge)

1. Throttle valve cable binding or misadjusted
2. Manual linkage misadjusted
3. Throttle valve binding
4. TV limit valve binding
5. Line bias valve binding
6. Reverse boost valve binding in pressure regulator bore
7. Reverse/third or low/reverse check ball missing or seat in spacer plate damaged
8. Reverse clutch
   (a) Piston cracked or missing inner or outer seals; clutch plates burned
   (b) Reverse oil seal in case missing or damaged
   (c) Missing clutch plate or wave plate
9. Center support
   (a) Center support attaching bolts loose or missing
   (b) Passages blocked or not drilled
   (c) Porosity
10. Direct clutch housing
    (a) Piston or housing cracked
    (b) Inner or outer piston seal missing or damaged
    (c) Check ball in either the direct clutch housing or the piston missing or damaged
    (d) Plates burned
11. Low/reverse overrun clutch orifice in spacer plate plugged

Drive in Neutral

1. Manual linkage
   (a) Misadjusted or disconnected
2. Forward clutch
   (a) Clutch does not release
   (b) Exhaust check ball sticking
   (c) Plates burned together
3. Case
   (a) Cross leakage to forward clutch passage (D4)

Slips 1-2 Shift

1. Low oil level
   (a) Correct oil level
2. Spacer plate and gaskets
   (a) Gaskets damaged or incorrectly installed
3. Accumulator valve
   (a) Valve sticking in valve body, causing low 1–2 accumulator pressure
   (b) Weak or missing spring
4. 1–2 accumulator piston
   (a) Seal leaking; spring broken or missing
   (b) Leak between piston and pin
   (c) 1–2 accumulator piston binding
   (d) 1–2 accumulator piston bore damaged
5. Intermediate band apply pin
   (a) Wrong selection of apply pin and case
   (b) Excessive leakage between apply pin and case
   (c) Apply pin feed hole not completely drilled
6. Intermediate servo assembly
   (a) Porosity in piston
   (b) Cover to servo oil seal ring damaged or missing
   (c) Leak between servo apply pin and case
7. Throttle valve cable not properly adjusted
8. Throttle valve binding, causing low TV pressure
9. TV limit valve binding
10. Line bias valve sticking, causing low line pressure
11. Intermediate band worn or burned
12. Case porosity in second-clutch passage

Rough 1-2 Shift

1. Throtle valve cable
   (a) Not adjusted properly
   (b) Binding
2. Throttle valve or TV plunger
   (a) Binding
3. TV limit valve binding

4. Accumulator valve binding
5. Line bias valve binding
6. Intermediate servo assembly
   (a) Wrong selection apply pin
   (b) Servo piston-to-case oil seal ring damaged or missing
   (c) Bleed cup plug missing in case
7. 1–2 accumulator
   (a) Oil ring damaged
   (b) Piston stuck
   (c) Broken or missing spring
   (d) Bore damaged
8. 1–2 shift check ball 8 missing or sticking

Slips 2–3 Shift

1. Oil level low
   (a) Correct oil level
2. Throttle valve cable not adjusted properly
3. Throttle valve binding
4. Spacer plate and gaskets
   (a) Direct clutch orifice partially blocked in spacer plate
   (b) Gaskets mispositioned or damaged
5. Intermediate servo assembly
   (a) Servo-to-case oil seal ring damaged or missing
   (b) Piston or servo bore damaged
   (c) Intermediate servo orifice bleed cup plug in case missing
   (d) Porosity in the case in the servo bore area
6. Direct clutch feed
   (a) Porosity in direct clutch feed channels in case
   (b) Case-to-support bolts not tight, causing leakage
   (c) Direct clutch piston or housing cracked
   (d) Piston seals cut or missing
   (e) Direct clutch plates burned
   (f) Check ball in piston or housing missing, damaged, or leaking
   (g) Check ball capsule damaged
   (h) Release spring guide mislocated, preventing check ball from seating in piston
7. Center support
   (a) Channels cross feeding, leaking, or restricted
   (b) Oil seal rings damaged or missing

Rough 2–3 Shift

1. Throttle valve cable mispositioned or missing
2. Throttle valve and plunger
   (a) TV plunger binding
   (b) Throttle valve binding
3. TV limit valve binding
4. Intermediate servo assembly
   (a) Exhaust hole undrilled or plugged between intermediate servo piston seals, not allowing intermediate servo piston to complete its stroke
5. 3–2 Exhaust check ball 4 missing or mispositioned
6. 3rd accumulator check ball 2 missing or mispositioned

Slips 3–4 Shift

1. Low oil level
2. Control valve assembly and spacer plate
   (a) Gaskets or spacer plate damaged or incorrectly installed
   (b) Accumulator valve sticking causing low 3–4 accumulator pressure
   (c) Weak or missing accumulator valve spring
3. 3–4 accumulator
   (a) Piston stuck
   (b) Bore damaged
   (c) Oil ring damaged
4. Center support
   (a) Porosity
   (b) Center support attaching bolts loose
   (c) Fourth-clutch piston surface damaged
   (d) Fourth-clutch piston seals damaged
   (e) Proper plate usage (see clutch plate usage chart)
   (f) Fourth-clutch plates burned
5. Case
   (a) Porosity
   (b) 1–2 accumulator housing bolts loose
   (c) 3–4 accumulator piston seal damaged
   (d) 3–4 accumulator leaking between the piston and pin
   (e) 3–4 accumulator bore damaged

Rough 3-4 Shift

1. Throttle valve cable mispositioned or missing
2. Throttle valve and plunger
   (a) TV plunger binding
   (b) Throttle valve binding
3. TV limit valve binding
4. 3–4 accumulator
   (a) Piston stuck
   (b) Bore damaged
5. Fourth-clutch piston binding

No Converter Clutch Apply

1. Electrical problem
   (a) 12 volts needed by the solenoid to engage
   (b) Defective solenoid
   (c) Damaged electrical connector
   (d) Defective pressure switch
   (e) Wire grounded
2. Control valve assembly and spacer plate
   (a) Converter clutch shift valve or throttle valve stuck
3. Pump assembly
   (a) Orifice plugged for converter signal oil in the pump
   (b) O-ring damaged or missing on the solenoid
   (c) Orificed cup plug missing in the oil cooler passage in the pump
   (d) Pump-to-case gasket damaged or mispositioned
   (e) Converter clutch apply valve stuck
   (f) Cup plug missing from the apply passage

Rough Converter Clutch Apply

1. Converter clutch pressure plate damaged
2. Check ball in end of turbine shaft damaged or missing

No Converter Clutch Release

1. Converter clutch apply valve stuck
2. Damaged converter
3. Cup plug in the release passage missing in the pump
4. Seal on the end of the turbine shaft missing or damaged
5. Hole not drilled through in the turbine shaft

First, Second, and Third Speed Only, No 3-4 Shift

Install pressure gauge to fourth-clutch pressure tap.

1. Control valve assembly and spacer plate
   (a) 3–4 shift valve or 3–4 throttle valve stuck
   (b) Orifice in spacer plate plugged
2. Center support
   (a) Oil passages plugged or not drilled
   (b) Center support attaching bolts loose or missing
   (c) Fourth-clutch piston cracked or damaged
   (d) Fourth-clutch piston seals damaged, missing, or improperly assembled
   (e) Proper plate usage (see clutch plate usage chart in manual)
   (f) Fourth-clutch plates burned
   (g) Overrun clutch plates binding
3. Case
   (a) Porosity
   (b) Orificed cup plug missing in 3–4 accumulator passage in the case
   (c) Leakage between the accumulator piston and pin
   (d) 3–4 accumulator bore damaged

No Engine Braking in Manual Low—First Gear

1. Control valve assembly and spacer plate
   (a) Manual linkage misadjusted
   (b) D–3 orifice in the spacer plate plugged
   (c) Valve body gaskets leaking, damaged, or incorrectly installed
   (d) D–2 oil pipe leaking or out of position
   (e) Low overrun clutch valve binding in valve body
   (f) Low/reverse check ball (10) mispositioned or missing
   (g) Low/detent check ball (9) mispositioned or missing
   (h) Low/reverse overrun clutch orifice in the spacer plate plugged
   (i) PT/D–3 check ball (3) mispositioned or missing
2. Turbine shaft and overrun clutch

No manual 3rd or 2nd should also be a complaint with the following:

(a) D–3 oil passage not drilled or plugged in turbine shaft

(b) D–3 oil passage not drilled through in overrun clutch hub

(c) Oil seals missing or damaged in the overrun clutch piston

(d) Overrun clutches burned

(e) Overrun clutch backing plate snap ring out of groove

3. Case

(a) Porosity

4. Low/reverse clutch assembly

No reverse should also be a complaint with any of the following conditions:

(a) Piston seals broken or missing

(b) Clutch housing snap ring out of case

(c) Piston or housing cracked or porous

(d) Cup plug or rubber seal missing or damaged between the case and low/reverse clutch housing

No Engine Braking in Manual Second—Second Gear

1. Control valve assembly and spacer plate

(a) Manual linkage misadjusted

(b) Valve body gaskets leaking, damaged, or incorrectly installed

(c) D–2 oil pipe leaking or out of position

(d) D–3 orifice in the spacer plate plugged

(e) PT/D–3 check ball (3) mispositioned or missing

2. Case

(a) Porosity

3. Intermediate servo assembly

(a) Servo cover-to-case oil seal ring missing or damaged

4. Intermediate band

(a) Off anchor pin

(b) Broken or burned

5. Turbine shaft and overrun clutch

(a) D–3 oil passage not drilled through in the overrun clutch hub

(b) Oil seals missing or damaged in the overrun clutch piston

(c) D–3 oil hole not drilled or plugged in the turbine shaft

(d) Overrun clutches burned

(e) Overrun clutch backing plate snap ring out of groove

No Engine Braking in Manual Third—Third Gear

1. Control valve assembly and spacer plate

(a) Manual linkage misadjusted

(b) D–3 orifice in the spacer plate plugged

(c) Valve body gaskets leaking, damaged, or incorrectly installed

(d) PT/D–3 check ball (3) mispositioned or missing

2. Turbine shaft and overrun clutch

(a) D–3 oil passage not drilled or plugged in turbine shaft

(b) D–3 oil hole not drilled through in overrun clutch hub

(c) Oil seals missing or damaged in the overrun clutch piston

(d) Overrun clutches burned

(e) Overrun clutch backing plate snap ring out of groove

Will Not Hold in Park

1. Manual linkage misadjusted

2. Internal linkage

(a) Parking pawl binding in the case

(b) Actuator rod, spring, or plunger damaged

(c) Parking pawl broken

(d) Parking bracket loose or damaged

(e) Manual shaft to case pin missing or mispositioned

3. Inside detent lever and pin assembly

(a) Nut loose

(b) Hole in the lever worn or damaged

4. Manual detent roller and spring assembly

(a) Loose bolt holding roller assembly to valve body

(b) Pin or roller damaged, mispositioned, or missing

No Part-Throttle Downshifts (Install Pressure Gauge)

1. Throttle valve binding

2. TV limit valve binding

3. Spacer plate hole plugged or undrilled

4. Valve body gaskets mispositioned or damaged
5. TV modulator downshift valve stuck
6. Throttle valve cable improperly set

No Part-Throttle 4–3 Downshift on Selected Models with a Part-Throttle Passage in the Throttle Plunger Bushing

1. Throttle plunger bushing passages not open
2. 3–4 throttle valve bushing passages not open
3. PT/D–3 check ball (3) mispositioned or missing
4. Valve body gaskets mispositioned or damaged
5. Throttle valve cable improperly set
6. TV limit valve binding

Low or High Shift Points (Install Pressure Gauge to Line Pressure Tap)

1. Throttle valve cable binding or misadjusted
2. TV limit valve binding
3. Throttle valve binding
4. TV modulator upshift valve binding
5. TV modulator downshift valve binding
6. Valve body gaskets mispositioned, leaking, or damaged
7. Throttle valve plunger binding
8. 1–2, 2–3, or 3–4 throttle valves binding in bushings
9. Pressure regulator valve binding
10. TV exhaust ball (5) and lifter mispositioned, unhooked, or missing
11. Throttle lever and bracket assembly
    (a) Binding, unhooked, or loose at mounting valve body bolt
    (b) Not positioned at the throttle valve plunger bushing pin locator
12. Governor shaft to cover seal ring broken or missing
13. Governor cover gasket broken or missing
14. Case porosity

Trouble Diagnosis

Refer to the diagnosis chart to determine a possible cause of a transmission problem. Additional diagnosis of a malfunction follows.

No Drive in Drive Range (Install Pressure Gauge)

- Low fluid level. Correct the level and check for external leaks or defective vacuum modulator (a leaking diaphragm will evacuate fluid from unit).
- Manual linkage misadjusted. Correct alignment to the manual lever shift quadrant is essential.
- Low fluid pressure. (Refer to the section on low line pressure).
- Forward clutch
  1. Forward clutch not applying—piston cracked; burned clutch plates.
  2. Pump feed circuit-to-forward clutch oil seal rings missing or broken on pump cover; leak in feed circuits, pump-to-case gasket mispositioned or damaged; clutch drum ball check stud or missing.
- Low and reverse roller clutch assembly. Broken spring, damaged cage, or installed backwards.

High or Low Fluid Pressure

Refer to the fluid pressure checks.

High Line Pressure

- Vacuum leak
  1. Vacuum line disconnected
  2. Leak in the line from the engine to the modulator
  3. Improper engine vacuum
  4. Leak in a vacuum-operated accessory (hoses, vacuum advance, etc.)
- Modulator
  1. Stuck modulator valve
  2. Water in the modulator
  3. Damaged, not operating properly
- Detent system—detent valve or cable stuck in the detent position
- Valve body
  1. Pressure regulator or boost valve stuck
  2. Boost valve sleeve broken or defective
  3. Incorrect pressure regulator valve spring

Low Line Pressure

- Low transmission fluid level
- Defective vacuum modulator assembly
- Strainer assembly
  1. Blocked or restricted
  2. Gasket omitted or damaged
- Oil pump

1. Gear clearance damaged, or worn; gear installed backward
2. Pump-to-case gasket mispositioned
3. Defective pump body or cover

- Valve body
  1. Pressure regulator or boost valve stuck
  2. Pressure regulator valve spring too weak
- Internal circuit leaks
  1. Forward clutch leak (pressure low in the drive range, pressure normal in neutral and reverse)
     a) Check pump oil seal rings
     b) Check forward clutch seals
  2. Direct clutch leak (pressure low in reverse, pressure normal in other ranges)
     a) Check the direct clutch outer seal
     b) Check 1–2 accumulator piston and ring for damage or if missing
- Case assembly
  1. Check ball missing from cored passages in the case face

1-2 Shift—Full Throttle Only

- Detent valve sticking or linkage misadjusted
- Vacuum leak; vacuum line or fittings leaking
- Control valve assembly
  1. Valve body gaskets leaking, damaged, or incorrectly installed
  2. Detent valve train stuck
  3. 1–2 valve stuck closed (in downshifted position)
- Case assembly (refer to case porosity repair)

First Speed Only, No 1-2 Shift

- Detent (downshift) cable binding
- Governor assembly
  1. Governor valve sticking
  2. Driven gear loose, damaged, or worn (check for the pin in the case and the length of the pin showing; also check the output shaft drive gear for nicks or a rough finish if the driven gear shows damage)
- Control valve assembly
  1. Valve body gaskets leaking, damaged, or incorrectly installed
  2. Governor feed channels blocked
  3. 1–2 shift valve train stuck closed (in downshifted position)
- Intermediate band
  1. Servo piston seals missing, improperly installed, or broken
  2. Band improperly adjusted; screw loose
  3. Servo apply rod broken
- Case
  1. Porosity between channels
  2. Governor feed channel blocked; governor bore scored or worn, allowing cross pressure leak

First and Second Speeds Only, No 2-3 Shift

- Control valve assembly
  1. Valve body gaskets leaking, damaged, or incorrectly installed
  2. 2–3 shift valve train stuck closed (in downshifted position)
- Direct clutch
  1. Pump hub and direct-clutch oil seal rings broken or missing
  2. Clutch piston seals missing, improperly assembled, or cut
  3. Clutch plates burned

No First-Speed Starts in Second Speed (Locks up to L Range)

- Intermediate band
  1. Band adjustment too tight
  2. 1–2 shift valve stuck in upshifted position

Drive in Neutral

- Manual linkage misadjusted (correct alignment in manual lever shaft quadrant is essential)
- Internal linkage—manual valve disconnected or end broken
- Oil pump—line pressure leaking into forward clutch apply passage
- Forward clutch—incorrect clutch plate usage or burned clutches

No Motion in Reverse or Slips in Reverse (Install Pressure Gauge)

- Low fluid level; add fluid
- Manual linkage misadjusted (correct alignment in manual lever shift quadrant is essential)

- Low oil pressure (refer to the section on low line pressure)
- Control valve assembly
  1. Valve body gaskets leaking, damaged, or incorrectly installed
  2. 2–3 shift valve train stuck open (in upshifted position)
- Intermediate servo piston or pin stuck so that intermediate band is applied
- Low and reverse clutch—piston outer seal damaged or missing
- Direct clutch
  1. Outer seal damaged or missing
  2. Clutch plates burned
- Forward clutch not releasing, causing drive in neutral

Slips in All Ranges or Slips on Start (Install Pressure Gauge)

- Low fluid level—add fluid
- Low fluid pressure (refer to the section on low line pressure)
- Forward clutch
  1. Clutch plates burned
  2. Pump cover oil seal rings broken or worn
- Case—cross leaks or porosity

Slipping 1–2 Shift (Install Pressure Gauge)

- Low fluid level—add fluid
- Low fluid pressure (refer to the section on low line pressure)
- 1–2 accumulator—oil ring damaged, missing, or case bore damaged
- Pump-to-case gasket mispositioned or damaged
- Intermediate band
  1. Piston seals damaged or missing
  2. Intermediate band burned
- Case—porosity between channels
- Improper intermediate band adjustment

Slipping 2–3 Shift (Install Pressure Gauge)

- Low fluid level—add fluid
- Low oil pressure (refer to the section on low line pressure)
- Direct clutch
  a) Piston seals leaks, damaged, or missing
  b) Clutch plates burned
  c) Inspect for proper number and type of clutch plates
- Case (refer to the section on case porosity repair)

Rough 1–2 Shift (Install Pressure Gauge)

- High fluid pressure (refer to the section on high line pressure)
- 1–2 accumulator
  a) Oil rings damaged
  b) Piston stuck
  c) Broken or improper spring
  d) Bore damaged
- Case
  a) Check for correct number and location of check balls
  b) Porosity between channels
- Intermediate band
  a) Improper adjustment
  b) Improper or broken servo spring

Rough 2–3 Shift (Install Pressure Gauge)

- High fluid pressure (refer to the section on high line pressure)
- 2–3 accumulator
  1. Oil ring damaged
  2. Piston stuck
  3. Broken or missing spring
  4. Piston bore damaged

No Engine Braking in I (Install Pressure Gauge)

- Low fluid pressure; pressure regulator or boost valve stuck
- Intermediate servo
  1. Servo oil rings or bores leaking or damaged
  2. Servo piston stuck or cocked
- Intermediate band broken or burned (look for the cause), not engaged on servo pin or improper adjustment

No Engine Braking in L (Install Pressure Gauge)

- Low fluid pressure; pressure regulator or boost valves stuck
- Manual low control valve assembly stuck
- Low and reverse clutch—piston inner seal damaged or missing

No Part-Throttle Downshift (Install Pressure Gauge)

- Fluid pressure in vacuum modulator assembly, modulator valve, or pressure regulator valve train (other malfunctions may also be noticed)
- Detent valve and linkage stuck, disconnected, or broken
- 2–3 shift valve stuck

No Detent (Wide-Open Throttle) Downshift

- Detent cable or retainer not adjusted properly
- Detent cable disconnected at transmission or throttle linkage
- Valve body
  1. Detent valve sticks
  2. Detent regulator valve sticks
  3. Incorrect spacer plate or gasket

High or Low Shift Points (Install Pressure Gauge)

- Fluid pressure
  1. Engine vacuum (check at the transmission end of the modulator pipe)
  2. Check vacuum line connections at the engine and transmission
  3. Vacuum modulator assembly and valve and pressure regulator valve train
- Governor
  1. Valve sticking
  2. Feed holes restricted or leaking
- Detent valve and linkage stuck open, causing high shift points
- Control valve assembly
  1. 1–2 Shift valve train sticking
  2. 2–3 shift valve train sticking
- Case (refer to the section on case porosity repair)

Won't Hold in Park (P)

- Manual linkage misadjusted (correct alignment in manual lever shift quadrant is essential)
- Internal linkage
  1. Inner lever and actuating rod assembly defective or improperly installed
  2. Parking pawl broken or inoperative
  3. Parking lock bracket loose, burred or rough edges, or incorrectly installed
  4. Parking pawl disengaging spring missing, broken, or incorrectly hooked

Transmission Noisy

Note: Before checking the transmission, make certain that noise is not coming from the water pump, generator, power steering, etc. These components can be isolated by removing the proper belt and running the engine not more than two minutes at one time.

Park (P), Neutral (N), and All Driving Ranges

- Pump cavitation
  1. Fluid level
  2. Plugged or restricted strainer
  3. Strainer-to-valve body gasket damaged
  4. Porosity in valve body intake area
  5. Water in fluid
  6. Porosity or voids at the transmission case (pump face) intake port
  7. Pump-to-case gasket off location
- Pump assembly
  1. Gears damaged
  2. Driving gear assembled backward
  3. Crescent interference
  4. Oil seal rings damaged or worn
- Converter
  1. Loose flexplate-to-converter bolts
  2. Converter damage
  3. Water in fluid, causing whine

First, Second, or Reverse Gear

- Planetary gearset
  1. Gears or thrust bearings damaged
  2. Input or output ring gear damaged

During Acceleration—Any Gear

- Transmission grounded to underbody
- Motor mounts loose or broken

Squeal at Low Vehicle Speed

- Speedometer-driven gear shaft seal requiring lubrication or replacement

GM 325-4L Diagnosis Procedure (Courtesy of General Motors Corporation)

1. Check transmission oil level.
2. Check and adjust TV cable.
3. Check outside manual linkage and correct.
4. Check engine tune.

5. Install an oil pressure gauge.
6. Connect the tachometer to the engine.

Check oil pressures in the following manner.

Minimum TV Line Pressure Check. Set the TV cable to specification; with the brakes applied, take the line pressure readings in the ranges and at the engine rpm's indicated in the following chart.

Full TV Line Pressure Check. Full TV line pressure readings are obtained by tying or holding the TV cable to the full extent of its travel; with the brakes applied, take the line pressure readings in the ranges and at the engine rpm's indicated in the following chart.

Note: Total running time for this combination must not exceed 2 minutes.

Caution: Brakes must be applied at all times.

Line pressure is basically controlled by pump output and the pressure regulator valve. In addition, line pressure is boosted in reverse, intermediate, and low by the reverse boost valve.

Also, in the neutral, drive, intermediate, and reverse positions of the selector lever, the line pressure should increase with throttle opening because of the TV system. The TV system is controlled by the TV cable, the throttle lever and bracket assembly, and the TV link, as well as the control valve assembly.

Check Fluid Level

1. Engine running
2. Vehicle on a level surface
3. Brakes applied
4. Selector lever moved through all ranges
5. Transmission placed in park
6. Oil level checked
7. If oil low, possible causes checked

The oil level, at room temperature of approximately 70°F (21°C), should be approximately 6 mm (1/4 inch) below the add mark. The oil level should be at the full mark at normal operating temperature, 200°F (90°C). This temperature is obtained after at least 15 miles (24 km) of expressway driving or equivalent city driving. Also, at normal operating temperature, the oil will heat the gauge end of the dipstick to a degree where the average person cannot grasp it firmly with a bare hand without discomfort.

Maintain the oil level at the full mark at normal operating temperature.

Note: Do not overfill the transmission, as this may cause foaming and loss of oil through the vent pipe.

Adjust Manual Linkage

The transmission manual linkage must be adjusted so that the indicator quadrant and stops correspond with the transmission detents. If the linkage is not adjusted properly, an internal leak could occur; this could cause a clutch or band to slip.

Refer to the car division shop manual for the manual linkage adjustment procedure.

Caution: If a manual linkage adjustment is made, the associated neutral start switch should be adjusted. If necessary, the neutral start switch should be adjusted so that the engine will start in park and neutral positions only. With the selector lever in the park position, the parking pawl should freely engage and prevent the car from rolling. Transmission, vehicle, or personal injury may occur if not properly adjusted.

Diagnosis Procedure

1. Check the transmission oil level and correct as required.

2. Be sure that the engine is operating properly and brakes are not dragging.

3. Check for correct TV cable, according to the parts catalog.

Model	AJ		AL, OK		AM, BE		OE	
Range	kPa	psi	kPa	psi	kPa	psi	kPa	psi
Park—neutral	—	—	—	—	—	—	—	—
N, D4, D3	438- 529	64- 77	438- 529	64- 77	438- 529	64- 77	380- 450	55- 65
D2, D1	875-1055	126-153	875-1055	126-153	1020-1231	148-179	885-1045	128-152
Reverse	625- 754	91-109	625- 754	91-109	871-1051	126-152	755- 892	109-129
Park—neutral	—	—	—	—	—	—	—	—
N, D4, D3	891-1066	129-155	891-1066	129-155	777- 931	113-135	833- 986	121-130
D2, D1	875-1055	126-153	875-1055	126-153	1020-1231	148-179	885-1045	128-152
Reverse	1270-1519	184-220	1270-1519	184-220	1543-1848	224-268	1654-1957	239-284

psi = kPa ÷ 6.895

4. Check that the TV cable is connected at both ends.

5. Check the TV cable for sticking or binding:

(a) The engine is running at idle speed.

(b) The transmission selector is in neutral with vehicle brakes set.

(c) Pull the TV cable to full-throttle position.

(d) Release the cable. The cable should return to the closed-throttle position. If the TV cable sticks, it may be caused by one or more of the following:

(1) Sharp bends or a damaged TV cable housing. Correct by rerouting the cable or replacing it if required.

(2) Sharp end or burr on the TV link, dragging in the TV cable housing. Correct by smoothing the end, using a file or stone.

(3) Bent TV link. Correct by straightening or replacing as required.

(4) Misalignment of the throttle lever and bracket assembly on the coiled pin in the control valve assembly.

(5) Damaged or binding throttle lever and bracket assembly. Correct by straightening or replacing as required.

(6) Throttle lever spring unhooked or damaged. Correct by assembling the spring properly or replacing the throttle lever and bracket assembly as required.

6. If the TV cable is adjusted too long, it may result in one of the two following conditions:

(a) Early and slipping shifts or no detent downshifts.

(b) Delayed or full-throttle shifts by causing the transmission to operate in the high pressure mode. The transmission senses a malfunction of the TV cable and associated parts; to prevent burning the clutches and band due to low line pressures, it will go into the high pressure mode. Line pressures checked under the minimum TV conditions in neutral and drive will be in the range of the full TV pressures, if the transmission is in the high pressure mode (see the oil pressure chart). The complaint could be described as no upshifts or delayed or sharp upshifts; the closed throttle 3–2 shifts may automatically occur as high as 45 mph (72 km/h).

7. If the TV cable is adjusted too short, it will result in raising the line pressure and shift points. It may also limit the carburetor opening or diesel fuel pump delivery to prevent full-throttle operation.

8. Readjust the TV cable. On models with manual TV cable adjustment, the cable housing will extend through the cable snap lock assembly about 1.57–7.95 mm (1/16–5/16 inch).

9. Road test the car. If delayed or only full-throttle shifts still occur, proceed with step 10.

10. Install an oil pressure gauge.

(a) The engine is running and at hot idle speed.

(b) Record oil pressures in neutral and park.

Note: Neutral pressure should be equal to, or no more than, 69 kPa (10 psi) above park pressure.

(c) If neutral pressure is more than 69 kPa (10 psi) above park pressure, proceed to step 11.

11. Remove the oil pan and inspect the throttle lever and bracket assembly on the control valve assembly. Check that the TV exhaust valve lifter rod is not distorted and not binding in the control valve assembly or spacer plate. The TV exhaust check ball must move up and down as the lifter does. Also, be sure that the lifter spring holds the lifter rod up against the bottom of the control valve assembly. Make sure that the TV plunger is not stuck.

TV Cable System

General Description

The TV cable used with the Turbo Hydra-Matic 325–4L transmission should not be thought of as a THM 350 (M–38) downshift cable. The TV cable used on the THM 325–4L controls line pressure, shift points, shift feel, part-throttle downshifts, and detent downshifts. The function of the cable is similar to the combined functions of the vacuum modulator and the downshift cable used in vehicles that use the THM 350 (M–38) transmission. The cable operates the throttle lever and bracket assembly.

The throttle lever and bracket assembly serves two basic functions.

1. The primary function of this assembly is to transfer the carburetor throttle plate movement (pump lever on the diesel) to the TV plunger in the control valve assembly as related by the TV cable and linkage. The TV pressure is then modulated to follow closely engine torque by the line bias valve. Modulated TV (MTV) pressure than acts on the pressure regulator valve. TV pressure is also directed to the (MTV) up and down valves to assist the valves with upshifts and downshifts. The proper adjustment of the TV cable is based on the TV plunger being fully

depressed to flush with the TV bushing at wide open throttle.

2. The second function of the assembly involves the TV exhaust valve lifter rod, spring, and TV exhaust ball. The function of this system is to prevent the transmission from operating at low (idle position) pressures, if the TV cable should become broken or disconnected. If the cable is connected, not broken or stretched, the TV lifter rod will not move from its normal, spring-loaded up position, which holds the TV exhaust check ball off its seat. The TV lifter rod will drop to allow the TV exhaust ball to seat only if the cable is broken, disconnected, or extremely out of adjustment. With the transmission pan removed, it should be possible to pull down on the TV exhaust valve lifter rod and the spring should return the rod to its normal up position. If the throttle lever and bracket assembly or lifter rod binds or sticks so that the TV lifter rod cannot lift the exhaust ball off its seat, high line pressures and delayed upshifts will result. The normal shape of the TV lifter rod must be maintained. The right angle leg must not be bent to any other angle, or it will not function properly.

Road Test Procedure

1. *Drive range.* Position the selector lever in the drive range and accelerate from a standing start. Check for 1–2, 2–3, and 3–4 upshifts. (Shift points will vary with throttle openings.) Check for part-throttle downshifts by depressing the throttle to a three-quarters open position. Check for detent downshifts by depressing the accelerator to the wide-open throttle position at various speeds.

2. *Drive 3.* At road speed in fourth gear, shift the transmission to D3. The transmission should shift to third immediately. It should not shift back to fourth. Check for detent downshifts by accelerating wide-open throttle.

3. *Drive 2.* While in the third range shift to D2, the transmission should downshift immediately. Test for a 2–1 detent downshift by depressing the accelerator pedal to the wide-open throttle position.

4. *Low range.* Position the selector lever to the low range and accelerate; a 1–2 upshift should not occur.

5. *Overrun braking.* Overrun braking can be checked by manually downshifting to the next lower range. Engine rpm's should increase and a braking effect should be noticed.

6. *Reverse.* Position the selector lever in reverse and check reverse operation.

GM 350 Diagnosis Procedure (Courtesy of General Motors Corporation)

Condition	Possible Cause	Correction
No drive in drive range (Install pressure gauge.)	1. Low oil level	1. Correct level—check for external leaks or vacuum modulator (leaking diaphragm will evacuate oil from unit).
	2. Manual linkage adjustment	2. Adjust linkage.
	3. Low oil pressure	3. a) Filter assembly blocked. b) Pump assembly-pressure regulator, pump drive gear, and tangs damaged by converter. c) Case-porosity in intake bore.
	4. Control valve assembly	4. Manual valve disconnected from inner lever.
	5. Forward clutch	5. a) Forward clutch does not apply—piston cracked; seals missing or damaged; clutch plates burned. b) Pump feed circuit to forward clutch oil seal rings missing or broken on pump cover; leak in feed circuits; pump to

GM 350 Diagnosis Procedure (Courtesy of General Motors Corporation) (*cont.*)

Condition	Possible Cause	Correction
		case gasket mispositioned or damaged. c) Clutch drum ball check stuck or missing.
	6. Roller clutch assembly	6. Broken spring or damaged cage.
Oil pressure high or low	1. High oil pressure	1. a) Vacuum line or fittings leaking. b) Vacuum modulator. c) Modulator valve. d) Pressure regulator. e) Oil pump.
	2. Low oil pressure	2. a) Vacuum line or fittings obstructed. b) Vacuum modulator. c) Modulator valve. d) Pressure regulator. e) Governor. f) Oil pump.
1-2 shift—full throttle only	1. Detent valve	1. Sticking or linkage misadjusted.
	2. Vacuum leak	2. Vacuum line or fittings leaking. Correct leaks.
	3. Control valve assembly	3. a) Valve body gaskets leaking, damaged, or incorrectly installed. b) Detent valve train stuck. c) 1-2 valve stuck.
	4. Case assembly	4. Porosity.
1st speed only—no 1-2 shift	1. Governor assembly	1. a) Governor valve sticking. Clean. b) Driven gear loose, damaged, or worn; check for pin in case and length of pin showing. Also check output shaft drive gear for nicks or rough finish if driven gear shows damage.
	2. Control valve assembly	2. a) 1-2 shift valve train stuck closed. b) Governor feed channels blocked. c) Valve body gaskets—leaking, damaged, incorrectly installed.
	3. Case	3. a) Porosity between channels. b) Governor feed channel blocked. Governor bore scored or worn, allowing cross pressure leak.
	4. Intermediate clutch	4. a) Clutch piston seals—missing, improperly assembled, or cut. b) Intermediate roller clutch. Broken spring or damaged cage.

(*continued*)

GM 350 Diagnosis Procedure (Courtesy of General Motors Corporation) (*cont.*)

Condition	Possible Cause	Correction
1st and 2nd speeds only—no 2-3 shift	1. Control valve assembly	1. a) 2-3 shift train stuck. b) Valve body gaskets leaking, damaged, or incorrectly installed.
	2. Direct clutch	2. a) Pump hub—direct clutch oil seal rings—broken or missing. b) Clutch piston seals missing, improperly assembled; cut piston; ball check stuck or missing.
Drive in neutral	1. Manual linkage	1. Maladjusted. Adjust.
	2. Forward clutch	2. Clutch does not release (this condition will also cause no reverse).
No motion in reverse or slips in reverse (Install pressure gauge.)	1. Low oil level	1. Add oil.
	2. Manual linkage	2. Maladjusted. Adjust.
	3. Oil pressure	3. a) Modulator valve stuck. b) Modulator and reverse boost valve stuck. c) Pump hub—direct clutch oil seal rings broken or missing. d) Direct clutch piston seal cut or missing. e) Low and reverse clutch piston seal cut or missing. f) Check ball 1 missing.
	4. Control valve assembly	4. a) Valve body gaskets leaking, damaged, or incorrectly installed (other malfunctions may also be indicated). b) 2-3 valve train stuck in upshifted position; this will also cause 1-3 upshift in drive range. c) 1-2 valve train stuck in upshifted position.
	5. Intermediate servo	5. Piston or pin stuck so intermediate overrun band is applied.
	6. Low and reverse clutch	6. Piston out or seal damaged or missing.
	7. Direct clutch	7. a) Outer seal damaged or missing. b) Clutch plates burned—may be caused by stuck ball check in piston.
	8. Forward clutch	8. Clutch does not release (will also cause drive in neutral).
Slips in all ranges, slips on takeoff (Install pressure gauge.)	1. Oil level low	1. Add oil.
	2. Oil pressure	2. a) Vacuum modulator inoperative.

GM 350 Diagnosis Procedure (Courtesy of General Motors Corporation) (*cont.*)

Condition	Possible Cause	Correction
		b) Vacuum modulator valve sticking. c) Filter assembly plugged or leaks.
	3. Case	3. a) Pressure regulator valve stuck. b) Pump-to-case gasket damaged or incorrectly installed.
	4. Forward clutch slipping	4. Case leaks, porosity. a) If burned, look for cause. b) Oil seal rings on pump cover broken or worn.
Slips 1–2 shift (Install pressure gauge.)	1. Oil level low	1. Add oil.
	2. Oil pressure	2. a) Vacuum modulator assembly inoperative. b) Modulator valve sticking. c) Pump pressure regulator valve.
	3. 2–3 accumulator	3. Oil ring damaged or missing.
	4. 1–2 accumulator	4. Oil ring missing or damaged; case bore damaged.
	5. Pump-to-case gasket	5. Mispositioned.
	6. Case	6. Porosity between channels.
	7. Intermediate clutch	7. Piston seals missing or damaged; clutch plates burned.
Rough 1–2 shift (Install pressure gauge.)	1. Oil pressure	1. a) Vacuum modulator—check for loose fittings, restrictions in line. b) Modulator valve stuck. c) Valve body—regulator or boost valve stuck. d) Pump-to-case gasket off-location or damaged.
	2. Case	2. Porosity between channels.
	3. 1–2 accumulator assembly	3. a) Oil rings damaged. b) Piston stuck. c) Broken or missing spring. d) Bore damaged. e) Check accumulator feed hold in valve body plate.
Slips 2–3 shift (Install pressure gauge.)	1. Oil level low	1. Add oil.
	2. Oil pressure low.	2. a) Modulator assembly. b) Modulator valve. c) Pump pressure regulator valve or boost valve pump-to-case gasket off location.
	3. Case	3. Porosity.
	4. Direct clutch	4. Piston seals leaking or ball check leak.

(*continued*)

GM 350 Diagnosis Procedure (Courtesy of General Motors Corporation) (*cont.*)

Condition	Possible Cause	Correction
Rough 2-3 shift (Install pressure gauge.)	1. Oil pressure high	1. a) Vacuum leak. b) Modulator valve sticking. c) Valve body—pressure regulator or boost valve inoperative.
	2. 2-3 accumulator assembly	2. a) 2-3 accumulator spring missing or broken. b) Accumulator piston stuck.
No engine braking—L2-2nd gear	1. Intermediate servo and 2-3 accumulator	1. a) Servo or accumulator oil rings or bores leaking. b) Servo piston stuck.
	2. Intermediate overrun band.	2. Intermediate overrun band broken or burned (check for cause).
	3. Oil pressure low	3. Pressure regulator or boost valve stuck.
No engine braking—L1-1st gear	1. Manual low control valve assembly	1. Stuck valve assembly.
	2. Oil pressure low	2. Pressure regulator or boost valve stuck.
	3. Low and reverse clutch	3. Piston inner seal damaged or missing.
No part-throttle downshift (Install pressure gauge.)	1. Oil pressure	1. Vacuum modulator assembly, modulator valve, pressure regulator valve train (other malfunctions may also be noticed).
	2. Detent valve and linkage	2. Sticks, disconnected, or broken.
	3. 2-3 shift valve	3. Stuck.
No detent downshifts	1. Control valve assembly	1. 2-3 valve stuck.
	2. Detent valve and linkage	2. Sticks, disconnected, or broken.
Low or high shift points (Install pressure gauge.)	1. Oil pressure	1. a) Engine vacuum—check at transmission end of the modulator pipe. b) Vacuum modulator assembly, vacuum line connections at engine and transmission, modulator valve, pressure regulator valve train.
	2. Governor	2. a) Valve sticking. b) Feed holes restricted or leaking, pipes damaged or mispositioned. c) Feed line plugged.
	3. Detent valve and linkage	3. Stuck open (will cause late shifts).
	4. Control valve assembly	4. a) 2-3 valve train sticking. b) 1-2 shift valve train sticking.
	5. Case	5. Porosity.

GM 350 Diagnosis Procedure (Courtesy of General Motors Corporation) (*cont.*)

Condition	Possible Cause	Correction
Won't hold in park	1. Manual linkage 2. Internal linkage	1. Misadjusted. 2. a) Parking brake lever and actuator assembly. Check for chamfer on actuator rod sleeve. b) Parking pawl broken or inoperative.
Locks up in manual low (usually hot only)	1. Converter pressure leaking into direct clutch thru stator shaft 2. Direct clutch 3. Low and reverse clutch	1. Check stator shaft index. 2. a) Direct clutch bore undersize or piston oversize. b) Direct clutch feed hole shy small chamber. 3. Low and reverse piston center seal leaks.
2nd-gear start or slips 2nd gear only	1. Intermediate clutch	1. Wrong number of clutch plates or wrong piston.
Locks up in reverse (usually hot only)	1. Forward clutch 2. Direct clutch	1. Bore undersize or piston oversize. 2. Direct clutch feeding forward clutch through stator shaft (check stator shaft index).
Locks in reverse from park to reverse only	1. Parking pawl	1. Parking pawl staying in due to a burr on leading edge.
Cold morning reverse, no drive till engine warms up	1. Pressure regulator bore or sleeve tight	1. Remove and repair.
Shifts cold but not warm	1. Governor assembly	1. Nylon gear roll pin missing.
No drive—but has manual low	1. Low and reverse roller clutch	1. Low and reverse roller clutch installed backward.
No 1-2 shift—makes 1-3 shift and 3-1 shift but has all shifts manually	1. Intermediate roller clutch	1. Intermediate roller clutch not locking.
Governor nylon gear stripped 360°	1. Case pin 2. Output shaft	1. Governor case pin missing. 2. Output shaft rough or worn.
Governor gear stripped one side	1. Governor siezing in bore	1. Repair or replace as necessary.
Slow reverse (cold only)	1. Low oil level	1. Adjust oil level.
Harsh 1-2 shift	1. 1-2 accumulator	1. a) Piston or spring. b) Accumulator feed hole in valve body plate.
Slow reverse (hot only)	1. Valve body 2. Shift selector lever	1. Leaking valve body support plate. 2. a) Bent or S-hook hole off location. b) S-hook bent. c) Detent roller spring hole off location. d) Manual valve S-hook hole off location.

Ford General Diagnosis—C3, C4, and FAOD (Courtesy of Ford Motor Company)

Diagnosis and Testing

Troubleshooting the automatic transmission is simplified by using the proven method of diagnosis. One of the most important things to remember is that there is a definite procedure to follow. Do not try to short cut it or take for granted that someone else has done the critical checks or adjustments. The following procedures are recommended for checking or verifying that the various components are adjusted and operating properly.

Automatic Transmission Tests

Kickdown Linkage Check

Check for wide-open carburetor and linkage travel at full throttle. The carburetor full-throttle stop must be contacted by the throttle linkage and there must be a slight amount of movement left in the downshift linkage. Be sure that the downshift linkage return spring is connected and the downshift lever returns to a closed position.

Manual Linkage Check

This is a *critical* adjustment. Be sure that the D detent in the transmission corresponds exactly with the stop in the steering column or console. Hydraulic leakage at the manual valve can cause delay in engagements or slipping while operating if the linkage is not correctly adjusted.

Transmission Fluid Level Check

The transmission is designed to operate with the oil level between the add and full mark on the dipstick, at an operating temperature of 66°–77°C (150°–170°F). Fluid level should be checked under these temperatures. Obtain the operating temperature by driving 24–32km (15–20 miles) of city driving with the outside temperature above 10°C (50°F).

Transmission at Operating Temperature, 66°-77°C (150°-170°F)

1. With the transmission in park, the engine at curb idle rpm, foot brakes applied, and the vehicle on a level surface, move the transmission selector lever through each range. Allow time in each range to engage the transmission, return to park, and apply the parking brake. Do not turn off the engine during the fluid level check.

2. Clean all dirt from the transmission fluid dipstick cap before removing the dipstick from the filler tube.

3. Pull the dipstick out of the tube, wipe it clean, and push all the way back into the tube. Be sure that it is fully seated.

4. Pull the dipstick out of the tube again and check the fluid level. Fluid level should be between the add and full mark. Before adding fluid, be sure that the correct type will be used. If in doubt, check the vehicle certification label affixed to the left front door lock face panel or door pillar for the transmission code. For C–3 automatic transmission (Code V), C–4 automatic transmission (Code W), and automatic overdrive transmission (Code T), use fluid that meets Ford specification ESP–M2C138–CJ or Dextron® II, Series D or the equivalent. *Caution: Use of a fluid other than that specified could result in transmission malfunction or failure.* The C5 uses Type H fluid. If necessary, add enough fluid through the filler tube to raise the level to the correct position. Do not overfill the transmission, as this will result in foaming, loss of fluid through the vent, and possible transmission malfunction.

5. Install the dipstick, making sure that it is fully seated in the tube.

Transmission Fluid Condition Check

1. Make the normal fluid check according to the preceding procedure.

2. Observe the color and odor of the fluid. It should be red, not brown or black. Odor can sometimes indicate that there is an overheating condition or clutch disc or band failure.

3. Use an absorbent white facial tissue to wipe the dipstick. Examine the stain for evidence of solids (specks of any kind) and for antifreeze signs (gum or varnish on the dipstick).

If specks are present in the oil or there is evidence of antifreeze, the transmission oil pan must be removed for further inspection. If fluid contamination or transmission failure is confirmed by further evidence of coolant or excessive solids in the oil pan, the transmission must be disassembled and completely cleaned and serviced. This includes cleaning the torque converter and transmission cooling system. It would be a waste of time to perform any further checks before cleaning and repairing the transmission. During disassembly and assembly, all overhaul checks and adjustments of clearances and end play must be made. After the transmission has been serviced, all diagnosis tests and adjustments listed in the diagnosis guide must be completed to ensure that the problem has been corrected.

FORD 3 SPD
AUTOMATIC TRANSMISSIONS DIAGNOSIS CHECK SHEET

R.O. ______________ Trans. ______________

Engine ______________

Code on Diagnosis Wheel	Check/Test	Remarks

☐ B — TRANSMISSION FLUID
1. Level ______________ ______________
2. Condition ______________ ______________

☐ C — ENGINE
Idle ______________ ______________
Power ______________ ______________

☐ D — EGR SYSTEM

☐ E — LINKAGE
Downshift ______________ ______________
Manual ______________ ______________

☐ F — SHIFT TESTS

Throttle Opening	Range	Shift	Shift Points (MPH) Record Actual	Record Spec.
Minimum (Above 12'' Vacuum)	D	1-2		
	D	2-3		
	D	3-1		
	1	2-1		
To Detent (Torque Demand)	D	1-2		
	D	2-3		
		3-2		
Thru Detent (wide open Throttle)	D	1-2		
	D	2-3		
	D	3-2		
	D	2-1 or 3-1		

☐ G — PRESSURE TEST

Engine RPM	Manifold Vacuum In-HG	Throttle	Range	PSI Record Actual	Record Spec.
Idle	Above 12	Closed	P		
			N		
			D		
			2		
			1		
			R		
As Required	10	As Required	D,2,1		
As Required	Below 3	Wide Open	D		
			2		
			1		
			R		

Results ______________________________

☐ H — STALL TEST

Range	Specified Engine RPM	Record Actual Engine RPM
D		
2		
1		
R		

Results ______________________________

☐ I — GOVERNOR TEST
Cutback Speed (C3, C4, C6, JATCO)
10'' Vacuum ______ MPH ______________
0-2'' Vacuum ______ MPH ______________
Pressure at MPH (FMX)
10 ______ PSI ______________
20 ______ PSI ______________
30 ______ PSI ______________

☐ J — LEAK TEST

CHECK THESE	OK	OIL/FLUID *(COLOR)
CONVERTER AREA		
OIL PAN GASKET		
FILLER TUBE/SEAL		
COOLER/CONNECTIONS		
LEVER SHAFT SEALS		
PRESSURE PORT PLUGS		
EXTENSION/CASE GASKET		
EXTENSION SEAL/BUSHING		
SPEEDOMETER ADAPTER		
SERVO COVERS		
AIR VENT		

*Color Codes		
	Auto. Trans.	Red
	Power Steering	Yellow-Green
	Engine Oil	Golden Brown

☐ K — VACUUM HOSE ROUTING
☐ L — BAND AND SERVO
1. Intermediate Band Adj.
2. Reverse Band Adj.
3. Polished, Glazed Band, Drum

☐ M — DRIVESHAFT, U-JOINTS, ENGINE MOUNTS
☐ P — VALVE BODY DIRTY, STICKING
☐ Q — INTERNAL LINKAGE
☐ R — VALVE BODY BOLT TORQUE
☐ S — AIR PRESSURE TEST
☐ T — MECHANICAL PARTS
☐ U — VERIFY PROBLEM
☐ V — VALVE BODY MOUNTING FACES
☐ W — SPEEDO DRIVEN GEAR
☐ X — VACUUM TO DIAPHRAGM
☐ Y — CHECK DIAPHRAGM FOR LEAKAGE

REFER TO DIAGNOSIS WHEEL OR TO CAR DIAGNOSIS MANUAL FOR ACTION TO TAKE ON ANY "NOT OK" CONDITION.

Figure 8-44.

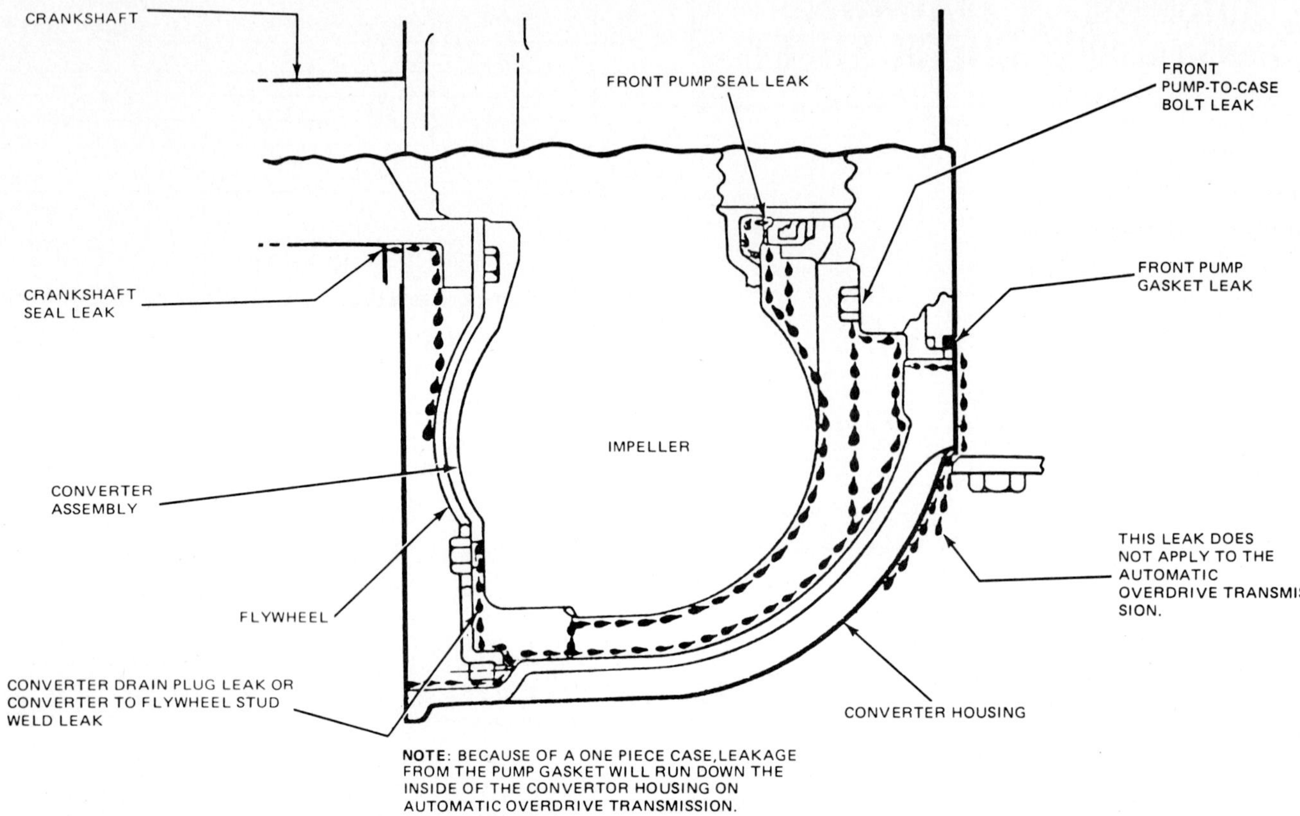

Figure 8-45. Fluid leakage in converter area.

High or Low Fluid Level

A fluid level that is too high will cause the fluid to become aerated. Aerated fluid will cause low control pressure, and the aerated fluid may be forced out the vent.

A fluid level that is too low can affect the operation of the transmission. A lower level may indicate fluid leaks that could cause transmission damage.

Fluid Level High before Starting Engine—OK during Normal Check (C3, C4, or AOD Transmissions)

1. Check for correct operation of the drainback valve in stator support.
2. Check the pump bushing.
3. Replace or repair the pump if required.

Transmission Fluid Leakage Checks

Check the speedometer cable connection at the transmission. Replace the rubber seal if necessary.

Leakage at the oil pan gasket often can be stopped by tightening the attaching bolts to specifications. If necessary, replace the gasket.

Check the fluid filler tube connection at the transmission case or pan. If leakage is found here, install a new O-ring or tighten the fitting to specifications.

Check the fluid lines and fittings between the transmission and the cooler in the radiator tank for looseness, wear, or damage. If leakage cannot be stopped by tightening a fluid line tube nut, replace the damaged parts. When oil is found to be leaking between the case and the cooler line fitting, tighten the fitting to maximum specification. Do not try to stop the oil leak by increasing the torque beyond specification. This may cause damage to the case threads. If the leak continues, replace the cooler line fitting and tighten to specifications. The same procedure should be followed for oil leaks between the radiator cooler and cooler line fittings.

Check the engine coolant in the radiator. If transmission fluid is present in the coolant, the cooler in the radiator is probably leaking.

The cooler can be further checked for leaks by disconnecting the lines from the cooler fittings and applying 345–517 kPa (50–75 psi) air pressure to the fittings. Remove the radiator cap to relieve the pressure buildup at the exterior of the oil cooler tank. If the cooler is leaking or will not hold pressure, the cooler must be replaced. If leakage is found at either the downshift control lever shaft or the manual lever shaft, replace either or both seals.

C3 and C4 Transmissions

Inspect the pipe plug on the left side of the transmission case at the front. If the plug shows leakage, tighten the plug to specifications, using Motorcraft sealing compound or the equivalent on the threads. If tightening does not stop the leaks, replace the plug.

Automatic Overdrive Transmission (AOD)

The automatic overdrive (AOD) transmission has four pipe plugs: three on the right side of the case and one on the left side. Inspect the plugs for leakage. Be sure that they are tightened to specification. If tightening does not stop the leak, replace the plug.

When a converter drain plug leaks, remove the drain plug with a sixpoint wrench. Coat the threads with Motorcraft sealing compound or the equivalent and install the plug. Tighten the drain plug to specification. Oil-soluble aniline or fluorescent dyes premixed at the rate of 1/2 teaspoon of dye powder to 1/2 pint of transmission fluid have proved helpful in locating the source of fluid leakage. Such dyes may be used to determine whether an engine oil or transmission fluid leak is present or if the fluid in the oil cooler leaks into the engine coolant system. A black light must be used with the fluorescent dye solution.

Fluid Leakage in Converter Area

In diagnosing and correcting fluid leaks in the front pump and converter area, use the following procedures to locate the exact cause of the leakage. Leakage at the front of transmission, as evidenced by fluid around the converter housing, may have several sources. By careful observation, it is possible, in many instances, to pinpoint the source of the leak before removing the transmission from the vehicle. The paths that the fluid takes to reach the bottom of the converter housing are shown in Figure 8–45.

1. Fluid leaking by the front pump seal lip will tend to move along the drive hub and onto the back of the impeller housing. Except in the case of a total seal failure, fluid leakage by the lip of the seal will be deposited on the inside of the converter housing only, near the outside diameter of the housing.

2. Fluid leakage by the outside diameter of the seal and front pump body will follow the same path that the leaks by the front pump seal follow.

3. Fluid that leaks by a front pump-to-case bolt will be deposited on the inside of the converter housing only. Fluid will not be deposited on the back of the converter.

4. Leakage by the front pump-to-case gasket may cause fluid to be deposited inside the converter housing, or it may seep down between the front of the case and converter housing (not on AOD). Fluid on the front of the case above the pan gasket on the C4 transmission is evidence that the front pump-to-case gasket or seal could be leaking.

5. Fluid leakage from the converter drain plugs will appear at the outside diameter of the converter on the back face of the flywheel and in the converter housing only near the flywheel.

6. With converter-to-flywheel-stud weld leak, engine oil leaks are sometimes improperly diagnosed as transmission front pump seal leaks. The following areas of possible leakage should also be checked to determine if engine oil leakage is causing the problem.

(a) Leakage at the rocker arm cover may allow oil to flow over the converter housing or seep down between the converter housing and cylinder block causing oil to be present in or at the bottom of the converter housing.

(b) Oil gallery plug leaks will allow oil to flow down the rear face of the block to the bottom of the converter housing.

(c) Leakage at the crankshaft seal will work back to the flywheel and then into the converter housing.

7. Fluid leakage from other areas, such as the power steering system forward of the transmission, could cause fluid to be present around the converter housing due to blow back or road draft. The following procedures should be used to determine the cause of the leakage before any repairs are made.

(a) Remove the transmission dipstick and note the color of the fluid. Original factory-fill fluid is dyed red to aid in determining if leakage is from the engine or transmission. Unless a considerable amount of makeup fluid has been added or the fluid has been changed, the red color should assist in pinpointing the leak. Fluid used in the power steering system has a yellowish green dye. Since road draft may cause leaking power steering fluid to be present on the transmission, this leakage should be eliminated before checking the transmission for fluid leakage.

(b) Remove the converter housing cover. Clean off any fluid from the top and bottom of the converter housing, front of the transmission case, and rear face of the engine and engine oil pan. Clean the converter area by washing with a suitable nonflammable solvent, and blow dry with compressed air.

(c) Wash out the converter housing, the front of the flywheel, and the converter drain plugs. The converter housing may be washed out using cleaning solvent and a squirt-type oil can. Blow all washed areas dry with compressed air.

(d) Start and run the engine until the transmission reaches its normal operating temperature. Observe the back of the block and top of the converter housing for evidence of fluid leakage. Raise the vehicle on a hoist and run the engine at fast idle, then at engine idle, occasionally shifting to the drive and reverse ranges to increase pressure within the transmission. Observe the front of the fly-wheel, back of the block (in as far as possible), and inside of the converter housing and front of the transmission case. Run the engine until fluid leakage is evident and the probable source of leakage can be determined.

Converter Leakage Check

If welds on the torque converter indicate leakage, remove the converter and make the following check. Assemble a Rotunda 72–0004 Torque Converter Leak Detector or equivalent. Test the converter for leaks, following the directions supplied with the detector kit.

Engine Idle Speed Check

If the idle speed is too low, the engine will run roughly. An idle speed that is too high will cause the vehicle to creep and have harsh engagements and harsh closed throttle downshifts. Check and, if necessary, adjust the engine idle speed.

Antistall Dashpot Clearance Check

After the engine idle speed has been properly adjusted, check the antistall dashpot clearance. Refer to the Ford manual for checking and adjusting this clearance.

Control Pressure Test

There are two methods of performing the control pressure test. One is to perform the test using the engine vacuum. The second method is to use a remote vacuum source such as the one provided in the distributor tester or a hand-operated vacuum pump.

Engine Vacuum Pressure (C3 and C4 Transmissions)

When the vacuum diaphragm unit is operating properly and the downshift linkage is adjusted properly, all the transmission shifts (automatic and kickdown) should occur within the road speed limits given in the Ford car performance specifications book.

If the shifts do not occur within limits, or the transmission slips during shift point, use the following procedure to determine whether the engine, transmission, linkage, vacuum diaphragm unit, or valve body is causing the condition.

1. Using the automatic transmission tester or equivalent, attach the tachometer to the engine, and the vacuum gauge to the transmission vacuum line at the manifold vacuum port.

2. Attach the pressure gauge to the control pressure outlet at the transmission.

3. Firmly apply the parking brake and start the engine. On a vehicle equipped with a vacuum brake release, apply the service brakes. The parking brake cannot be used because the brake automatically releases when the transmission selector lever is moved to drive.

4. Adjust the engine idle speed to the specified rpm, using the carburetor idle adjustment screw. If the engine idle speed cannot be brought within limits, check the throttle and downshift linkage for a binding condition. If linkage is satisfactory, check for vacuum leaks in the transmission diaphragm unit and its connecting tubes and hoses. Check all other vacuum operated units (such as the power brake) for vacuum leaks.

Vacuum Pump Procedure (C3 and C4 Transmissions)

Install the automatic transmission tester. Disconnect and temporarily plug the vacuum line at the vacuum diaphragm unit. Apply both the parking and service brakes. Start the engine and vacuum pump. Set the vacuum at 15 inches; read and record the control pressure in all selector positions. The two charts in Figure 8–61 show what components are inoperative when the control pressure test is not within specifications. Do not proceed with the main diagnosis guide until you have completed any required service and the control pressure is within specification.

Vacuum Supply Test (C3 and C4 Transmissions)

The vacuum supply to the vacuum diaphragm unit and the diaphragm itself must be checked. To check the supply, disconnect the vacuum line at the diaphragm unit and connect it to a vacuum gauge. With the engine idling, the gauge must have a steady acceptable vacuum reading for the altitude at which the test is being performed. If the vacuum reading is low, check for a vacuum leak or poor engine vacuum. If the vacuum reading is okay, rapidly accelerate the engine momentarily. The vacuum reading must drop rapidly at acceleration and return immediately upon release of the accelerator. If the vacuum reading does not change or changes slowly, the transmission vacuum line is plugged, restricted, or connected to a reservoir supply. Correct as required.

Vacuum Diaphragm Test (C3 and C4 Transmissions)

On Vehicle

To check the vacuum diaphragm unit, start the vacuum pump and set the regulator knob so the vacuum gauge reads 18 inches with the end of the vacuum hose blocked off. Then connect the vacuum hose to the diaphragm unit. If the reading does not remain at 18 inches but drops, the vacuum unit diaphragm is leaking. Replace the vacuum diaphragm unit. Also, if automatic transmission fluid is present in the vacuum side of the diaphragm or in the vacuum hose, the diaphragm is leaking and must be replaced.

Off Vehicle

To check the vacuum unit for diaphragm leakage, remove the unit from the transmission. Use a distributor tester equipped with a vacuum pump. Set the regulator knob so the vacuum gauge reads 18 inches with the end of the vacuum hose blocked off.

Connect the vacuum hose to the manifold vacuum port. If the gauge still reads 18 inches, the vacuum unit diaphragm *is not leaking.*

A second leakage check can be made as the hose is removed from the transmission vacuum unit. Hold a finger over the end of the control rod. When the hose is removed, the internal spring of the vacuum unit should push the control rod outward. If the vacuum diaphragm needs replacing, install a new unit that has been released for service. Vacuum diaphragm assembly identification is given in the specifications section of the Ford service manual.

Stall Test

Start the engine to allow it to reach its normal temperature. Apply both the parking and service brakes while testing.

The stall test is made in drive, 2, 1 (first) or reverse (automatic overdrive-overdrive, 3, L, or reverse) at full throttle to check engine performance, converter clutch operation or installation, and the holding ability of the forward clutch, reverse-high clutch and low-reverse band and the gear train one-way clutch. While making this test, do not hold the throttle open for more than five seconds at a time.

After the test, move the selector lever to neutral and run the engine at 1,000 rpm for about 15 seconds in order to cool the converter before the next test. If the engine speed recorded by the tachometer exceeds the maximum limits given in specifications, release the accelerator immediately because clutch or band slippage is indicated.

Governor Check (C3 and C4 Transmissions)

The governor can be checked at the same time as the control pressure test is performed and in the same manner. Raise the vehicle with an axle or frame hoist so that the rear wheels are clear of the floor. Disconnect and plug the vacuum line to the vacuum diaphragm unit. Connect the line from the distributor tester to the vacuum diaphragm unit. Install the automatic transmission tester.

Caution: Never exceed 97 km/h (60 mph) speedometer speed.

Place the transmission in manual 2, no load on the engine, and apply 10 inches of vacuum to the vacuum diaphragm unit. Increase the speed slowly and watch the speedometer. Check the km/h (mph) at which the control pressure cutback occurs. It should occur between 16 and 32 km/h (10–20 mph).

Note: After each test, move the selector to neutral, run the engine at 1000 rpm to cool the transmission.

Decrease the vacuum at the vacuum diaphragm to 0 to 2 inches. Repeat the check. Control pressure cutback should occur between 48 and 80 km/h (30 and 50 mph).

The governor is good if the cutback occurs within these specifications. If the cutback does not occur within specifications, check shift speeds to verify that it is the governor and not a stuck cutback valve; then service or replace the governor.

Automatic Overdrive Transmission

Accelerate the vehicle to 40 km/h (25 mph) then back off throttle completely. The transmission should shift to third gear.

Transmission Fluid Cooler Flow Check

The linkage, fluid, and control pressure must be within specifications before performing this flow check.

Remove the transmission dipstick from the filler tube. Place a funnel in the transmission filler tube. Raise the vehicle; remove the cooler return line from its fitting in the case. Attach a hose to the cooler return line and fasten the free end of the hose in the funnel installed in the filler tube. Start the engine and set idle speed at 1,000 rpm with the transmission in neutral.

Observe the fluid flow at the funnel. When the flow is solid (air bleeding has been completed), the flow should be liberal. If there is not a liberal flow at 1,000 rpm in neutral, low pump capacity, main circuit system leakage, or cooler system restriction is indicated.

To separate transmission trouble from cooler system trouble, observe the flow at the transmission case converter outfitting.

Shift Point Checks—Road Test (C3 and C4 Transmissions)

Check the minimum throttle upshifts in drive. The transmission should start in first gear, shift to second, and then shift to third, within the shift points given in the Ford car performance specifications book.

With the transmission in third gear, depress the accelerator pedal through the detent (to the floor). The transmission should shift from third to second or third to first, depending on vehicle speed.

Check the closed-throttle downshift from third to first by coasting down from about 48 km/hr (30 mph) in third gear. The shift should occur within the limits given in the Ford car performance specifications book.

When the selector lever is at 2 (second), the transmission can operate only in second gear.

With the transmission in third gear and road speed over 48 km/h (30 mph), the transmission should shift to second gear when the selector lever is moved from drive to 2 (second) to 1 (first). The transmission will downshift from second or third to first gear when this same manual shift is made below approximately 40 km/h (25 mph) with a C3, C4, and AOD transmission.

This check will determine if the governor pressure and shift control valves are functioning properly. During the shift point check operation, if the transmission does not shift within specifications or certain gear ratios cannot be obtained, refer to the diagnosis guide to resolve the problem.

Altitude Compensating Modulator (C3 and C4 Transmissions)

The altitude-compensating modulator assembly controls shift spacing and timing so that these will not be adversely affected by changes in engine performance at altitudes, as in the Denver area.

The modulator assembly exerts force on the throttle valve as a function of engine intake manifold vacuum and local atmospheric pressure. The modulator assembly is composed of a flexible diaphragm, aneroid bellows assembly, calibration spring, and case. The calibration spring and evacuated bellows assembly supply a nominal (maximum) throttle valve load determined by initial calibration and local atmospheric pressure. The bellows assembly varies nominal load based on the effective bellows area and the pressure differential between variable atmospheric pressure and constant (approximately zero psia) internal bellows pressure. Engine intake manifold vacuum acting on the effective diaphragm area decreases the nominal load according to vacuum magnitude.

Automatic Overdrive Transmission

Choke on will affect shift points minimum throttle upshifts. Check minimum throttle upshifts in the fourth gear (overdrive). The transmission should start in first gear, shift to second gear, then shift to third gear and finally shift to fourth gear, within the shift points given in the Ford car performance specifications book.

Forced Downshifts

With the transmission in fourth gear (overdrive), depress the accelerator pedal to the floor. The transmission should downshift to third gear or to second gear, depending on vehicle road speed. Refer to the car performance specifications book for the vehicle shift specification.

Closed-Throttle Downshifts

Closed-throttle downshifts should be extemely difficult to detect. It may be necessary to attach pressure gauges to the forward and direct clutch pressure taps in order to detect 4–3 and 3–2 coastdown shifts.

With gauges attached a 4–3 coast (closed-throttle) downshift is signified by the application of the forward clutch, and 3–2 coast downshift is signified by the release of the direct clutch. A 2–1 coast downshift should be imperceptible. The coast downshifts should occur within the limits set forth in the car performance specifications book.

When the shift selector is in manual low (1), the transmission can operate only in first gear.

When the shift selector is moved from either the overdrive or direct drive (3) ranges to the manual low (1) position, the transmission will downshift into second gear if the vehicle is moving over approximately 25 mph and into first gear if the vehicle is moving less than approximately 25 mph.

The overdrive band is applied when the transmission is in manual second. The low-reverse band is applied when the transmission is in manual low.

The automatic overdrive transmission will not shift into fourth gear (overdrive) at wide-open throttle. The automatic overdrive transmission will not make a 4–1 downshift.

Shift Test—In Shop

A shift test can be performed in the shop to check shift valve operation, governor circuits, shift delay pressures, throttle boost, and downshift valve action.

Raise the vehicle with an axle or frame hoist so

that the rear wheels are clear of the floor. Disconnect and plug the vacuum line to the vacuum diaphragm unit. Connect the line from the distributor tester vacuum pump to the vacuum diaphragm unit (except AOD).

Caution: Never exceed 97 km/h (60 mph) speedometer speed.

1. To check the shift valves and governor circuits, apply 18 inches of vacuum to the transmission vacuum diaphragm unit. The automatic overdrive transmission is not equipped with a vacuum-regulated throttle pressure system. Eighteen inches of vacuum is the equivalent to a closed throttle for the automatic overdrive transmission. Place the transmission in drive and make a minimum throttle 1–2, 2–3, and 3–4 on automatic overdrive transmission shift test. When the shift occurs, you will see the speedometer needle make a momentary surge and feel the driveline bump. If the shift points are within specifications, the 1–2 and 2–3 shift valves and governor are OK.

If the shift points are not within specifications, perform a governor check to isolate the problem.

Note: After each test, move the selector lever to neutral; run the engine at 1,000 rpm to cool the transmission.

2. To check the shift delay pressures and throttle boost (do not attempt on automatic overdrive transmission), decrease the vacuum at the vacuum diaphragm to 0 to 2 inches. Make a 1–2 shift test. If the shift point raises to the specifications, the throttle boost and shift delay systems are functioning.

Note: After each test, move the selector lever to neutral; run the engine at 1,000 rpm to cool the transmission.

3. To check downshift valve action (do not attempt on automatic overdrive transmission), leave the vacuum to the vacuum diaphragm at 0 to 2 inches. Position the downshift linkage in the wide-open throttle position (through the detent) and repeat the 1–2 shift test. The speed at the shift point should be higher.

Air Pressure Checks

A no-drive condition can exist, even with the correct transmission fluid pressure, because of inoperative clutches or bands. Erratic shifts could be caused by a stuck governor valve. The inoperative units can be located through a series of checks by substituting air

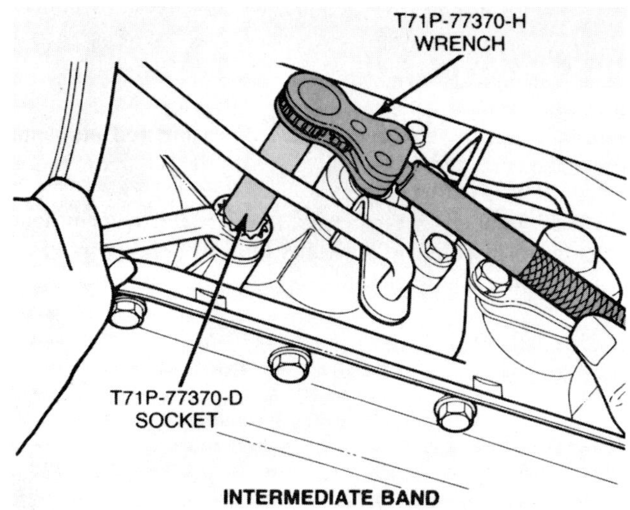

Figure 8-46. Intermediate band adjustment on C5 transmission. (*Courtesy of Ford Motor Company*)

pressure for the fluid pressure to determine the location of the malfunction.

When the selector lever is at 2 (C3 and C4), a no-drive condition may be caused by an inoperative forward clutch. A no-drive condition in drive (D) or overdrive may be caused by an inoperative forward clutch or one-way clutch. When there is no drive in 1, the difficulty could be caused by improper functioning of the forward clutch or low-reverse band and the one-way clutch. Failure to drive in reverse range could be caused by a malfunction of the reverse-high clutch or low-reverse band or clutch.

When you have a slip problem but do not know whether it is in the valve body or in the hydraulic

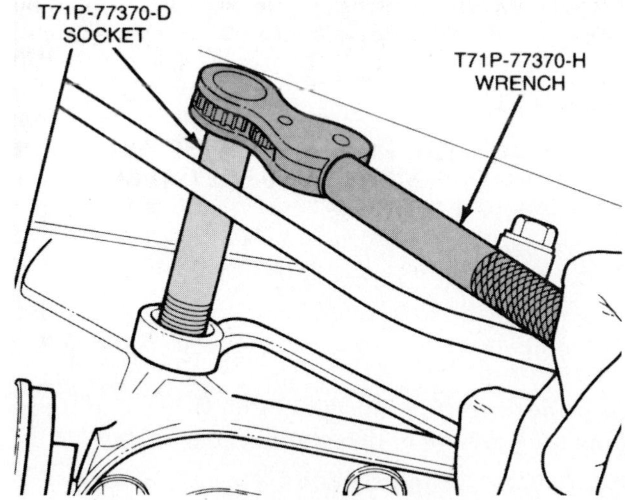

Figure 8-47. Adjusting low and reverse band on C5 transmission. (*Courtesy of Ford Motor Company*)

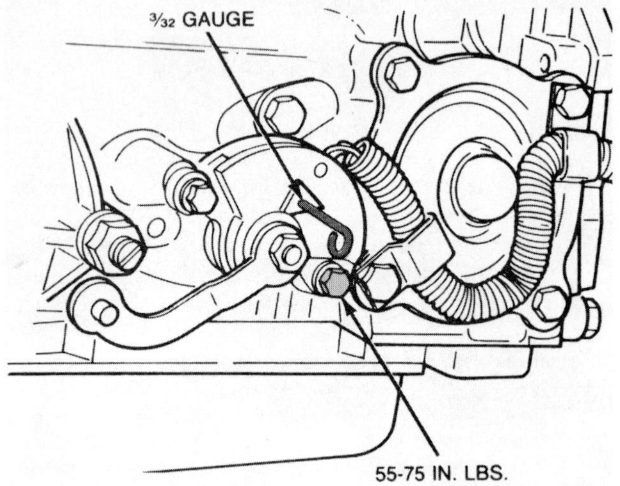

Figure 8-48. Neutral start switch adjustment on Ford C5 transmission. (*Courtesy of Ford Motor Company*)

system beyond the valve body, the air pressure tests can be valuable.

To make the air pressure checks, loosen the oil pan bolts and lower one edge of the oil pan to drain the transmission fluid. Remove the fluid pan and the control valve body assembly. The inoperative units can be located by introducing air pressure into the transmission case passages leading to the clutches, servos, and governor.

Forward Clutch (C3, C4, and AOD)

Apply air pressure to the transmission case forward clutch or front clutch passages. A dull thud can be heard when the clutch piston is applied. If no noise is heard, place the fingertips on the input shell and again apply air pressure to the forward or front clutch passage. Movement of the piston can be felt as the clutch is applied.

Governor

Apply air pressure to the control pressure to to governor passage, and listen for a sharp clicking or whistling noise. The noise indicates governor valve movement.

Reverse High Clutch—
C3, C4 Reverse Clutch—AOD

Apply air pressure to the reverse-high clutch. A dull thud indicates that the reverse-high clutch piston has moved to the applied position. If no noise is heard, place the fingertips on the clutch drum and again apply air pressure to detect movement of the piston.

Intermediate Servo (C3 and C4)

Hold the air nozzle in the front servo apply tube or the intermediate servo apply passages. Operation of the servo is indicated by a tightening of the front or intermediate band around the drum on C3 and C4 transmissions. Continue to apply air pressure to the servo apply tube or passage, and introduce air pressure into the front release tube or the intermediate servo release pressure. The front or intermediate should release the band against the apply pressure.

Low-Reverse Servo (C3, C4, and AOD)

Apply air pressure to the low-reverse or rear servo apply passage. On C3, C4, and AOD, the low-reverse band should tighten around the drum if the servo is operating properly.

If the servos do not operate, disassemble, clean, and inspect them in order to locate the source of the trouble. If air pressure applied to either of the clutch passages fails to operate a clutch or operates both clutches at once, remove and, with air pressure, check the fluid passages in the case and front pump to detect obstructions.

Direct clutch (AOD)

Apply air pressure to the direct clutch passages. A dull thud should be heard if the clutch is operating properly.

Overdrive Servo (AOD)

Apply air pressure to the overdrive servo apply passage in the case. Operation of the band is indicated by the tightening of the band around the reverse clutch drum. The servo will return to the release position as a result of spring force from the release spring. The band should be relaxed. (*Note:* It may be necessary to use service tool T80L–77030–B or the equivalent in order to seal the apply or release passages.)

Intermediate Clutch (AOD)

Apply air pressure to the intermediate clutch apply passage in the case. A dull thud should be heard if the clutch is operating properly.

Ford ATX Transaxle Diagnosis Procedure (Courtesy of Ford Motor Company)

1. Fluid level and condition.
2. Movement of TV linkage and adjustment check.
3. Manual linkage.
4. Road test and engine performance.
5. Engine idle—TV adjustments.
6. Stall test.
7. Line pressure test.
8. Case air pressure tests.

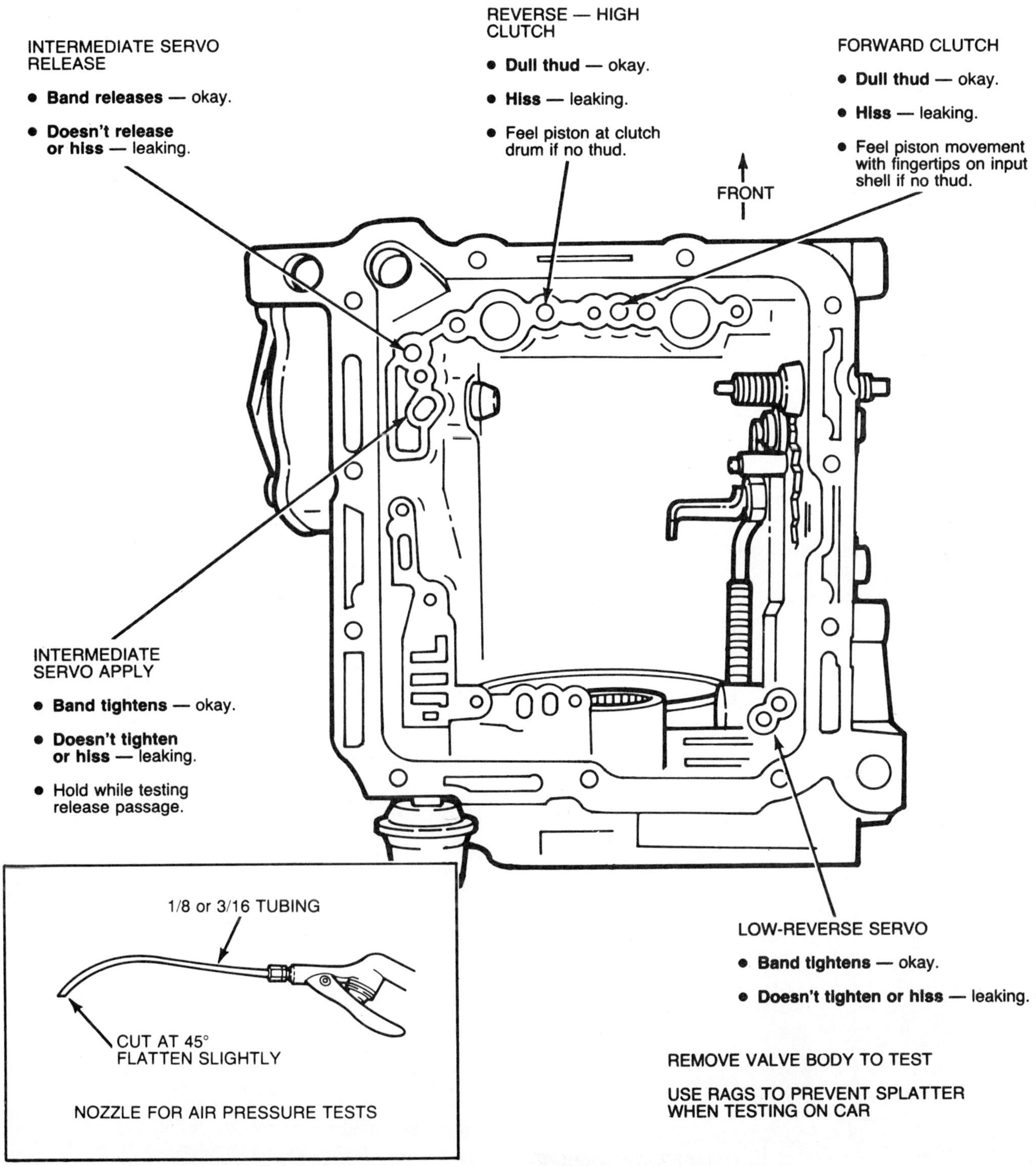

Figure 8-49. Air pressure testing clutches and bands (typical). (*Courtesy of Ford Motor Company*)

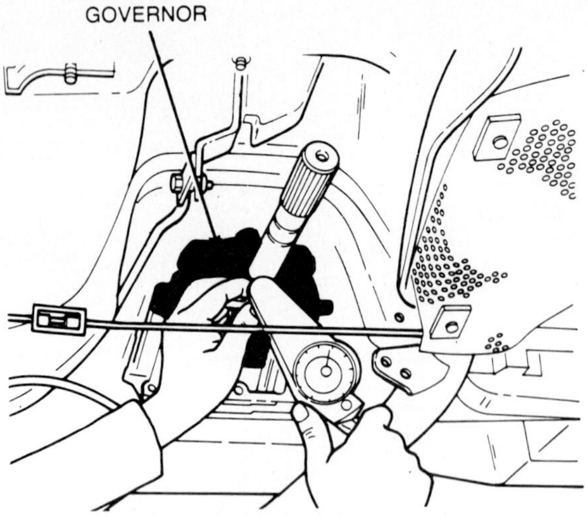

Figure 8-50. Checking governor bolt torque to correct governor pressure leakage. (*Courtesy of Ford Motor Company*)

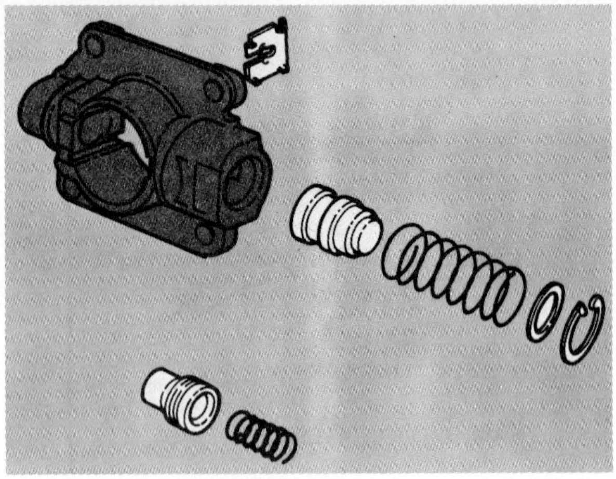

Figure 8-52. Disassembled governor. (*Courtesy of Ford Motor Company*)

Diagnosis Sequence

It is generally recommended that the first three checks—fluid, TV linkage, and manual linkage—be made before the road test. Never drive the car unless the fluid level is correct, the manual selector is in synchronization with the detents, and the TV linkage moves free of all obstructions. The sequence of tests 5 through 8 can vary, then, depending on the conditions encountered on the road test.

For example, conditions related to engagement or shift feel would generally lead to checking and possibly readjusting the TV linkage properly. Slipping conditions that occur in only certain gears might call for a stall test or line pressure test. Case air pressure tests, of course, require removing the valve body and should be used after the other checks have been done in order to isolate the problem.

TV adjustment is critical. On the road test, be alert for soft or mushy shifting or engagement. If this occurs, do *not* drive the vehicle at heavy throttle. Low TV pressure could cause damage. Check the TV and set it up properly; then give the car a full road test. Most shifting problems will be related to TV linkage. This should always be checked and properly adjusted before removing the governor or valve body.

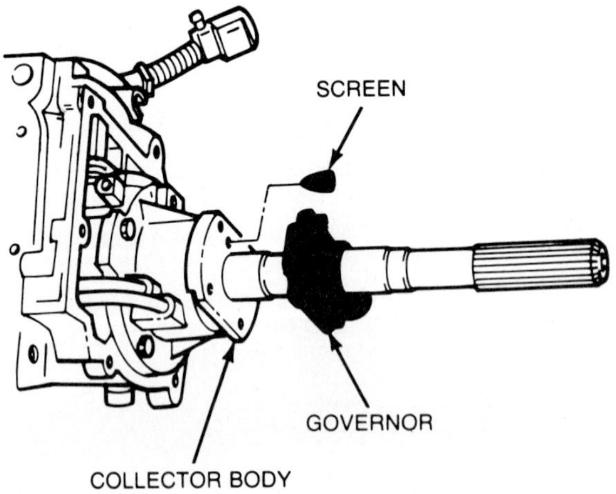

Figure 8-51. Removing governor and screen from output shaft for cleaning and inspection. (*Courtesy of Ford Motor Company*)

1. Pre-Road Test Check

The fluid level should ideally be checked with the transaxle at operating temperature. Test conditions are:

- Engine at hot curb idle and
- Shift selector in park (after moving through all the ranges to fill clutch and servo cavities).

For road testing, a cold check can be made. The fluid level should be between the inner holes. If the level is at or below the lowest hole, the vehicle should *not* be driven without adding fluid.

Hot Fill

After the vehicle is driven some 20 minutes, the fluid temperature should be in the operating range of 130°–180°F (55°–85°C). The final fluid fill should always be between the arrows at operating temperature.

Note: Do not overfill. The space between the arrows is slightly less than one pint. Do not fill to the arrows unless the transmission is hot.

DIAGNOSIS – AUTOMATIC TRANSMISSION

CONDITION	POSSIBLE SOURCE	RESOLUTION
• Slow initial engagement.	• Improper fluid level. • Damaged or improperly adjusted linkage. • Contaminated fluid. • Improper clutch and band application, or oil control pressure. • Damaged 2-3 accumulator valve (AOD).	• Add fluid as required. • Service or adjust linkage. • Perform fluid level check. • Perform control pressure test. • Service as required.
• Rough initial engagement in either forward or reverse. • Rough initial engagement (Automatic Overdrive Transmission)	• Improper fluid level. • High engine idle. • Automatic choke on (warm temp.) • Looseness in the driveshaft, U-joints or engine mounts. • Incorrect linkage adjustment. • Improper clutch or band application, or oil control pressure. • Damaged 2-3 accumulator valve (AOD). • Sticking or dirty valve body.	• Perform fluid level check. • Adjust idle to specifications. • Disengage choke • Service as required. • Service or adjust linkage. • Perform control pressure test. • Service as required. • Clean, service or replace valve body.
• No drive in any gear	• Improper fluid level. • Damaged or improperly adjusted linkage. • Improper clutch or band application, or oil control pressure. • Internal leakage. • Valve body loose. • Damaged or worn clutches. • Sticking or dirty valve body.	• Perform fluid level check. • Service or adjust linkage. • Perform control pressure test. • Check and repair as required. • Tighten to specification. • Perform air pressure test. • Clean, service or replace valve body.
• No drive forward – reverse OK.	• Improper fluid level. • Damaged or improperly adjusted linkage. • Improper clutch or band application, or oil pressure control system. • Damaged 2-3 accumulator valve (AOD). • Damaged or worn forward clutch or governor. • Valve body loose. • Dirty or sticking valve body.	• Perform fluid level check. • Service or adjust linkage. • Perform control pressure test. • Service as required. • Perform air pressure test. • Tighten to specification. • Clean, service or replace valve body.
• No drive, slips or chatters in reverse – forward OK.	• Improper fluid level. • Damaged or improperly adjusted linkage. • Looseness in the driveshaft, U-joints or engine mounts. • Reverse bands out of adjustment or damaged reverse clutches. • Improper oil pressure control. • Damaged or worn reverse clutch or servo. • Valve body loose. • Dirty or sticking valve body.	• Perform fluid level check. • Service or adjust linkage. • Service as required. • For C-3 or C-4 – adjust reverse band. (Automatic Overdrive has no adjustment) service clutches as required. • Perform control pressure test. • Perform air pressure test. • Tighten to specifications. • Clean, service or replace valve body.

Figure 8-53. Ford C3, C4, and AOD.

DIAGNOSIS – AUTOMATIC TRANSMISSION

CONDITION	POSSIBLE SOURCE	RESOLUTION
• No drive, slips or chatters in first gear in D. All other gears normal. • First gear in Ⓓ or 3 Automatic Overdrive transmission.	• Damaged or worn one-way clutch.	• Service or replace one-way clutch.
• No drive, slips or chatters in second gear.	• Improper fluid level. • Damaged or improperly adjusted linkage. • Intermediate band out of adjustment. (Except AOD). • Intermediate friction clutch or one way clutch (for Automatic Overdrive transmission only). • Improper band or clutch application, or oil pressure control. • Damaged or worn servo and/or internal leaks. • Dirty or sticking valve body. • Polished, glazed intermediate band or drum. (Except AOD).	• Perform fluid level check. • Service or adjust linkage. • Adjust intermediate band. • Service as required. • Perform control pressure test. • Perform air pressure test. • Clean, service or replace valve body. • Replace or service as required.
• Starts up in 2nd or 3rd	• Improper fluid level. • Damaged or improperly adjusted linkage. • Improper band and/or clutch application, or oil pressure control system. • Damaged or worn governor. Sticking governor. • Valve body loose. • Dirty or sticking valve body. • Cross leaks between valve body and case mating surface.	• Perform fluid level check. • Service or adjust linkage. • Perform control pressure test. • Perform governor check. Replace or service governor, clean screen. • Tighten to specification. • Clean, service or replace valve body. • Replace valve body and/or case as required.

Figure 8-54. Ford C3, C4, and AOD.

Check for Leaks

If the fluid level checks low, be sure to leak test the transaxle after the road test. Locate and repair leaks the same as on any other Ford automatic.

Fluid Specification

The ATX transaxle uses 9.8 quarts of type CJ transmission fluid for a complete fill. (This is the same fluid used in late model C6 transmissions and in the automatic overdrive transmission.) The use of incorrect fluid, as noted on the dipstick, will void the warranty.

Never add Type A or Type F where CJ is specified. In an emergency, Dexron® II may be used to fill to a safe driving level.

Fluid Condition

If the fluid is burnt or discolored or if it has band or clutch residue, the transaxle should be overhauled. During overhaul, be sure to flush the torque converter and cooler. Also determine the cause of the condition and fix it so it doesn't happen again.

2. TV Linkage Check and Adjustment

Check before Road Test

Before road testing the vehicle, make these two checks of the TV linkage:

1. Examine the fine adjustment screw with the throttle closed at hot idle. There should be *no* gap

DIAGNOSIS – AUTOMATIC TRANSMISSION

CONDITION	POSSIBLE SOURCE	RESOLUTION
• Shift points incorrect.	• Improper fluid level.	• Perform fluid level check.
	• Improper vacuum hose routing or leaks. (Except A.O.D.)	• Correct hose routing.
	• Improper operation of EGR system. (Except A.O.D.)	• Service or replace as required.
	• Linkage out of adjustment.	• Service or adjust linkage.
	• Throttle linkage out of adjustment.	
	• Improper speedometer gear installed.	• Replace gear.
	• Improper clutch or band application, or oil pressure control system.	• Perform shift test and control pressure test.
	• Damaged or worn governor.	• Service or replace governor – clean screen.
	• Dirty or sticking valve body.	• Clean, service or replace valve body.
• No upshift at any speed in D. 3 or Ⓓ for (A.O.D.)	• Improper fluid level. (Except A.O.D.)	• Perform fluid level check.
	• Vacuum leak to diaphragm unit.	• Service vacuum line or hose.
	• Linkage out of adjustment.	• Service or adjust linkage.
	• Throttle linkage A.O.D.	• Service as required.
	• Improper band or clutch application, or oil pressure control system.	• Perform control pressure test.
	• Damaged or worn governor.	• Service or replace governor, clean screen.
	• Dirty or sticking valve body.	• Clean, service or replace valve body.
• Shifts 1-3 in D. (Ⓓ or 3 range for A.O.D.)	• Improper fluid level.	• Perform fluid level check.
	• Intermediate band out of adjustment.	• Adjust band.
	• Intermediate friction or one way clutch A.O.D.	• Service.
	• Damaged intermediate servo and/or internal leaks. (Except A.O.D.)	• Perform air pressure test. Service front servo and/or internal leaks.
	• Polished, glazed band or drum. (Except A.O.D.)	• Service or replace band or drum.
	• Improper band or clutch application, or oil pressure control system.	• Perform control pressure test.
	• Dirty or sticking valve body.	• Clean, service or replace valve body.
• Engine over-speeds on 2-3 shift.	• Improper fluid level.	• Perform fluid level check.
	• Linkage out of adjustment.	• Service or adjust linkage.
	• Throttle linkage A.O.D.	• Service or adjust.
	• Improper band or clutch application, or oil pressure control system.	• Perform control pressure test.
	• Damaged or worn high clutch and/or intermediate servo.	• Perform air pressure test. Service as required.
	• Dirty or sticking valve body.	• Clean, service or replace valve body.

Figure 8-55. Ford C3, C4, and AOD.

DIAGNOSIS – AUTOMATIC TRANSMISSION

CONDITION	POSSIBLE SOURCE	RESOLUTION
• Mushy 1-2 shift.	• Improper fluid level	• Perform fluid level check.
	• Incorrect engine performance.	• Tune adjust engine idle as required
	• Improper linkage adjustment.	• Service or adjust linkage.
	• Intermediate band out of adjustment. (Except A.O.D.)	• Adjust intermediate band.
	• Improper band or clutch application, or oil pressure control system.	• Perform control pressure test.
	• Damaged high clutch and/or intermediate servo or band.	• Perform air pressure test. Service as required.
	• Polished, glazed band or drum.	• Service or replace as required.
	• Dirty or sticking valve body.	• Clean, service or replace valve body.
• Rough 1-2 shift.	• Improper fluid level.	• Perform fluid level check.
	• Incorrect engine idle or performance.	• Tune, and adjust engine idle.
	• Intermediate band out of adjustment. (Except A.O.D.)	• Adjust intermediate band.
	• Improper band or clutch application, or oil pressure control system.	• Perform control pressure test.
	• Damaged intermediate servo.	• Air pressure check intermediate servo.
	• Dirty or sticking valve body.	• Clean, service or replace valve body.
• Rough 2-3 shift.	• Improper fluid level.	• Perform fluid level check.
	• Incorrect engine performance.	• Tune and adjust engine idle.
	• Damaged 2-3 accumulator valve.	• Service as required.
	• Improper band or clutch application, or oil control pressure system.	• Perform control pressure test.
	• Damaged or worn intermediate servo release and high clutch piston check ball.	• Air pressure test the intermediate servo apply and release and the high clutch piston check ball. Service as required.
	• Dirty or sticking valve body.	• Clean, service or replace valve body
• Rough 3-1 shift at closed throttle in D.	• Improper fluid level.	• Perform fluid level check.
	• Incorrect engine idle or performance.	• Tune, and adjust engine idle.
	• Improper linkage adjustment.	• Service or adjust linkage.
	• Improper clutch or band application, or oil pressure control system.	• Perform control pressure test.
	• Improper governor operation.	• Perform governor test. Service as required.
	• Dirty or sticking valve body.	• Clean, service or replace valve body.

Figure 8-56. Ford C3, C4, and AOD.

DIAGNOSIS AUTOMATIC TRANSMISSION – Continued

CONDITION	POSSIBLE SOURCE	RESOLUTION
• Mushy 3-4 or rough 3-4 shift (Automatic Overdrive Transmission Only)	• Improper fluid level • Incorrect engine performance • Improper linkage adjustment • Improper band application • Damaged or glazed reverse clutch drum • Polished or glazed band • Dirty or sticking valve body • Damaged 2-3 accumulator valve	• Perform fluid level check. • Tune, adjust engine idle as required. • Service or adjust linkage. • Perform control pressure test. • Service as required. • Service as required. • Clean, service or replace valve body. • Service as required.
• Rough or mushy 4-2 or 3-1 shift (Automatic Overdrive Transmission Only)	• Improper fluid level • Incorrect engine performance • Improper throttle or manual linkage adjustment • Improper application of intermediate friction and one way clutch • Dirty or sticking valve body	• Perform fluid level check. • Tune, adjust engine idle as required • Service or adjust linkage. • Service as required. • Clean, service or replace valve body.
• Rough or mushy 4-3 shift (Automatic Overdrive Transmission Only)	• Improper fluid level • Incorrect engine performance • Improper throttle linkage adjustment • Improper overdrive servo release • Dirty or sticking valve body or governor	• Perform fluid level check. • Tune, adjust engine idle as required. • Service or adjust linkage. • Service as required. • Clean, service or replace valve body.

Figure 8–57. Ford C3, C4, and AOD.

between the end of the screw and the end of the bell crank rod. If there is, TV pressure can be too low. The linkage should be adjusted before the vehicle is driven.

2. Push the TV linkage rod down until the plunger is bottomed against its stop inside the transaxle, then release it. The rod should move freely both ways. If there is any binding, it will cause high TV pressure. Correct the binding by removing the cause of interference or repairing the linkage as required. (*Note: If the rod doesn't return by spring force, something is disconnected. Make the necessary repair before driving the vehicle.*)

To adjust the TV linkage:

1. Turn the ignition off, put the selector in park, and set the parking linkage.

2. If the curb idle needs setting, adjust it *before* setting the TV linkage.

3. Remove the air cleaner and be sure that the throttle is *off* the fast idle cam.

4. With all slack removed from the TV linkage (rod pulled up), set the fine adjustment screw to its midposition.

5. While holding the linkage in position to avoid damage, loosen the trunnion nut.

6. Check that the bell crank is not binding and is bottomed against the fine adjustment screw by the pull of the bell crank tension spring.

7. Move the TV linkage arm upward to its stop position and hold it.

8. Tighten the trunnion nut.

9. Turn the fine adjustment screw out until there is a gap between its tip and the end of the bell crank rod.

10. Turn the screw back in until it just touches the rod. Always reset this adjustment anytime the curb idle speed is adjusted.

3. Manual Linkage Check and Adjustment

Pre-road Test Check

Before driving the vehicle, you can check the selector linkage adjustment very easily as follows:

1. Move the selector lever through all the ranges.

2. Feel the detents in the transaxle. Are they synchronized with the shift selector markings?

3. Shift to the D position and release the gate button.

4. Check if the gate stop is against the D position by trying to pull the lever toward 2 without pressing the gate button.

DIAGNOSIS – AUTOMATIC TRANSMISSION

CONDITION	POSSIBLE SOURCE	RESOLUTION
• No forced downshifts.	• Improper fluid level.	• Perform fluid level check.
	• Linkage out of adjustment (external).	• Service or adjust linkage.
	• Improper clutch or band application, or oil pressure control system.	• Perform control pressure test.
	• Damaged internal kickdown linkage.	• Service internal kickdown linkage.
	• Dirty or sticking valve body.	• Clean, service or replace valve body.
	• Dirty or sticking governor	
• Runaway engine on 3-2 downshift.	• Improper fluid level.	• Perform fluid level check.
	• Linkage out of adjustment.	• Service or adjust linkage.
	• Intermediate band out of adjustment.	• Adjust intermediate band.
	• Improper band or clutch application, and one way clutch, or oil pressure control system.	• Perform control pressure test service clutch.
	• Damaged or worn intermediate servo.	• Air pressure test check the intermediate servo. Service servo and/or seals.
	• Polished, glazed band or drum.	• Service or replace as required.
	• Dirty or sticking valve body.	• Clean, service or replace valve body.
• No engine braking in manual first gear.	• Improper fluid level.	• Perform fluid level check.
	• Linkage out of adjustment.	• Service or adjust linkage.
	• Bands out of adjustment.	• On C-3 or C-4 – adjust reverse band. (A.O.D. has no band adjustment)
	• Oil pressure control system.	• Perform control pressure test.
	• Damaged or worn reverse servo.	• Perform air pressure test of reverse servo. Service reverse clutch or rear servo as required.
	• Polished, glazed band or drum.	• Service or replace as required.

Figure 8–58. Ford C3, C4, and AOD.

The adjustment is okay if the lever is right on the stop. If it is over the stop or there is some movement to the stop, adjustment is required. Do not drive the car if the adjustment is not correct.

Adjustment

The adjustment procedure is:

1. Move the selector into D and against the gate stop. (The lever must be held against the stop during the adjustment.)

2. Raise the vehicle and loosen the nut that retains the control cable to the manual lever.

3. Move the lever rearward to the last detent, then forward *two* steps to the D detent.

4. Tighten the nut to 10 to 15 foot pounds (14 to 20 N.m.) torque.

5. Lower the vehicle.

6. Perform the pre-road test check.

4. Road Test

Figure 8–24 shows the items that should be checked on a road test. Refer to your specifications for the shift points in D range.

Minimum throttle shifts should be within specification with just enough throttle opening to accelerate the car. To check WOT shift points, the accelerator should be pressed firmly to the floor.

PROBLEM DIAGNOSIS – AUTOMATIC TRANSMISSION

CONDITION	POSSIBLE SOURCE	RESOLUTION
• No engine braking in manual second gear.	• Improper fluid level.	• Perform fluid level check.
	• Linkage out of adjustment	• Service or adjust linkage.
	• Intermediate band out of adjustment.	• Adjust intermediate band.
	• Improper band or clutch application, or oil pressure control system.	• Perform control pressure test
	• Intermediate servo leaking.	• Perform air pressure test of intermediate servo for leakage. Service as required.
NOTE: No manual second gear A.O.D.	• Polished or glazed band or drum.	• Service or replace as required.
• Transmission noisy – valve resonance.	• Improper fluid level.	• Perform fluid level check.
	• Linkage out of adjustment.	• Service or adjust linkage.
	• Improper band or clutch application, or oil pressure control system.	• Perform control pressure test.
NOTE: Gauges may aggravate any hydraulic resonance. Remove gauge and check for resonance level.	• Cooler lines grounding.	• Free up cooler lines.
	• Dirty sticking valve body.	• Clean, service or replace valve body.
	• Internal leakage or pump cavitation.	• Service as required.
• Transmission overheats.	• Improper fluid level.	• Perform fluid level check.
	• Incorrect engine idle, or performance.	• Tune, or adjust engine idle.
	• Improper clutch or band application, or oil pressure control system	• Perform control pressure test.
	• Restriction in cooler or lines.	• Service restriction.
	• Seized converter one-way clutch.	• Replace one-way clutch.
	• Dirty or sticking valve body.	• Clean, service or replace valve body.
• Transmission fluid leaks.	• Improper fluid level.	• Perform fluid level check.
	• Leakage at gasket, seals, etc.	• Remove all traces of lube on exposed surfaces of transmission. Check the vent for free breathing. Operate transmission at normal temperatures and inspect for leakage. Service as required.
	• Vacuum diaphragm unit leaking (except A.O.D.).	• Replace diaphragm.

Figure 8–59. Ford C3, C4, and AOD.

Here are some tips on diagnosing problems with specific components.

Slipping Conditions or No Drive

Low-Intermediate Band. If the band slips, there will be slip in first gear in any range.

Note: There is no external band adjustment. Band application can be diagnosed by air pressure testing. Adjustment is accomplished by selective servo piston rods. Refer to the ATX overhaul flipchart and shop manual for overhaul procedures.

Turbine Shaft One-Way Clutch. A slipping one-way clutch will allow slip in first gear in ranges D and 2. (Drive should be normal in manual low, since the direct clutch is applied.)

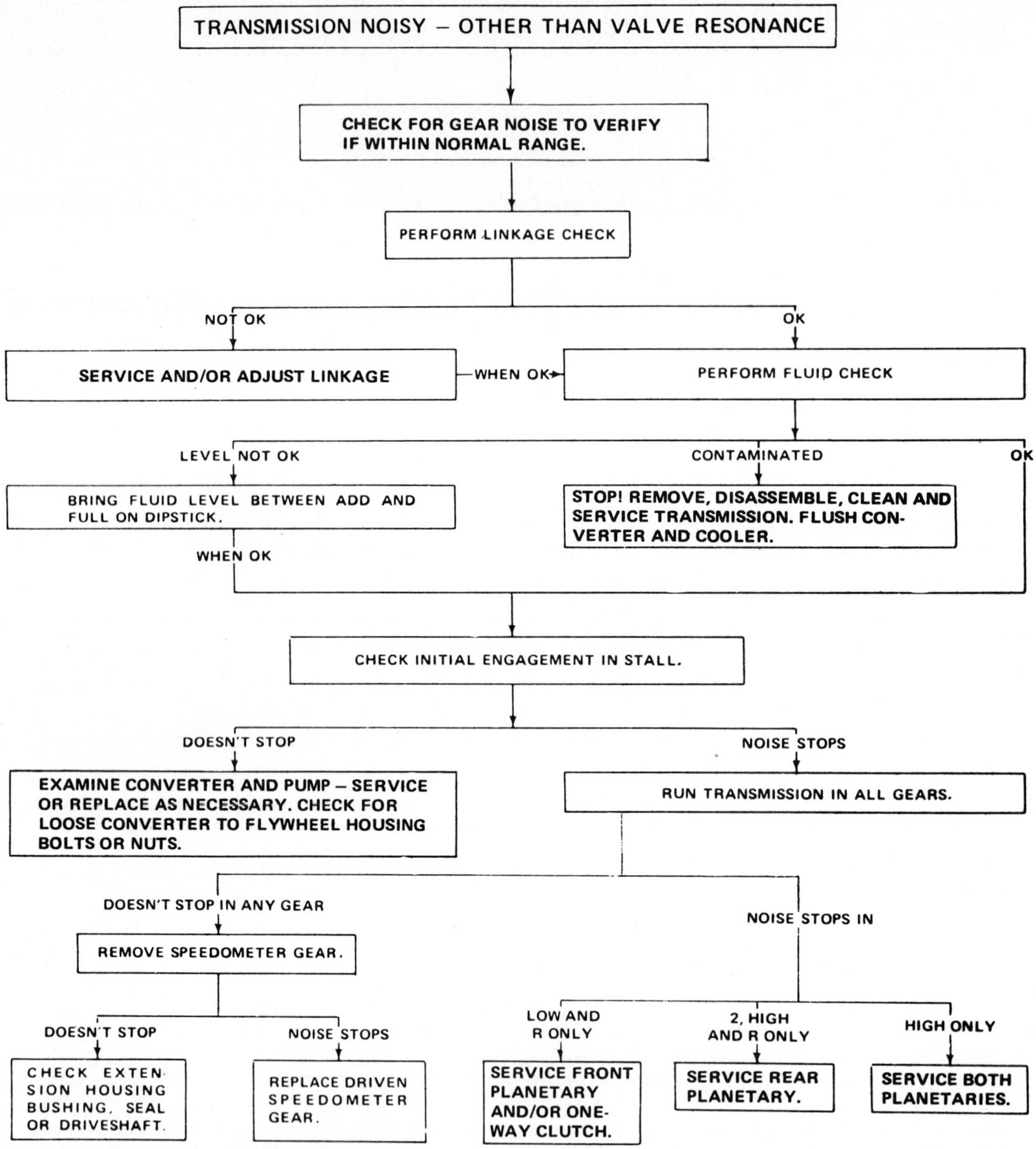

NOTE: A.O.D. TRANSMISSION HAS ONLY ONE GEAR SET.
GEAR NOISE REQUIRES THE REPLACEMENT OF THE PLANETARY CARRIER ASSEMBLY.

Figure 8-60. Ford C3, C4, and AOD.

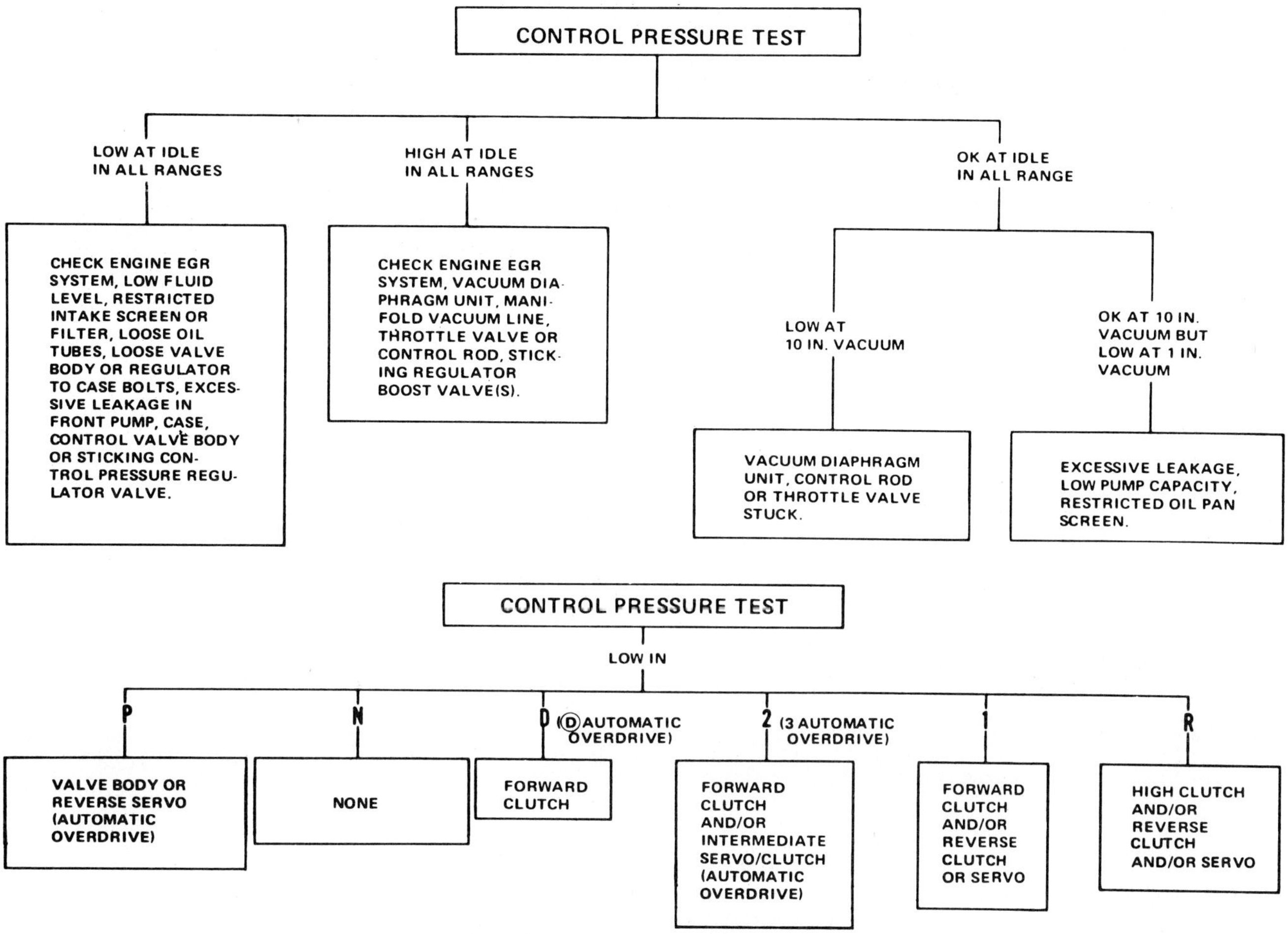

Figure 8–61. Ford C3, C4, and AOD.

Reverse Clutch. If this clutch slips, there will be slip or no drive in reverse.

Intermediate Clutch. A slipping intermediate clutch will cause slip in second and high gear. First gear should be normal.

Direct Clutch. A direct clutch condition would show up in high gear. Also, if this clutch slips, there will be no engine braking in manual low or reverse.

Erratic Shift or Shift Points

Incorrect shift points or erratic shifting can be caused by a stuck governor valve or problems in the TV linkage. Always check the TV adjustment first because it is readily accessible in the engine compartment.

Control or Air Pressure Testing

Slipping conditions normally call for control pressure testing to isolate the cause to a holding member or the valve body. Air pressure testing with the valve body removed also will help isolate the cause and can be used to check the governor as well.

5. Stall Test

The stall test checks for clutch or band slippage, for engine performance, and for torque converter operation. It should be done only with the engine coolant and transmission fluid at proper levels and at operating temperature and with the TV linkage set properly. Apply the service and parking brakes firmly for each stall test.

The chart in Figure 8–44 has spaces for writing in the specified stall speed and results of the test. Proceed as follows.

1. Look up the specified stall for the vehicle. Use a grease pencil to mark the rpm on the dial of a tachometer.

2. Connect the tachometer to the engine and position it for easy reading from the driver seat.

3. In each of the driving ranges, press the accelerator to the floor and hold it just long enough to let the engine get to full rpm.

Note: If the needle goes past the specified rpm, something is slipping. Release the accelerator immediately.

4. Record the results in each range (high, low, or OK).

5. Run the engine at fast idle, in neutral, between each test in order to cool the fluid.

6. Line (Control) Pressure Test

Line pressure on the automatic transaxle is tested in the idle position (minimum TV) and wide-open-throttle position (full TV). It is suggested that the test specifications and results be recorded on paper on a chart similar to the one in Figure 8–44. (In each of the two modes, the reverse specifications will be higher than the others.)

Test Procedure

1. Connect a 300-psi gauge to the line pressure port on the case.

2.. Run the engine until it is hot. Idle pressure *must* be read with the throttle off the fast idle cam.

3. Be sure that the TV linkage is properly adjusted.

4. Apply the service and parking brakes firmly and shift through all the ranges. Record the line pressure and compare it with specifications.

Note: WOT readings are to be made at full stall. However, be sure to run the engine at fast idle in neutral for cooling between tests. To avoid operating at stall, you may have a helper manually force the TV linkage to the kickdown position, which will simulate WOT.

5. Refer to the control pressure test results chart (Figure 8–20) for possible conditions disclosed by out-of-specification pressure readings.

Line Pressure Test Results

After making the control pressure tests, analyze the results to relate to the conditions in Figure 8–28.

Keep in mind that clutch and servo leakage may not show up on the control pressure test. This is because (1) the pump has a high output volume and the leak may not be severe enough to cause a pressure drop and (2) orifices between the pump and pressure chamber may maintain pressure at the source, even with a leak downstream. Pressure loss caused by a less-than-major leak is more likely to show up at idle than at WOT, where the pump is delivering full volume.

Conversely, if you are manipulating the TV linkage to simulate WOT but actually testing at idle, the leak is more likely to cause a pressure loss in the WOT position. To isolate leakage in a clutch or servo circuit further, it is necessary to remove the oil pan (to drain the fluid) and the valve body and to perform case air pressure tests.

7. Air Pressure Tests

The clutch, servo, and governor circuits can be air tested through the passages identified in Figure 8–49. A nozzle should be used similar to that shown in the inset view. A rubber tip is helpful when the nozzle is inserted into a fluid supply hole. Where there is not a tight-fitting hole in the passage, stuff rags around the nozzle.

Air pressure should be regulated to about 25 psi. The compressed air should be filtered and dry to avoid contaminating the fluid.

Clutches

The clutch should apply with a dull thud when compressed air is blown into the pressure ports. A hissing sound and the absence of application noise indicate a leak.

Servo

The servo application should be heard as the piston bottoms with air pressure at the forward servo apply passage. When the air is removed, you should be able to feel the piston move back against the cover. (The band application and release can be seen with the pump assembly removed.)

The servo release circuit also can be checked by holding the servo applied and pressurizing the release passage. The servo and band should *release* with pressure held at both the apply and release passages.

Governor

Air blown into the control-pressure-to-governor passage should cause the governor to whistle or make a sharp clicking noise. If there is no noise, the governor may be stuck.

8. Diagnosis of Transaxle Noise

The automatic transaxle has only one planet carrier assembly in the gear train. Normally planetary gear noise requires that the planet carrier assembly be replaced. However, it is important to verify first that a noise condition is gear noise.

Ford C5 Diagnosis Procedure (Courtesy of Ford Motor Company)

General Diagnosis Sequence

A good, general diagnosis sequence to follow on C5 transmission problems is:

1. Fluid level and condition,
2. Free movement of downshift linkage,
3. Vacuum hose connected to TV diaphragm,

Condition	Possible Cause	Resolution
Slow initial engagment.	• Improper fluid level • Damaged or improperly adjusted manual linkage • Incorrect TV linkage adjustment • Contaminated fluid • Improper clutch and band application or oil control pressure • Dirty valve body	• Add fluid as required. • Service or adjust linkage. • Change fluid and filter. • Perform control pressure test. • Clean, repair, or replace valve body.
Rough initial engagement in either forward or reverse.	• Improper fluid level • High engine idle • Automatic choke on (warm temperature) • Looseness in halfshafts, CV joints, or engine mounts • Improper clutch or band application or oil control pressure • Incorrect TV linkage adjustment • Sticky or dirty valve body	• Perform fluid check. • Adjust idle to specs. • Disengage choke. • Service as required. • Perform control pressure test. • Service or adjust linkage. • Clean, repair, or replace valve body.
No drive in any gear.	• Improper fluid level • Damaged or improperly adjusted manual linkage • Improper clutch or band application or oil control pressure • Internal leakage • Valve body loose • Damaged or worn clutches or band • Sticking or dirty valve body	• Perform fluid check. • Service or adjust linkage. • Perform control pressure test. • Check and repair as required. • Tighten to specs. • Perform air pressure test. • Clean, repair, or replace valve body.
No forward drive—reverse OK.	• Improper fluid level • Damaged or improperly adjusted manual linkage • Improper one-way clutch, band application, or oil pressure control system • Damaged or worn band, servo, or clutches • Valve body loose • Dirty or sticking valve body	• Perform fluid level check. • Service or adjust linkage. • Perform control pressure test. • Perform air pressure test. • Tighten to specs. • Clean, service, or repair valve body.
No drive, slips, or chatters in reverse—forward OK.	• Improper fluid level • Damaged or improperly adjusted manual linkage • Looseness in half-shafts, CV joints, or engine mounts • Improper oil pressure control • Damaged or worn reverse clutch • Valve body loose • Dirty or sticking valve body	• Perform fluid level check. • Service or adjust linkage. • Service as required. • Perform control pressure test. • Perform air pressure test. • Tighten to specs. • Clean, service, or replace valve body.
Car will not start in neutral or park.	• Neutral start switch improperly adjusted • Neutral start wire disconnected/damaged • Manual linkage improperly adjusted	• Service or adjust neutral start switch. • Replace or repair. • Service or adjust linkage.
No drive, slips, or chatters in 1st gear in D.	• Damaged or worn one-way clutch • Improper fluid level • Damaged or worn band • Incorrect TV linkage adjustment	• Service or replace one-way clutch. • Perform fluid level check. • Service or replace band assembly. • Service or adjust linkage.

(*continued*)

Condition	Possible Cause	Resolution
No drive, slips, or chatters in 2nd gear.	• Improper fluid level • Incorrect TV linkage adjustment • Intermediate friction clutch • Improper clutch application • Internal leakage • Dirty or sticking valve body • Polished, glazed band or drum	• Perform fluid level check. • Service or adjust linkage. • Service clutch. • Perform control pressure test. • Perform air pressure test. • Clean, service, or replace valve body. • Replace or service as required.
Starts up in 2nd or 3rd.	• Improper fluid level • Damaged or improperly adjusted manual linkage • Improper band or clutch application or oil pressure control system • Damaged or worn governor • Valve body loose • Dirty or sticking valve body • Cross leaks between valve body and case mating surface	• Perform fluid level check. • Service or adjust linkage. • Perform control pressure test. • Perform governor check. Replace or service governor. • Tighten to specs. • Clean, service, or replace valve body. • Replace valve body or case as required.
Shift points incorrect.	• Improper fluid level • TV linkage out of adjustment • Improper clutch or band application or oil pressure system • Damaged or worn governor • Dirty or sticking valve body	• Perform fluid level check. • Service or adjust linkage. • Perform shift test and control pressure test. • Service or replace governor. • Clean, service, or replace valve body.
No upshift at any speed in D.	• Improper fluid level • TV linkage out of adjustment • Improper band or clutch application or oil pressure control system • Damaged or worn governor • Dirty or sticking valve body	• Perform fluid level check. • Service or adjust linkage. • Perform control pressure test. • Service or replace governor. Clean screen. • Clean, service, or replace valve body.
Shifts 1–3 in D.	• Improper fluid level • Intermediate friction clutch • Improper clutch application or oil pressure control system • Dirty or sticking valve body	• Perform fluid level check. • Service. • Perform control pressure test. • Clean, service, or replace valve body.
Engine overspeeds on 2–3 shift.	• Improper fluid level • Improper band or clutch application or oil pressure control system • Damaged or worn direct clutch or servo • Dirty or sticking valve body	• Perform fluid level check. • Perform control pressure test. • Perform air pressure test. Service as required. • Clean, service, or replace valve body.
Mushy 1–2 shift.	• Improper fluid level • Incorrect engine performance • Improper TV linkage adjustment • Improper intermediate clutch application or oil pressure control system • Damaged intermediate clutch • Dirty or sticking valve body	• Perform fluid level check. • Tune and adjust engine idle as required. • Service or adjust. • Perform control pressure test. • Perform air pressure test. Service as required. • Clean, service, or replace valve body.
Rough 1–2 shift.	• Improper fluid level • Improper TV linkage adjustment • Incorrect engine idle or performance • Improper intermediate clutch application or oil pressure control system • Dirty or sticking valve body	• Perform fluid level check. • Service or adjust linkage. • Tune and adjust engine idle. • Perform control pressure test. • Clean, service, or repair valve body.

Condition	Possible Cause	Resolution
Rough 2-3 shift.	• Improper fluid level • Incorrect engine performance • Improper band release or direct clutch application, or oil control pressure system • Damaged or worn servo release and direct clutch piston check ball • Improper TV linkage adjustment • Dirty or sticking valve body	• Perform fluid level check. • Tune and adjust engine idle. • Perform control pressure test. • Air pressure test the servo apply and release and the direct clutch piston check ball. Service as required. • Service or adjust linkage. • Clean, service, or replace valve body.
Rough 3-2 shift at closed throttle in D.	• Improper fluid level • Incorrect engine idle or performance • Improper TV linkage adjustment • Improper band or clutch application or oil pressure control system • Improper governor operation • Dirty or sticking valve body	• Perform fluid level check. • Tune and adjust engine idle. • Service or adjust linkage. • Perform control pressure test. • Perform governor test. Service as required. • Clean, service, or replace valve body.
No forced downshifts.	• Improper fluid level • Improper clutch or band application or oil pressure control system • Damaged internal kickdown linkage • TV linkage out of adjustment • Dirty or sticking valve body • Dirty or sticking governor	• Perform fluid level check. • Perform control pressure test. • Service internal kickdown linkage. • Service or adjust TV kickdown. • Clean, service, or replace valve body. • Clean or replace governor.
Runaway engine on 3-2 or 3-1 downshift.	• Improper fluid level • TV linkage out of adjustment • Band out of adjustment • Improper band or clutch application or oil pressure control system • Damaged or worn servo • Polished, glazed band or drum • Dirty or sticking valve body	• Perform fluid level check. • Service or adjust TV linkage. • Check and adjust servo rod travel. • Perform control pressure test. • Air pressure test check the servo. Service servo or seals. • Service or replace as required. • Clean, service, or replace valve body.
No engine braking in manual 1st gear.	• Improper fluid level • TV linkage out of adjustment • Damaged or improperly adjusted manual linkage • Band or clutch out of adjustment • Oil pressure control system • Polished, glazed band or drum • Dirty or sticking valve body	• Perform fluid level check. • Service or adjust linkage. • Check direct clutch and service as required. Check servo rod travel. • Perform control pressure test. • Service or replace as required. • Clean, service, or replace valve body.
No engine braking in manual 2nd gear.	• Improper fluid level • TV linkage out of adjustment • Damaged or improperly adjusted manual linkage • Improper band or clutch application or oil pressure control system • Servo leaking • Polished, glazed band or drum	• Perform fluid level check. • Service or adjust linkage. • Perform control pressure test. • Perform air pressure test of servo for leakage. Service as required. • Service or replace as required.

(*continued*)

Condition	Possible Cause	Resolution
Transaxle noisy—valve resonance. *Note: Gauges may aggravate any hydraulic resonance. Remove gauge and check for resonance level.*	• Improper fluid level • TV linkage out of adjustment • Improper band or clutch application or oil pressure control system • Cooler lines grounding • Dirty or sticking valve body • Internal leakage or pump cavitation	• Perform fluid level check. • Service or adjust linkage. • Perform control pressure test. • Free cooler lines. • Clean, service, or replace valve body. • Service as required.
Transaxle overheats.	• Excessive tow loads • Improper fluid level • Incorrect engine idle or performance • Improper band or clutch application or oil pressure control system • Restriction in cooler or lines • Seized converter one-way clutch • Dirty or sticking valve body	• Check owner's manual for tow restriction. • Perform fluid level check. • Tune or adjust engine idle. • Perform control pressure test. • Service restriction. • Replace converter. • Clean, service, or replace valve body.
Transaxle fluid leaks.	• Improper fluid level • Leakage at gaskets, seals, etc.	• Perform fluid level check. • Remove all traces of lube on exposed surfaces of transaxle. Check the vent for free-breathing. Operate transaxle at normal temperatures and inspect for leakage. Service as required.

4. Manual linkage synchronized,
5. Road test (shift speeds and engine performance),
6. Stall test, and
7. Line pressure test.

In addition to these, the following are covered in this section:

8. Governor test and inspection
9. Case air pressure tests
10. Band adjustments

Items 1, 2, 3, and 4 can be checked easily and quickly in the shop. They should be done on *all* problems before driving the vehicle. Further testing would depend on the results of the road test. The cause may be found before road testing, in which case the road test will confirm the repair.

Fluid Level and Condition

Pre-Road Test Check

The fluid level should ideally be checked with the transmission at operating temperatures. Test conditions are:

- Engine at hot curb idle and
- Selector in park (after moving through all the ranges to fill clutch and servo cavities).

For road testing, a cold check can be made. The fluid level should be between the inner holes. If the level is at or below the lowest hole, the vehicle should *not* be driven without adding fluid. If it's overfilled, the vehicle should not be driven without removing fluid.

Hot Fill

After the vehicle is driven some 20 minutes, the fluid temperature should be in the operating range of 150°–170°F (66°–77°C). The final fluid fill should always be between the arrows at operating temperature.

Note: Do not overfill. The space between the arrows is less than one pint. Do not fill to the arrows unless the transmission is hot.

Check for Leaks

If the fluid level checks low, be sure to leak test the transmission after the road test. Locate and repair leaks as described in the shop manual.

Fluid Specification

Use only Type H fluid as specified on the dipstick. Use of any other fluid can cause damage and will void the warranty.

Check Fluid Condition

The fluid should be clean and normally not discolored. Fluid used with the C5 automatic transmission contains a detergent, which retains in suspension particles generated during normal transmission use. This characteristic may result in a dark coloration of the fluid and does not by itself indicate malfunction or need for repair.

Smell the Fluid

A burnt smell or rotten egg odor indicates a major failure. Overhaul is required.

Look for These Conditions

- Clean and normal color—OK.
- Burnt—burnt clutch or band. Overhaul the transmission, replace the converter, and flush the cooler lines.
- Solid residue—clutch or band material. Overhaul the transmission, replace the converter, and flush the cooler lines.
- Varnish on the dipstick—overhaul the transmission. Check for engine overheating and for coolant leak into the transmission. (There would be evidence of oil in the radiator, indicating leakage between the radiator core and transmission cooler.)

Diagnosis—Throttle Valve and Diaphragm

Before Road Test

Take a quick look at the diaphragm to be sure that the vacuum line is connected. If the vacuum isn't there, the upshifts will be delayed and harsh.

If these symtoms show up on the road test, check out the vacuum system and leak test the diaphragm.

Manifold Vacuum Check

To test the vacuum to the diaphragm:

1. Remove the vacuum hose and connect it to a vacuum gauge.
2. Run the engine at hot curb idle. Check for a steady idle vacuum.
3. Goose the engine and let it return to idle. The vacuum should drop; then return to a steady high vacuum.

If there is not *manifold* vacuum to the diaphragm, inspect the hoses and check for proper routing with the vacuum schematic.

Diaphragm Inspection

Check at the vacuum port of the diaphragm for transmission fluid or water.

- Transmission fluid means the diaphragm is a leaker. Replace it.
- Water may have leaked through the vacuum system causing the diaphragm to rust or freeze during cold weather and to stick. This also would cause shift problems; a new diaphragm will fix the condition.

Vacuum Diaphragm Leak Test

Leak test the diaphragm if shift points are delayed or shifts are harsh. Also leak test it if the fluid is low or the engine appears to be burning oil. (The fluid may be getting sucked into the intake manifold through a ruptured vacuum diaphragm.)

The leak test is simple and can be done on or off the vehicle.

1. Remove the manifold vacuum hose and connect a vacuum source to the diaphragm.
2. Apply 18 inches of vacuum to the diaphragm and see if it holds.

The diaphragm is probably OK if it will hold vacuum. Replace it if it fails to hold vacuum or if the rod doesn't move with vacuum changes.

Is the Diaphragm Sticking?

Check the diaphragm and rod for any binding or sticking.

- Insert the throttle valve operating rod and push it back and forth. If the diaphragm binds, install a new one.
- Install a new rod if the rod is bent.

Throttle Valve Inspection

Do this before you install the diaphragm:

1. Check the throttle valve for free operation in its bore in the case.
2. If the valve sticks, remove it from the bore and inspect it for burrs.

A sticking throttle valve can often be deburred by rolling it on a flat Arkansas stone and quickly correct a shifting problem.

Manual Linkage

Before Road Test

This is a simple check to make at the beginning of the road test.

1. Pull the shift lever toward you (or depress the thumb button for the floor shift) to "ungate" it. Move it through all the ranges.
2. Feel the detents in the transmission. Are they synchronized with the markings on the shift selector?
3. Shift to D and let the lever drop into position.
4. Check if the shift gate pawl is against the D stop by trying to move the lever toward 2 without "ungating" it.

If there is free movement to the D stop or if the pawl is up on the 2 stop, an adjustment is required.

Note: Do not drive the vehicle if the adjustment is not correct.

Linkage Adjustment

1. Locate the slotted rod in the linkage. It will usually be at the transmission manual lever but may be elsewhere. (Refer to the shop manual if necessary.) Loosen the nut or screw.

2. Put the selector in D firmly against the gate stop.

3. Shift the manual lever on the transmission to the D position. It will be three detents away from park.

4. Tighten the nut securely.

5. Repeat the linkage check with the shift lever and gate stops before driving the vehicle.

Downshift Linkage

Check before Road Test

Here is a quick check to eliminate downshift rod problems before the road test.

1. Check for binding. Move the rod down by hand as far as it will go. Check that the valve spring moves it back when you release it.

2. Check for proper setting. The rod must not move until the throttle is almost wide-open. At wide-open throttle, the rod must still have some slight travel (0.050–0.070 inch).

Note: Binding of the rod, a sticking valve, or rod misadjustment must be corrected before road testing.

Downshift Linkage Adjustment

To adjust the downshift linkage:

1. Hold the throttle wide-open against its stop.

2. Push the rod down in order to force the downshift valve to bottom in the valve body.

3. Measure the clearance between the tip of the adjusting screw and the throttle lever. It should be 0.050 to 0.070 inch.

4. Turn the screw as required to obtain this clearance.

Quick Checks

The bands can be quickly checked before the road test as follows:

1. Be sure that the oil level is correct.

2. Shift the selector to positions R and 2, and check for firm engagement.

- If engagement in R is delayed or mushy, adjust the rear (low-reverse) band.
- If engagement in 2 is delayed or mushy, adjust the front (intermediate) band.

Adjustment

To adjust either band:

1. Loosen the adjuster stop.

2. Remove and discard the locknut. Install a new nut loosely on the stop.

3. Tighten the screw to 10 foot-pounds torque or until the adjuster wrench overruns. (All the special adjuster tools click and overrun at 10 foot pounds.)

4. Back the screw off exactly the specified number of turns—intermediate band, $4\frac{1}{4}$ turns; low-reverse band, 3 turns.

5. Hold the adjustment and tighten the new locknut 35 to 45 foot-pounds.

6. Road test the car. If the complaint isn't corrected, air test the servo.

Neutral Start Switch

Adjust the neutral start or backup lamp (truck only) switch if the engine won't crank in N and P, if it cranks in any other selector position (passenger car), or the backup lamps will not light (truck).

1. Put the selector in neutral and hold it there.

2. Loosen the switch bolts.

3. Insert a 3/32 gauge pin or a drill through the hole in the switch. Wiggle the switch until the drill seats in the case.

4. Tighten the bolts 55 to 75 inch-pounds and remove the drill.

Road Test Shift Speeds

Road Test Diagnosis

A properly conducted road test is really your most valuable diagnosis tool if you know your transmission and apply this knowledge. It will tell you if any of the drive train holding members are slipping and will usually give a clear indication of valve body and governor functioning.

Figure 8–24 is a chart of what to check for on the road test. You can use it to record test results.

Shift Speeds

Specifications are published in the service manual for automatic shift speeds in three operating conditions in D range:

1. The *minimum throttle* shifts check the operation of the shift valves and governor circuit.

2. The *to-detent* or *torque demand* shifts check the throttle valve systems delay of upshifts and boost

of control pressure. (To-detent is almost full throttle just *before* the downshift rod starts to move.)

3. The *through-detent* or *wide-open* throttle shifts check the downshift (kickdown) system, with the downshift valve actuated.

Test Preparations

Note: Shift speed testing can be done in the shop but will give no indication of shift "feel." A road test is preferable.

To prepare for the shift speed test:

1. Look up the specifications for shift speeds for the particular engine, transmission, and axle combination. Write them on the check sheet. (Record the actual shift speeds during the test for analysis following the test.)

2. Depress the accelerator a time or two to get the feel for where kickdown is actuated. (Some technicians put an extra spring or rubber band on the downshift rod to help feel when it moves.)

Be sure to observe the following procedures and analysis for each of the tests. Always pay attention to the feel of the shifts as well as the shift speeds.

Minimum Throttle Upshifts

To check the minimum throttle upshifts, feed just enough gas to accelerate the vehicle barely. Specifications call for above 12 inches of vacuum for all minimum throttle shifts. That means that the throttle pressure should be low—not enough to raise control pressure or to cause modulated throttle pressure to act on the 2–3 shift valve to delay upshifts.

When you move the selector to D, the forward clutch and governor supply system is charged with control pressure. The clutch applies; with the one-way clutch holding, you are in first gear. Control pressure to the governor is dead-ended at the governor valve so that there is no governor pressure.

As you drive off in first gear, you can be fairly sure that the governor valve is not stuck in the maximum pressure position. If it were, the governor pressure would build up immediately, causing an immediate shift to second.

Minimum Throttle 1-2

When you reach about 10 mph, governor pressure should cause a shift. Since control pressure opposing the governor is low, the 1–2 shift valve moves to charge the servo apply system. A smooth 1–2 shift at 12 mph tells you that the governor and 1–2 shift valve are free and that the servo is applying the band. If the band doesn't apply, you'll stay in first gear and upshift to third at a higher speed.

Minimum Throttle 2-3

Without increasing the throttle setting, check for the 2–3 upshift at specified speed. Now governor pressure has risen high enough to overcome the 2–3 shift valve spring. (There is still no modulated throttle pressure.) The 2–3 shift valve moves and charges the reverse-and-high clutch system. The clutch is applied and at the same time release pressure on the servo releases the band. The governor is still doing its job, the clutch circuit is hydraulically tight, and the 2–3 shift valve is free.

Coasting Downshifts

Let the vehicle coast down in the D range. When the governor primary valve moves (about 10 mph), governor pressure goes to zero. The shift valves move simultaneously to downshift 3–2 and 2–1. In first gear, the car should be freewheeling.

To-Detent Shifts

The to-detent upshifts are full-throttle shifts. Throttle pressure is maximum and acts on the modulator valve so that maximum modulated throttle pressure is available to delay the 2–3 upshift and high control pressure to delay the 1–2 shift. There is no downshift valve input to the shift valves.

Upshifts

If the upshift points are within specification and the shifts are firm, the throttle valve and throttle modulator valve are operating freely. Also, the governor still is doing a good job.

3-2 Shift

At about 25 mph in high gear, push the accelerator to the detent. This raises modulated throttle pressure enough to force the 3–2 shift.

Through-Detent Shifts

Check the through-detent upshifts by pushing the accelerator to wide-open throttle in D and holding it there. Check the kickdown shift points by pushing the accelerator through the detent at the appropriate road speed. These shift points are a check on the downshift valve and on the governor pressure output above 10 mph.

Test Results—Conclusions

After the road test, determine what repairs are needed by analyzing what's going on during the problem shift. Here are some tips on road test analysis.

Governor. It is okay if shift speeds are all okay. If they aren't, refer to the section on governor test and inspections in this section.

Valve body. The problem may be correctable by cleaning the valve body or tightening its bolts to the specified torque.

Throttle valve and diaphragm. Perform the checks in this section if the shifts are early and mushy or harsh and delayed.

Bands and clutches. Use the application summary chart to analyze whether any holding member is slipping. If a band slips, adjust it and drive the vehicle again. Further diagnosis can be made by the line pressure test and case air pressure tests.

Stall Test

The stall test checks for clutch or band slippage, for engine performance, and for torque converter operation. It should be done only with the engine coolant and transmission fluid at proper levels and at operating temperature. Apply the service and parking brakes firmly for each stall test.

1. Look up the specified stall rpm for the vehicle. Use a grease pencil to mark the rpm on the dial of a tachometer.

2. Connect the tachometer to the engine and position it for easy reading from the driver seat.

3. In each of the following ranges, press the accelerator to the floor and hold it just long enough to let the engine get to full rpm.

Range	Specified rpm	High, low, or OK
D		
2		
1		
R		

Note: If the needle goes past the mark, something is slipping. Release the accelerator immediately.

4. Record the results in each range (high, low, or OK).

5. Run the engine at fast idle, in neutral, between each test to cool the fluid.

Line Pressure Test

Line (control) pressure testing is often called for on shift complaints and may be useful in diagnosing slip conditions. Where ordinary checks of the TV system do not disclose the cause of the condition, the pressure test will pinpoint it to other parts of the transmission such as the pump or valve body.

Preparation

1. Attach a 400-psi gauge with a long hose to the pressure takeoff at the left front of the case. Put the gauge where you can see it from the driver's seat.

2. Look up the control pressure specifications and write them on your diagnosis guide sheet or a chart as shown in Figure 8–44.

3. If a vacuum bleed valve is available, tee it into the diaphragm vacuum line. This will let you simulate heavy throttle conditions without operating at stall. If a bleed is not available, tee in an ordinary vacuum gauge, such as on the automatic transmission tester.

Testing

The stall test can be performed at the same time as the pressure test. The procedure following is for the stall method. If you use a vacuum bleed, run the engine at 1,000 rpm for the 10 inches and WOT readings.

1. At idle, with vacuum above 12 inches, record the control pressure in all selector positions.

Note: Apply the parking and service brakes firmly for each test. Do not operate at stall longer than five seconds; just long enough to get a reading. Operate the engine in neutral at fast idle between tests for cooling.

2. Perform the test at 10 inches of vacuum in the forward ranges and record the results.

3. Perform the wide-open throttle tests in D, 2, 1, and R and record the results.

Test Results—Conclusions

The road-map chart shows the likeliest causes of out-of-specification pressure readings in various ranges. In some cases, low pressure can be caused by internal leakage in the clutch or servo circuits. To verify the leakage, remove the valve body and air test the circuits.

Here are some additional tips on analysis of the control pressure test.

Low Pressure at Idle

Low pressure at idle only is usually caused by low pump output, an internal leak, or a sticking regulator valve. Check the pump inlet screen, oil pan filter, and valve body bolt torque. Other causes require major repair procedures.

High Pressure at Idle

High pressure at idle means that there is probably TV input to the pressure regulating system. Check out the throttle valve, diaphragm, and vacuum system. If they're okay, the valve body may need to be serviced.

High or Low in All Tests

Very rarely the pressure will test high or low in all conditions. This can be corrected by replacing the TV diaphragm control rod with a longer or shorter selective rod. Changing the rod length changes the regulating pressure of the throttle valve.

Note: See the shop manual for this procedure. Never change the setting just to soften the shift feel, only to meet specifications.

Pressure OK in All Tests

It may be time to stop the diagnosis process. If the fluid level, linkage, shift speeds, and pressure tests check OK, the transmission is OK. Certain shift-feel conditions then are due to drivability characteristics; no further attempt should be made to diagnose the transmission.

Governor Test

The governor is normally okayed on the road test if all the shift points are to specification. If they are not, a governor test can be performed either in the shop or on the road to determine if the governor is functioning. The test is based on the fact that governor pressure acts on a cutback valve in the valve body to reduce control pressure with increased road speed. If the pressure cutback doesn't happen, the governor or valve body may be at fault.

Preparation

1. Disconnect and plug the TV diaphragm vacuum hose. Attach a hand vacuum tester with a long hose to the diaphragm.
2. In the shop, raise the vehicle slightly. Support it on jack stands to permit the rear wheels to turn.
3. Look up the cutback road speed specifications.

Testing

1. Apply 10 inches of vacuum.
2. Slowly accelerate in range 2.
3. Watch for a pressure cutback between 10 and 13 mph.
4. Record the road speed at which the cutback occurs.
    - Cutback speeds okay. The governor is okay.
    - Cutback early or late. Inspect the governor and screen. If okay, check the cutback valve in the valve body.

Governor Inspections

To inspect the governor and screen, remove the extension housing. Refer to the shop manual for the correct procedure.

Governor In-Car Inspection

1. Manipulate the valves with a thin rod to see if either is sticking.
2. Check the mounting bolt torque. It should be 80–120 inch-pounds. If not, tighten the bolts correctly and recheck for sticking.

Governor Screen

1. Unbolt the governor and remove it from the car.
2. Remove the screen. Check it for contamination. If it's severely contaminated, so is the forward clutch circuit. Remove the oil pan for fluid inspection.
3. Clean and replace the screen.

Governor Bench Inspection

1. Disassemble and inspect the governor. Check valves for free movement in bores when dry.
2. Check sticking valves for burrs.
3. Remove burrs by rolling on an Arkansas stone.
4. Deburr the valve bore with rolled crocus cloth.
5. Assemble and install the governor.

Case Air Pressure Tests

Testing the clutch and servo circuits with air pressure tells you whether slipping is caused by internal leakage or by a mechanical failure. Use 25-psi shop air and a nozzle. If the circuit leaks, locate the cause by inspecting the seals during disassembly. (The servos can be repaired on-car; clutches require bench overhaul.) If the circuit doesn't leak, the problem is mechanical. Refer to the service manual for a detailed procedure for these tests.

Chrysler A404, A413, and A470 Transaxle Diagnosis Procedure (Courtesy of Chrysler Corporation)

Diagnosis—General

Automatic transaxle malfunctions may be caused by four general conditions: poor engine performance, improper adjustments, hydraulic malfunctions, and mechanical malfunctions. Diagnosis of these problems should always begin by checking the easily accessible variables: fluid level and condition, manual cable adjustment, and throttle cable adjustment. Then perform a road test to determine if the problem has been corrected or more diagnosis is necessary. If the problem exists after the preliminary tests and corrections are completed, hydraulic pressure tests should be performed.

Fluid Level and Condition

Before removing the dipstick, wipe all dirt off of the protective cap and top of the dipstick. Since this torque converter fills in both the P (park) and N (neutral) positions, place the selector lever in P to be sure that the fluid level check is accurate. The engine should be running at idle speed. The fluid should be at normal operating temperature (approximately 82°C or 180°F). The fluid level is correct if it is between the full and add marks on the dipstick.

Low fluid level can cause a variety of conditions because it allows the pump to take in air along with the fluid. As in any hydraulic system, air bubbles make the fluid spongy; therefore, pressures will be low and build up slowly.

Improper filling can also raise the fluid level too high. When the transaxle has too much fluid, the gears churn up foam and cause the same conditions that occur with a low fluid level.

In either case, the air bubbles can cause overheating, fluid oxidation, and varnishing, which can interfere with normal valve, clutch, and servo operation. Foaming can also result in fluid escaping from the transaxle vent, where it may be mistaken for a leak.

Along with fluid level, it is important to check the condition of the fluid. When the fluid smells burned and is contaminated with metal or friction material particles, a complete transaxle overhaul is needed. Be sure to examine the fluid on the dipstick closely. If there is any doubt about its condition, drain out a sample for a double-check.

After the fluid has been checked, seat the dipstick fully to seal out water and dirt.

Manual Linkage

Normal operation of the neutral safety switch provides a quick check to confirm proper manual linkage adjustment.

Move the selector lever slowly upward until it clicks into the P (park) notch in the selector gate. If the starter will operate, the P position is correct.

After checking the P position, move the selector slowly toward the N (neutral) position until the lever drops at the end of the N stop in the selector gate. If the starter will also operate at this point, the manual linkage is properly adjusted. If adjustment is required, refer to gearshift linkage adjustment in Maintenance and Adjustments in the service manual.

Throttle Cable

The throttle cable adjustment is important to proper transaxle operation. This adjustment positions a valve that controls shift speed, shift quality, and part-throttle downshift sensitivity. If the setting is too short, early shifts and slippage between shifts may occur. If the setting is too long, shifts may be delayed and part-throttle downshifts may be very sensitive.

Road Test

Prior to performing a road test, be certain that the fluid level and condition and the control cable adjustments have been checked and approved.

During the road test the transaxle should be operated in each position to check for slipping and any variation in shifting. Note whether the shifts are harsh or spongy and check the speeds where the upshifts and downshifts occur. Approximate shift speeds for the various modes of operation are shown in the automatic shift speeds and governor pressure chart in the service manual.

Observe closely for slipping or engine speed flare-up. Slipping or flare-up in any gear usually indicates clutch, band, or overrunning clutch problems. If the condition is far advanced, an overhaul will probably be necessary to restore normal operation.

In most cases, the clutch or band that is slipping can be determined by noting the transaxle operation in all selector positions and by comparing which internal units are applied in those positions. The elements in use chart provides a basis for road test analysis.

By observing that the rear clutch is applied in both the D first-gear and 1 first-gear positions but that the overrunning clutch is applied in D first and the low and reverse band is applied in 1 first, if the transaxle slips in D-range first-gear but does not slip in 1 first gear, the overrunning clutch must be the unit that is slipping. Similarly, if the transaxle slips in any two forward gears, the rear clutch is the slipping unit.

Using the same procedure, the rear clutch and front clutch are applied in D third gear. If the transaxle slips in third gear, either the front clutch or the rear clutch is slipping. By selecting another gear that does not use one of those units, the unit that is slipping can be determined. If the transaxle also slips in reverse, the front clutch is slipping. If the transaxle does not slip in reverse, the rear clutch is slipping.

This process of elimination can be used to detect any unit that slips and to confirm proper operation of good units. However, although road test analysis can usually diagnose slipping units, the actual cause of the malfunction usually cannot be decided. Practically any condition can be caused by leaking hydraulic circuits or sticking valves.

Therefore, unless the condition is obvious, like no drive in D-range first gear only, the transaxle should never be disassembled until hydraulic pressure tests have been performed.

Hydraulic Pressure Tests

Pressure testing is an important step in the diagnostic procedure. These tests usually reveal the cause of most transaxle problems.

Before performing pressure tests, be certain that fluid level and condition and the control cable adjustments have been checked and approved. Fluid must be at operating temperature (150° to 200°F).

Install an engine tachometer, raise the vehicle on a hoist that allows front wheels to turn, and position the tachometer so that it can be read. Disconnect the throttle cable and shift cable from transaxle levers so that they can be controlled from outside the vehicle. Attach 150-psi gauges to ports required for test being conducted. A 300-psi gauge (C-3293) is required for the "reverse" pressure test at the rear servo.

Test 1 (Selector in 1)

1. Attach gauges to the line and rear servo ports.
2. Operate the engine at 1,000 rpm for the test.
3. Move the selector lever on the transaxle all the way forward (1 position).
4. Read pressures on both gauges as the throttle lever on the transaxle is moved from the full forward position to the full rearward position.
5. Line pressure should read 52 to 58 psi with the throttle lever forward and gradually increase, as the lever is moved rearward, to 80 to 88 psi.
6. Rear servo pressure should read the same as line pressure within 3 psi.
7. This tests pump output, pressure regulation, and condition of rear clutch and rear servo hydraulic circuits.

Test 2 (Selector in 2)

1. Attach the gauge to the line pressure port and tee into lower cooler line fitting to read lubrication pressure.
2. Operate the engine at 1,000 rpm for the test.
3. Move the selector lever on the transaxle one detent rearward from full forward position. This is the selector 2 position.
4. Read pressures on both gauges as the throttle lever on the transaxle is moved from the full forward position to the full rearward position.
5. Line pressure should read 52 to 58 psi with the throttle lever forward and gradually increase, as the lever is moved rearward, to 80 to 88 psi.
6. Lubrication pressure should be 10 to 25 psi with the lever forward and 10 to 35 psi with the lever rearward.
7. This tests pump output, pressure regulation, and condition of rear clutch and lubrication hydraulic circuits.

Test 3 (Selector in D)

1. Attach gauges to the line and front servo release ports.
2. Operate the engine at 1,600 rpm for the test.
3. Move the selector lever on the transaxle two detents rearward from the full forward position. This is the selector D position.
4. Read pressures on both gauges as the throttle lever on the transaxle is moved from the full forward position to the full rearward position.
5. Line pressure should read 52 to 58 psi with the throttle lever forward and gradually increase, as the lever is moved rearward.
6. Front servo release is pressurized only in direct drive and should be the same as line pressure within 3 psi, up to kickdown point.
7. This tests pump output, pressure regulation, and condition of rear clutch, front clutch, and hydraulic circuits.

Test 4 (Selector in Reverse)

1. Attach 300-psi gauge to the rear servo apply port.
2. Operate the engine at 1,600 rpm for the test.
3. Move the selector lever on the transaxle four detents rearward from the full forward position. This is the selector R position.
4. Rear servo pressure should read 160 to 180 psi with the throttle lever forward and gradually increase, as the lever is moved rearward, to 250 to 300 psi.
5. This tests pump output, pressure regulation, and condition of front clutch and rear servo hydraulic circuits.
6. Move the selector lever on the transaxle to the D position to check that rear servo pressure drops to zero.

7. This tests for leakage into the rear servo due to case porosity, which can cause reverse band burnout.

Test Result Indications

1. If proper line pressure, minimum to maximum, is found in any one test, the pump and pressure regulator are working properly.

2. Low pressure in D, 1, and 2 but correct pressure in R indicates rear clutch circuit leakage.

3. Low pressure in D and R but correct pressure in 1 indicates front clutch circuit leakage.

4. Low pressure in R and 1 but correct pressure in 2 indicates rear servo circuit leakage.

5. Low line pressure in all positions indicates a defective pump, a clogged filter, or a stuck pressure regulator valve.

Governor Pressure

Test only if the transaxle shifts at wrong vehicle speeds when the throttle cable is correctly adjusted.

1. Connect a 0- to 150-psi pressure gauge to the governor pressure takeoff point, located at the lower right side of case, below the differential cover.

2. Operate the transaxle in third gear to read pressures and compare speeds shown in the chart.

If governor pressures are incorrect at the given vehicle speeds, the governor valves are probably sticking. The governor pressure should respond smoothly to changes in mph and should return to 0 to 3 psi when the vehicle is stopped. High pressure at standstill (above 3 psi) will prevent the transaxle from downshifting.

Throttle Pressure

No gauge port is provided for throttle pressure. Incorrect throttle pressure should be suspected only if part-throttle upshift speeds are either delayed or occur too early in relation to vehicle speeds. Engine runaway on either upshifts or downshifts can also be an indicator of incorrect (low) throttle pressure setting. In no case should throttle pressure be adjusted until the transaxle throttle cable adjustment has been verified as correct.

Converter Stall Test

Caution: During the test no one should stand in front of the vehicle.

The stall test consists of determining the engine speed obtained at full throttle in the D position only. This test checks the torque converter stator clutch operation and the holding ability of the transaxle clutches. The transmission oil level should be checked and the engine brought to normal operating temperature before stall operation. Both the parking and service brakes must be fully applied and rear wheels blocked during this test.

Do not hold the throttle open any longer than necessary to obtain a maximum engine speed reading and never longer than 5 seconds at a time. If more than one stall check is required, operate the engine at approximately 1,000 rpm in neutral for 20 seconds to cool the transmission fluid between runs. If engine speed exceeds the maximum limits shown, release the accelerator immediately since transaxle clutch slippage is indicated.

Stall Speed above Specification

If stall speed exceeds the maximum specified in the chart by more than 200 rpm, transaxle clutch slippage is indicated. Follow the transaxle oil pressure and air pressure checks outlined in this section to determine the cause of slippage.

Stall Speed below Specification

Low stall speeds with a properly tuned engine indicate torque converter stator clutch problems. A road test will be necessary to identify the exact problem.

If stall speeds are 250–350 rpm below specification and the vehicle operates properly at highway speeds but has poor through-gear acceleration, the stator overrunning clutch is slipping.

If stall speed and acceleration are normal but abnormally high trottle opening is required to maintain highway speeds, the stator clutch has seized.

Both of these stator defects require replacement of the torque converter.

Noise

A whining or sirenlike noise due to fluid flow is normal during stall operation with some converters; however, loud metallic noises from loose parts or interference within the assembly indicate a defective torque converter. To confirm that the noise originates within the converter, operate the vehicle at light throttle in D and N on a hoist and listen under the transaxle bell housing.

Clutch and Servo Air Pressure Tests

A no-drive condition might exist even with correct fluid pressure because of inoperative clutches or bands. The inoperative units, clutches, bands, and servos can be located through a series of tests by substituting air pressure for fluid pressure.

The front and rear clutches, kickdown servo, and low-reverse servo may be tested by applying air pressure to their respective passages after the valve body assembly has been removed. To make air pressure tests, proceed as follows. Compressed air supply must be free of all dirt or moisture. Use a pressure of 30 psi.

Front Clutch

Apply air pressure to the front clutch apply passage and listen for a dull thud, which indicates that the front clutch is operating. Hold air pressure on for a few seconds and inspect the system for excessive oil leaks.

Rear Clutch

Apply air pressure to the rear clutch apply passage and listen for a dull thud, which indicates that the rear clutch is operating. Also inspect for excessive oil leaks. If a dull thud cannot be heard in the clutches, place finger tips on the clutch housing and again apply air pressure. Movement of the piston can be felt as the clutch is applied.

Kickdown Servo (Front)

Operation of the servo is indicated by a tightening of the front band. Spring tension of the servo piston should release the band.

Low and Reverse Servo (Rear)

Direct air pressure into the rear servo apply passage. Operation of the servo is indicated by a tightening of the rear band. Spring tension on the servo piston should release the band.

If clutches and servos operate properly, no upshift or erratic shift conditions indicate that malfunctions exist in the valve body.

Fluid Leakage— Transaxle Converter Housing Area

1. Check for the source of leakage. Since fluid leakage at or around the converter area may originate from an engine oil leak, the area should be examined closely. Factory-fill fluid is dyed red and therefore can be distinguished from engine oil.

2. Prior to removing the transaxle, perform the following checks. When leakage is determined as originating from the transaxle, check fluid level prior to removal of the transmission and torque converter. High oil level can result in oil leakage out the vent located in the dipstick. If the fluid is high, adjust to the proper level.

After performing this operation, inspect for leakage. If a leak persists, perform the following operation on the vehicle to determine if it is the converter or transaxle that is leaking.

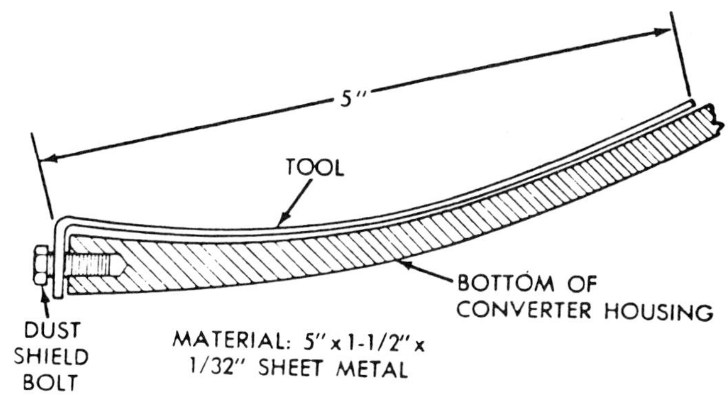

Figure 8-62.

Leakage Test Probe

1. Remove the converter housing dust shield.

2. Wipe the bottom interior of the converter housing as dry as possible. A solvent spray followed by compressed air drying is preferable.

3. Fabricate and fasten the test probe (Figure 8–62) securely to the convenient dust shield bolt hole. Make certain that the converter is cleared by the test probe. The tool must be clean and dry.

4. Run the engine at approximately 2,500 rpm with the transmission in neutral for about two minutes. The transaxle must be at operating temperature.

5. Stop the engine and carefully remove the tool.

6. If the upper surface of the test probe is dry, there is no converter leak. A path of fluid across the probe indicates a converter leak. Oil leaking under the probe is coming from the transaxle converter area.

7. Remove the transaxle and torque converter assembly from the vehicle for further investigation. The fluid should be drained from the transaxle. Install the oil pan (with R.T.V. sealant) at specified torque.

Possible sources of transaxle converter area fluid leakage are:

1. Converter hub seal:

(a) Seal lip cut; check converter hub finish,

(b) Bushing moved or worn,

(c) Oil return hole in front pump housing plugged or omitted, and

(d) Seal worn out (high-mileage cars);

2. Fluid leakage at the outside diameter from pump housing O-ring seal;

3. Fluid leakage at the front pump-to-case bolts; and

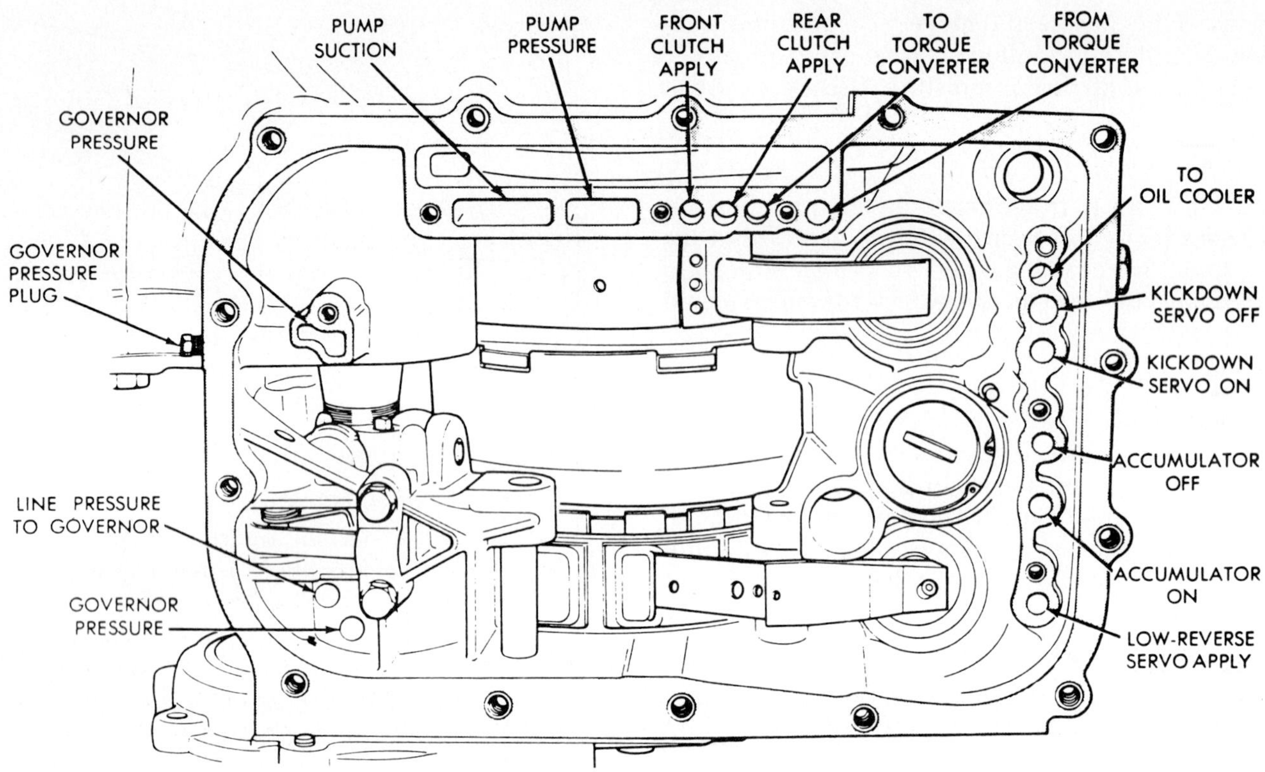

Figure 8-63. Air pressure test locations on Chrysler transaxle. (*Courtesy of Chrysler Corporation*)

4. Fluid leakage due to case or front pump housing porosity.

Torque Converter Leakage

Possible sources of converter leakage are:

1. Torque converter weld leaks at the outside diameter (peripheral) weld and

2. Torque converter hub weld.

The hub weld is inside and not visible. Do not attempt to repair. Replace the torque converter.

Air Pressure Test of Transaxle

Fabricate equipment needed for the test as shown in Figures 8–64 and 8–65.

The transaxle should be prepared for pressure tests as follows after removal of the torque converter.

1. Install a dipstick bore plug and plug oil cooler line fitting (lower fitting).

2. With a rotary motion, install the converter hub seal cup over the input shaft and through the converter hub seal until the cup bottoms against the pump gear lugs. Secure with the cup retainer strap using the converter housing to the engine block retaining bolt for the tapped hole.

Note: Use the lower starter hole and front dowel hole.

3. Attach and clamp hose from nozzle of tool C–4080 to the upper cooler line fitting position in case.

4. Pressurize the transaxle using tool C–4080 until the pressure gauge reads 8 psi. Position the transaxle so that the pump housing and case front may be covered with a soapy solution of water. Leaks are sometimes caused by porosity in the case or pump housing.

Caution: Do not under any circumstances pressurize a transaxle to more than 10 psi.

Note: Air leakage at the differential vent indicates leakage in the area of the transfer shaft bearing retainer.

If a leak source is located, that part and all associated seals, O-rings, and gaskets should be replaced with new parts.

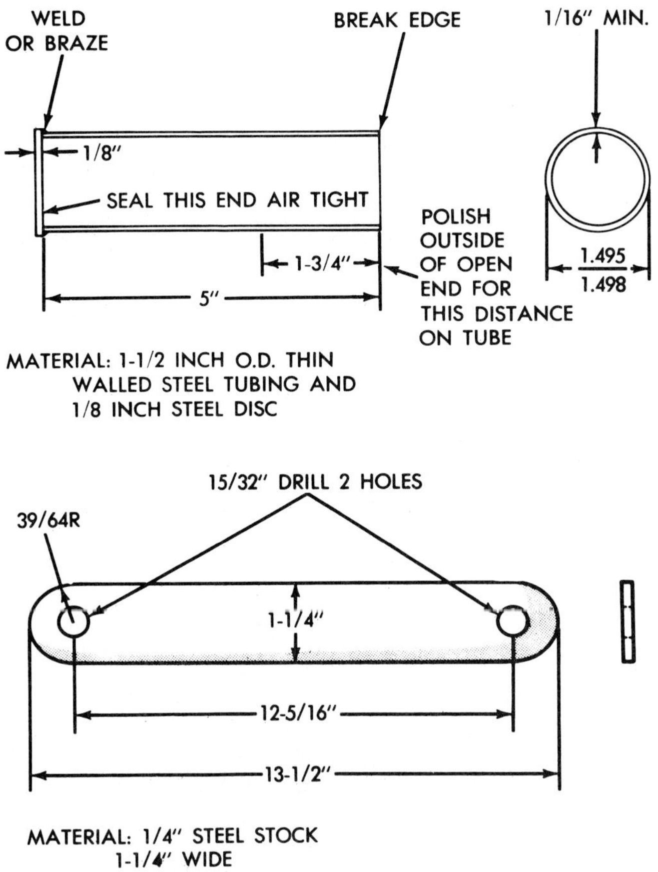

Figure 8-64. Fabrication of converter hub seal cup and retainer for air pressure testing transmission case. (*Courtesy of Chrysler Corporation*)

Chrysler A904 and A727 Diagnosis Procedure (Courtesy of Chrysler Corporation)

Diagnosis—General

Automatic transmission malfunctions may be caused by four general conditions: poor engine performance, improper adjustments, hydraulic malfunctions, and mechanical malfunctions. Diagnosis of these problems should always begin by checking the easily accessible variables: fluid level and condition, manual linkage adjustment, and throttle linkage adjustment. Then perform a road test to determine whether the problem has been corrected or that more diagnosis is necessary. If the problem exists after the preliminary tests and corrections are completed, hydraulic pressure tests should be performed.

Fluid Level and Condition

Before removing the dipstick, wipe all dirt off the protective cap and top of the filler tube. Since the torque converter fills more slowly in the P (park) position, place the selector lever in N (neutral) to be sure that the fluid level check is accurate. The engine should be running at idle speed. The fluid should be at a normal operating temperature (approximately 175°F). The fluid level is correct if it is between the full and add-one-pint marks on the dipstick.

Low fluid level can cause a variety of conditions because it allows the pump to take in air along with the fluid. As in any hydraulic system, air bubbles

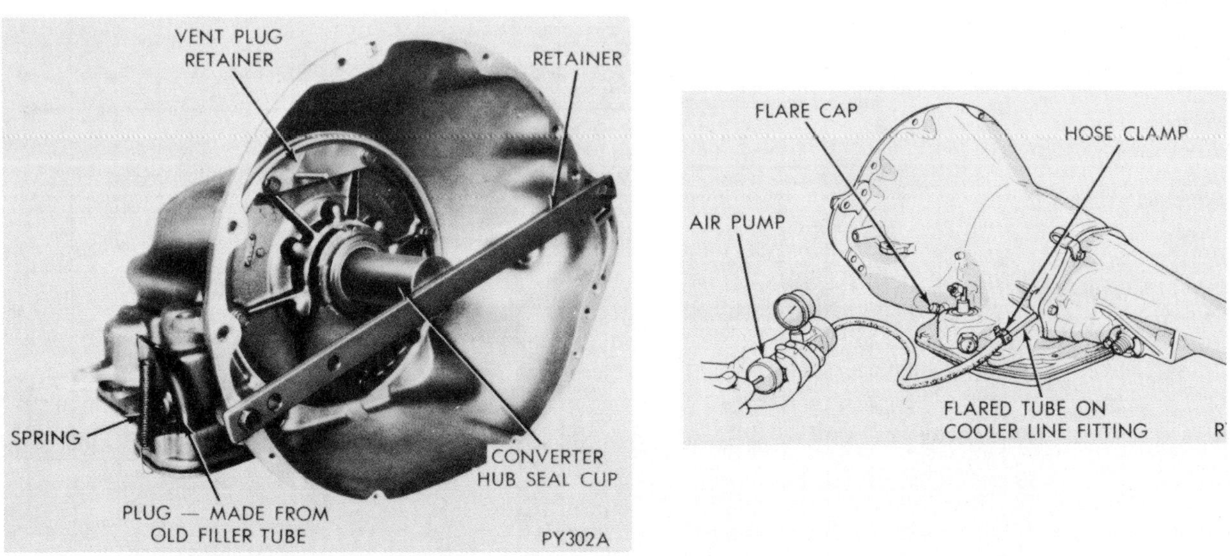

Figure 8-65. Air pressure testing transmission case for leaks. (*Courtesy of Chrysler Corporation*)

No.	Possible Cause	Harsh engagement from neutral to D	Harsh engagement R	Delayed engagement from neutral to D	Delayed engagement R	Runaway upshift	No upshift	3-2 kickdown runaway	No kickdown or normal downshift	Shifts erratic	Slips in forward drive positions	Slips in reverse only	Slips in all positions	No drive in any position	No drive in forward drive positions	No drive in reverse	Drives in neutral	Drags or locks	Grating, scraping growling noise	Buzzing noise	Hard to fill, oil blows out filler hole	Transaxle overheats	Harsh upshift	Delayed upshift
33	Overrunning clutch inner race damaged.																			X				
32	Overrunning clutch worn, broken or seized.										X				X			X	X					
31	Planetary gear sets broken or seized.													X	X	X		X	X					
30	Rear Clutch dragging.																X							
29	Worn or faulty rear clutch.	X	X	X	X						X				X	X	X							
28	Insufficient clutch plate clearance.																X					X		
27	Faulty cooling system.																					X		
26	Kickdown band adjustment too tight.																	X				X		
25	Hydraulic pressure too high.	X	X																				X	
24	High fluid level.																				X			
23	Worn or faulty front clutch.				X	X	X	X		X		X				X								X
22	Kickdown servo band or linkage malfunction.					X	X	X	X	X														X
21	Governor malfunction.						X		X	X														X
20	Worn or broken reaction shaft support seal rings.				X	X	X			X		X				X								X
19	Governor support seal rings broken or worn.						X	X		X														X
18	Drive shaft(s) bushing(s) damaged.																		X					
17	Overrunning clutch not holding.										X				X									
16	Kickdown band out of adjustment.							X											X				X	X
15	Incorrect throttle linkage adjustment.					X	X	X	X	X	X												X	X
14	Engine idle speed too low.			X	X																			
13	Aerated fluid.			X	X	X		X		X	X	X	X							X	X			
12	Worn or broken input shaft seal rings.			X	X						X		X		X									
11	Faulty oil pump.			X	X					X	X	X	X	X								X		
10	Oil filter clogged.			X	X	X				X	X		X	X							X			
9	Incorrect gearshift control linkage adjustment.			X	X		X			X	X	X				X	X					X		
8	Low fluid level.			X	X	X	X	X		X	X	X	X	X	X					X		X		
7	Low-reverse servo, band or linkage malfunction.		X		X							X				X								
6	Valve body malfunction or leakage.	X		X	X	X	X	X	X	X	X	X	X	X	X	X	X			X				
5	Low-reverse band worn out.		X		X							X				X		X	X					
4	Hydraulic pressures too low.			X	X	X	X	X		X	X	X	X	X	X	X						X	X	
3	Engine idle speed too high.	X	X																			X		
2	Stuck switch valve.																					X		
1	Low-reverse band misadjusted (A-413 & A-470).		X		X							X										X		

Figure 8-66. Chrysler transaxle.

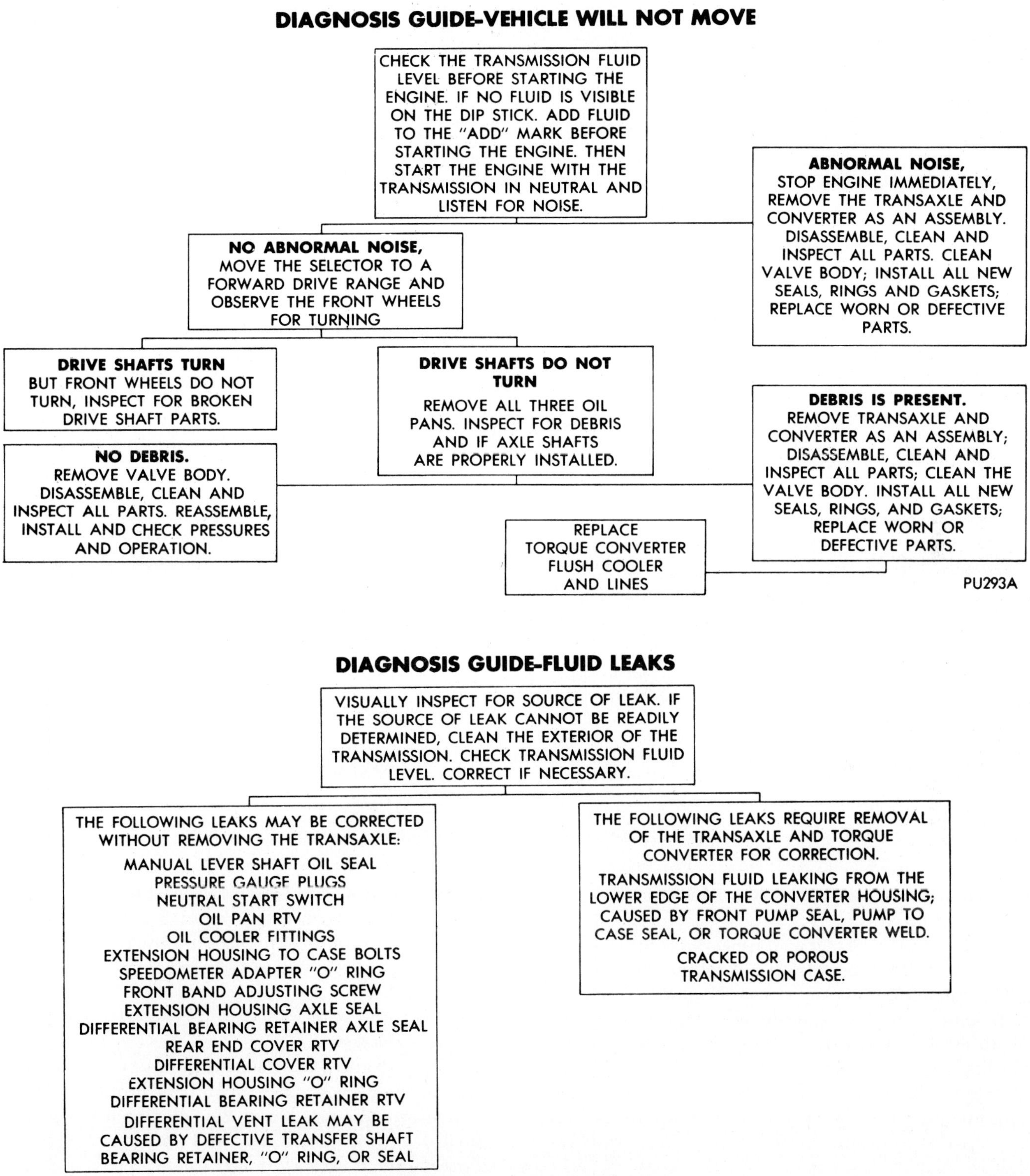

Figure 8-67. Chrysler transaxle.

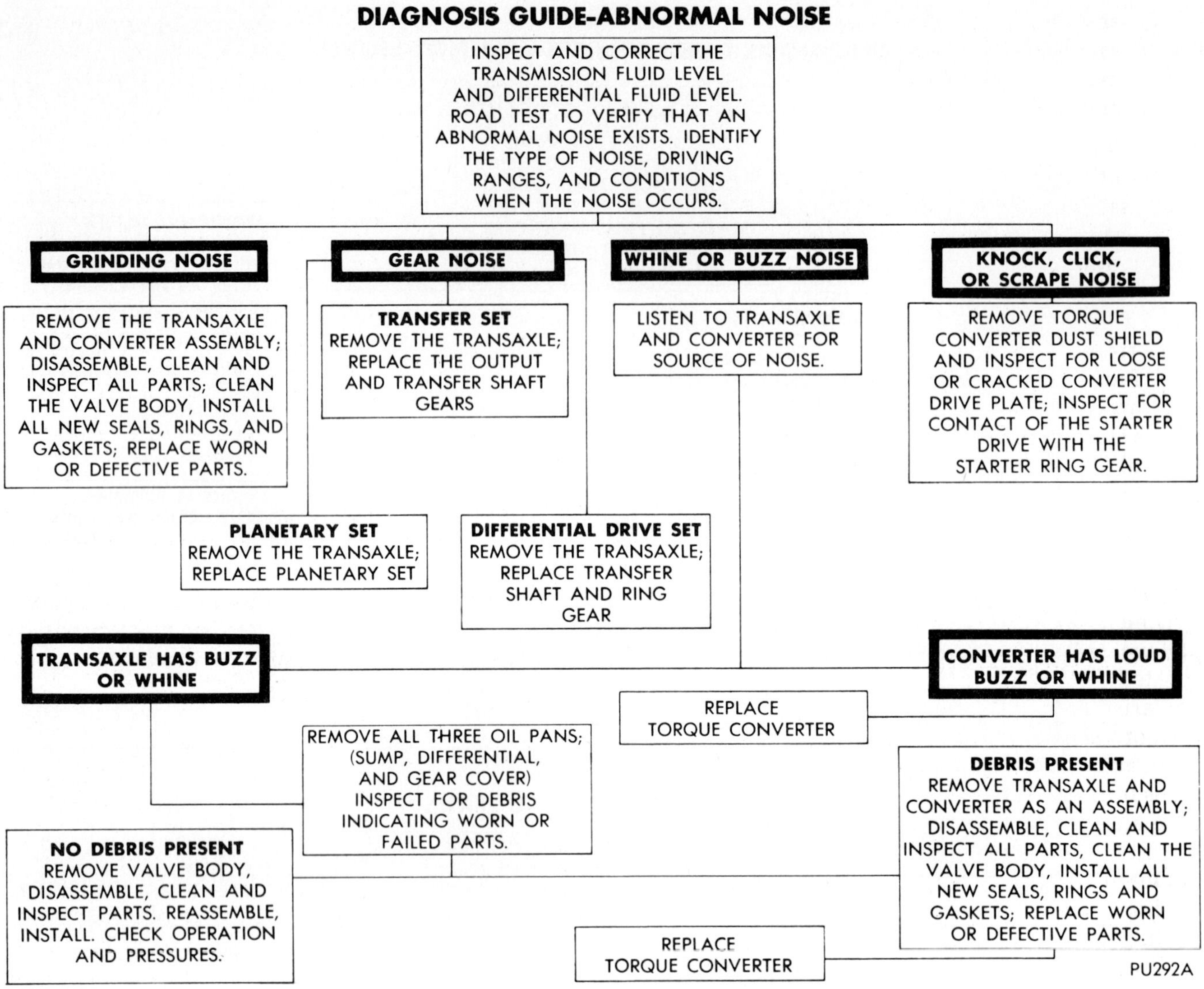

Figure 8-68. Chrysler transaxle.

make the fluid spongy; therefore pressures will be low and build up slowly.

Improper filling can also raise the fluid level too high. When the transmission has too much fluid, the gears churn up foam and cause the same conditions that occur with a low fluid level.

In either case, the air bubbles can cause overheating, fluid oxidation, and varnishing, which can interfere with normal valve, clutch, and servo operation. Foaming can also result in fluid escaping from the transmission vent, where it may be mistaken for a leak.

Along with fluid level, it is important to check the condition of the fluid. When the fluid smells burned and is contaminated with metal or friction material particles, a complete transmission overhaul is needed. Be sure to examine the fluid on the dipstick closely. If there is any doubt about its condition, drain a sample for a double check. After the fluid has been checked, seat the dipstick fully to seal out water and dirt.

Manual Linkage

Normal operation of the neutral safety switch provides a quick check to confirm proper manual linkage adjustment. Move the selector lever slowly upward until it clicks into the P (park) notch in the selector gate. If the starter will operate, the P position is correct.

After checking the P position, move the selector slowly toward the N (neutral) position until the lever drops at the end of the N stop in the selector gate. If the starter will also operate at this point, the manual linkage is properly adjusted. If adjustment is required, refer to gearshift linkage adjustment in maintenance and adjustments in the shop manual.

Throttle Linkage

The throttle rod adjustment is important to proper transmission operation. This adjustment positions a valve that controls shift speed, shift quality, and part-throttle downshift sensitivity. If the setting is too short, early shifts and slippage between shifts may occur. If the setting is too long, shifts may be delayed and part-throttle downshifts may be very sensitive.

In fact, this adjustment is so critical that the use of a throttle lever holding spring is necessary to remove slack in the linkage during adjustment. Refer to throttle rod adjustment in maintenance and adjustments in the shop manual.

Road Test

Prior to performing a road test, be certain that the fluid level and condition and the control linkage adjustments have been checked and approved.

During the road test the transmission should be operated in each position to check for slipping and any variation in shifting. Note whether the shifts are harsh or spongy and check the speeds where the upshifts and downshifts occur. Approximate shift speeds for the various modes of operation are shown in the automatic shift speeds and governor pressure chart in the service manual.

Observe closely for slipping or engine speed flare-up. Slipping or flare-up in any gear usually indicates clutch, band, or overrunning clutch problems. If the condition is far advanced, an overhaul will probably be necessary to restore normal operation.

In most cases, the clutch or band that is slipping can be determined by noting the transmission operation in all selector positions and by comparing which internal units are applied in those positions. The elements-in-use chart provides a basis for road test analysis.

By observing that the rear clutch is applied in both the D first-gear and 1 first-gear positions but that the overrunning clutch is applied in D first and the low and reverse band is applied in 1 first, if the transmission slips in D-range first gear but does not slip in 1 first gear, the overrunning clutch must be the unit that is slipping. Similarly, if the transmission slips in any two forward gears, the rear clutch is the slipping unit.

Using the same procedure, the rear clutch and front clutch are applied in D third gear. If the transmission slips in third gear, either the front clutch or the rear clutch is slipping. By selecting another gear that does not use one of those units, the unit that is slipping can be determined. If the transmission also slips in reverse, the front clutch is slipping. If the transmission does not slip in reverse, the rear clutch is slipping.

This process of elimination can be used to detect any unit that slips and to confirm proper operation of good units. However, although road test analysis can usually diagnose slipping units, the actual cause of the malfunction usually cannot be determined. Practically any condition can be caused by leaking hydraulic circuits or sticking valves. Therefore, unless the condition is obvious, like no drive in D-range first gear only, the transmission should never be disassembled until hydraulic pressure tests have been performed.

An engine tachometer can be used to determine if the lockup clutch in the torque converter is functioning. An instantaneous rise in engine speed of more than 250 rpm at 45 miles per hour when the throttle is opened just short of kickdown, indicates that the lockup clutch is slipping more than normal. Slippage less than 250 rpm is normal.

Hydraulic Pressure Tests

Pressure testing is an important step in the diagnostic procedure. These tests usually reveal the cause of most transmission problems.

Before performing pressure tests, be certain that the fluid level and condition and the control linkage adjustments have been checked and approved. Fluid must be at operating temperature (150° to 200°F).

Install an engine tachometer, raise the vehicle on a hoist that allows rear wheels to turn, and position the tachometer so that it can be read under the vehicle.

Disconnect the throttle rod and shift rod from transmission levers so that they can be controlled under the vehicle.

Attach 100-psi gauges to ports required for the test being conducted. A 300-psi gauge is required for the reverse pressure test at rear servo.

Test 1 (Selector in 1)

1. Attach gauges to the line and rear servo ports.
2. Operate the engine at 1,000 rpm for the test.
3. Move the selector lever on the transmission all the way forward (1 position).
4. Read pressures on both gauges as the throttle lever on the transmission is moved from the full forward position to the full rearward position.

5. Line pressure should read 54 to 60 psi with the throttle lever forward and gradually increase, as the lever is moved rearward, to 90 to 96 psi.

6. Rear servo pressure should read the same as line pressure within 3 psi.

7. This tests pump output, pressure regulation, and condition of rear clutch and rear servo hydraulic circuits.

Test 2 (Selector in 2)

1. Attach the gauge to the line pressure port and tee into the rear cooler line fitting to read lubrication pressure.

2. Operate the engine at 1,000 rpm for the test.

3. Move the selector lever on the transmission one detent rearward from the full forward position. This is selector 2 position.

4. Read pressures on both gauges as the throttle lever on the transmission is moved from the full forward position to the full rearward position.

5. Line pressure should read 54 to 60 psi with the throttle lever forward and gradually increase, as the lever is moved rearward, to 90 to 96 psi.

6. Lubrication pressure should be 5 to 15 psi with the lever forward and 10 to 30 psi with the lever rearward.

7. This tests pump output, pressure regulation, and condition of rear clutch and lubrication hydraulic circuits.

Test 3 (Selector in D)

1. Attach gauges to the line and front servo release ports.

2. Operate the engine at 1,600 rpm for the test.

3. Move the selector lever on the transmission two detents rearward from the full forward position. This is selector D position.

4. Read pressures on both gauges as the throttle lever on the transmission is moved from the full forward position to the full rearward position.

5. Line pressure should read 54 to 60 psi with the throttle lever forward and gradually increase, as the lever is moved rearward.

6. Front servo release is pressurized only in direct drive and should be the same as line pressure within 3 psi, up to the downshift point.

7. This tests pump output, pressure regulation, and condition of rear clutch, front clutch, and lockup clutch hydraulic circuits.

Test 4 (Selector in Reverse)

1. Attach a 300-psi gauge to the rear servo apply port.

2. Operate the engine at 1,600 rpm for the test.

3. Move the selector lever on the transmission four detents rearward from the full forward position. This is selector R position.

4. Rear servo pressure should read 160 to 270 psi.

5. This tests pump output, pressure regulation, and condition of front clutch and rear servo hydraulic circuits.

6. Move the selector lever on the transmission to the D position to check that rear servo pressure drops to zero.

7. This tests for leakage into the rear servo due to case porosity, which can cause reverse band burnout.

Test Result Indications

1. If proper line pressure, minimum to maximum, is found in any one test, the pump and pressure regulator are working properly.

2. Low pressure in D, 1, and 2 but correct pressure in R indicates rear clutch circuit leakage.

3. Low pressure in D and R but correct pressure in 1 indicates front clutch circuit leakage.

4. Low pressure in R and 1 but correct pressure in 2 indicates rear servo circuit leakage.

5. Low line pressure in all positions indicates a defective pump, a clogged filter, or a stuck pressure regulator valve.

Governor Pressure

Test only if the transmission shifts at wrong vehicle speeds when the throttle rod is correctly adjusted.

1. Connect a 0- to 100-psi pressure gauge to the governor pressure takeoff point, located at the lower left side of extension near the mounting flange.

2. Operate the transmission in third gear to read pressures and compare speeds shown in the chart.

If governor pressures are incorrect at the given vehicle speeds, the governor valve or weights are probably sticking. The governor pressure should respond smoothly to changes in mph and should return to 0 to $1\frac{1}{2}$ psi when the vehicle is stopped. High pressure at standstill (above 2 psi) will prevent the transmission from downshifting.

Throttle Pressure

No gauge port is provided for throttle pressure. Incorrect throttle pressure should be suspected only if part-throttle upshift speeds are either delayed or occur too early in relation to vehicle speeds. Engine runaway on either upshifts or downshifts can also be an indicator of incorrect (low) throttle pressure setting. In no case should throttle pressure be adjusted until the transmission throttle linkage adjustment has been verified as correct.

Converter Stall Test

Note: During the test no one should stand in front of the vehicle.

The stall test consists of determining the engine speed obtained at full throttle in D position only. This test checks the torque converter stator clutch operation and the holding ability of the transmission clutches. The transmission oil level should be checked and the engine brought to the normal operating temperature before stall operation. Both the parking and service brakes must be fully applied and front wheels blocked during this test.

Do not hold the throttle open any longer than necessary to obtain a maximum engine speed reading and never longer than 5 seconds at a time. If more than one stall check is required, operate the engine at approximately 1,000 rpm in neutral for 20 seconds to cool the transmission fluid between runs. If engine speed exceeds the maximum limits shown, release the accelerator immediately since transmission clutch slippage is indicated.

Stall Speed above Specification

If stall speed exceeds the maximum specified in the chart by more than 200 rpm, transmission clutch slippage is indicated. Follow the transmission oil pressure and air pressure checks outlined in this section to determine the cause of slippage.

Stall Speed below Specification

Low stall speeds with a properly tuned engine indicate torque converter stator clutch problems. A road test will be necessary to identify the exact problem.

If stall speeds are 250–350 rpm below specification and the vehicle operates properly at highway speeds but has poor through-gear acceleration, the stator overrunning clutch is slipping (lockup and nonlockup torque converters). If stall speed and acceleration are normal but abnormally high throttle opening is required to maintain highway speeds, the stator clutch has seized (nonlockup torque converter only). Both of these stator defects require replacement of the torque converter.

Noise

A whining or sirenlike noise due to fluid flow is normal during stall operation with some converters; however, loud metallic noises from loose parts or interference within the assembly indicates a defective torque converter. To confirm that the noise originates within the converter, operate the vehicle at light throttle in D and N on a hoist and listen under the transmission bell housing.

Clutch and Servo Air Pressure Tests

A no-drive condition might exist even with correct fluid pressure because of inoperative clutches or bands. The inoperative units, clutches, bands, and servos can be located through a series of tests by substituting air pressure for fluid pressure.

The front and rear clutches, kickdown servo, and low-reverse servo may be tested by applying air pressure to their respective passages after the valve body assembly has been removed. To make air pressure tests, proceed as follows. Compressed air supply must be free of all dirt or moisture. Use a pressure of 30 psi.

Front Clutch. Apply air pressure to the front clutch apply passage and listen for a dull thud, which indicates that the front clutch is operating. Hold air pressure on for a few seconds and inspect the system for excessive oil leaks.

Rear Clutch. Apply air pressure to the rear clutch apply passage and listen for a dull thud, which indicates that the rear clutch is operating. Also inspect for excessive oil leaks. If a dull thud cannot be heard in the clutches, place finger tips on the clutch housing and again apply air pressure. Movement of the piston can be felt as the clutch is applied.

Kickdown Servo (Front). Direct air pressure into the front servo apply passage. Operation of the servo is indicated by a tightening of the front band. Spring tension on the servo piston should release the band.

Low and Reverse Servo (Rear). Direct air pressure into the rear servo apply passage. Operation of the servo is indicated by a tightening of the rear band. Spring tension on the servo piston should release the band.

If clutches and servos operate properly, no upshift or erratic shift conditions indicate that malfunctions exist in the valve body.

Fluid Leakage— Transmission Converter Housing Area

1. Check for the source of leakage. Since fluid leakage at or around the converter area may originate from an engine oil leak, the area should be examined

closely. Factory-fill fluid is dyed red and therefore can be distinguished from engine oil.

2. Prior to removing the transmission, perform the following checks. When leakage is determined to originate from the transmission, check fluid level prior to removal of the transmission and torque converter. High oil level can result in oil leakage out the vent located at the top of the front pump housing. If the fluid is high, adjusted to proper level.

After performing this operation, inspect for leakage. If a leak persists, perform the following operation on the vehicle to determine whether it is the converter or transmission that is leaking.

Leakage Test Probe

1. Remove the converter housing dust shield.

2. Position the vehicle with the front lower that the back so that accumulated fluid in the converter housing will drain out. Wipe the bottom interior of the converter housing as dry as possible. A solvent spray followed by compressed air drying is preferable.

3. Fabricate and fasten the test probe ($5\frac{1}{2}$ inches long) securely to the convenient dust shield bolt hole. Make certain that the converter is cleared by the test probe. The tool must be clean and dry.

4. Run the engine at approximately 2,500 rpm with the transmission in neutral for about two minutes. The transmission must be at operating temperature.

5. Stop the engine and carefully remove the tool.

6. If the upper surface of the test probe is dry, there is no converter leak. Oil leaking under the probe is coming from the transmission converter area.

7. Remove the transmission and torque converter assembly from the vehicle for further investigation. The fluid should be drained from the transmission. Reinstall the oil pan (with a new gasket) at specified torque.

Possible sources of transmission converter area fluid leakage are:

1. Converter hub seal:

(a) Seal lip cut. Check the converter hub finish,

(b) Bushing moved or worn,

(c) Oil return hole in front of the pump housing plugged or omitted, and

(d) Seal worn-out (high-mileage cars);

2. Fluid leakage at the outside diameter from the pump housing O-ring seal;

3. Fluid leakage at the front pump to case bolts;

4. Fluid leakage due to case or front pump housing porosity;

5. Oil leakage out the vent; and

6. Kickdown lever shaft access plug.

Converter Leakage

Possible sources of converter leakage are:

1. Torque converter weld leaks at the outside diameter (peripheral) weld and

2. Front pump hub weld.

Air Pressure Test of Transmission

Fabricate equipment needed for the test as shown in the shop manual.

The transmission should be prepared for the pressure test as follows after removal of the torque converter.

1. Install the filler tube bore plug, propeller shaft yoke (tie in with cord or wire), flared tube fitting cap (on front cooler line fitting), and pipe nipple (in case at rear cooler line fitting).

2. Remove necessary front pump housing bolts. Install the vent plug (rubber stopper) and vent plug retainer, preferably using longer bolts than those removed.

3. With a rotary motion, install the converter hub seal cup over the input shaft and through the converter hub seal until the cup bottoms against the pump rotor lugs. Secure with the cup retainer strap, using the converter housing to the engine block retaining bolts.

4. Attach and clamp the hose from the nozzle of tool C–4080 to the pipe nipple, which is in the rear cooler line fitting position in the case.

5. Pressurize the transmission, using tool C–4080 until the pressure gauge reads 8 psi. Position the transmission so that the pump housing and case front may be covered with soapy solution or water. Leaks are sometimes caused by porosity in the case or pump housing.

Caution: Do not under any circumstances pressurize a transmission to more than 10 psi.

If a leak source is located, that part and all associated seals and gaskets should be replaced with new parts.

TORQUEFLITE DIAGNOSIS CHART—GENERAL

No.	POSSIBLE CAUSE	HARSH ENGAGEMENT FROM NEUTRAL TO D OR R	DELAYED ENGAGEMENT FROM NEUTRAL TO D OR R	RUNAWAY UPSHIFT	NO UPSHIFT	3-2 KICKDOWN RUNAWAY	NO KICKDOWN OR NORMAL DOWNSHIFT	SHIFTS ERRATIC	SLIPS IN FORWARD DRIVE POSITIONS	SLIPS IN REVERSE ONLY	SLIPS IN ALL POSITIONS	NO DRIVE IN ANY POSITION	NO DRIVE IN FORWARD DRIVE POSITIONS	NO DRIVE IN REVERSE	DRIVES IN NEUTRAL	DRAGS OR LOCKS	GRATING, SCRAPING GROWLING NOISE	BUZZING NOISE	HARD TO FILL, OIL BLOWS OUT FILLER TUBE	TRANSMISSION OVERHEATS	HARSH UPSHIFT	DELAYED UPSHIFT	SLIPS IN REVERSE OR MANUAL LOW
35	Faulty lock-up clutch.	X																			X		
34	Overrunning clutch inner race damaged.																	X					
33	Overrunning clutch worn, broken or seized.								X				X			X	X						
32	Planetary gear sets broken or seized.											X	X	X		X	X						
31	Rear clutch dragging.														X								
30	Worn or faulty rear clutch.	X	X						X				X	X	X								
29	Insufficient clutch plate clearance.														X					X			
28	Faulty cooling system.																			X			
27	Kickdown band adjustment too tight.															X				X			
26	Hydraulic pressure too high.	X																			X		
25	Breather clogged.																		X				
24	High fluid level.																		X				
23	Worn or faulty front clutch.		X	X	X	X		X		X				X								X	
22	Kickdown servo band or linkage malfunction.			X	X	X	X	X														X	
21	Governor malfunction.				X		X	X														X	
20	Worn or broken reaction shaft support seal rings.		X	X	X	X		X		X				X								X	
19	Governor support seal rings broken or worn.				X			X														X	
18	Output shaft bearing and/or bushing damaged.																X						
17	Overrunning clutch not holding.								X				X										
16	Kickdown band out of adjustment.					X											X				X	X	
15	Incorrect throttle linkage adjustment.			X	X	X	X	X	X												X	X	
14	Engine idle speed too low.		X																				
13	Aerated fluid.		X	X		X		X	X	X	X							X	X				
12	Worn or broken input shaft seal rings.		X						X		X		X										
11	Faulty oil pump.		X					X	X	X	X	X				X		X		X			X
10	Oil filter clogged.		X	X				X	X		X	X							X				
9	Incorrect gearshift control linkage adjustment.		X		X			X	X	X				X	X					X			
8	Low fluid level.		X	X	X	X		X	X	X	X	X	X					X		X			
7	Low-reverse servo, band or linkage malfunction.		X							X				X									
6	Valve body malfunction or leakage.	X	X	X	X	X	X	X	X	X	X	X	X	X	X			X					
5	Low-reverse band out of adjustment.									X				X		X	X						X
4	Hydraulic pressures too low.		X	X	X	X		X	X	X	X	X	X	X						X	X		
3	Engine idle speed too high.	X																		X			
2	Stuck lock-up valve.															X							
1	Stuck switch valve.																			X			

CONDITION

Figure 8-69.

DIAGNOSIS CHART—LOCK-UP TORQUE CONVERTER

POSSIBLE CAUSE / CONDITION	NO LOCK-UP	WILL NOT UNLOCK	STAYS LOCKED UP TO TOO LOW A SPEED IN DIRECT	LOCKS UP OR DRAGS IN LOW OR SECOND	STALLS OR IS SLUGGISH IN REVERSE	LOUD CHATTER DURING LOCK-UP ENGAGEMENT—(COLD)	VIBRATION OR SHUDDER DURING LOCK-UP ENGAGEMENT	VIBRATIONS AFTER LOCK-UP ENGAGEMENT	VIBRATION WHEN "REVED" IN NEUTRAL	OVERHEATING; OIL BLOWING OUT DIPSTICK OR PUMP SEAL	SHUDDER AFTER LOCK-UP ENGAGEMENT
FAULTY OIL PUMP	X			X	X		X				X
STICKING GOVERNOR VALVE	X	X	X								
PLUGGED COOLER, LINES OR FITTINGS					X					X	X
VALVE BODY MALFUNCTION	X	X	X	X	X		X				X
STUCK SWITCH VALVE	X	X	X	X	X					X	
STUCK LOCK-UP VALVE	X	X	X								
STUCK FAIL-SAFE VALVE	X	X	X	X							
FAULTY TORQUE CONVERTER:	X					X	X	X			X
OUT OF BALANCE									X		
FAILED LOCKING CLUTCH	X					X					X
LEAKING TURBINE HUB SEAL	X					X					
ALIGN EXHAUST SYSTEM								X			X
TUNE ENGINE							X	X			X
FAULTY INPUT SHAFT OR SEAL RING	X				X						
THROTTLE LINKAGE MISADJUSTED								X			X

Figure 8-70.

American Motors 904 and 998 Diagnosis Procedure (Courtesy of American Motors Corporation)

General

Automatic transmission malfunctions are generally the result of poor engine performance, incorrect fluid level, incorrect linkage adjustment, band or hydraulic control pressure adjustments, hydraulic system malfunctions, or mechanical component malfunctions.

In all automatic transmission repair, the logical and proper procedure is diagnosis before disassembly. A systematic diagnosis procedure is important in avoiding repair delays caused by incorrect or unnecessary repairs. The diagnosis and test procedures outlined in this section should be performed in the following sequence to realize maximum effect: (1) preliminary diagnosis, (2) road test, (3) hydraulic pressure test, (4) air pressure test, (5) stall test, (6) analysis of test results, in consultation with diagnosis guides and charts for the cause of malfunction.

Preliminary Diagnosis

Two basic procedures are required: one procedure for automobiles that are drivable and an alternate procedure for automobiles that will not back up or move forward.

Automobile Is Drivable

1. Check fluid level and condition.

2. Adjust throttle and gearshift linkage before road testing if the complaint is based on delayed, erratic, or harsh shifts.

3. Road test the automobile and note transmission operating characteristics. The road test of six-cylinder automobiles must also include the test of lockup torque converter operation.

4. Perform the stall test if the complaint is based on sluggish, low-speed acceleration or abnormal

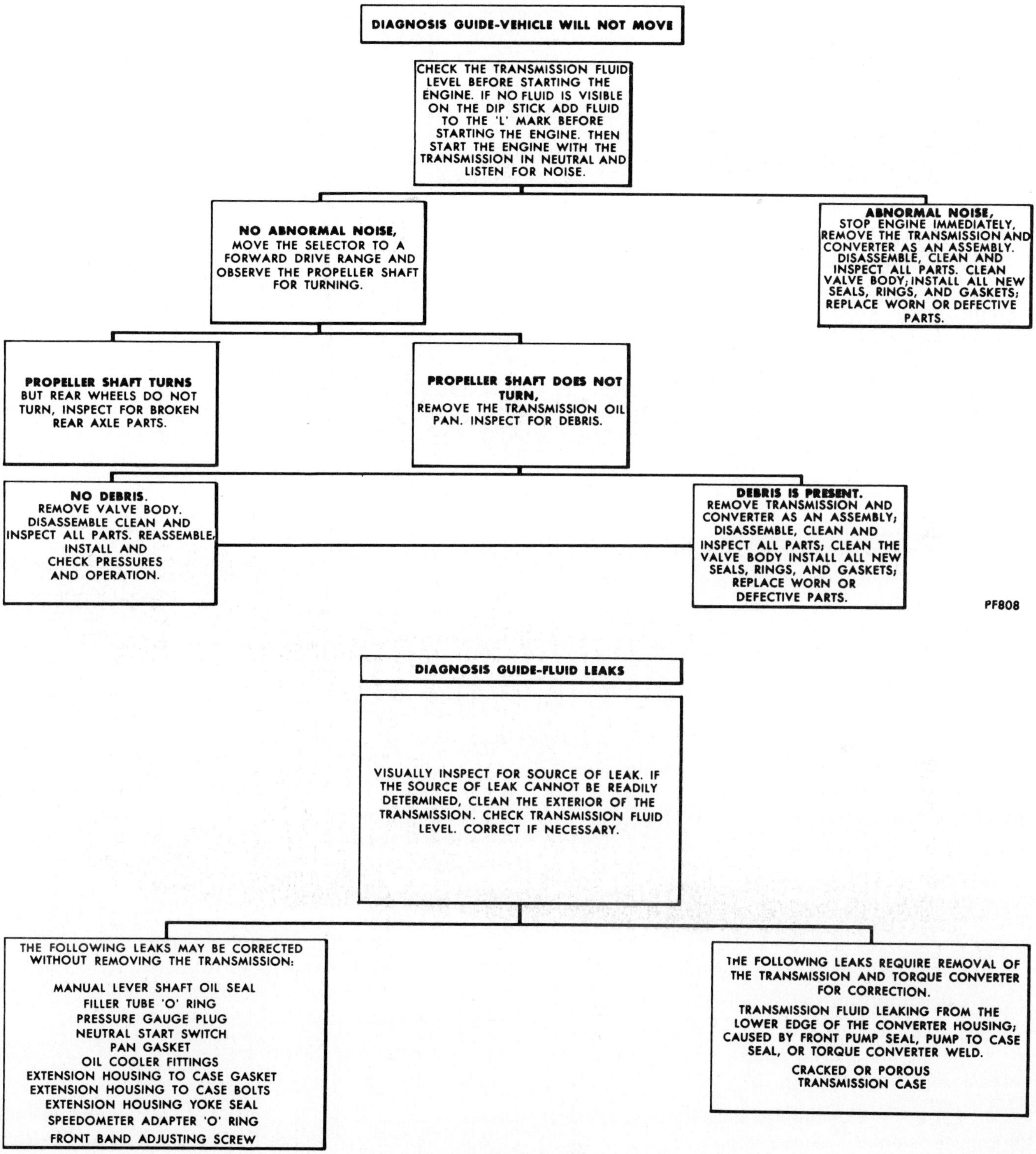
DIAGNOSIS GUIDE-VEHICLE WILL NOT MOVE
CHECK THE TRANSMISSION FLUID LEVEL BEFORE STARTING THE ENGINE. IF NO FLUID IS VISIBLE ON THE DIP STICK ADD FLUID TO THE 'L' MARK BEFORE STARTING THE ENGINE. THEN START THE ENGINE WITH THE TRANSMISSION IN NEUTRAL AND LISTEN FOR NOISE.
NO ABNORMAL NOISE, MOVE THE SELECTOR TO A FORWARD DRIVE RANGE AND OBSERVE THE PROPELLER SHAFT FOR TURNING.
ABNORMAL NOISE, STOP ENGINE IMMEDIATELY, REMOVE THE TRANSMISSION AND CONVERTER AS AN ASSEMBLY. DISASSEMBLE, CLEAN AND INSPECT ALL PARTS. CLEAN VALVE BODY; INSTALL ALL NEW SEALS, RINGS, AND GASKETS; REPLACE WORN OR DEFECTIVE PARTS.
PROPELLER SHAFT TURNS BUT REAR WHEELS DO NOT TURN, INSPECT FOR BROKEN REAR AXLE PARTS.
PROPELLER SHAFT DOES NOT TURN, REMOVE THE TRANSMISSION OIL PAN. INSPECT FOR DEBRIS.
NO DEBRIS. REMOVE VALVE BODY. DISASSEMBLE CLEAN AND INSPECT ALL PARTS. REASSEMBLE, INSTALL AND CHECK PRESSURES AND OPERATION.
DEBRIS IS PRESENT. REMOVE TRANSMISSION AND CONVERTER AS AN ASSEMBLY; DISASSEMBLE, CLEAN AND INSPECT ALL PARTS; CLEAN THE VALVE BODY INSTALL ALL NEW SEALS, RINGS, AND GASKETS; REPLACE WORN OR DEFECTIVE PARTS.
PF808
DIAGNOSIS GUIDE-FLUID LEAKS
VISUALLY INSPECT FOR SOURCE OF LEAK. IF THE SOURCE OF LEAK CANNOT BE READILY DETERMINED, CLEAN THE EXTERIOR OF THE TRANSMISSION. CHECK TRANSMISSION FLUID LEVEL. CORRECT IF NECESSARY.
THE FOLLOWING LEAKS MAY BE CORRECTED WITHOUT REMOVING THE TRANSMISSION:
MANUAL LEVER SHAFT OIL SEAL
FILLER TUBE 'O' RING
PRESSURE GAUGE PLUG
NEUTRAL START SWITCH
PAN GASKET
OIL COOLER FITTINGS
EXTENSION HOUSING TO CASE GASKET
EXTENSION HOUSING TO CASE BOLTS
EXTENSION HOUSING YOKE SEAL
SPEEDOMETER ADAPTER 'O' RING
FRONT BAND ADJUSTING SCREW
THE FOLLOWING LEAKS REQUIRE REMOVAL OF THE TRANSMISSION AND TORQUE CONVERTER FOR CORRECTION.
TRANSMISSION FLUID LEAKING FROM THE LOWER EDGE OF THE CONVERTER HOUSING; CAUSED BY FRONT PUMP SEAL, PUMP TO CASE SEAL, OR TORQUE CONVERTER WELD.
CRACKED OR POROUS TRANSMISSION CASE

Figure 8–71.

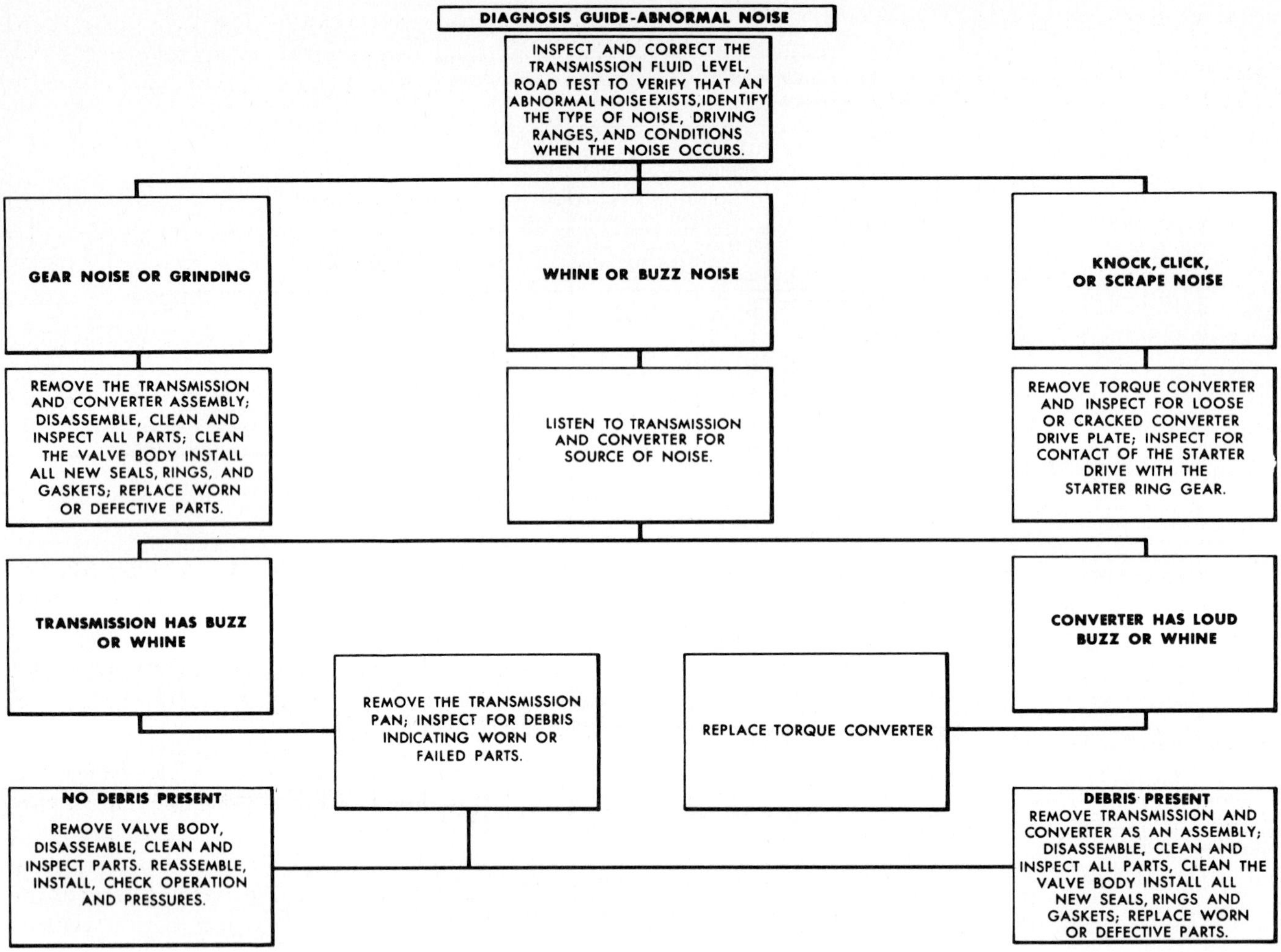

Figure 8–72.

throttle-opening requirements in order to maintain highway speeds with engine in good tune.

5. Perform hydraulic pressure tests.

6. Perform the air pressure test of clutch and band operation.

Automobile Will Not Back up or Move Forward

1. Check fluid level and condition.

2. Check for broken or disconnected throttle linkage.

3. Check for broken cooler lines and loose or missing pressure port plugs.

4. Raise the automobile, start the engine, shift the transmission into gear, and note the following:

(a) If propeller shafts turn but axle wheels do not, the problem is in differential or axle shafts.

(b) If propeller shafts do not turn and the transmission is noisy, stop the engine, remove the oil pan, and check for debris. If debris is not found, remove the transmission and check for broken drive plate or drive plate-to-converter bolts, broken converter hub, broken input or output shaft, broken oil pump, or stripped torque converter turbine hub splines.

(c) If propeller shafts do not turn and the transmission is not noisy, perform hydraulic pressure test to determine if the problem is malfunction of hydraulic or mechanical component.

Fluid Level and Condition

Check fluid level and condition as follows:

1. Drive the automobile until transmission fluid is at the normal operating temperature of approximately 175°F (79.4°C).

Note: To avoid false readings, which could result in an over- or underfill condition, do not check the fluid level until the fluid is at operating temperature.

2. Shift the transmission into neutral.

Note: The transmission fluid level is checked in neutral because the converter fills more rapidly in this position.

3. Apply the parking brake.
4. Operate the engine at idle speed.

Note: When checking fluid level or performing other underhood operations with the engine running, observe the following precautions to avoid possible personal injury. Keep hands away from hot or rotating engine components or accessories. Do not wear loose articles of clothing that could become entangled in rotating engine components or accessories.

5. Wipe dirt from the filler cap and tube before removing the dipstick.

6. Remove the dipstick and check fluid level. Fluid level is correct when between the add-one-pint and full marks on the dipstick.

7. Add or drain fluid as necessary to bring fluid to the correct level. If the level is low, check the transmission for leaks.

8. Check condition of fluid.

(a) Fluid should be dark red in color and free of dirt or debris.

(b) If fluid is discolored or smells burnt but the transmission operates properly, replace fluid and filter.

(c) If fluid is badly discolored, smells burnt, contains metal or friction material particles and transmission problems were experienced, transmission may require overhaul.

A low fluid level allows the pump to take in air along with the fluid. As in any hydraulic system, aerated fluid will cause hydraulic pressures to be low and develop slower than normal.

If the transmission is overfilled, the gears churn the fluid into foam, aerating the fluid and causing the same conditions that occur with a low fluid level. In either case, air bubbles cause fluid overheating, oxidation, and varnish buildup, which interferes with valve, clutch, and servo operation. Foaming can also result in fluid escaping from the transmission vent, where it may be mistaken for a leak.

Along with the fluid level, it is also important to check fluid condition. When the fluid is dark, smells burnt, or is contaminated with metal or friciton material particles, an overhaul may be necessary. Examine the fluid on the dipstick closely. If there is any doubt about its condition, drain a sample for further inspection. After checking fluid level and condition, seat the dipstick fully in order to seal out water and dirt.

Flushing Oil Cooler and Cooler Lines

If transmission or lockup clutch failure has contaminated fluid, the oil cooler should be reverse flushed to ensure that metal particles or sludged oil are not later transferred back into reconditioned transmission.

1. Disconnect both cooler lines at the radiator.

2. Dislodge any foreign material at the inlet side of the cooler with a small screwdriver.

3. Reverse the flush cooler with a combination of mineral spirits and pulsating air under pressure (shop air).

4. Treat cooler lines separately and ensure that they are clear by flowing mineral spirits or automatic transmission fluid through them.

5. Remove leftover mineral spirits from the cooler and cooler lines by flowing automatic transmission fluid through them.

6. Cooler flow should now be checked by connecting cooler tubes and placing the rear cooler tube into a 1-quart container. Overfill transmission by 1 quart. Watching a clock, start the engine (run at curb idle) and run it in neutral for exactly 20 seconds. If cooler flow is less than 1 quart in 20 seconds, replace the radiator or have the radiator bottom cooler professionally reconditioned.

Note: After completing any repairs that require draining the transmission fluid, add 6 quarts (5.7 liters) of AMC, Dexron II®, or equivalent automatic transmission fluid to the transmission before starting the engine.

Throttle Linkage

The throttle linkage adjustment is important to proper operation. This adjustment positions a valve that controls shift speed, shift quality, and part-throttle downshift sensitivity. If the setting is too short, early shifts and slippage between shifts may occur. If the linkage setting is too long, shifts may be delayed and part-throttle downshifts may be very sensitive.

This adjustment is so important that the use of throttle lever holding spring is necessary to remove slack in the linkage during adjustment. Refer to the throttle linkage adjustment in the in-automobile service and adjustment section of the service manual.

Gearshift Linkage

The gearshift linkage adjustment is important because the linkage positions the manual valve in the valve body. Incorrect adjustment will result in creeping in neutral, premature clutch wear, delayed engagement in any gear, or a no-start in park or neutral.

Proper operation of the neutral start switch will provide a quick check of linkage adjustment as follows:

1. Insert the key in the ignition lock and turn the lock to the on position in order to unlock the column and gearshift lever.

2. Move the gearshift lever slowly upward until it clicks into park detent in the shift selector gate.

3. Turn the ignition lock cylinder to the start position and start the engine. If the starter operates, the park position is correct.

4. Stop the engine.

5. Move the gearshift lever slowly toward neutral until the lever engages in edge of neutral detent in the shift selector gate.

6. Turn the ignition lock cylinder to the start position and start the engine. If the starter operates, the neutral position is correct and linkage is properly adjusted.

7. If the starter failed to operate in park or neutral or if the gearshift lever had to be moved back and forth to achieve start in either position, linkage adjustment is required. Refer to gearshift linkage adjustment in the in-automobile service and adjustment section of the service manual.

Lockup Torque Converter Service Diagnosis

Preliminary diagnosis of a transmission malfunction in six-cylinder Spirit, Concord, and Eagle models must include a test of the lockup torque converter. Since it is possible for transmission or lockup torque converter problems to be mistaken for each other, attention must be paid to torque converter operation during a comprehensive road test. The lockup torque converter service diagnosis chart provides a basis for analyzing road test results.

Road Test

Before road testing, be sure that fluid level and throttle and gearshift linkage adjustments have been checked and corrected if necessary.

Observe engine performance during the test. On Eagle models, be sure that the transfer case and front axle malfunction will have an adverse effect on transmission operation. In addition, operate the transmission in all gearshift positions to check for slippage and shift variations. Note whether the shifts are harsh, spongy, delayed or early or part-throttle downshifts are sensitive.

Watch closely for slippage or engine speed flare-up, which usually indicates clutch, band, or overrunning clutch problems. If the condition is advanced, an overhaul may be necessary to restore normal operation.

In most cases, a slipping clutch or band can be determined by noting transmission operation in all gearshift lever positions and by comparing which internal units are applied in those positions. The clutch and band application chart provides a basis for analyzing road test results.

Analyzing the Road Test

Refer to the clutch and band application chart and note which elements are in use in the various gear ranges. The rear clutch is applied in all forward ranges (D, 2, and 1). The overrunning clutch is applied in all forward ranges (D, 2, and 1). The overrunning clutch is applied in first-gear D and 2 range only and the front band is applied in 1 and R range only.

For example, if slippage occurs in D- and 2-range first gear but not in the 1 range, the overrunning clutch is slipping. Similarly, if slippage occurs in any two forward gears, the rear clutch is slipping.

Applying the same method of analysis, note that both clutches are applied in D third gear only. If the transmission slips in third gear, either the front clutch or the rear clutch is slipping. By selecting another gear that does not use one of these units, the slipping clutch can be determined. For example, if the transmission also slips in reverse, the front clutch is slipping. If the transmission does not slip in reverse, the rear clutch is slipping.

This process of elimination can be used to determine the slipping unit and to check operation. The key is proper use of the clutch and band application chart.

Although road test analysis will help determine the slipping unit, the actual cause of a malfunction usually cannot be ascertained until hydraulic and air pressure tests are performed. Practically any condition can be caused by leaking hydraulic circuits or sticking valves. Unless the condition is obvious, such as no drive in D-range first gear only, do not disassemble the transmission until hydraulic and air pressure tests have been performed.

Hydraulic Pressure Tests

The hydraulic pressure tests are performed, using test set J–240027. The set consists of five color-coded pressure hoses, a 400-psi (2,758-kPa) capacity pressure gauge, and a 100-psi (689-kPa) capacity pressure gauge. The high-pressure gauge is used to record rear servo pressure in R and 1 positions only; the low pressure gauge is used for all other readings. The test set permits simultaneous connection to all the pressure ports and allows sequential or independent pressure readings as desired.

Pressure Test Port Location

The accumulator line pressure port is located on the right side of the case between the front and rear servo castings just above the oil pan mating surface. The front servo release pressure port is located on the right side of the case, just behind the filler tube opening just above the oil pan mating surface. The rear servo apply pressure port is on the right rear side of the case, facing rearward and just above the oil pan mating surface.

The governor pressure port faces to the left side in the front lower center section of the extension housing between the speedometer cable adapter and case mating surface. Lubrication pressure is measured by installing a T-fitting in the fluid cooler return line on the left side at the rear edge of the case halfway to the top of the case.

Hydraulic Pressure Test Procedure

Before the test, check and correct fluid level and control linkage adjustments and connect a tachometer to the engine. Raise the automobile on a hoist that will allow the rear wheels to rotate freely and connect the test gauge hoses to the appropriate transmission pressure ports. Compare the test results with the hydraulic pressure test specifications and diagnosis charts in the service manual.

Test 1

This test checks pump output, pressure regulator valve operation, front/rear clutch and oil filter condition by measuring transmission operating pressure at the accumulator and front servo pressure ports in the D range.

1. Turn the test gauge selector handle to the line pressure position.

2. Start and operate the engine at idle speed, move the gearshift lever to the D range, and note line pressure.

3. At idle, pressure should be 54 to 60 psi (372 to 414 kPa).

4. Increase engine rpm to 1,600 and slowly move the valve body throttle lever fully forward, then fully rearward. With the lever in the forward position, pressure should be 54 to 60 psi (372 to 414 kPa) and gradually increase to 94 psi (648 kPa) when the lever is moved rearward.

5. Turn the test gauge selector handle to the front servo apply position.

6. Operate the engine at 1,000 rpm, allow the transmission to upshift into third gear, and note the front servo pressure. Pressure should not vary from previously noted line pressure by more than 3 psi (21 kPa).

7. Slowly move the throttle linkage to the full throttle position and note front servo pressures just prior to and after 3–2 downshift. Servo pressure should increase, then drop to 0 psi after downshift.

Test 2

This test checks pump output, pressure regulation, and front clutch/rear servo condition by measuring and comparing pressure at the rear servo and accumulator pressure ports in gear ranges 1 and 2.

1. Turn the test gauge selector handle to the line pressure position.

2. Move the gearshift lever to the 1 range, operate the engine at 1,000 rpm, and slowly move the valve body throttle lever fully forward, then fully rearward. With the lever in the forward position, pressure should be 54 to 60 psi (372 to 414 kPa) and gradually increase to 90 to 96 psi (621 to 662 kPa) as the lever is moved rearward.

3. Turn the test gauge selector handle to the rear servo apply position.

4. Operate the engine at 1,000 rpm. Slowly move the valve body throttle lever fully forward, then fully rearward. Compare pressure at the servo with pressure at the accumulator port. Servo pressure should not vary from accumulator port pressure by more than 3 psi (21 kPa).

5. Turn the test gauge selector handle to the line pressure position.

6. Place the gearshift lever in the 2 range and operate the engine at 1,000 rpm.

7. Slowly move the valve body throttle lever fully forward, then fully rearward, and note pressures. With the lever in the forward position, pressure should be 54 to 60 psi (372 to 414 kPa) and gradually increase to 90 to 96 psi (621 to 662 kPa) as lever is moved rearward.

Test 3

This test checks pump output, pressure regulation, and front clutch/rear servo condition by measuring and comparing operating pressure at the rear servo pressure port in reverse.

Note: For this test, only the high pressure gauge in the test set is used. Since the gauge hose is connected directly to the rear servo, it is not necessary to move the test gauge selector handle to any specific position.

1. Operate the engine at idle speed, move the gearshift lever to reverse, and note the reading on high pressure gauge. Pressure should be approximately 160 psi (1,103 kPa) at idle.

2. Slowly increase engine speed to 1,600 rpm and note the reading at the high pressure gauge again. As engine speed is increased, pressure should also increase to approximately 270 psi (1,862 kPa).

3. Operate engine at idle speed, move the gearshift lever from reverse to drive, and note the reading at high pressure gauge. Pressure should drop to 0 psi when the gearshift lever is moved to drive. This test checks for leakage into the rear servo, which can cause rear band burnout.

Test 4

This test checks governor operation by measuring governor pressure response to changes in engine speed. It is usually not necessary to check governor operation unless shift speeds are incorrect or if the transmission will not downshift.

1. Turn the test gauge selector handle to the governor pressure position.

2. Move the gearshift lever to the D range, operate the engine at idle speed, and note pressure. Pressure should be 0 to $1\frac{1}{2}$ psi (10.3 kPa) maximum. If pressure exceeds the maximum, the governor valve or weights are sticking open.

3. Slowly increase the engine speed and observe speedometer and test gauge pressure. Governor pressure should increase in proportion to automobile speed (approximately 1 psi per 1 mph). Pressure rise should be smooth and drop back to 0 to $1\frac{1}{2}$ psi (10.3 kPa) when the throttle is closed.

Test 5

This test checks lubrication pressure, which is measured by connecting the white test gauge hose to a T-fitting that has been inserted in the left-side cooler line. This test is especially important when the transmission is noisy, unexplained rear clutch failure has occurred, or fluid is forced out the fill tube and fluid level was known to be correct.

1. Install a T-fitting in the transmission cooler line.

2. Connect the white test gauge hose to the T-fitting and turn the test gauge selector handle to the lubrication pressure position.

3. Operate the engine at idle speed, move the gearshift lever to the D range, and note pressure.

4. Pressure at idle should be 5 to 15 psi (35 to 103 kPa).

5. Increase engine speed to 1,000 rpm and slowly move the valve body throttle lever fully forward and then fully rearward. With the lever forward, pressure should be 5 to 15 psi (35 to 103 kPa) and 10 to 30 psi (68 to 207 kPa) with the lever rearward.

Air Pressure Test

Air pressure testing is used as a diagnostic tool before transmission removal and also as a method of confirming proper clutch and band servo operation after repair. The tests involve substituting air pressure for fluid pressure by applying air pressure to the appropriate case passages after the valve body has been removed.

Caution: Use dry, filtered compressed air only when performing air pressure tests. Pressures of 30 to 100 psi (207 to 689 kPa) are required to perform the tests.

Front Clutch Test

Place one or two fingers on the clutch housing and apply air pressure to the front clutch apply passage. Movement of the piston can be felt and a soft thud may be heard as the clutch applies. While air pressure is applied, check for excessive air leakage.

Rear Clutch Test

Place one or two fingers on the clutch housing and apply air pressure to the rear clutch apply passage. Movement of the piston can be felt and a soft thud may be heard as the clutch applies. While air pressure is applied, check for excessive or unusual air leakage.

Front Servo Test

Apply air pressure to the front servo apply passage. The servo rod should extend and cause the band to tighten around the drum. While air pressure is applied, check for excessive air leakage. Spring tension should release the servo when air pressure is removed.

Rear Servo Test

Apply air pressure to the rear servo apply passages. The servo rod should extend and cause the band to tighten around the drum. While air pressure is applied, check for excessive air leakage. Spring tension should release the servo when air pressure is removed.

Converter Housing Fluid Leak Diagnosis

When diagnosing converter housing fluid leaks, two facts must be established before attempting repair. First, it must be verified that a leak condition does actually exist; second, the real source of the leak must be determined. Failure to establish these facts beforehand can result in incorrect and unnecessary repairs.

In some cases, suspected converter housing fluid leaks may not be leaks at all. They may be the result of residual fluid in the converter housing or excess fluid spilled during factory filling or initial transmission operation. These conditions may be incorrectly diagnosed as fluid leaks.

Converter housing area leaks may have several sources. Through careful observation, it is possible to locate the leak source before removing the transmission. The paths that various types of fluid leaks follow are described next.

- Oil pump seal. Leaks past the seal lip tend to move along the drive hub and onto the rear of the converter housing. However, if total seal failure occurs, fluid will be deposited inside the converter housing only, near the outside diameter of the housing.
- Oil pump body. Leaks past the pump body follow the same path as an oil pump seal leak, or fluid may travel down the pump face into the converter housing.
- Oil pump-to-case bolt. Leaks past any one of these bolts are deposited on the inside of the converter housing only and not on the converter itself.
- Oil pump-to-case gasket. Leaks past the gasket are deposited inside the converter housing only.
- Front band lever pin plug. Leaks past the plug threads are deposited inside the housing and not on the converter.

Leak Diagnosis Procedure

1. Check the fluid level and condition. Refer to the section on fluid level and condition.

2. Raise and support the automobile.

3. Inspect the transmission and correct any external leaks from oil pan gasket, fill tube, cooler line fittings, pressure test port plugs, and cast-to-extension or adapter housing gasket.

4. Remove the converter housing spacer plate.

5. Wipe all fluid from the converter housing area.

6. Operate the engine at 2,000 rpm for two minutes and observe the converter housing for fluid accumulation pattern.

7. If fluid accumulation pattern is not evident, proceed to the next step.

8. If a circular pattern develops, it indicates a defective or damaged torque converter. Correct the leak by replacing the converter.

9. If a trickle develops, it indicates an oil pump leak caused by one or more of the following conditions.

- Pump drainback hole obstructed
- Pump housing vent obstructed
- Pump bushing or converter hub scored, nicked, pitted, or burred
- Oil pump O-ring, gasket, or seal leak
- Front band level pin plug loose or plug threads in case stripped

10. Correct these conditions as outlined in following steps.

(a) Remove the transmission and converter.

(b) Tighten the front band adjusting screw until band is tight around the front clutch retainer. This will prevent the front clutch assembly from also coming out when the oil pump is removed and will prevent damaging clutch discs.

(c) Remove the oil pump and oil seal.

(d) Inspect the pump housing drainback and vent holes for obstructions. If the drainback hole cannot be opened using thin wire, replace the pump housing. Check the vent hole by blowing solvent through the vent. If the vent cannot be opened, replace the pump housing.

(e) Inspect the condition of the pump housing and converter hub. If the bushing is scored, replace it and polish the hub, using fine sandpaper.

(f) Install the replacement oil pump seal, O-ring, gasket, and oil pump.

(g) Loosen the kickdown lever pin plug two turns.

(h) Apply a small quantity of No. 2 Permatex, or equivalent sealer, to plug threads and tighten the plug to 150 inch-pounds (17 N.m) torque.

(i) Adjust the front band.

11. Install the transmission and converter.

12. Install the converter housing spacer plate.

13. Remove supports and lower the automobile.

Governor

Governor operational problems should be diagnosed, using the road test and hydraulic pressure test procedures.

Stall Test

Stall testing determines the maximum engine rpm obtainable at full throttle with the rear wheels locked and the transmission in drive. Stall testing also checks the holding ability of the converter-stator overrunning clutch and both transmission clutches. When stall testing is completed, refer to the stall speed specifications chart at the end of this section and the stall speed diagnosis guides.

Note: Never allow anyone to stand in front of the automobile when performing a stall test. In addition, always block the front wheels and have the parking and service brakes fully applied during the test.

1. Connect the tachometer to the engine.
2. Check and adjust the transmission fluid level. Refer to the section on fluid level and condition.
3. Operate the engine until the transmission fluid reaches the operating temperature.
4. Block the front wheels.
5. Fully apply parking brakes.
6. Fully apply service brakes.
7. Open the throttle completely and record maximum engine rpm registered on the tachometer.

Note: Do not hold the throttle open any longer than necessary and never longer than 5 seconds at a time. If more than one stall test is required, operate the engine at 1,000 rpm with the transmission in neutral for at least 20 seconds to cool the transmission fluid.

8. If engine speed exceeds the maximum shown in the stall speed chart, release the accelerator immediately. This indicates that transmission clutch slippage is occurring.
9. Shift the transmission into neutral, operate the engine for 20 seconds, stop the engine, shift the transmission into park, and release the brakes.
10. Compare test results with the stall speed chart and refer to the stall test diagnosis.

Stall Test Diagnosis

Stall Speed Too High

If stall speed exceeds specifications by more than 200 rpm, transmission clutch slippage is indicated.

Stall Speed Too Low

Low stall speeds with a properly tuned engine indicate a torque converter stator clutch problem. The condition should be confirmed by road testing prior to converter replacement. If stall speeds are 250 to 350 rpm below the minimum specified in the chart and the automobile operates properly at highway speeds but has poor low-speed acceleration, the stator overrunning clutch is slipping and the torque converter should be replaced.

Stall Speed Normal

If stall speeds are normal but road testing shows that abnormally high throttle opening is required to maintain highway speeds even though low-speed acceleration is normal, the stator overrunning clutch is seized and the torque converter should be replaced.

Noise

A whining or sirenlike noise caused by fluid flow is normal during a stall test. Loud metallic noises from loose internal parts or interference within the assembly indicates a defective torque converter. To confirm that a noise is originating from within the converter, operate the automobile at light throttle in drive and neutral on a hoist and listen for noise coming from the torque converter housing. When the stall test is completed, compare the test speeds recorded with those listed in the stall speed specifications chart.

Service Diagnosis

General

The diagnosis charts provide a quick reference for transmission diagnosis. A step-by-step approach to diagnosing and correcting transmission malfunctions is used.

The in-automobile procedures chart lists problems that can be corrected with the transmission in the automobile. The out-of-automobile procedures chart lists problems that require transmission removal and disassembly.

The in-automobile procedures should always be performed first. Do not remove the transmission unless the in-automobile procedures fail to correct the problem.

How to Use the Charts

The condition columns in each chart describe the most frequently encountered malfunctions. Each problem condition is cross-referenced to the necessary service procedures.

The code letters in the boxes at the top of each chart identify the individual service procedures. These code letters correspond to descriptions of the various procedures, which appear on the pages immediately following the charts.

Captial letters *A* through *T* denote in-automobile

procedures. Lower-case letters *a* through *j* denote out-of-automobile procedures. The numbers in the boxes adjacent to the condition columns cross-reference each problem condition to the necessary service procedures. These numbers also show the order in which the various service procedures should be performed.

As an example, assume that the problem condition is harsh engagement in R–D–2–1. First, locate the problem description in the condition column; then note which service procedures are indicated. As shown in the chart, procedures B, D, E, G, N, and Q are required. Next, note the numbers that indicate the sequence in which these procedures are to be performed. In this case, the correct order will be D, B, E, G, Q, and N. Finally, refer to the service procedure descriptions, which appear on the pages immediately following the charts, for details of each procedure.

Become familiar with both charts and the procedures required. Some conditions require in-automobile service only, others require out-of-automobile service only, and some require a combination of both.

In-Automobile Procedures

A. Fluid level and condition. Fluid should be at the full mark with the engine idling. Fluid should not be milky, full of bubbles, or dark and burnt smelling. Use AMC, Dexron®, or equivalent transmission fluid only.

B. Throttle linkage. Check for smooth travel. Clean, but do not lube, linkage pivot points as necessary, then adjust to specifications.

C. Gearshift linkage. Adjust to specifications.

D. Engine idle adjustment. Set to specifications.

E. Hydraulic pressures. Perform the hydraulic pressure test to determine if operating pressures are within specifications. Repair hydraulic components as necessary. Check and correct throttle and line pressure settings if required.

F. Front band. Adjust to specifications.

G. Rear band. Adjust to specifications.

H. Neutral start switch. Check wires and connections. Test the switch. See if the valve body manual lever grounds switch in the P and N positions. If not OK, check the ground strip at the valve body manual lever. If OK, check the starting circuit.

I. Park lock. Check the condition of the lock rod, lock rod ball, sprag reaction plug, governor support, and sprag shaft. Replace parts as required.

J. Transmission oil cooler. Check lines and cooler for obstructions or leaks (look for transmission fluid in radiator coolant or milky colored transmission fluid, which indicates coolant in the fluid).

K. Output shaft bearing, bushing, or seal. Remove the extension housing, inspect parts, and replace parts as necessary.

L. Governor valve. Clean and inspect all parts. Check weights, shaft, and valve for burrs, nicks, scores, or binding. Check the spring for collapsed or distorted coils and snap rings for distortion. Check the filter for dirt and debris. Inspect the body for cracks or warpage. Check the torque on the governor and output shaft support bolts.

M. Oil filter. Inspect and replace if clogged.

N. Valve body. Remove, disassemble, clean thoroughly, and inspect valves and plugs for nicks, scratches, burrs, and rounded edges on valve lands. Check bores for scratches, springs for collapsed coils, and all mating surfaces for nicks, burrs, or warpage. Reassemble and install, tightening all screws to exact specifications.

O. Front servo and linkage. Inspect piston for wear, cracks, and worn or broken seal rings. Check springs for collapsed or broken coils. Check servo bore for scratches, nicks, or wear. Check lever, strut, and band for damage. Check the lever shaft for wear, looseness in the case, or leaking O-ring.

P. Rear servo and linkage. Inspect the piston for wear, cracks, worn or broken seal ring, or damaged seal. Check springs for collapsed or broken coils. Check the servo bore for scratches, nicks, or wear. Check the lever and band for damage. Check the lever shaft for wear or looseness in the case.

Q. Accumulator. Clean and inspect for broken seal rings, scratched bore, or broken or collapsed spring. Check the piston for cracks or evidence of piston cocking in the bore.

R. Air pressure test. Remove the valve body and use air pressure to apply clutches and bands to check operation.

S. Engine tune and performance. Verify proper engine operation. Be sure that compression meets specifications and fuel and ignition systems are functioning properly.

T. Stall test. Perform the stall test to check the holding ability of the converter and transmission clutches.

Out-of-Automobile Procedures

a. Converter drive plate. Check the plate for flatness, cracks at mounting bolt holes, loose attaching bolts, or damaged ring gear teeth. A broken

IN AUTOMOBILE

Service Diagnosis

CONDITIONS	A	B	C	D	E	F	G	H	I	J	K	L	M	N	O	P	Q	R	S	T
HARSH ENGAGEMENT R, D, 2, 1		2		1	3		4							6			5			
SLOW TO ENGAGE N, R, D, 2, 1	1		2	3	4		6						5	9			8	7		
NO UPSHIFT, STUCK IN LOW GEAR	1	2	3		4	7						5		6	8			9		
NO LOW GEAR MOVES IN 2ND OR 3RD GEAR												1		2						
NO KICKDOWN OR NORMAL DOWNSHIFT	1	2	3		5	4						6		7	8			9		
DELAYED, ERRATIC SHIFTS (HARSH AT TIMES)	1	2	3		4	5						6	7	8	9	10				
SLIPS IN FORWARD DRIVE RANGES	1	2	3		4	5								6	7	8	9	10		
SLIPS IN REVERSE ONLY	1		2		3		4							5		6		7		
WILL NOT MOVE IN FORWARD OR REVERSE	1		2		3								4	5				6		
REVERSE OK, WILL NOT MOVE FORWARD IN D, 2, 1			1		2									3				4		
NO REVERSE			1		2		3									4		5		
MOVES IN NEUTRAL POSITION (CREEPS IN "N")			1											2						
DRAGS OR LOCKS UP					1	2	3		4					5	6	7	8			
GROWLING, GRATING OR SCRAPING NOISES	1								2		3		4							
BUZZING NOISE	1											2		3				4		
OIL BLOWS OUT FILLER TUBE	1									2			3	4						
OVERHEATS	1			2	5	4	7			3			6	8						
WILL NOT START IN N OR P POSITION			1					2						4					3	
SLUGGISH ACCELERATION, EXCESSIVE THROTTLE NEEDED TO MAINTAIN SPEED	1	3			5														2	4

Figure 8–73. AMC Torque Command.

drive plate may indicate engine-to-transmission misalignment caused by loose, missing, or misaligned dowels.

b. Oil pump. Clean the pump and check all clearances. Inspect rotors for scoring and the seal and bushings for wear. Inspect the pump housing and reaction shaft support mating surfaces for flatness.

c. Transmission vent. Make sure that the vent is open and not obstructed.

d. Front clutch. Clean and inspect all parts. Examine the retainer and piston for scores and scratches, discs and plates for wear, return springs for collapsed coils, and the seal rings for damage. The vent check ball in the retainer must operate freely.

e. Rear clutch. Inspect all rear clutch parts as outlined under the front clutch procedure.

f. Planetary gear set. Clean and inspect the annulus gear, planet pinion carrier assembly, and sun gear for worn thrust washers, damaged gear teeth, and excessive pinion and clearance. Examine the bushings in the sun gear for excessive wear.

g. Rear band. Inspect the band for wear and for good bond of lining to band. Inspect the lining for burn marks, glazing, uneven wear patterns, flaking,

OUT OF AUTOMOBILE PROCEDURES

CONDITIONS	a	b	c	d	e	f	g	h	i	j
HARSH ENGAGEMENT R, D, 2, 1				1	2					
SLOW TO ENGAGE N, R, D, 2, 1		1		2	3					4
NO UPSHIFT, STUCK IN LOW GEAR		1		2						
DELAYED, ERRATIC SHIFTS (HARSH AT TIMES)		1								
SLIPS IN FORWARD DRIVE RANGES		1		2	3		4			5
SLIPS IN REVERSE ONLY		1		2			3			
SLIPS IN LOW GEAR "D" ONLY BUT NOT IN "1" POSITION								1		
WILL NOT MOVE IN FORWARD OR REVERSE	1	2				3				
REVERSE OK, WILL NOT MOVE FORWARD IN D, 2, 1					1					
NO REVERSE				1			2			
MOVES IN NEUTRAL POSITION (CREEPS IN "N")					1					
DRAGS OR LOCKS UP				1	2	3	4	5		
GROWLING, GRATING OR SCRAPING NOISES	1	2		3		4		5	6	
BUZZING NOISE		1							2	
OIL BLOWS OUT FILLER TUBE		1	2							
OVERHEATS	1									
SLUGGISH ACCELERATION, EXCESSIVE THROTTLE NEEDED TO MAINTAIN SPEED					2				1	

Figure 8-74. AMC Torque Command.

or band grooves worn away at any portion of the band. Replace the band if it exhibits any of these conditions.

h. Overrunning clutch. Clean and inspect clutch parts for brinelled clutch rollers or cam or improperly assembled rollers or springs. Check for collapsed springs and bent spring retainer tabs.

i. Torque converter. If the converter hub seal surface or drive slots are damaged or if the converter contains foreign material, burnt-oxidized fluid, or debris, replace the converter. Do not attempt to clean or flush the converter.

j. Seal rings. Inspect seal rings on the reaction shaft support and governor support for wear, cracks, or breakage. Inspect ring grooves on both support assemblies for nicks, burrs, or distortion. Inspect bores in the front clutch retainer and output shaft support for nicks, grooves, wear, cracks, or scratches.

Stall Speed Specifications

Engine	Transmission Model	Engine rpm
Four-cylinder	904	2,750–3,000
Six-cylinder	904	1,850–2,150
Six-cylinder	998	1,850–2,150

Hydraulic Pressure Test Diagnosis

Condition	Indication
1. Line Pressure OK during any one test	1. Pump and regulator valve OK
2. Line pressure OK in R but low in D, 2, and 1	2. Leakage in rear clutch area (servo, clutch seals, governor support seal rings)
3. Pressure OK in 1 and 2 but low in D3 and R.	3. Leakage in front clutch area (servo, clutch seals, retainer bore, pump seal rings)
4. Pressure OK in 2 but low in R and 1	4. Leakage in rear servo
5. Front servo pressure in 2	5. Leakage in servo; broken servo ring or cracked servo piston
6. Pressure low in all positions	6. Clogged filter, stuck pressure regulator valve, worn or defective pump
7. Governor pressure high at idle (over 0 to $1\frac{1}{2}$ psi)	7. Governor valve sticking open
8. Governor pressure zero or below specifications at all mph figures	8. Governor valve sticking in closed position
9. Lubrication pressure low at all throttle positions	9. Clogged oil cooler or lines, seal rings leaking, output shaft plugged with debris, worn bushings in pump or clutch retainer

Hydraulic Pressure Test Specifications

Lube pressure	Closed throttle	5-30 psi (34-207 kPa)
	Full throttle	5-30 psi (34-207 kPa)
Line pressure	Closed throttle	54-60 psi (372-414 kPa)
	1,000 rpm	57-94 psi (393-648 kPa)
Front servo release	3rd gear only	No more than 3 psi (21 kPa) lower than line pressure.
Rear servo release	1 range	No more than 3 psi (21 kPa) lower than line pressure.
	R range	160 psi (1,103 kPa) at idle, builds to 270 psi (1,862 kPa) at 1,600 rpm.
Governor	D range	Pressure should respond smoothly to changes in mph and return to 0 to 11/2 psi (0-7 kPa) when stopped with transmission in D, 1, 2.
	Closed throttle	Pressure above 11/2 psi (7 kPa) at standstill will prevent transmission from downshifting.

Clutch and Band Application Chart

Drive position	P	R	N	D			2		1
				1	2	3	1	2	
Front clutch		●				●			
Front band					●				
Rear clutch				●	●	●	●	●	●
Rear band		●							●
Overrunning clutch				●			●		●
Converter lockup clutch						●			

Lockup Torque Converter Service Diagnosis

Condition	Possible Cause	Correction
No lockup	1. Faulty oil pump 2. Sticking governor valve 3. Valve body malfunction a) Stuck switch valve b) Stuck lockup valve c) Stuck fail-safe valve 4. Faulty torque converter 5. Failed locking clutch 6. Leaking turbine hub seal 7. Faulty input shaft or seal ring	1. Replace oil pump. 2. Repair or replace as necessary. 3. Repair or replace valve body or its internal components as necessary. 4. Replace torque converter. 5. Replace torque converter. 6. Replace torque converter. 7. Repair or replace as necessary.
Will not unlock	1. Sticking governor valve 2. Valve body malfunction a) Stuck switch valve b) Stuck lockup valve c) Stuck fail-safe valve	1. Repair or replace as necessary. 2. Repair or replace valve body or its internal components as necessary.
Stays locked up at too low a speed in direct	1. Sticking governor valve 2. Valve body malfunction a) Stuck switch valve b) Stuck lockup valve c) Stuck fail-safe valve	1. Repair or replace as necessary. 2. Repair or replace valve body or its internal components as necessary.
Locks up or drags in low or 2nd	1. Faulty oil pump 2. Valve body malfunction a) Stuck switch valve b) Stuck fail-safe valve	1. Replace oil pump. 2. Repair or replace valve body or its internal components as necessary.
Sluggish or stalls in reverse	1. Faulty oil pump 2. Plugged cooler, cooler lines, or fittings 3. Valve body malfunction a) Stuck switch valve b) Faulty input shaft or seal ring	1. Replace oil pump as necessary. 2. Flush or replace cooler and flush lines and fittings. 3. Repair or replace valve body or its internal components as necessary.
Loud chatter during lockup engagement (cold)	1. Faulty torque converter 2. Failed locking clutch 3. Leaking turbine hub seal	1. Replace torque converter. 2. Replace torque converter. 3. Replace torque converter.
Vibration or shudder during lockup engagement	1. Faulty oil pump 2. Valve body malfunction 3. Faulty torque converter 4. Engine needing tuneup	1. Repair or replace oil pump as necessary. 2. Repair or replace valve body or its internal components as necessary. 3. Replace torque converter. 4. Tune engine.
Vibration after lockup engagement	1. Faulty torque converter. 2. Exhaust system striking underbody 3. Engine needing tuneup 4. Throttle linkage misadjusted	1. Replace torque converter. 2. Align exhaust system. 3. Tune engine. 4. Adjust throttle linkage.
Vibration when revved in neutral	1. Torque converter out of balance	1. Replace torque converter.
Overheating oil blows out of dipstick tube or pump seal	1. Plugged cooler, cooler lines or fittings 2. Stuck switch valve	1. Flush or replace cooler and flush lines and fittings. 2. Repair switch valve in valve body or replace valve body.
Shudder after lockup engagement	1. Faulty oil pump 2. Plugged cooler, cooler lines or fittings 3. Valve body malfunction 4. Faulty torque converter 5. Failed locking clutch 6. Exhaust system striking underbody 7. Engine needing tuneup 8. Throttle linkage maladjusted.	1. Replace oil pump. 2. Flush or replace cooler and flush lines and fittings. 3. Repair or replace valve body or its internal components as necessary. 4. Replace torque converter. 5. Replace torque converter. 6. Align exhaust system. 7. Tune engine. 8. Adjust throttle linkage.

Part 8–5 Worksheet

Perform any or all of the following tasks as required by your instructor on vehicles specified. Perform all tasks to 100 percent accuracy as to procedures and specifications given by your instructor, the appropriate service repair manual, and this text.

Record the results of all tests and measurements.

1. Check the fluid level, type, and condition.
2. Adjust the manual shift linkage.
3. Adjust the throttle linkage.
4. Check and adjust the vacuum modulator system and vacuum lines.
5. Check and adjust engine idle speed.
6. Perform the stall test and interpret results.
7. Perform the road test and interpret results.
8. Perform hydraulic pressure tests and interpret results.
9. Change the oil and filter.
10. Remove the valve body, perform air pressure checks, and interpret results.
11. Service the hydraulic servos.
12. Adjust the bands.
13. Service the governor.

PART 8-6 SELF-CHECK

1. Why is a systematic approach to problem diagnosis necessary?
2. Describe how to check the fluid level and condition in a vehicle of your choice.
3. What problems can maladjusted manual shift linkage cause?
4. What effect does throttle linkage adjustment have on shift timing and on shift quality?
5. What effect does a vacuum leak in the modulator diaphragm have on transmission operation?
6. What is the purpose of a stall test?
7. Why should fluid level and linkage adjustments be checked and adjusted before road testing?
8. What is the purpose of air pressure checks?
9. What effect does improper band adjustment have on transmission operation?
10. Describe how to perform a stall test.
11. Slippage in all forward gears could be caused by __________.
12. A malfunctioning governor could cause ________ ____________________.

PART 8-7 PERFORMANCE EVALUATION

Did you complete the following tasks as required to 100 percent accuracy as to procedures and specifications given by your instructor, the appropriate service repair manual, and this text and did you accurately record and interpret the results?

1. Check the fluid level, type, and condition.
2. Adjust the manual shift linkage.
3. Adjust the throttle linkage.
4. Check and adjust the vacuum modulator system and vacuum lines.
5. Check and adjust engine idle speed.
6. Perform the stall test and interpret results.
7. Perform the road test and interpret results.
8. Perform hydraulic pressure tests and interpret results.
9. Change the oil and filter.
10. Remove the valve body, perform air pressure checks, and interpret results.
11. Service the hydraulic servos.
12. Adjust the bands.
13. Service the governor.
14. Prepare the vehicle properly for customer acceptance.

Chapter 9

Overhaul Procedure

Performance Objectives

After sufficient study of this text, the appropriate training models, the instructor's demonstrations, and the proper service repair manuals, the student should be able to perform the following tasks.

1. Follow all the required safety procedures and general precautions required for the safe and successful removal, overhaul, and installation of automatic transmissions.
2. Clean, disassemble, inspect, determine the serviceability of, overhaul, assemble, install, and adjust all automatic transmission subassemblies and components designated by the instructor.
3. After overhaul, leave the vehicle in a clean and presentable condition for customer acceptance.

PART 9-1 INTRODUCTION

This chapter deals with the general procedures required for proper removal, cleaning, disassembly, overhaul, reassembly, installation, adjustment, and road testing of automatic transmissions. Although the sequence of operations varies somewhat between different makes and models of transmissions, the general procedure is quite similar for all of them.

Some differences in removal procedure exist between rear-wheel-drive cars and front-wheel-drive cars, and there are also differences in disassembly and measuring procedures. However, all transmissions consist of the same types of operating assemblies, which require the same type of service procedures.

The procedures given in this chapter can be used as guidelines that will make the job easier. The manufacturer's service repair manual as well as any superseding service bulletins, of course, take precedence over this text. These manuals and bulletins are required in any case for specific procedures, specifications, and modifications that apply to specific transmissions.

Before Transmission Removal

No automatic transmission should be removed unless a thorough systematic diagnosis has definitely established that removal and overhaul are required. Automatic transmissions are designed to provide 100,000 miles (160,00 km) or more of normal service life. Malfunctions may require only minor in-vehicle adjustment or service. Removing and overhauling a transmission unnecessarily is costly and unfair to the customer as well as damaging to the reputation of the technician and the repair shop.

Severe automatic transmission service may be defined as any one or more of the following:

1. Driving in heavy city traffic
2. Operating regularly in ambient temperatures of 90°F (32°C) or higher
3. Driving in hilly or mountain areas
4. Frequent or heavy trailer towing
5. Use as a taxi, police car, or delivery service vehicle

More frequent service may be required under these operating conditons.

Transmission Identification

In order to obtain the correct parts and the proper specifications for any automatic transmission, it is necessary to identify the unit positively.

Transmission identification includes the following:

1. Make of the transmission (vehicle make)
2. Model year of the vehicle
3. Transmission number

The transmission number may be stamped on the side of the pan mounting flange on either the right or left side on some units. On others, it is found on a tag attached with one of the valve body attaching screws or it may be stamped or painted on the transmission converter housing. Refer to the appropriate manufacturer's service repair manual to help locate and identify the transmission number.

PART 9-2 GENERAL PRECAUTIONS

1. Be careful of hot transmission fluid and hot exhaust parts. (Fluids may reach 350°F [177°C]).

2. Follow all normal personal safety habits: no loose clothing; no smoking; no jewelry; cleanliness; face mask; and the like.

3. Use only a soft hammer and light tapping blows in order to prevent damage to transmission parts.

4. Follow all normal shop safety procedures: proper jacking and hoisting of vehicle; clean floor and tools; proper routing of electrical cords and air hoses; no jack handles sticking out; proper shop ventilation; proper use of transmission jack; and the like.

A good practice to develop and to follow is to anticipate hazardous conditions and to prevent accident and injury from happening. The competent technician has developed the skill of anticipating the consequences of actions into a sixth sense, which avoids unsafe and dangerous habits almost without conscious effort.

5. Parts that are machined to very close tolerances and are highly polished are used in automatic transmissions. Parts should be handled so as to prevent nicking and scoring. Do not drop any parts. Do not allow parts to bump and scrape against each other. Do not place parts where they can roll off and drop on the floor with the slightest jar.

6. Never remove the transmission attaching bolts until both the engine and the transmission are properly supported.

7. Always use a block of wood between the transmission jack and the oil pan in a manner that will prevent any damage to the pan or the transmission.

8. It is good practice to leave the hood open during removal and installation of the transmission in order to prevent damage to it when raising the engine slightly.

9. Always disconnect the battery to prevent anyone from cranking the engine accidentally and to prevent electrical shorts during transmission overhaul.

These are a few reminders to help avoid injury and damage. It is possible for individuals not thinking of the consequences of their actions to proceed in any number of unpredictable and unsafe practices. It is critical for you as a technician to avoid this kind of action.

PART 9-3 TRANSMISSION REMOVAL AND INSTALLATION

1. Open the hood and place fender protectors on both sides.

2. Disconnect the negative battery cable for safety reasons.

3. Remove the air cleaner if clearance to the fire wall is limited. Lowering the rear of the engine during transmission removal may cause interference or damage if the air cleaner is not removed.

4. Remove the transmission dipstick. If left in, damage could result due to limited clearance during transmission removal.

5. Disconnect the throttle valve linkage or cable as needed.

6. Raise the car on a hoist.

7. Mark the drive shaft with scribe marks or spray paint for assembly in the same position. Remove the drive shaft.

8. Remove the starter if required. The starter must be removed if any starter mounting bolts are attached to the converter housing.

9. Remove any steering linkage or drive axles (transaxles) that may interfere with transmission removal.

10. Disconnect the manual shift linkage and throttle linkage from the transmission. Use pliers to push rods out of plastic grommets.

11. Disconnect the speedometer cable at the transmission.

12. Disconnect the vacuum line at the vacuum modulator if so equipped.

13. Disconnect all electrical connections at the transmission.

14. Disconnect or remove any exhaust system components that may interfere with transmission removal and lowering.

15. Remove transmission rear mount bolts. Re-

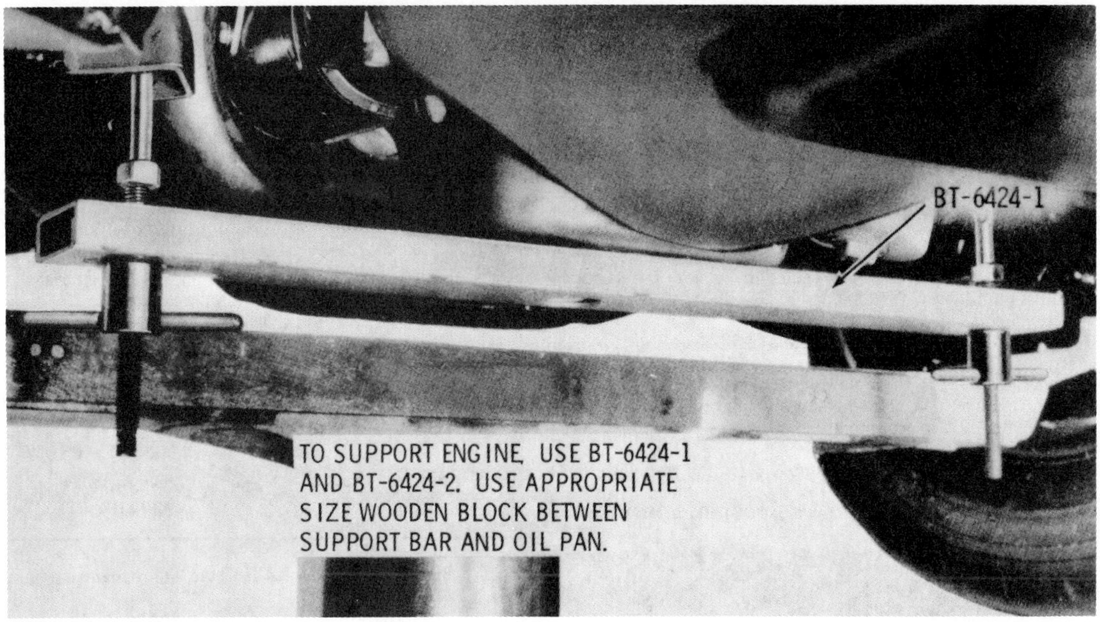

Figure 9-1. Typical engine support bracket required for automatic transmission removal. Similar bracket can be fabricated. (*Courtesy of General Motors Corporation*)

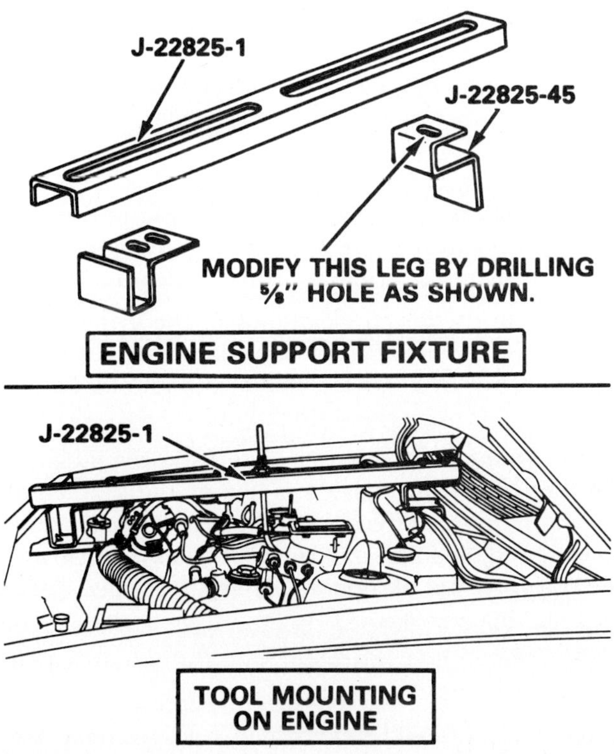

Figure 9-2. Engine support fixture used for transaxle removal. (*Courtesy of General Motors Corporation*)

move crossmember bolts if the crossmember is to be moved for transmission clearance.

16. Remove the converter housing cover plate and mark the converter to drive plate position with scribe marks or spray paint for proper assembly later.

17. Remove the torque converter to the flexplate attaching bolts or nuts. You may have to use a socket wrench at the crankshaft pulley bolt at the front of the engine in order to turn the converter for access to all bolts. The converter and transmission must always be removed or installed as an assembly. This

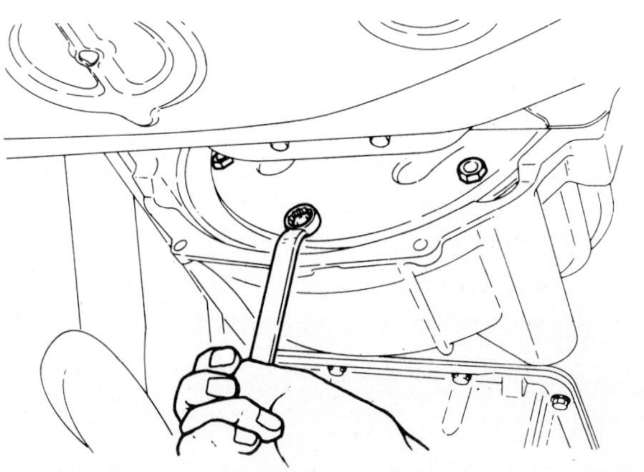

Figure 9-3. Removing torque converter drive bolts prior to transmission removal. (*Courtesy of Chrysler Corporation*)

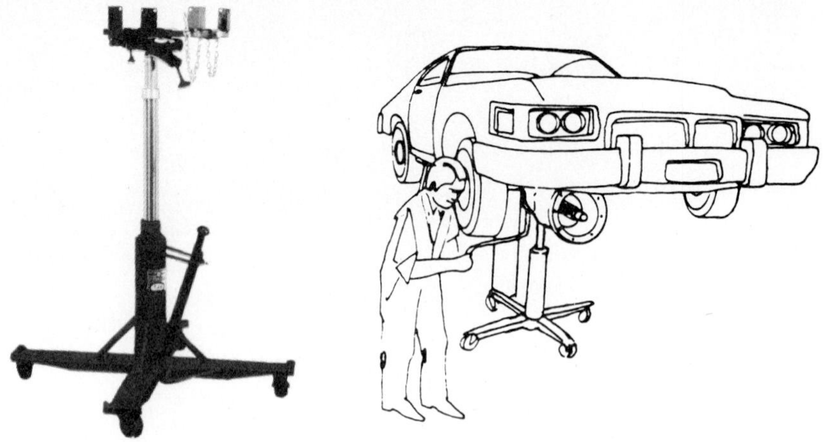

Figure 9-4. Hydraulic transmission jack with adjustable platform and hold-down chain assures proper transmission alignment and avoids heavy lifting.

prevents converter or drive plate damage. The weight of the transmission or converter must never hang on the drive plate.

18. Disconnect the oil cooler lines from the transmission. Cap all openings.

19. Install the engine holding fixture if required. Some engines must be properly supported with a holding fixture to prevent damage to the radiator, shroud, radiator hoses, and the like during transmission removal or installation. In many cases the engine is balanced well enough on its mounts not to require a holding fixture. In both cases, however, due care and attention must be given to prevent the fan from damaging the radiator or fan shroud and the radiator hoses from being overstretched.

20. Position a transmission jack under the transmission and install the hold-down chain to secure the transmission to the jack. Tilt the jack contact pad to fit the transmission properly. Raise the jack enough to raise the engine slightly.

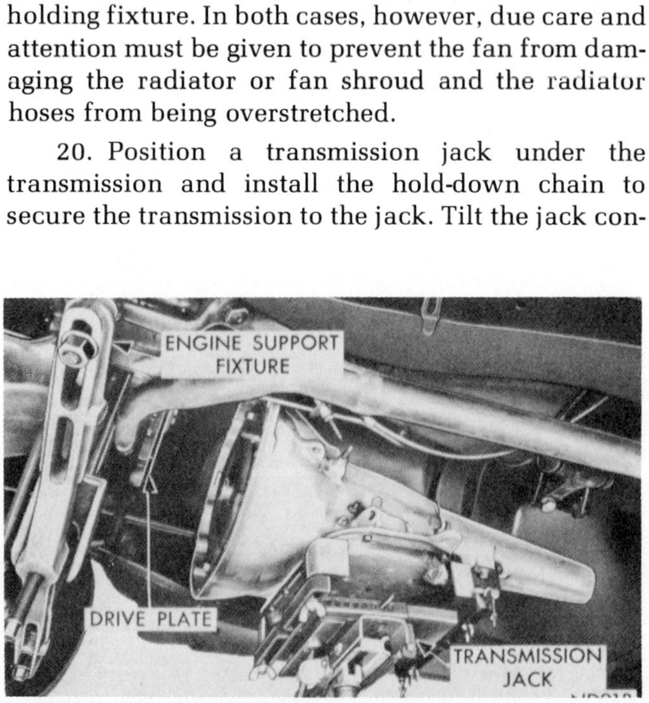

Figure 9-5. Removing or installing transmission and converter assembly (*Courtesy of Chrysler Corporation*)

21. Remove the previously disconnected cross-member (if necessary), keeping all alignment shims in place for proper reassembly.

22. Remove the converter housing to engine attaching bolts.

23. Install the converter retaining tool to prevent the converter from moving forward during transmission removal.

24. Carefully lower the transmission.
Caution: It may be necessary to wiggle the transmission from side to side at the rear to break it loose from the dowel pins. Do this carefully so as not to upset the jack when the transmission lets go. This requires having the jack at the correct height for enough support on the transmission, without raising the engine.

25. Carefully and slowly lower the transmission jack to maneuver the transmission clear of any wiring, tubing, or linkage. Do not damage or bend any of these during transmission removal.

26. Remove the transmission from the jack.
Caution: The transmission is heavy and you should have help for this. Place the transmission in the clean-up area and clean all exterior dirt from the transmission. Be sure that all openings remain capped during this cleaning process.

27. Place the transmission on the repair bench.

28. At this point, a careful inspection of the converter flex plate will determine if replacement is required. Inspect for starter gear tooth wear or damage, cracks, worn converter attaching holes, and worn or damaged converter hub pilot hole. If transmission

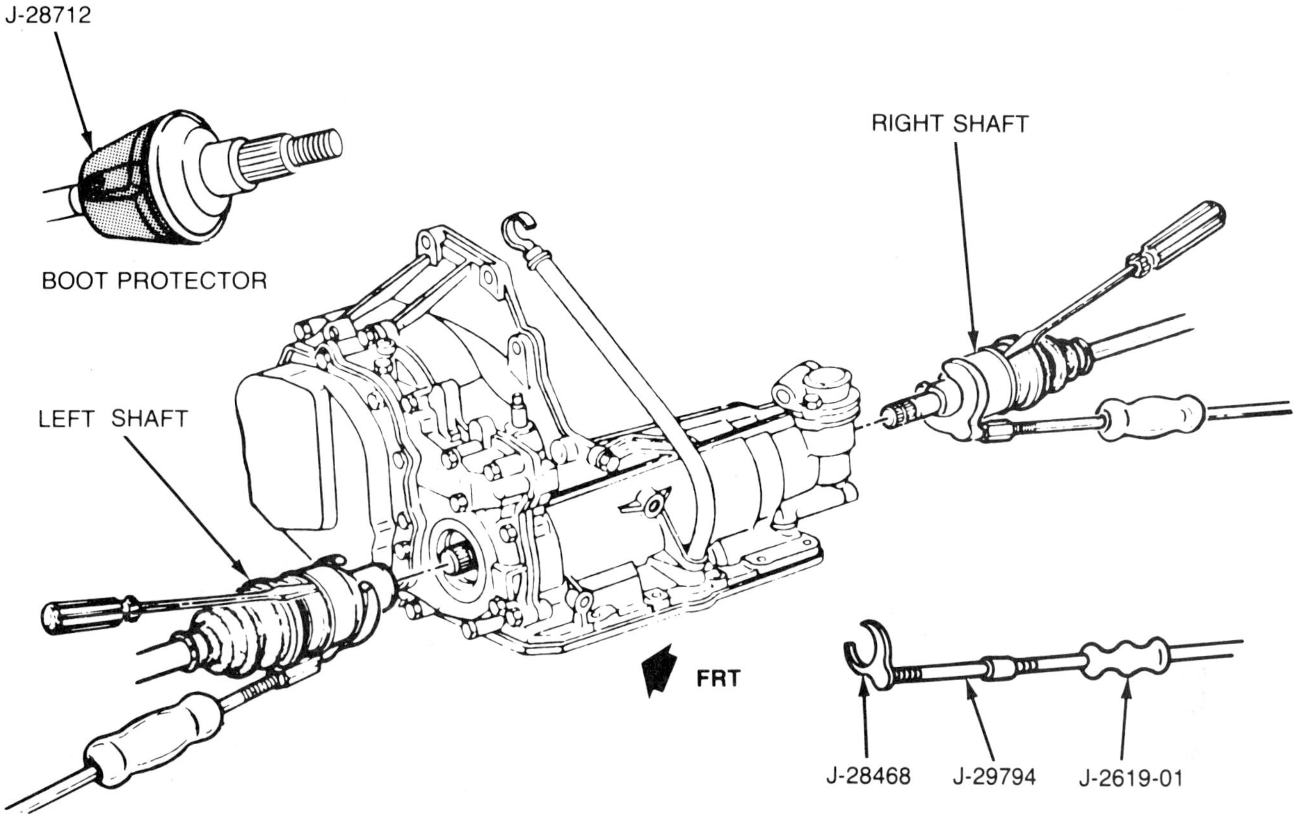

Figure 9-6. Removing drive axles from front-wheel-drive transaxle. (*Courtesy of General Motors Corporation*)

diagnosis established that there is a vibration problem in the converter area or if there is front seal leakage, it is good practice to check the flex plate runout if otherwise in good condition. A flex plate that does not run true can cause vibration and rapid front seal and pump bushing wear, which results in oil leakage. Check the flex plate runout with a dial indicator and compare to specifications. Maximum al-

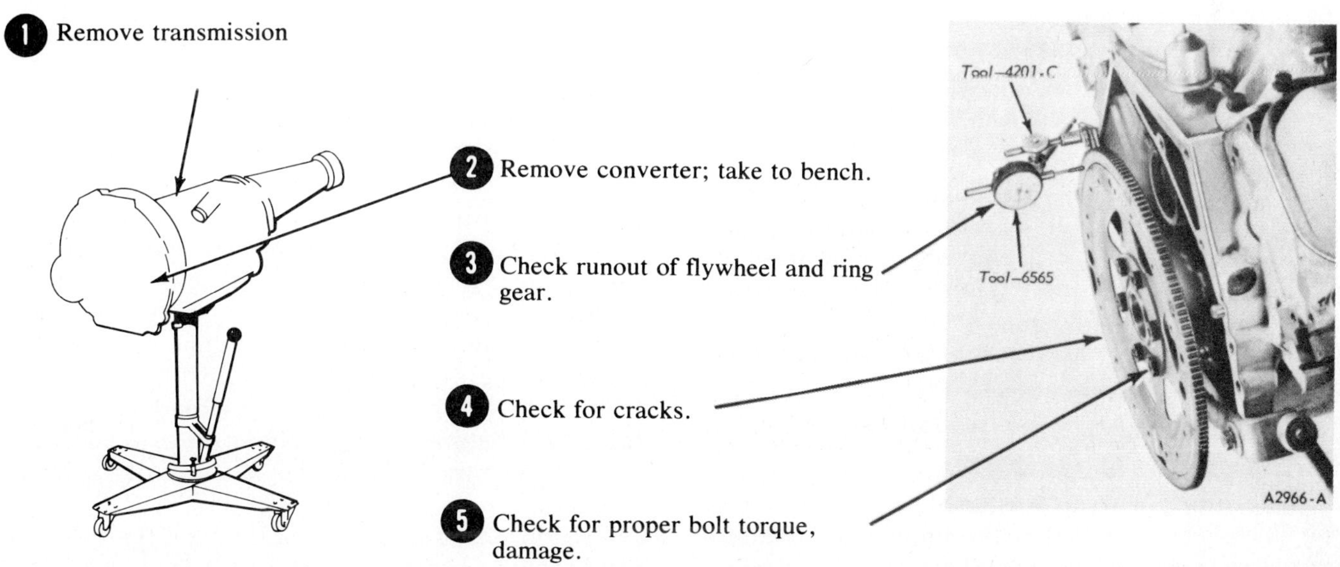

Figure 9-7. (*Courtesy of Ford Motor Company*)

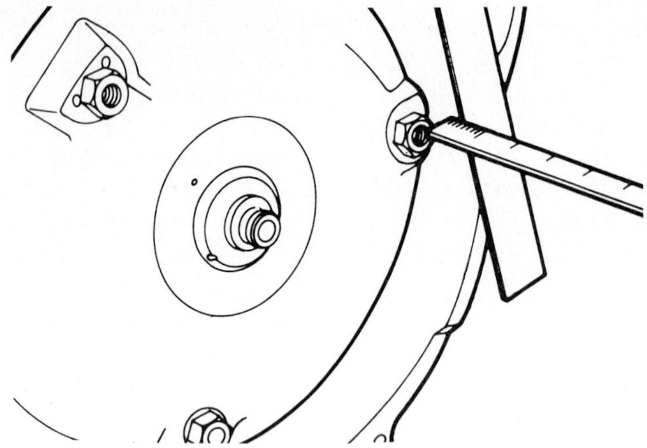

Figure 9-8. Measuring converter for full engagement in transmission. (*Courtesy of Chrysler Corporation*)

lowable runout is approximately 0.060 inch (1.50 mm).

Cooler and Cooler Line Flushing

In a major transmission failure, where particles of metal have been carried with the oil throughout the units of the transmission, it will be necessary to flush out the oil cooler and connecting lines. To flush the oil cooler and lines, use the following procedures.

1. With both cooler lines from the transmission disconnected:

2. Place a hose over the end of the cooler inlet line (from the bottom of the cooler) and insert the hose into an empty container.

3. Flush clean solvent through the return line (from the top of the cooler) using an oil suction gun until clean solvent comes out of the hose. This will back-flush the cooler.

4. Remove the hose from the inlet cooler line and place it on the return line.

5. Flush clean solvent through the inlet line until clean solvent comes out the return line. Remove the remaining solvent from cooler with compressed air applied to the return line and flush with transmission fluid.

Caution: Maximum air pressure should not exceed 50 psi.

6. Reconnect oil cooler lines and torque nuts to specified torque.

This procedure can also be done with converter flushing equipment. Follow the equipment manufacturer's instructions for procedures and pressures to use.

Oil Cooler Lines

If replacement of transmission steel tubing cooler lines is required, use only double-wrapped and brazed steel tubing. Under no condition use copper or aluminum tubing to replace steel tubing. Those materials do not have satisfactory fatigue durability to withstand normal car vibrations. Steel tubing should be flared using the double-flare method and the proper fittings.

Transmission Installation

The transmission and torque converter must be installed as an assembly. This requires that the torque converter fully engage the drive gear of the hydraulic pump. If full engagement of the torque converter is not assured, pump damage will occur when the transmission is installed and the mounting bolts tightened.

To assure full engagement, a measurement is taken between the mounting surface of the converter bell housing and the converter to the flex plate mounting surface. If this dimension is less than specified, the converter is not fully engaged. Once the converter is fully engaged, the converter holding bracket (you can make your own from a suitable piece of flat iron) should be installed to maintain converter engagement during transmission installation.

To install the transmission, lubricate the converter pilot hub. Then install the transmission by reversing the removal procedure. Clean and lubricate all bolts and tighten to specified torque. Bolts not lubricated may destroy aluminum threads.

When the transmission is raised into position, guide it over the dowel pins and push it toward the engine until mating surfaces are flush. Check to see that there is no interference between the converter and the converter drive plate. If there is, the converter may not be fully engaged with the hydraulic pump. Proceeding under these circumstances would damage the pump. If the converter is properly installed and the correct dimensions obtained between the converter face and the bell housing face, there should be no problem. Install all the mounting bolts by hand, then tighten to specified torque. Install the cross member, remove the jack and engine support, and complete the assembly. Make sure that the torque converter is aligned with marks made during removal. Tighten attaching bolts or nuts to specifications. Lower the vehicle. Apply the parking brake securely.

Put in 2, 3, or 4 quarts (or liters) of the specified fluid in the transmission and start the engine. Let the engine idle; do not accelerate. With the engine idling, add the remaining fluid to the cold level or add-oil level on the dipstick.

Caution: Do not fill above this level at this time. Move the selector lever through all gear ranges and place it in the specified position (P or N); add oil if needed to bring fluid level to cold level mark or add-oil mark on the dipstick.

During this procedure note if the transmission has the proper engagement feel.

Run the engine until transmission reaches a normal operating temperature (about 175°F, 80°C), recheck the fluid level, and add fluid if necessary.

Caution: Do not overfill. If accidentally overfilled, the fluid level must be corrected by draining until the proper level is reached (Chapter 8). Road test the vehicle as described in Chapter 8 to determine success of transmission service.

PART 9-4 DISASSEMBLY AND ASSEMBLY GUIDELINES

After the transmission has been removed, the exterior cleaned and placed on the work bench, the overhaul should proceed in an organized and systematic manner.

The sequence of operations includes the following.

First all subassemblies are removed from the transmission, generally in the following order:

- Torque converter
- Oil pan and filter
- Valve body
- Hydraulic pump
- Clutch units, gear train, and bands
- Extension housing
- Governor
- Output shaft

End-play measurements are taken and recorded before and during disassembly. These measurements are used to determine selection of proper thrust washer thickness during assembly. Thrust washer thickness determines end-play requirements for the transmission. Excessive end play indicates excessive wear internally and allows clutch drums to move back and forth excessively in the transmission case. Assembled end-play measurements should be between minimum and maximum specifications with preference being at the low end of specifications since end play will increase as the transmission is put in service. Refer to the appropriate shop manual for specifications and for information on the proper procedure for end-play measurements.

After end-play measurements have been recorded and all subassemblies removed, the procedure is to disassemble, clean, inspect, measure (where required), repair, and assemble the various units. When all units have been properly reconditioned, the transmission is assembled and installed. Although there are differences in the number of clutches and bands and the design of clutches, bands, servos, valve bodies, governors, and planetary gears, the general procedure for servicing these units is similar for all transmissions.

Disassembly, Cleaning, and Inspection Guidelines

1. Follow the disassembly sequence and procedure given in the appropriate service repair manual and this text.

2. Remove all old gasket material from metal surfaces with a suitable scraper. Avoid scoring or gouging the metal since this can cause fluid leakage.

3. Replace all old gaskets and seals with those supplied in the gasket and seal kit.

4. Replace oil transfer rings and metal seal rings if worn. Gasket and seal kits do not normally contain replacement oil transfer rings. Overhaul kits (which are more expensive) usually contain all gaskets, seals, oil transfer rings, check balls, and in some cases servo pistons.

5. Use only filtered, moisturefree compressed air regulated at 50 psi (345 kPa) maximum for cleaning passages, etc.

6. Do not use solvents, detergents, or vapor degreasers on composition clutch plates, bands, synthetic seals, and check balls.

7. Use only lintfree cloth for any wiping of parts. Lint can cause valves to stick and oil to leak.

8. Do not overexpand snap rings. Damaged snap rings must be replaced.

9. Do not overstretch seals and seal rings. Fluid leaks and malfunction are caused by stretched and distorted seals and seal rings.

10. Ends of angle-cut (scarf-cut) seal rings must be installed with angled cuts facing each other.

11. Hooked cast-iron seal rings must be properly hooked and flush with no step in hooked ends when installed.

12. Use only the recommended type of transmission fluid for lubrication during assembly. Petroleum-

jelly (Vaseline) type of lubricant can also be used. In some cases a lubricant such as Door Ease® may be useful in keeping a clutch seal in place for assembly. Never use any high temperature lubricant such as white grease. It will clog oil passages since it is not compatible with transmission fluids.

13. Every passage in every component must be free and clear of all obstructions.

14. Where fasteners of unequal length are used, be sure that each fastener is installed in the correct location. Installing a shorter fastener in place of the normal length will destroy the threads when it is tightened. Tighten all fasteners to specifications with a torque wrench.

15. Install all paper gaskets dry. The cork pan gasket should also be installed dry with dry mating surfaces. Oily, greasy, or slippery sealant surfaces cause cork gaskets to crack at bolt holes and squish out as fasteners are tightened to specified torque. A little petroleum jelly may be used on paper gaskets to help keep the gasket in place during assembly.

16. Always maintain the original position of parts to be reused.

PART 9-5 UNIT OVERHAUL PROCEDURE

This section deals wih the overhaul of the transmission unit by unit.

Torque Converter

The following procedure based on Ford C3, C4, and AOD automatic transmissions is typical of the checks required to determine the serviceability of a torque converter. Some manufacturers recommend flushing converters by the agitation method while others recommend flushing by machine only. The manufacturer's recommendations should be followed. It may not be recommended to flush some lockup converters.

Converter and Oil Cooler Service

When internal wear or damage has occurred in the transmission, metal particles, clutch plate material, or band material may have been carried into the converter and oil cooler. These contaminants are a major cause of recurring transmission troubles and *must* be removed from the system before the transmission is put back into service.

Whenever a transmission has been disassembled to replace worn or damaged parts or because the valve body sticks from foreign material, the converter and oil cooler *must* be cleaned by using a mechanically agitated cleaner, such as the Rotunda 14–0028 or equivalent. Under *no* circumstances should an attempt be made to clean converters or oil coolers by hand agitation with solvent.

Figure 9-9. After transmission is drained, transmission is mounted in holding fixture. (*Courtesy of General Motors Corporation*)

Converter End Play and One-Way Clutch Check

The tool shown in Figure 9–14 is used to check the converter end play. The tools shown in Figure 9–15 are used to check the converter one-way clutch.

End-Play Check

1. Insert the end-play checking tool into the converter pump drive hub until it bottoms.

Figure 9-10. Removing torque converter. (*Courtesy of General Motors Corporation*)

INSPECT DRIVE STUDS

- Studs pilot converter to run true with flywheel.

NOTE: C-3 converters have welded drive nuts in place of drive studs. Check these for damage and for cross threading. Replace drive bolts after each disassembly.

1 Check STUDS for:
- Tightness
- Good threads

2 Stud shoulders must not be damaged.

3 Raised or lowered shoulder causes:
- Misalignment
- Pump drive hub eccentric
- Pump bushing damage

DRIVE PLATE

RESULTS
- **Studs or welded drive nuts damaged or loose** — replace converter.
- **Shoulder damaged** — clean up burrs; inspect pump body bushing.
- **Okay** — check drive hub (step E).

Figure 9-11. *(Courtesy of Ford Motor Company)*

INSPECT DRIVE HUB AND PILOT HUB

- Scoring may not be as deep as it appears to the eye.

CHECK FOR DAMAGE
- **Burrs or nicks** — clean up.

CHECK WITH TIP OF FINGERNAIL.
- **Light scoring** — remove as shown below.
- **Deeply scored** — replace converter.
- Cover impeller hub to prevent dirt from entering.

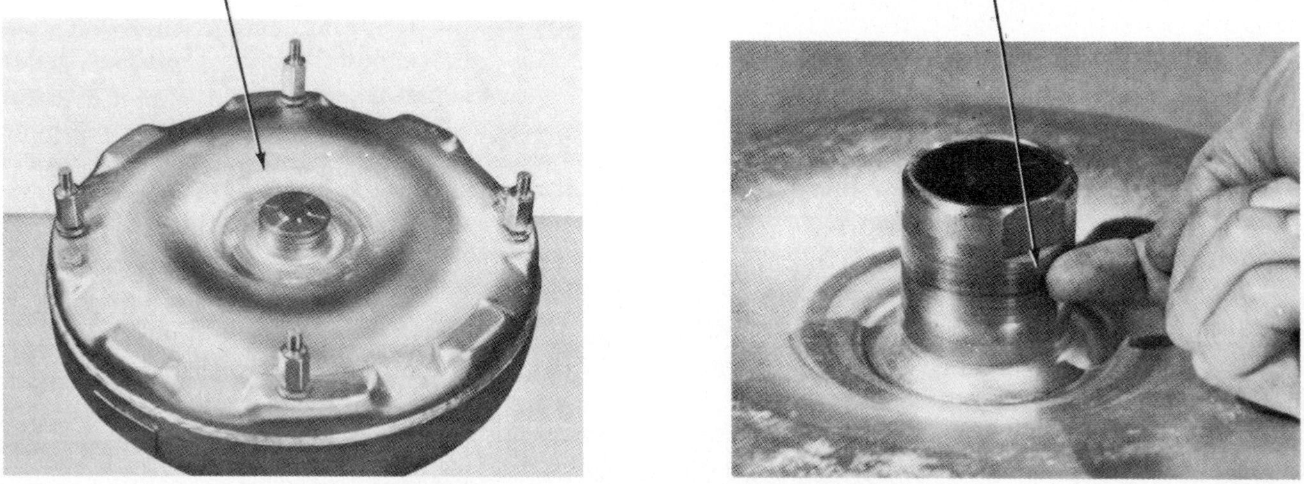

Figure 9-12. *(Courtesy of Ford Motor Company)*

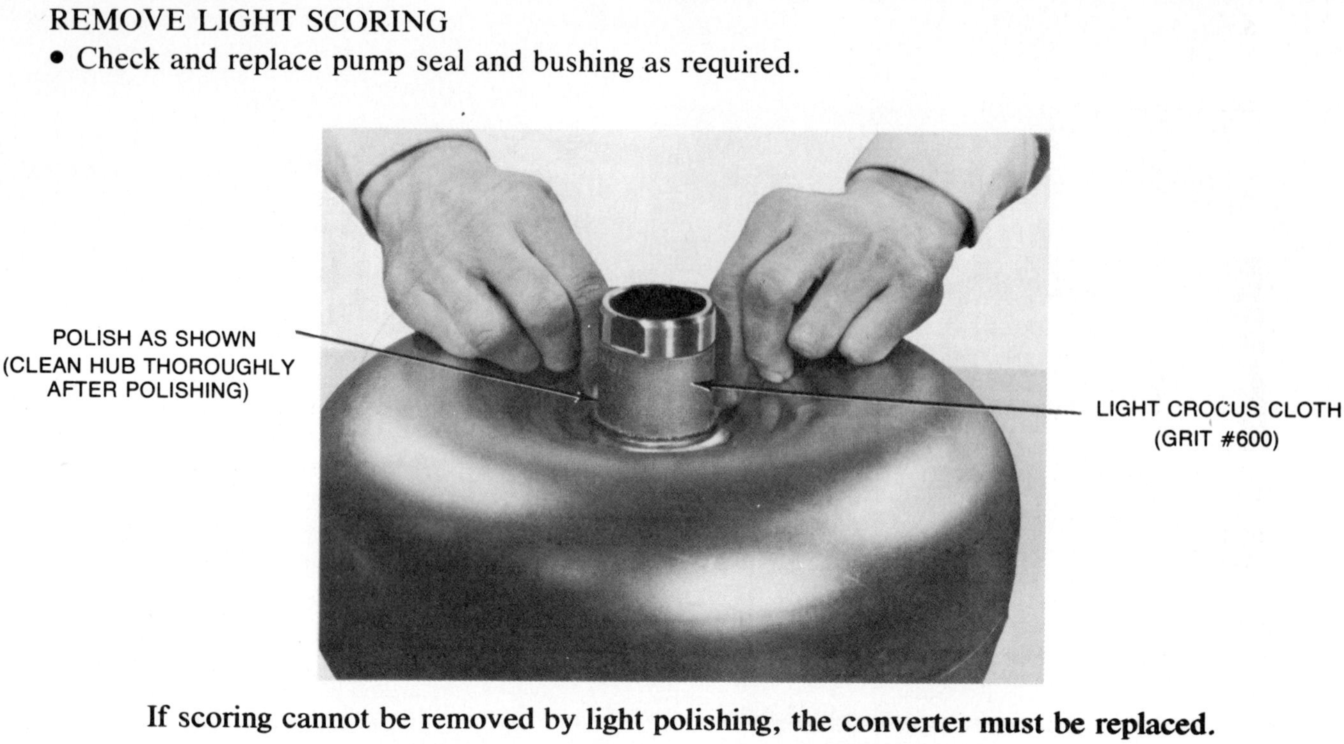

Figure 9-13. *(Courtesy of Ford Motor Company)*

2. Expand the sleeve in the turbine spline by tightening the threaded inner post until the tool is securely locked into the spline.

3. Attach a dial indicator to the tool. Position the indicator button on the converter pump drive hub, and set the dial face at zero.

4. Lift the tool upward as far as it will go and note the indicator reading. The indicator reading is the total end play that the turbine and stator share. Replace the converter unit if the total end play exceeds the limits indicated in the specifications in the manual.

5. Loosen the threaded inner post to free the tool and then remove the tool from the converter.

Converter One-Way Clutch Check

1. Insert the one-way clutch holding tool in one of the grooves in the stator thrust washer.

2. Insert the one-way clutch tightening tool for C3 automatic transmission (or the equivalent for automatic overdrive C5 and C4 automatic transmissions) in the converter pump drive hub so as to engage the one-way clutch inner race.

3. Attach a torque wrench to the one-way clutch tightening tool. With the one-way clutch holding tool held stationary, turn the torque wrench counterclockwise. The converter one-way clutch should lock up and hold a 13 N.m (10 pound-feet) force. The converter one-way clutch should rotate freely in a clockwise direction. Try the clutch for lockup and hold in at least five different locations around the converter.

4. If the clutch fails to lock up and hold a 13 N.m (10 pound-feet) torque, replace the converter unit.

Stator to Impeller Interference Check

1. Position the front pump assembly on a bench with the spline end of the stator shaft pointing up.

2. Mount a converter on the pump with the splines on the one-way clutch inner race engaging the mating splines of the stator support. The converter hub will then engage the pump drive gear.

3. Hold the pump stationary, and try to rotate the converter counterclockwise. The converter should rotate freely without any signs of interference or scraping within the converter assembly.

4. If there is an indication of scraping, the trailing edges of the stator blades may be interfering with the leading edges of the impeller blades. In such cases, replace the converter.

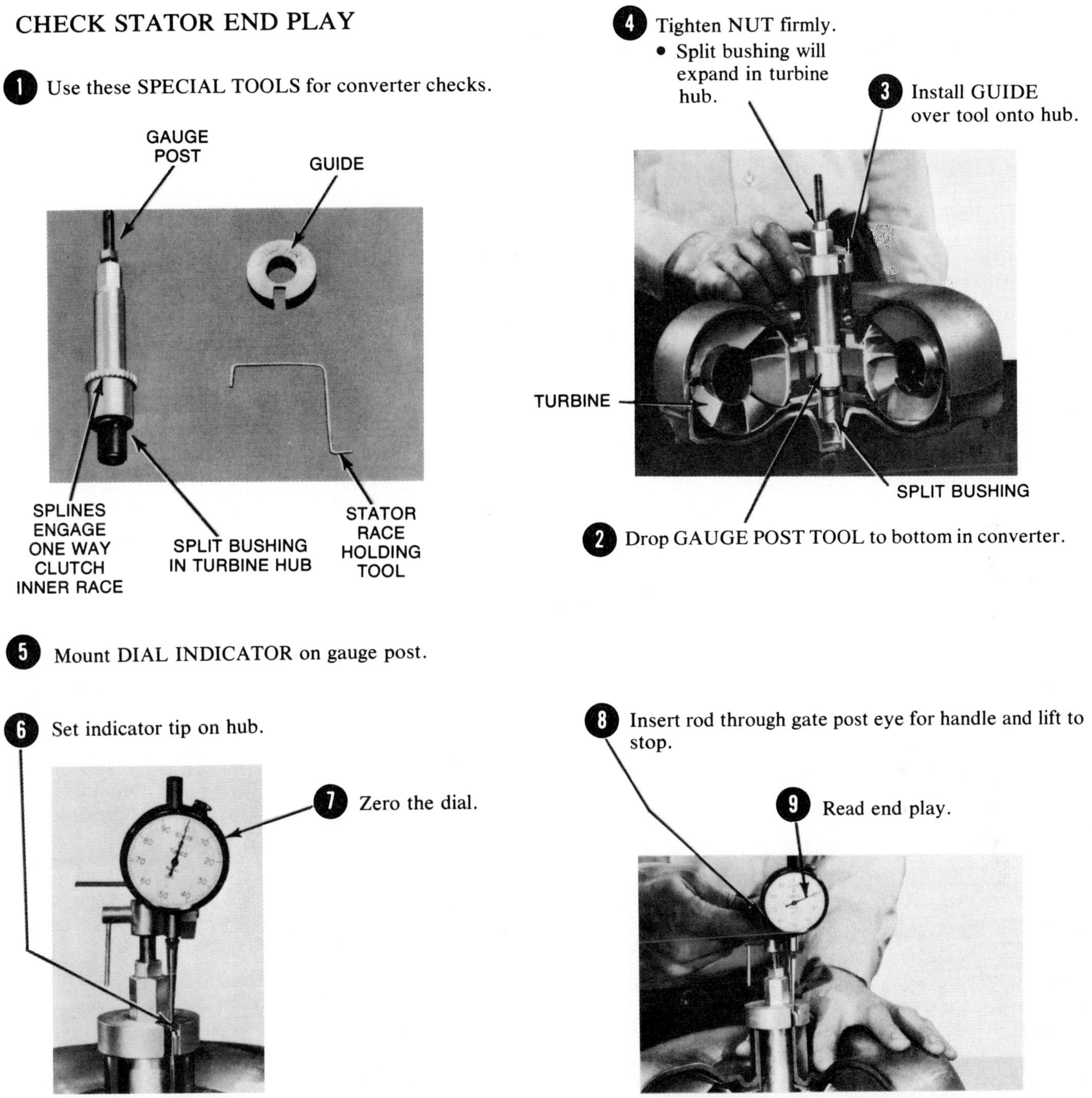

Figure 9-14. (*Courtesy of Ford Motor Company*)

Stator to Turbine Interference Check

1. Position the converter on the bench frontside down.

2. Install a front pump assembly to engage the mating splines of the stator support and stator, and pump drive gear lugs.

3. Install the input shaft, engaging the splines with the turbine hub.

4. Hold the pump stationary and attempt to rotate the turbine with the input shaft. The turbine should rotate freely in both directions without any signs of interference or scraping noise.

CHECK ONE-WAY CLUTCH

- Must hold stator for torque multiplication at low speeds.
- Must "free-wheel" at high speeds (coupling phase).

1 Use these SPECIAL TOOLS for converter checks.

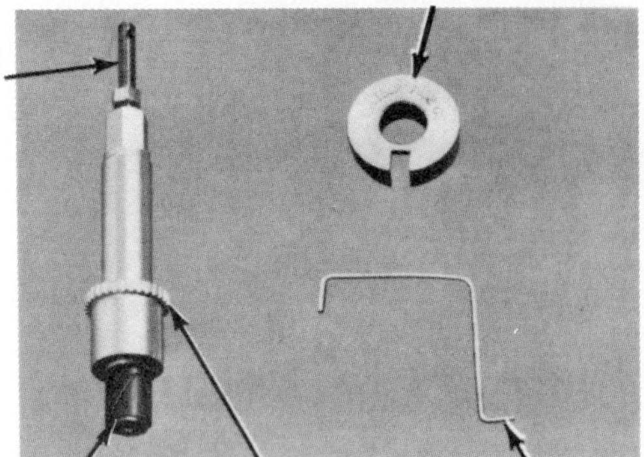

2 Insert HOLDING TOOL in groove in stator.
- Prevents stator from turning.

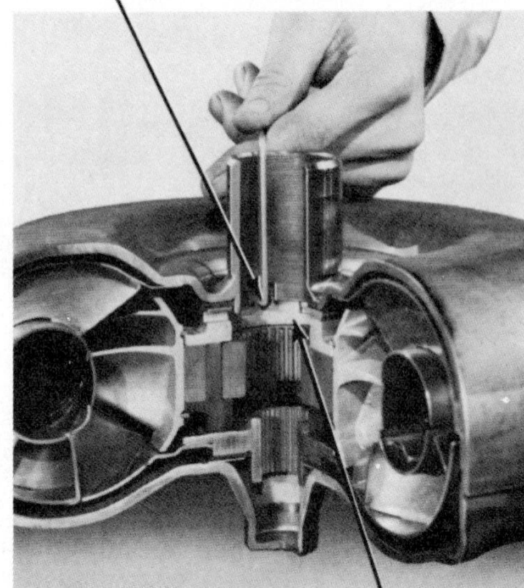

3 Drop GAUGE POST into converter.
- Use to turn inner race.

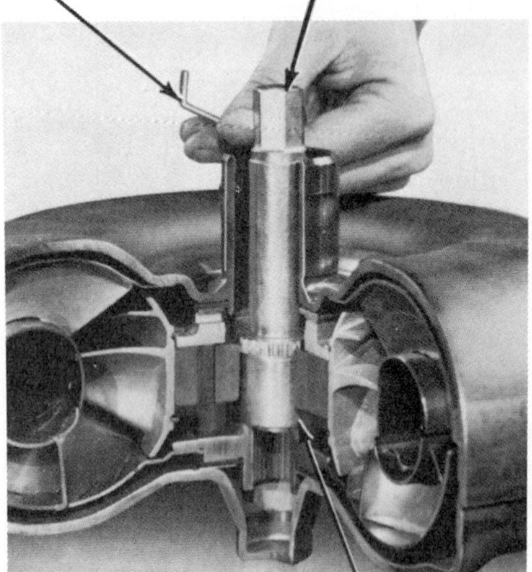

4 Install GUIDE to lock holding tool.

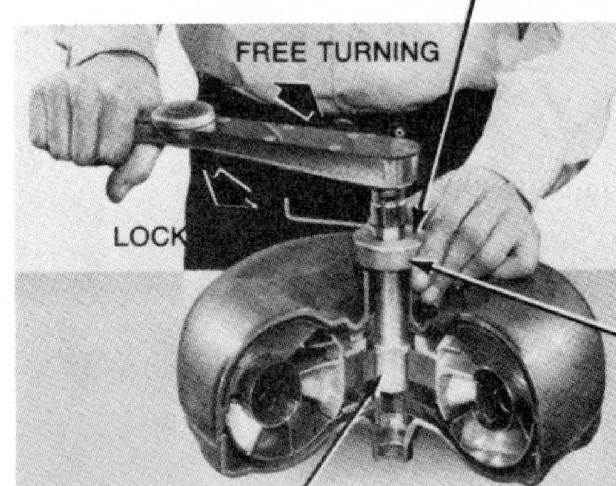

5 Turn with torque wrench.
- Check in five positions.

RESULTS REQUIRED

- **Clockwise** — turn freely.
- **Counter-clockwise** — lock up with a 10 foot-pound pull.

REPLACE CONVERTER IF NOT AS REQUIRED.

Figure 9-15. (*Courtesy of Ford Motor Company*)

CHECK FOR STATOR-TO-IMPELLER INTERFERENCE

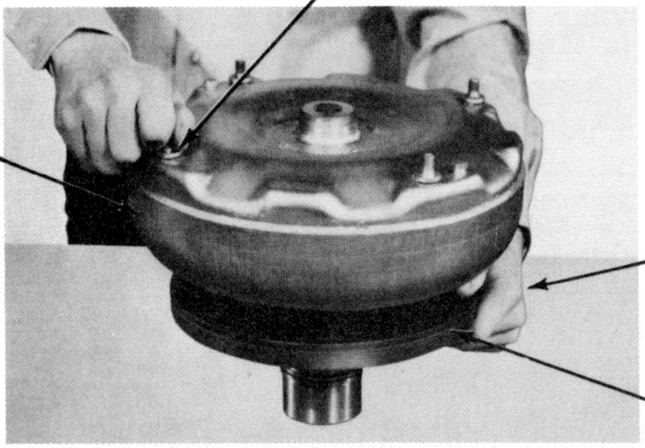

RESULTS
- **Converter turns freely** — no interference.
- **Slight rubbing noise** — okay.
- **Binding or loud scraping** — replace converter.

Figure 9-16. (*Courtesy of Ford Motor Company*)

CHECK FOR TURBINE INTERFERENCE

RESULTS
- **Turns freely** — no interference.
- **Slight rubbing noise** — okay.
- **Binding or loud scraping** — replace converter.

Figure 9-17. (*Courtesy of Ford Motor Company*)

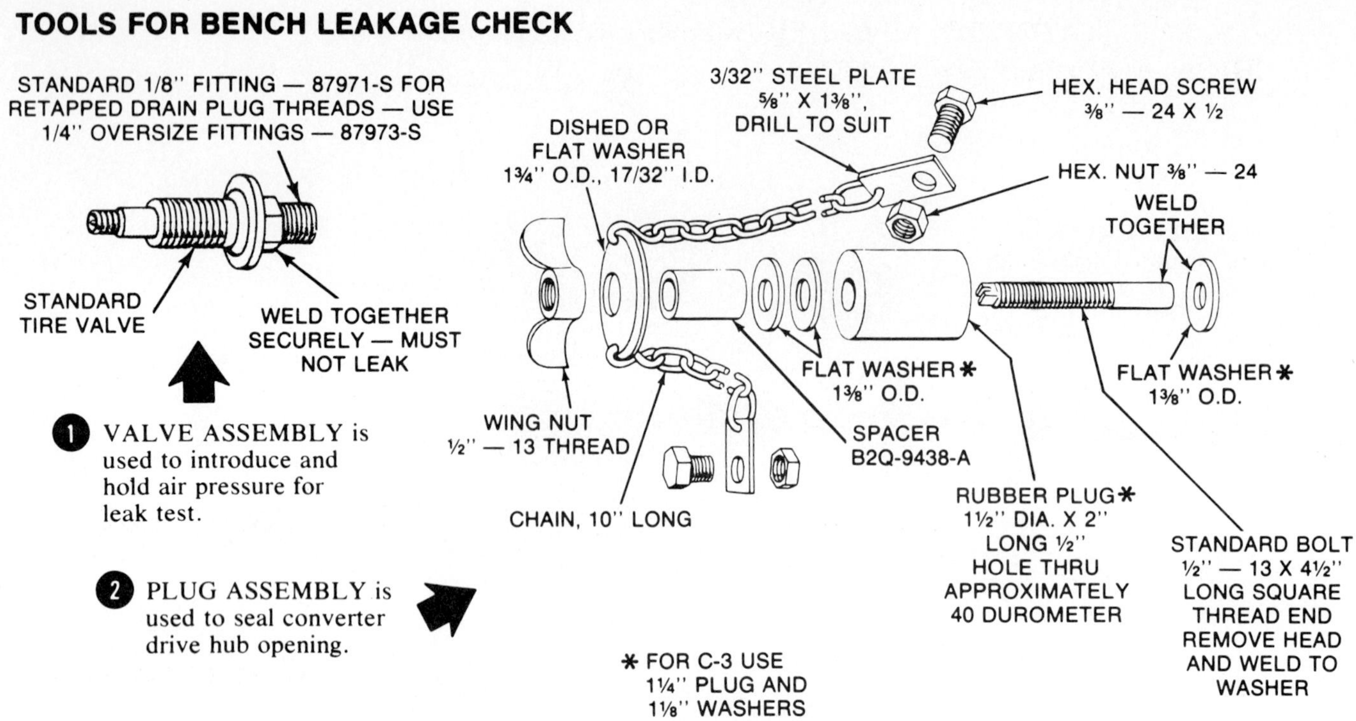

Figure 9-18. *(Courtesy of Ford Motor Company)*

5. If interference exists, the stator front thrust washer may be worn, allowing the stator to hit the turbine. In such cases, the converter must be replaced. The converter crankshaft pilot should be checked for nicks or damaged surfaces that could cause interference when installing the converter into the crankshaft. Check the converter front pump drive hub for nicks or sharp edges that would damage the pump seal.

Transmission Case and Thread Repair (Based on Ford C3, C4, and AOD Transmissions)

Inspect the case for cracks and stripped threads. Inspect the gasket surfaces and mating surfaces for burrs. Check the vent for obstructions and check all fluid passages for obstructions and leakage. Check servo piston bores, accumulator piston bores, and throttle valve bores for damage or scoring. Use crocus cloth to remove minor scratches and scoring.

Inspect the case bushing for scores. Check linkage parts for wear or damage. Replace any seals or damaged parts.

If a transmission case thread is damaged, service kits may be purchased from local jobbers. To service a damaged thread, the following procedures should be carefully followed.

1. Drill out the damaged threads, using the same drill size as the thread OD. For example, use a 5/16-inch drill for a 5/16,18 thread.

2. Select the proper special tap and tap the drilled hole. The tap is marked for the size of the thread being serviced. Thus, the special tap marked 5/16,18 will not cut the same thread as a standard 5/16,18 tap. It does cut a thread large enough to accommodate the insert; after the insert is installed, the original thread size (5/16,18) is restored.

3. Select the proper coil inserting tool. These tools are marked with the thread size being serviced. Place the insert on the tool and adjust the sleeve to the length of the insert being used. Press the insert against the face of the tapped hole. Turn the tool clockwise and wind the insert into the hole until the insert is one-half turn below the face.

4. Working through the insert, bend the insert tang straight up and down until it breaks off at the notch.

5. Improperly installed inserts can be removed with the extractor tool. Place the extractor tool in the insert so that the blade rests against the top coil one-quarter to one-half turn away from the end of the coil. Tap the tool sharply with a hammer so that the blade cuts into the insert. Exert downward pressure on the tool and turn it counterclockwise until the insert is

B. CHECK FOR LEAKAGE ON BENCH

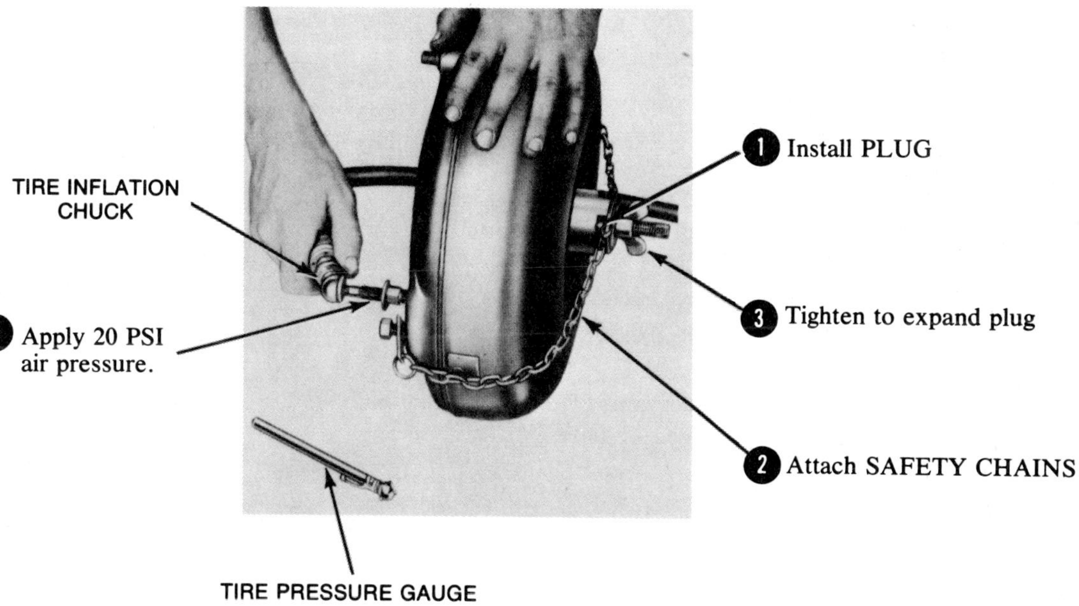

5 Check for bubbles at possible leak points —

- Wait 5-10 minutes for small leaks
- May not show with converter cold

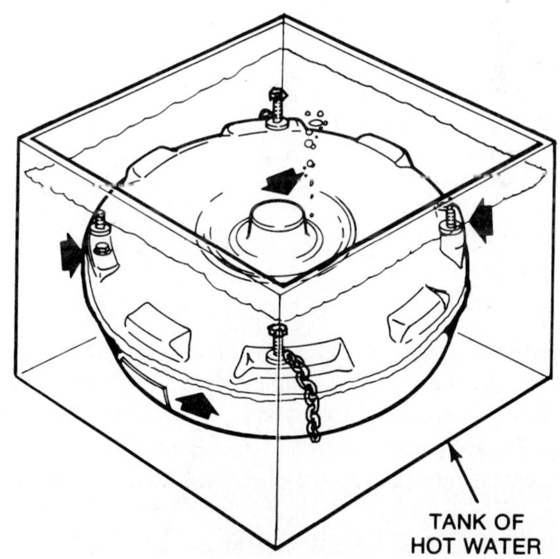

RESULTS

- **Leakage at Weld** — replace converter.
- **Leakage at drain plug** — tighten.
- **No leakage** — check pump and other causes in converter area

Figure 9-19. *(Courtesy of Ford Motor Company)*

INSPECT FOR TYPE OF CONTAMINATION

- DO **NOT** REPLACE CONVERTER BECAUSE OF CONTAMINATION CAUSED BY **TRANSMISSION** FAILURE:
  1. Clutch or band material.
  2. Sludge from overheating.
- **DO** REPLACE CONVERTER BECAUSE OF CONTAMINATION FROM CONVERTER THRUST WASHER FAILURE.
  1. Look for bronze, steel, aluminum (thrust washer) particles.
  2. Verify by interference checks and end play after **cleaning converter**

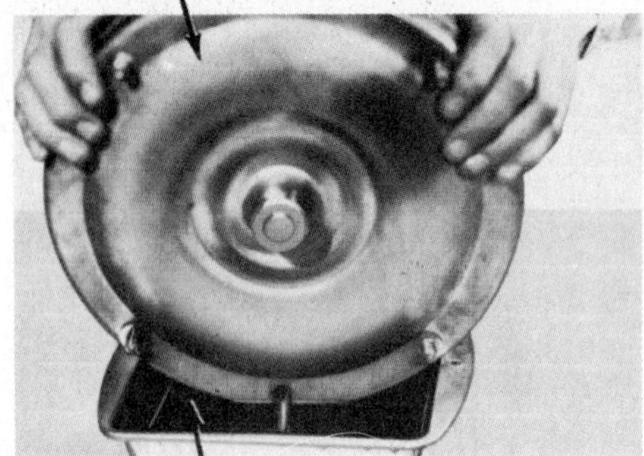

CLEAN CONVERTER

- If contaminated from transmission failure, clean and reuse.
- Use Rotunda Converter Cleaner
- Follow instructions with Cleaner.

Figure 9-20. *(Courtesy of Ford Motor Company)*

removed. The tool is marked with the thread size being serviced. Place the insert on the tool and adjust the sleeve to the length of the insert being used. Press the insert against the face of the tapped hole.Turn the total clockwise and wind the insert into the hole until the insert is one-half turn below the face.

Servicing the Clutch Units

Service procedures for multiple-disc clutches are generally quite similar for both driving and holding clutches.

Disassembly

Remove the clutch pack retaining snap ring and all clutch plates. Keep the clutch plates and pressure plates in order whether they are to be used again or not. This helps keep the order of assembly intact for later reference.

Position the clutch piston spring compressor on the clutch. Position the compressor so that it will not slip when tightened but will push on the spring retainer only, not on the snap ring. Tighten the compressor enough to allow snap ring removal with the appropriate snap ring pliers. Be careful not to push against the compressor and cause it to slip out of position. Do not overtighten.

With the snap ring removed, loosen the spring compressor carefully and watch that the spring retainer does not get caught in the snap ring groove during release. With the spring retainer and spring removed, remove the compressor and the clutch piston. You may have to use compressed air directed

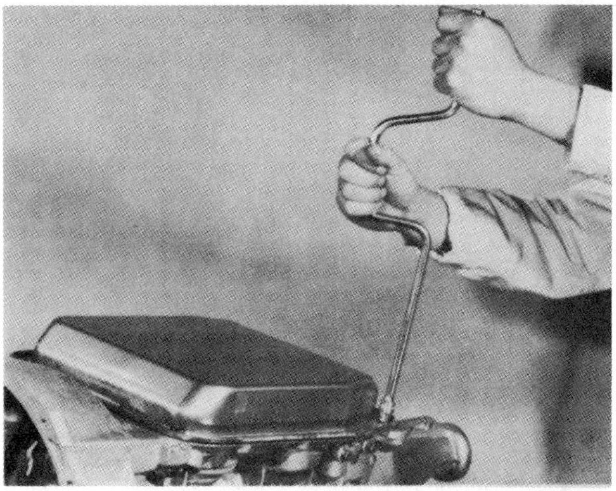

Figure 9-21. Removing oil pan bolts for pan removal. (*Courtesy of Ford Motor Company*)

Figure 9-22. Removing fluid filter from valve body. (*Courtesy of Ford Motor Company*)

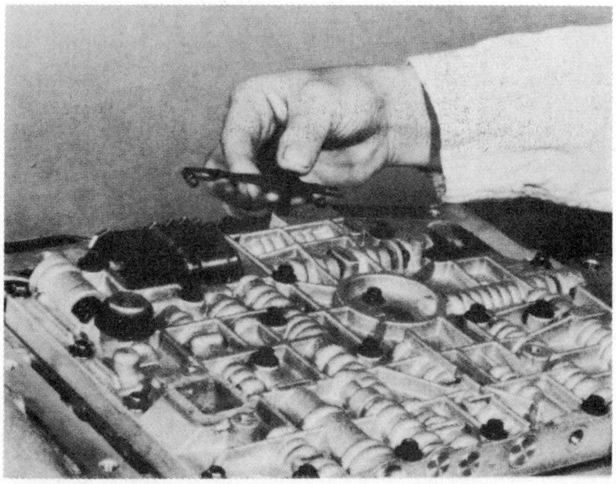

Figure 9-23. Removing detent spring from valve body. (*Courtesy of Ford Motor Company*)

Figure 9-24. Removing valve body attaching bolts. (*Courtesy of Ford Motor Company*)

Figure 9-25. Removing valve body and separator. (*Courtesy of Ford Motor Company*)

Figure 9-26. Use a screwdriver to remove snap ring from low-reverse band servo cover. The cover may pop out due to spring pressure. If not, tap cover with screwdriver handle to release seal. (*Courtesy of Ford Motor Company*)

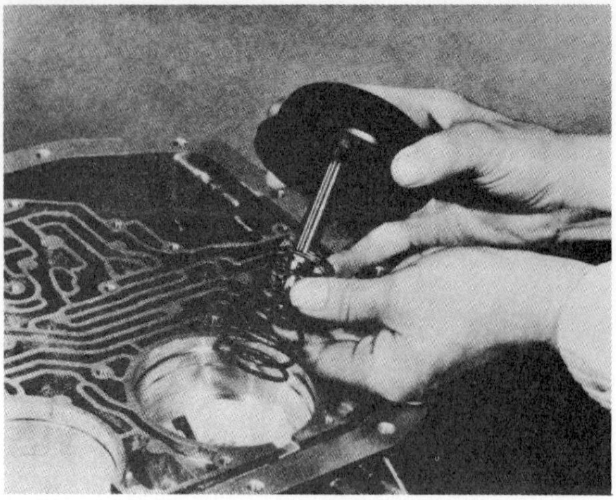

Figure 9-27. Remove low-reverse band piston and spring from transmission case. (*Courtesy of Ford Motor Company*)

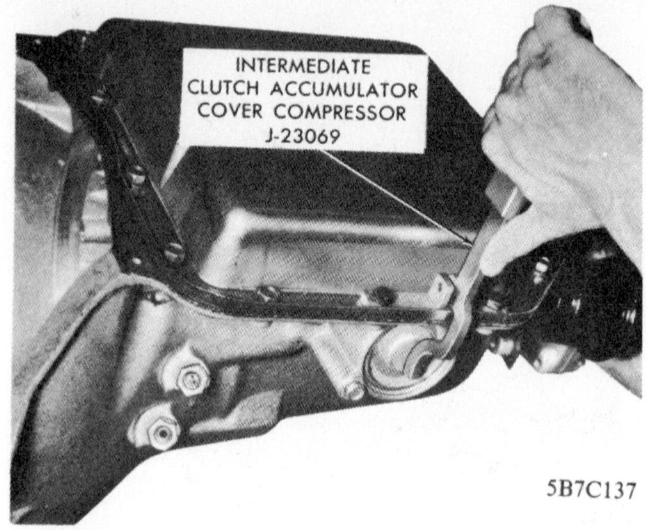

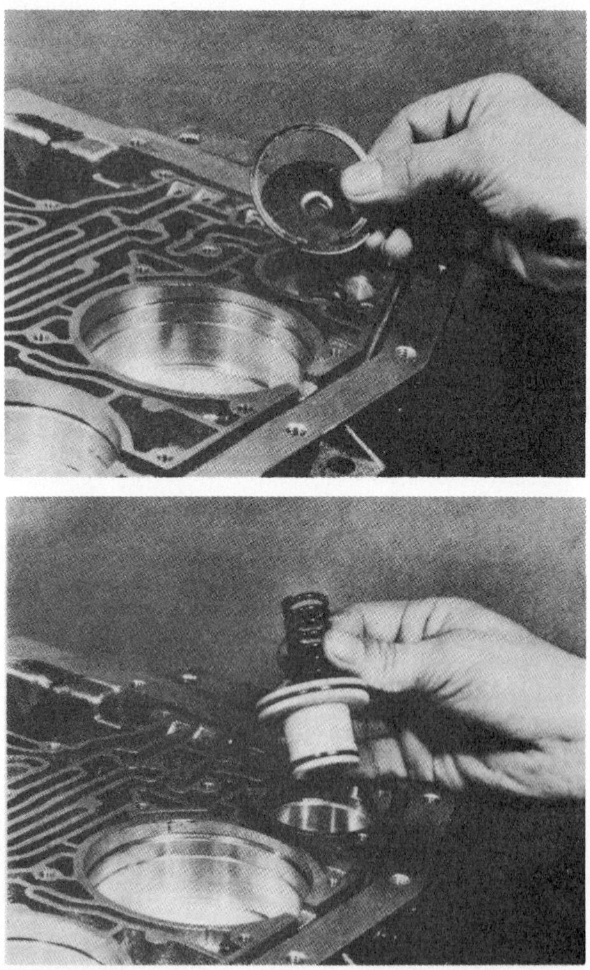

Figure 9-28. Remove snap ring from 2-3 accumulator piston cover. Remove cover, accumulator spring, and piston from transmission case. (*Courtesy of Ford Motor Company*)

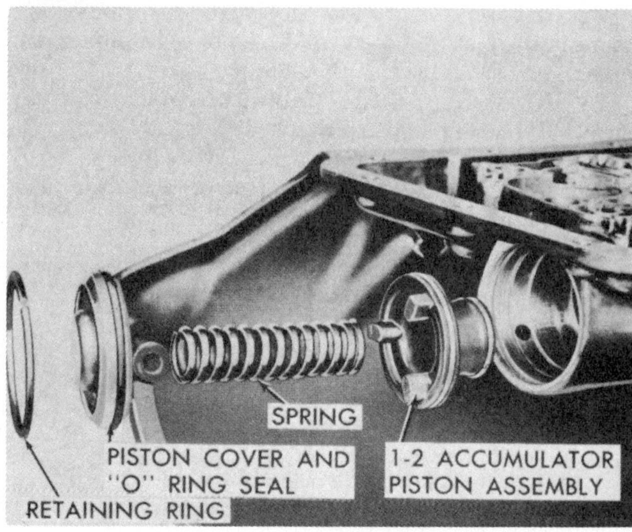

Figure 9-29. (*Courtesy of General Motors Corporation*)

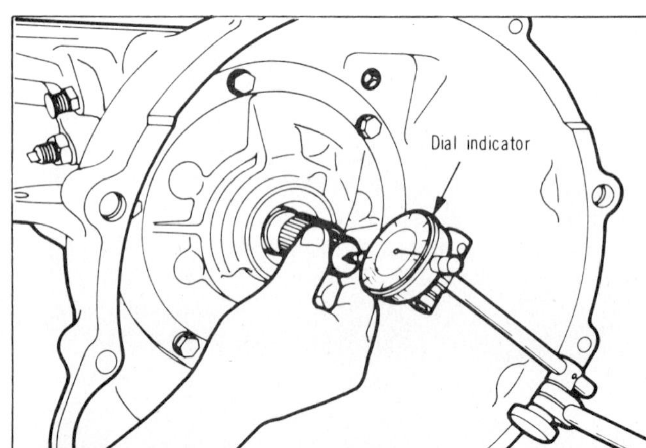

Figure 9-30. Measuring turbine shaft end play. (*Courtesy of Chrysler Corporation*)

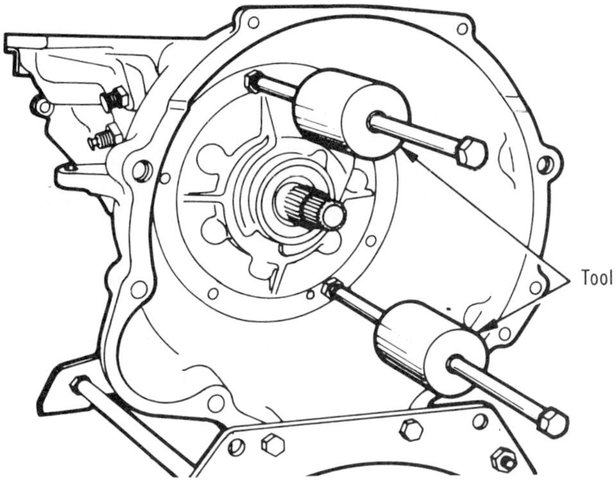

Figure 9-31. Using slide hammer pullers to remove hydraulic pump after pump attaching bolts have been removed. (*Courtesy of Chrysler Corporation*)

into the clutch apply passage to help remove the piston. Remove all seal rings from the clutch piston and clutch drum hub. Wash in appropriate cleaning fluid and blow dry.

Inspection

Clutch Plates

1. Compositioned plates

(a). Dry the plates and inspect for pitting, flaking, wear, glazing, cracking, charring, and chips or metal particles imbedded in lining. Check for flatness. Plates must not be dished.

(b). If a compositioned plate shows any of the preceding conditions, replacement is required.

2. Steel plates and pressure plates

(a). Wipe plates dry and check for heat discoloration. If the surfaces are smooth and even in color, the plate could be reused. If severe heat spot discoloration or surface scuffing is indicated, the plate must be replaced.

3. Clutch release springs

(a). Extreme heat or burning in the area of the clutch may have caused the springs to take a heat set and would require replacement of the springs.

Figure 9-32. Rotate transmission to horizontal position. Grasp turbine shaft and pull out assembly, consisting of intermediate clutch pack; overdrive band; forward and reverse clutch assembly; forward and reverse sun gear shell assembly from Ford AOD transmission. (*Courtesy of Ford Motor Company*)

Figure 9-33. Grasp center support low-reverse planetary assembly and pull it out of the transmission case. Typical of Ford AOD transmission. (*Courtesy of Ford Motor Company*)

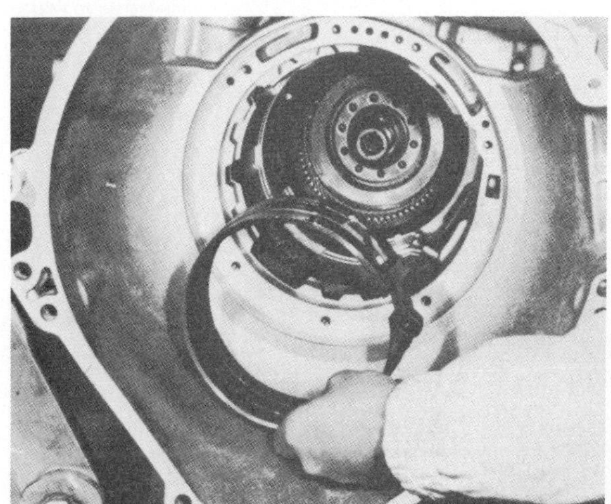

Figure 9-34. Remove low-reverse band from transmission case (typical). (*Courtesy of Ford Motor Company*)

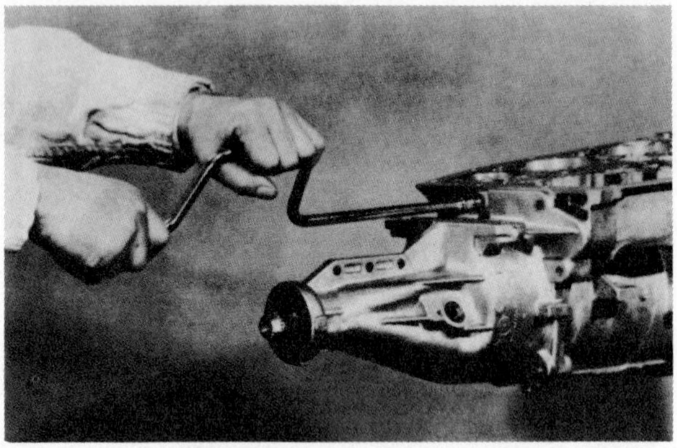

Figure 9-35. Remove bolts from extension housing. (*Courtesy of Ford Motor Company*)

Figure 9-37. Remove governor retaining snap ring with retaining ring plier. (*Courtesy of Ford Motor Company*)

Causes of Burned Clutch Plates

Be sure to correct the cause of clutch failure to prevent recurrence of the problem:

1. Check ball in clutch damaged, stuck, or missing
2. Clutch piston cracked, seals damaged or missing
3. Low line pressure
4. Oil seal rings missing, broken, or undersize; ring groove oversize (worn)
5. Case valve body face not flat or porosity between channels

Burned clutch plates can be caused by using incorrect clutch plates. Engine coolant in transmission fluid can cause severe damage, such as large pieces of clutch plate material peeling off.

On transmissions equipped with a vacuum modulator, an inoperative exhaust gas recirculation (EGR) system will raise engine vacuum, causing lower transmission oil pressure, which will result in burned clutch plates. Make sure that the EGR system is working properly. On transmissions with a linkage-operated throttle valve system, incorrect linkage adjustment can have the same effect.

Figure 9-36. Remove extension housing and gasket from transmission case. Remove extension housing seal (typical). (*Courtesy of Ford Motor Company*)

Figure 9-38. Remove governor assembly from output shaft. If necessary, carefully tap the assembly with a soft head hammer. Remove drive ball. Typical of Ford AOD transmission. (*Courtesy of Ford Motor Company*)

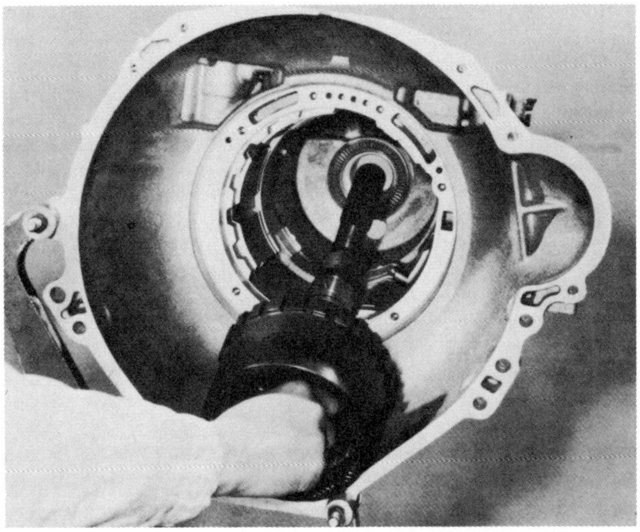

Figure 9-39. Grasp and remove output shaft, ring gear and direct clutch assembly. *Caution:* Take care to avoid damage to seal ring grooves and mating surface in transmission case. (*Courtesy of Ford Motor Company*)

Clutch Cylinder (Piston Bore)

Inspect the thrust surfaces and bushing for minor burrs or scores. Remove with crocus cloth. Replace worn bushings. Inspect the piston bore sliding surfaces in the same manner. If the clutch check ball is located in the clutch cylinder, make sure that it is clear of any obstruction and operates freely. A badly scored clutch cylinder should be replaced. If the clutch cylinder is the type on which a band operates, examine the outer circumference for discoloration and heat checks caused by overheating due to band slippage. If the drum is overheated badly or worn badly, it should be replaced. Minor scuffing and scoring can be removed with crocus cloth. Inspect drum and hub clutch grooves. Clutch plates must move freely in all grooves.

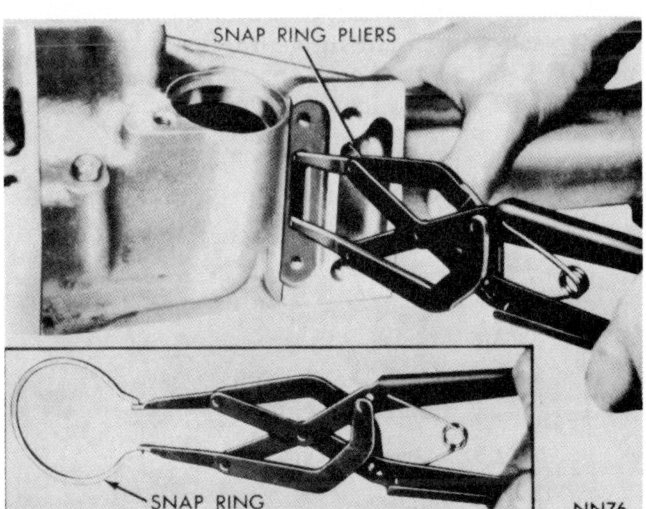

Figure 9-40. Expanding snap ring at extension housing front bearing to allow housing removal. (*Courtesy of Chrysler Corporation*)

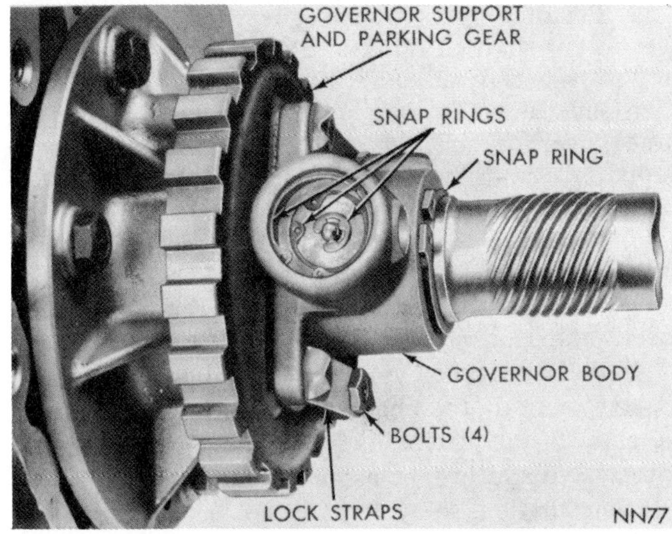

Figure 9-41. Details of Chrysler TorqueFlite governor mounting method. (*Courtesy of Chrysler Corporation*)

Clutch Piston and Springs

Inspect the clutch piston inner and outer circumferences for damage such as nicks, burrs, and scoring. Remove minor damage with a crocus cloth. If the check ball is located in the piston, make sure that it operates freely and is clear of any obstruction. If the clutch is badly overheated, the clutch springs should be replaced.

Figure 9-42. Removing General Motors-type governor cover. (*Courtesy of General Motors Corporation*)

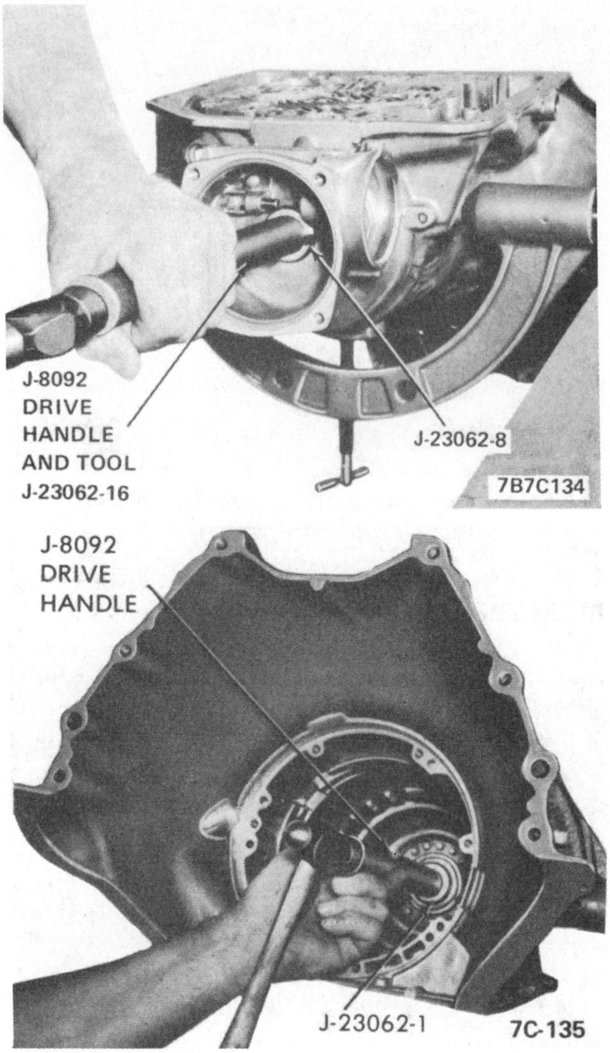

Figure 9-43. Removing (top) and installing (bottom) transmission case bushing, 350 THM. (*Courtesy of General Motors Corporation*)

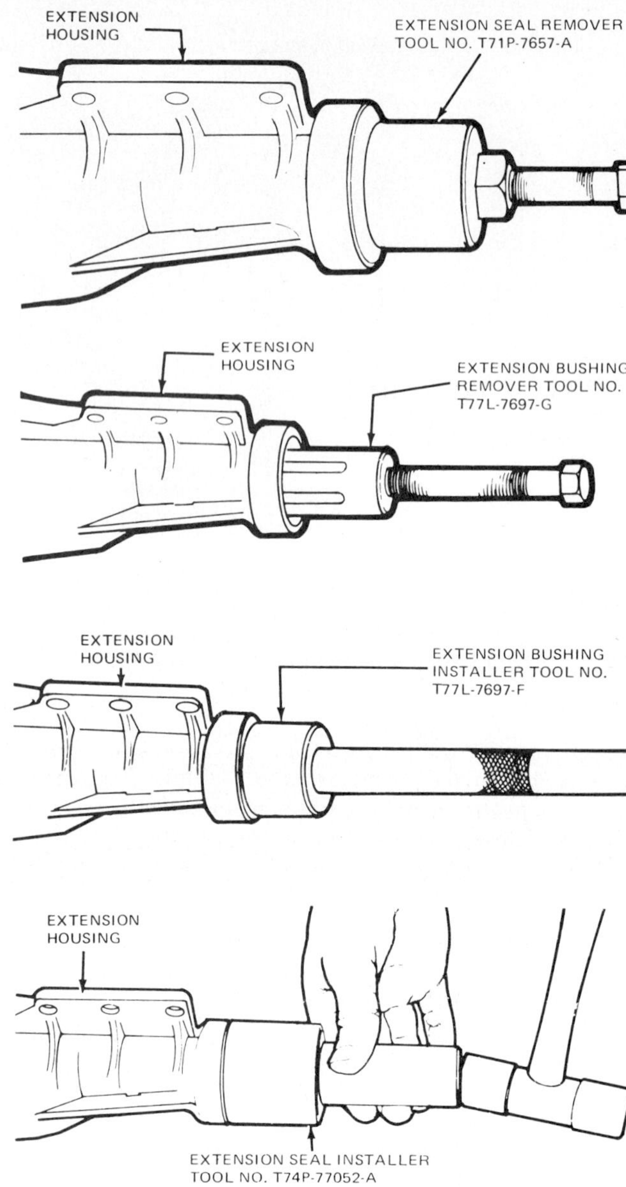

Figure 9-44. (*Courtesy of Ford Motor Company*)

Make sure that all snap rings, snap ring grooves, and spring retainers are not damaged or distorted. Replace any damaged items.

Clutch Assembly

Install all new clutch seal rings. Make sure that they are the correct type and size. It is possible to stretch a smaller-diameter seal ring and install it; however, this would reduce the cross-sectional size of the seal ring and cause leakage. In other words, overstretching a smaller seal ring causes it to be thinner in the installed position than a proper sized seal would be in the groove. Be sure that the lip of a lip-type seal is installed with the lip edge facing the hydraulic pressure side of piston or clutch cylinder. There may be two or

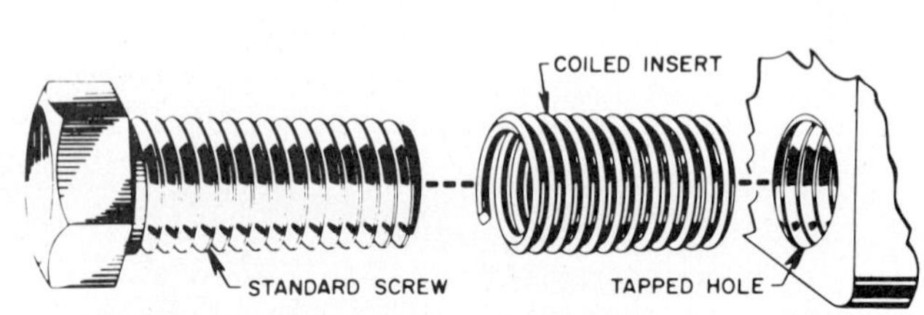

Figure 9-45. Damaged threads can be restored in a threaded part by the use of helically coiled insert. The damaged hole is drilled to a precise oversize, tapped, and a coiled insert installed. This provides new threads of original diameter and type.

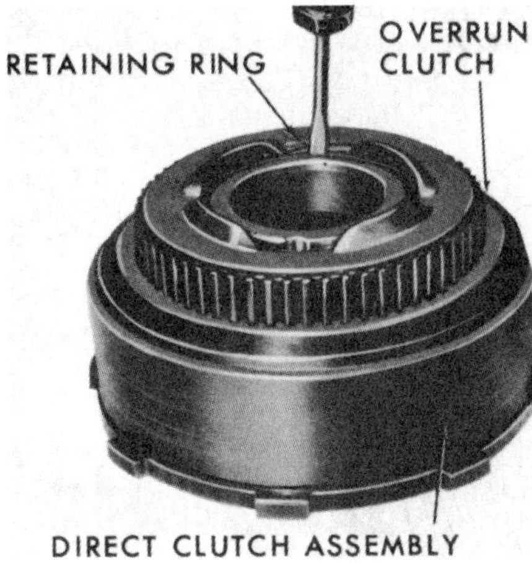

Figure 9-46. Removing retaining ring from THM 350 direct clutch assembly. (*Courtesy of General Motors Corporation*)

Figure 9-47. Removing roller clutch outer race from direct clutch, THM 350. (*Courtesy of General Motors Corporation*)

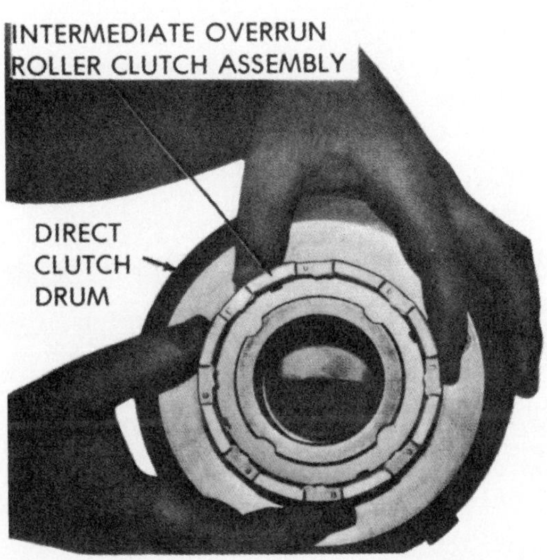

Figure 9-48. Removing roller clutch from TH 350 direct clutch. (*Courtesy of General Motors Corporation*)

Figure 9-49. Removing needle bearing, THM 350. (*Courtesy of General Motors Corporation*)

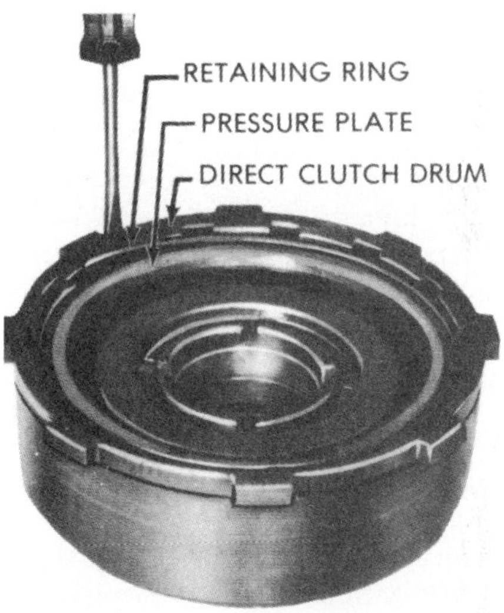

Figure 9-50. Direct clutch snap ring and pressure plate removal, THM 350. (*Courtesy of General Motors Corporation*)

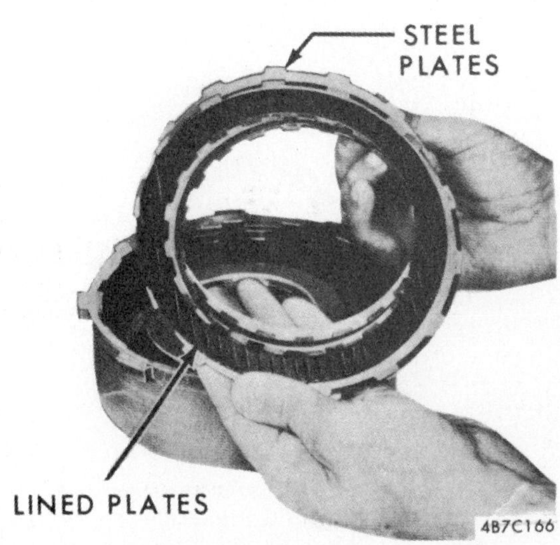

Figure 9-51. Removing clutch plates from direct clutch, THM 350. (*Courtesy of General Motors Corporation*)

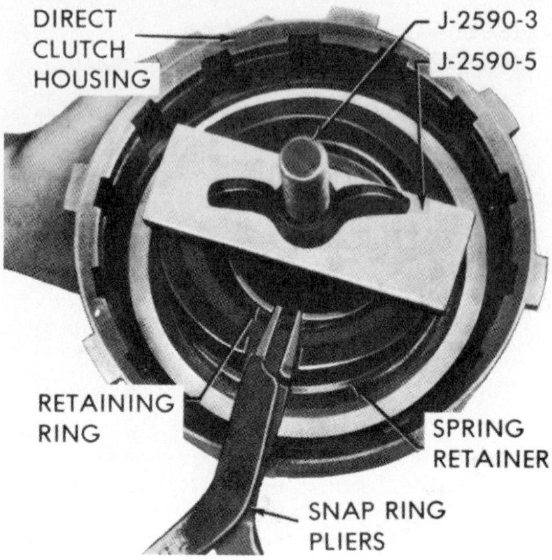

Figure 9-52. Using clutch spring compressor to allow removal of snap ring and spring retainer, THM 350. (*Courtesy of General Motors Corporation*)

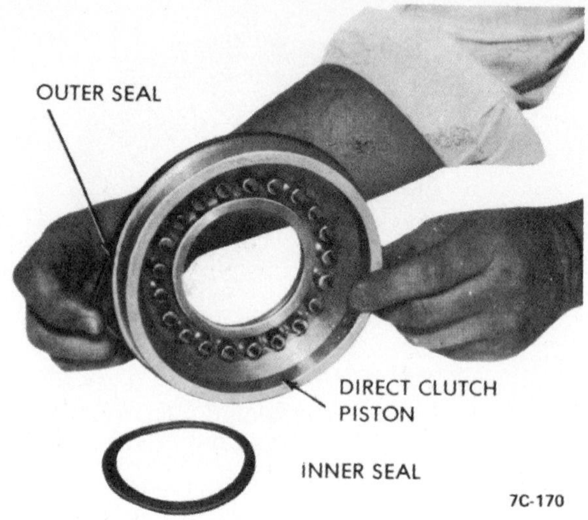

Figure 9-54. Replacing outer and inner clutch piston seals, THM 350 direct clutch. (*Courtesy of General Motors Corporation*)

three seal rings, depending on clutch design. One of these may be located on the clutch cylinder hub. Make sure that all seals are properly installed, are not twisted, and are lubricated for assembly with the appropriate transmission fluid. Be sure to soak all composition clutch plates in the appropriate fluid prior to assembly. Clutch slippage and clutch failure (burn out) can result from neglecting to do this.

Install the piston in the clutch cylinder with a slight turning motion to assist in assembly. Some units may require a special sleeve-type ring to prevent the seal from being damaged or the seal lip from folding back. In some cases all that is needed is a smooth-edge (no sharp cutting edges) feeler gauge to help install the piston. A very useful tool for this purpose can be made from a 3/16-inch (4.75-mm) diameter tubing about 6 inches (about 152 mm) long and two pieces of 0.06- to 0.12-inch (1.5- to 3-mm) diameter music or piano wire 3 inches (76.2 mm) long. Form the wire into a loop, insert the ends into the tubing as shown, and clamp the tubing securely to retain the wire loops—one in each end. Insert the tool between the piston seal and the cylinder. Slide the tool around the cylinder to feed the seal into the cylinder while pressing down slightly on the piston.

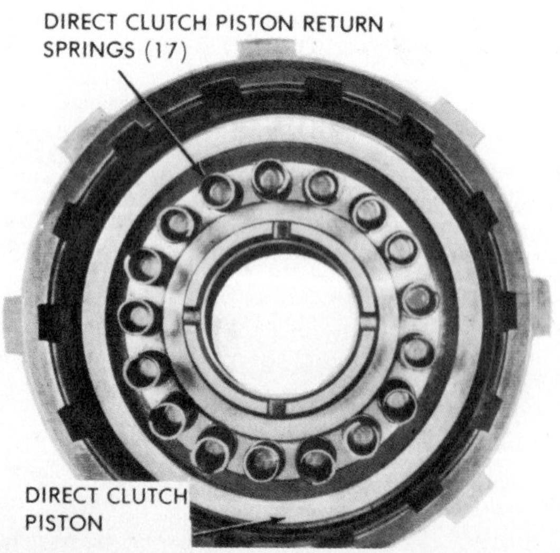

Figure 9-53. Location of clutch return springs, THM 350 direct clutch. (*Courtesy of General Motors Corporation*)

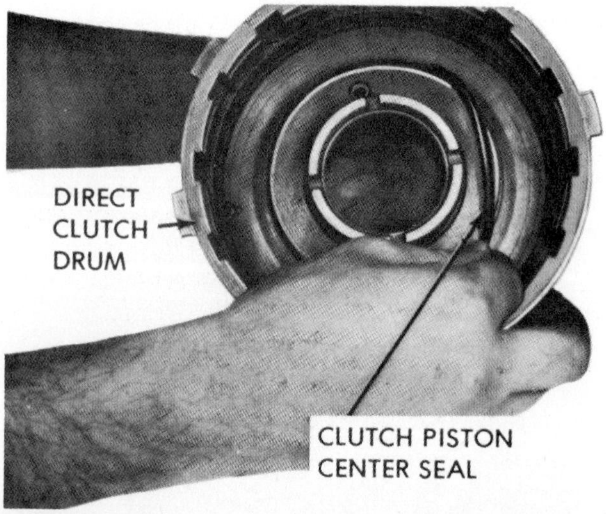

Figure 9-55. Replacing direct clutch piston center seal, THM 350. (*Courtesy of General Motors Corporation*)

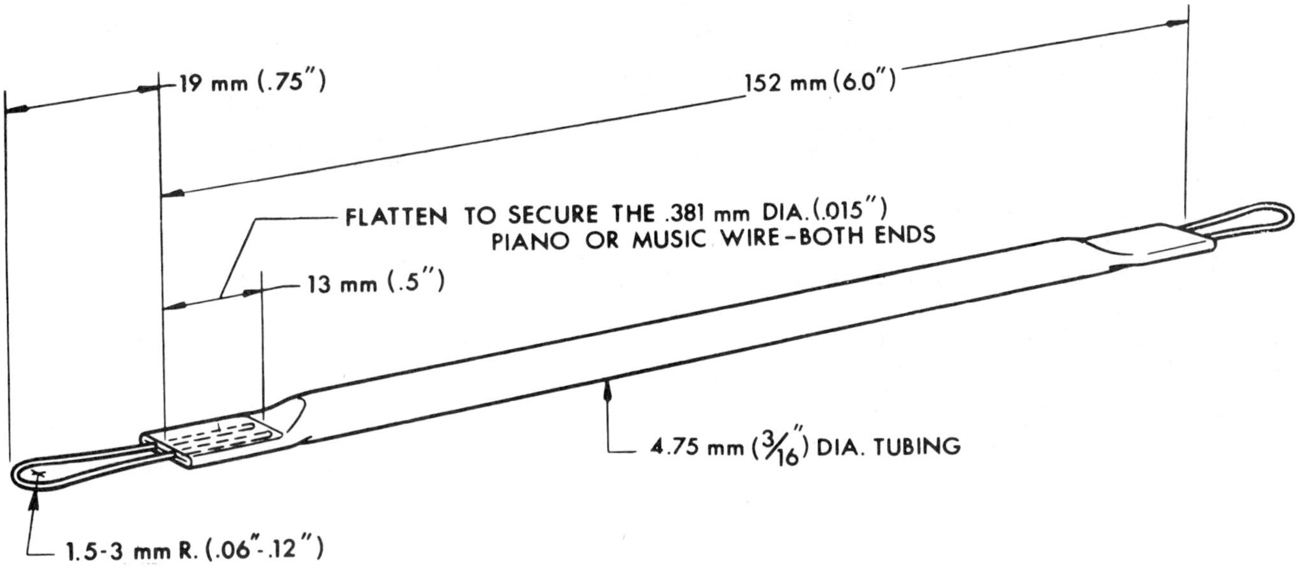

Figure 9-56. Piston and seal installing tool can be fabricated as above. (*Courtesy of General Motors Corporation*)

Be sure not to nick or cut the seal in any way and be sure that lip of seal is not allowed to fold back during piston installation. Nicks, cuts, or foldback will cause pressure loss and clutch failure.

With the piston in place, install the piston return spring or springs, the spring retainer, and snap ring, using the same spring compressor used for disassembly. If a number of return springs are used, be sure to use the number of springs specified in the manual and be sure to place them in the positions specified for the transmission. Make sure that the snap ring is fully seated before removing the spring compressor.

The Belleville or diaphragm spring type of clutch requires no spring compressor for assembly. Install the spring with the dished side away from the piston, then install the wave spring or snap ring, depending on design. Make sure that it is fully seated in the groove.

Next install the clutch pack as specified in the service manual. The Belleville spring type of clutch requires two pressure plates, one on each side of the clutch pack. Other types require only one pressure plate installed after all clutch plates are in place.

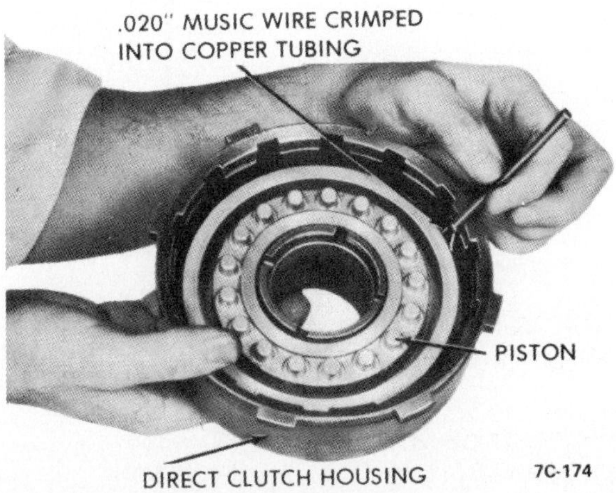

Figure 9-57. Using tool to install clutch piston. Lip of seals must not fold back or be damaged during installation. (*Courtesy of General Motors Corporation*)

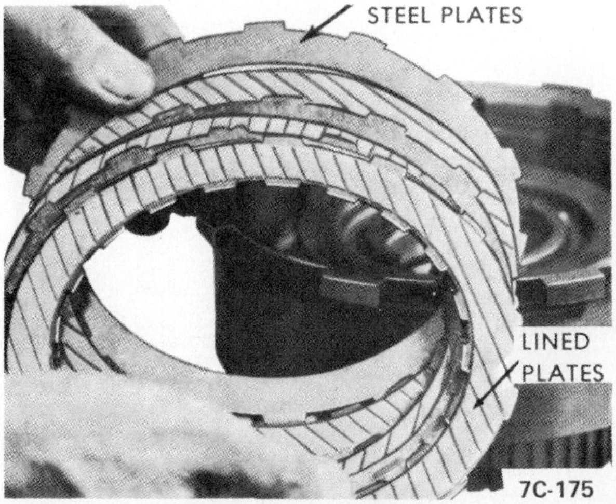

Figure 9-58. Installing steel and lined clutch plates alternately in clutch drum after piston installation. (*Courtesy of General Motors Corporation*)

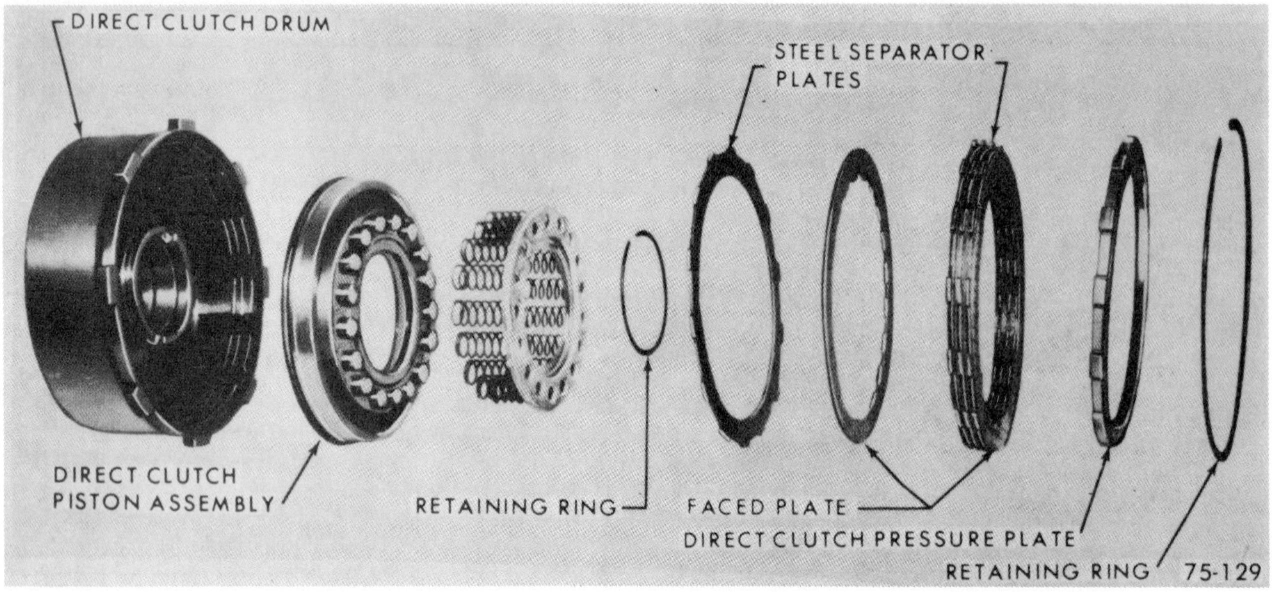

Figure 9-59. Exploded view of THM 350 direct clutch components. (*Courtesy of General Motors Corporation*)

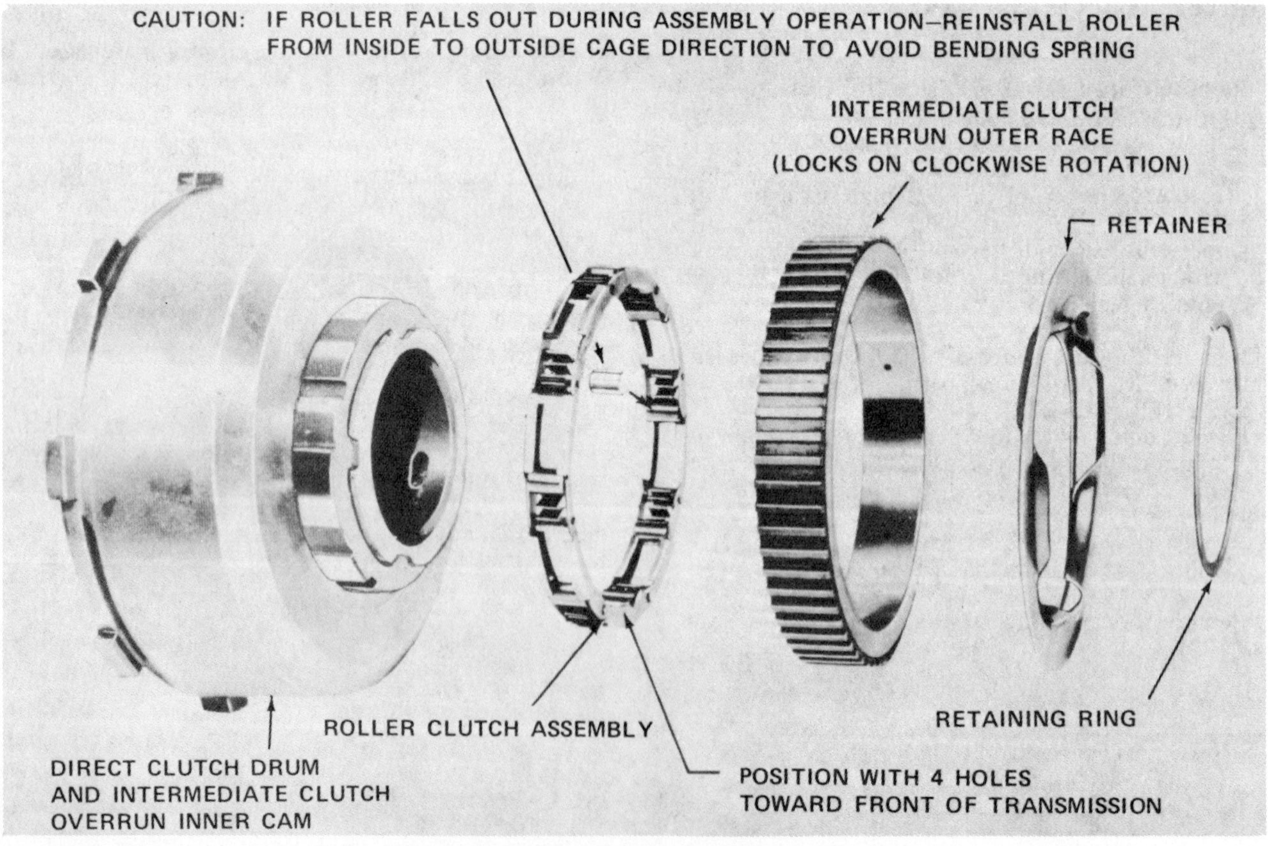

Figure 9-60. (*Courtesy of General Motors Corporation*)

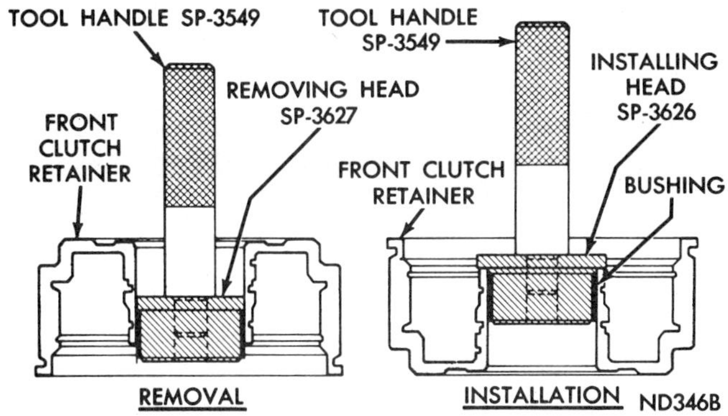

Figure 9-61. Replacing front clutch retainer bushing in TorqueFlite transmission. (*Courtesy of Chrysler Corporation*)

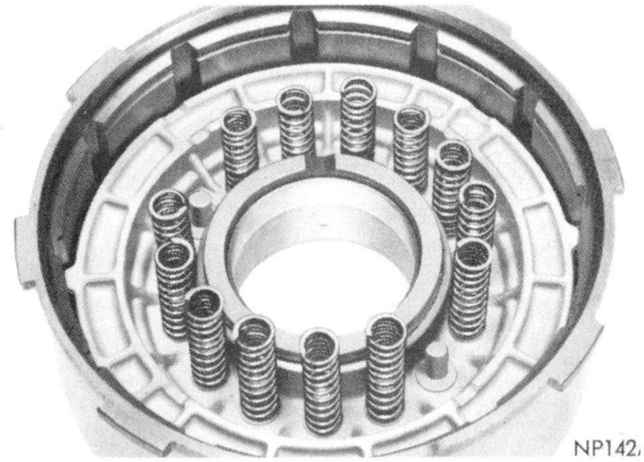

Figure 9-63. Location of front clutch piston springs in 727 TorqueFlite transmission. (*Courtesy of Chrysler Corporation*)

Be sure to install the *correct number* of steel and composition plates and be sure to install them in the *correct sequence*. (Be sure that all composition plates have been soaked in the appropriate fluid prior to assembly as stated earlier.)

If indexing the clutch plates with the clutch cylinder is required, follow the service manual to obtain proper indexing. Indexing may be required to aid oil flow or clutch balance.

In the Belleville spring type of clutch, the normal sequence is to start with the pressure plate first, then a composition plate and a steel plate alternately, ending with the outer pressure plate and snap ring. Other clutch units start with a steel plate, then a composition plate and so on alternately, finishing with the pressure plate and snap ring.

Clutch Pack Clearance

For a multiple-disc clutch to operate properly, the clutch pack must have sufficient clearance to provide for full clutch release and must not have excessive clearance, which could prevent full clutch application. Insufficient clearance would cause plates to rub against each other when the clutch is released. The clutch would attempt to drive when it should be released. The resulting slippage would cause over-

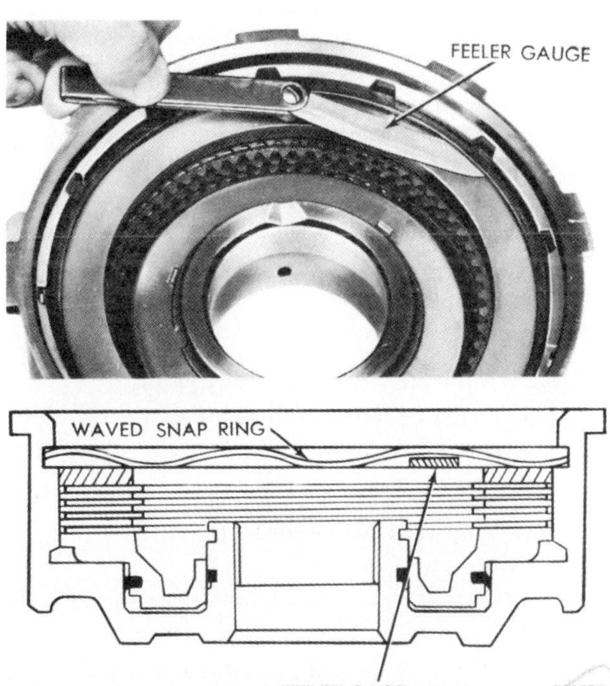

Figure 9-62. Measuring front clutch pack clearance on TorqueFlite transmission. (*Courtesy of Chrysler Corporation*)

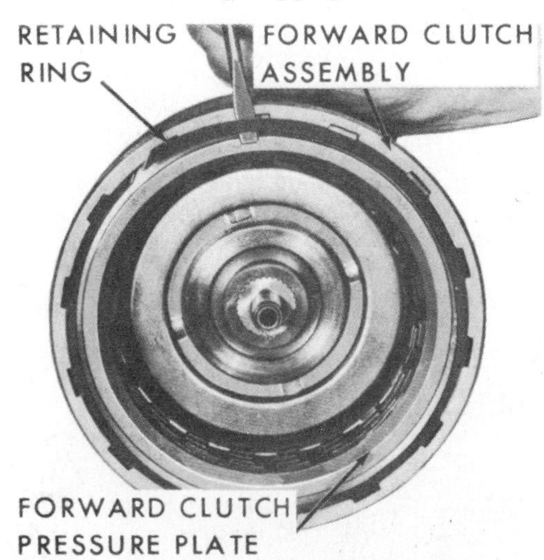

Figure 9-64. Removing forward clutch snap ring, THM 350. (*Courtesy of General Motors Corporation*)

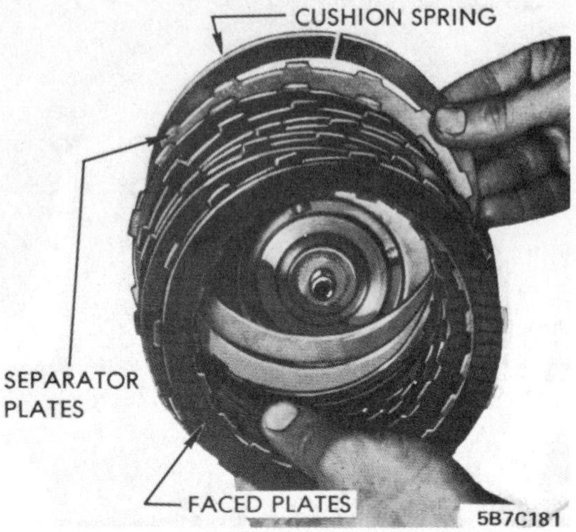

Figure 9-65. Removing clutch plates and cushion spring from forward clutch, THM 350. (*Courtesy of General Motors Corporation*)

heating and damage as well as affect the shift pattern of the transmission. Excessive clearance could allow the piston to reach its travel limit (bottom cut) before the clutch is fully applied. This would result in clutch slippage and clutch failure due to burnout. Therefore it is extremely important that assembled clutch pack clearances be maintained. The objective in adjusting clutch pack clearance is to achieve clearances in the low end of specifications since clearance will increase as the clutch is put into service.

The methods of adjusting clearance include selective thickness snap rings—thinner snap rings provide increased clearance, selective thickness pressure plates, or steel plates; thicker plates reduce clearance and thinner plates increase clearance.

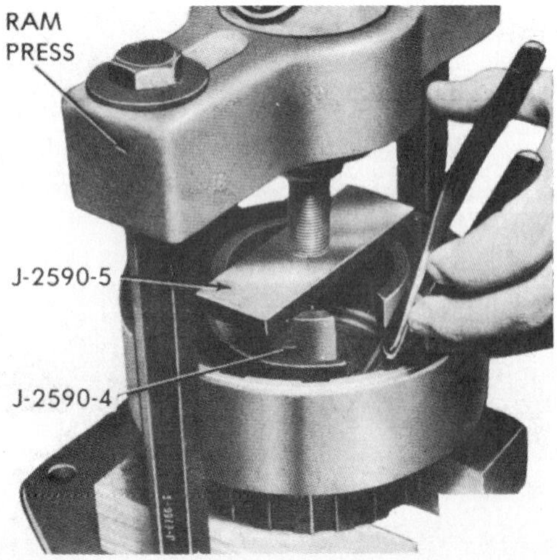

Figure 9-66. Compressing clutch piston springs to allow snap ring removal from forward clutch, THM 350. (*Courtesy of General Motors Corporation*)

Follow the manufacturer's service manual for the method of measuring clutch pack clearance, method of correcting clutch pack clearance, and clearance specifications. Some typical examples are illustrated for clarification.

In some cases, notably holding-type clutches assembled in the transmission case, no clearance specifications are given. In this instance assembling the proper number and type of clutch plates will provide the proper clearance since there is sufficient piston travel to provide full clutch application. Typical clutch pack clearances range from about 0.025 inch (0.635 mm) to 0.165 inch (4.191 mm).

After assembly, check clutch operation by using air pressure directed to the clutch apply passage. The clutch should apply with an audible clunk and release when air pressure is removed.

Servicing Gears and Shafts

Inspect planetary gear units, sun gears, planet carriers and pinions, and ring gears for damage. Inspect all thrust surfaces for minor burrs or scores and remove with crocus cloth if required. Badly damaged units or units showing excessively worn thrust surfaces (stepped wear) must be replaced. Measure pinion gear end play with a feeler gauge and compare to specifications. If excessive, replace the assembly. Check pinion bearings for roughness while turning them by hand. Check for looseness by trying to tilt them on their shaft. If wear is excessive, replace the carrier assembly. Inspect all gears for excessive tooth wear (stepped wear) and replace if present. Inspect bushings in sun gears, ring gears, or carriers for wear. If wear is excessive, replace bushings, using suitably sized bushing drivers. Make sure that all oil passages are properly aligned when installing new bushings.

Check park gear for wear or damage.

Inspect turbine shaft, direct drive shaft, overdrive shaft, output shaft, and sun gear shaft at the following points:

- Bushing wear areas for stepped wear, scoring, pitting, and the like.
- Splines for cracks, stepped wear on drive side of splines, or other damage.
- Seal ring grooves, oil transfer ring grooves, and snap ring grooves for wear or damage.
- All oil passages for obstruction or damage; use compressed air to blow through all passages to ensure that they are free and clear of all ob-

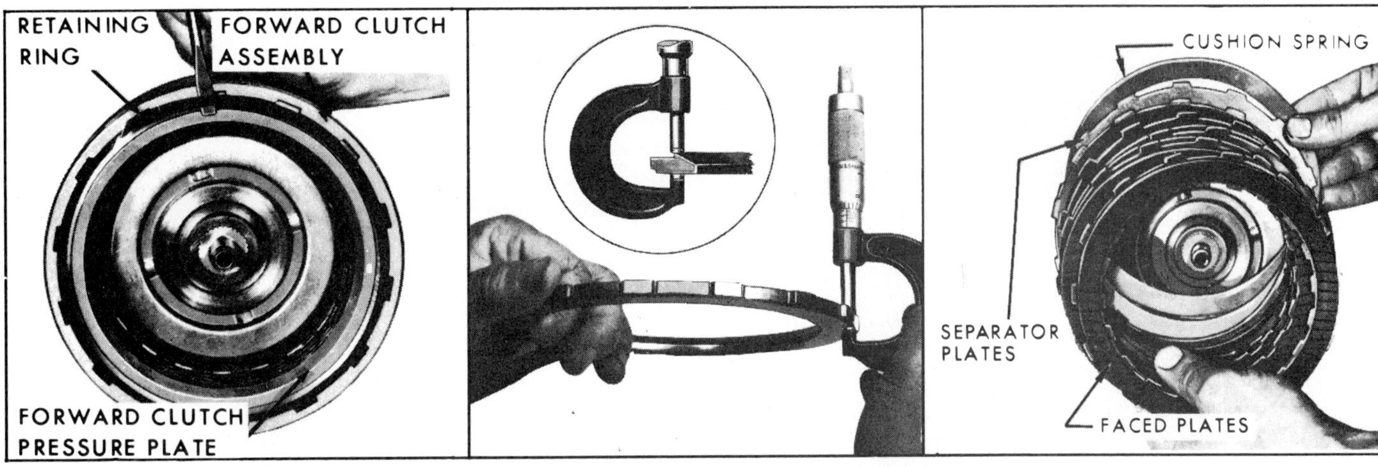

Figure 9-67. Removal and inspection of forward clutch pressure plate and clutch pack, THM 350. (*Courtesy of General Motors Corporation*)

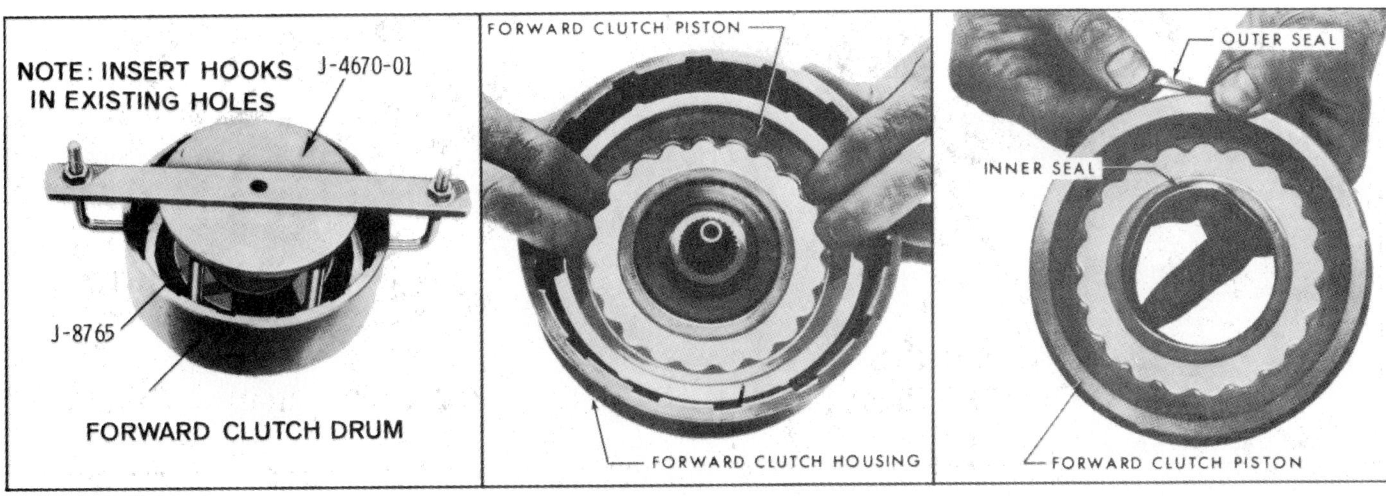

Figure 9-68. Removal of forward clutch piston and piston seals, THM 350. (*Courtesy of General Motors Corporation*)

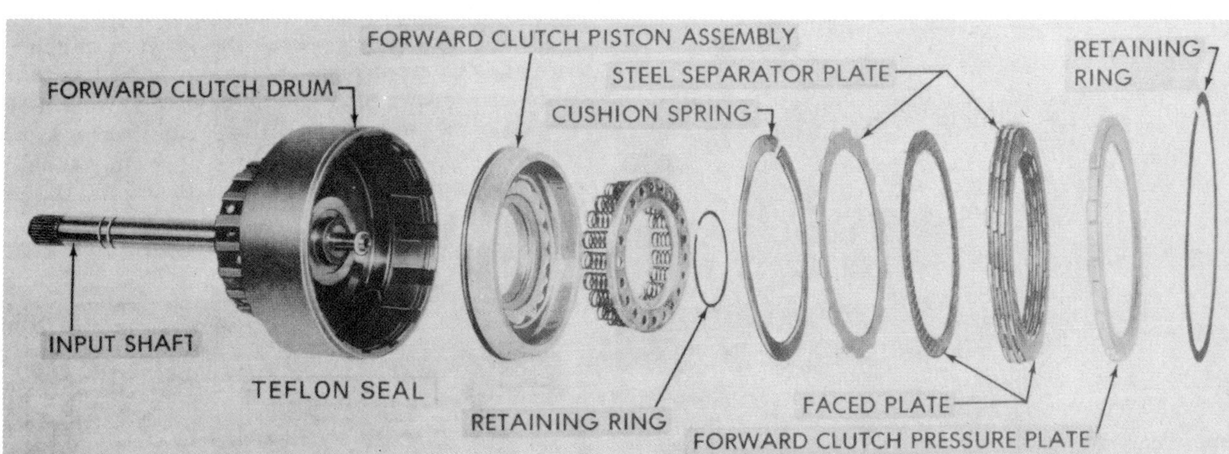

Figure 9-69. Exploded view of forward clutch, THM 350. (*Courtesy of General Motors Corporation*)

Figure 9-70. Forward clutch piston removal, THM 350. (*Courtesy of General Motors Corporation*)

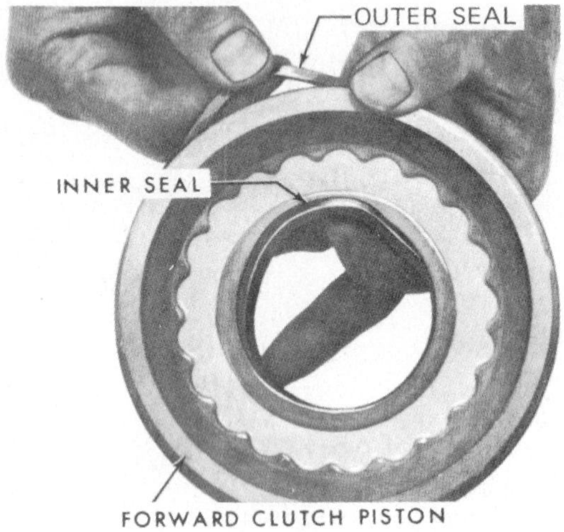

Figure 9-71. Replacing forward clutch piston seals. (*Courtesy of General Motors Corporation*)

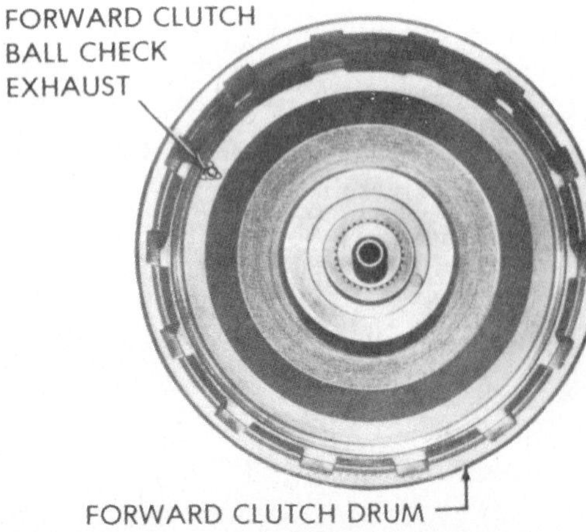

Figure 9-72. Exhaust check ball must be checked to assure operation. (*Courtesy of General Motors Corporation*)

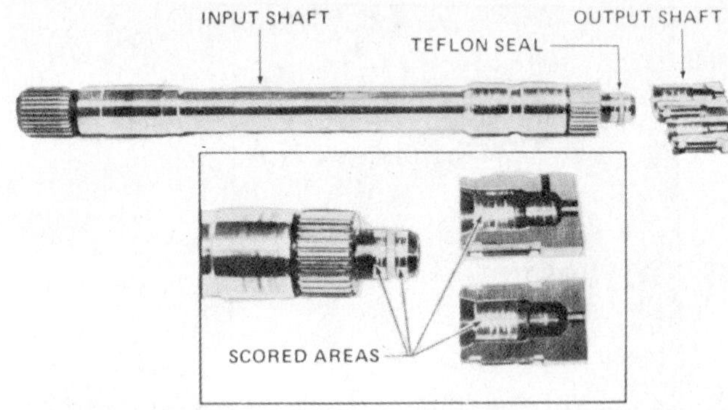

Figure 9-73. Turbine shaft must be inspected for scoring or damage, THM 350. (*Courtesy of General Motors Corporation*)

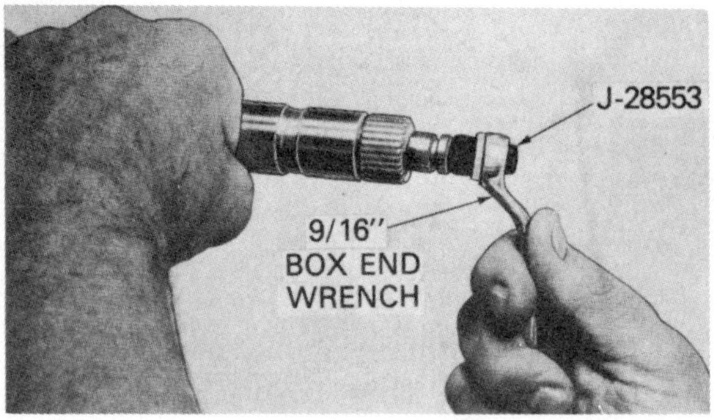

Figure 9-74. Installing input shaft seal, THM 350. (*Courtesy of General Motors Corporation*)

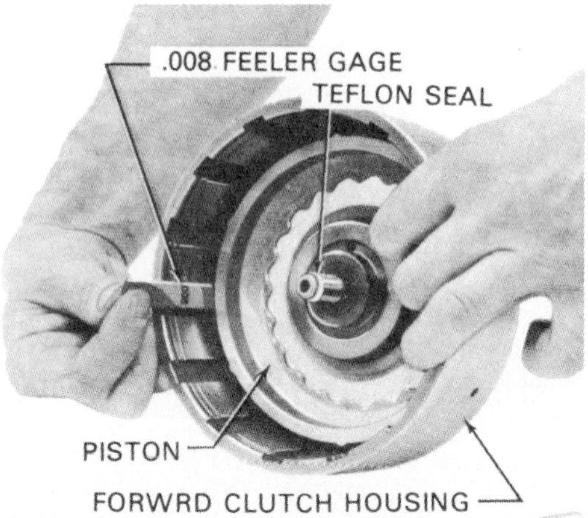

Figure 9-75. Installing forward clutch piston, THM 350. (*Courtesy of General Motors Corporation*)

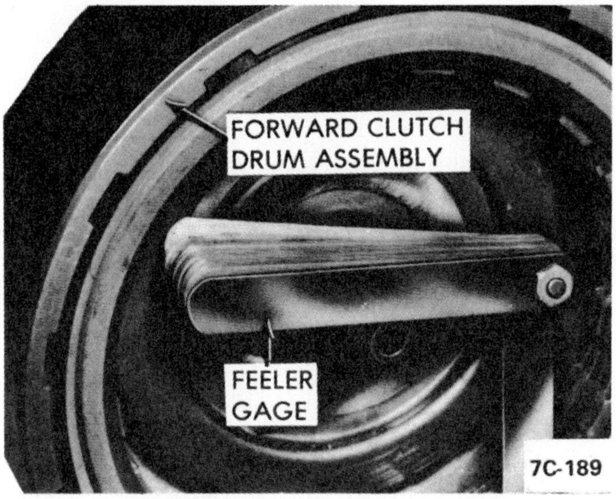

Figure 9-76. Measuring forward clutch pressure plate clearance, THM 350. (*Courtesy of General Motors Corporation*)

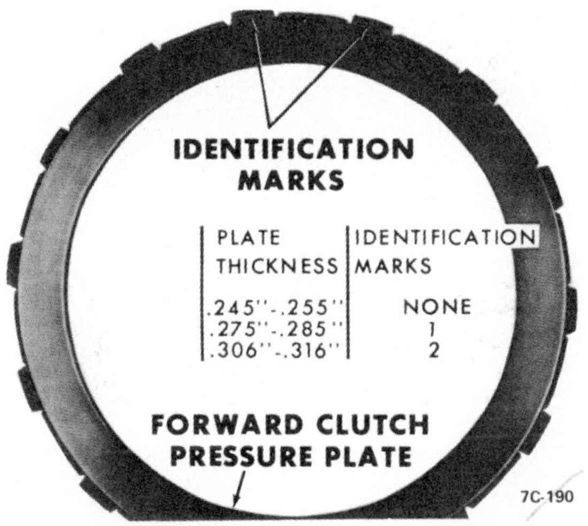

Figure 9-77. Identification of selective thickness forward clutch pressure plate. (*Courtesy of General Motors Corporation*)

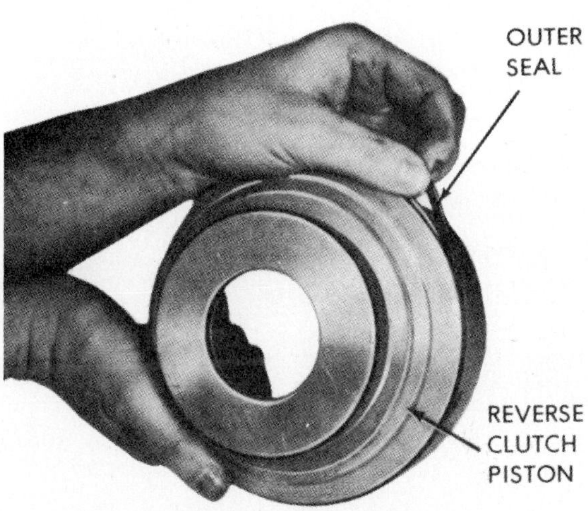

Figure 9-78. Replacing reverse clutch piston outer seal, THM 350. (*Courtesy of General Motors Corporation*)

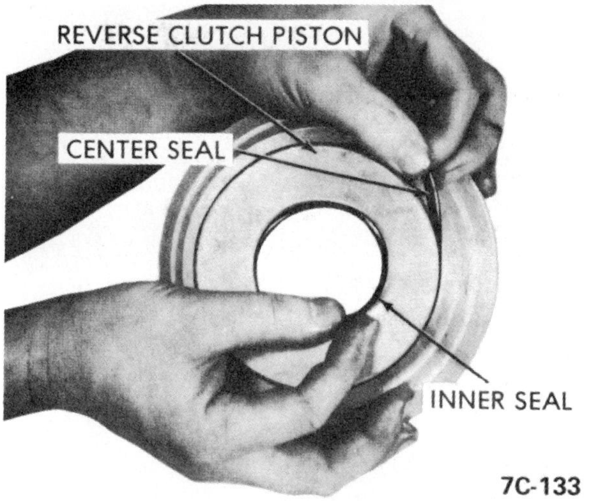

Figure 9-79. Replacing inner and center reverse clutch piston seals, THM 350 (*Courtesy of General Motors Corporation*)

Figure 9-80. Installing low and reverse clutch piston, THM 350. (*Courtesy of General Motors Corporation*)

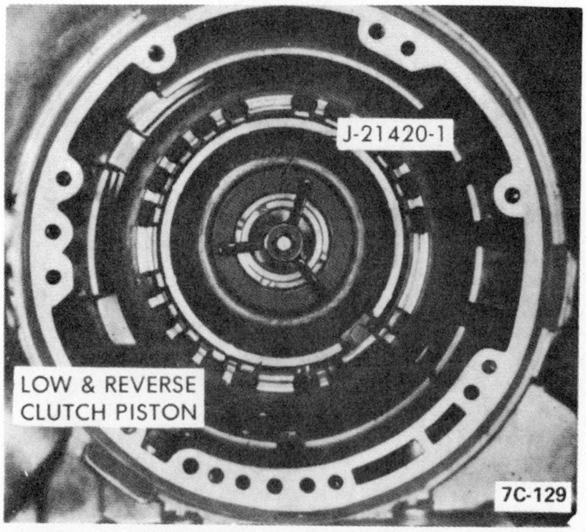

Figure 9-81. Low and reverse clutch piston snap ring installation, THM 350. (*Courtesy of General Motors Corporation*)

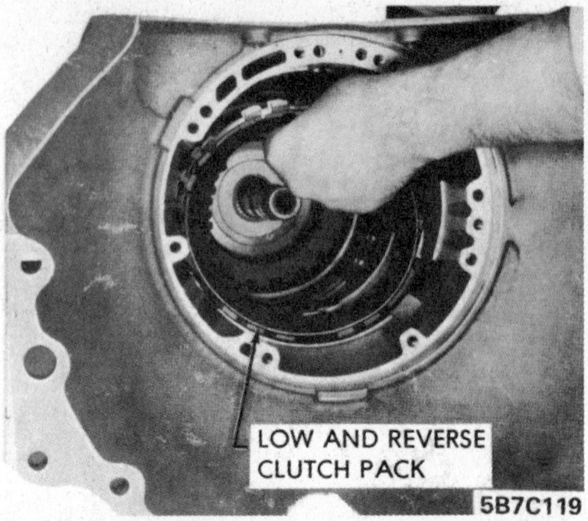

Figure 9-82. Installing low and reverse clutch pack, THM 350. (*Courtesy of General Motors Corporation*)

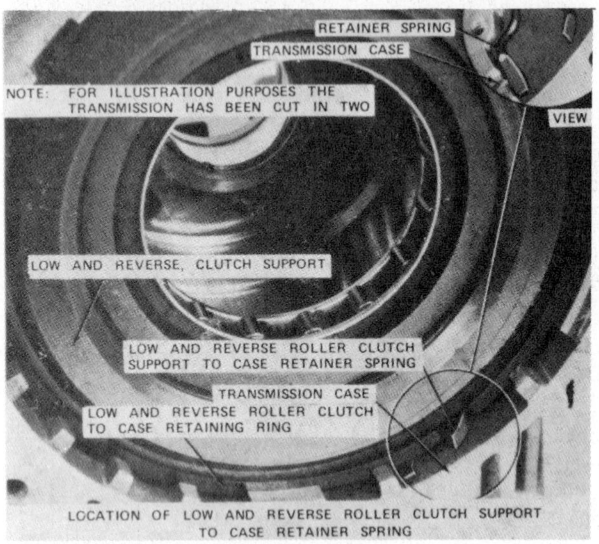

Figure 9-83. THM 350. (*Courtesy of General Motors Corporation*)

Figure 9-84. Removing sun gear retaining ring, THM 350. (*Courtesy of General Motors Corporation*)

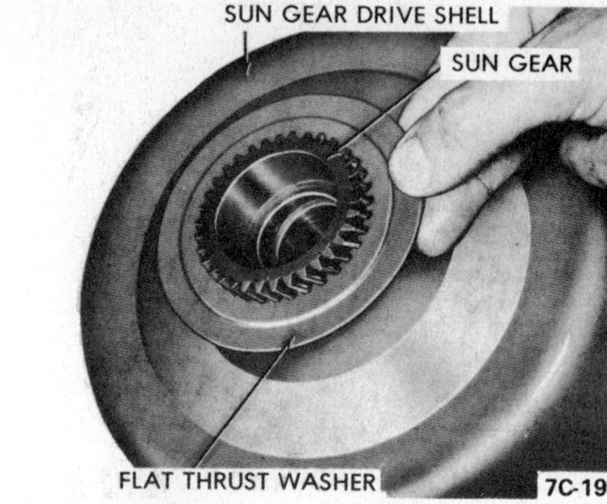

Figure 9-85. Removing sun gear drive shell flat washer, THM 350. (*Courtesy of General Motors Corporation*)

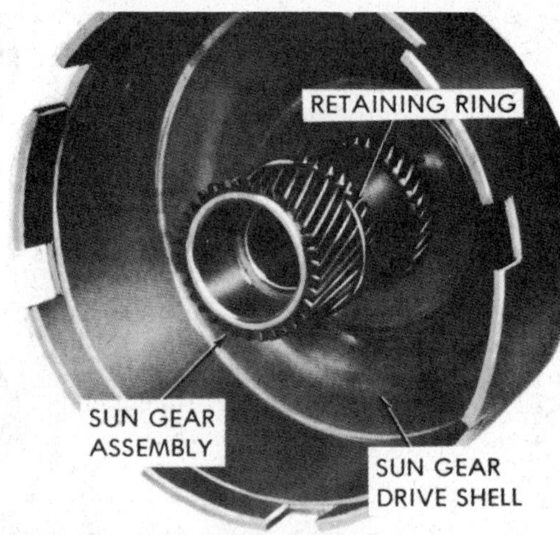

Figure 9-86. Removing front sun gear retaining rinc THM 350. (*Courtesy of General Motors Corporation*)

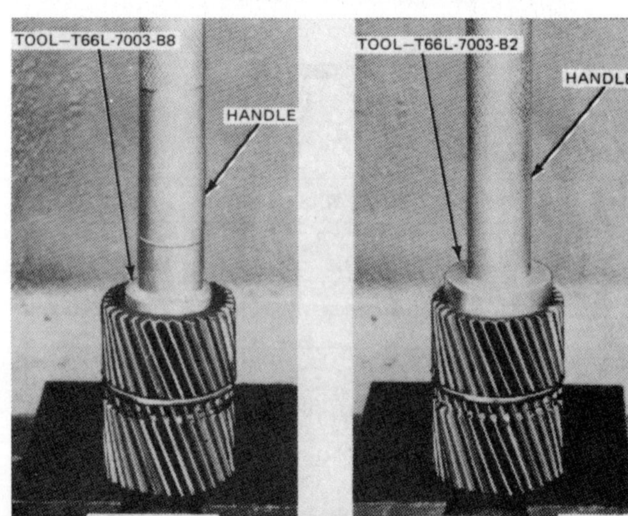

Figure 9-87. Replacing sun gear bushing on Ford C4 transmission. (*Courtesy of Ford Motor Company*)

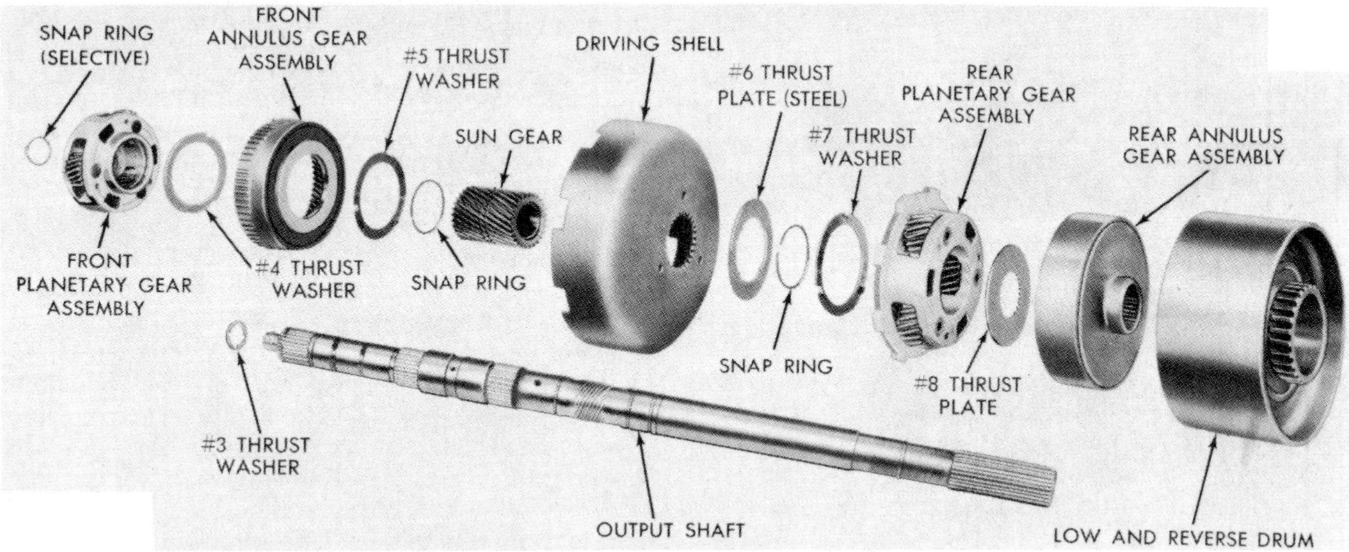

Figure 9-88. TorqueFlite planetary gear train exploded view. (*Courtesy of Chrysler Corporation*)

struction. Make sure that the shaft is not bent. Roll the shaft on a good flat metal surface to check if the shaft is bent.

Replace shafts that do not meet specifications. If the shaft assembly includes a bushing, inspect the bushing for wear or damage and replace if necessary. Be sure to align all oil passages when installing new bushings.

If required, make the necessary output shaft and gear train assembly end-play check, and correct if necessary to meet specifications. End-play corrections are made by selective thickness snap rings or selective thrust washers, depending on transmission design.

One-Way Clutch Service

Two types of one-way clutches are used in automatic transmissions: the roller type and the sprag type. A transmission may have one or more one-way clutches.

Generally inspection procedures are similar. The inner and outer races should be inspected for wear and damage to the friction surfaces. The springs, rollers or sprags, and cage should also be inspected for wear and damage. Most one-way clutches are subjected to full engine torque application and should therefore be in good condition to withstand this treatment.

Damaged parts should be replaced. Some clutch races are riveted to the transmission case. In this case the rivets must be marked in the exact center with a center punch, then drilled out carefully with the exact drill size specified in the service manual. The replacement clutch race is installed with cap screws provided with the replacement parts.

Assembly of one-way clutches during transmission assembly requires placing the springs and rollers or sprags in proper position in relation to each other to ensure clutch lockup in one direction but not in the other. However, it is possible to assemble or install many one-way clutches incorrectly without causing any assembly problems but when the transmission is put into operation, it will not perform properly.

It is important to remember that all one-way clutches designed to perform a reaction-holding function during engine torque application must be installed so that they will freewheel when the gear train turns in a clockwise direction and must lock up in a counterclockwise direction when viewed from the

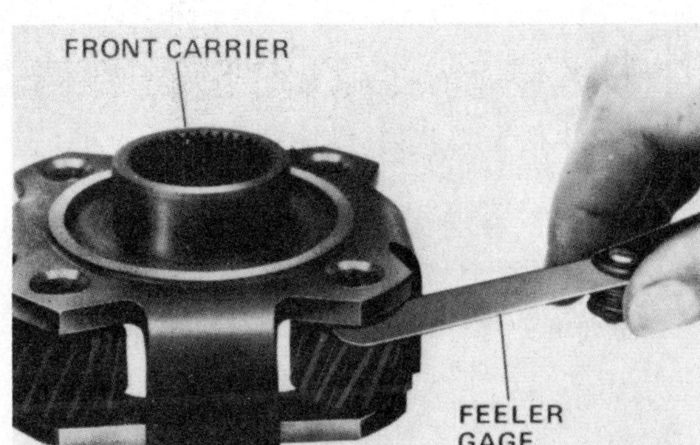

Figure 9-89. Measuring pinion end play on GM 200 transmission. (*Courtesy of General Motors Corporation*)

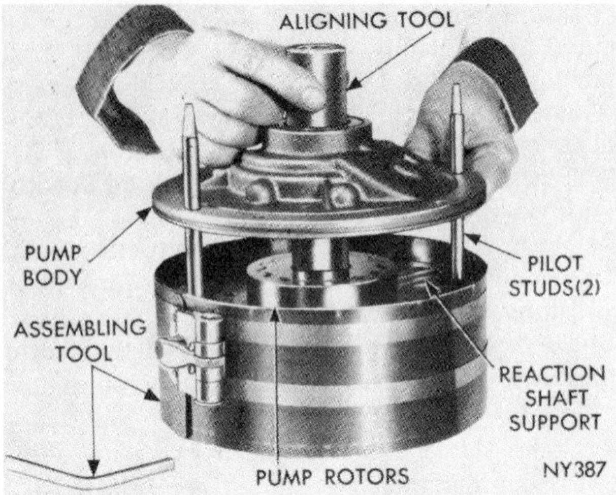

Figure 9-90. Measuring planetary gear train and play on TorqueFlite transmission. (*Courtesy of Chrysler Corporation*)

torque converter or input shaft end of the transmission. See Part 3, Chapter 4, for one-way clutch operation. Be sure to lubricate the one-way clutch with transmission fluid during assembly. Assembly is easier if the unit is turned during installation.

Servo and Band Service

Clean and inspect servo pistons, piston seal rings, push rods, springs, covers, and snap rings for damage and distortion. Rubber seal rings should be replaced. Cast-iron or Teflon seal rings should be inspected for sliding surface damage, cracks, or other damage; if present they should be replaced. Be sure to replace any damaged snap rings or servo covers.

Inspect the servo piston bores for surface damage such as scoring or scratches. Minor scoring or scratches should be removed with crocus cloth.

Examine the bands for damage to the friction material such as cracks, flaking, excessive wear, or overheating. Overheating is also usually evidenced by dark spots on the corresponding drum surface. Inspect band ends for excessive wear at the anchor pin or strut engagement area.

Inspect any band operating linkage for wear or damage as well. Replace bands that do not pass inspection.

Lubricate servo pistons with transmission fluid and assemble, making sure that all springs, snap rings, and seals are in the proper position and fully seated. Some servo piston springs must be compressed with a special tool similar to an engine valve spring compressor for assembly to be possible. Make sure that the servo piston cover, cover seal, and snap ring are properly installed. If the servo cover is of the bolt-on type, be sure that the gasket and seal are in place. Tighten bolts evenly a few turns at a time to prevent damage to the cover. Tighten bolts to specified torque. Check assembled servo operation with air pressure to ensure proper action.

During assembly some transmissions require a band adjustment to be performed to provide proper clearance between band and drum. Some models have screw-type adjusters while others have a selective length push-rod adjustment. Still others have no provision for adjustment. Follow service manual procedures and specifications for any specific transmission.

Hydraulic Pump Service

Gear- and rotor-type pumps require similar inspection, measuring, and overhaul procedures. The pump

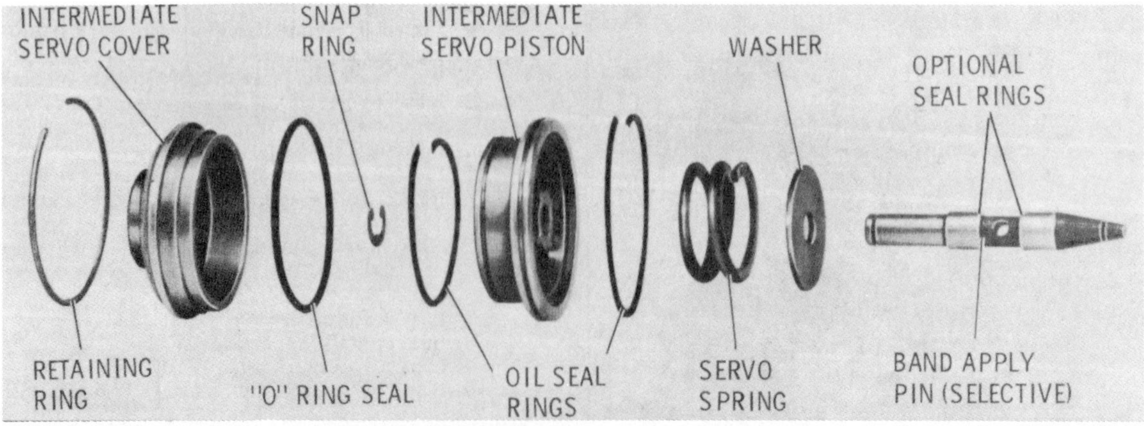

Figure 9-91. Intermediate servo piston detail, GM 200 transmission. (*Courtesy of General Motors Corporation*)

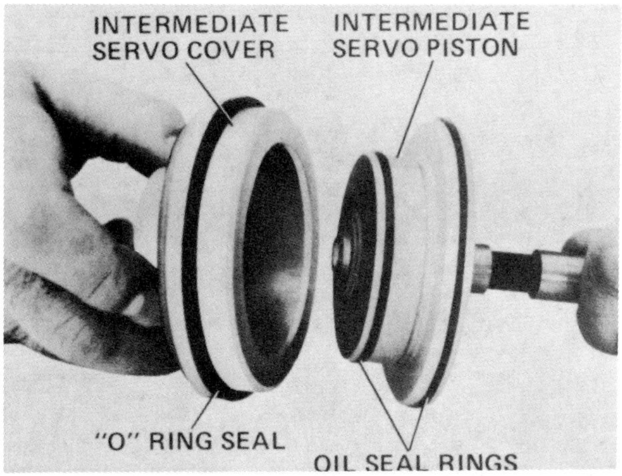

Figure 9-92. Installing intermediate servo piston into servo cover. (*Courtesy of General Motors Corporation*)

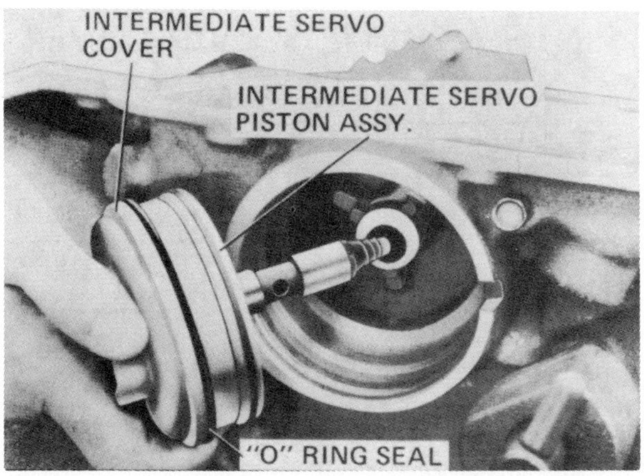

Figure 9-93. Installing servo piston and cover assembly, GM 200 transmission. (*Courtesy of General Motors Corporation*)

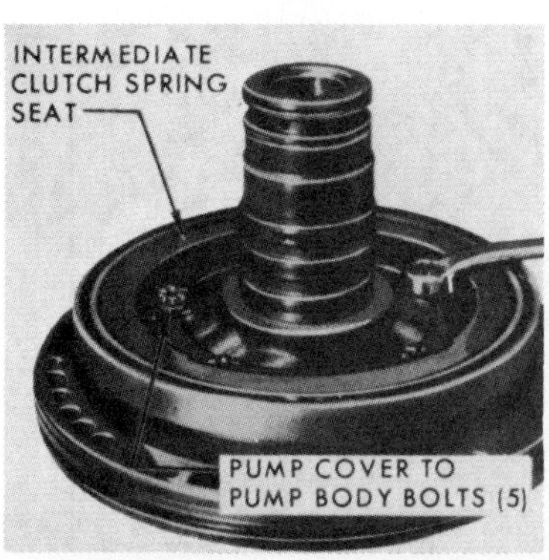

Figure 9-94. Removing clutch spring seat from pump, THM 350. (*Courtesy of General Motors Corporation*)

Figure 9-95. Details of steady rest ring and steel sleeve as used on Buick 350 transmission. (*Courtesy of General Motors Corporation*)

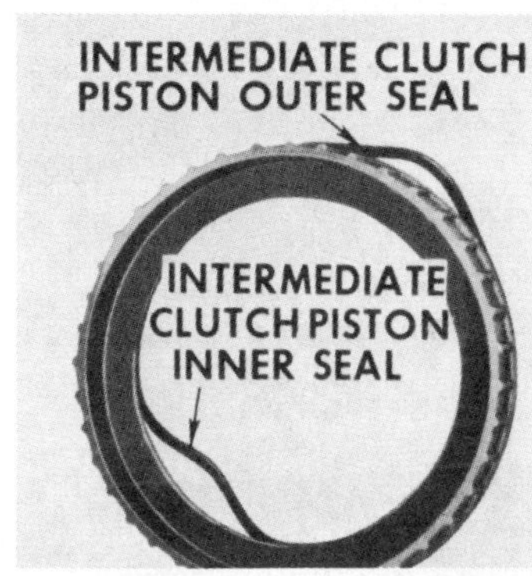

Figure 9-96. Replacing intermediate clutch piston seals, GM 350. (*Courtesy of General Motors Corporation*)

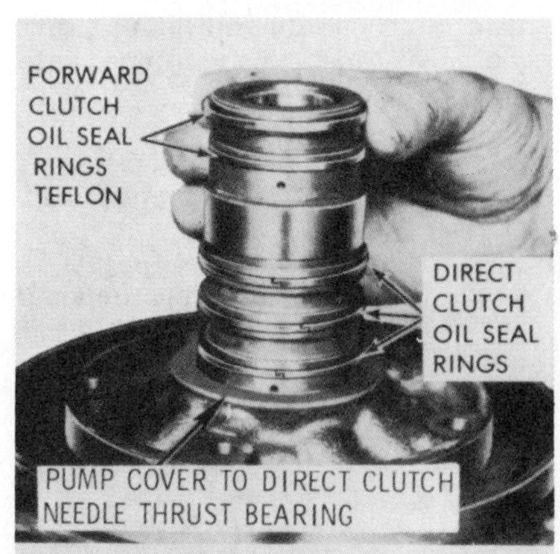

Figure 9-97. Replacing forward clutch and direct clutch oil seal rings, GM 350. (*Courtesy of General Motors Corporation*)

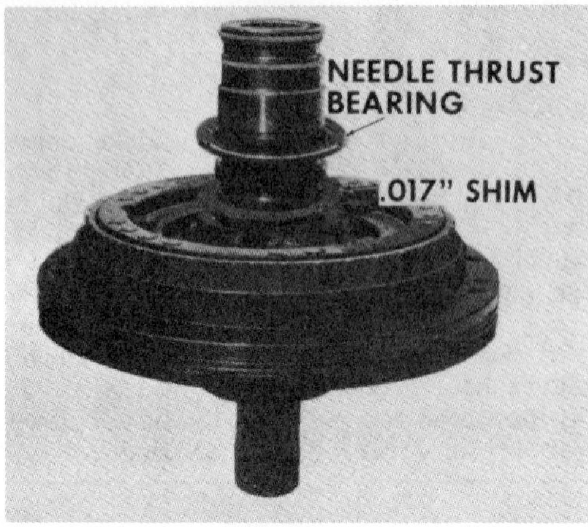

Figure 9-98. Replacing needle thrust bearing and shim. (*Courtesy of General Motors Corporation*)

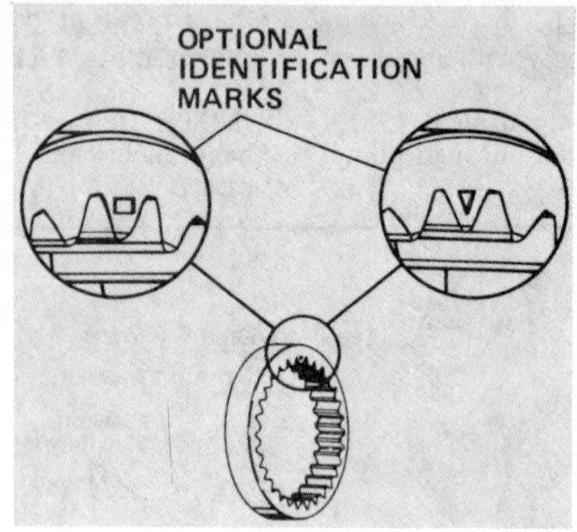

Figure 9-101. Pump driven gear identification marks, GM 200 (typical). (*Courtesy of General Motors Corporation*)

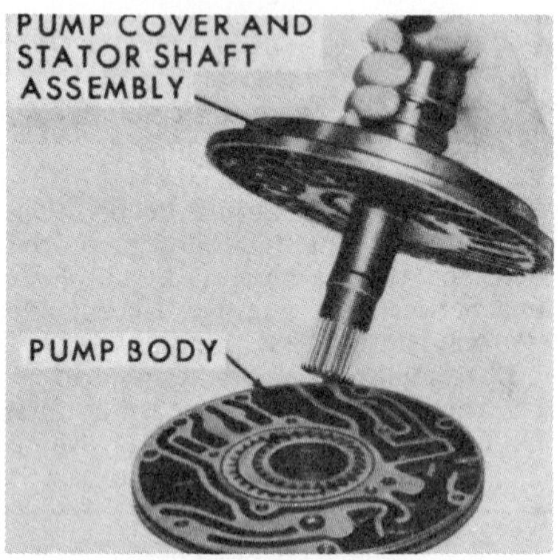

Figure 9-99. Separating pump cover and pump body, GM 350. (*Courtesy of General Motors Corporation*)

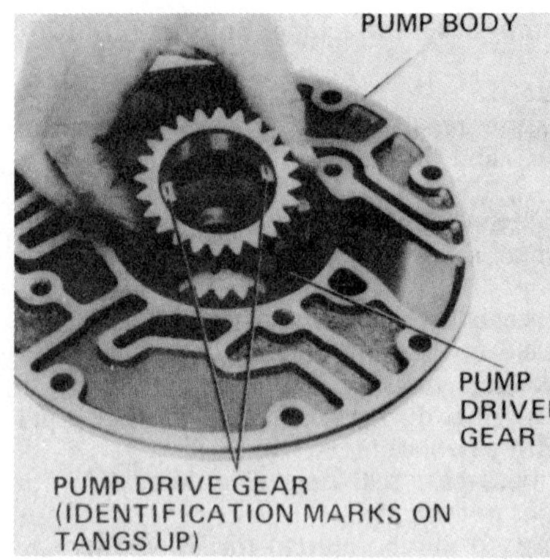

Figure 9-102. Replacing pump drive gear, GM 200 (typical) (*Courtesy of General Motors Corporation*)

Figure 9-100. Checking pump body to gear face clearance, GM 200. (*Courtesy of General Motors Corporation*)

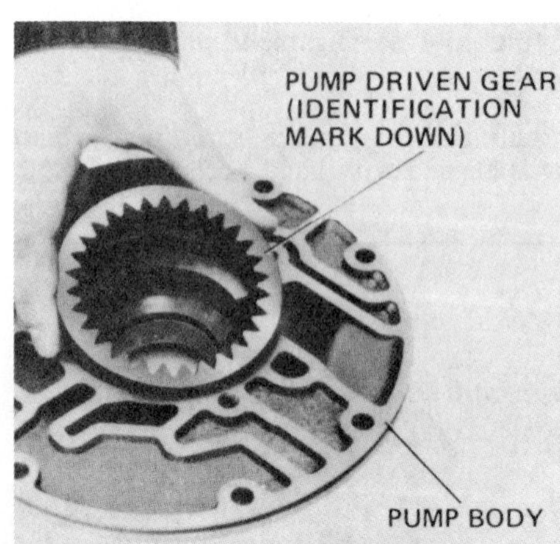

Figure 9-103. Replacing pump-driven gear GM 200 (typical). (*Courtesy of General Motors Corporation*)

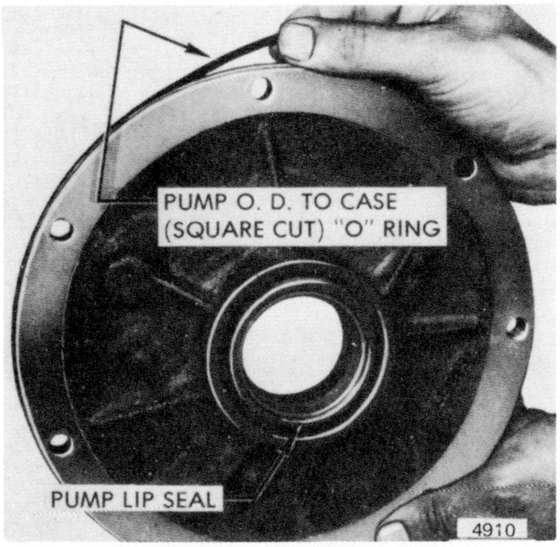

Figure 9–104. Replacing pump outer seal, GM 350 (typical). (*Courtesy of General Motors Corporation*)

assembly may include a check valve or the pressure regulator valve. The pump body contains the gears or rotors, and the pump cover includes the stator shaft (reaction shaft). The vane-type pump includes the pump body, which contains the rotor and vanes, and the pump cover, which contains several valves and the stator shaft.

Disassemble the pump by removing the outer seal ring, the pump cover to pump body bolts, the pump hub to clutch seal rings, and the thrust washer. Measure and record the thrust washer thickness for later reference.

Be sure to maintain the pump drive and driven-gear position to pump body relationship. A small dab

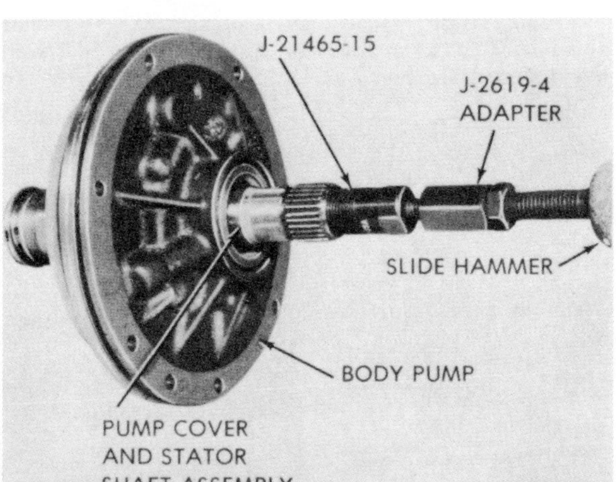

Figure 9–105. Removing front stator shaft bushing, GM 350. (*Courtesy of General Motors Corporation*)

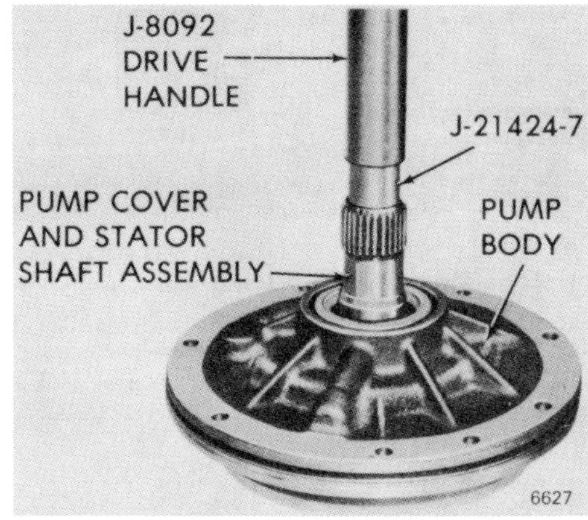

Figure 9–106. Installing front stator shaft bushing GM 350. (*Courtesy of General Motors Corporation*)

of paint on each gear or rotor can be used to do this. Do not punch-mark gears or rotors. Some gears and rotors have factory markings and do not need to be marked.

Inspect the gears or rotors and pump body for excessive wear, breakage, or scoring. Some manufacturers recommend only visual inspection for pump wear while others require clearance measurements. Feeler gauges can be used to determine gear tooth or rotor tip clearance and gear- or rotor-to-pump-body clearance. A feeler gauge and straight-edge can be used to measure gear or rotor to pump cover clearance. A pump not meeting service manual specifications must be replaced. Pump gears and rotors are not available separately but are purchased with the pump body as an assembly. The pump body and cover can be checked with a feeler gauge and straight-edge to determine if it is warped. Warped parts must be replaced.

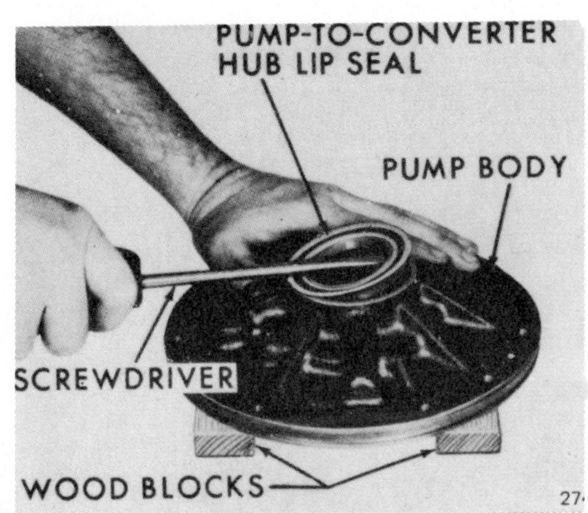

Figure 9–107. Removing pump to converter seal, GM 350. (*Courtesy of General Motors Corporation*)

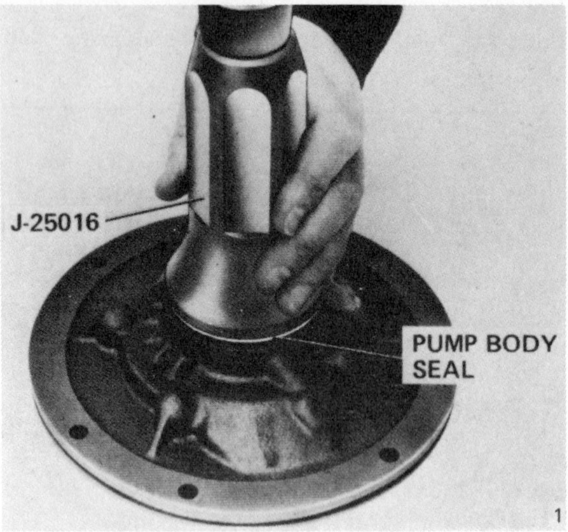

Figure 9-108. Installing pump body seal, GM 200. (*Courtesy of General Motors Corporation*)

Inspect the seal ring grooves for broken or damaged edges or stepped wear. Inspect the stator support splines for wear or damage. Inspect pump bushings for wear. Replace worn bushings with proper size bushing driver. Align oil holes when replacing bushings.

Remove and inspect any spool-type valves,

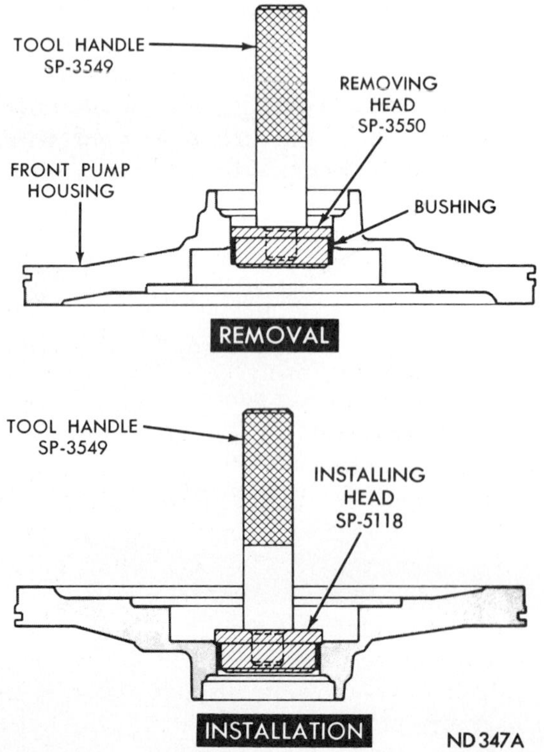

Figure 9-109. Replacing pump bushing on TorqueFlite transmission. (*Courtesy of Chrysler Corporation*)

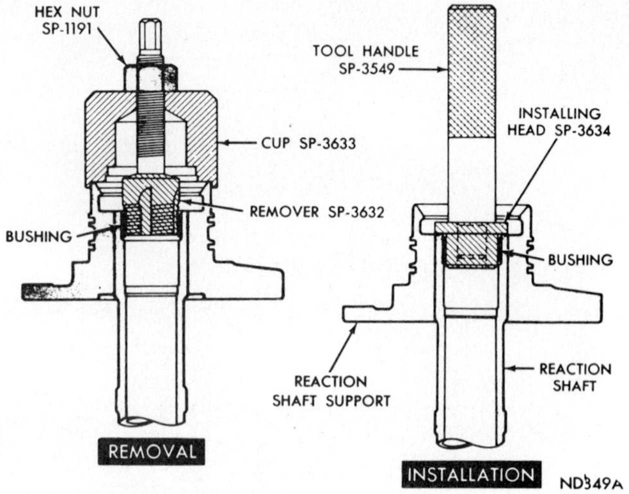

Figure 9-110. Replacing TorqueFlite reaction shaft bushing. (*Courtesy of Chrysler Corporation*)

springs, and valve bores for scoring or damage. Minor scoring can be touched up with crocus cloth. The valve should move freely in its bore. Remove the front pump seal.

Wash all parts in approved cleaning fluid and blow dry. Make sure that all oil passages are free and clear. Install a new pump seal.

Lubricate valves and gears or rotors with proper type of automatic transmission fluid and assemble the pump. Refer to marks and place gears or rotors in pump body right side up. Assemble the pump and tighten bolts to specified torque.

It is essential on some hydraulic pumps that assembly of the pump cover and pump body result in perfect alignment of the two pieces before tightening the bolts. If the pump body and cover are not properly aligned, the total diameter of the pump assembly is in-

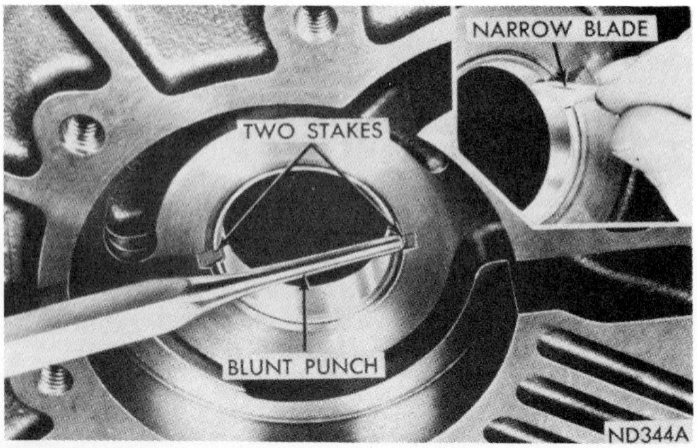

Figure 9-111. Staking pump bushing on TorqueFlite A904. (*Courtesy of Chrysler Corporation*)

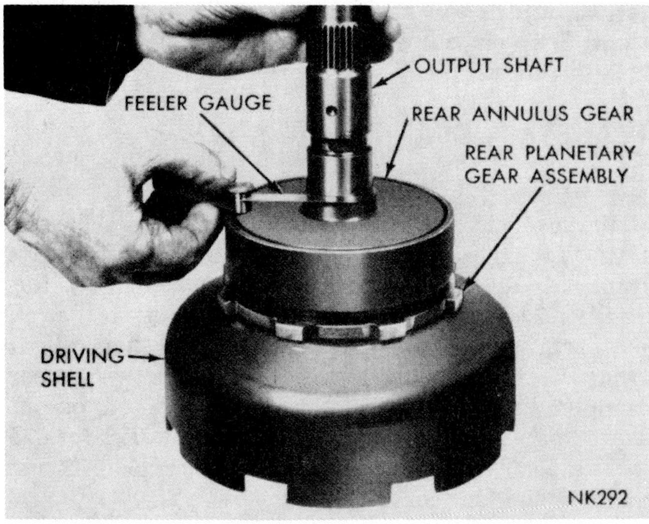

Figure 9-112. Assembling pump with aligning tool, TorqueFlite. (*Courtesy of Chrysler Corporation*)

creased and it will not fit in the transmission case. A steel band type of clamping tool is used to achieve alignment during pump assembly as illustrated.

Pump gears or rotors installed incorrectly will cause pump and torque converter damage. In some cases the pump gears will be seized if incorrectly installed.

If front unit end-play measurement during disassembly of the transmission was excessive, the proper thrust washer selection must be made to correct end play.

For example, assuming the following conditions—

- End play originally—0.105 inch (2.66 mm)
- End play required—0.020 to .090 (0.5 mm to 2.28 mm)
- Present thrust washer thickness—0.100 inch (2.54 mm)
- Available selective size thrust washers—
  0.080 inch
  0.100 inch
  0.120 inch
  0.140 inch
  0.160 inch
  0.180 inch

which thrust washer should be used to correct the end play? To decide, simply select a thicker thrust washer from those available, which will reduce end play the most but will still keep it at or above the lowest limit (in this case, 0.020 inch or 0.5 mm). A 0.180-inch thrust washer is 0.080 inch thicker than the present thrust washer and would therefore reduce end play by 0.080 inch, which would be 0.105–0.080 = 0.25 inch. This is just over the specified lower limit. Use the appropriate service manual specifications and the existing thrust washer measurement to select the thrust washer that will provide the proper end play for any specific transmission.

Lubricate the thrust washer and install it on the pump hub. Then lubricate and install all the pump hub to clutch seal rings. Make sure that seal rings are not overstretched during installation. Hooked rings must be properly hooked. Taper-cut ends must have tapered ends face and overlap each other. Install a new pump to the transmission case seal ring. Refer to the appropriate service manual for specific transmissions.

Vane-Type Pump Service

The vane-type pump is used on some rear-wheel-drive vehicle transmissions as well as on some transaxles. The pump is part of the control valve body assembly on the 125C General Motors transaxle. Regardless of the application of the pump, inspection and service of the pump itself are similar.

After disassembly and cleaning of the pump, it should be inspected for wear and damage. Inspect the pump body and cover for cracks, wear, and scoring. Inspect the pump rotor, vanes, vane rings, and slide for cracks, scoring, wear, or damage. Inspect the bearing for wear or damage.

Replace the pump seal. Replace any parts that are cracked or show excessive wear or damage. Lubricate all parts with the appropriate transmission fluid and assemble the pump as outlined in the appropriate service manual.

Governor Service

The governor on most Ford and Chrysler transmissions is mounted on the output shaft and requires removal of a snap ring from the output shaft for removal. A second snap ring must be removed from the valve shaft; the valve shaft that passes through a hole in the output shaft must also be removed. The valves and valve body can then be removed, disassembled, cleaned, and inspected. Minor nicks, burrs, and scratches can be removed from the valves and valve bores with crocus cloth. Valves must move freely in their bores to prevent sticking during operation, which causes erratic shifts. Lubricate the valves with transmission fluid, assemble the governor, and install it during transmission assembly. Make sure that snap rings are not distorted and are fully seated in the grooves.

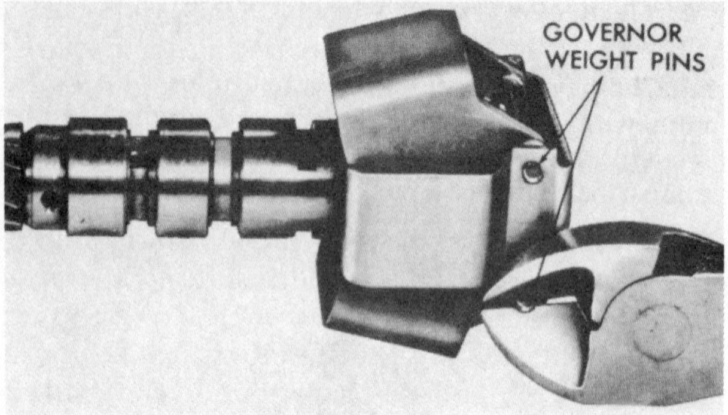

Figure 9-113. Removing governor weight pins, GM type. (*Courtesy of General Motors Corporation*)

The General Motors governor is output-shaft driven but is mounted in the transmission case under a governor cover. The cover may be bolted in place or held in place by a retaining ring or wire, depending on the transmission model.

Be careful not to damage the cover during removal or installation. A damaged cover must be replaced. Removing the cover allows removal of the governor. The governor may even come out as the cover is removed. Do not use pliers on the governor. The output shaft may have to be turned counterclockwise to help remove the governor.

The following procedures are typical for the 250C and 350C transmissions; although others are similar, be sure to follow shop manual procedures.

Disassembly

All parts of the governor assembly, with the exception of the driven gear, are a select fit, and each assembly is calibrated. The governor, including the driven gear, is serviced as a complete assembly. However, the driven gear can also be serviced separately.

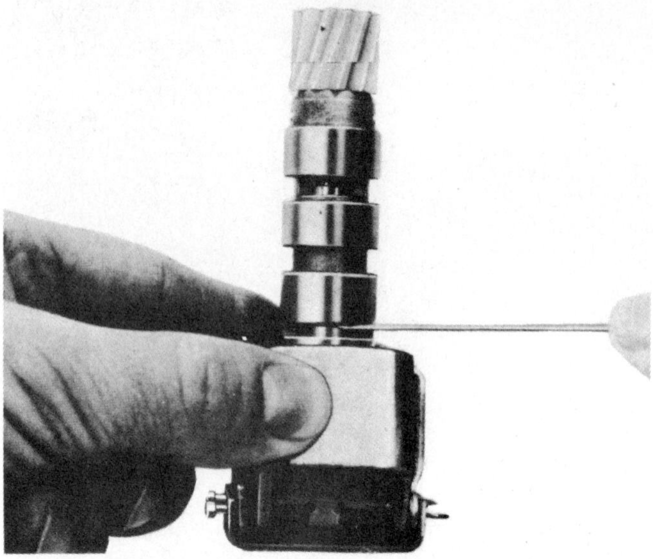

Figure 9-115. Measuring check valve opening (exhaust), GM. (*Courtesy of General Motors Corporation*)

It is necessary to disassemble the governor in order to replace the driven gear. Disassembly may also be necessary due to foreign material causing improper operation. In such cases, proceed as follows.

1. Cut off one end of each governor weight pin and remove pins, governor thrust cap, governor

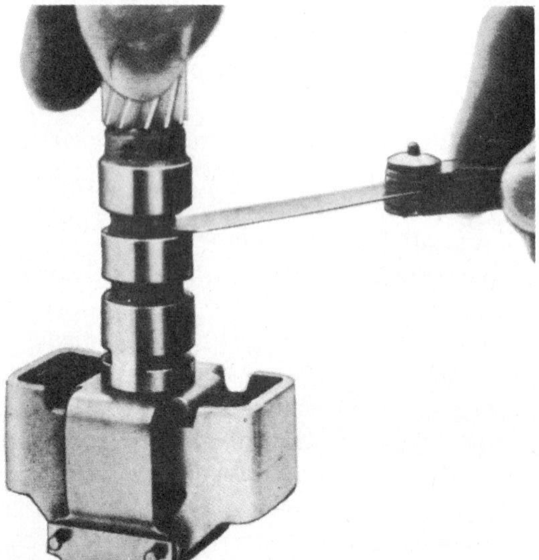

Figure 9-114. Measuring check valve opening (entry), GM. (*Courtesy of General Motors Corporation*)

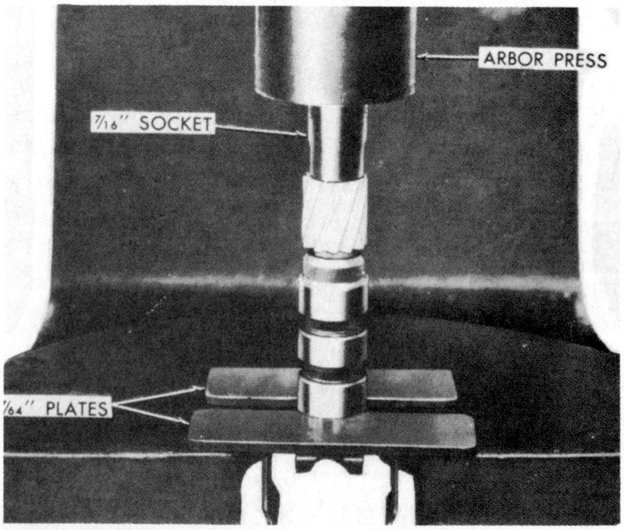

Figure 9-116. Installing governor-driven gear, GM. (*Courtesy of General Motors Corporation*)

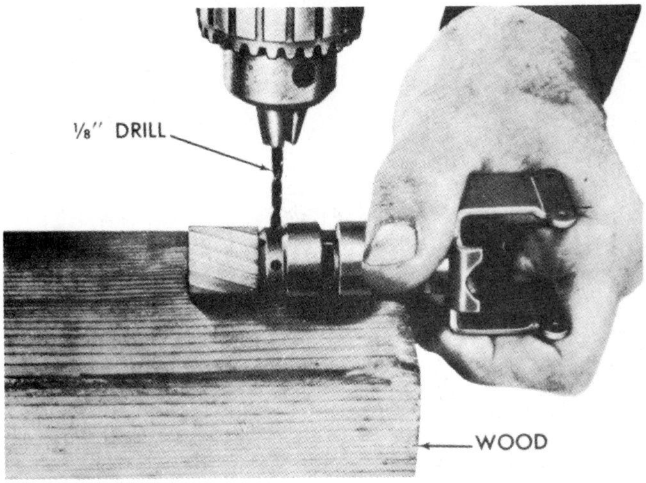

Figure 9-117. Drilling governor-driven gear after replacement, GM. (*Courtesy of General Motors Corporation*)

Figure 9-118. Positioning governor bushing in case, GM 350. (*Courtesy of General Motors Corporation*)

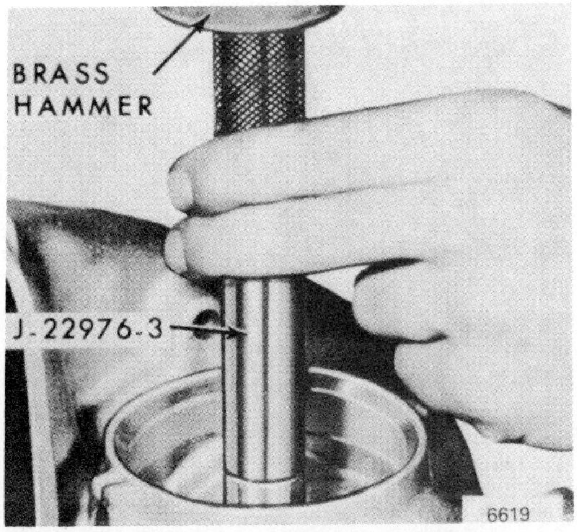

Figure 9-119. Installing governor bushing in case, GM. (*Courtesy of General Motors Corporation*)

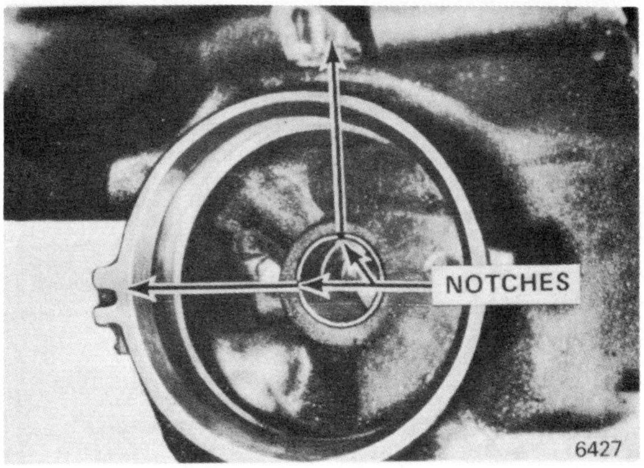

Figure 9-120. Bushing installed correctly in case, GM 350. (*Courtesy of General Motors Corporation*)

Figure 9-121. Installing 350 GM governor. (*Courtesy of General Motors Corporation*)

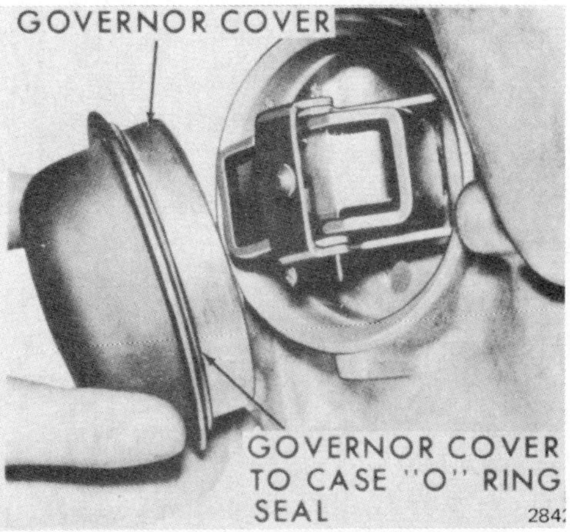

Figure 9-122. Installing governor cover, GM 350. (*Courtesy of General Motors Corporation*)

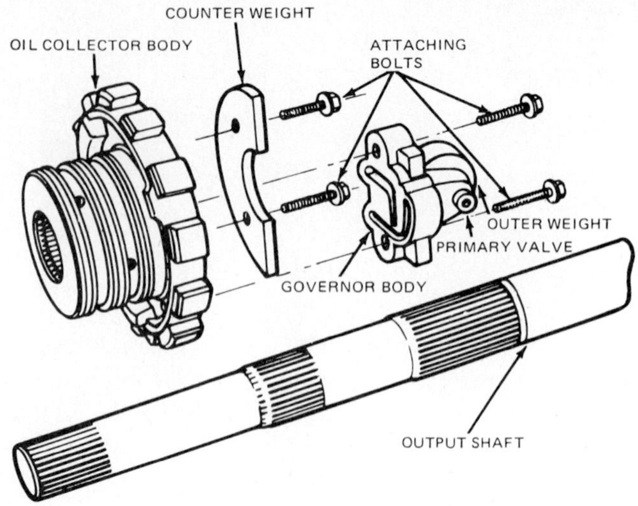

Figure 9-123. Ford C3 governor components. *(Courtesy of Ford Motor Company)*

weights, and springs. Governor weights are interchangeable from side to side and need not be identified.

2. Remove the governor valve from the governor sleeve. Be careful not to damage the valve.

3. Perform the following inspections and replace the governor driven gear, if necessary.

Inspection

1. Wash all parts in cleaning solvent; air dry and blow out all passages.

2. Inspect the governor sleeve for nicks, burrs, scoring, or galling.

3. Check the governor sleeve for free operations in the bore of the transmission case.

4. Inspect the governor valve for nicks, burrs, scoring, or galling.

5. Check the governor valve for free operation in the bore of the governor sleeve.

6. Inspect the governor-driven gear for nicks, burrs, or damage.

7. Check the governor-driven gear for looseness on the governor sleeve.

8. Inspect governor weight springs for distortion or damage.

9. Check governor weights for free operation in their retainers.

10. Check the valve opening at entry to specifications with a feeler gauge, holding the governor with governor weights extended completely outward.

11. Check the valve opening at exhaust to specifications with a feeler gauge, holding the governor with governor weights completely inward. The sleeve must be tight in the carrier.

12. Replace the gear if necessary. Follow the procedure in the service manual for gear replacement.

Assembly

1. Install the governor valve in the bore of the governor sleeve.

2. Install governor weights and springs and thrust cap on the governor sleeve.

3. Align pin holes in the thrust cap, governor weight assemblies, and governor sleeve; install new pins. Crimp both ends of the pin to prevent them from falling out.

4. Check governor weight assemblies for free operation on pins and the governor valve for free operation in the governor sleeve.

Installation

1. Install the governor; rotate the output shaft if necessary to help installation.

2. Apply the recommended gasket or sealant (such as Loctite Cup Plug Sealer 2 or equivalent if specified in the shop manual) to the cover. Then install the governor cover and cover bolts, retaining ring, or wire. Be sure that the cover is not distorted during installation. In some cases tapping around the outer edges with a brass drift may be required.

Speedometer Drive Gear and Pinion Service

The speedometer drive gear may be part of the output shaft on some transmissions. On others it is a separate gear mounted on the output shaft or final drive unit. In all cases inspect the condition of the drive gear. Check for wear, cracked, or broken teeth. If wear is excessive or teeth are broken or cracked, replacement is required. Inspect the speedometer-driven gear or pinion assembly for wear or damage. Replace damaged parts and install new seals.

Valve Body Service

The valve body may be serviced on most transmissions without transmission removal. A complete overhaul of the transmission should include disassembly, cleaning, inspection, repair, and assembly of the valve body.

Valve body service includes the following steps:

1. Removal
2. Preliminary cleaning of assembly

3. Disassembly
4. Cleaning of the valve body parts
5. Inspection of all parts
6. Removal of nicks and burrs from valves and valve bores and checking that all valves move freely in their bores
7. Lubrication and assembly
8. Installation

It is important to remember some general precautions regarding valve body service procedures in order to ensure success.

1. Never use excessive force during disassembly or assembly, particularly when trying to remove or install valves. If a valve will not move, other methods must be used rather than increased force.

2. Be careful of springs behind valves when removing cover plates in order to prevent popping out.

3. Never use a magnet on any valve body part.

4. Do not attempt to straighten any valve body part; a bent or warped part requires valve body replacement. Damaging one small valve body part due to careless handling or improper service procedures may require replacement of the entire valve body assembly. Individual parts are not available.

5. Use only crocus cloth to remove nicks or burrs from valves and valve bores. Abrasive stones of the type specified in the appropriate service manual only are also used for this purpose but cannot be used in valve bores.

6. Never damage or round off the sharp edges of spool valves since they serve an important shearing and cleaning function during operation.

7. Use the cleaning method recommended in the appropriate service manual since some ordinary cleaning solvents leave varnish deposits, which can cause valves to stick. Mineral spirits can be used.

8. Be sure that you have a clean workbench large enough to lay all parts out in an orderly fashion by referring to the shop manual.

9. Be absolutely sure that all valves and springs are assembled correctly.

10. Tighten all screws and bolts to specified torque only.

11. Use only clean, lintfree cloth spread on the workbench to lay out all valve body parts in order.

Removal

Valve body removal may require disconnecting electrical connections and linkage to the valve body before or during valve body removal. Make sure that you have completely disconnected the valve body and remove only the attaching screws. Such items as kickdown solenoids, converter lockup solenoids, park pawl linkage, downshift linkage, throttle linkage, and manual linkage are involved.

Preliminary Cleaning

If the valve body does not show any serious signs of sludge or varnish buildup, preliminary cleaning may not be required before disassembly. If, however, there are deposits or grit, sludge, or varnish evident on the assembly, one or more of the following procedures will be required.

1. Using a brush, wash the assembly in mineral spirits and let it soak for a while. This may be all that is required for some valve bodies unless there are valves that are stuck during disassembly.

2. Use a spray type of varnish and gum remover and spray all sides of the assembly. Allow the spray to soak for a while, then wash the assembly in mineral spirits. If it still looks dirty, repeat the procedure.

3. If recommended in the service manual, the assembly may be soaked in carburetor cleaning fluid for a while. Leaving it in too long may allow some cleaners to attack the aluminum valve body parts. Follow the manufacturer's instructions only for time and method of using carburetor cleaner.

Disassembly

Disassemble the valve body in the sequence specified in the service manual. The general procedure is to separate the two halves of the valve body and then remove any other plates or smaller parts. Be sure to hold spring loaded plates tightly against the valve body while removing the screws in order to prevent damage to the plate and the valves. When all screws have been removed from the plate, slowly release the pressure while you hold the plate side of the valve body up.

Be careful when removing roll-pin–type spring retainers that you do not damage springs or the soft aluminum valve body.

Carefully remove all springs and valves in succession from each valve bore and keep them in the same order as in the valve body. This helps considerably during assembly later. In some cases springs from different valve bores may be almost identical in size but not in tension. Interchanged springs or valves cause serious problems.

OVERHAUL AND CLEAN
VALVE BODY IF:

1. Contaminated from another transmission failure.
   - Make sure converter, cooler and cooler lines are cleaned.
2. Diagnosis procedure indicates valve body malfunction (such as sticking valve, etc.)

NOTE: Inspect all parts carefully to determine cause of malfunction.

Figure 9-124. *(Courtesy of Ford Motor Company)*

If valves are stuck, a small screwdriver may be used to help pry the valve out. However, this must be done with great care not to scratch the sliding surfaces of any valves or the valve bores. It may be necessary at this time to spray the valve again with varnish remover to help remove it.

Another method that can be used is to rap the valve body vigorously flat side down onto a clean cloth-covered, flat area of the workbench several times. This may help jar the valve loose so that the screwdriver method will complete the job. If fine metal particles have been circulating in the fluid, these may have caused the valves to be jammed tightly in their bores.

These methods may have to be repeated several times in order to remove some of the more stubborn valves.

Cleaning Valve Body Parts

Valves, springs, plugs, and valve bodies can be sprayed with a varnish remover and washed in mineral spirits. Another method is to soak them in automatic transmission cleaning fluid or carburetor cleaner, then rinse them with hot water; then they are washed in mineral spirits to remove all traces of water. Water will collect in the bottom of the mineral spirits cleaning pan.

Inspection of Parts

Mating surfaces of the valve body must be flat and true. Warpage can be checked with a straight-edge and feeler gauge. A warped valve body must be replaced.

If the valve body is not warped, valves and valve bores should be inspected for nicks and scratches. Nicks and scratches that cause valves to stick even slightly in their bores must be polished with crocus cloth. Check the freedom of valve movement in the bore dry.

To remove scratches from valves, lay a sheet of crocus cloth on a flat-machined surface (not a rough-surfaced wood or metal workbench) and slide the valve across the crocus cloth while you turn the valve in the opposite direction. Never round off sharp valve edges during this procedure.

To remove scratches from a valve bore, roll up a piece of crocus cloth (abrasive side out) small enough to fit in the valve bore. Insert it in the bore and turn it in the direction that tries to unwind the crocus cloth. Do this for 10 or 20 revolutions. Then check the previously polished valve for freedom of movement in the valve bore. The valve should move within its normal travel limits by its own weight as you tip the valve body back and forth. If it does not do this, the procedure may have to be repeated several times.

After all valves have been inspected and move freely in their bores, all parts should again be washed to remove abrasives before assembly.

Assembly

1. Use the appropriate shop manual for reference during assembly. The illustrations will help determine correct assembly. Read the text to assure proper assembly sequence.

2. Lubricate all parts with the appropriate transmission fluid.

EQUIPMENT FOR CLEANING

- Valve body is a precision "hydraulic computer."
- **Clinical** cleanliness is necessary for proper operation.
- All items shown are **essential.**

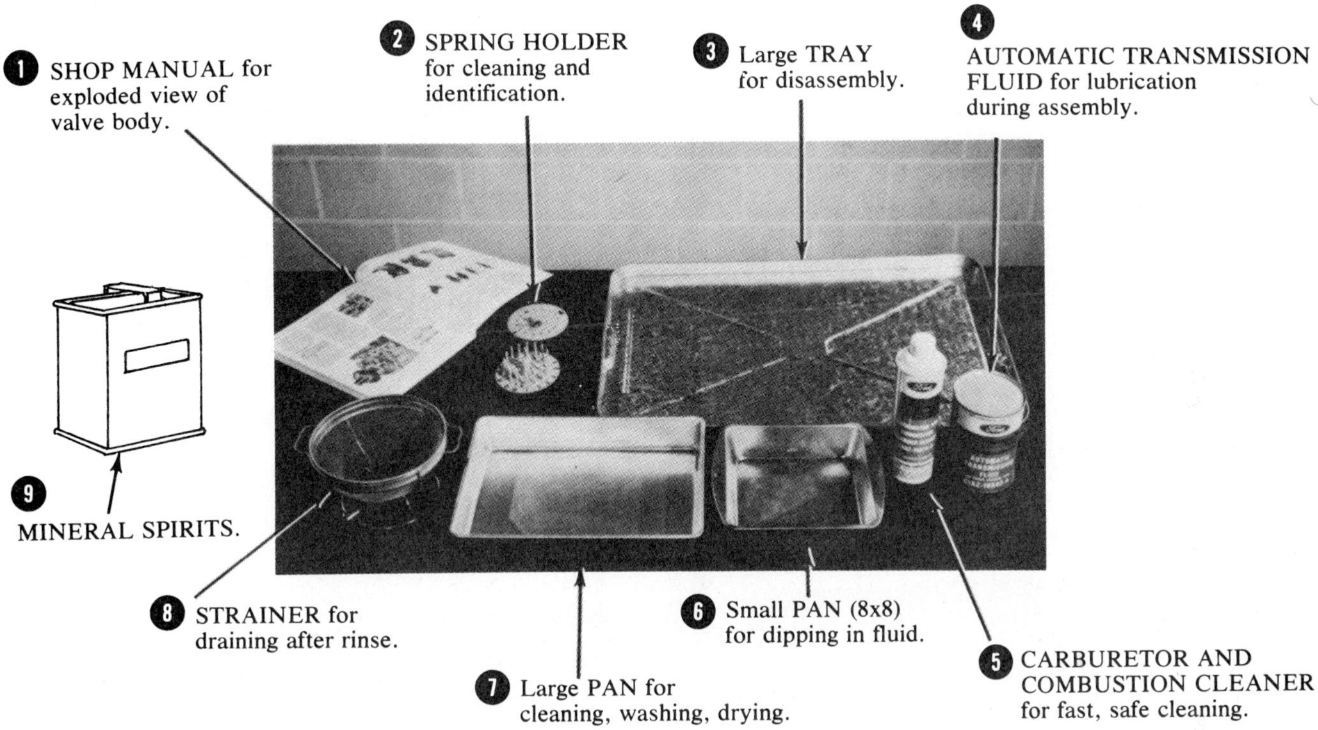

Figure 9-125. *(Courtesy of Ford Motor Company)*

3. If a valve body gasket is involved, be certain that it is the correct gasket. All holes should match the old gasket.

4. Two-piece valve bodies are usually bolted together before valves are installed to prevent valves from sticking as bolts or screws are tightened. Follow the sequence outlined in the service manual.

5. Always start all screws and bolts by hand and install all of them before tightening any of them. Thread damage, stripped threads, and false torque readings are the result of not following this procedure.

6. Always tighten all screws and bolts in the order specified in the service manual and to specified torque.

7. If any valve body adjustment is required (such as TorqueFlite's pressure adjustments), follow the appropriate shop manual procedures and specifications to assure accuracy and success.

Installation

Install the valve body at the proper stage of transmission assembly, making all the necessary linkage and electrical connections required. Tighten all valve body attaching bolts to specified torque.

Transmission Assembly (Subassembly Installation)

After all subassemblies have been reocnditioned, they are installed in the transmission case. The necessary checks and adjustments must be made during subassembly installation. These include such checks as end play, band adjustment, and air pressure checks. Follow the assembly sequence prescribed in the appropriate manufacturer's service manual. Be sure

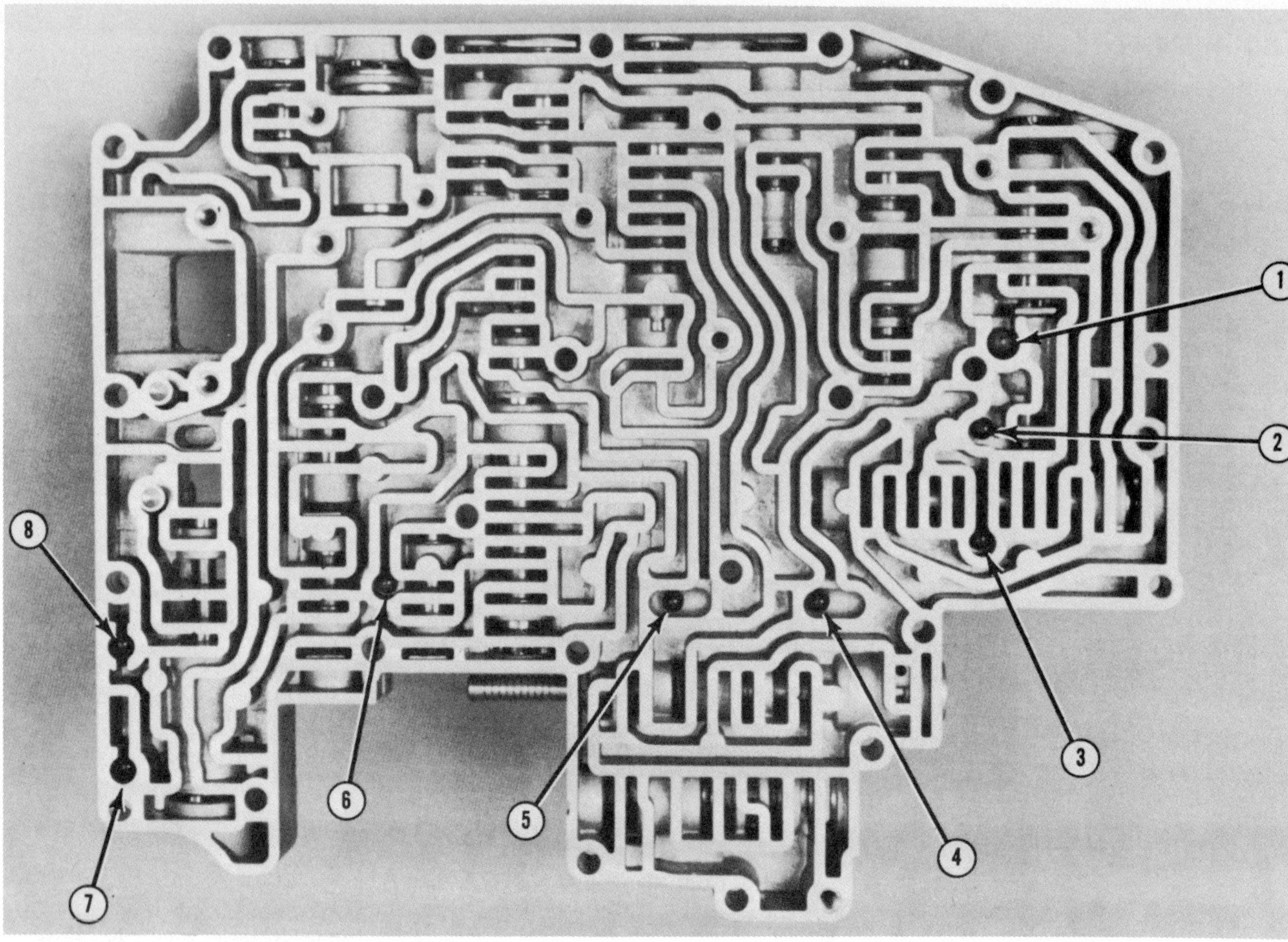

Figure 9-126. Check ball locations, Ford AOD. (*Courtesy of Ford Motor Company*)

to adhere to all specifications in the manual for the transmission being serviced.

Here are some points to remember when assembling the automatic transmission.

1. Absolute cleanliness is essential. Work in a clean area on a clean workbench with clean hands and clean tools.

2. Follow the prescribed sequence of steps for assembly given in the appropriate service manual.

3. Perform all checks and measurements required during the assembly process. Correct any problems discovered.

4. Never use force during transmission assembly. The ball peen hammer has no place in transmission assembly.

5. Dip all bolts in transmission fluid before installing to prevent damage to threads, especially in aluminum parts.

6. Never use bolts to draw parts together. Serious damage can be done to the transmission if this practice is used. All parts must assemble fully by hand before tightening any bolts.

7. Be sure that thrust washers remain in place during the assembly procedure. A thrust washer moved out of place slightly by even a light bumping of parts will prevent assembly of the transmission. Use petroleum jelly (petrolatum or vaseline) on one side only to stick thrust washers in place. Never use any other type of grease since it will clog hydraulic passages.

8. Be sure to lubricate all parts where there is sliding friction with the appropriate type of transmission fluid.

WASHING AND DRYING

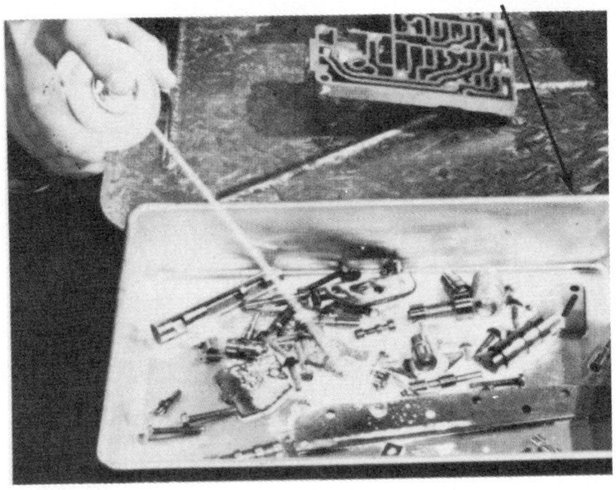

1 Spray all small parts with cleaner.

- Cover all parts with foam.
- Spray spring holder also.

2 Clean valve bodies and separator plate.

- Use cleaner liberally.
- Don't leave bristles in body.

3 Rinse thoroughly.

- Use <u>hot</u> tap water.
- Cleaner will rinse away readily.

4 Drip dry small parts.

5 Dip <u>all</u> parts in mineral spirits for final water separation.

Figure 9-127. (*Courtesy of Ford Motor Company*)

9. Handle clutch units with care. Friction disc internal teeth are easily bent or damaged if assembled carelessly.

10. Be sure that clutch units are fully engaged on splines (all discs engaged on splines) before proceeding to the next assembly. If not fully engaged or if a thrust washer is out of place, the assembly will bind and prevent shafts from rotating freely or even lock up. Bands must be presoaked in transmission fluid before installation. Make all band adjustments as required. Transmission assembly is made easier if the case is held in a fixture that allows the output shaft to point downward. This allows "loading" the case from

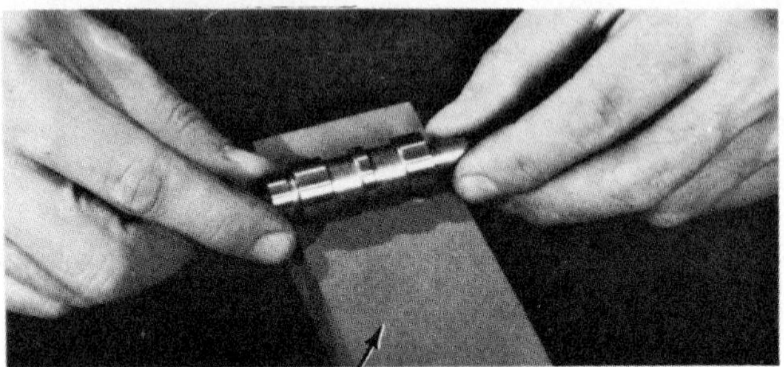

Figure 9-128. (*Courtesy of Ford Motor Company*)

the top and avoids the tendency of installed subassemblies, particularly clutch units, to fall out of place. Clutch assemblies should be installed with a slight turning of the unit to help align disc teeth with

Figure 9-129. (*Courtesy of Ford Motor Company*)

hub splines. Slight wiggling of the unit or the turbine shaft may also help. A solid metal-to-metal bump is evident when full engagement is achieved.

11. When the hydraulic pump is being bolted into place, make sure to keep checking that the turbine shaft rotates freely at all times. If the turning effort suddenly increases, stop tightening immediately;

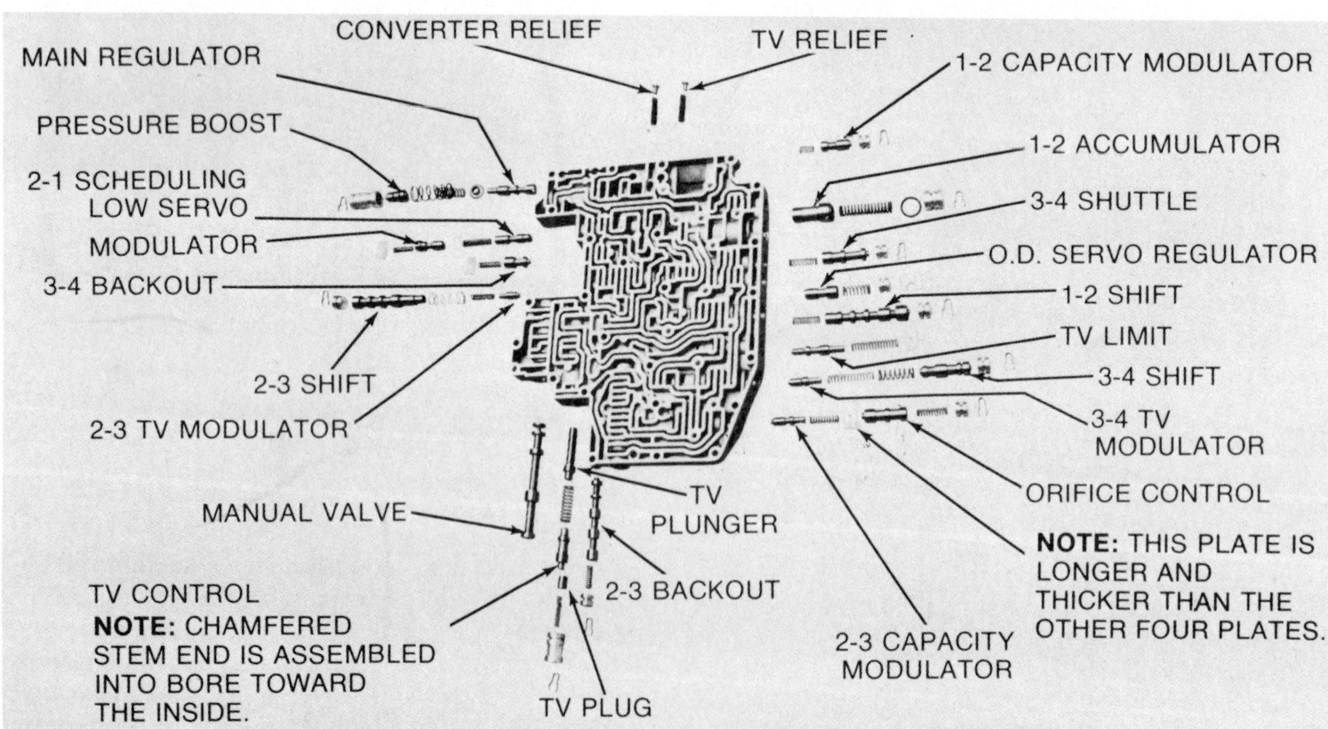

Figure 9-130. Exploded view of Ford AOD valve body. (*Courtesy of Ford Motor Company*)

Figure 9–131. Torqueing valve body bolts. (*Courtesy of Ford Motor Company*)

determine the cause and correct the problem. The cause will be either a thrust washer out of place or one or more clutch discs not engaged with the splines on the hub. Correction requires removing the pump and the clutch assemblies, correcting the problem, and reassembly.

12. The pump should be installed with a new gasket and pump-to-case O-ring seal. Guide pins are helpful in aligning the pump-to-case position. Lubricate the O-ring seal with transmission fluid or petroleum jelly. The gasket may also be stuck in place with petroleum jelly. Use it sparingly so as not to restrict hydraulic flow even momentarily.

13. Use a torque wrench to tighten all bolts to specified torque. Uneven tightening, undertightening, or overtightening of bolts can cause hydraulic fluid leaks.

14. Before valve body installation, perform air pressure checks as specified in the manual to ensure that clutches and servos operate properly. A tube-type air chuck nozzle may be required to obtain access to apply passages. In some cases an irregular-shaped apply passage may require sealing around the nozzle to achieve clutch or servo apply. In many cases the governor circuit is also air pressure checked.

15. Always use a new vacuum modulator on transmissions equipped with this device when the transmission is overhauled. Using a used modulator is not advisable since its life expectancy is unpredictable. A new modulator will ensure troublefree operation for a longer time.

16. Install the valve body with a new filter, tightening all bolts evenly to specifications with a torque wrench.

17. Straighten the oil pan mounting flange if necessary and install the pan with a new gasket. Cork gaskets are best installed dry to prevent "squishing" out at the bolts or cracking at the bolt holes.

18. Install the converter. Make sure that the converter is fully engaged with stator splines, turbine splines, and pump-drive gear. Measure converter engagement depth to ensure full engagement.

Install the transmission as outlined in Part 9–4. Follow the fluid fill recommendations and the road test procedure outlined in Chapter 8. Be sure to inspect the transmission for leaks after the road test and correct the fluid level if required. Remove all evidence of fingerprints, oil, or grease from any part of the vehicle inside or out to complete the job.

Part 9–6 Worksheet

1. Complete the diagnostic chart provided in this section.
2. Perform the following checks on a transmission specified by your instructor and record the results on the sheets provided:
   a) Fluid checks
   b) Linkage checks
   c) Vacuum modulator checks
   d) Hydraulic pressure tests
3. Perform a stall test on a vehicle specified by your instructor and record the results on the sheet provided.
4. Perform a road test with a vehicle specified by your instructor and record the results on the sheet provided.
5. Remove, overhaul, and install an automatic transmission on a vehicle specified by your instructor. Record the information required on the sheets provided during the overhaul. Follow the instructions given by your instructor, this text, and the appropriate service manual. Road test the vehicle to determine the success of the overhaul.

Automatic Transmission Diagnosis

Problem	Cause	Correction
Harsh engagement.	1.	1.
	2.	2.
	3.	3.
	4.	4.
Delayed engagement.	1.	1.
	2.	2.
	3.	3.
	4.	4.
	5.	5.
	6.	6.
	7.	7.
	8.	8.
No upshift.	1.	1.
	2.	2.
	3.	3.
	4.	4.
	5.	5.
	6.	6.
Downshift abnormal.	1.	1.
	2.	2.
	3.	3.
	4.	4.
	5.	5.
	6.	6.
Slips in forward only.	1.	1.
	2.	2.
	3.	3.
	4.	4.
	5.	5.
Slips in reverse only.	1.	1.
	2.	2.
	3.	3.
	4.	4.
	5.	5.
Slips in all positions.	1.	1.
	2.	2.
	3.	3.
	4.	4.
	5.	5.
Transmission overheats.	1.	1.
	2.	2.
	3.	3.
	4.	4.
	5.	5.
Delayed upshift.	1.	1.
	2.	2.
	3.	3.
	4.	4.
	5.	5.

Worksheet—Automatic Transmission Service

Fluid			
Level	OK ________	Not OK ________	Explain ________________
Color	OK ________	Not OK ________	Explain ________________
Smell	OK ________	Not OK ________	Explain ________________
Contamination	OK ________	Not OK ________	Explain ________________
Leaks	OK ________	Not OK ________	Explain ________________
Cooler	OK ________	Not OK ________	Explain ________________
Linkage			
Manual shift	OK ________	Not OK ________	Explain ________________
Throttle linkage	OK ________	Not OK ________	Explain ________________
Downshift linkage	OK ________	Not OK ________	Explain ________________
Vacuum Modulator			
Vacuum source	OK ________	Not OK ________	Explain ________________
Vacuum lines	OK ________	Not OK ________	Explain ________________
Modulator	OK ________	Not OK ________	Explain ________________

Hydraulic pressure tests

Engine rpm	Manifold Vacuum	Range	Actual	Specified
		P		
		N		
		R		
		D		
		2		
		1		
		O.D.		

Explain and interpret the results of these tests. ________________

____

____

Stall Test

	Stall Speed Data		Diagnosis	
Selector Lever Position	Specified Engine rpm	Actual Engine rpm	Above Specified Engine rpm	Below Specified Engine rpm
R				
D				
2				
1				

Explain the results of these tests. ________________

____

Road Test

Throttle Opening	Range	Shift	Shift Points mph	
			Actual	Specified
Light throttle	D	1–2		
		2–3		
		O.D.		
Full throttle	D	1–2		
		2–3		
		O.D.		
Through detent (WOT)	D	1–2		
		2–3		
		O.D.		

Explain results of road test.____

____

____

Worksheet—Automatic Transmission Overhaul ____________________

Make of vehicle ____________________ Year ____________________

Engine C.I.D. ____________________ Transmission type ____________________

After removal of the transmission from the vehicle, remove necessary assemblies and check end play.

Specified____________________ Actual____________________

Disassemble the transmission according to procedures outlined in the service manual and check and record the condition of the following units or subassemblies.

	OK	Not OK	Explain
Converter and front pump	___	___	___
• Bushing	___	___	___
• Seal	___	___	___
• Turbine end play	___	___	___
• Drive lugs	___	___	___
Forward clutch			
• Drum	___	___	___
• Bushings	___	___	___
• Piston	___	___	___
• Steel plates	___	___	___
• Friction plates	___	___	___
• Release springs	___	___	___
• Pressure plates	___	___	___
• Snap rings	___	___	___
Reverse and high clutch			
• Drum	___	___	___
• Bushings	___	___	___
• Piston	___	___	___
• Steel plates	___	___	___
• Friction plates	___	___	___
• Pressure plates	___	___	___
• Snap rings	___	___	___
Intermediate holding clutch			
• Case	___	___	___
• Piston	___	___	___
• Steel plates	___	___	___
• Friction plates	___	___	___
• Pressure plates	___	___	___
• Snap rings	___	___	___
Reverse holding clutch			
• Case	___	___	___
• Piston	___	___	___
• Steel plates	___	___	___
• Friction plates	___	___	___
• Pressure plates	___	___	___
• Snap rings	___	___	___
O.D. clutch			
• Drum	___	___	___
• Piston	___	___	___
• Steel plates	___	___	___

	OK	Not OK	Explain
• Friction plates	____	______	______
• Pressure plates	____	______	______
• Snap rings	____	______	______
Forward planetary gear set			
• Carrier	____	______	______
• Ring	____	______	______
• Sun	____	______	______
• Pinions	____	______	______
Reverse planetary gear set			
• Carrier	____	______	______
• Ring	____	______	______
• Sun	____	______	______
• Pinions	____	______	______
O.D. planetary gear set			
• Carrier	____	______	______
• Ring	____	______	______
• Sun	____	______	______
• Pinions	____	______	______
Low and reverse band	____	______	______
Intermediate band	____	______	______
Servos and pins	____	______	______
Struts or links	____	______	______
Input shaft	____	______	______
Intermediate shaft	____	______	______
Output shaft	____	______	______
• Pressure plates	____	______	______
• Snap rings	____	______	______
Over-running clutches	____	______	______
Pump	____	______	______
Valve body	____	______	______
Governor	____	______	______
Vacuum modulator	____	______	______
Neutral switch	____	______	______
Oil pick up screen or filter	____	______	______
Linkages within transmission	____	______	______
Transmission case	____	______	______

Identify those thrust washers, snap rings, bearings, or bushings that require replacement.

Thrust washers	____	______	______
Snap rings	____	______	______
Bearings	____	______	______
Bushings	____	______	______

PART 9-7 SELF-CHECK

1. Define severe automatic transmission service.
2. List as many general precautions as you can that apply to automatic transmission overhaul procedures.
3. Why should you not remove the transmission first and then the torque converter?
4. Why is full engagement of the torque converter required prior to and during transmission installation?
5. What is the importance of taking end-play measurements during transmission disassembly?
6. List the checks required to determine the serviceability of a torque converter.
7. Why must clutch pack clearance not be over or under specifications?
8. List four causes of burned clutch discs.
9. How is clutch pack clearance corrected?
10. What happens if the low-gear one-way clutch is installed backward?
11. Excessive pump rotor or gear clearance will cause reduced pump __________ and __________.
12. How is front unit (turbine shaft) end play corrected?
13. How should minor nicks and scratches be removed from spool valves?
14. Spool valves are available as replacement parts for rebuilding valve bodies. True or False?
15. State two causes for a turbine shaft not able to be turned after tightening hydraulic pump mounting bolts.

PART 9-8 PERFORMANCE EVALUATION

1. Did you safely and successfully remove, overhaul, and install one or more automatic transmissions specified by your instructor as outlined in this chapter, in the appropriate service repair manual, and by your instructor?

Index

V